International Conference on Nutrient Recovery from Wastewater Streams

International Conference on Nutrient Recovery from Wastewater Streams

May 10–13, 2009
The Westin Bayshore Hotel and Resort,
Vancouver, British Columbia, Canada

Editors
Ken Ashley,
Don Mavinic
and Fred Koch

Publishing
London • New York

Published by   **IWA Publishing**
Alliance House
12 Caxton Street
London SW1H 0QS, UK
Telephone: +44 (0)20 7654 5500
Fax: +44 (0)20 654 5555
Email: publications@iwap.co.uk
Web: www.iwapublishing.com

First published 2009
© 2009 IWA Publishing and the Authors

Printed by TJ International, Padstow, Cornwall, UK.
Cover design by www.designforpublishing.co.uk
Typeset in India by Alden Prepress Services Private Limited.
Index provided by Alden Prepress Services Private Limited.

FSC

Mixed Sources
Product group from well-managed
forests and other controlled sources

Cert no. SGS-COC-2482
www.fsc.org
© 1996 Forest Stewardship Council

British Library Cataloguing in Publication Data
A CIP catalogue record for this book is available from the British Library

Library of Congress Cataloging-in-Publication Data
A catalog record for this book is available from the Library of Congress

ISBN: 1843392321
ISBN13: 9781843392323

Contents

xi   Preface
K. Ashley

1   Elimination of eutrophication through resource recovery
J.L. Barnard

23   Preferred future phosphorus scenarios: a framework for meeting long-term phosphorus needs for global food demand
D. Cordell, Dr T. Schmid-Neset, Dr S. White and Dr J.-O. Drangert

45   Impact of supply and demand on the price development of phosphate (fertilizer)
J. von Horn and C. Sartorius

55   Wastewater treatment and the green revolution
P.M. Sutton, A.P. Togna and O.J. Schraa

69   A review of struvite nucleation studies
S.C. Galbraith and P.A. Schneider

79   A quantitative method analyzing the content of struvite in phosphate-based precipitates
X.-D. Hao, C.-C. Wang, L. Lan and M.C.M. van Loosdrecht

89   Phosphorus removal from an industrial wastewater by struvite crystallization into an airlift reactor
A. Sánchez, S. Barros, R. Méndez and J.M. Garrido

99   Quantifying phosphorus recovery potentials by full-scale process analysis and modelling
M. Beier, R. Pikula, V. Spering and K.-H. Rosenwinkel

111   Validation of a comprehensive chemical equilibrium model for predicting struvite precipitation
S. Gadekar, P. Pullammanappallil and A. Varshovi

121   A thermochemical approach for struvite precipitation modelling from wastewater
M. Hanhoun, C. Azzaro-Pantel, B. Biscans, M. Frèche, L. Montastruc, L. Pibouleau and S. Domenech

131   Numerical investigations of the hydrodynamics of the UBC MAP
        fluidized bed crystallizer
        **M.S. Rahaman, D.S. Mavinic, A.T. Briton, M. Zhang, K.P. Fattah
        and F.A. Koch**

145   About the economy of phosphorus recovery
        T. Dockhorn

159   Different strategies for recovering phosphorus: technologies and costs
        D. Montag, K. Gethke and J. Pinnekamp

169   Social and economic feasibility of struvite recovery from Urine at the
        community level in Nepal
        E. Tilley, B. Gantenbein, R. Khadka, C. Zurbrügg and K.M. Udert

179   Induced struvite precipitation in an airlift reactor for phosphorus recovery
        D. Stumpf, B. Heinzmann, R.-J. Schwarz, R. Gnirss and M. Kraume

193   Pilot testing and economic evaluation of struvite recovery from
        dewatering centrate at HRSD's Nansemond WWTP
        A. Britton, R. Prasad, B. Balzer and L. Cubbage

203   Standardizing the struvite solubility product for field trial optimization
        A.L. Forrest, K.P. Fattah, D.S. Mavinic and F.A. Koch

215   Plant availability of P fertilizers recycled from sewage sludge and
        meat-and-bone meal in field and pot experiments
        R.C. Pérez, B. Steingrobe, W. Römer and N. Claassen

225   Ecological testing of products from phosphorus recovery
        processes – first results
        K. Weinfurtner, S.A. Gäth, W. Kördel and C. Waida

235   Strategy for separation of manure P through flocculation
        M. Hjorth and M.L. Christensen

245   Phosphate removal in agro-industry: pilot and full-scale operational
        considerations of struvite crystallisation
        **W. Moerman, M. Carballa, A. Vandekerckhove, D. Derycke and
        W. Verstraete**

257   Development of a process control system for online monitoring and
        control of a struvite crystallization process
        K.P. Fattah, D.S. Mavinic, M.S. Rahaman and F.A. Koch

269   Increasing cost efficiency of struvite precipitation by using alternative
        precipitants and P-remobilization from sewage sludge
        T. Esemen, W. Rand, T. Dockhorn and N. Dichtl

281   Temperature dependence of electrical conductivity and its relationship
       with ionic strength for struvite precipitation system
       M. Iqbal, H. Bhuiyan and D.S. Mavinic

291   Study on phosphorus recovery by calcium phosphate precipitation
       from wastewater treatment plants
       W. Hui-Zhen, Z. Ya-jun, F. Cui-Min, X. Ping and W. Shao-Gui

301   Phosphorus removal and recovery from sewage sludge as calcium
       phosphate by addition of calcium silicate hydrate compounds (CSH)
       S. Petzet and P. Cornel

317   Field application methods for the liquid fraction of separated animal
       slurry in growing cereal crops
       T. Nyord

327   Research on nutrient removal and recovery from swine wastewater
       in china
       Y-h. Song, P. Yuan, G.-l. Qiu, J-f. Peng, X-y. Cui and P. Zeng

339   Chemical recycling of phosphorus from piggery wastewater
       M.-L. Daumer, F. Béline and S.A. Parsons

351   Struvite harvesting to reduce ammonia and phosphorus recycle
       R. Baur, R. Prasad and A. Britton

361   The application of process systems engineering to the development
       of struvite recovery systems
       P.A. Schneider and Md. I. Ali

371   Membrane EBPR for phosphorus removal and recovery using a
       sidestream flow system: preliminary assessment
       H. Srinivas, F.A. Koch, A. Monti, D.S. Mavinic, and E. Hall

389   Phosphorus recovery from eluated sewage sludge ashes by
       nanofiltration
       **C. Niewersch, S. Petzet, J. Henkel, T. Wintgens, T. Melin and
       P. Cornel**

405   P-recovery from sewage sludge ash – technology transfer from
       prototype to industrial manufacturing facilities
       L. Hermann

417   Phosphorus recovery by thermo-chemical treatment of sewage
       sludge ash – results of the European FP6-project SUSAN
       C. Adam, C. Vogel, S. Wellendorf, J. Schick, S. Kratz and E. Schnug

431   Remediation of phosphorus from animal slurry
       A.M. Thygesen, E. Skou, O. Wernberg and S.G. Sommer

443   Affecting corn processing nutrients using membrane separation and
        biological extraction and conversion
        **K.D. Rausch, R.L. Belyea, L.M. Raskin, V. Singh, D.B. Johnston,
        T.E. Clevenger, M.E. Tumbleson and E. Morgenroth**

459   Technology for recovery of phosphorus from animal wastewater
        through calcium phosphate precipitation
        M. Vanotti and A. Szogi

469   Determining the operational conditions required for homogeneous
        struvite precipitation from belt press supernatant
        **B. Lew, M. Kummel, C. Sheindorf, S. Phalah, M. Rebhum and
        O. Lahav**

479   Involvement of filamentous bacteria in the phosphorus recovery cycle
        J. Suschka, E. Kowalski and K Grübel

489   Carbon and struvite recovery from centrate at a biological nutrient
        removal plant
        B.G. Dirk, A. Gibb, H. Kelly, F. Koch and D.S. Mavinic

503   Recovery of phosphorus from sewage sludge incineration ash by
        combined bioleaching and bioaccumulation
        J. Zimmermann and W. Dott

511   Energy efficient nutrient recovery from household wastewater using
        struvite precipitation and zeolite adsorption techniques. A pilot study in
        Sweden.
        Z. Ganrot, J. Broberg and S. Bydén

521   Crystallisation of calcium phosphate from sewage: efficiency of batch
        mode technology and quality of the generated products
        **A. Ehbrecht, D. Patzig, S. Schönauer, M. Schwotzer and
        R. Schuhmann**

531   Effect of osmotic pressure and substrate resistance on transmembrane
        flux during the concentration of pretreated swine manure with reverse
        osmosis membranes
        L. Masse, D.I. Massé and Y. Pellerin

543   P and N in solids from manure separation: separation efficiency and
        particle size distribution
        K. Jorgensen, M. Hjorth and J. Magid

551   Treating solid dairy manure by using the microwave-enhanced
        advanced oxidation process
        A.A. Kenge, P.H. Liao and K.V. Lo

Contents

563 Profitable recovery of phosphorus from sewage sludge and meat &
bone meal by the Mephrec process – a new means of thermal sludge
and ash treatment
K. Scheidig, M. Schaaf and J. Mallon

567 Empirical evaluation of nutrient recovery using seaborne technology at
the wastewater treatment plant Gifhorn
**L.C. Phan, D. Weichgrebe, I. Urban, K.H. Rosenwinkel, L. Günther,
T. Dockhorn, N. Dichtl, J. Müller and N. Bayerle**

579 Sewage treatment to remove ammonium ions by struvite precipitation
S. Lobanov

591 Full-scale plant test using sewage sludge ash as raw material for
phosphorus production
W.J. Schipper and L. Korving

599 Phosphorous recovery and nitrogen removal from wastewater using
BioIronTech process
V. Ivanov, C.H. Guo, S.L. Kuang, and V. Stabnikov

609 Phosphorus speciation of sewage sludge ashes and potential for
fertilizer production
**S. Nanzer, M. Janousch, T. Huthwelker, U. Eggenberger,
L. Hermann, A. Oberson and E. Frossard**

615 Savings from integration of centrate ammonia reduction with
BNR operation: simulation of single-sludge and two-sludge plant
operation
M. Orentlicher, A. Fassbender and G. Grey

623 The use of phosphorus-saturated ochre as a fertiliser
S.T.D. Carr, K.E. Dobbie, K.V. Heal and K.A. Smith

635 Volatile Fatty Acid (VFA) and nutrient recovery from biomass
fermentation
Q. Yuan, F. Zurzolo and J.Oleszkiewicz

645 Phosphorus recovery from sewage sludge ash by a wet-chemical
process
C. Dittrich, W. Rath, D. Montag and J. Pinnekamp

659 Phosphorus recovery from sewage sludge ash: possibilities and
limitations of wet chemical technologies
C. Schaum, P. Cornel and N. Jardin

671 Phosphate adsorption from sewage sludge filtrate using Zinc-Aluminium
layered double hydroxides
X. Cheng, X. Huang, X. Wang, B. Zhao, A. Chen and D. Sun

687   Urine reuse as fertilizer for bamboo plantations
J.E. Ndzana and R. Otterpohl

697   Ammonium absorption in reject water using vermiculite
N. Åkerback, S. Engblom and K. Sahlén

707   Alternating anoxic-aerobic process for nitrogen recovery from
wastewater in a biofilm reactor
M.F. Hamoda and R.A. Bin-Fahad

719   Air stripping of ammonia from anaerobic digestate
F. Wäger, T. Wirthensohn, A. Corcoba and W. Fuchs

737   Effect of air temperature and air humidity on mass transfer coefficient
for volume reduction and urine concentration
P.M. Masoom, R. Ito and N. Funamizu

753   Phosphorus cycling by using biomass ashes
B. Eichler-Loebermann and S. Bachmann

763   Phosphorus recovery from high-phosphorus containing excess sludge
in an anaerobic-oxic-anoxic process by using the combination of
ozonation and phosphorus adsorbent
T. Kondo, Y. Ebie, S. Tsuneda, Y. Inamori and K.-Q. Xu

773   Struvite control techniques in an enhanced biological phosphorus
removal plant
R. Baur

781   A novel waste sludge operation to minimize uncontrolled phosphorus
precipitation and maximize the phosphorus recovery: a case study in
Tarragona, Spain
R. Barat, M. Abella, P. Castella, J. Ferrer and A. Seco

791   Study of uncontrolled precipitation problems in Tarragona WWTP (Spain)
R. Barat, M. Abella, J. Roig, J. Ferrer and A. Seco

801   Phosphorus recovery in EBPR systems by struvite crystallization
N. Martí, L. Pastor, A. Bouzas, J. Ferrer and A. Seco

813   Index

Preface:
"The Philosopher's Stone"

Phosphorus has been a defining element throughout modern human history. The elemental form was discovered by the German alchemist Hennig Brand in 1669 by heating urine to high temperatures. Brand thought he had discovered the *Philosopher's Stone*, which turned lead into gold. He actually generated income entertaining European nobility by demonstrating his mysterious source of light: phosphorus Greek: *phôs* meaning "light", and *phoros* meaning "bearer" – a reaction with oxygen takes place at the surface of the solid (or liquid) phosphorus, forming short-lived molecules HPO and P_2O_2 which both emit visible light. During the Edo era (1603–1868) in Japan, the residents of early Tokyo supported a population of over 500,000 residents with no external supplies of nutrients, by carefully recycling their waste nutrients, including phosphorus, back to their agricultural fields.

However, following the Aug 28, 1854 localized outbreak of cholera in London – the "Broad Street Pump" incident, Dr. John Snow and Rev. Henry Whitehead demonstrated the source of infection was a well contaminated by wastewater, which lead to the London sewers being built between 1859 and 1865. Consequently, water based disposal of human wastes fundamentally changed modern civilization from a P recycling society to a P through-put society. P was now mined, used once, then disposed in one pass and "night soil" was no longer returned to the land in most Western cities, as the tonnages and distances became too large, and public health concerns mandated disposal, rather than re-use. The Industrial Revolution accelerated the expansion of cities, and the resulting problems of water born pollution. The most famous incident being "The Big Stink" when British Parliament was suspended in the summer of 1858 due to the stench from the Thames River.

The engineering community responded by developing wastewater treatment processes to treat human waste, while animal waste treatment remained at medieval levels of sophistication. Eutrophication of lakes and rivers then mirrored the development of modern societies through Europe and North America: the Wisconsin lakes – Mendota and Monona in 1882, Lake of Zurich, Switzerland in 1896, Lake Erie in 1930 and Lake Washington in the 1950s, to name a few. Creative engineering once again solved the problem by developing more advanced treatment processes, including tertiary treatment, to remove nutrients from the effluent stream. Initially by chemical treatment, then by the more elegant and sustainable Biological Nutrient Removal process, and the problem was solved – or was it?

© 2009 The Authors, *International Conference on Nutrient Recovery from Wastewater Streams.* Edited by Ken Ashley, Don Mavinic and Fred Koch. ISBN: 9781843392323. Published by IWA Publishing, London, UK.

As predicted in 1798 by Thomas Malthus, and repeated in the early 1970s publications "The Limits to Growth" and "The Population Bomb", concerns were expressed that global population demands would eventually outrun food supply as certain elements were of finite supply on planet earth, and that one day they could be depleted. While modern society has been preoccupied with concerns about international terrorism and climate change, which is an enormous social and environmental problem intimately linked to combustion of fossil fuels, the real show stopper of "Peak Phosphorus" has attracted little attention. However, as eloquently stated by Isaac Asimov:

"We may be able to substitute nuclear power for coal, and plastics for wood, and yeast for meat, and friendliness for isolation – but for phosphorus there is neither substitute nor replacement".

Ominously, food riots were a common scene in many developing countries throughout most of 2008.

This is the new reality, and the impetus for this conference, the 4th in a series of forward looking conferences sponsored by the Global Phosphate Forum and CIWEM, and other gracious supporting agencies. The importance of developing techniques, technologies and processes to capture and recycle phosphorus, and to lesser degrees, nitrogen and other useful molecules (e.g. methane), are critical to the future of mankind. While the global threat of climate change must be addressed, the issue of looming phosphorus shortages, exacerbated by a global population of 6.7 billion humans, 63 billion livestock, and increasing demands from the biofuels sector, has substantially raised the stakes on this issue.

We hope that this conference will stimulate various regulators, operators, politicians and the scientific community to examine the challenges, recognize opportunities and develop solutions to recover nutrients from wastewater streams, or as they should be correctly termed "resource streams". It is therefore appropriate that this 4th international conference on nutrient recovery be held in Vancouver, British Columbia, where the magnificent Pacific salmon, though their selfless life history characteristic of recycling nutrients after spawning, have sustained themselves for millennia in low nutrient streams and rivers. Hopefully, we are intelligent and observant enough to learn from nature and follow their lead towards an environmentally sustainable future.

Ken Ashley
Final Program Chair and Co-Chair
International Technical Scientific Committee
Department of Civil Engineering
University of British Columbia
Vancouver, B.C., Canada

Committees

Local Organizing Committee

Don Mavinic – Conference Chair, UBC, Canada
Ken Ashley – Final Program Chair, UBC, Canada
Chris Thornton – Coordinator, Global Phosphate Forum, France
Joe McHugh – Director, Ostara Inc., Canada
Michelle Gock – Scientific Program Manager, Venue West Ltd., Canada

Final Program Committee

Don Mavinic – Canada
Ken Ashley – Canada
Chris Thornton – France
Peggy Shepard – Canada
Michelle Gock – Canada
Rob Simm – USA

International Technical Scientific Committee

Rob Simm (Chair) – Stantec Consultants Ltd., USA
Ken Ashley (Co-Chair) – UBC, Canada
Mark van Loosdrecht (Co-Chair) – Delft University, The Netherlands
Fred Koch – UBC, Canada
Ahren Britton – Ostara Inc., Canada
Simon Parsons – Cranfield University, United Kingdom
Harlan Kelly – Dayton and Knight Ltd., Canada
Peter Cornell – Institute WAR, Germany
T.D. Evans – Tim Evans Environment, United Kingdom
Norbert Jardin – Ruhrverband, Planungsabteilung, Germany
Bill Oldham – UBC, Canada
Rao Surampalli – EPA, USA
Bernd Heinzmann – Berliner Wasserbetriebe, Germany

Acknowledgements

Sponsored by:

- *Bonneville Power Administration*
- *Dayton and Knight Ltd.*
- *Global Phosphate Forum*
- *Government of Canada and Province of British Columbia*
- *Metro Vancouver*
- *NORAM Engineering and Constructors Ltd.*
- *Ostara Nutrient Recovery Technologies, Inc. (ONRTI)*
- *Stantec Consulting Ltd.*
- *The Chartered Institution of Water and Environmental Management (CIWEM)*
- *The University of British Columbia (UBC)*

Supported by:

- *The United States Environmental Protection Agency (EPA)*
- *The British Columbia Water and Wastewater Association (BCWWA)*
- *The Canadian Society for Civil Engineering (CSCE)*
- *The Ostara Research Foundation (ORF)*
- *Water Environment Federation (WEF)*

Elimination of eutrophication through resource recovery

James L Barnard

Black & Veatch 8400 Ward Parkway, Kansas City MO 64114

Abstract The undiminished growth in the world population and the spread of industrialization to former agricultural societies have put a relentless pressure on resources in terms of not only the supplies of food and fuel but also of the rejected energy that can cause serious pollution of receiving waters. This paper will look at the causes of deterioration of the water environment and how this resource can be recovered while solving many of the pollution problems. Proteins can be recovered from wastewater to augment the food supplies, urine can be separated and used as fertilizer, phosphorus can be recovered and used in fertilizers or incorporated in compost, algae can be grown and harvested for converting to bio-diesel fuel, wastewater biosolids can be turned into organic fertilizers and water can be reclaimed for re-use which would prevent deterioration of receiving water quality.

INTRODUCTION

Over the past 35 years the population of the world has doubled, from roughly 3 billion to about 6 billion. Since land is at a premium in most high-growth countries, most of this growth was in the cities. It is expected that by 2050, the world population will stabilize at around 10 billion. The growth will not be spread uniformly around the globe but will be concentrated in countries that are already disadvantaged. The concentration of people leads to concentration of pollution and the need for treatment of the wastewater generated in the urban areas occupied by this population. As an example, the population of Mexico City now exceeds 30 million, but the wastewater produced there receives virtually no treatment. It is simply used to irrigate land to produce food. Increase in population means increases in the demand for food and for fertilizers to grow this food. The Green Revolution spanning the period from 1967–68 to 1977–78 changed India from a starving nation to one of the world's leading agricultural producers. This change came as a result of harvesting two crops a year from the same land, developing high-yield grains and using more water, more fertilizer, more pesticides, fungicides, and certain other chemicals. The US and Canada produce more than 60% of the surplus food in the world, to make up for the shortfall in other countries. Crop yields in the US have been increased by intensified agriculture, a massive increase in fertilizer application, and installation of under-drainage to leach out the build-up of minerals in the soil.

The leachate containing nitrates and phosphorus was discharged to streams and eventually began to enrich the receiving waters. McCarty (1969) described methods used for reducing the discharges of nitrates to San Francisco Bay. The intensified production of meat led to the use of large feed-lots for raising cattle, which resulted in massive discharges of nutrients, mainly nitrogen and phosphorus, to receiving streams.

Discharges of wastewater effluent containing excess nitrogen and phosphorus can contribute to the growth of algae in receiving water, which results in eutrophication.

What is eutrophication?

The term "eutrophication" comes from the Greek word "eutrophos", meaning well-nourished. It referred to the natural ageing of water bodies through the natural addition of nutrients. However in modern terms it means the enrichment of receiving waters with excess nutrients. It is not unusual to find lakes and rivers which have become rich in the main nutrients such as carbon, silicon, nitrogen, and phosphorus, as a result of erosion or runoff from adjacent soils. Other nutrient sources include drainage and wash down of excess nutrients from applied fertilizers, from agricultural feed lots, and domestic and industrial wastes. In the receiving water, these nutrients support for the growth of phytoplankton (algae) which are the first link in the food chain and, hence, the basis of all aquatic life. In surface waters, particularly in oceans, this "primary production" speeds the diffusion of carbon dioxide from the atmosphere to the oceans, the largest sink for carbon dioxide.

The food chain itself consists of many links; each with complex interactions but, in the simplest terms can be described as follows: phytoplanktons are consumed by zooplankton such as daphnia (water fleas) – the food for many species of small fish. These fish are consumed by larger predatory fish which, together with their prey, are food for birds and mammals and, indeed, man. A healthy and well-nourished water body (river, lake, or sea) sustains a rich and diverse aquatic life with all components of the food chain existing in a dynamic equilibrium of production and consumption. A healthy food chain can often survive large changes in nutrient load or climatic conditions with remarkable resilience without any long-term changes in water quality or species diversity.

However, the pressures of expanding population, urbanization, industrialization, and agricultural intensification in many regions have resulted in a massive increase in the loadings of not just nutrients, but also of untreated or secondary treated sewage into the rivers, lakes, and estuaries. Industrial discharges, pesticides, animal wastes, and countless other pollutants can have a direct and devastating effect on the functioning of the food chain.

The combination of greatly increased nutrient input and a wide range of other, potentially ecotoxic, inorganic and organic products that reach the water can have serious effects on the aquatic ecosystem. While primary production of algae is promoted by the increased nutrient supply, the ability of the zooplankton (usually the most pollution-sensitive organisms in the food chain) to respond to this increased food supply is impaired by the presence of other kinds of pollutants. The result is often that the balance of production and consumption in the food chain is disturbed which, in most cases, leads to algae becoming the dominant form of life in the water.

In the worst case, algae will proliferate in a way that can no longer be controlled at the higher levels in the food chain. This may lead to the decline in the populations of other water plants, particularly the bottom-growing plants which fail to obtain adequate light in the turbid-water column. In the most extreme cases, toxic algal scum may be formed and water may become deoxygenated, which will result in fish kills. There are many lakes and reservoirs where elevated nutrient levels have not caused the water quality problems associated with high algal biomass, while other lakes, with similar nutrient loads exhibit signs of algal domination.

An example of such imbalance can be found in the deteriorating condition of Lake Erie in the early 70's which was of particular concern (Knud-Hanson, 1994). The approximately 20,000 pounds of phosphorus per day being discharge into the lake resulted in an about 2,600 square-mile area of the lake with no oxygen within ten feet from the bottom (Beeton, 1971). As of 1967, mats of attached algae covered Lake Erie's shoreline, and the populations of desirable fish such as whitefish, blue pike, and walleye had either severely declined or disappeared altogether.

This was a great concern at the time and even led to a poem:

> You're glumping the pond where the Humming-Fish hummed
> No more can they hum for their gills are all gummed.
> So I'm sending them off. Oh, their future is dreary
> They'll walk on their fins and get woefully weary
> In search of some water that isn't so smeary
> **I hear things are just as bad up in Lake Erie**
> – The Lorax, by Dr. Seuss

The so-called Gulf Anoxia, is caused by excessive amounts of nitrogen being discharged into the gulf from the Mississippi River, which enhances the growth of algae in the gulf. When the algae die, oxygen is consumed, leading to development of a zone where the dissolved oxygen is too low to support fish and other aquatic life, while causing large swings in the pH value as CO_2 is extracted or returned to the water. Midsummer coastal hypoxia in the northern Gulf of Mexico was first recorded in the 1970s. From 1993 to 1999 the extent of the bottom-water hypoxia covered between 16,000 and 20,000 km^2 and was twice the size of the Chesapeake Bay, rivaling the extensive hypoxia of the Baltic and the Black Sea (Rabelais *et al.*, 2001). This area exceeds the area of the states of Connecticut and Rhode Island.

Much of the initial work on nutrient removal was sparked by the situation around Johannesburg, now known as the Province of Gauteng, South Africa, with close to 10 million people. Because of the discovery of gold, the city is on the continental divide and water is pumped to the city from long distances. In spite of rigid effluent standards, by the early 1970s eutrophication of the reservoirs to the north and south of the city became severe enough to resemble pea soup. Inevitably, as the urban areas grew, recycled wastewater effluent began to constitute an ever higher percentage of the flow to these reservoirs, which in turn supplied water to downstream users. With the salinity of the drinking water reaching a concentration of 800 mg/L during years of drought, the addition of chemicals for the removal of phosphorus was not considered an option. Activated carbon was used to remove tastes and odors at potable water treatment plants.

The concept of a limiting nutrient

Algae, the lowest link in the food chain, need a number of conditions to sustain their growth: sunlight for photosynthesis; an elevated temperature, certain water conditions (turbulence), and nutrients, in particular carbon, nitrogen, and phosphorus which, broadly speaking, are required in the ratio 100:10:1 respectively, as well as a wide range of trace elements. This "primary production" is the foundation of the food chain that sustains all higher life forms: invertebrates, fish, birds, and mammals. If any one of these essential conditions is removed, the primary production ceases and with it, all higher life forms.

In a healthy ecosystem, the ability of the food chain to adapt to variations in nutrient load can be quite remarkable. In its simplest sense, the water body is capable of sustaining a richer and more productive food chain. This can also be achieved without any deterioration in water quality. While the availability of nutrients causes the production of algae to increase, this is balanced by the increase in the consumption of algae by the organisms higher in the food chain that prosper on the increased food supply. Such a situation can often be beneficial by supporting a productive sport or commercial fishery and wildlife.

Since carbon dioxide is freely available from the atmosphere, reducing either nitrogen of phosphorus in discharges to the receiving water to below the values that correspond with the ratio required for growth will limit the growth of the algae. In inland water, phosphorus is mostly the limiting nutrient while in bays and estuaries, nitrogen is predominantly the limiting nutrient, and in some environments both nutrients may be limiting.

Where do the main nutrients come from?

Secondary treatment of wastewater is widely practiced in the USA and Europe, Japan, Australia, and South Africa. Wastewater is treated by exposing it under aerobic conditions to organisms that promote the breakdown of carbonaceous compounds – protein, sugars, soaps, etc. – to carbon dioxide and water. Typically the nitrogen and phosphorus compounds in domestic wastewater are in excess of that required for growth of the organisms in the treatment plant. The excess nutrients are not removed but merely converted from the organic to the inorganic form. The process is depicted on Figure 1. Protein is a combination of many elements but mainly carbon, hydrogen, nitrogen phosphorus, and sulfur. The bound ammonia is converted to ammonia or, when it is further oxidized, to nitrites and nitrate. The phosphorus radical, PO_4^{-3} is discharged with the effluent.

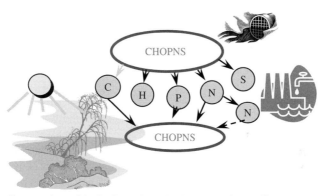

Figure 1. Conversion of typical wastewater to secondary effluent.

Until 1974, phosphorus used in detergents made up about 50% of the phosphorus in wastewater effluent. After a heated battle between environmentalists and the detergent industry, a number of states outlawed the use of phosphorus in detergents, which resulted in the reduction the phosphorus in treated wastewater effluent from around 11 mg/L to between 5 and 7 mg/L. Phosphorus compounds are being added to some water supplies in older communities where lead pipes are still in use in the water supply system to prevent leaching of lead into the water. In a recent article in Environmental Science and Technology (October 1, 2001) a concern was raised that phosphorus compounds may become too expensive and even be limiting for use for this purpose and the community is urged to recycle and re-use phosphorus. The City of Winnipeg adds 1 mg/L P to the water supply for lead control.

The sources of nutrients that are discharged to water bodies vary from one place to another. Figure 2 shows the nutrient sources in the European Union. Even there, the contribution from human sources equals the contribution from livestock. Those two sources, combined with industry, can be considered "point sources", meaning that nutrients are discharged at discrete points as opposed to dispersed runoff from crop fields, lawns, and parks where fertilizers are applied.

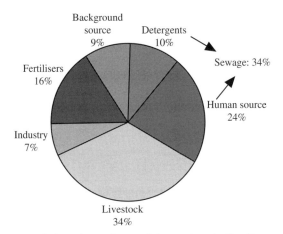

Figure 2. Sources of phosphorus in receiving waters in the European Union.

Nitrogen is more often the limiting nutrient for the growth of algae in bays and estuaries and is the main cause of the anoxic conditions in the Gulf of Mexico. Sources contributing nitrogen to the Gulf anoxia are indicated on Figure 3.

Only 11% of the total nitrogen in the Mississippi comes from municipal and industrial point sources in the catchment of the river system which include cities such as Chicago, St Louis, Minneapolis, Kansas City, Nashville, Memphis, and New Orleans, while 15% originates in animal manure and more than 50% comes from fertilizers and mineralized soil through tile drainage systems from crop fields. (Hey *et al.,* 2005) With the expected increase in crop production for the bio-fuel program, the problem will increase in severity. In the Long Island Sound in which the algae growth is also nitrogen-limited, more than 50% of

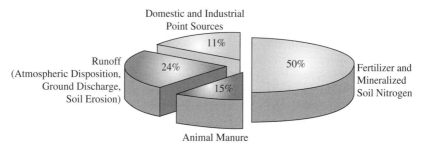

Domestic and Industrial
Point Sources

11%

Runoff
(Atmospheric Disposition,   24%   50%
Ground Discharge,   Fertilizer and
Soil Erosion)   15%   Mineralized
                                                    Soil Nitrogen

Animal Manure

Figure 3. Sources of nitrogen in the Mississippi River.

the nitrogen originates in wastewater treatment plants or City of New York, Westchester County and the state of Connecticut. Many US cities have begun removing nitrogen from municipal wastes, but little effort has gone into tackling the bigger problem of removing the nitrogen from agricultural and other diffused sources.

The Las Vegas metropolitan area discharges all wastewater effluent to the Las Vegas Wash and into the arm of Lake Mead that also serves as the area's water supply. Because of the low rainfall in the area, most of the flow into that arm consists of municipal effluent, and the contribution of point sources can be more than 90% of the total. Thus all nutrients contained in the effluent will make up a very high percentage of the total going into that arm of the lake. At present there is sufficient exchange of flow between the Las Vegas Wash and the main body of Lake Mead but as lake levels are dropping, the exchange will be reduced and the effluent of the treatment plants will make up a larger proportion of the domestic water supply.

How do we deal with this surplus of nutrients?

Multiple strategies will be required to reduce the nutrients that affect water bodies. The first line of attack is the removal of nutrients from point sources such as domestic and industrial wastewater treatment plants.

Phosphorus removal: At domestic or industrial waste treatment plants, phosphorus can be removed either by precipitation with metal salts such as aluminum and iron salts which form an insoluble precipitate of aluminum or ferric phosphate, or by incorporation into biological cells. The solids containing the phosphorus can be removed by settling and disposed of with the treated excess biosolids.

In response to public concern about the eutrophication of Lake Erie, pollution control laws were adopted in both countries to deal with water quality problems, including phosphorus loadings to the lakes. In 1972, Canada and the United States signed the Great Lakes Water Quality Agreement to begin a bi-national Great Lakes cleanup that emphasized the reduction of phosphorus entering the lakes. Iron or aluminum salts were added in the wastewater treatment plants to precipitate phosphorus.

The later development of processes for the biological removal of phosphorus is of particular significance for sustainable practice since it also facilitates recycling and re-use of the phosphorus as discussed further below. All living things take up phosphorus to grow, but certain micro-organisms can store phosphorus very much in excess of their biological needs. To trigger the mechanism by which these organisms take up phosphorus in activated sludge plants, they are continuously recycled from an aerobic to an anaerobic environment where they come in contact with the influent wastewater. While these bacteria are strictly aerobic, they can store phosphorus in the form of high-energy polyphosphate bonds when dissolved oxygen is available, forming a large pool of polyphosphates within the cells. When they are recycled and contacted in the absence of dissolved oxygen or nitrates, with short-chain Volatile Fatty Acids (VFA) which are natural constituents of "older" waste-water, these pools of high-energy phosphorus can supply the energy to take up and store these acids as an intermediate product, breaking the phosphorus bonds and releasing phosphorus. When they are passed back to the aerobic environment, they have a surplus of stored food which they can use to provide energy to take up all the phosphorus they released during the anaerobic phase, plus all the surplus phosphorus in the influent. If there is a shortage of VFA in the influent, primary or secondary solids can be acid fermented and VFA decanted to the anaerobic zone as was first done during the design of the Kelowna WWTP. The Phosphorus Accumulating Organisms (PAO) are very efficient and can take up the soluble phosphorus to levels as low as 0.03 mg/L. When the solids are separated from the liquid, up to 98% of the phosphorus can be removed. This plays an important part in the recovery of phosphorus as discussed below.

On Figure 4, the polyphosphate within the bacteria stains black on the electron-microscope picture. The right-hand picture shows the aerated zone where virtually all the phosphorus has been taken up into the bacterial cells. A portion of the phosphate-rich solids are continuously removed from the bacterial mass, thus removing the phosphorus from the main stream.

Nitrogen removal: Urea, free ammonia, and complex nitrogen compounds are converted to free ammonia compounds in solution by biological wastewater treatment processes. The ammonia is converted to nitrates and nitrites (nitrified) by specialized soil bacteria in the presence of surplus dissolved oxygen and then reduced by another set of organisms to nitrogen gas in the absence

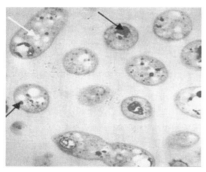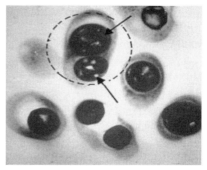

Figure 4. Release and uptake of phosphorus in biological phosphorus removal process.

of oxygen. These "anoxic" bacteria need a supply of organic carbon to grow while they get their energy from the reduction of nitrates to nitrogen gas. Most of my work concerned the discovery and development of processes for simultaneous biological removal of nitrogen and phosphorus using the internal carbon sources.

While there are many treatment plants in the country that remove more than 90% of the nitrogen and more than 95% of the phosphorus, much still needs to be done, and billions of dollars are being spent, to remove nutrients from discharges to sensitive water bodies such as Chesapeake Bay, other estuaries on the East Coast and Florida, the Great Lakes and many inland reservoirs. In some sensitive areas, such as Las Vegas and the water supply reservoirs in New York and Washington, phosphorus in the effluent needs to be removed to less than 0.02 mg/L. Although the power consumption for these processes is high, it is a tiny fraction of one percent of the carbon footprint of the average person, amounting to less than 30 kWh per person per year, but efforts at sustainability demand that we take a new look at the resources in wastewater that are not used beneficially.

Protein recovery: This is an area with huge potential from an environmental view point. De Villiers (2000) demonstrated, at full-scale, that 65% protein recovery was possible at an abattoir treating the waste of 2000 cattle units per day. This was after the normal blood and fat recovery taking place in the industry itself. The value of the protein recovered exceeded $1 million per annum. Such processes are applicable to many food wastes such as beer brewery waste and whey treatment. The Alaska seafood industry uses bio-technology to recover marine oils and proteins for a global marketplace that is in short supply of such products. Both phosphorus and nitrogen will be recovered in this way.

Phosphorus recovery

"*The phosphorus content of our land, following generations of cultivation, has greatly diminished. It needs replenishing. I cannot over-emphasize the importance of phosphorus not only to agriculture and soil conservation but also the physical health and economic security of the people of the nation. Many of our soil deposits are deficient in phosphorus, thus causing low yield and poor quality of crops and pastures ...*" President Franklin D. Roosevelt 1938. Thus the President underscored the importance of phosphorus for food production.

Unlike oil, which is lost once used, phosphates can be recovered and recycled. However, phosphorus cannot be replaced once resources begin to be depleted: there is no substitute. At present, 80% of phosphate mined is lost in fertilizer production, field application, food processing, and does not reach the food we consume. Worldwide, mankind emits 3 million tonnes of phosphorus annually in faeces and urine, with even more in animal manures. Human emissions represent more than 10% of phosphate rock production. Recovery and recycling of phosphorus offer an important opportunity to reduce dependency on mined phosphates and make food production more sustainable.

Although the United States currently produces more than 25% of the world's supply of phosphorus, Morocco has more than 6 times the reserves of the US. A war has already been fought over Morocco's phosphorus deposits. Some of the richest lodes of phosphorus in the US lie under cities such as Jacksonville, FL and will probably never be exploited. Production is limited to a few countries, including Russia, Tunisia, Jordan, Brazil, and South Africa, and to a lesser extent, some other countries. Predictions as to when the reserves will run out vary, but it is generally accepted that in less than 50 years there will be fewer producers, and thus possibilities of severe competition, and that the known reserves may be depleted within around 200 years if nothing is done to recover and recycle phosphorus. The highest-quality deposits are being depleted rapidly. Lower-grade ore will require more energy for processing. The price of fertilizer will increase. This does not accord with the notions of sustainability. The world will eventually be faced with a shortage, and if even at this point the poorer countries cannot afford the price of fertilizers, the future scenario seems bleak.

In a report (1999), the European Environmental Agency indicated that cleaner rivers as a result of phosphorus removal at wastewater treatment plants were one of the few environmental success stories. Almost ten more years have passed and the next ten years will see a considerable increase in the conversion of phosphorus from the liquid to the solid state at treatment plants in the USA, where it will be present in the surplus biosolids produced. At the same time, recycling of phosphorus for agricultural use is reduced, since there is pressure to reduce the application of biosolids on land. As a result of alternative disposal methods, more of the phosphorus is dispersed in the environment and thus lost to future generations.

However, the concept of biological phosphorus removal lends itself to recovery of phosphorus from wastewater as demonstrated on Figure 5. The phosphorus- accumulating organisms concentrate it in their cells in the aerated section of the main plant and release it in the anaerobic section. The solids are separated by sedimentation or more recently, by membranes. The separated surplus solids containing all the phosphorus are then held under anaerobic conditions which cause them to release the phosphorus to solution, and the phosphorus-rich supernatant can be decanted. When treated in an upflow column with lime, the phosphorus precipitates on small calcium phosphate granules in a form that is readily accessible to plant growth. While the excess biosolids are subjected to anaerobic conditions for releasing phosphorus, acid fermentation takes place to form volatile fatty acids which are the sole substrate used by the phosphorus accumulating organisms for growth and which then allows them to take up the phosphorus from the main stream. One could almost compare this with a sponge that soaks up the phosphorus and can then be squeezed to release it as a concentrate.

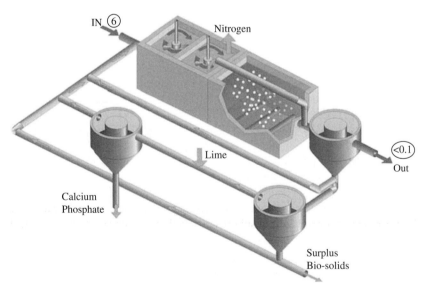

Figure 5. Process for removing nitrogen and recovering phosphorus from wastewater.

Phosphorus recovery from wastewater has enjoyed great interest over the last 20 years. Eggers *et al.* (1991) described experiences with a full-scale Crystal-lactor Process operating at pH between 10 and 10.5 and succeeded in reducing the effluent phosphorus to less than 0.5 mg/L through precipitation of calcium phosphate crystals. Momberg & Oellermann (1992) described laboratory experiments with struvite and hydroxylapatite precipitation. Momberg (1993) precipitated calcium phosphate on fine sand particles in a fluidized bed reactor at the Daspoort WWTP (the same site where biological phosphorus removal in BNR plants was discovered 20 years earlier). Pilot plant developments of struvite recovery at an Osaka City WWTP had the recovered struvite analyzed for content of magnesium, ammonium and phosphate, showing levels very close to those expected in pure struvite. Analysis of heavy metals and other contaminants were below levels fixed by the Fertilizer Control Act in Japan, so that the recovered struvite could be used as a fertilizer or in fertilizer manufacture. (Nakamura 2008).

Recent work at the University of British Columbia has perfected a way of precipitating struvite or magnesium ammonia phosphate from the concentrated stream (Huang *et al.*, 2006). Successful trials at a number treatment plants have demonstrated the viability and economics of this process. Phosphorus can be released and concentrated in a sidestream as shown on Figure 5 or it can be released during anaerobic digestion of the biosolids. Magnesium is taken up in the biological cells during phosphorus uptake and released when phosphorus is released. Ammonia is released in the digestion process and is normally returned to the plant for nitrification and denitrification, consuming energy in the process. With excess ammonia available, magnesium is usually in short supply and more may need to be added for crystal formation. The cost is relatively low since most of the nutrients are present in the side-stream from the wastewater treatment plant.

Processes like these are presently operating in Holland, Germany, and Japan, producing substantial amounts of fertilizer. With the potential to remove 5 mg/L of phosphorus from the wastewater being treated in a 100 mgd plant – a typical size for smaller cities – 50,000 tons of superphosphate fertilizer can be produced annually.

Other methods of recovering phosphorus include simply drying the biosolids containing the phosphorus into pellets that could be applied to land. However, this is energy-intensive. The move to restrict land application of biosolids has led to large-scale incineration, especially in Europe. Incineration does not destroy the phosphorus, and it can be extracted from the residue. Matsuo (1996) was able to extract more than 80% of the phosphorus from incinerated Bio-P biosolids using water at 86°C. Veeken *et al.* (1999) used citric acid to elutriate metal salts and between 80 and 90% of the phosphorus from sludge, making land application possible while limiting the application of phosphorus on the one hand and capturing phosphorus on the other hand.

Composting

An alternative method of recovering phosphorus is to concentrate it in the solids as above, and then compost the solids to destroy pathogens, and to apply the compost to land. The Kelowna Pollution Control Center serves a population of 65,000 produces 27.5 ML/day of wastewater. Placed in service in 1982, the Kelowna treatment plant was the first in Canada and the second in North America to use the new technology of biological nutrient removal which incorporated the phosphorus in the biosolids without chemical addition. The plant produces 13,000 wet tonnes of biosolids each year. The biosolids contain pathogens and require further treatment to make them safe to handle. Composting uses the natural heat generated during aerobic decomposition to kill pathogens, and active decomposition stabilizes the biosolids, transforming them into a valuable, environmentally safe and economically beneficial resource. The phosphorus from the plant wastewater influent is concentrated into the composted material and recycled. This method of the removal of nutrients and incorporating them into compost is widely used in western Canada.

Recovery of nitrogen

The history of nitrogen as a fertilizer is in some ways similar to the future situation facing phosphorus. Before the industrial revolution, animal manure was the main, if not the only, source of both nitrogen and phosphorus as fertilizers. The industrial revolution led to reduced reliance on animals while also contributing to the growth of, especially urban, population. With many European nations near the point of starving, guano, discovered on the islands off the Pacific Ocean coast of South America was imported and used as fertilizer. After the guano resources were exhausted, only limited supplies of crystalline nitrogen were available and became the cause of wars between Chile and Bolivia (Leigh, 2004). In 1798, Malthus wrote his historical thesis, "Essay on the principle of population" on the eventual fate of man due to quadratic growth of the population while food production was increasing only linearly. The development of the Haber-Bosch process in the 19th century for industrial fixation of nitrogen and hydrogen atoms to form ammonia saved the world from nitrogen deficiency. The process requires high pressure and high temperature and metal catalysts using natural gas as feedstock. When natural gas is depleted, a new challenge or a price increase may result, and the process may be switched back to using coal gas as the feedstock. According to Leigh (2004), the secret of how plants and algae fix nitrogen still needs to be discovered. While nitrogen could be "fixed" from the huge reservoir in the atmosphere, there is no such reservoir of phosphorus other than deposits of ore.

Nitrogen can be recovered from wastewater, but the cost of recovery far exceeds that of fixing nitrogen from the atmosphere. Attempts were made more than 30 years ago to remove ammonia and phosphorus by adding lime to the effluent containing ammonia, and raising the pH above 11.5. The phosphorus could be precipitated in the form of calcium phosphate (hydroxy-apatite) and the ammonia was stripped from the liquid in a closed loop reactor, and dissolved in sulfuric acid to form ammonia sulfate which could be recovered. This is method was not economically viable and is no longer practiced. However, as mentioned above, some ammonia can be precipitated with phosphorus in struvite at low cost.

The main problem with nitrogen today is not a shortage but rather an overabundance of it resulting form excessive use of fertilizers as indicated by from the pie-chart on Figure 3, showing the contribution to the anoxia in the Gulf of Mexico. Hey *et al.* (2005) suggested that huge wetlands be created to remove the nitrogen from the streams such as the Illinois River. Barnard *et al.* (2006) proposed a biological method whereby a source of organic carbon is added for denitrification of the nitrates to nitrogen gas in a way similar to the nutrient removal wastewater treatment plants. This would require the consumption of ethanol that produces greenhouse gases.

Urine separation

This revolutionary, yet age-old approach to the control and recovery of nutrients is vigorously promoted by the Stockholm Environment Institution with large exhibits during the Stockholm Water Week. Urine contains approximately 70% of the nitrogen and more than 50% of the phosphorus and potassium in all household wastewater, while its volume is less than 1% of the total daily flow. Using urine-separating toilets, the urine could be "harvested" and used as fertilizer (Kvarnström *et al.,* 2006). The urine is already fairly pathogen-free but when kept for a few days, its pH increases as a result of hydrolyzation of ammonia which then contributes to the further disinfection. Urine diversion is a complementary sanitation technology that has been implemented in several countries. Technical and organizational experience gained in Sweden and elsewhere is now applied widely for meeting especially Goal 7 of the Millennium Development Goals of the WHO. See box below.

A urine-diverting toilet has two outlets and two collection systems: one for urine and one for the faeces. Other than that, the system consists only of conventional technical construction materials/devices, even though they might be used in completely or partly new ways. The urine-diverting toilets can be either water-flushed or dry depending on cultural and economic differences. There are ways of achieving urine diversion in both rural settings and urban areas. Research and experience show that the systems function in all these different settings, provided that they are properly installed, operated, and maintained. Urine diversion in itself should be seen as a complementary technology, since the other wastewater flows (faeces fraction, grey water, and stormwater) also need to be handled and treated. The faecal fraction will, due to its possibly high content of pathogens constitute the main hygienic risk. Effective treatment to reduce the pathogen content, combined with safe handling procedures, is of importance to manage health risks.

A new-up scale development near Stockholm has installed urine separating toilets with a two-pipe system. The urine is collected in strategically placed basins for use by farmers. The toilets were so designed that during flushing a small quantity of water spills over to wash the urine side without substantially increasing the volume. During a visit to such a housing development, during the Stockholm Water Week in 2005, the owners had no complaints, except that it was necessary for the men to sit down. There is obviously some difficulty in persuading the farmers to use this as fertilizer when the cost of fertilizer is still very low, but some progress is being made. (Kvarnstöm et al., 2006) Source-separated human urine is a complete, well balanced fertilizer with its nutrients readily available to plants. The nitrogen effect was found to be close to that of chemical fertilizer ($\sim$90%). It varied between 70% and more than 100% between different years. The phosphorus effect was equal to that of chemical fertilizer. (Jönsson, 2001; Johansson et al., 2001)

In the developing countries, where a low percentage of people have access to water, not to mention toilets, the concept is more appealing. This is especially so in light of the cost of fertilizers which is mostly beyond reach for the ordinary people. The results of demonstration projects about increasing crop production for people living in poverty are astounding and have created great interest in using this technology to increase food production on small plots. According to Kvarnström et al. (2006), urban agriculture in Kampala, Uganda, supplies the city with a substantial percentage of its food. An example of such a garden is shown on Figure 6.

One of the main unknowns associated with the use of diluted urine as fertilizer is the fate of the pharmaceuticals and endocrine disruptors. The consensus is that they should have less impact when applied to land than to

Figure 6. Urban agriculture in Kampala, Uganda. Photo: Margaret Azuba.

water because of the much longer time available for their breakdown by natural decay than provided in a wastewater treatment plant.

Studies in Switzerland looked at using separated urine for struvite recovery.

A significant source of nutrients in the Chesapeake Bay catchment is the contribution from the septic tanks and tile fields of households that are not connected to the municipal sewer systems. The separation and application of urine as fertilizer needs to be promoted in these areas.

A substantial volume of knowledge is available on this subject, as well as a great willingness by the population to protect water bodies and reduce the carbon footprint. A good start would be to install urine-separating toilets in housing developments around golf courses to supply fertilizer for the fairways. Swedish studies have shown that when the material is applied correctly, it could be quite acceptable.

Use of algae as bio-fuels

According to the statistics of U.S. Department of Energy, 60 billion gallons of petroleum-based diesel fuel and 120 billion gallons of gasoline are consumed annually in the United States. The economic strain on the US resulting from the $100–150 billion spent every year to buy oil from other nations makes the development of alternatives to oil one of the highest priorities. In the United States, roughly two-thirds of all oil is used primarily for transportation. Developing an alternative means of powering our cars, trucks, and buses would go a long way towards weaning us, and the rest of the world, off oil. While hydrogen is receiving a lot of attention in the media as fuel for automobiles, the best alternative at present, according to Briggs (2004), is bio-diesel that can be used without modifications in existing diesel engines. Bio-diesel can be produced from vegetable oils or animal fats, or even biosolids from wastewater treatment plants. Micro-algae offer the best option for producing bio-diesel in quantities sufficient to replace petroleum. While traditional crops can yields around 50–150 gallons of bio-diesel per acre per year, Briggs (2004) estimates that algae can yield 5,000–20,000 gallons per acre per year. Nutrients derived from waste streams, drainage fields, and human and animal wastes are well-balanced for optimal production of algae.

The Office of Fuels Development, a division of the U.S. Department of Energy, funded during the period of 1978 through 1996 the National Renewable Energy Laboratory program known as the "Aquatic Species Program" (Sheehan *et al.,* 1998). The focus of this program was to investigate high-oil algae that could be grown specifically for the purpose of reducing the emissions of carbon dioxide (CO_2) from the stacks of power generation plants. Noticing that some algae have very high oil content, the project shifted its focus to growing algae for another purpose – the production of bio-diesel fuel. Some species of algae are ideally suited for bio-diesel production because of their high oil content (up to well over 50%), and extremely fast growth rates. The research discovered over 300 species of algae that have a suitable oil content. One of the advantages of biofuel is that it fixes the CO_2 from the air, and could even be used to remove it from the stack emissions.

Algae can be grown in shallow ponds, with perhaps some circulation to ensure uptake of CO_2 from the air. There are more than 10,000 pond systems in the US used for treating domestic and industrial wastes. The algae, growing on the nutrients present in the wastewater, supply the oxygen to enable other bacteria to break down the wastewater compounds. Middlebrooks *et al.* (1974) described several ways of harvesting algae from the ponds, such as centrifugation, microscreening, coagulation and sedimentation, the dissolved aeration process, skimming off the algae, or some other innovative method such as cross-flow screening. Durand-Chastel (1980) described the long-existing method of cultivating Spirulina in Mexico as a food source with great health benefits. He described the culture of the alga *Dunaliella* as a potential raw material for glycerol in semi-saline waters such as, for example, around the Dead Sea in Israel, and predicted that by the year 2020, 30% of the petro-chemicals will be obtained from agricultural products. Pretorius *et al.* (1984) described a method of selecting desirable algae for cultivation. He used a mechanically circulated looped pond with a retention time of 10 days. For each unit volume of liquid added to the pond, a similar volume is removed and passed through a 200 µm cross-flow screen. Algae retained on the screen are recycled into the pond while those that pass through the screen are washed out. Eventually, through this selection process, only the algae that can be retained on the screens will grow in the ponds. Pretorius retained a culture of *Stigeoclonium* in the pond. The surplus production could be removed mechanically and compressed to 20% solids content. It is not clear whether this strain of algae had a high oil content, since the work was not related to bio-fuel production but it demonstrates how algae could be grown and harvested. Pretorius also developed a system for growing high-rate activated sludge biosolids that could be harvested in a similar way to produce protein for animal fodder.

Growing biomass on the scale required for the production of bio-fuels will entail massive use of plant nutrients, so why not look at nutrients that are presently waste products that must be removed from many rivers and streams? The bar chart on Figure 7 shows the increase in nitrogen and phosphorus in the Illinois River at Valley City. The Illinois River contributes to the anoxia problems in the Gulf of Mexico. The USGS National Water Quality Assessment Program (NAWQA) of the highly-agricultural Lower Illinois River Basin (LIRB) concluded that nitrate concentrations in this basin were among the highest in the country. Agriculture is the predominant land use in the area – typically corn and soybean row crops. It accounts for the use of 88% of the overall land, whereas forests account for 7% and urban areas about 2%. The remaining land, about 3%, is mostly grassland, wetlands, and water (Groschen, *et al.,* 2000). The annual discharge of nitrogen in the Illinois River, mostly in the form of nitrate nitrogen, was estimated at around 120,000 tonnes, and the discharge of phosphorus at around 10,000 tonnes. Hey *et al.* (2005) suggested that farmers in the Mississippi flood plain be paid for turning their farms into wetlands to reduce the discharges of nitrates. Barnard & Andrews (2006) suggested that a small-footprint biological process with ethanol as feed be used as a complementary process to reduce nitrates. However, this process would produce, rather than reduce, CO_2.

It can be expected that every piece of land in the areas now contributing nutrients to the Gulf Anoxia will be planted with crops for producing ethanol, thus increasing rather that decreasing the discharge of nutrients, in an effort to produce the biofuel to replace the petroleum-based fuel. It is suggested that instead, most of the farms in the lower flood plains be converted to ponds for utilizing the nutrients already contained in the main rivers to propagate algae for conversion to bio-diesel. Processing the algae will release a large portion of the nutrients which could be used again to grow more algae, making this a very

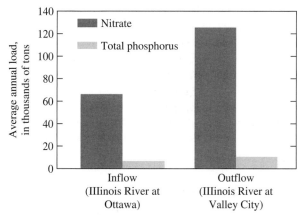

Figure 7. The loads for the upper Illinois River Basin upstream from Ottawa – are about one-half of the respective loads for the lower Illinois River Basin.

sustainable process for producing fuel. With all these preconditions, the cost of producing the algae will be very low, since all the energy comes from the sun.

If, in addition, the runoff from farms and feedlots, which contributes more nitrogen to the Gulf than all the urban areas, could be collected and treated through bacterial breakdown and algae production in scientifically designed pond systems, a secondary industry for fuel production could eliminate another large source of nutrients to the Gulf. In existing pond systems for treatment of domestic wastewater, the algae-rich biosolids that are captured need further treatment. In some instances the removed algae are returned to the anaerobic section of the pond system to undergo anaerobic digestion with the release of methane, a highly active greenhouse gas. Meiring (1992) and Meiring *et al.* (1995) proposed an elemental system for upgrading of algae-rich ponds in the Petro process which in its simplest form consists of a series of ponds, the first of which is anaerobic, allowing the sedimentation and digestion of solid organic matter, and followed by more ponds for abundant growth of algae. The ponds are followed by a bed or rocks or plastic media with sprinklers to distribute pond effluent over the media. About 20% of the flow from the anaerobic pond is then mixed with the algae pond effluent. The readily biodegradable substance in the anaerobic pond effluent provides a bacterial layer on the filter media which traps the algae in the effluent very efficiently. The attached algae change their energy supply from phototrophic to chemotrophic and help in the purification process. The excess biosolids mixture sloughs off from the media and can be settled and used for the production of bio-fuels.

The method proposed by Meiring (1992) and Meiring *et al.* (1995) was used to upgrade the large pond treatment system on the Werribee Farm, treating 480,000 m³/d of the wastewater from the city of Melbourne, Australia. In this case, the captured algae were returned to covered anaerobic ponds where they were fermented to produce methane gas which in turn was used for power generation. During fermentation, the nitrogen is again released to the liquid phase where it stimulates algae growth in the following ponds. The follow-up activated sludge process includes a step for conversion of nitrogen in the effluent to nitrogen gas while capturing the solids. "As a result of the upgrade, a large supply of high-quality recycled water can be used for agricultural, horticultural, and other applications. The upgrade is also opening up opportunities for renewable energy generation and new agricultural land uses at the plant. The Victorian Coastal Award for Excellence 2006 in Water Quality was awarded to Melbourne Water in recognition of the environmental improvements achieved through the upgrade. The award recognizes excellence in developing, adopting, or implementing practices that have improved water quality in the marine environment" (Melbourne Water Web-site).

Concluding remarks

We need new visions, we need paradigm shifts, and we need young professionals at a time when we see an alarming decline in the number of graduates at our universities. In order to improve or even just to maintain the quality of receiving reservoirs, estuaries and bays, we need massive investments in sustainable solutions. Resource recovery and water reclamation to hitherto unthinkable proportions must be considered.

REFERENCES

Barnard, J.L. and Andrews, H.O. (2006). An innovative approach to nitrate removal: making a dent in the Gulf of Mexico hypoxia. Proceedings WEFTEC 2006.

Beeton, A.M. 1971. Eutrophication of the St. Lawrence Great Lakes. In: *Man's Impact on Environment*, T.R. Detwyler (ed.), McGraw-Hill Book Co., New York, pp. 233–245.

Briggs (2004). http://www.unh.edu/p2/biodiesel/article_alge.html

de Villiers G.H. (2000). Abattoir effluent treatment and protein production: Full-scale application, Water SA, Vol. 26, No. 4, Oct. 2000, 559.

Durand-Chastel, H. (1980). Production and use of Spirulina in Mexico. In: *Algae Biomass Production and Use*. G. Shelef and C.J. Soeder (eds.), Elsevier/North-Holland Biomedical Press, 51–64.

Eggers E., Dirkzwager A.H. and van der Honing H. (1991). Full-scale experience with phosphate crystallization in a Crystallactor. *Water Science & Technology*, **23**, 819–824.

Groschen, G.E., Harris, M.A., King, R.B., Terrio, P.J. and Warner, K.L. (2000). Water Quality in the Lower Illinois River Basin, Illinois, 1995–98, U.S. Geological Survey, 2000.

Hey, D.L., Kostel, J.A., Hunter, A.P. and Kadlec, R.H. (2005). Metropolitan Water Reclamation District of Greater Chicago. Nutrient Farming and Traditional Removal: An Economic Comparison, WERF Report 03-WSM-6CO. Water Environment Research Foundation, Alexandria, Virginia, 2005.

Huang, H., Mavinic, D.S., Lo, K.V. and Koch, F.A. (2006) Production and Basic Morphology of Struvite Crystals from a Pilot-Scale Crystallization Process Environmental Technology, Vol. 27, No. 3, Mar. 2006, pp. 233–245(13).

Johansson, M., Jönsson, H., Höglund, C., Richert Stintzing, A. and Rodhe, L. (2001). Urine separation – closing the nutrient cycle (English version of report originally published in Swedish). Stockholm Water Company. Stockholm., Sweden.

Jönsson, H. (2001). Urine separation – Swedish experiences. EcoEng Newsletter 1, Oct. 2001 (http://www.iees.ch/EcoEng011).

Knud-Hansen, C. (1994). Congressional Report HR 91-1004. Apr. 14, 1970. "Phosphates in Detergents and the Eutrophication of America's Waters" Committee on Government Operations. Conflict Research Consortium. Working Paper 94–54, Feb. 1994 (www.colorado.edu/conflict).

Kvarnström, E., Emilsson, K., Stintzing, A.R., Johansson, M., Jönsson, H., af Petersens, E. Schönning, C. Christensen, J., Hellström, D., Qvarnström, L., Ridderstolpe, P. and Drangert, J. (2006). Urine Diversion: One step towards sustainable sanitation. Report 2006-1, Stockholm Environment Institute. Downloadable from www.ecosanres.org

Leigh, G.J. (2004). The world's greatest fix – A history of nitrogen and agriculture, Oxford University Press ISBN 0-19-516582-9.

Matsuo, Y. (1996). Release of phosphorus from ash produced by incinerating waste activated sludge from enhanced biological phosphorus removal. *Wat. Sci. Tech.*, **34**(1–2), 407–415.

McCarty, P.L., Beck, L. and Amant, P.S. (1969). Biological denitrification of wastewaters by addition of organic materials. Proceedings of the 24th Purdue Ind. Waste Conf., Purdue University, Lafayette, Indiana.

Meiring, P.G.J. (1992). Introducing the PETRO process. Proceedings 3rd SA Anaerobic Digestion Symposium. 13–16 July 1992. Pietermaritzburg, South Africa.

Meiring, P.G.J. and Oellermann, R.K. (1995). Biological removal of algae in an integrated pond system. *Wat. Sci. and Tech.*, **31**(12), 21–31.

Middlebrooks, E.J., Porcella, D.B., Gearheart, R.A., Marshall, G.R., Reynolds, J.H. and Grenney, W.J. (1974). Techniques for algae removal from wastewater stabilization ponds. *J. Wat. Pollut. Control Fed.*, **46**(12), 2676–2695.

Momberg, G.A. and Oellermann, R.A. (1992). The removal of phosphorus by hydroxyl-apatite and struvite crystallization in South Africa. *Water Science and Technology*, **26**, 987–996.

Momberg G.A. (1993). Phosphate crystallization in activated sludge systems. WRC Report 215/1/93 available from Librarian Water Research Commission. P.O. Box 824, Pretoria South Africa.

Nakamura, T. (2008). Experiment on phosphorus recovery from digested sludge using struvite crystyllization method. Unitaka Ltd Japan (quoted from Scope Newsletter Feb. 2008).

Pretorius, W.A. and Hensman, L.C. (1984). The selective cultivation of easily harvestable Algae using crossflow-microscreening. *Wat. Sci. Tech.*, **17**, 791–802.

Rabelais, N.N., Turner, E.E. and Wiseman, Jr. W.J. (2001). Hypoxia in the Gulf of Mexico. *Environ. J. Qual.*, **30**, 320–329. 7. RTI (Research Triangle Institute).

Sheehan, J., Dunahay, T., Benemann, J. and Roessler, P. (1998). A Look Back at the U.S. Department of Energy's Aquatic Species Program—Biodiesel from Algae. U.S. Department of Energy's Office of Fuels Development Prepared by: the National Renewable Energy Laboratory 1617 Cole Boulevard Golden, Colorado 80401-3393

A National laboratory of the U.S. Department of Energy Operated by Midwest Research Institute under Contract No. DE-AC36-83CH10093 (http://www1.eere.energy.gov/biomass/pdfs/biodiesel_from_algae.pdf).

Veeken, A. and Hamelers, H. (1999). Removal of heavy metals from sewage sludge by extraction with organic acids. *Wat. Sci. Tech.*, **40**(1), 129–136.

Preferred future phosphorus scenarios: A framework for meeting long-term phosphorus needs for global food demand

Dana Cordell[*,a,b], Dr Tina Schmid-Neset[a], Dr Stuart White[b] and Dr Jan-Olof Drangert[a]

[a]Department of Water and Environmental Studies, Linköping University, Sweden
[b]Institute for Sustainable Futures, University of Technology, Sydney, Australia

[*]Corresponding author: Dana.Cordell@uts.edu.au

Abstract This paper puts phosphorus recovery in a global sustainability context, with particular reference to future phosphate rock scarcity and global food security. While phosphorus fertilizers are essential for sustaining high crop yields, all modern agricultural systems currently rely on constant input of mined phosphate rock. However, phosphate rock, like oil, is a finite resource, and global production of high quality phosphate rock is estimated to peak by 2033, after which demand for phosphorus fertilizers will increasingly exceed supply. Phosphorus cannot be manufactured; though fortunately there are a number of technologies and practices that together could potentially meet long-term future phosphate fertilizer needs for global food demand. This paper develops probable, possible and preferred long-term scenarios for supply and demand-side measures. The preferred scenarios together demonstrate how substantial reduction in demand for phosphorus can be achieved, and how the remaining demand can be met through high recovery and reuse of organic sources like human and animal excreta (e.g. direct reuse, struvite crystals), crop residues, food waste and 'new' sources like seaweed, ash, bonemeal and some phosphate rock.

INTRODUCTION

Phosphorus, together with nitrogen and potassium, is an essential nutrient in fertilizers. As an element, phosphorus has no substitutes for plant and animal growth. Hence there will always be a global demand for phosphorus to produce fertilizers to in turn feed a growing world population. While historically agriculture relied on organic sources of phosphorus, the past century has witnessed a dramatic increase in the dependence on phosphate rock to achieve high crop yields. Yet environmental, economic and geopolitical concerns could mean the end of an era for phosphate rock this century. The fertilizer industry acknowledges that cheap fertilizers are a thing of the past and it is widely accepted that quality of remaining reserves is decreasing

© 2009 The Authors, *International Conference on Nutrient Recovery from Wastewater Streams.* Edited by Ken Ashley, Don Mavinic and Fred Koch. ISBN: 9781843392323. Published by IWA Publishing, London, UK.

(Stewart *et al.*, 2005; IFA, 2006; Smil, 2002). This paper presents an integrated approach to how we use and source phosphorus that looks beyond the current focus on agricultural efficiency (largely driven by concern of phosphate pollution in rivers and lakes). It addresses inefficiencies in the food commodity chain, consumer demand for phosphorus-intensive diets, and reuse opportunities from organic sources.

METHODOLOGY: SCENARIO DEVELOPMENT IN THE FACE OF UNCERTAINTY

This paper presents qualitative and quantitative scenarios for meeting global phosphorus fertilizer demand, while accounting for substantial uncertainty about future food demand, lack of consensus about the key issues and limited data regarding the availability of phosphate rock. The purposes of the scenarios are, firstly, to allow consistent analysis of a disparate group of options within a single framework (there are currently numerous options under investigation by different groups). Secondly, to trigger debate among scientists and policy-makers about preferred phosphorus futures, alternative pathways and what is feasible. Finally, to support future decision-making.

Forecasting and backcasting approaches are combined to provide three scenarios of the future. While forecasting projects a present point into the future, backcasting[i] works backwards from a specified preferred future to the present (Dreborg, 1996). The preferred future in this case is based on global food security (no under- or over-nutrition), since this is considered the greatest global significance of phosphorus resources for humanity, in addition to optimal soil fertility and minimum environmental impacts (Cordell, 2008).

A 'probable' scenario considers '*where are we heading?*' by forecasting business-as-usual. A 'possible' scenario considers '*where could we go?*' by backcasting from a maximum achievable scenario while a 'preferred' scenario considers '*where do we want to go?*' (Gidley *et al.*, 2004) by backcasting from a desired future situation taking into consideration what is possible and the criteria in box 1.

Best available data[ii] and knowledge are used to determine future supply and demand side measures for meeting phosphorus needs, presented in million tonnes of phosphorus in 2005, 2050 and 2100 and supported by a qualitative 'storyline'. Demand-side measures include changing diets, food chain efficiency and agricultural efficiency, while the supply-side measures consider sourcing phosphorus from mined rock, manure, human excreta, food waste and crop residues. All scenarios use UN medium world population estimates of 6.1 billion in 2005, 9.19 billion in 2050 and 9.1 billion in 2100 (UN, 2007; UN, 2003). Combining qualitative and quantitative scenarios has been used extensively in long-term global studies on energy, climate, water and global change (Royal Dutch Shell, 2008; Mitchell and White, 2003; Netherlands Environmental Assessment Agency, 2006; Pacala and Socolow, 2004).

The preferred scenarios are aggregated together in an iterative manner, taking into account potential physical quantities of phosphorus and other important environmental, economic, institutional and social criteria (Box 1).

Box 1: 10 *PROVISIONAL* CRITERIA FOR PHOSPHORUS SUSTAINABILITY IN THE CONTEXT OF GLOBAL FOOD SECURITY

1. Availability in the **long-term** (50–100 years).

2. **Equitably** distributed, **accessible** and **affordable** to all farmers – either fertilizer markets are accessible, or access to non-market fertilizers such as manure and excreta; more locally and renewable sources.

3. **Cost-effective** from a whole-of-society perspective (i.e. Not just from a single stakeholder perspective).

4. Sufficient quantity and quality (i.e. Future **demand** can be met by **supply**).

5. **Minimises adverse environment impacts**, including at all key life-cycle phases (e.g. cadmium levels and radium-phosphogypsum management at the mine, energy intensity of production and transport, prioritise renewable rather than non-renewable sources where possible, minimises losses to waterways where eutrophication is a problem).

6. **Minimises losses** in the entire food production and consumption system.

7. **Ethical** – not supporting and trading with a country illegally occupying regions with phosphate reserves.

8. Potential synergies and/or **value-adding** to other systems (e.g. Water, energy, sanitation, poverty reduction, environmental health).

9. Independent **monitoring** of phosphorous resources and future trends, data and analysis **transparent** and publicly available.

10. System has **adaptive capacity** to adapt in a timely manner to changes, to ensure annual availability.

The data points for the present (2005) are sourced from the authors' flows analysis of phosphorus through the global food production and consumption system (Cordell *et al.*, in press), with estimates of the major flows in mining, fertilizer application, harvest, food processing, consumption and excretion (and phosphorus losses and recovery from the chain at each of these stages).

AN INTEGRATED FRAMEWORK FOR ANALYSING FUTURE SUPPLY AND DEMAND OPTIONS

There are few integrated analyses or frameworks covering the spectrum of options available to meet future global phosphorus demand in a sustainable way[iii]. Figures 1 and 2 provide an integrated classification of supply and demand-side measures which are then analysed in the subsequent sections.

Supply measures range from *used* sources to *new* sources and vary in terms of the *process* by which they are sourced, treated and applied in agriculture.

SOURCE:	PROCESS:					
	i. source separation, composting & reuse	*ii.* wastewater mixing & reuse	*iii.* recovery & reuse of byproducts/ residuals	*iv.* struvite generation & reuse	*v.* virgin extraction, processing & use	*vi.* incineration/ burning & reuse
type A: USED SOURCES A1. Human excreta	*e.g. urine (storage and direct reuse), composted dry faeces*	*e.g. direct use of diluted wastewater; use of treated effluent as irrigation water*	*e.g. activated sewage sludge from wastewater treatment plant; sludge from biogas/biofuel digester; composted filter cake from sugar factories*	*e.g. from mixed wastewater at the treatment plant*		*e.g. incinerating toilet*
A2. Greywater	*e.g. minimum treatment and non-portable reuse*					
A3. Animal manure	*e.g. direct application of manure*			*e.g. from dairy waste*		
A4. Other industrial waste						
A5. Animal meal	*e.g. ground bonemeal, meatmeal, bloodmeal*		*e.g. ground bonemeal, meatmeal, bloodmeal*			
A6. Food waste	*e.g. composted food waste*	*e.g. ground in-sink-orator*	*e.g. composted residues from food processing.*			
A7. Crop residues	*e.g. crop residues ploughed back in to field*		*e.g. oil cakes and press mud from oil and sugar industry.*			
type B: NEW SOURCES A8. Crops	*e.g. Green manure*		*e.g. sludge from anaerobic digestion of virgin crops*			*e.g. slash and burn*
A9. Phosphate rock			*e.g. extracting P from phosphogypsum stockpiles*		*e.g. mining existing and potential reserves; seabed phosphate*	
A10. Aquatic vegetation, sediments and seawater					*e.g. seaweeds as liquid fertilizer, algae, seawater*	

Figure 1. A classification matrix of supply-side measures to meet future global phosphorus needs for food security.

DEMAND MEASURES:	STAGE IN FOOD PRODUCTION & CONSUMPTION PROCESS:			
	i. Extracting and processing raw fertilizer materials	*ii.* agricultural fertilizer use and crop growth	*iii.* food processing and retailing	*iv.* food consumption
A1. Minimising P losses	*e.g. reducing storage and distribution losses of phosphate rock and other sources*	*e.g. reducing erosion and runoff*	*e.g. minimise crop and food losses from food processing, storage and transport; minimise losses from retailers (e.g. supermarket dumping of expired, excess or unpresentable food)*	*e.g. Minimising household food waste*
A2. Reducing P demand		*e.g. Soil testing, precision agriculture (e.g. using remote sensing)*		*e.g. Reducing demand for P-intensive diets like meat and dairy; reducing over-eating*
A3. Increasing P uptake	*e.g. Breading, selection and manipulation of seeds and livestock to improve P uptake*	*e.g. Biostimulants (inoculants) at root zone to increase soil P availability; optimising soil biochemistry (pH, carbon, moisture); organic farming techniques, permaculture*		*e.g. Ensuring consumers have healthy bodies (e.g. free of diseases like diarrhoea) to maximise P uptake*

Figure 2. A classification matrix of demand-side measures to meet future global phosphorus needs for food security.

Demand measures include minimising phosphorus losses in each stage of the food production and consumption process, reducing the overall demand for phosphorus, and increasing phosphorus uptake by plants, humans and animals.

FUTURE GLOBAL PHOSPHORUS DEMAND: A MOVING TARGET

There are many interlinked and dynamic variables affecting the overall demand for phosphorus, making informed long-term forecasts challenging (FAO, 2000). Phosphorus fertilizer demand is strongly linked with population and crop demand, in addition to price of food and non-food commodities, consumer preferences for meat and dairy diets, farmer' soil knowledge, price of phosphate fertilizer commodities, farmer purchasing power and demand for alternative sources of phosphorus (Cordell, forthcoming). Approximately 90% of the global demand for phosphorus is for fertilizers, animal feed and food additives (Smil, 2000). However the future fraction of fertilizers that is for food versus non-food crops is difficult to estimate, given the recent dramatic increase in biofuel crop production.

Steen's (1998) 'most likely' 2% annual growth estimate for phosphate consumption (estimated a decade ago) did not foresee the increased demand for biofuel crops that require fertilizer application or growing demand for meat and dairy foods. The most recent official forecasts suggest a growing annual demand at around 3% until 2010–12 (FAO, 2007a; Heffer and Prud'homme, 2008). In the short-term, biofuel demand, lack of investment and possible investor speculation, resulted in the price of phosphate rock increasing 700% within a 14-month period (IFA, 2008).

While fertilizer demand has stabilised or is decreasing in the developed world because of previous overfertilisation (FAO, 2006; European Fertilizer Manufacturers Association, 2000), demand continues to increase sharply in developing and emerging economies (especially China and India), and Africa in particular will need decades of high application before its agricultural soils reach the so-called critical soil P point. Koning *et al.* (2008) suggest approximately 13 times current global P consumption would be required to reach the critical soil P point in the world's phosphorus-poor agricultural soils.

If demand for meat/dairy and biofuels continues to increase, along with population growth, total phosphorus demand could grow annually by 3% until 2050 (99 MT P) and by 4% until 2100 (707 MT P). However, if every person in the world received precisely the required amount of nutrients, i.e. the Recommended Daily Intake (RDI) of 1.2g P/person/day (European Fertilizer Manufacturers Association, 2000) and system losses are minimised to 20%, this would equate to a demand of just 4.79 MT P in 2050 and 4.78 MT P in 2100.

Our 'probable' scenario estimates demand will rise by 2% annually until 2050 (64 MT P) due to increases in population, meat and dairy demand, biofuel crops, fertilizer application on P-deficient soils to attain critical P levels (and a modest level of agricultural efficiency). It is probable that the annual growth rate will reduce to 0.5% until 2100 (82 MT P) due to a stabilisation of world population, soils reaching critical P levels (hence only what is lost in harvest needs replacing) while increased per capita demand remains due to demand for biofuel crops and meat and dairy demand. This probable scenario provides the baseline demand in Figure 11.

FUTURE PHOSPHORUS DEMAND REDUCTION MEASURES

The current food production and consumption system is extremely inefficient: while globally we mine 15 million tonnes of P annually for food production, 80% of this never reaches the food on our dinner table (Cordell *et al.*, in press). Much of the lost phosphorus can be physically recovered and reused, however preventing losses first is typically more energy and economically efficient.

Substantial opportunities exist for not only increasing the efficiency of the food production and consumption system, but to also rethink the way we use of phosphorus to achieve nutritional security for the global population. This broader view (conceptualised in the demand matrix in Figure 2) facilitates the exploration of options that reduce the overall demand for phosphorus.

Changing diets

Changing diets in this context refers to influencing consumer preferences away from phosphorus-intensive diets or overeating (there are currently more overweight people than undernourished in the world (Lundqvist *et al.*, 2008; WHO, 2006b)). Meat and dairy-based diets require up to three times as much phosphorus as a vegetarian diet, in addition to requiring substantially more water, energy and nitrogen (Fraiture, 2007; Cordell *et al.*, in press). FAO predicts a doubling of global meat, dairy and fish consumption by 2050 (particularly in the rapidly developing world) (WWF, 2004). A deliberate reversal of the current trend (in addition to a reduction of the already high meat demand in the developed world) could therefore drastically reduce the global demand for phosphorus in addition to other environmental impacts. Smil (2007) calls for a 'smart vegetarian' diet that prioritises lower P-intensive foods, while the Food Ethics Council is encouraging supermarkets to reduce consumer choice and availability of environmentally damaging goods (including meat products) (Food Ethics Council, 2008). Figure 3 provides 3 scenarios.

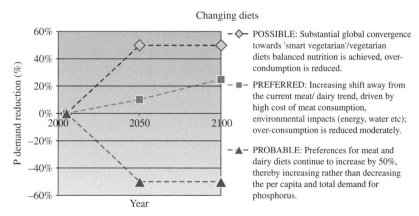

Figure 3. Scenarios for reducing phosphorus demand through changing diets.

Food chain efficiency

Food chain efficiency involves reducing food losses throughout the food chain, indicated in the scenarios in Figure 4. The globalised food commodity chain has resulted in more players, more processes, further distances and increased trade of commodities. Longer production chains contribute to more food losses in transport, production, storage and retail (Lundqvist *et al.*, 2008). In developed countries, the percentage of food lost from farm to fork can be as much as 50%, while in developing countries this is significantly less (IWMI, 2006). In Britain alone, householders throw out £10 billion worth of food (equal to a third of the food that is purchased). Approximately 60% of this is unused edible food and hence avoidable waste (WRAP, 2008).

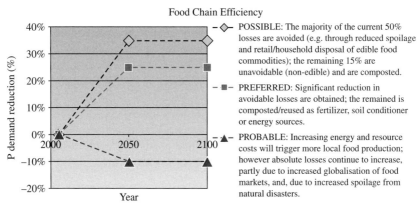

Figure 4. Scenarios for reducing phosphorus demand through increasing food chain efficiency.

Producing food closer to the point of demand – mostly from cities – would reduce food waste as well as energy, water and other resources. Ongoing urban and peri-urban agriculture (e.g. growing food on roofs, gardens, public spaces, agroparks) (FAO, 1999) are examples of more sustainable food chains.

Agricultural efficiency

Agricultural efficiency in this paper refers to increasing crop yields per unit input of phosphorus. The current level of efficiency varies widely between regions, and typically only 15–30% of the applied fertilizer P reaches the crop[iv] (FAO, 2006), hence there is still potential for new innovations to increase crop phosphorus use efficiency. For over a decade, the FAO and fertilizer industry have called for more integrated nutrient management (INM), that ensures crop productivity through optimising soil fertility and meeting nutrient needs from a range of organic and inorganic sources (FAO, 2006; FAO, 2008; IFA, 2007). Efficiency measures include: appropriate timing and application rate of fertilizers (e.g. precision agriculture); improving chemical and physical properties of soil (e.g. pH, moisture, aeration, root-penetrability); and addition of microbial inoculants (e.g. Mycorrhizae fungi) to the root zone to increase the uptake of nutrients (FAO, 2006). Organic farming, permaculture and conservation farming all aim to minimise nutrient losses and to close on-farm nutrient cycles, thereby requiring zero or minimal external fertilizer inputs (FAO, 2007b; Holmgren, 2003; García-Torres *et al.*, 2003). Figure 5 provides 3 scenarios.

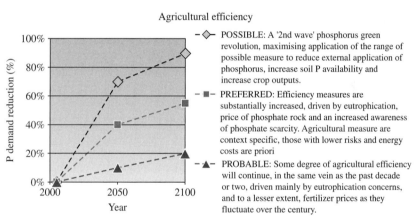

Figure 5. Scenarios for reducing phosphorus demand through increasing agricultural efficiency.

FUTURE PHOSPHORUS SUPPLY OPTIONS

This section introduces a range of organic and inorganic sources that can be used as phosphorus fertilizers. These include recovering organic phosphorus from the food production and consumption chain (manure, crop residues, food waste, human excreta), and virgin sources (such as seaweed, algae, phosphate rock). These vary widely in terms of phosphorus concentration, chemical form, and state (solid, liquid or sludge). Our preferred scenarios build on important sustainability criteria such as total energy consumption of sourcing and using phosphorus, level of contaminants, phosphorus concentration, chemical use, long-term availability and accessibility to farmers, and reliability of quality and quantity.

Phosphate rock

Phosphate rock has been the dominant source of phosphorus fertilizers for the past century (Buckingham and Jasinski, 2004). While initially perceived as an abundant source of highly concentrated and easily accessible phosphorus, numerous environmental, economic, geopolitical and social concerns about remaining phosphate rock reserves could mean the end of the phosphate rock era this century (Cordell *et al.*, in press).

US Geological Survey data (the most comprehensive publicly available data sets on global phosphate reserves) suggest remaining economically and technically feasible global reserves are around 18,000 million tonnes of phosphate rock (Jasinski, 2008). At the present rate of consumption, these reserves will be depleted within 50–100 years (Steen, 1998; Smil, 2000; Gunther, 2005), leaving behind lower quality and less accessible rock. More concerning however is that a peak phosphorus analysis based on industry data suggests a peak in maximum production could occur within 30 years (Cordell *et al.*, in press). While the timeline of the peak may be disputed[v], the fertilizer industry recognises that the quality of remaining reserves are decreasing and cheap phosphate rock is a thing of the past (Stewart *et al.*, 2005; IFA, 2006; Smil, 2002).

Remaining phosphate rock reserves are under the control of only a handful of countries, mainly China, the US and Morocco (which occupies Western Sahara and it's reserves) (WSRW, 2007). The US, historically the world's largest producer and exporter of phosphate rock, now has 25 years remaining of their reserves (Stewart *et al.*, 2005; Jasinski, 2008). China has recently imposed an export tariff that effectively bans phosphate exports[vi].

Each tonne of processed phosphate also generates five tonnes of phosphogypsum with radium levels too high for reuse (Wissa, 2003). Cadmium and other heavy metals are increasingly present in low-grade phosphate rock (even 'cadmium-free' phosphorus contains cadmium hundredfold the level of excreta) (Steen, 1998; Driver, 1998; Jönsson *et al.*, 2004). Processing and transporting phosphate fertilizers from the mine to the farm gate bears an increasingly significant energy cost, which currently relies on cheap fossil fuel energy.

For the reasons outlined above, it is probable that phosphate rock will constitute an even smaller proportion of global phosphate demand in the first half of this century (see scenarios in Figure 6).

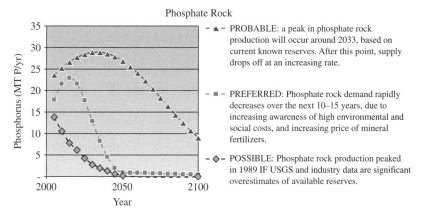

Figure 6. Scenarios for the contribution of phosphate rock to meeting future phosphorus demand.

Manure

Animal manure has always been widely used as a source of fertilizer in most regions of the world. Its phosphorus content is easily available for plants, however the concentration varies from 2.9% P_2O_5 in poultry manure, to 0.1% P_2O_5 in cattle dung (FAO, 2006). Livestock manures can also be mixed and composted with other solid farm organic matter such as bedding (known as Farm Yard Manure), food waste and human excreta. The resultant compost also has good soil conditioning properties that does not occur with direct application of organic wastes. While composted urban material can contain as much as 1% P_2O_5 (compared to 0.2% P_2O_5 in rural organic matter), it is more likely to contain heavy metals if industrial wastes and sewage sludge are mixed in (FAO, 2006). Another common source is sludge from biogas digesters[vii] (1.1–1.7% P_2O_5) designed to convert organic wastes into methane and hydrogen for cooking and lighting. Over 21 million small-scale biogas digesters are already in use in China and other regions (UNDP, 2007).

Smil estimates 40–50% of the annual 15MT of phosphorus generated in manure is recirculated to agriculture (Smil, 2000). It would be difficult to recirculate 100% of manure if the livestock industry remains geographically separate from crop production. Further, in some parts of the world manure supply exceeds demand (e.g. North America, The Netherlands), while in other regions demand exceeds supply (e.g. Australia, Africa). Future scenarios are presented in Figure 7.

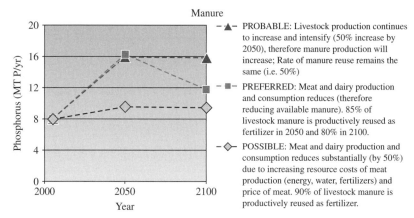

Figure 7. Scenarios for the contribution of manure to meeting future phosphorus demand.

Human excreta

Reusing phosphorus in human excreta as a fertilizer can occur through direct use (e.g. of urine), following composting, mixing with municipal wastewater, incineration, struvite crystallisation and sludge reuse, with phosphorus concentrations ranging from 0.16% (urine) to 3.2% P_2O_5 (activated sewage sludge) (Kvarnström *et al.*, 2006; FAO, 2006; Raschid-Sally and Jayakody, 2008; Reindl, 2007; SCOPE, 2004).

Human excreta has been reused as fertilizer by the Chinese for the past 5,000 years. Because urine is essentially sterile and contains plant available nutrients (N,P,K), it can be used directly as a fertilizer in a safe way if it is not mixed with faeces in toilets and by taking simple precautions. Urine from one person alone provides more than half the per capita phosphorus required to fertilize cereal crops, yet its potential as a fertilizer is often overlooked (Drangert, 1998) and should be considered along side other phosphorus recovery options.

Some 200 million (mostly poor) farmers today divert wastewater from cities to agricultural fields because it is a cheap and reliable source of water and nutrients (Raschid-Sally and Jayakody, 2008). In aquaculture, two thirds of farmed fish globally are fertilized by wastewater (World Bank, 2005). However it is essential that minimum precautionary measures are adhered to in order to avert serious health risks. Indeed, the World Health Organisation has now published extensive risk-based guidelines on the safe reuse of human excreta in agriculture and aquaculture (WHO, 2006a).

Water bodies are often polluted by high anthropogenic nutrient loads. Capturing urine at source (at the toilet) can be much more energy efficient and cost-effective than removing high levels of phosphorus at the wastewater treatment plant. At the same time this avoids heavy metal contamination (like Cadmium) from the mixed wastewater. Struvite recovery can also be more cost effective than chemical and biological removal (Shua *et al.*, 2005).

Humans produce 1–1.5 g P/person/day in human excreta[viii]. This means globally, we produce around 3 million tonnes of P in our excreta each year. Approximately 10% is currently returned to agriculture as sludge or direct wastewater reuse (Cordell *et al.*, in press). Future scenarios are presented in Figure 8.

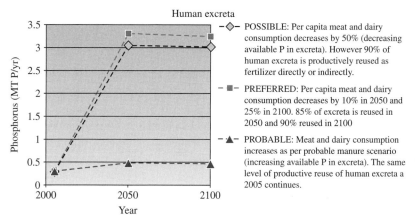

Figure 8. Scenarios for the contribution of human excreta to meeting future phosphorus demand.

Food Waste

For the purpose of this analysis, food waste constitutes all organic matter byproducts from post-harvest food processing through to consumption waste. For example, the residual byproduct 'oil cakes' following oil extraction from oilseeds contain at least 0.9–2.9% P_2O_5 which is significantly higher than crop residues (FAO, 2006). Approximately 2 million tonnes of phosphorus in post-harvest and food waste is currently lost and not recirculated (Cordell *et al.*, in press). While food chain efficiency could reduce avoidable losses substantially, the remainder can be composted or digested and reused. Figure 9 provides three scenarios.

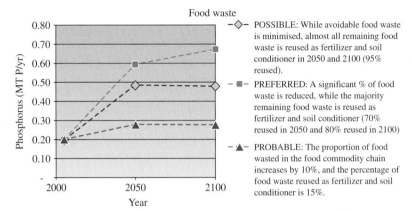

Figure 9. Scenarios for the contribution of food waste to meeting future phosphorus demand.

Crop Residues

Crop residues such as straw, husks, and stalks can be ploughed back into the soils after harvest, for their soil conditioning and fertilizer value (0.05–0.75% P_2O_5) (FAO, 2006). Around 40% of the 5 MT P in crop residues generated annually are currently reused as fertilizers (Smil, 2002). The remainder are used for feed, fuel, roofing bedding, sold or disposed of through burning or other means. Figure 10 presents future scenarios.

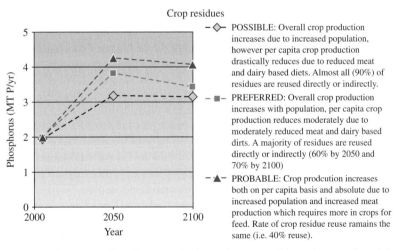

Figure 10. Scenarios for the contribution of crop residues to meeting future phosphorus demand.

Other sources

Other concentrated sources of phosphorus include: commercial organic fertilizers (processed to ensure consistent characteristics) (FAO, 2006); ash (e.g. from slash-and-burn techniques); animal meal (ground bone, meat and blood) (Mårald, 1998); guano (bat and bird droppings) (Cordell *et al.*, in press); aquatic vegetation and sediments, (Koning *et al.*, 2008, p.34).

It is assumed that these sources together could provide the remaining 1–1.2 MT P required in 2050 and 2100 to meet the new phosphorus demand following implementation of preferred demand measures outlined in section 1.5.

A LONG-TERM PERSPECTIVE: HISTORICAL AND FUTURE SCENARIOS

A synthesis of the *preferred* future supply and demand scenarios together with historical sources of phosphorus fertilizers (Cordell *et al.*, in press) are provided in Figure 11.

Figure 11 illustrates how if unchecked, demand for phosphorus will continue to rise over the remainder of the 21st century, reaching 83 MT P in 2100. The figure also demonstrates how this demand can be substantially reduced through measures such as changing diets, food chain efficiency, and most substantially through improved agricultural efficiency. However there will likely be lag time of at least a decade (i.e. until 2020) before significant results from these policy and technical measures are realised. After 2020, total demand decreases and can be met through multiple sources of phosphorus, shifting away from relying on phosphate rock by around 2015–2020 (assuming a lag time for the shift). In this case, the majority (~65–90%) of human excreta, manure, crop residues and remaining food waste will need to be recovered and reused in an efficient and ethical manner in order to meet future food needs. The remaining gap between demand and supply can be supplemented through new sources such as seaweed, algae and ash.

INSTITUTIONAL CHALLENGES AND POLICY IMPLICATIONS

Phosphorus recovery and demand management are more than just technical solutions. Significant institutional changes will also be required at the

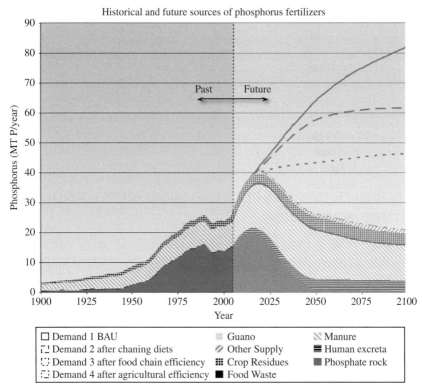

Figure 11. Long-term historical and future sources of phosphorus fertilizers for global food demand, based on aggregated preferred future scenarios.

international and national level in order to achieve the preferred scenarios in Figure 11. At minimum, this will require:

- Global governance that considers: roles and responsibilities for managing phosphorus use; integrating the phosphorus issue into global food security policies; independent and transparent monitoring and analysis of phosphorus source and use data; and, a deliberate paradigm shift to decouple sanitation's current association with water (e.g. 'WatSan' programs) towards an institutional link with food programs (e.g. 'FoodSan').
- Local and regional physical infrastructure to collect, treat and use phosphorus effectively and efficiently from multiple sources (food waste, excreta etc).

- National and local institutions and markets to manage and finance the new systems of reuse (e.g. new entrepreneurs).
- New national and local dialogues on community preferences and perceptions on phosphorus and food (from global scarcity to excreta reuse) and preferred pathways to achieving a sustainable situation.

CONCLUSIONS

The scenarios developed in this paper are designed to facilitate discussions among international and national policy makers, scientists, industry and community groups, aiming for a common framework for future phosphate recovery, demand management and governance structures to meet future global food demand.

Four key messages can be inferred from this analysis. Firstly, achieving food security in a sustainable way will likely mean the end of an era of agriculture's dependence on phosphate rock. Secondly, while there is no single quick fix solution to replacing the dependence on phosphate rock, a number of different supply and demand-side measures can together meet the growing demand for phosphorus in the long-term. Thirdly, in order to reach a desired pathway that is equitable and sustainable, significant changes will be required to both physical and institutional infrastructure. Finally, due to the serious lack of data, analysis and discussions at the international and national level regarding policies to achieve phosphorus security for food production, there is a significant and pressing need to act now, in order to avert a potentially serious shortage of phosphorus for food production.

ACKNOWLEDGEMENTS

The authors would like to thank Tom Lindström for his modelling and graphical assistance in the iterative aggregation of the preferred scenarios in Figure 11.

REFERENCES

Buckingham, D. and Jasinski, S. (2004). *Phosphate Rock Statistics 1900–2002*, US Geological Survey.

Cordell, D. (2008). *Phosphorus, food and 'messy' problems: A systemic inquiring into the management of a critical global resource*, paper submitted to the Australia New Zealand Systems Conference (ANZSYS), December 1st–3rd 2008, Perth.

Cordell, D. (forthcoming) *Phosphorus: A nutrient with no home – Multiple stakeholder perspectives on a critical global resource*, Institute for Sustainable Futures, University of Technology, Sydney (UTS) Sydney.

Cordell, D., Drangert, J.-O. and White, S. The Story of Phosphorus: Global food security and food for thought. *Global Environmental Change,* in press.

Drangert, J.-O. (1998). Fighting the urine blindness to provide more sanitation options. *Water SA,* **24**.

Dreborg, K. (1996). Essense of Backcasting. *Futures,* **28,** 813–828.

Driver, J. (1998). Phosphates recovery for recycling from sewage and animal waste. *Phosphorus and Potassium,* **216,** 17–21.

European Fertilizer Manufacturers Association (2000). *Phosphorus: Essential Element for Food Production*, European Fertilizer Manufacturers Association (EFMA) Brussels.

FAO (1999). *Urban and peri-urban agriculture*, Food and Agriculture Organisation of the United Nations, Rome.

FAO (2000). *Fertilizer Requirements in 2015 and 2030*, Food and Agriculture Organisation of the United Nations Rome.

FAO (2006). *Plant nutrition for food security: A guide for integrated nutrient management, FAO Fertilizer And Plant Nutrition Bulletin 16*, Food And Agriculture Organization Of The United Nations Rome.

FAO (2007a). *Current world fertilizer trends and outlook to 2010/11*, Food and Agriculture Organisation of the United Nations Rome.

FAO (2007b). *Organic agriculture and food availability*, International Conference on organic agriculture and food security, 3–5th May, 2007, Food and Agriculture Organisation of the United Nations Rome.

FAO (2008). *Efficiency of soil and fertilizer phosphorus use: Reconciling changing concepts of soils phosphorus behaviour with agronomic information*, FAO Fertilizer and Plant Nutrition Bulletin 18, Food and Agriculture Organization of the United Nations Rome.

Food Ethics Council (2008). *Food Distribution: An ethical agenda*, Food Ethics Council, October 2008 Brighton.

Fraiture, C.d. (2007). *Future Water Requirements for Food – Three Scenarios, International Water Management Institute (IWMI)*, SIWI Seminar: Water for Food, Bio-fuels or Ecosystems? in World Water Week 2007, August 12th–18th 2007 Stockholm.

García-Torres, L., Benites, J., Martinez-Vilela, A. and Holgado-Cabrera, A. (2003). *Conservation Agriculture: Environment, Farmers Experiences, Innovations, Socio-Economy, Policy,* Springer.

Gidley, J., Batemen, d. and Smith, C. (2004). *Futures in Education: Principles, practice and potential*, Monograph Series 2004, No. 5, Australian Foresight Institute, Swinburne University, Melbourne.

Gunther, F. (2005). *A solution to the heap problem: The doubly balanced agriculture: integration with population*, Available: http://www.holon.se/folke/kurs/Distans/Ekofys/Recirk/Eng/balanced.shtml

Heffer, P. and Prud'homme, M. (2008). *Medium-Term Outlook for Global Fertilizer Demand, Supply and Trade 2008–2012, Summary Report*, International Fertilizer Industry Association (IFA), Paris.

Holmgren, D. (2003). *Permaculture: Principles and Pathways Beyond Sustainability*, Holmgren Design Services.

IFA (2006). *Production and International Trade Statistics*, International Fertilizer Industry Association Paris, available: http://www.fertilizer.org/ifa/statistics/pit_public/pit_public_statistics.asp (accessed 20/8/07).

IFA (2007). *Fertilizer Best Management Practices: General Principles, Strategy for their Adpotion and Voluntary Initatives vs Regulations*, International Fertilizer Industry Association, Paris.

IFA (2008). *Feeding the Earth: Fertilizers and Global Food Security, Market Drivers and Fertilizer Economics* International Fertilizer Industry Association Paris.

IWMI (2006). *Comprehensive Assessment of water management in agriculture*, Co-sponsers: FAO, CGIAR, CBD, Ramsar, www.iwmi.cgiar.org/assessment

Jasinski, S.M. (2008). *Phosphate Rock, Mineral Commodity Summaries*, January 2008 <minerals.usgs.gov/minerals/pubs/commodity/phosphate_rock/>.

Jönsson, H., Stintzing, A.R., Vinnerås, B. and Salomon, E. (2004). *Guidelines on the Use of Urine and Faeces in Crop Production*, EcoSanRes, Stockholm Environment Institute Stockholm.

Koning, N.B.J., Ittersum, M.K.v., Becx, G.A., Boekel, M.A.J.S.v., Brandenburg, W.A., Broek, J.A.v.d., Goudriaan, J., Hofwegen, G.v., Jongeneel, R.A., Schiere, J.B. and Smies, M. (2008). Long-term global availability of food: continued abundance or new scarcity? . *NJAS Wageningen Journal of Life Sciences* **55**, 229–292.

Kvarnström, E., Emilsson, K., Stintzing, A.R., Johansson, M., Jönsson, H., Petersens, E.a., Schönning, C., Christensen, J., Hellström, D., Qvarnström, L., Ridderstolpe, P. and Drangert, J.-O. (2006). *Urine Diversion: One Step Towards Sustainable Sanitation*, EcoSanRes programme, Stockholm Environment Institute Stockholm.

Lundqvist, J., Fraiture, C.d. and Molden, D. (2008). *Saving Water: From Field to Fork – Curbing Losses and Wastage in the Food Chain, SIWI Policy Brief*, Stockholm International Water Institute, Stockholm.

Mårald, E. (1998). *I mötet mellan jordbruk och kemi: agrikulturkemins framväxt på Lantbruksakademiens experimentalfält 1850–1907*, Institutionen för idéhistoria, Univ Umeå.

Mitchell, C. and White, S. (2003). Forecasting and backcasting for sustainable urban water futures. *Water Policy*, **30**, 25–28.

Netherlands Environmental Assessment Agency (2006). *Integrated modelling of global environmental change. An overview of IMAGE 2.4*, Netherlands Environmental Assessment Agency (MNP), MNP publication number 500110002/2006 October 2006 Bilthoven.

Pacala, S. and Socolow, R. (2004). Stabilization Wedges: Solving the Climate Problem for the Next 50 Years with Current Technologies. *Science*, **305**, 968–972.

Raschid-Sally, L. and Jayakody, P. (2008). *Drivers and characteristics of wastewater agriculture in developing countries – results from a global assessment*, Comprehensive Assessment of water management in agriculture, International Water Management Institute.

Reindl, J. (2007). *Phosphorus Removal from Wastewater and Manure Through Struvite Formation: an annotated bibliography*, Dane County Dept of Highway, Transportation and Public Works.

Royal Dutch Shell (2008). *Shell Energy Scenarios 2050*, Shell International BV. Available: www.shell.com/scenarios

SCOPE (2004). Struvite: its role in phosphorus recovery and recycling,. *International conference 17th–18th June 2004, Cranfield University, Great Britain. Summary report in SCOPE 57.*

Shua, L., Schneidera, P., Jegatheesana, V. and Johnsonb, J. (2005). An economic evaluation of phosphorus recovery as struvite from digester supernatant. *Bioresource Technology,* **97**(17), 2211–2216.

Smil, V. (2000). Phosphorus in the Environment: Natural Flows and Human Interferences *Annual Review of Energy and the Environment,* **25**, 53–88.

Smil, V. (2002). Phosphorus: Global Transfers. IN DOUGLAS, P.I. (Ed.) *Encyclopedia of Global Environmental Change.* Chichester, John Wiley & Sons.

Smil, V. (2007). *Policy for Improved Efficiency in the Food Chain*, SIWI Seminar: Water for Food, Bio-fuels or Ecosystems? in World Water Week 2007, August 12th–18th 2007 Stockholm.

Steen, I. (1998). Phosphorus availability in the 21st Century: Management of a non-renewable resource. *Phosphorus and Potassium,* **217**, 25–31.

Stewart, W., Hammond, L. and Kauwenbergh, S.J.V. (2005). Phosphorus as a Natural Resource. *Phosphorus: Agriculture and the Environment, Agronomy Monograph No.46.* Madison, American Society of Agronomy, Crop Science Society of America, Soil Science Society of America.

UN (2003). *World Population in 2300: Highlights*, draft, United Nations Department of Economic and Social Affairs, Population Division, New York.

UN (2007). *World Population Prospects: The 2006 Revision*, United Nations Department of Economic and Social Affairs, Population Division New York.

UNDP (2007). *Overcoming vulnerability to rising oil prices: Options for Asia and the Pacific*, United Nations Development Programme (UNDP), Regional Energy Programme for Poverty Reduction, Bangkok, available: http://www.energyand environment.undp.org/undp/indexAction.cfm?module=Library&action=GetFile& DocumentAttachmentID=2327

WHO (2006a). *Guidelines for the safe use of wastewater, excreta and greywater, Vol. 4: Excreta and greywater use in agriculture*, World Health Organisation.

WHO (2006b). *Obesity and overweight, Fact sheet number 311*, World Health Organisation Global Strategy on Diet, Physical Activity and Health, Geneva, available: http://www.who.int/mediacentre/factsheets/fs311/en/index.html

Wissa, A.E.Z. (2003). *Phosphogypsum Disposal and The Environment* Ardaman & Associates, Inc. Florida, available: http://www.fipr.state.fl.us/pondwatercd/phospho-gypsum_disposal.htm

World Bank (2005). *Water Resources And Environment Technical Note F.3 Wastewater Reuse* Series Editors Richard Davis, Rafik Hirji Washington, DC.

WRAP (2008). *The food we waste, Food waste report v2*, Waste & Resources Action Programme.

WSRW (2007). *The phosphate exports*, Western Sahara Resource Watch, http://www.wsrw.org/index.php?cat=117&art=521

WWF (2004). *Living Planet Report*, World Wide Fund For Nature, Cambridge. Available: http://assets.panda.org/downloads/lpr2004.pdf

Notes

[i] Backcasting is ideal to address complex, long-term, solutions-oriented future studies with a high degree of uncertainty. See Robinson (1990), Dreborg (1996), Mitchell and White (2003).

[ii] Little data is available on such sources, particularly organic phosphorus sources, due to lack of previous monitoring and research. The figures thus present best available data and should be considered for their order of magnitudes. In this way, they can be used as a framework to stimulate discussion and further data collection to increase accuracy of assumptions.

[iii] The FAO's Integrated Nutrient Management (INM) approach does provides an integrated approach to agricultural efficiency and multiple source of plant nutrients, however this does not address reducing demand beyond the field.

[iv] The rest remains chemically unavailable in the soil, or washed off to waterways.

[v] However these data sets have been heavily criticised by some scientists (Michael Lardelli pers comm 9/8/08; Ward, 2008) claiming USGS assumptions behind reserves are over-estimates and may be biased towards industry interests. On the other hand, these estimates are also viewed as under-estimates by others. Table 2 in Cordell et al (submitted) outlines factors leading to potential over- or under-estimates of phosphate rock reserves and the timeline of peak phosphorus.

[vi] Phosphate rock deposits exist in other countries and offshore on the seabed, however these are difficult to access and of lower quality and hence uneconomical to mine.

[vii] Nearly all the P in the input material is retained in the slurry, since the phosphorus cycle does not have a significant atmospheric phase.

[viii] This varies with diet, for example, excreta from a person eating meat can contain twice as much phosphorus as a vegetarian's excreta.

Impact of supply and demand on the price development of phosphate (fertilizer)

J. von Horn and C. Sartorius

Fraunhofer Institute Systems and Innovation Research, Breslauer Str. 48, 76139 Karlsruhe, Germany

Abstract In this paper the development of the phosphate price is investigated. The historical price development is analyzed and a prospect to the future development is given.

Generally the price is determined by the supply and demand situation. The phosphate reserves minable for the production of fertilizer are diminishing in quality and quantity. The mining of deeper soil layers leads to higher production costs and the phosphate rock is increasingly contaminated with heavy metals like cadmium and uranium. Consequently the quantity of phosphate economically minable at actual costs is decreasing. Additionally substantial costs for environment protection and transportation increase the phosphate price. Since 2007 the supply situation is tight due to production capacity shortage. For this reason the phosphate price increased rapidly. New mines and processing plants are supposed to be in operation by 2012. Most of the reserves are located in Africa and in the Middle East. Asian Countries, particularly India, are investing in processing plants to satisfy their increasing demand. The demand for phosphate is increasing due to population growth and the necessity of nurturing that population. The growth of energy plants for the production of agro-fuels is another factor that increases the phosphate demand. Since the development of agro-fuel production is difficult to predict, two scenarios – business as usual and increased agrofuel production – are distinguished. The phosphate price estimated for 2030 is about US$ 100 and 120 per tonne respectively in both scenarios.

INTRODUCTION

Phosphate reserves are diminishing in quantity and quality throughout the world. At the same time the demand for phosphate fertilizer grows due to the increasing food demand of a growing world population and, more recently, the demand for agro-fuels as renewable alternatives to fossil fuels. To ensure future supply of these agro-products it is important to take adequate measures now. Even though phosphate recycling from wastewater is one option to ensure future phosphate supply, it does not seem economically feasible yet. To strengthen research in this field a prospect of its economy and impact on the future supply will be helpful. This paper shows the development of the supply and demand situation for phosphates and from this makes conclusions for the future phosphate prices.

SUPPLY: P-RESERVES

The world's phosphate resources are hard to specify. The deposit of currently explored phosphate mines sum up to about 6,370 million tonnes [CRU, 2003]. An estimation of phosphate reserves that can be mined at costs lower than 40 $/t showed an amount of 12,000 million tonnes of phosphate rock in 2001. This is about one third of the world's total reserves [US Bureau of Mines, 2001 in FAO, 2004]. Some deposits might be used in the future with new mining technologies [FAO, 2004]. Other deposits are offshore and their mining is not likely.

Most of the profitably minable phosphate reserves are located in politically unstable countries in Northern Africa and the Middle East. Political decisions of local governments leading to the restriction of phosphate exports could therefore lead to a severe shortage in phosphate rock – including a strong impact on the phosphate price.

The increasing oil price affects the price of fertilizers in several ways. Rising transportation costs increase the price for phosphate imports. Furthermore, the production of phosphate fertilizer from phosphate rock requires ammonia and sulphuric acid. Especially the cost of ammonia production is strongly affected by increasing energy costs.

An upcoming problem with phosphate rock is heavy metals (e.g. cadmium, uranium) that contaminate the raw material. With a further exploitation of the mines the heavy metal content will tend to increase. The usual processing of phosphate rock to fertilizer does not eliminate heavy metals. Upgrading of the processing plants is technically feasible, but it increases the processing costs by more than 10 $/t. The German fertilizer ordinance, for instance, sets a maximum limit for cadmium of 1.5 mg/kg fertilizer and 50 mg/kg P_2O_5 for fertilizers with more than 5% P_2O_5. Table 1 shows that most mines can comply with this limit but there are some that could exceed the limit in the future when the mining proceeds to deeper soil layers.

In some countries like USA and Canada, environment protection requires cost intensive measures. Renaturation of the surface mining fields is an obligation for the mines. Costs can be very high and can even lead to the closure of mines.

In the phosphate business a general trend towards integrating further parts of the production chain exists. This includes the further processing of phosphate rock to phosphoric acid or P-containing fertilizers. The reason for this is the need

Table 1. Reserves and Heavy Metal contents of phosphate rock [Data from FAO (2004) and Duley (2001)].

Country	Deposit	P$_2$O$_5$ (wt %)	As	Cd (ppm)	U	Reserves (1000 tonnes)	Reserve base (1000 tonnes)
Israel		32	5	25	150	180,000	180,000
Jordan		32	8	5	78	900,000	1,700,000
Morocco						5,700,000	21,000,000
	Bu Craa	35.1		37.5	75		
	Louribga	32.6	13.3	15.1	88		
	Youssoufia	31.2	9.2	29.2	97		
Togo		36.7	10	58.4	94	30,000	60,000
USA						1,000,000	4,000,000
	Florida	31.9	11.3	9.1	141		
	Idaho	31.7	23.7	92.3	107		
	N. Carolina	29.9	11.2	38.2	65		
S. Africa		39.5	11	<2	9	1,500,000	2,500,000
Tunisia		29.3	4.5	39.5	44	100,000	600,000
Senegal		35.9	17.4	86.7	67	150,000	1,000,000
China		31	26	2.5	22.8	500,000	1,200,000
Russia						150,000	1,000,000
others						1,200,000	4,000,000
World total						12,000,000	37,000,000

to satisfy a rapidly increasing fertilizer demand or to realize the higher profits associated with the higher value added by further processing phosphate products.

Phosphate supply in 2030

Today about 140 million tonnes of phosphate rock are mined. The amount of mined phosphate is increasing due to increasing fertilizer demand. With strong agro-fuel growth the amount of phosphate rock that needs to be mined will rise to about 171 million tonnes per annum in 2030. That means that in the time period until 2030 another 3,400 million tonnes of phosphate rock would have to be mined. This is about half the amount of the reserves of the currently operated mines.

DEMAND: P-CONSUMPTION

Current situation

In the CRU (2003) analysis of phosphate mines the total mining of P-Rock sums up to 126 million tonnes per year. From 2003 to 2006 an increase of the phosphate consumption of 10 million tonnes can be seen due to an increasing consumption in South and East Asia and Oceania. This increase exceeds the decreasing consumption in Western Europe and North America. More than 90% of all phosphate is used in agriculture.

Factors affecting the future fertilizer demand

Population growth: According to the statistics of the UN Population Division, the world population (2005 to 2050) grows by at least 1.3 billion in the low variant and by up to 4.2 billion in the high variant scenario. Even the low variant scenario represents a population increase by about 20%. In 2030 the world population will be about 9 billion people.

Nutrition: At the same time as the population increases the average income in developing countries is rising. This will lead to a better nourished population and an increasing need for further processed food (milk products, meat). For this reason the need for fertilizer is expected to increase more strongly than the population (see Figure 1).

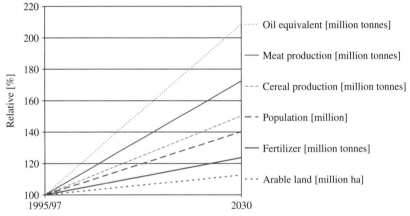

Figure 1. Development of factors affecting future fertilizer demand [FAO 2002, FAO 2000].

Farming technologies: The increase of the agricultural production has to be achieved mainly on the existing agricultural land as most of the arable land is already used. Only in South America and Sub-Saharan Africa further land can be made arable by clearing. The most important factors for crop increase on existing agricultural land are improved irrigation and fertilization. Like water, phosphate is essential for the growth of plants and can not be substituted.

Modern farming technology like precision farming can help to reduce the consumption of phosphate in agriculture by enabling a more precise adjustment of fertilizer supply to the specific local demand. Additionally, the use of precision farming can lead to an at least temporary decrease of the mineral fertilizer consumption in countries like Germany where agricultural land has partly been over-fertilized over many years by applying large amounts of manure to the cultured land.

Energy plants: The production of agro-fuels is increasing rapidly. The European Unions target is to increase the amount of agro-fuels used in transportation to 10% by 2020. At the same time there are ongoing debate as to whether the climate impact of agro-fuels is altogether positive if issues like land use change, GHG emissions from agriculture itself and water demand are included into the balance. With regard to these controversial discussions the future development of agro-fuel production is hard to predict. If however policy continues to consider the production of agro-fuels as a way out of the dependence on fossil fuels and the adverse effects on climate change, the increase in energy plant production will certainly result in a strong increase in fertilizer demand.

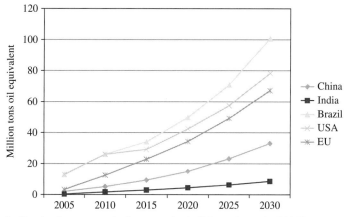

Figure 2. Production scenario for agro-fuels [Msangi *et al.*, 2007].

The production of biodiesel requires 100 kg NPK-Fertilizer per hektar. With a crop yield of 1.4 tonnes of biodiesel per hektar the fertilizer consumption for biodiesel production is 0.071 t NPK/t biodiesel. The agro-fuel scenario of Msangi *et al.* (2007) considers an aggressive agro-fuel growth (20% of transportation fuels by 2020) and no crop productivity change. Due to this scenario, the agro-fuel production is about 300 million tonnes in 2030. This would require about 21 million tonnes of NPK fertilizer. That means about 13% of the total fertilizer demand of 166 million tonnes in 2030 (FAO, 2000) will be used for energy plant production. According to FAO (2008) the fertilizer demand varies from 1% to 27.6% in different studies.

Phosphate demand in 2030

The FAO (2000) fertilizer forecast includes already the rising per capita income that leads to a higher fertilizer demand for nutrition, the increasing world population and the progress in farming technology that leads to a better fertilizer per crop yield. The increasing production of agro-fuels is not included in the forecast. Considering the terms shown in the energy plant chapter, an additional amount of 21 million tonnes of fertilizer are needed in 2030. Altogether the fertilizer demand will then be about 187 Million tonnes.

PRICE DEVELOPMENT

Actual and historical development

Phosphate is not traded globally. Instead fertilizer companies usually make long term contracts with locally specific phosphate mines. Therefore the phosphate rock market has an oligopolic structure. The phosphate rock price has been constant at about 30 $/t since 1990. In 2006 the price increased by 4.6 $/t, in 2007 an even greater increase in P-rock price of 12 $/t was recognized by CRU (2007). In 2008 the phosphate rock price jumped dramatically up to 200 $/t (see Figure 3).

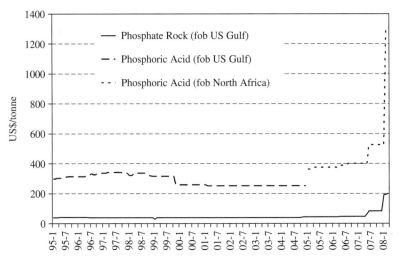

Figure 3. Phosphate price development (Fertilizer International, 2008).

The actual phosphate price is mainly influenced by the increasing demand for phosphate fertilizer. Since the year 2007, the current production capacity of mines, beneficiation and fertilizer plants are working at their maximum capacity. On the other hand, there is no substitute for phosphate in agriculture and, accordingly, the price elasticity of phosphate demand is very low. So, phosphate prices are expected to increase as long as demand increases and no new plant capacity will be in operation. The lack of production capacity is the main reason for the price increase.

For the last few years, the price for energy increased due to the high oil price. Although the conversion of phosphate is an energy intensive process the energy costs are less than 10% of the production costs in the phosphate fertilizer production. The high oil price however affects the transportations costs for phosphate. The transportation costs for coal increased by 100% in 2007. Phosphate transportation costs are supposed to respond to an increasing energy price in a similar way. Accordingly, the expected price increase for transportation from 10 $/t to 20 $/t may result in a price increase of more than 20% for Phosphate rock imports to the EU. At the same time, the high oil price is an important argument for processing plants are even more dependent on supply from the nearest phosphate mines. This further decreases the price elasticity and increases the rents actually realized by the phosphate mines.

To emphasize it again, the price jump is not caused by a shortage of phosphate reserves. Estimations show that the reserves minable at less than 40 $/t are available for the next 70 years at the current consumption level (FAO, 2004). The diminishing quantity and quality of phosphate reserves would not lead to a significant price increase yet.

Future development

Since in the past the producers of phosphate never undertook efforts to restrict their output below the actual demand and thereby realize substantial monopoly rents, we assume that they will respond also to this challenge by expanding the capacity of their facilities. Since capacities are actually being expanded for the extraction of a wide variety of resources and capacities find it difficult to even buy the necessary equipment, we further assume that the process of adjusting the capacity to the higher level will take at least 5 years starting from the beginning of the capacity shortage in 2007. For this adjustment period (until 2012), it will not be possible to determine a reasonable phosphate price, because the price elasticity of the demand is essentially unknown. The main reason for this is that the economic rule that suppliers will first satisfy the demand with the highest willingness-to-pay will not hold because policy may not allow poor farmers (with low willingness to pay) to be excluded from the access to fertilizers. So it is essentially unclear which demand will actually be satisfied in the case of a shortage of supply and who is going to pay which price for the required phosphate.

For the time after 2012, by contrast, we assume that a price determination on the basis of demand and supply will again be possible. On the supply side, we start our analysis with the actual supply curve showing in increasing order the total cost of phosphate production over all worldwide production capacities (CRU, 2003). The cost figures at the left and right end of the resulting steadily increasing graph respectively represent the lowest and highest costs of phosphate production at the time of investigation (2002). This supply curve is then extrapolated into subsequent years by assuming that the present cost curve is valid only for a static period of exploitation of 25 years and, accordingly, more costly production capacities have to replace the least costly exploitation opportunities running out every year. As a result, a cost range is yielded of which the lower and upper costs increase steadily over time. Figure 4 shows the estimation of the phosphate price until 2030 in a business-as-usual (BAU) scenario, that is, under the assumption that agricultural production and the respective fertilizer needs will develop as estimated by the FAO (2002).

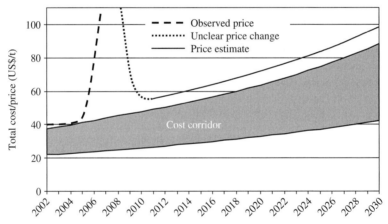

Figure 4. Prediction of the price development of phosphate rock in the business-as-usual scenario.

According to this picture it is unclear how the phosphate price will develop in the short term (by 2012) – which peak level it may reach and when exactly it will return. In the medium and long run, we are however confident that under the assumption of a profit margin of 10% a phosphate price of about 100 $ per tonne may be reached in 2030.

The move of the longer-term price curve at a margin of 10% above the upper cost limit goes with the assumption that in the future phosphate rock production facilities will always work close to their production capacity. This will cause more intense competition on the phosphate market which will lead to the formation of a more uniform phosphate price and the realisation of a moderate profit.

While in the business-as-usual scenario the fertilizer demand was expected to increase by 1% annually, we assumed an increase by up to 2% annually in the agro-fuel scenario, that is when agro-fuels will gain a substantial share (>20%) of the future agricultural production. In this case, the phosphate fertilizer demand would be similar to the BAU scenario until 2012, but up to 22% higher than in the BAU scenario in 2030. The phosphate price is estimated to eventually reach about 120 US$ per tonne.

CONCLUSIONS

The world's phosphate reserves are estimated to about 12,000 million tonnes. The currently operated mines have about 6,370 million tonnes of reserves. By 2030 these reserves will be diminished by half. Currently the phosphate price has jumped dramatically due to capacity shortage. New production capacity is under construction to satisfy the increasing fertilizer demand.

The phosphate demand is increasing due to population growth and a rising income in developing countries that leads to more sophisticated food preparation including more input per kilocalorie of taken up food. Furthermore increasing oil prices and concerns about the safety of oil supply enhances the cultivation of energy plants for agro-fuel production. In developed countries the fertilizer per crop yield improves due to precision farming. In total, the fertilizer demand will increase significantly to about 187 million tonnes per annum in 2030 due to these influences and energy plant growth.

In the phosphate price development, a long term slow price increase can be forecasted on the basis of the increasing fertilizer demand and the increasing efforts needed on the supply side to meet this demand. The more difficult accessibility and the decreasing quality of the reserves leads to higher processing costs that enhance the price increase – leading to phosphate prices of about 100 US$/t (without energy plants) or 120 US$/t (with aggressive energy plant cultivation) in the year 2030.

Phosphate recycling from wastewater will be self sufficient at costs of about 100 $/t. Especially for European countries without phosphate reserves even slightly higher prices might be acceptable to reduce the dependence on phosphate imports.

REFERENCES

CRU (2003). The Competitiveness of Phosphate Fertilizer Producers to 2020: Coping with Changing Production Costs, Resource Availability and Market Structures. CRU Group international of British Sulphur Consultants. Confidential.

CRU (11/2007). Homepage CRU Nov. 2007 Issue. www.cru.de

DüMV (2003). Verordnung über das Inverkehrbringen von Düngemitteln, Bodenhilfsstoffen, Kultursubstraten und Pflanzenhilfsmitteln (Düngemittelverordnung – DüMV). Vom 26. November 2003, BGBl. I S. 2373, zuletzt geändert am 23. Jul. 2008, BGBl. I S. 1410.

Duley, B. (2001). Recycling Phosphorus by Recovery From Sewage, Rhodia Consumer Specialties UK Ltd. (for Centre Europeen d'Etudes des Polyphosphates).

FAO (2000). Fertilizer requirements in 2015 and 2030. Food and Agriculture Organization of the United Nations (FAO), Rome 2000.

FAO (2002) World agriculture: towards 2015/2030. Summary report. Food and Agriculture Organization of the United Nations (FAO), Rome 2002.

FAO (2004). Use of phosphate rocks for sustainable Agriculture. FAO Fertilizer and Plant Nutrition Bulletin 13, 2004.http://www.fao.org/docrep/007/y5053e/y5053e06.htm#TopOfPage Zugriff am 11.03.2008

Fernandez, Ivan (2002). Diminishing Phosphate Reserves: Is the Fertilizer Industry Prepared? Frost & Sullivan Market Insight. 18 Jan. 2002. http://www.frost.com/prod/servlet/market-insight-top.pag?docid=JEVS-5N2JS7. Zugriff am 11.03.2008

Heffer und Prud, Çhomme (2005). Medium-Term Outlook for Global Fertilizer Demand, Supply and Trade. Summary Report. International Fertilizer Industry Association (ifa).

Ifa (2007). Production and International Trade Statistics. Production and international trade statistics covering nitrogen, phosphates, potash and sulphur products: production, exports, imports by region from 1999 to 2006. http://www.fertilizer.org/ifa/statistics/pit_public/pit_public_statistics.asp. Updated: 27 Sept. 2007.

Integer (200x). The Biofuels Boom and Fertilizers. Integer Research Ltd., London, U.K. http://www.integer-research.com/Products/Services/?ServiceID=159&ckIndustryID=4

IVA (2008). Industrieverband Agrar. Homepage. http://www.iva.de/presse_news/content/Kuhlmann_Folien_060508.pdf

Msangi, S., Sulser, T., Rosegrant, M., Valmonte-Santos, R. and Ringler, C. (2007). Global Scenarios for Biofuels: Impacts and Implications. International Food Policy Research Institute (IFPRI).

OECD/FAO (2007). OECD-FAO Agricultural Outlook 2007-2016. Paris: OECD.

Schrödter, K., Bettermenn, G., Staffel, T., Wahl, F., Klein, T. and Hofmann, T. (2008). Phosphoric Acid and Phosphates. Ullmann's Encyclopedia of Industrial Chemistry, Wiley.

USGS (U.S. Geological Survey) (2007): Phosphate rock statistics, in Kelly, T.D. and Matos, G.R., comps., Historical statistics for mineral and material commodities in the United States: U.S. Geological Survey Data Series 140, available online at http://pubs.usgs.gov/ds/2005/140/ (Zugriff am 11.08.2008).

Wastewater treatment and the green revolution

Paul M. Sutton[a], A.P. Togna[b] and O.J. Schraa[c]

[a]P.M. Sutton & Assoc., Inc., Enfield, NH, USA
[b]Basin Water, Lawrenceville, NJ, USA
[c]Hydromantis, Inc., Cambridge, ON, Canada

Abstract There is growing belief that in the future the objective should be to treat wastewater not as a waste but as a resource for water, energy and nutrients while reducing the impact of wastewater treatment on greenhouse gas (GHG) emissions. With this objective in mind, a municipal wastewater treatment flowsheet has been developed which achieves the goals of energy sustainability and water and nutrient recovery, while minimizing residual solids production (i.e., biomass) and the release of GHGs. This new flowsheet in part accomplishes these goals by abandoning the conventional norms of municipal wastewater treatment stating, the organic carbon fraction in the wastewater is best removed through biological oxidation to carbon dioxide, and the nitrogen and phosphorus components are best managed by biological nutrient removal processes (i.e., nitrification, denitrification and bioP removal). The new flowsheet integrates established biological and physical-chemical process systems in a unique fashion which allows for conversion of the organic carbon in the wastewater to methane, the removal and recovery of phosphorus and nitrogen from the wastewater, and the production of water suitable for reuse. Computer modelling results provide confidence in the validity of many of the claimed advantages of the new flowsheet, implying 80% of the wastewater COD can be shunted to the anaerobic digestion step of the flowsheet, and the flowsheet can reduce residual solids production and energy consumption by respectively, 40 and 80% in comparison to conventional methods.

INTRODUCTION

A municipal wastewater treatment flowsheet has been developed which achieves the goals of energy sustainability and water and nutrient recovery, while minimizing residual solids production (i.e., biomass) and the release of greenhouse gases (GHGs). This "new flowsheet" in part accomplishes these goals by abandoning the conventional norms of municipal wastewater treatment stating, the organic carbon fraction in the wastewater is best removed through biological oxidation to carbon dioxide, and the nitrogen and phosphorus components are best managed by biological nutrient removal (BNR) processes.

The new flowsheet derives energy from the wastewater by first shunting a large fraction of the organic carbon measured as COD, to a particulate or solids slurry form and then ultimately treating the solids via anaerobic digestion. The aerobic membrane bioreactor (MBR) configuration plays a key role in the new flowsheet in achieving the organic carbon shunt. Furthermore, use of the MBR and uniquely designed physical-chemical process systems allow for the production of water for reuse, and the capture of phosphorus and nitrogen from the wastewater.

The purpose of this paper is to discuss the treatment principles and mechanisms dictating the performance and operation of the various unit processes making up the new flowsheet, allowing achievement of the stated green revolution goals. In addition, the paper includes the following;

- an evaluation of the new flowsheet relative to current conventional methods for achieving complete treatment of wastewater,
- examples of attractive, commercially available, system embodiments representing each of the new flowsheet unit processes,
- information in regard to how the independent, physical-chemical phosphorus and nitrogen unit processes can be designed with the intent of nutrient recovery,
- the use of existing computer models to provide confidence in the validity of the claimed advantages of the new flowsheet, and
- a discussion of the technical and economic hurdles associated with each of the principle unit processes comprising the new flowsheet which must be overcome in order for it to be viewed as attractive.

THE NEW FLOWSHEET

Principles and treatment mechanisms

Biological treatment is generally considered the most cost-effective method to achieve secondary treatment of municipal wastewater. Figure 1 represents a simplified schematic of the typical, conventional biological treatment plant designed to achieve secondary treatment of smaller municipal wastewater flows.

The flowsheet purposely consists of only a few unit processes each requiring minimal instrumentation and control translating to a relatively low capital cost. To upgrade this system to produce an effluent for reuse, the flowsheet (Figure 1) would typically be altered as follows.

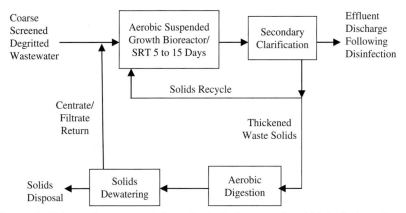

Figure 1. Schematic representation of a typical conventional biological treatment plant flowsheet to achieve secondary treatment of smaller municipal wastewater flows.

- Tankage would be added and/or the existing biological reactor(s) modified to increase the biomass inventory to achieve ammonia oxidation to nitrate (i.e., nitrification).
- Tankage, equipment components and chemicals would be added, and the existing biological reactor(s) modified to achieve nitrate reduction to nitrogen gas (denitrification).
- Tankage and equipment components would be added to achieve phosphorus removal by biological methods (i.e., enhanced phosphorus uptake by phosphorus accumulating organisms or PAOs) and/or by physical-chemical methods with chemical addition.
- Tankage, equipment components and/or chemicals would be added to achieve near complete TSS removal.
- Membrane systems (e.g., microfiltration or ultrafiltration plus reverse osmosis (RO)) would be added in order to produce effluent for reuse.

The resulting system would consist of a large number of treatment steps each with its own instrumentation and control requirements, and an elaborate piping and pumping network with a number of locations where chemicals are added. These characteristics translate to high capital costs, and high operating costs related to aeration energy requirements to achieve wastewater nitrification and waste solids aerobic digestion. Additional high operating costs are incurred related to the disposal of captured and generated wastewater solids, and the addition of chemicals and disposal of the resulting chemical solids. The complexity of the system and the variety and number of equipment components results in high manpower costs for operation and maintenance. With the exception of water recovery, the system can be categorized as a resource consumer, destroying the value associated with the wastewater components and requiring a substantial energy input.

The new flowsheet (Figure 2) represents a less complex treatment system in terms of the number of unit processes, and their operational requirements and interdependency.

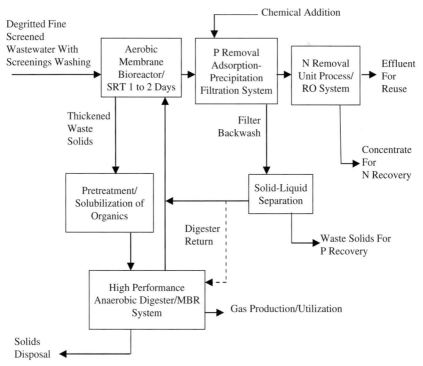

Figure 2. Schematic representation of the new flowsheet to achieve complete treatment of smaller municipal wastewater flows with resource recovery.

Principal treatment steps and process systems

The new flowsheet is made up of uniquely designed but commercially available process systems, consisting of four principal treatment steps that when combined translate to a high-performance, energy efficient flowsheet characterized by resource recovery and minimal residual solids generation. The four principal treatment steps and corresponding process systems constituting the flowsheet (Figure 2) with the goal of achieving water reuse are;

- an aerobic MBR system operated at a low reactor solids retention time (SRT) designed to capture the wastewater particulate organics, and maximize the wastewater organics conversion to cell mass and/or sorption of the organics onto the TSS present in the reactor,
- an anaerobic MBR system representing a high performance anaerobic digester designed to maximize the conversion of the organics in the form of waste solids from the aerobic MBR to methane,
- an adsorption-precipitation filtration system designed to maximize phosphorus removal with minimal chemical addition and allow for its recovery, and
- an RO system designed to remove ammonia and allow for N recovery, and produce an effluent for reuse.

A more detailed discussion of each of the process systems follows together with examples of corresponding commercial embodiments of the systems where applicable.

Aerobic MBR

The MBR process system consists of a suspended growth bioreactor operated under aerobic conditions and at a low SRT controlled in the range from 1.0 to 2.0 days, coupled to a membrane separation step designed to capture and remove soluble, particulate and colloidal organics, and colloidal inorganics.

The bioreactor is designed at a high biomass concentration (e.g., 7 g/l measured as volatile suspended solids (VSS)). The mechanisms responsible for CBOD removal across the MBR system are the storage or conversion of organic carbon into cell mass, sorption of organic carbon onto the TSS present in the reactor, and organic carbon oxidation. The bioreactor is purposely operated/controlled near the minimum SRT to achieve growth of organisms only responsible for CBOD removal. Operation of the reactor at low SRT, high TSS and completely under aerobic conditions, achieves the following.

- The low SRT and high TSS conditions maximize the mass of organic matter present in the wastewater that is stored in the cells or converted to cell mass (i.e., high cell yield), and/or sorbed onto the TSS in the reactor. The result is a much larger fraction of the organic matter present in the wastewater is transferred to the anaerobic digester through solids wasting (Figure 2) than would be characteristic of the bioreactors designed for use in more conventional treatment systems.

- The low SRT and high TSS conditions minimize the size of the required bioreactor translating to lower equipment and installation costs for new plant construction.
- Operation at a low SRT prevents the growth of organisms responsible for nitrification (i.e., nitrifiers) and enhanced biological phosphorus uptake (i.e., PAOs). Relative to the organisms responsible for CBOD removal, nitrifiers and PAOs are much fewer in number and require more specific conditions to ensure their growth.
- Operation at a low SRT minimizes the mass of organic matter oxidized to carbon dioxide versus stored in the cells or converted to cell mass, and eliminates the oxygen requirements associated with ammonia oxidation or nitrification.
- Operation of the bioreactor under completely aerobic conditions simplifies design of the reactor vessel, and eliminates the need for components and/or additional vessels required to achieve the alternating aerobic, anoxic and anaerobic zones necessary for nitrification, denitrification and biological P removal. Preventing nitrification eliminates the potential generation of nitrous oxide a powerful GHG, as previously noted.
- Operation at a low SRT minimizes the production of soluble microbial products (SMPs) resulting from cell decay. The presence of a higher concentration of SMPs generally reduces the efficiency (i.e., flux per unit of applied pressure) of membrane separation.

Precise control of the SRT is important when the reactor is designed to achieve only CBOD removal. The MBR system allows precise SRT control as wasting is done directly from the suspended growth reactor (Figure 2).

Calculations related to the bioreactor component of the aerobic MBR, imply in this application the most attractive MBR system will be characterized by membranes with a high packing density, immersed directly in an aerated membrane tank. With these considerations in mind, an MBR designed around the use of a General Electric (GE) ZeeWeed membrane system or another commercial membrane system embodiment with similar characteristics, is an attractive choice in this application for the following reasons. This conclusion is based on treating "typical" U.S. municipal wastewater as defined in a subsequent section of this paper.

1. The volume of the tank required to house the membrane cassettes is approximately equal to that required to operate the tank as the aerobic bioreactor at an SRT in the 1 to 2 day range, assuming a mixed-liquor TSS concentration of approximately 8 to 9 g/l.

2. Specifying the membrane/bioreactor tank at a liquid depth to ensure submergence of the cassettes and with sufficient coarse bubble aeration to maximize the efficiency of the membranes, translates to an airflow and corresponding oxygen transfer sufficient to meet the bioprocess oxygen requirements.
3. The amount of membrane area contained in the membrane/bioreactor tank is sufficient for operation of the membrane component at a realistic net flux (i.e., approximately 25 $l/m^2/h$).

Anaerobic MBR digestion system

The solids from the aerobic MBR system containing a large fraction of the wastewater organic carbon, are routed to a volumetrically efficient, high performance, anaerobic MBR digestion system (Figure 2). Choice of the MBR configuration for the anaerobic reactor provides the following benefits;

1. maximizes the volumetric efficiency of the digestion step recognizing the need to operate at a long reactor SRT (e.g., 30 to 50 days), provided the membrane system selected affords operation at a high reactor solids concentration (e.g., TS 3 to 5%),
2. maximizes performance of the digester (i.e., organics conversion to methane) as the membrane step ensures retention of the reactor biomass and all other particulates, colloidal and other higher molecular weight compounds,
3. maximizes digester stability as allows operation at a precise SRT, and
4. provides for a single point for accumulation/collection of low value, residual solids from the new flowsheet (Figure 2; solids disposal stream).

Various design considerations in regard to selection of the specific anaerobic MBR system configuration in this application, imply the use of a membrane system designed as a skidded assembly located external to the anaerobic reactor may be favoured.

To maximize the conversion of the organics to methane and minimize the accumulation or production of solids in the anaerobic MBR digestion system, a pretreatment step (i.e., solubilisation of particulate organics and solids) precedes the digester (Figure 2). To minimize the energy required to operate the pretreatment step, the waste solids from the aerobic MBR are first thickened. Designing the membrane system of the aerobic MBR to allow for membrane thickening of the waste solids in a separate membrane compartment or tank, appears to be an attractive approach to achieving this need. The focused pulsed (FP) or pulsed electric field technology, commercialized for this application by

OpenCel into a small, mechanically simple FP system, appears to be an attractive approach for achieving pretreatment of the thickened solids. The technology discussed in detail elsewhere (Salerno *et al.*, 2008), uses high voltage electrical micropulses to lyse the biomass cellular membranes releasing organic material trapped in the thickened waste solids from the aerobic MBR.

Phosphorus and nitrogen removal

An objective of the new flowsheet is nutrient removal, and the opportunity for phosphorus and nitrogen recovery. In the flowsheet physical-chemical process systems are used to achieve this objective while allowing for the production of water for reuse.

The phosphorus removal and recovery step is a physical-chemical unit process designed to provide near complete removal of the total P present in the effluent from the MBR system and convert it to a product for reuse (e.g., agricultural fertilizer product or product component). A number of physical-chemical treatment systems could be selected for this step including adsorption and ion exchange (IX) systems. Evaluations to-date imply phosphorus removal through an adsorption-precipitation filtration process such as the Blue Cat system offered by Blue Water Inc., represents an attractive alternative. Details regarding the removal mechanisms involved in this reactive filtration process are provided elsewhere (Newcombe *et al.*, 2008a). The Blue Cat system is a continuous backwash filter utilizing a reactive filter bed media (i.e., hydrous ferric oxide coated sand) with relatively small chemical additions (i.e., ferric chloride and ozone) to achieve P removal and advanced oxidation of any residual organics not removed in the low SRT based MBR system. The reactive filtration process is attractive in this application as it will result in the need for a low Fe chemical addition rate to achieve a low effluent total P (TP) concentration (e.g., less than 0.5 mg/l) even when the TP in the feed to the process system is greater than 4 mg/l, according to Blue Water (Newcombe 2008b). The fact the feed to the process system contains no suspended solids is likely to further minimize chemical addition requirements and the size of the filter.

The nitrogen in the effluent from the aerobic MBR of the new flowsheet is in the form of ammonia plus a small fraction of soluble or dissolved organic nitrogen. This is a result of the biochemical reactions occurring in the aerobic MBR, the waste solids from the reactor being treated anaerobically and removal of the non-solubilized, inert, particulate nitrogen by the membrane component of the aerobic or anaerobic MBR system. Assuming little or no change across the phosphorus removal step, the nitrogen to the final treatment step of the new flowsheet is essentially in the form of ammonia. If nitrogen removal and recovery are the treatment goals, a number of physical-chemical systems could be selected for this step including IX and air stripping-absorption systems. With the additional objective of producing an effluent for reuse, the use of a RO system designed specifically to remove ammonia is viewed as attractive.

Nutrient recovery

Essentially all the phosphorus and nitrogen contained in the degritted and fine screened wastewater feed to the new flowsheet is captured respectively, as waste solids from the solid-liquid separator receiving the backwash from the phosphorus removal reactive filtration process and in the concentrate from the RO step (Figure 2). This statement assumes only a small mass of N and P will be present in the solids wasted from the anaerobic digester and a small mass of N will be present in the digester gas stream.

The underflow waste solids from the solid-liquid separation step receiving the backwash from the Fe based reactive filtration process, has the potential to represent a slow release, high P containing fertilizer product following dewatering and drying. The fact Fe is found to strongly fix P, and P release from the solids matrix could be controlled by adjustment of the soil pH (Silveria *et al.*, 2006), support this conjecture.

If a treatment goal of the new flowsheet is producing an effluent for reuse, RO is favoured for ammonia removal and as the final treatment step. Although the concentrate from the RO system represents recovered ammonia, it may not be practical to process the concentrate to produce an ammonia containing product of value. If nitrogen recovery has particular benefit and less stringent waster reuse criteria apply, alternative ammonia removal technologies may be more attractive, as noted previously. For example, an IX system can be designed with the regeneration system consisting of pH adjustment and air stripping to release the ammonia from the media and then recovery of the ammonia by absorption, resulting in an aqueous ammonia solution.

New flowsheet modelling

A model is under development using the GPS-X wastewater treatment process simulator to help quantify the attractiveness of the new flowsheet in the treatment of municipal wastewater relative to conventional methods. The model is based on mass balances for the various wastewater constituents and includes bio-kinetic rate expressions for the biological processes occurring within the treatment system. Energy calculations have been added to the model to quantify the energy requirements of the flowsheet. To-date modelling efforts have focused on the aerobic MBR system, the anaerobic MBR digestion system and the treatment steps associated with these systems. The plant flowsheet as represented in GPS-X is shown in Figure 3. The model configuration and simulation results based on treatment of wastewater with characteristics as defined in Table 1 (i.e., typical U.S. municipal wastewater), are detailed in a report prepared by Schraa (2008). Summary information from the report follows.

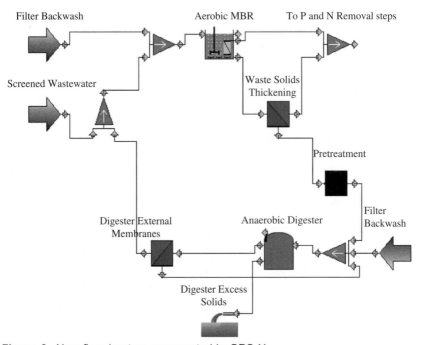

Figure 3. New flowsheet as represented in GPS-X.

The modelling and simulation results were developed based on the following.

1. The wastewater COD was fractioned into various components.
2. The reactor of the aerobic MBR was modelled as completely-mixed and the ASM3 model was used to model biological growth. This model was selected as it accounts for storage of soluble substrate. The default kinetic and stoichiometric parameters in ASM3 were used with the exception of the growth and storage yield values, and the particulate COD to VSS ratio. These values were marginally increased supported by reference to the work of others considering this specific reactor configuration/operating conditions.

Table 1. Characteristics of wastewater used in GPS-X simulation.

Characteristic	Value
Flow rate	18,925 m^3/day
COD	280 mg/l
CBOD	150 mg/l
TSS	175 mg/l
VSS	137 mg/l
TN	30 mg/l
TKN	30 mg/l
Soluble TKN	24 mg/l
NH4-N	19 mg/l
TP	6 mg/l
Soluble TP	5 mg/l

3. A customized mass balance model was used for the digester pretreatment step based on application of the OpenCel FP technology. The model was developed based on the input of Rittman (2008).
4. The MantisAD biological model was used for the anaerobic MBR digester with certain additions to account for phosphorus release and uptake, and the effect of the FP technology pretreatment step.
5. A membrane separator was used to achieve thickening of the waste solids from the aerobic MBR. This membrane step and those associated with the aerobic and anaerobic MBRs, were modelled using a membrane filter model.
6. Energy balance calculations associated with the anaerobic digester were comparable to those associated with a conventional digester with modifications to account for the effect of the pretreatment step and its configuration as an MBR. Energy requirements for the other new flowsheet treatment steps were based on input provided by the system suppliers (i.e., GE and OpenCel).

The results from the modelling and simulation efforts are detailed in the report referenced previously. Results of particular significance follow.

1. Results indicate approximately 80% of the wastewater COD is shunted to the anaerobic MBR digestion system.
2. The mass of dry solids produced requiring disposal (i.e., Figure 2, Solids Disposal) is calculated at 1480 kg/day or 78 kg per 1000 m^3 of wastewater treated. This value represents approximately 40% of the value characteristic

of a conventional activated sludge system operated without primary treatment.

3. The total energy required for operation of the new flowsheet excluding that associated with the phosphorus and nitrogen removal steps, equates to 32 kWh per 1000 m^3 of wastewater treated. A conventional activated sludge system with anaerobic digestion would use approximately 176 kWh per 1000 m^3 of wastewater treated.

Technical and economic hurdles

Despite the perceived advantages of the new flowsheet relative to current conventional methods for achieving complete treatment of wastewater, there are a number of technical and economic hurdles associated with each of the principle unit processes which must be overcome in order for the new flowsheet to be viewed as an attractive alternative, including the following.

- A lower specific flux typically characterizes the membrane component of an aerobic MBR system operated at short versus long SRTs translating into the need for more membrane area.
- There is little current data to support the anticipated attractiveness of the MBR technology for anaerobic treatment of municipal wastewater solids.
- The attractiveness of the reactive filtration process for P removal depends on achieving a low effluent TP concentration at a low Fe chemical addition rate. There is little current data to support this contention.
- Current physical and/or chemical methods for ammonia removal are generally not considered attractive for post-treatment of municipal wastewater. Little data exists to support the economic attractiveness of RO to achieve ammonia removal together with other contaminants, in order to produce an effluent for reuse.

SUMMARY

A municipal wastewater treatment flowsheet has been developed which achieves the goals of energy sustainability and water and nutrient recovery, while minimizing residual solids production and the release of GHGs. The new flowsheet integrates established process systems in a unique fashion which allows for conversion of the organic carbon in the wastewater to methane, the removal and recovery of phosphorus and nitrogen from the wastewater, and the production of water suitable for reuse. Although computer modelling completed to-date supports the major claimed advantages of the new flowsheet, there are various technical and economic hurdles to overcome in order for the new flowsheet to be viewed as an attractive alternative to the conventional approach for achieving complete treatment of wastewater.

REFERENCES

Newcombe, R.L., Strawn, D.G., Grant, T.M., Childress, S.E. and Moller, G. (2008). Phosphorus removal from municipal wastewater by hydrous ferric oxide reactive filtration and coupled chemically enhanced secondary treatment: Part II-mechanism. *Wat. Env. Research*, **80**, 248–256.

Newcombe, R.L. (2008). Personal communication to P.M. Sutton, Blue Water Inc., May.

Rittman, B.E. (2008). Personal communication to O.J. Schraa, Arizona State Univ., August.

Salerno, M.B., Lee, H-S, Parameswaran, P. and Rittman, B.E. (2008). Using a pulsed electric field as a pretreatment for improved biosolids digestion and methanogenesis. *Proc. Of the 81st Annual Wat. Env. Tech. Exhibition and Conf.*, Chicago, 18–22, October.

Schraa, O.J. (2008). Low SRT flowsheet modelling report: part I. Prepared for P.M. Sutton & Assoc. and Basin Water, Hydromantis, Inc., October.

Silverira, M.L., Miyittah, M.K. and O'Conner, G.A. (2006). Phosphorus release from a manure-impacted spodosol: effects of water treatment residual. *Jour. of Env. Quality*, **35**, 529–541.

A review of struvite nucleation studies

S.C. Galbraith and P.A. Schneider[*]

School of Engineering, James Cook University, Townsville, Queensland, 4811, Australia

[*]Corresponding author

Abstract A review of struvite nucleation studies is given. Low supersaturation induction time experiments are presented. A thermodynamic solver was developed in Engineering Equation Solver (EES) which was validated against PHREEQc. This solver allowed meaningful comparisons to be made by processing the raw data under consistent thermodynamic conditions. Differences in the results that still remained are attributed to hydrodynamic variations. An image capture and analysis experimental method proved successful for the low supersaturation induction time experiments. Previous studies were unable to predict induction times at low supersaturation.

INTRODUCTION

Intentional struvite crystallisation, formed via Equation (1), is an attractive processing route for nutrient recovery from a variety of wastewater streams because of its potential as a sustainable fertiliser (Driver *et al.*, 1999; Li and Zhao, 2003). However, effective process design and optimisation requires the investigation of crystallisation mechanisms. This paper focuses its attention on struvite crystal nucleation by reviewing contributions to this field and conducting a nucleation study.

$$Mg^{2+} + NH_4^+ + PO_4^{3-} + 6H_2O \leftrightarrow MgNH_4PO_4 \cdot 6H_2O \tag{1}$$

BACKGROUND

Before the analysis of previous and current work it is important to address the solution thermodynamics that govern struvite precipitation. Some background theory to nucleation will also be given as well as in introduction to the previous research.

Solution thermodynamics

The formation of struvite is not as simple as Equation (1) suggests. There are complicated thermodynamics influencing the precipitation reaction. Free magnesium, ammonium and phosphate ions that react to form struvite are subject to a range of speciations that are pH dependant. It is therefore important to understand these equilibria in order to calculate the free ion concentrations and other thermodynamic properties of the system.

Thermodynamic equilibria

In water chemistry equilibrium reactions take place between various dissolved species in solution (Snoeyink and Jenkins, 1980). These equilibria form the basis for all subsequent thermodynamic calculations and measurements, the most important of these being supersaturation which is the most significant variable in any nucleation study (Mullin, 1993).

An early investigation of struvite thermodynamics conducted by Snoeyink and Jenkins (1980) based their calculations on the following species; Mg^{2+}, $MgOH^+$, NH_4^+, NH_3, PO_4^{3-}, HPO_4^{2-}, $H_2PO_4^-$ and H_3PO_4. As computing improved the list of species included in thermodynamic calculation expanded in subsequent research (Ohlinger et al., 1998; Ali and Schneider, 2005). The species included for this work were taken from Ohlinger et al. (1998) which were used in the nucleation investigations by Ohlinger et al. (1999). The equations and constants corresponding to the equilibrium reactions are located in Table 1.

Some other species that have been included in previous research are $MgNH_3^{2+}$, $Mg(NH_3)_2^{2+}$ and $Mg(NH_3)_2^{2+}$ (Bouropoulos and Koutsoukos, 2000). However, these complexes have little influence due to their low equilibrium constants, compared with other species (Ali and Schneider, 2005).

The various equilibria require activity coefficients to be calculated in order to determine ion species activities. These calculations are performed with Equations (2), (3) and (4).

$$I = \frac{1}{2} \sum C_i Z_i^2 \tag{2}$$

$$-\log \gamma_i = A Z_i^2 \left(\left[\frac{\sqrt{I}}{1 + \sqrt{I}} \right] - 0.3I \right) \tag{3}$$

$$\{i\} = \gamma_i C_i \tag{4}$$

where I = Ionic Strength, C_i = Concentration of species i, Z_i = Valency of species i, γ_i = Activity coefficient of species i, A = DeBye-Hückel constant (0.509 at 25°C, (Mullin, 1993)). Equation (3) is the DeBye-Hückel equation

Table 1. Thermodynamic equilibria and their governing equations.

Compound	Equilibrium Equation	Equilibrium Constant (K_i)	Reference
HPO_4^{2-}	$\dfrac{\{H^+\}\{PO_4^{3-}\}}{\{HPO_4^{2-}\}} =$	$10^{-12.35}$	(Morel and Hering, 1993)
$H_2PO_4^-$	$\dfrac{\{H^+\}\{HPO_4^{2-}\}}{\{H_2PO_4^-\}} =$	$10^{-7.20}$	(Morel and Hering, 1993)
H_3PO_4	$\dfrac{\{H^+\}\{H_2PO_4^-\}}{\{H_3PO_4\}} =$	$10^{-2.15}$	(Martell and Smith, 1989)
$MgPO_4^-$	$\dfrac{\{Mg^{2+}\}\{PO_4^{3-}\}}{\{MgPO_4^-\}} =$	$10^{-4.80}$	(Martell and Smith, 1989)
$MgHPO_4$	$\dfrac{\{Mg^{2+}\}\{HPO_4^{2-}\}}{\{MgHPO_4\}} =$	$10^{-2.91}$	(Martell and Smith, 1989)
$MgH_2PO_4^+$	$\dfrac{\{Mg^{2+}\}\{H_2PO_4^{3-}\}}{\{MgH_2PO_4^+\}} =$	$10^{-0.45}$	(Martell and Smith, 1989)
$MgOH^+$	$\dfrac{\{Mg^{2+}\}\{OH^-\}}{\{MgOH^+\}} =$	$10^{-2.56}$	(Childs, 1970)
NH_4^+	$\dfrac{\{H^+\}\{NH_3\}}{\{NH_4^+\}} =$	$10^{-9.25}$	(Taylor *et al.*, 1963)
H_2O	$\dfrac{\{H^+\}\{OH^-\}}{\{H_2O\}} =$	10^{-14}	(Harris, 2003)

with Davies approximation and was chosen due to its simplicity and accuracy at moderate ionic strengths (Mullin, 1993).

Supersaturation

Supersaturation is a term used to describe a solution where the solute concentration is greater than its value at equilibrium; precipitation is a direct result of supersaturation. As struvite contains three constituent ions, its solute concentration is defined by the ion activity product (IAP), see Equation (4). When the IAP is greater than the minimum solubility product (K_{so}) the system is supersaturated and struvite may nucleate and grow, returning the system to equilibrium.

When dealing with non-ideal, multi-component systems a more complex expression for supersaturation is required, such as supersaturation index, SI (Allison *et al.*, 1991; Parkhurst, 1999; Ali, 2005; Bhuiyan *et al.*, 2008), or supersaturation ratio, S_a (Snoeyink and Jenkins, 1980; Ohlinger *et al.*, 1999; Bouropoulos and Koutsoukos, 2000). Both expressions incorporate the ion activity product and the minimum solubility product. The equations for SI and S_a are given below in Equations (6) and (7), respectively.

$$IAP = \{Mg^{2+}\}\{NH_4^+\}\{PO_4^{3-}\} \tag{5}$$

$$SI = \log\left(\frac{IAP}{K_{SO}}\right) \tag{6}$$

$$S_a = \left(\frac{IAP}{K_{SO}}\right)^{1/3} \tag{7}$$

The minimum solubility product of struvite used in this research was that published by Ohlinger et al. (1998) where $pK_{so} = 13.26$. This value was used by Ohlinger et al. (1999) and Bouropoulos and Koutsoukos (2000). Bhuyian et al. (2008) used a Kso value of 13.36 (Bhuyian et al., 2007).

Thermodynamic solvers

In order to solve these equations simultaneously a computer program is required. Some common packages used in struvite research are PHREEQC (Bhuyian et al., 2008), MINTEQA2 or VisualMINTEQ (Ali, 2005), MINEQL+ (Ohlinger et al., 1999) and ChemEQL (Bouropoulos and Koutsoukos, 2000). These programs use the initial conditions of the system and a database of equilibrium information to calculate the concentration of all species in solution. For the purposes of understanding struvite thermodynamics more thoroughly, Engineering Equation Solver (EES) was used to solve the relevant nonlinear equation set that specifies the solution thermodynamics. The output from the EES solver was validated by comparing the ionic concentration of Mg^{2+}, NH_4^+ and PO_4^{3-} against the outputs of PHREEQC as seen in Figure 1. The outputs from both solvers are almost identical showing the EES solver to be numerically robust in handling these complex thermodynamic calculations.

Nucleation

Nucleation is the first step in the crystallisation process. It occurs when solute molecules come together in clusters and grow by accretion. They then coalesce to form large amounts of a new phase (Mullin, 1993). A period of time usually passes between the achievement of supersaturation and the appearance of crystal nuclei; this is called the induction time.

A common approach in nucleation studies and the one taken by the three studies being reviewed is measuring the induction time and applying classical nucleation theory which is based on homogeneous nucleation (Mullin, 1993).

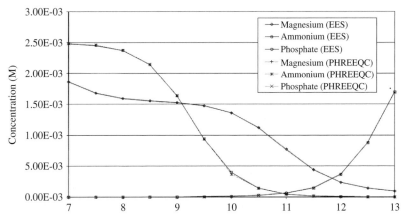

Figure 1. EES thermodynamic solver validated against PHREEQC at 0.0025M.

This produces a relationship between induction time and supersaturation shown in Equation (8), the derivation of this equation can be found in Mullin (2003). Homogeneous nucleation is the precipitation of nuclei in a solution free from foreign particles which is practically impossible to achieve in real solutions. However, the assumption of homogeneous nucleation is valid under conditions of high supersaturation (Mullin, 1993).

$$\log t_{ind} = \frac{A}{(\log S_a)^2} - B \tag{8}$$

A problem emerges in the relevance of this data to struvite reactor design as suspension reactors designed for crystal growth operate at low supersaturation in the metastable zone where crystal growth is favoured over nucleation. This is important as struvite reactors can encounter loss of product, fouling of process equipment, poor product quality and handling difficulties due to fines resulting from excess nucleation (von Munch and Barr, 2001, Battistoni *et al.*, 2005, Adnan *et al.*, 2003). Therefore it is a major aim of this work to gather nucleation data at low levels of supersaturation, which is relevant to reactor design and to compare these results with the predictions made by the other studies.

Previous work

Because of the complex nature of struvite thermodynamics it can be difficult to compare the experimental results of previous work. It has already been established in thermodynamic equilibria that Ohlinger *et al.* (1999) and Bouropoulos and Koutsoukos (2000) included different species in the thermodynamics and the species included by Bhuyian *et al.* (2008) are not specified. This means that the same initial condition inputs would give three different supersaturation outputs. In order to perform a valid comparison the raw data from the three previous studies was processed using the EES solver. A summary of the experimental conditions and techniques used by all studies is detailed in Table 2.

Table 2. Summary of experimental conditions and techniques.

	Ohlinger *et al.* (1999)	Bouropoulos and Koustoukus (2000)	Bhuyian *et al.* (2008)	This study (2008)
Concentration	4.0–20.0 mM	2.75–4.0 mM	56, 70 mg/L at 1:1:10 $Mg:PO_4:NH_4$	1.0, 2.5 mM
pH range	6.3–7.9	8.5	8.2–8.51	7.8–9.2
Temperature	22°C	25°C	25°C	22°C
Detection method	Laser scintilla-tions	pH change	pH change	Laser scintilla-tions
Thermodynamic solver	MINEQL+	ChemEQL v2.0	PHREEQC	EES
Induction time (sec)	13–2280	360–7500	12–500	999–26289

EXPERIMENTAL METHOD

Experiments were conducted at 25°C with 250 mL solutions at two levels of equimolar concentration (0.001 M and 0.0025 M) of Mg, NH_4 and PO_4. Supersaturation was established by adjusting solution pH with sodium hydroxide. Induction times were determined by monitoring light scintillations from a HeNe laser directed through the supersaturated solution and recorded with a low-light CCD camera placed perpendicular to the laser.

Prior to nucleation the solution is clear and therefore no light will be reflected. The induction time is determined taking images of the solution at fixed time intervals and measuring the time until scintillations are detected. JPEG images from the CCD camera were automatically archived to a high capacity disk drive at a suitable sampling frequency. Experiments could therefore be conducted unsupervised over extended periods, enabling the investigation of induction at low solution supersaturation.

Image files were subsequently processed using a custom MATLAB script, yielding the time course of average red light intensity versus time. Regression of the rate of change of red light intensity was used to determine induction times. A typical plot of light intensity and pictures captured are shown in Figure 2.

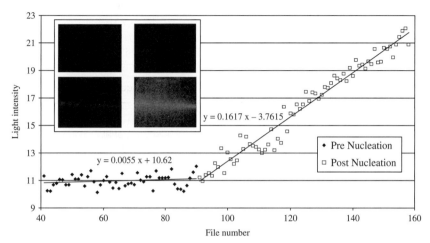

Figure 2. Light intensity versus the time stamped file number; the subfigure shows the light scintillations as nucleation onsets.

RESULTS AND DISCUSSION

The results from processing the raw data from previous studies through the EES thermodynamic solver are demonstrated in Figure 3. The differences between the studies are apparent and none of the studies have overlapping results. This is to be expected due to the different detection methods used and varying hydrodynamic conditions. The results also support the assumption that homogeneous nucleation is valid at high levels of supersaturation. This is demonstrated by R^2 values of the linear regression.

The idea that this assumption is no longer valid at lower levels of supersaturation is also supported by the results. Bouropoulos and Koustsoukos (2000) decreased supersaturation to see where the assumption of homogeneous nucleation no longer applies. This can be clearly seen in Figure 3 by the change in gradient. Furthermore it can be seen that the Ohlinger *et al.* (1999) data points begin lie further away from the regressed line as supersaturation decreases.

The experiments conducted for this work investigated nucleation behavior at low supersaturation; the results can be seen in Figure 4. It can be clearly seen that concentration influences induction time at low supersaturation, this is contrary to classical theory which states induction time is only a function of

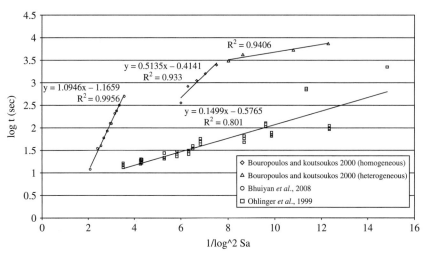

Figure 3. Logarithmic plots of induction time versus the inversed square of supersaturation ratio for the previous studies.

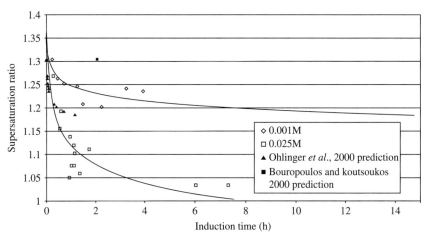

Figure 4. Induction time curves obtained by this study and predictions using previous studies.

supersaturation. This means that nucleation rates found at high supersaturation cannot be used for valid reactor design at low supersaturation.

Further evidence supporting this can be found by using the results of previous studies to predict the induction times at the levels of supersaturation used in this study. These predictions can be made by using Equation 8 which can be solved using the information in Figure 3. The results from these predictions were overlaid the results from this research in Figure 4.

It can be seen that the Ohlinger *et al.* (1999) predictions are in reasonable agreement at high supersaturation. Below $S_a = 1.2$ the induction time quickly approaches infinity though it is apparent in Figure 4 that nucleation does onset between 1 to 8 hours below $S_a = 1.2$. The Bouropoulos and Koutsoukos (2000) predictions only appear on Figure 4 at the highest level of supersaturation and the Bhuyian *et al.* (2008) predictions do not appear at all demonstrating predictions of infinite induction time at higher supersaturation levels than Ohlinger *et al.* (1999). The induction time curves themselves show clearly that, even at very low supersaturation, nucleation cannot be avoided indefinitely.

CONCLUSIONS

The method of image capture and analysis can be used to determine induction times at low supersaturation and the results produced with this method indicated that nucleation of struvite could not be avoided indefinitely at low super-saturation. The comparison of previous studies in a common thermodynamic solver demonstrated that hydrodynamic conditions and detection methods may account for variations in observed induction times. Comparing the nucleation information from previous studies to this work showed that they cannot be used to design low supersaturation crystallisers. As nucleation is practically unavoidable in struvite crystallization it is also concluded that a means of dealing with fine nuclei is not an option but a necessity in the design of a struvite recovery unit.

REFERENCES

Adnan, A., Koch, F.A. and Mavinic, D.S. (2003). Pilot-scale study of phosphorus recovery through struvite crystallization – II: Applying in-reactor supersaturation ratio as a process control parameter. *Journal of Environmental Engineering and Science*, **2**, 473–483.

Ali, M.I. (2005). Struvite Crystallisation from Nutrient Rich Wastewater. *Chemical Engineering*. Townsville, James Cook University.

Ali, M.I. and Schneider, P.A. (2005). Crystallization of struvite from metastable region with different types of seed crystal. *Journal of Non-Equilibrium Thermodynamics*, **30**, 95–111.

Battistoni, P., Paci, B., Fatone, F. and Pavan, P. (2005). Phosphorus removal from supernatants at low concentration using packed and fluidized-bed reactors. *Ind. Eng. Chem. Res.*, **44**, 6701–6707.

Bhuiyan, M.I.H., Mavinic, D.S. and Beckie, R.D. (2007). A solubility and thermo-dynamic study of struvite. *Environ. Technol.*, **28** (2007) 1015.

Bhuiyan, M.I.H., Mavinic, D.S. and Beckie, R.D. (2008). Nucleation and growth kinetics of struvite in a fluidized bed reactor. *J. Crystal Growth.*, **310**, 1187–1194.

Bouropoulos, N.C. and Koutsoukos, P.G. (2000). Spontaneous precipitation of struvite from aqueous solutions. *Journal of Crystal Growth.*, **213**, 381–388.

Childs, C.W. (1970). A potentiometric study of equilibria in aqueous divalent metal orthophosphate solutions. *Inorganic Chemistry*, **11**, 2465–2469.

Driver, J., Lijmbach, D. and Steen, I. (1999). Why recover phosphorus for recycling, and how? *Environmental Technology*, **20**, 651–662.

Harris, D.C. (2003). *Quantitative Chemical Analysis,* New York, W.H. Freeman and Company.

Li, X.Z. and Zhao Q.L. (2003). Recovery of ammonium-nitrogen from landfill leachate as a multi-nutrient fertilizer, *Ecological Engineering.*, **20**, 171–181.

Martell, A.E. and Smith, R.M. (1989). *Critical Stability Constants,* New York, Plenum Press.

Morel, F.M.M. and Hering, J.G. (1993). *Principles and Applications of Aquatic Chemistry,* New York, John Wiley and Sons.

Mullin, J.W. (1993). *Crystallization,* Oxford, Butterworth-Heinemann.

Munch, E.V. and Barr, K. (2001). Controlled struvite crystallisation for removing phosphorus from anaerobic digester sidestreams. *Water Res.*, **35**, 151–159.

Ohlinger, K.N., Young, T.M. and Schroeder, E.D. (1998). Predicting struvite formation in digestion. *Water Research*, **32**, 3607–3614.

Ohlinger, K.N., Young, T.M. and Schroeder, E.D. (1999). Kinetics effects on preferential struvite accumulation in wastewater. *J. Environ. Eng.*, **125**, 730–737.

Snoeyink, V.L. and Jenkins, D. (1980). *Water Chemistry,* New York, John Wiley & Sons.

Taylor, A.W., Frazier, A.W. and Gurney, E.L. (1963). Solubility products of di and trimagnesium phosphates and the dissociation of magnesium phosphate solutions. *Trans. Faraday Soc.*, 1585–1589.

A quantitative method analyzing the content of struvite in phosphate-based precipitates

X.-D. Hao[a], C.-C. Wang[a], L. Lan[a] and M.C.M. van Loosdrecht[b]

[a]The R & D Centre of Sustainable Environmental Biotechnology, Beijing University of Civil Engineering and Architecture, Beijing 100044. P. R. of China 7(E-mail: xdhao@hotmail.com)
[b]Dept. of Biochemical Engineering, Delft University of Technology, Julianalaan 67, 2628 BC Delft, the Netherlands (E-mail: m.c.m.vanloosdrecht@tnw.tudelft.nl)

Abstract Although X-ray Diffraction (XRD) can be applied to ascertain the existence of struvite (MAP: $MgNH_4PO_4 \cdot 6H_2O$) in phosphate-based precipitates, it is not an easy way to quantitatively determine the exact content of struvite. For this reason, a method to quantitatively determine the content of struvite was developed, in which acid dissolution was applied to analyze the elements in precipitates. Based on the NH_4^+-N content in precipitates, the content of struvite can calculated. This method was testified with both purchased pure struvite and formed pure struvite. At the same time, the effect of pH and Ca^{2+} on formation and crystallization of struvite was evaluated. The developed method was effective enough to determine the content of struvite, which could be an easy method than XRD.

INTRODUCTION

Chemical precipitation of struvite ($MgNH_4PO_4 \cdot 6H_2O$) from phosphate-rich wastewater (anaerobic supernatant/urine and/or animal wastes) is increasingly getting global attention for resource recovery and closing nutrient cycles. Some processes recovering phosphates as struvite have been theoretically and experimentally investigated by chemists, biochemists and civil engineers (Pastor *et al.*, 2008; Ronteltap *et al.*, 2007; Wilsenach *et al.*, 2007; Hao *et al.*, 2006; Ohlinger *et al.*, 1998; Abbona *et al.*, 1982; Taylor *et al.*, 1963).

The struvite in precipitates from phosphate recovery was traditionally characterized by X-ray diffraction (XRD) (Le Corre *et al.*, 2005; Stratful *et al.*, 2001; Doyle *et al.*, 2003; Sundaramoorthi and Kalaninathan, 2007; Kim *et al.*, 2007; Pastor *et al.*, 2008) and by scanning electron microscopy coupled with energy dispersive X-ray analysis (SEM-EDS) (Le Corre *et al.*, 2005). However, XRD might be only a qualitative method to judge if struvite is present in precipitates, simply by comparing the positions and intensities of XRD peaks, as described in literatures (Pastor *et al.*, 2008; Sundaramoorthi and Kalaninathan, 2007; Kim *et al.*, 2007; Le Corre *et al.*, 2005; Doyle *et al.*, 2003). No one has yet stated that the exact contents of struvite in precipitates were determined by XRD. In principle, detailed calculations based on the results of XRD could determine the contents of struvite. In practice, however, the calculating work is not so easy due to complicated and uncertain compositions in precipitates.

For this reason, it is necessary to develop an effective method to measure the contents of struvite in precipitates. The method to be developed might also contribute to determining the optimal reaction environments (pH, *etc.*) forming pure struvite or high struvite contents in precipitates. For this purpose, a method analyzing elements in precipitating crystals was developed with experiments, which includes pre-dissolution of precipitates by acid (like HCl) solution followed by element analyses.

MATERIALS AND METHOD

Analytical grade chemicals were used as received without further purification. A relatively pure struvite crystal (99.0% in the content) was used as a reference compound, which was commercially purchased from Alfa-Aesar (US).

Formation of single crystal of struvite

Reactant solutions:
I: 5 mmol $NaH_2PO_4 \cdot 2H_2O$ was dissolved in 0.5-l ultra pure water;
II: 6 mmol $MgSO_4 \cdot 7H_2O$ and 15 mmol NH_4Cl were dissolved in 0.5-l ultra pure water.

Solution II was moved into a 1.5-l plastic beaker, and then Solution I was gradually poured in the beaker. In the mixed solution, the optimal ratio of Mg:N:P was maintained at 1.2:3:1 (Wang *et al.*, 2007). A magnetic stirrer at the bottom of the beaker was used for stirring. After mixture, the pH of the solution was at 6.78. Precipitate was immediately filtered out, and the clear solution was transferred to a clean beaker with a filter-paper's cover. During the natural evaporation, some colourless rod-like crystals (maximal dimension: L2.0 × W1.5 × H0.2 cm) were formed.

Formation of phosphate-based precipitates

With the same solutions and method as above, the solutions were mixed with magnetic stirrers for 30 min. And then two NaOH solutions (a supersaturated solution followed by 1−M solution) were gradually added into the mixed solutions to reach to a fixed pH point from 6.5 to 11.5. After pH was adjusted each time, the solutions were continuously mixed for 30 min with a very low speed (to prevent formed crystals from crushing), and then aged for an hour to obtain big precipitates' crystals. Collected precipitates were finally filtered with filter papers and washed with ultra pure water for 2 times to remove potential soluble impurities like ammonium compounds. Filtrated precipitates were naturally dried and stored in desiccators also at the ambient temperature around 25°C.

With reference to the above ultra pure water's solutions, tap water was also used to make solutions and acquire phosphate-based precipitates with the identical methods mentioned above. Tap water mainly consisted of ground water, with a high mineral content: $c(Ca^{2+}) = 2.17$ mM and $c(Mg^{2+}) = 1.34$ mM.

Crystal characterization and image analyses

Crystal characterization of the single crystal of struvite and phosphate-based precipitates was performed by X-ray diffraction (Rigaku D/max IIIA). Micro-scopic image were analysed under a high-resolution colour digital camera (Zeiss: AxioCam MRc5).

Element analyses

A dissolution method was testified to implement the element analyses for all the acquired samples. Clear solutions were prepared by dissolving 40 mg of each sample with a little volume of concentrated HCl (pH < 1), and then diluted to 250 ml with ultra pure water. Clear solutions should be adjusted to a suitable pH point before the element analyses. $PO_4^{3-} - P$ was analyzed using Inductively Coupled Plasma Atomic Emission Spectrometer (He and Zhang, 2002), and Mg^{2+}, Ca^{2+} and $NH_4^+ - N$ were determined using Ion Chromatograph (Thomas et al., 2002).

Calculating the struvite content

The molar ratio of N, P and Mg in struvite is 1:1:1, as defined in the molecular formula of struvite, $MgNH_4PO_4 \cdot 6H_2O$. At high pHs, P and Mg tend to be bound into some impure compounds rather than struvite when Ca is existent. For this reason, the $NH_4^+ - N$ content could become the only reference element to calculate the struvite content in precipitates. In other words, 1 mole of $NH_4^+ - N$

could be equivalent to 1 mole of struvite. Therefore, the struvite content in precipitates can be calculated as below:

$$\text{Struvite content } (\%) = \frac{n_{NH_4^+-N} \times M_{struvite}}{m_{precipitates}} \times 100\% \qquad (1)$$

where, $m_{precipitates} = $ mass of the precipitates;
$\quad\quad n_{NH_4^+-N} = $ molar amounts of NH_4^+;
$\quad\quad M_{struvite} = $ molar mass of struvite.

RESULTS AND DISCUSSION

XRD and image analyses

At pH < 7.5, no visible precipitates were observed due to the low concentrations of Mg, N and P compounds. At pH > 7.5, white precipitates gradually emerged in the two solutions. XRD and microscopic images were applied to characterize the morphology of formed precipitates, in which a purchased pure struvite was characterized for reference, as shown in Figure 1.

No clear differences (i.e. position and intensity) can be observed by comparing the XRD patterns of formed precipitates, both from the solution with ultra pure water in the pH range of 7.5–9.0 and from that with tap water in the pH range of 7.0–8.5. As shown in Figure 1, the XRD images of these precipitates are quite similar to those of the purchased struvite and the formed single crystal struvite. Figure 1 also illustrates that precipitates' crystals are transparent and bigger at lower pHs. At pH(9.0 with ultra pure water, more impurities such as $Mg(OH)_2$ ($K_{sp} = 5.1 \times 10^{-12}$) and $Mg_3(PO_4)_2$ ($K_{sp} = 1.0 \times 10^{-24}$) probably emerged in precipitates (Wang et al., 2005), which resulted in more noisy XRD patterns with reduction in peak intensities and with change in peak positions. The existence of Ca^{2+} in tap water at high pHs would contribute to more impurities like $Ca_3(PO_4)_2$ ($K_{sp} = 2.1 \times 10^{-33}$) and even $CaHPO_4$ ($K_{sp} = 1.8 \times 10^{-7}$) (Yigit and Mazlum, 2007), besides $Mg(OH)_2$ and $Mg_3(PO_4)_2$ found with ultra pure water, which resulted in a very narrow pH range for forming transparent crystals.

The XRD patterns shown in Figure 1 reveal that increasing pH in the two solutions would result in deviating from the XRD patterns of the purchased struvite and formed single crystal struvite, which implies that a higher pH would impose a negative impact on forming pure struvite. Regretfully, it is almost impossible to directly determine the content of pure struvite in precipitates with only comparing peak intensities and positions of XRD images.

The microscope images (Figure 1) further reveal that transparent rod-like crystals could be only observed in the pH range of 7.5–9.0 with ultra pure water

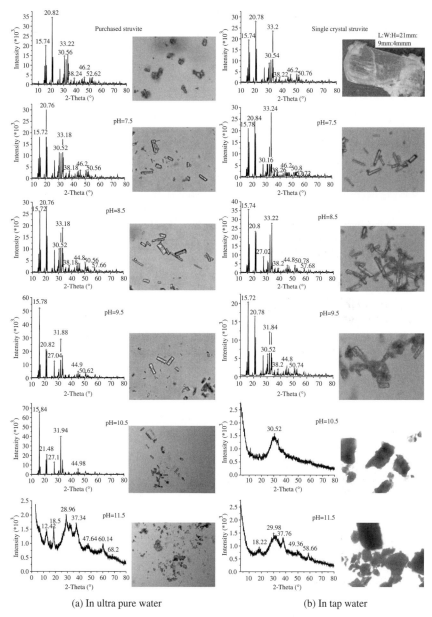

Figure 1. XRD and microscopic images (morphology) of formed precipitates at different pHs.

and in the pH range of 7.0–8.5 with tap water. At higher pHs with both solutions, precipitates became white and even colourful amorphous powders (no crystals at all), which implies that it was almost impossible to form pure struvite at higher pHs.

Element analyses

The element analyses indicated two sets of the measured molar concentrations for the purchased struvite and the single crystal struvite, respectively at N = 0.6456 mM, P = 0.6464 mM and Mg = 0.6532 mM and at N = 0.6510 mM, P = 0.6514 mM, and Mg = 0.6522 mM. These measured molar concentrations are very close to the calculated molar ones of 0.6520 mM for each element (N/P/Mg according to the struvite formula ($MgNH_4PO_4 \cdot 6H_2O$). On the one hand, this reveals that the formed single crystal struvite is really pure. On the other hand, this means that the element analyses can be reliably applied to determine the content of struvite from precipitates. For this reason, the element analyses of all precipitates were conducted and the results are shown in Figure 2 and Figure 3.

As shown in Figure 2, the $NH_4^+ - N$ content in precipitates decreased gradually at pH $\leq$ 10.5 and decreased sharply at pH $>$ 10.5, along with increasing pH in the original solution with ultra pure water. In principle, the NH_4^+ ratio in solution depends on the equilibrium with NH_3, as shown in Equation 2. As a result, the NH_4^+ ratio tends to decrease and NH_3 tends to volatilize from solution at higher pHs (Andrade and Schuiling, 2001). Therefore, pure struvite is hardly formed at a high pH.

$$NH_4^+ + OH^- \rightleftharpoons NH_3 + H_2O \qquad (2)$$

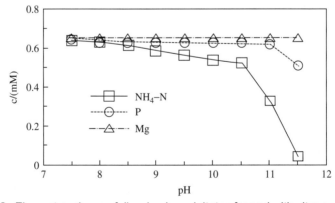

Figure 2. Element analyses of dissolved precipitates formed with ultra pure water.

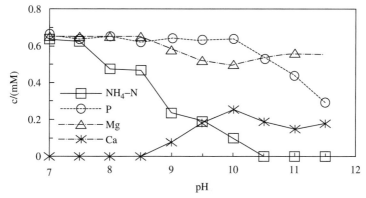

Figure 3. Element analyses of dissolved precipitates formed with tap water.

Figure 2 indicates that the P content did not change until pH $= 11.0$ and that the content of Mg did not change over all the pH range. In theory, precipitates of impurities like $Mg_3(PO_4)_2$ and $Mg(OH)_2$ might occur at high pH. The element analyses (Figure 2) and XRD analyses (Figure 1) testified indeed the presence of such impurities as $Mg_3(PO_4)_2$ and $Mg(OH)_2$.

Figure 3 indicates that the $NH_4^+ - N$ content in precipitates decreased rapidly along with increasing pH in the original solution with tap water. At pH ≥ 10.5, $NH_4^+ - N$ in precipitates disappeared at all, which implies impossibility for the existence of struvite in the precipitates. At pH ≤ 8.5, Ca^{2+} could not be detected in precipitates. At pH ≤ 8.5, however, Ca^{2+} emerged suddenly, as described in the literature (Le Corre *et al.*, 2005). This means that formation of struvite was seriously limited by Ca^{2+} compounds like $Ca_3(PO_4)_2$, $CaHPO_4$, *etc.* at pH > 8.5. Moreover, other impurities like $Mg_3(PO_4)_2$ and $Mg(OH)_2$ might coexist at pH > 8.5. Ca^{2+} present in the original solutions even affected the P content in precipitates at pH ≥ 10.0, which also resulted in a negative effect on formation of struvite.

Calculating the struvite content in precipitates

It is essential to acquire a pure struvite for testifying the accuracy of the calculating method of the struvite content in precipitates as shown in Equation 1. Besides the purchased pure struvite, both XRD and element analyses indicate that the colorless rod-like crystal formed with ultra pure water at pH 6.78 is really pure struvite. According to Equation 1, the contents of both purchased and formed pure struvite were thus calculated at 98.9% (the labelled content at

99.0% and 99.8% respectively. Obviously, Equation 1 can be reliably used to determine the content of struvite in precipitates, as shown in Figure 4.

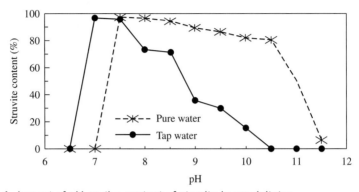

Figure 4. Impact of pH on the content of struvite in precipitates.

Figure 4 indicates that the content of struvite in precipitates decreased gradually in the pH range of 7.5 to 10.5 with ultra pure water being the solution, and then decreased sharply at pH > 10.5. Figure 4 also indicates that the content of struvite with tap water being the solution was controlled not only by pH, but also by the Ca^{2+} content in the original solution. The content of struvite was 96.8% and 95.7% at pH = 7.0 and 7.5 respectively. At pH > 7.5, the content of struvite decreased sharply, down to 15.5% at pH = 10.5. An even higher pH (> 10.5) resulted in complete disappearance of struvite in precipitates.

In principle, $CaNH_4PO_4 \cdot 7H_2O$ is probably formed in precipitates when Ca^{2+} exists in reaction solution (Takagi et al., 1984). If this situation happens, calculating the content of struvite based on $NH_4^+ - N$ could result in an error. However, the element analyses of precipitates formed with tap water being the solution reveal that no calcium precipitates were formed at pH ≤ 8.5 (Figure 3), which means that the contents of struvite calculated above ≤ 8.5 are accurate enough. A minor error might emerge at pH > 8.5 due to the existence of $CaNH_4PO_4 \cdot 7H_2O$.

A key procedure analyzing and calculating the exact content of struvite in precipitates is to determine the $NH_4^+ - N$ concentration, which was performed using Ion Chromatograph in the experiments. In fact, other methods like Nessler's reagent-colorimetry, salicylate-hypochlorous acid ameliorated method, flow injection analysis and ammonia gas-sensitive electrode method, could be also applied.

CONCLUSIONS

Some conclusions can be drawn as follow:

(i)   Although XRD can be applied to analyze the presence of struvite in precipitates, it is not an easy way to analyze the exact content of struvite in precipitates;
(ii) The element analyses could be a quantitative and efficient method to analyze the elements in precipitates;
(iii) Calculations based on the $NH_4^+ - N$ content can quantitatively determine the content of struvite in precipitates;
(iv) The content of struvite in precipitates was controlled by both pH and Ca^{2+} in the original solutions.

ACKNOWLEDGEMENTS

The study was financially supported by the China's High-tech R & D (863) Program (2006AA06Z320), Funding Project for Academic Human Resources Development in Institutions of Higher Learning Under the Jurisdiction of Beijing Municipality (BJE10016200611), and Scientific Research Common Program of Beijing Municipal Commission of Education(KM200910016009).

REFERENCES

Abbona, F., Boistelle, R. and Lundager, H. (1982). Crystallization of two magnesium phosphates, struvite and newberyite: effect of pH and concentration. *J. Cryst. Growth*, **57**, 6–14.
Andrade, A. and Schuiling, R. (2001). The chemistry of struvite crystallization. *Miner. J.*, **23**(5/6), 37–46.
Doyle, J., Oldring, K., Churchley, J., Price, C. and Parsons, S. (2003). Chemical control of struvite precipitation. *J. Environ. Eng.*, **129**(5), 419–426.
Hao, X.D. and van Loosdrecht M.C.M. (2006). Model-based evaluation of struvite recovery from P-released supernatant in a BNR process. *Water Sci. Technol.*, **53**(3), 191–198.
He, H. and Zhang, P. (2002). Determination of trace phosphorus in water by ICP-AES. *Chin. J. Spectrosc. Lab.*, **2**, 244–246.
Kim, D., Ryu, H., Kim, M., Kim, J. and Lee S. (2007). Enhancing struvite precipitation potential for ammonia nitrogen removal in municipal landfill leachate. *J. Hazard. Mater.*, **146**(1–2), 81–85.
Le Corre, K., Valsami-Jones, E., Hobbs, P. and Parsons, S. (2005). Impact of calcium on struvite crystal size, shape and purity. *J. Cryst. Growth*, **283**(3–4), 514–522.

Ohlinger, K., Young, T. and Schroeder, E. (1998). Predicting struvite formation in digestion. *Water Res.*, **26**(3), 2229–2232.

Pastor, L., Mangin, D., Barat, R. and Seco, A. (2008). A pilot-scale study of struvite precipitation in a stirred tank reactor: conditions influencing the process. *Bioresour. Technol.*, **99**(14), 6285–6291.

Ronteltap, M., Maurer, M. and Gujer, W. (2007). Struvite precipitation thermodynamics in source-separated urine. *Water Res.*, **41**, 977–984.

Stratful, I., Scrimshaw, M. and Lester, J. (2001). Conditions influencing the precipitation of magnesium ammonium phosphate. *Water Res.*, **35**(17), 4191–4199.

Sundaramoorthi, P. and Kalaninathan, S. (2007). Growth and characterization of struvite crystals in silica gel medium and its nucleation reduction process. *Asian J. Chem.*, **19**(4), 2783–2791.

Takagi, S., Mathew, M. and Brown, W. (1984). Structure of ammonium calcium phosphate heptahydrate, $Ca(NH_4)PO_4 \cdot 7H_2O$. *Acta. Crystallogr.*, **C40**, 1111–1113.

Taylor, A., Frazier, A. and Gurney, E. (1963). Solubility products of magnesium ammonium phosphate. *Trans. Faraday Soc.*, **59**, 1580–1584.

Thomas, D., Rey, M. and Jackson, P. (2002). Determination of inorganic cations and ammonium in environmental waters by ion chromatography with a high-capacity cation-exchange column. *J. Chromatogr. A.*, **956**(1–2), 181–186.

Wang, J., Burken, G.L., Zhang, X. and Surampalli, R. (2005). Engineered struvite precipitation: Impacts of component-ion molar ratios and pH. *J. Environ. Eng.*, **131**(10), 1433–1440.

Wang, Y., Cao, X., Meng, X., Li, D. and Zhang, J. (2007). Effect of Mg^{2+}, PO_4^{3-} and NH_4^+ in sludge dewatering filtrate on struvite formation. *China Water & Wastewater*, **23**(19), 6–9.

Wilsenach, J., Schuurbiers, C. and van, Loosdrecht, M. (2007). Phosphate and potassium recovery from source separated urine through struvite precipitation. *Water Res.*, **41**, 458–466.

Yigit, N. and Mazlum, S. (2007). Phosphate recovery potential from wastewater by chemical precipitation at batch conditions. *Enviro. Technol.*, **28**, 83–93.

Phosphorus removal from an industrial wastewater by struvite crystallization into an airlift reactor

A. Sánchez[a], S. Barros[b], R. Méndez[a] and J.M. Garrido[a]

[a]Chemical Engineering Department, School of Engineering, University of Santiago de Compostela, Campus Sur, E-15782 Santiago de Compostela, Spain.
[b]3R Ingeniería Ambiental, Vía Ptolomeo 3, E-15890, Santiago de Compostela, Spain.

Abstract The crystallization of struvite (magnesium ammonium phosphate, MAP, $NH_4MgPO_4 \cdot 6H_2O$) represents an interesting technique to recover this element from wastewater. The product recovered, struvite, is easy to hand and free of sludge-handling problems. Struvite crystallization has been demonstrated to be an economical way to remove phosphorous of wastewater effluents, especially at high phosphorous concentrations. Traditional systems based on orthophosphate precipitation with metal salts (Al or Fe) are expensive and generates a large amount of sludge that must be treated. In contrast, struvite has a commercial value as a fertilizer, so is supposed to be really beneficial in all senses for nutrient recovery.

In this work the performance of a three-phase reactor was studied. An airlift reactor was used to precipitate phosphate as struvite. Secondary treated industrial wastewater with a high phosphorus concentration was fed to the reactor. Wastewaters were generated in a fish-canning industry located nearby the shore of the sea. Magnesium should be added to the wastewater in order to precipitate struvite. For this reason, during an experimental period, seawater was used as magnesium source. Crystallization of struvite was observed. The main parameter that could affect the phosphorus removal was pH. The best periods of operation took place when pH is above 8.5 and close to 9. Amorphous calcium formation was observed during the period in which seawater is used as magnesium source. The formation of calcium phosphate had two different effects, one positive as acts as glue that promotes the agglomeration of struvite crystals and one negative as some fraction of this phosphate was washed out with the effluent. Seawater, for those facilities located near the seashore, is a reliable and inexpensive magnesium source for MAP crystallization.

INTRODUCTION

Chemical phosphorus removal from wastewaters is currently achieved by precipitation with Fe and Al salts or lime. These precipitation processes are widely used for removing phosphorus during the treatment of wastewater from diluted wastewaters. However the precipitation could be disadvantageous when the concentration of phosphorus, and the load treated, is very high as occurs for some industrial wastewaters. In such a case, a high amount of precipitates should be managed as a waste material. Thus, the operating cost due to the chemicals consumption in the plant and sludge management could be a drawback that could made uneconomic the precipitation of phosphates. Moreover, considering sustainability, the phosphorus recovery process from these precipitates is very difficult and expensive. Due to the above mentioned drawbacks, the precipitation of phosphate by crystallization of struvite (magnesium ammonium phosphate, MAP, $NH_4MgPO_4 \cdot 6H_2O$) could be an alternative for the sustainable and economical recovery of this compound from concentrated wastewaters.

Struvite is a white, crystalline substance of distinctive orthorhombic crystal structure (Abbona et al., 1984). Its density is 1710 kg/m^3. The crystallization of struvite represents an interesting technique to recover this element from wastewater. The product recovered, struvite, is easy to hand and free of sludge-handling problems. Struvite crystallization has been demonstrated to be an economical way to remove phosphorous in wastewater effluents, especially at high phosphorus concentrations. Additionally, struvite has a commercial value as a fertiliser, so is supposed to be really beneficial in all senses for nutrient recovery (Ghosh et al., 1996). Struvite usually precipitates as result of a reversible reaction, producing stable white orthorhombic crystals in a 1:1:1 molar ratio according equation 1:

$$Mg^{2+} + NH_4^+ + H_nPO_4^{n-3} + 6H_2O \leftrightarrow MgNH_4PO_4 \cdot 6H_2O + nH^+ \qquad (1)$$

Struvite has a solubility product, K_{sp}, of around $5.37 \cdot 10^{-14}$ (Ohlinger, 1998). As suggested by equation 1, struvite solubility is conditioned by the pH of the solution. It is highly soluble at acidic pH and highly insoluble at alkaline pH. Thus, its precipitation could be controlled by pH adjustment, as well as by alteration of the concentration of the ions indicated in equation 1. The pH range within which struvite may precipitate is between 7 and 11. This is closely related to the decrease in struvite's solubility when pH increases (Battistoni et al., 1997; Matynia et al., 2002).

Struvite crystallization from wastewaters may be achieved with different plant configurations: Complete Stirred Tank Reactors (Regy et al., 2001) and either Two-phase (Bhuiyan et al., 2008) or Three-phase Fluidized bed reactors (Battistoni et al., 1997; Matsumiya et al., 2000). The main aim of this research was to study the performance of a three-phase airlift reactor for recovering MAP from industrial wastewaters with a high phosphorus concentration. Wastewaters were generated in a frozen seafood factory located near the seashore. Magnesium concentration in the wastewater is low, and thus should be externally added in order to precipitate struvite in the reactor.

MATERIALS AND METHODS

A 2.7 L volume airlift reactor with a three-phase separator at the top of the system was used during the experiments of MAP crystallization (Figure 1). The system has a three phase separator at the top of the reactor. This device was designed for separating the treated water from the gas and the estruvite crystals. An external settler was coupled in series to the airlift reactor in order to avoid the outflow of fine particles with the effluent. Two different wastewaters with a similar P concentration, were fed to the reactor: A synthetic solution, and industrial wastewater with high P concentration. The objective of this strategy was to study and compare the origin of the influent in the efficiency of the MAP crystallization system.

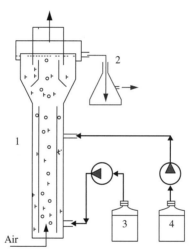

Figure 1. Schematic diagram of the Lab-scale airlift reactor for struvite crystallization. (1) Airlift reactor with a three-phase separator at the top of the unit; (2) External settler; (3) Wastewater and (4) Magnesium source.

The synthetic wastewater had a phosphate concentration similar to that observed in the industrial wastewater (average 630 mg $P-PO_4^{3-}/L$), and was prepared by diluting sodium phosphate in tap water. The industrial wastewater was a stream generated in a frozen seafood factory located in Galicia (NW Spain) with a very high phosphorus concentration. The wastewater used, have a previous secondary treatment in an aerobic biological pilot plant. This was done in order to reduce the COD concentration and hydrolyze the organic phosphorous and polyphosphates present in the raw influent.

Two different magnesium sources were used during the precipitation assays: A concentrated magnesium chloride solution and seawater. The seawater was collected at the industrial facility nearby the shore and contained around 1250 mg/L of Magnesium and 400 mg/L of Calcium. The reactor was operated during 128 days and the operation could be divided in three different stages:

Stage I: Days 0 to 34, Start up period: The Airlift reactor was operated semi-continuously and fed during 6 hours per day, feeding it with the synthetic medium. Ammonium chloride and magnesium chloride solutions were used as Magnesium and ammonium sources. During the operation the Hydraulic Retention Time (HRT) of the reactor was fixed at 6 h.

Stage II: Days 43 to 91, Operation with industrial wastewaters: During this period the reactor was operated continuously. The industrial wastewater secondary treated was fed to the reactor. Seawater was added as magnesium source. HRT ranged between 1.1 and 6 h.

Stage III: Days 103 to 128, Synthetic wastewater and the concentrated ammonia media were fed continuously. The differences with regard to stage I were the change in the strategy of operation of the reactor, from a semi-continuous to a continuous operation, and the use of seawater as magnesium source. HRT was maintained in 1.1 h.

Most of the analyses, including ammonia and phosphate, were done accordingly to the Standard Methods. Phosphate and ammonia concentration, Suspended solids, Conductivity, alkalinity, TOC and IC were measured in the influent and effluent from the reactor. Magnesium and Calcium were punctually determined in the effluent. Suspended solids concentration, temperature and pH were determined in the reactor. X-ray fluorescence and ICP-OES were used for determine the elemental composition of the precipitates and the effluent, respectively. Samples of struvite crystals were taken in the reactor or the settler for observation by using SEM with Energy Dispersive X-ray analysis (SEM-EDS). The nature of the crystals was identified by using X-ray diffraction (XRD).

RESULTS AND DISCUSSION

Figure 2 shows pH evolution all over the operation time. Reactor pH was controlled by using a NaOH solution, in order to maintain it above 7.5. Air stripping of a fraction of the inorganic carbon present in the influent was observed. The efficiency of the reactor depended on the pH of operation; in fact the better results in terms of low fine particles outflow and phosphate concentration in the effluent, were achieved when pH values were between 8.5

and 9.0 (Figures 2 and 3). These were independently of either the magnesium or phosphate source used. pH was manipulated either by adding NaOH into the reactor or by using free ammonia solution instead a solution with ammonium salts. Some authors referred optimum pH for MAP crystallization of 9.0 (Wang *et al.*, 2006) or even at 10 if seawater is used as magnesium source (Lee *et al.*, 2003). This was especially observed during period II in which the secondary treated wastewater was fed.

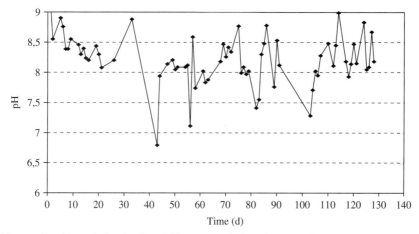

Figure 2. pH evolution in the airlift reactor during the experiments.

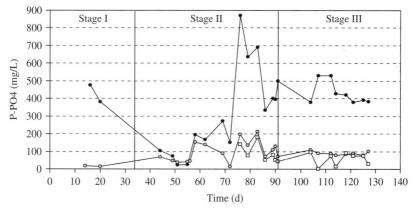

Figure 3. Evolution of phosphate concentration in the influent (●), soluble phosphate in the effluent (□) and total phosphate in the effluent (◓) of the airlift reactor.

Phosphorus recovery, through struvite precipitation, from either the synthetic or the industrial wastewaters was achieved during the three experimental periods in which the reactor was fed with different sources of phosphate and magnesium. Primary nucleation (crystal birth) was induced during the first experimental of the stage I, by raising the pH of the reactor, and later secondary nucleation (crystal growth) was observed. Struvite crystals concentration in the reactor ranged from 20 to 90 g/L during the operation. Obviously, the best results were obtained during the first operating period, with HRT of 6 h. Nevertheless, a high phosphate separation was also achieved under different feeding conditions, even with HRT of 1.1 h, by maintaining an adequate range of pH. Recovery efficiency (Phosphate precipitates that were retained into the reactor) was above 80% during most of the research.

The operation of the system during stage I, was very stable in terms of both P removal efficiency and concentration at the effluent. During the first days of stage 2 the P concentration of the industrial wastewater was lower than expected and pH was not controlled or manipulated in the reactor. Around day 43 and 80, pH drops in the reactor were observed, caused by insufficient alkali addition during the formation of struvite. The P removal efficiency increased when pH was higher than 8–8.5. One of the main differences of stage 2 with regard stage 1 was the higher P content in the effluent. Two different causes were identified: the washout of fines with the effluent and the operation during some days of this stage at pH below 8.0. During stage 2 seawater was employed as magnesium source and calcium present in it formed an amorphous precipitate of $Ca_3(PO_4)_2$ (Figures 4A and 4B) that tends to be washout easily. Some authors found that Struvite formation can be inhibited by the interaction of calcium and magnesium, depending on their relative concentrations (Battistoni et al., 1997). During this study the struvite formation was not inhibited by Ca present in the seawater, but the precipitation reactions of both struvite and an amorphous Ca phosphate occurred simultaneously. This fact also occurred during stage 3, in which the same synthetic wastewater used during stage 1 was fed. However, during this stage seawater, instead of MgCl solution, was used as Mg source.

Microscopical observations of precipitates using SEM showed different kinds of precipitates depending on the period of operation (Figure 4). During stage 1 in which synthetic wastewater was fed, small amounts of fine particles were detected in the effluent. Needle-form struvite crystals grew as monocrystrals, and did not agglomerate. Kinetics studies demonstrated that during this period magnesium excess was not a key factor in the formation of struvite, as the P efficiency was similar at molar ratios higher than 1:1 of Mg and P.

The precipitation of an amorphous calcium phosphate was identified when seawater was used, as this contained around 400 mg/L of Calcium. XRD

Figure 4. Microphotography of the struvite crystals during stages (A) 1; (B) 2; (C) 3 and (D) from the fines washed out from the reactor and retained in the external settler.

analysis demonstrated the amorphous nature of this precipitate. The most important feature of this Calcium phosphate [$Ca_3(PO_4)_2$] precipitate is that also acted like a glue for the struvite crystals in the reactor. The presence of agglomerates of struvite with amorphous calcium was observed by using SEM-EDS. Le Corre *et al.* (2005) also observed an increase of the struvite particle size formed, that was caused by the addition of Calcium during batch assays. In this sense the Calcium precipitate had a positive effect, as promotes the growth of aggregates of struvite.

Phosphate concentration in the effluent was influenced more by the operation pH than for the magnesium source. In this sense, seawater provides inexhaustible and inexpensive source for MAP precipitation in facilities near the seashore. Related to this fact, the cost of adding magnesium salts could be a major economic constrain to application of MAP crystallization for nutrient recovery. Nevertheless, as indicated, the addition of seawater also had a negative effect on Phosphorous recovery efficiency. Most of the P precipitates detected in the effluent were composed by amorphous calcium phosphate (Figure 4).

CONCLUSIONS

Crystallization of struvite was observed during the three periods studied and was not influenced by either the nature of the phosphorus or magnesium source used. The main parameter that could affect the phosphorus removal was pH. Regardless of the feed synthetic or real wastewater, best periods of operation took place when pH is above 8.5 and close to 9. pH may be controlled by adding NaOH or other suitable alkali.

Amorphous calcium formation was observed during the period in which seawater is used as magnesium source. The formation of calcium phosphate had two different effects, one positive as acts as glue that promotes the agglomeration of struvite crystals, and one negative as some fraction of this phosphate was washed out with the effluent. Seawater, for those facilities located near the seashore, is a reliable and inexpensive magnesium source for MAP crystallization.

ACKNOWLEDGEMENTS

To the Ministry of Education and Science through the Novedar-Consolider project (CSD2007-00055) and to the Consellería de Innovación e Industria, Xunta de Galicia (PGIDIT06TAM008E) that funded this project.

REFERENCES

Abbona, F., Lundager Madsen, H.E. and Boistelle, R. (1982). Crystallization of two magnesium phosphates, struvite and newberyite: Effect of pH and concentration. *Journal of Crystal Growth*, **57**(1), 6–14.
Battistoni, P., Fava, G., Pavan, P., Musacco, A. and Cecchi, F. (1997). Phosphate removal in anaerobic liquors by struvite crystallization without addition of chemicals: Preliminary results. *Water Research*, **31**(11), 2925–2929.
Bhuiyan, M.I.H., Mavinic, D.S. and Koch, F.A. (2008) Phosphorus recovery from wastewater through struvite formation in fluidized bed reactors: a sustainable approach. *Wat. Sci. Technol.*, **57**(2), 175–181.
Ghosh, G.K., Mohan, K.S. and Sarkar, A.K. (1996). Characterization of soil-fertilizer P reaction products and their evaluation as sources of P for gram (Cicer arietinum L.) *Nutrient Cycling in Agroecosystems*, **46**(1), 71–79.
Lee, S.I., Weon, S.Y., Lee, C.W. and Koopman, B. (2003) Removal of nitrogen and phosphate from wastewater by addition of bittern. *Chemosphere*, **51**(4), 265–271.

Le Corre, K.S., Valsami-Jones, E., Hobbs, P. and Parsons, S.A. (2005) Impact of calcium on struvite crystal size, shape and purity. *Journal of Crystal Growth*, **283**(3–4), 514–522.

Matsumiya, Y., Yamasita, T. and Nawamura, Y. (2000). Phosphorus removal from sidestreams by crystallisation of magnesium-ammonium-phosphate using seawater. *Journal of the Chartered Institution of Water and Environmental Management*, **14**(4), 291–296.

Matynia, A., Koralewska, J., Wierzbowska, B. and Piotrowski, K. (2006). The influence of process parameters on struvite continuous crystallization kinetics. *Chemical Engineering*, **193**(2), 160–176.

Ohlinger, K.N., Young, T.M. and Schroeder, E.D. (1998). Predicting struvite formation in digestion. *Water Research*, **32**(12), 3607–3614.

Regy, S. *et al.* (2001). Phosphate Recovery by struvite precipitation in a stirred reactor. CEEP-LAGEP report.

Wang, J., Song, Y., Yuan, P., Peng, J. and Fan, M. (2006). Modeling the crystallization of magnesium ammonium phosphate for phosphorus recovery. *Chemosphere*, **65**(7), 1182–1187.

Quantifying phosphorus recovery potentials by full-scale process analysis and modelling

M. Beier, R. Pikula, V. Spering and K.-H. Rosenwinkel

Institute for Water Quality and Waste Management (ISAH), Leibniz University of Hanover, Welfengarten 1, 30167 Hanover, Germany (beier@isah.uni-hannover.de)

Abstract The objective of a German joint research project of the universities of Darmstadt, Hannover and Karlsruhe (integrated in a superordinate research association supported by the Federal Ministry of Research) is to investigate and optimize the options of phosphorus recovery in municipal waste water treatment plants. The cooperation allows the participants to focus on the phosphorus recovery capacity in the liquid phase, the digester system and a possible elution from ashes. Additionally, different precipitation processes (struvite and calcium phosphate) are tested, and finally different concepts will be evaluated for specific facility constellations on the basis of the collected data. Within this project the evaluation tool – a new plant-wide modelling system for continuous phosphorus flows – will be created. In this paper you will find information about the basic approach, the enhanced model and first data from the measurements carried out in full scale plants and during laboratory and pilot plant experiments.

INTRODUCTION

There are currently two main ways to remove phosphorus from municipal wastewater:

1. Enhancing the phosphorus binding capacity of the sludge by the enrichment of phosphorus-accumulating bacteria in the system (enhanced biological phosphorus removal)
2. Via the precipitation of dissolved phosphates using Fe, Al precipitation, high- or low-lime precipitation or struvit precipitation, by adding different kinds of chemicals.

There are three main flows in the WWTP where precipitation can be carried out for P recovery as an additional goal of the treatment concept. First, "post-precipitation" in the effluent of a WWTP, secondly, "side-stream precipitation" – enhancing phosphorous release by installing anaerobic tanks and stripping units, and finally "sludge liquor precipitation" in the effluent of the digester. The last process step can be optimized by integrating a disintegration unit.

© 2009 The Authors, *International Conference on Nutrient Recovery from Wastewater Streams.* Edited by Ken Ashley, Don Mavinic and Fred Koch. ISBN: 9781843392323. Published by IWA Publishing, London, UK.

Which substream and which process combination will be the best for phosphorus recovery, considering the economic and ecological aspects, depends, among other things, on the process schema of the WWTP and on the particular composition of the wastewater. Due to the various links between the different process steps and the interaction between the process parameters it is nearly impossible to identify the best P recovery process for a specific plant situation just by comparing the different P recovery processes available. This is where the research project described below comes in. One target of the project is to create a basic model that includes the main processes for P removal and P recovery. This basic P-recovery-model can be used as a decision support system for the purpose of evaluating the different options, taking the specific situation into consideration.

Before the model can be used to identify the best process concept for phosphorus recovery and/or to determine the optimal operating point that matches the boundary conditions in each individual case, the following prerequisites have to be fulfilled:

1. Creation of a consistent flow model for all relevant P fractions encompassing all process steps of the WWTP (plant-wide-model)
2. Calibration and determination of the main kinetic parameters and operands used in the biological and/or chem./phys. models employed
3. Integration of the various phosphorus recovery process modules into the WWTP-model
4. Verification of the basic model by applying it to characteristic plants
5. If necessary, extension of the model for evaluation models (costs, energy, benchmarks) to compute the major parameters.

This model, with appropriate interfaces for integrating the various phosphorus fractions and with specific parameters to describe the various phosphorus displacements, has already been designed. Currently, data are being collected from laboratory-scale, semi-scale and full-scale plants to verify the process parameters in the model used and to develope additional conversion functions for process steps not taken into account in previous models.

DESCRIPTION OF RELEVANT FLOWS

Due to the shift from precipitation to biological enhanced phosphorus elimination in the 1990s, data on the quantities and bonding forms of the phosphorus in full-scale plants are hardly available. The research project will close this gap by literature research, extensive measurements, and a detailed balance sheet, recorded on at least four representative full-scale WWTPs. The different phosphorus fractions are determined distinguished by bonding type and process step.

Relevant flows and fractions

One advantage mapping the relevant phosphorus flows in WWTPs is that phosphorus is a non-volatile element therefore it is possible to set up a balance of solid and liquid phases for P. Based on the four most important partial flows in a WWTP as shown in Figure 1 phosphorus recovery potential can be quantified.

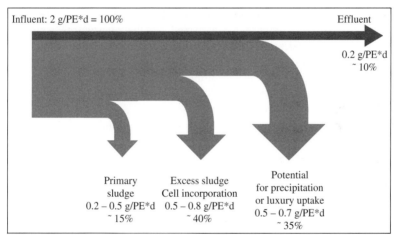

Figure 1. Relevant partial flows for the P recovery.

It is obvious that due to a very low content of solids it is easier to implement a post-precipitation process, however, only a maximum of 45% is recoverable, which depends on the preceding treatment step. In contrast thereto, all recovery processes following the digester or even a burning step (elution of ashes) allow a potential recovery of up to 90% however with a considerable increasing complexity of the phosphorus elution process from the sludge/ashes. And this complexity again depends on the applied P elimination concept of the WWTP, as this process step influences the bonding type, just as the different processing steps of sludge treatment do.

Correspondingly, it does not suffice anymore to determine only the total phosphorus load (P_{tot}) in order to estimate the suitability of various processes, or to investigate the optimum locations for the respective recovery procedures. The same applies to additionally measured phosphate concentrations in the liquid phase in order to quantify the phosphorus recovery potential. In fact, the rearrangement processes of the phosphorus are to be described along the process chain by means of the determination of the different P-fractions of the measured P_{tot} concentration for each process step.

Regarding the phosphorus recovery, four different phosphorus fractions of the total phosphorus content (P_{tot}) are to be distinguished:

- **Phosphate (PO_4-P):** The dissolved and precipitable part of P_{tot}. This part corresponds to the direct recoverable potential by precipitation.
- **Physiological fraction (P_{phys}):** The phosphorus fraction which is used for building up the cell structure – usually estimated to be 3% of the biomass. This part of P_{tot} can only be activated for the phosphorus recovery by disintegration processes. Around 50% of P_{phys} will usually redissolve during the mineralization process in the digester.
- **Luxury uptake (P_{lux}):** Additionally accumulated phosphorus by phosphorus-accumulating bacteria for energy supply under anaerobic conditions. The maximum luxury uptake can be estimated to be up to 4% of the biomass. P_{lux} will be redissolved under anaerobic conditions. One of the most common phosphorus recovery concept (Phostrip) is based on the use of this effect. In the digester 100% of the P_{lux} will be released by the bacteria.
- **Mineral fraction (P_{min}):** This is the minerally bonded fraction of the total phosphorus content – this part can usually be traced back to precipitation reactions, it is technically induced or results from a change in the chemical equilibrium e.g. in the digester.

Figure 2 shows exemplarily the rearrangement of the phosphorus fractions in a typical WWTP with mechanical treatment, biological enhanced phosphorus elimination and anaerobic sludge digestion. To extend the existing models (e.g. ASM2a, ADM) to these fractions is an integral part of the new created model-based assessment tools.

Comparison of literature and measured data

Based on the measurements a balance sheet model for different WWTPs was created. The calibration of the models is based on long term data collected from full-scale plants and data from a pilot plant for phosphorus recovery from surplus sludge. As a basis for the plant-wide dynamic simulation system for phosphorus recovery processes a stationary model of the complete plant was set up (see Figure 3). After the successful calculation of the mass balances the dynamic model was designed. The obtained results are in good agreement with the published standard assumptions.

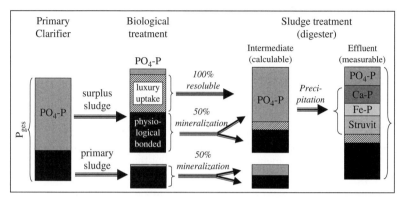

Figure 2. P fractions to be differentiated at the different process steps of a WWTP.

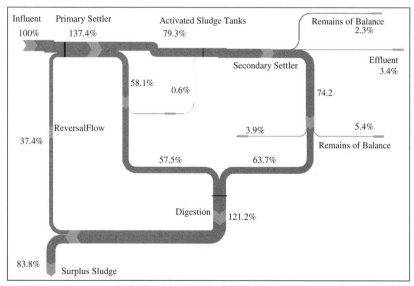

Figure 3. Mass balance for a WWTP (120.000 PE) using a stationary model approach.

MODELLING APPROACH

The current status in modelling wastewater treatment plants (WWTPs) is to look either at the activated sludge process or the processes of digestion. If the activated sludge system and the sludge digestion processes influence each other to a very great extent, this procedure has reached its limits. In this case, the approach of plant-wide modelling, which covers all important components and material flows of the WWTP, provides a way to solve these problems ([Jeppsson *et al.*, 2007]).

The evaluation of P recovery methods and processes in the liquid phase of the WWTP is one area of application for these kinds of models. For this purpose different techniques are being developed. Up to now, all of them have been implemented at an experimental or pilot-plant level. Models are used in order to upscale these processes to full-scale plants. The question of fixation processes during digestion and the choice of location for the recovery plant are of major interest.

In the existing sludge digestion models, phosphorus is not taken into account. As regards the example mentioned, it is essential that a phosphorus monitoring system be implemented in order to receive information about the dilution of any stored phosphorus and parallel precipitation and to be able to estimate phosphorus concentrations in the effluent and in reverse flows. However, the direct implementation of additional processes in the digestion models is difficult due to its complexity. Therefore in this script a conceptual approach involving a combination of digestion models in a plant-wide system is presented. This procedure is based on a modular approach which models all processes dealing with phosphorus separately, taking only the stated variables from the digestion model. The following figure 4 shows the concept behind the approach.

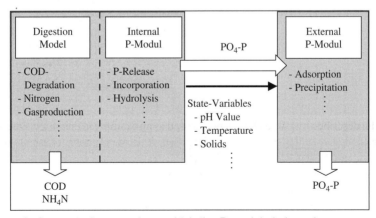

Figure 4. Conceptual approach on which the P module is based.

This method is only valid because the phosphorus concentration in the digester doesn't affect the anaerobic processes. The incorporation of phosphorus can be considered by calculating the sludge production. Several advantages of this method can be stated:

- The P process module can be coupled with any digestion model
- Calibration can be performed independently
- Processes can be defined more precisely

The particular purpose of modelling the phosphorus transport and degradation mechanisms is to gain information about internal processes which lead to changing concentrations in the effluent of the digestion tank. In the case study described in this paper the phosphorus concentration downstream of the digestion process rises to 160 mg/l PO_4-P. This leads to uncontrolled struvite precipitation in the pipe system upstream of the reactor due to CO_2 removal and shifts in the pH-value resulting from the latter.

In a plant-wide modelling system the biological stage is displayed and calibrated first. The luxury uptake of the phosphate-accumulating organisms (PAOs) is adapted to the phosphorus concentration in the effluent of the anaerobic reactor. The sludge production and the sludge retention time are displayed on the basis of the real system if the remaining processes of the activated sludge process are calibrated correctly. Based on these figures, the total phosphorus potential, i.e. the quantity that is recovered during digestion, can be calculated by using modelling values. For this purpose, additional processes are incorporated into the digestion models. Firstly, the internally-stored phosphate of the PAOs and the soluble phosphate are added together in the form of additional fractions. In addition, magnesium monitoring is incorporated to display the amount of magnesium that is released parallel to P release. The phosphate fixation processes are calculated externally with the exception of the quantities absorbed by the biomass. Hence, the additional processes only display the total quantity of phosphorus that can be potentially released during digestion. These processes have to be accounted for when defining the interfaces.

In order to describe the P fixation process in the digestion tank, the stationary model described by [Wild et al., 1997] is used as an example in this study. Due to the long sludge retention time during digestion, it is appropriate to use stationary approaches. In this model the individual precipitation processes are divided up into partial steps. They are defined as unspecified first-order reactions. The total phosphorus load that is released is calculated according to the organic degradation rate. Concerning the precipitation rate, four processes are accounted for, representing the fixation of iron, calcium, aluminium silicates, and struvite.

Furthermore, adsorption to solids is calculated. The values for the released phosphate and magnesium, calculated in the digestion model, are transferred to the P-fixation model. The reduction of the dissolved phosphate is calculated by means of the fixation processes. It is agreed, as regards the P fractions, that no re-dissolution of the precipitation products takes place and that the precipitation products can be defined as solids. Figure 5 shows the communication between the models.

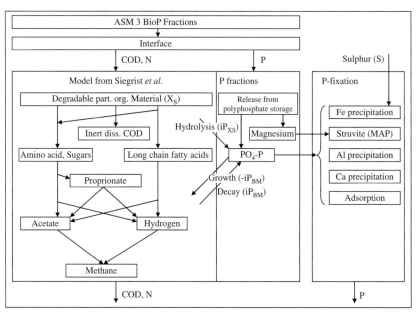

Figure 5. Method of coupling the digestion model ([Siegrist *et al.*, 2002]) with the P-fixation model (derived from [Wild *et al.*, 1997]).

Based on the results of this study, the following conclusions can be established:

1. For the calculation of continuous phosphorus flows it is necessary to incorporate P-fixation processes into the digestion models used. A method of combining different models of digestion and P fixation with each other has been presented. In a first step, a stationary P-fixation model was used.
2. This study has introduced a method of combining different models. The example shows the coupling of differently-scaled models. Dynamic P-fixation models, like kinetic-based equilibrium models, are much more complex if direct implementation is excluded. Therefore, the next step in the development of a fully-dynamic modelling system is the implementation of complex P-fixation models following the pattern presented.

RESULTS OF LABORATORY-, SEMI- AND FULL-SCALE MEASUREMENTS

The extensive data collection will be used among others for the improvement of the model regarding the following aspects:

1. Extension of the usually applied ASM2/ADM1 matrix for enhanced biological phosphorous removal (EBPR)
   - Influence of the amount and quality of the substrate
   - Influence of the treated part of return sludge
   - Amount of redissolved phosphorus from redissoluble phosphorus
   - Influence of nitrate in the return sludge
2. Differentiation in terms of process engineering, integration, dimensioning/ plant engineering
   - Influence of the anaerobic contact time
   - Degree of disintegration of different disintegration processes
   - Efficiency of separation procedures

Check of kinetic parameters of the enhanced biological phosphorous removal

Analysis shows that the allocated amount of substrate primarily influences the speed of the P redissolution. Further examination suggests that the influence of the substrate supply is being superposed by the quality of the substrate. Utilizable, organic acids like acetate and propionate are presumably used first as an energy source for the P-release. A substrate dose of 2:1 already leads to good results. Figure 6 shows exemplary the determined influence of the substrate on the redissolving behaviour. Redissolution of the increasingly retained phosphorus in the anaerobic basin shows a clear correlation to the existing amount of substrate.

This analysis is intended to improve the model. It provides data for a detailed representation of the rearrangement processes within the biological enhanced phosphorus elimination. These results may be integrated e.g. by an inhibiting factor of the reaction activities in dependence of the substrate supply.

Model enhancement

Figure 7 clearly shows the correlation between the redissolution of ortho-phosphate and magnesium under anaerobic conditions in WWTPs with biological enhanced phosphorus elimination. This correlation also allows the determination of the magnesium concentration present in other parts of the system like e.g. the digestion or in the redissolution basin in a bypass flow. Based on these data it is possible to calculate the required amount of precipitant.

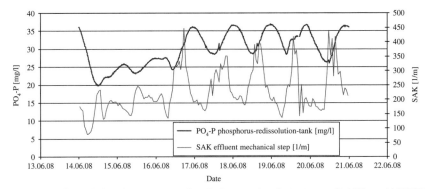

Figure 6. Correlation between redissolution and substrate availability – WWTP Husum.

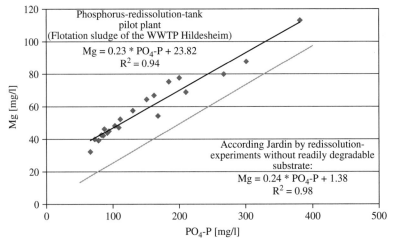

Figure 7. Correlation between phosphorus and magnesium redissolution.

Module integration "P recovery via struvite"

The process of phosphorus recovery is carried by first transforming the soluble PO_4-P into a solid phase and then separating it from the liquid. Suitable for this procedure is the struvite precipitation. Relevant operational parameters are the pH-value and the stoichiometric relation of the reactants magnesium, ammonium and phosphate. There is no existing process module for this process. Thus, it has to be developed, and the influence of the relevant operational parameters have to be determined. Figure 8 shows exemplary the influence of the pH-value on the struvite precipitation during labscale operation.

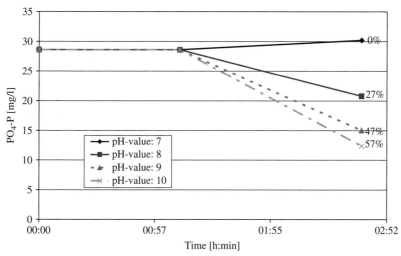

Figure 8. Struvite precipitation efficiency depending on the pH-value (increase after 60 minutes) in the presence of stoichiometric excess magnesium and ammonium at t = 0.

Analysis of the level of precipitation showed a minimum phosphorus concentration of approx. 30 mg/l. This value can only be reduced by increasing the pH-value (>8.5) through an extensive dosage of chemicals. Therefore, the precipitation procedures in the liquid phase become economical, if the phosphorus concentration after P-release exceeds 100 mg/l. This can be achieved through the use of concentrated sewage sludge. By using already settled surplus sludge (TSS > 6 g/l) P-concentrations of 600 mg/l could be realized without any additional hydrolysis. If the phosphorus concentration is that high, no additional increase of the pH-value is necessary. Furthermore, the results prove that a pH-value >8 led to a shift of the precipitation products from struvite to calcium phosphate.

FUTURE PROSPECTS

Among optimisation of precipitation the optimisation of phosphorus-redissolution plays a crucial role regarding to utilise as much as possible potential of phosphor-recovery. The data basis obtained in full-scale plants is currently enlarged and compacted. Measurements are carried out in summer and winter in order to integrate seasonal fluctuations in the model due to varying inflow characteristics and temperature influences.

In addition, disintegration procedures (high-pressure homogenisation, ultrasonic disintegration and electroporation) are examined in order to integrate them as additional modules into the model. Also a hydrocyclone as a separation unit for CaP and struvite will be included. As regards this aspect, some experiments have already been carried out which will be completed by further test series.

Data are collected for all examined processes and procedures to determine their efficiency, costs and energy consumptions. The thus created DSSM (decision support system model) will allow to identify the best procedures and their phosphorus recovery potential for the specific situation of each WWTP. It will also provide corresponding evaluation parameters like energy demand and costs per recoverable amount of phosphorus.

REFERENCES

Jardin, N. (1995). Untersuchungen zum Einfluss der erhöhten biologischen Phosphorelimination auf die Phosphordynamik bei der Schlammbehandlung Schriftenreihe WAR, Band 87.

Jeppsson, U., *et al.* (2007). "Benchmark simulation model no 2: general protocol and exploratory case studies." *Water Science and Technology*, **56**(8): 67–78.

Rosenwinkel, K.-H., Beier, M., Pabst, M. (2005). Optimierung der Phosphorrückgewinnung aus der wässrigen Phase mittels MAP, 75. Darmstädter Seminar, Schriftenreihe WAR, Band 167.

Siegrist, H., *et al.* (2002). "Mathematical model for meso- and thermophilic anaerobic sewage sludge digestion." *Environmental Science and Technology*, **36**(5): 1113–1123.

Wild, D., *et al.* (1997). "Prediction of recycle phosphorus loads from anaerobic digestion" *Water Research*, **31**(9): 2300–2308.

Validation of a comprehensive chemical equilibrium model for predicting struvite precipitation

Sachin Gadekar[a], Pratap Pullammanappallil[a] and Amir Varshovi[b]

[a]Department of Agricultural and Biological Engineering University of Florida Gainesville, Fl 32611-0570 USA
[b]GreenTechnologies LLC, Gainesville, FL 32635-7905, USA

Abstract Nutrients like ammonia and phosphate in wastewater can be recovered by precipitating these as struvite, a magnesium ammonium phosphate compound. Predicting struvite precipitation potential, yield and purity is important for the design and operation of reactors for struvite precipitation. In this paper a mathematical model of this precipitation process is developed for a closed system using physicochemical equilibrium expressions, mass balance equations for nitrogen, phosphorous and magnesium, and charge balance. Dissolved and ionic species modeled are NH_3, NH_4^+, PO_4^{3-}, HPO_4^{2-}, $H_2PO_4^-$, H_3PO_4, $MgOH^+$, $MgH_2PO_4^+$, $MgPO_4^-$, $MgHPO_4$(dissolved), liquid phase inorganic carbon as CO_3^{2-}, HCO_3^- and H_2CO_3, liquid phase organic carbon as CH_3COOH and CH_3COO^-, Ca^{2+}, Na^+, K^+, Mg^{2+}, H^+, Cl^- and OH^-. Fifteen different precipitates are also modeled: $MgNH_4PO_4 \cdot 6H_2O$, $Mg_3(PO_4)_2 \cdot 8H_2O$, $Mg_3(PO_4)_2 \cdot 22H_2O$, $Mg(OH)_2$, $MgHPO_4$, $CaMg(CO_3)_2$, $CaMg_3(CO_3)_4$, $MgCO_3$, $MgCO_3 \cdot 3H_2O$, $CaCO_3$, $Ca_5(PO_4)_3OH$, $Ca_3(PO_4)_2$, $Ca_8(HPO_4)_2(PO_4)_4 \cdot 5H_2O$, $CaHPO_4$ and $CaHPO_4 \cdot 5H_2O$. The model was simulated to explicitly solve for equilibrium concentrations of all the dissolved, ionic and solid species for a given set of initial concentrations of ammonium, magnesium, calcium, total inorganic carbon, organic carbon, phosphate, and pH. The model predictions were validated against our experimental data using synthetic wastewater and also data from literature. The model satisfactorily predicted all data. The purity of struvite in the precipitate was dependent on the initial concentrations of ammonium, calcium, total inorganic carbon, magnesium and phosphate. Struvite fraction in the precipitate increased as magnesium became limiting, or as ammonia to phosphate ratio increased and magnesium to phosphate ratio decreased. In presence of magnesium, phosphate, calcium and carbonate species, the struvite component is always a fraction of the total solids, hence it is important to clearly distinguish between struvite and other solid components in the mixture.

INTRODUCTION

Struvite is a magnesium ammonium phosphate mineral with the chemical formula $(NH_4)MgPO_4 \cdot 6H_2O$. It is a premium grade slow releasing fertilizer because it is sparingly soluble in water and also finds uses as a raw material in

© 2009 The Authors, *International Conference on Nutrient Recovery from Wastewater Streams.* Edited by Ken Ashley, Don Mavinic and Fred Koch. ISBN: 9781843392323. Published by IWA Publishing, London, UK.

the phosphate industry, for making fire resistant panels and as a binding material in cements (Sarkar, 1990; Schuiling and Andrade, 1999). Struvite precipitation can also serve as means for removal of nutrients like nitrogen and phosphorus in wastewater. Nutrient removal from wastewater is becoming an increasing challenge for operators, as regulatory authorities tighten discharge standards to avoid eutrophication problems in receiving waters. Significant costs are associated with the extra treatment processes required to meet the discharge standards (Wang *et al.*, 2006). The most widely used technologies for nutrient removal include biological nitrification/denitrification for nitrogen removal and metal salt precipitation or biological treatment for phosphorus removal. Both approaches result in the nutrient being made unrecoverable as a useful product. An alternative to these conventional technologies would be the precipitation of both nitrogen (in the form of ammonia) and phosphate as struvite which is a valuable product. Struvite recovery helps to meet legal requirements imposed on wastewater disposal and reduce the area needed for land application.

Predicting struvite precipitation potential is important to designers and operators for the development and operation of reactors for struvite precipitation. For process control, it essential to know the conditions under which struvite precipitation is likely to occur. Upon mixing salts of magnesium, ammonium and phosphate several ionic and dissolved species are formed in addition to precipitates including struvite. Various researchers have developed mathematical models for struvite precipitation (Harada *et al.*, 2006; Loewenthal *et al.*, 1994; Ohlinger *et al.*, 1998; Scott *et al.*, 1991; Wang *et al.*, 2006). These models are based on physico-chemical equilibrium of the various ionic, dissolved and solid species. A struvite precipitation model at least requires the incorporation of concentrations of ionic species: NH_4^+, PO_4^{3-} and Mg^{2+}, dissolved species: NH_3 and H_3PO_4 and solid species $MgNH_4PO_4$. However, a number of other ionic species (e.g. HPO_4^{2-}, $H_2PO_4^{2-}$, $MgOH^+$, $MgPO_4^-$ $MgH_2PO_4^+$, dissolved species (e.g. H_3PO_4, $MgHPO_{4(dissolved)}$) and solid species (e.g. $Mg_3(PO_4)_2 \cdot 8H_2O$, $Mg_3(PO_4)_2 \cdot 22H_2O$, $Mg(OH)_2$, $MgHPO_{4(solid)}$) exist in equilibrium. The complexity of models depends on the number of soluble and solid species considered.

Loewenthal *et al.* (1994) predicted struvite precipitation potential in synthetically prepared solutions that mimicked anaerobic digester effluents. This was a simple model that considered struvite as the only solid species, ionic species considered were Mg^{2+}, NH_4^+, PO_4^{3-}, HPO_4^{2-}, $H_2PO_4^{2-}$, and dissolved species were NH_3 and H_3PO_4. In addition to the above H_2CO_3, CH_3COO^-, CH_3COOH, carbonate and bicarbonate were also considered. Ohlinger *et al.* (1998) also considered only struvite as the solid species in their model. But, in addition to species modeled by Loewenthal *et al.* (1994), they also included $MgH_2PO_4^+$ and

$MgPO_4^-$ as these complexes exert a strong influence on equilibrium conditions. The model also included ionic strength effects. Wang *et al.* (2006) included formation of $Mg(OH)_2$ solid in addition to struvite and also considered $MgHPO_4$ as a dissolved species. Scott *et al.* (1991) modeled five solid species; struvite, $Mg_3(PO_4)_2 \cdot 8H_2O$, $Mg_3(PO_4)_2 \cdot 22H_2O$, $Mg(OH)_2 \cdot 6H_2O$ and $MgHPO_4$ and all the dissolved and ionic species considered by Wang *et al.* (2006). Harada *et al.* (2006) developed a model to predict struvite formation in urine. In all eight solid species were modeled; calcium precipitates: $Ca_3(PO_4)_2$, $CaHPO_4$, $Ca(OH)_2$, $CaCO_3$ and $CaMg(CO_3)_2$ and precipitates containing Mg, namely struvite, $Mg(OH)_2$ and $MgCO_3$. As the number of solid and soluble species considered increases, complexity of model also increases so analytical solutions are no longer possible and hence numerical solution is needed. Loewenthal *et al.* (1994) and Ohlinger *et al.* (1998) employed an iterative technique to converge one concentration value to an experimentally measured value while the other concentrations were calculated from equilibrium expressions. Scott *et al.* (1991) did not explicitly determine the concentrations of dissolved, ionic and solid species but compiled operating curves that related pH with total concentrations of magnesium, nitrogen and phosphorus in liquid under struvite recovery conditions. Only the mole fractions of the solids were computed. Harada *et al.* (2006) simplified the solution procedure by limiting the formation of solid mixtures to 14 patterns that was generated *a priori* by including or excluding various solid species. The concentrations of dissolved and ionic species were calculated for each pattern and the concentrations that gave reasonable values were chosen.

The more complex models described above are not easily amenable for developing dynamic models to predict struvite precipitation in dedicated reactors. In this paper a comprehensive model is used to explicitly determine the concentrations of all species (dissolved, ionic and solid) to enable investigation of the purity and yield of struvite for various operating conditions (pH and ratios of initial NH_4^+, Mg^{2+} and PO_4^{3-}). It uses mass and charge balances in addition to the physico-chemical equilibrium equations. The model described here considers 15 different solid species which is the maximum number reported in the literature (Çelen *et al.*, 2007). The model was validated by comparing it to experimental data from the literature and data obtained from our experiments.

MODEL FORMULATION

Assumptions

The model describes the evolution of a closed system in the presence of ionic species like ammonium (ammonium chloride), magnesium (magnesium chloride) and phosphate (potassium phosphate). It was formulated based on the following assumptions:

- Dissolved and ionic species present in the system are NH_3, NH_4^+, PO_4^{3-}, HPO_4^{2-}, $H_2PO_4^-$, H_3PO_4, $MgOH^+$, $MgH_2PO_4^+$, $MgPO_4^-$, $MgHPO_4$ (dissolved), CO_3^{2-}, HCO_3^{2-}, H_2CO_3, CH_3COOH, CH_3COO^-, Ca^{2+}, Na^+, K^+, Mg^{2+}, H^+, Cl^- and OH^-.
- Fifteen different precipitates are produced which are listed in Table 1.
- pH was kept constant by addition of NaOH or HCl. This was simulated by adjusting the variable 'exions', which was set equal to $\sum[$cations modeled$] - \sum[$anions modeled$]$.
- Reactions are at equilibrium.
- Reactions proceed in a closed system or a batch reactor.
- Reactions occur at room temperature (25°C).
- Activity coefficients were assumed to be unity.
- Effect of ionic strength on activity was neglected.

Table 1. List of solids included in the comprehensive model.

Number	Chemical name/Commercial name	Chemical formula
1	Magnesium ammonium phosphate, Struvite	$MgNH_4PO_4 \cdot 6H_2O$
2	Magnesium hydrogen phosphate, Newberyite [MHP]	$MgHPO_4$
3	Magnesium phosphate, Bobierrite [MP8]	$Mg_3(PO_4)_2 \cdot 8H_2O$
4	Magnesium phosphate, Cattiite [MP22]	$Mg_3(PO_4)_2 \cdot 22H_2O$
5	Hydroxyapatite [HAP]	$Ca_5(PO_4)_3OH$
6	Tricalcium phosphate, Whitlockite [TCP]	$Ca_3(PO_4)_2$
7	Monenite [DCP]	$CaHPO_4$
8	Octacalcium phosphate, [OCP]	$Ca_8(HPO_4)_2(PO_4)_4 \cdot 5H_2O$
9	Dicalcium phosphate dihydrate, Brushite [DCPD]	$CaHPO_4 \cdot 2H_2O$
10	Calcium carbonate, Calcite	$CaCO_3$
11	Magnesium carbonate, Magnesite	$MgCO_3$
12	Nesquehonite	$MgCO_3 \cdot 3H_2O$
13	Dolomite	$CaMg(CO_3)_2$
14	Huntite	$CaMg_3(CO_3)_4$
15	Magnesium hydroxide, Brucite	$Mg(OH)_2$

Equations

The model considers overall mass balance for magnesium, nitrogen and phosphorus, electro-neutrality, and physico-chemical and solubility equilibrium equations to describe the system. Values of equilibrium constants and solubility products, at 250C are taken from Scott (2001), Harada *et al.* (2006) and Moon *et al.* (2007). Of particular importance is the solubility product for struvite. Various values have been reported ranging between 12.6–13.26 (Ohlinger *et al.,* 1998). In the present model, a value of 12.7 was used unless stated otherwise.

Polymath Educational Version 6.1 was used to solve the model equations. Initial conditions which include pH, total concentrations and mass balance equations for nitrogen, magnesium, phosphorus, total inorganic carbon and calcium are input along with all equilibrium constants. Initial guesses for Mg^{2+}, NH_4^+, $PO_4^{3-} \cdot CO_3^{2-}$ and Ca^{2+} are needed. The Polymath program solves the expressions explicitly and gives concentrations of dissolved and ionic species and concentrations of solid components. Using the charge balance equation the appropriate exion concentration was determined. This gives the acid or base requirement for the given pH.

RESULTS AND DISCUSSION

First the model was validated by comparing its predictions to twelve struvite precipitation studies collected from literature that were carried out using synthetic and real wastewater at different pH values and initial concentrations of magnesium, phosphate and ammonium. The initial concentrations of these ions in the solution mixtures were equal to the total concentrations of magnesium, nitrogen and phosphorus species, namely Mg_T, N_T and P_T in the system. Data from model simulations is shown in Table 2. Five solid precipitates containing magnesium are listed in the table but the model considers a total of fifteen solids mentioned earlier.

The model was found to satisfactorily predict struvite formation for all literature data. Error between model predictions and experimental measurements varied between 1.2% and 24.5%. Except for two data sets the rest of the model predictions were less than 7.5%. Five of these data sets marked by (*) in Table 2 need further explanation. Closer inspection of experimental methods described in these references revealed that the struvite concentration reported was actually total solids concentration. For example, Loewenthal *et al.* (1994) estimated struvite concentration from differences in average molar concentration of Mg^{2+}, NH_4^+ and PO_4^{3-} between initial and final values in solution of these species. Since Mg^{2+} and PO_4^{3-} can form insoluble compounds other than struvite, for example MP8, MP22, MHP and $Mg(OH)_2$, differences in initial and final concentrations does not truly represent concentration of struvite but represents concentration of solids. Using experimental conditions of Loewenthal *et al.* [9]

Table 2. Comparison of model predictions with experimental measurements.

Ref.	Type of wastewater	Initial concentrations mM				Exptl struvite (mg/L)	Model predictions						Predicted struvite fraction in solids (%)	Error$ in model prediction of struvite (%)
		pH	Mg_T	P_{T-}	N_T		Struvite (mg/L)	MP8 (mg/L)	MP22 (mg/L)	$Mg(OH)_2$ (mg/L)	MHP (mg/L)	Total solids (mg/L)	(%)	(%)
*Loewenthal et al. (1994)	Solutions prepared by adding NH_4Cl, K_2HPO_4, $MgCl_2$, carbonate and acetate	6.8	8.23	12.9	21.43	601*	300.2	0.001	0	0	432.9	733	40.95	18
*Harada et al. (2006)	Synthetic urine containing PO_4, NH_4, Na, Mg, K, Ca, Cl, citrate, carbonate	8	20	13.45	20.18	1685*	1317	88	11.54	17.72	222.9	1687.7	78	1.23
Wilsenach et al. (2007)	Synthetic urine containing PO_4, NH_4, Na, Mg, K, Ca, Cl, citrate, carbonate	9.4	7.415	14.83	18.7	1045	987.07	0.001	0	0.061	14.75	1001.89	98.52	5.87
Wilsenach et al. (2007)	Synthetic urine containing PO_4, NH_4, Na, Mg, K, Ca, Cl, citrate, carbonate	9.4	14.83	14.83	18.7	2011	1845.03	7.03	0.922	11.625	80.25	1944.86	94.87	9
Çelen et al. (2007)	Liquid swine manure	8.5	2.39	5.51	80	338	322.2	0	0	0	2.7	324.9	99.17	4.9
Much and Barr (2001)	Supernatant from anaerobically digested sludge dewatering centrifuge	8.5	1.51	1.9677	43.88	195	200.86	0	0	0.003	3.08	203.95	98.49	2.92

Reference	Sample													
Yoshono et al. (2003)	Anaerobic digester effluent supernatant	8.5	7.025	6.387	24.5	805	818.23	1.006	0.132	12.157	29.68	883.02	92.66	1.62
*Tünay et al. (1997)	Synthetic samples prepared by using $MgCl_2$, NaH_2PO_4, NH_4Cl	9	14.26	14.26	14.26	1714*	1017.7	44.44	5.82	284.9	4.75	2273	44.75	24.5
*Altinbas et al. (2002)	Domestic Wastewater + 2% landfill leachate (DWL3 sample)	9.2	7.785	7.785	7.785	1420	1495	7.83	1.02	7.7	104.1	1616.1	92.5	5.01
Battistoni et al. (1998)	Supernatant from sludge centrifuges in a biological nutrient removal plant	8.12	1.54	2	44.5	210.98	198.21	0	0	0.001	6.56	204.77	96.79	6.44
Burns et al. (2001)	Swine Waste	9	9.736	6.085	12	758.6	705.47	44.963	5.896	161.05	54.53	971.91	72.59	7.53
*Stratful et al. (2001)	Deionized water with varying concentration of Mg^{2+}, NH_4^+ and PO_4^{3-}	10	7.692	7.81	14.77	1629	1757	0.67	0.04	17.45	15	1790.5	98	7.2

$ Error in struvite concentration predicted by model and that measured experimentally.

* For these data the error between solids concentration predicted by model and the struvite reported in experiments was calculated.

the model predicted total solids concentration of 733 mg/L of which struvite was only 300 mg/L with MHP being other dominant solid. Experimentally reported struvite was 601 mg/L which in reality could be the total solids concentration and was close in agreement to the total solids predicted by model. No data on calcium concentration was reported in this paper.

Harada et al. (2006) reported that most phosphate in their experiments was precipitated in the form of struvite. Initial total calcium concentration reported is 1.21 mmol/L which was input for validation. Experimental data indicated that 12.3 mM of PO_4^{3-} was precipitated as struvite(1685 mg/L). This value, however, was approximately equal to the total solids predicted (1706.2 mg/L) by the model. Concentrations of NH_4^+ and Mg^{2+} used here were higher than concentration of phosphate. The model showed that other significant components that could make up the solid phase included MP8 (88 mg/L), MHP (222.9 mg/L) and $CaHPO_4$ (15.6 mg/L) and $CaHPO_4 \cdot 5H_2O$ (7 mg/L). Hence under phosphate limiting conditions, struvite is not the only component in the precipitate.

Tunay et al. (1997) used a 1:1:1 stoichiometric proportion of ammonia: magnesium phosphate for experiments. At the conclusion of experiments the solution was filtered and concentrations of total ammonia and phosphate measured in the filtrate. The difference in the initial and final concentrations of ammonia was assumed to have been precipitated as struvite. In an experiment with initial pH of 9 and magnesium, total ammonia and phosphate concentrations of 14.26 mM each, 1714 mg/L struvite was precipitated. However, this concentration corresponded to the concentration of total solids predicted by the model. Struvite produced was only 1017.7 mg/L about 44.75% of total solids produced. Formation of two calcium containing solids $Ca_3(PO_4)_2$ (495.8 mg/L) and $CaHPO_4$ (384.4 mg/L) was also predicted by the model. A 1:1:1 molar ratio of ammonia, phosphate and magnesium does not yield a predominantly struvite precipitate. Similar observations are applicable to data by Stratful et al. (2001).

Depending on the experimental conditions like pH and initial ratios of magnesium, ammonium and phosphate, the solid phase can consist of several components not just struvite. Therefore, equating total solids concentration to struvite can lead to erroneous results. The usual approach to accurately determining struvite component in the solid phase involves redissolving the solid fraction after filtration by digesting in acid and measuring the ammonia concentration (Çelen et al., 2007; Munch and Brar, 2001; Wilsenach et al., 2007; Yoshino et al., 2003). In this approach, it is assumed that ammonia is only precipitated as struvite and the molar concentration of ammonia in solids is equal to molar concentration of struvite. The model agreed very well with experimental struvite data presented using this approach; the error in model predictions ranging from 1.6% to 9%. The error in prediction of struvite was much less than error in prediction of total solids.

Experiments using defined solutions were also carried out in our laboratory to validate the model. A solution of 10 mM of PO_4^{3-} was prepared by dissolving 1320 mg of ammonium phosphate $(NH_4)_2PO_4$ in 1 liter of distilled water. The total ammonia concentration was made up to 50 mM by adding 1605 mg of ammonium chloride. A 200 mM magnesium chloride $(MgCl_2 \cdot 6H_2O)$ stock solution was used as a magnesium source. Precipitation experiments were carried out in 500 ml Erlenmeyer flask in which 475 ml of $(NH_4)_2PO_4$ and NH_4Cl solution was taken. To this solution 25 ml of magnesium chloride was added. This gave an initial $NH_4^+:Mg^{2+}:PO_4^{3-}$ ratio of 48.6:5:9.5 or ~5:0.5:1. A magnetic stirrer was used for mixing. The pH of the solution was continuously monitored with a pH probe. pH of the solution was adjusted to 9.6 by adding 10 N NaOH. Precipitation was found to occur instantaneously after addition of magnesium chloride solution which increased with addition of NaOH. After reaching a pH of 9.6 the solution was stirred for another 5 minutes. Then the whole 500 ml solution was filtered using 0.45 μm Whatman filter paper to recover precipitate. Precipitate was dried overnight in oven at 104°C and weighed. Experiment was repeated three times. The average concentration of precipitate measured in the experiments was 653 mg/L $\pm$ 55 mg/L. Model predicted total solids of 682 mg/l. The error in model prediction was about 4%. This was consistent with errors in model predictions for data in Table 2. The model was then applied to study the effect of various experimental conditions on struvite concentration and purity.

CONCLUSIONS

- A model was developed for predicting precipitation in closed systems containing solutions of ammonium, magnesium and phosphate. The model incorporates fifteen different precipitates and explicitly solves for precipitate, residual ion and dissolved species concentrations using mass balance equations for magnesium, phosphorus and nitrogen along with chemical equilibrium and charge balance equations.
- The model was validated against data collected from literature for synthetic and real wastewaters. The model was able to predict struvite to within 1.6% to 9% and total precipitate prediction errors ranged from 1 to 24.5%.
- Equimolar stoichiometric ratio of magnesium, ammonium and phosphate (i.e. the ratio of their occurrence in struvite) was not ideal for struvite precipitation.
- To obtain pure struvite it was necessary to have excess ammonia in the solution with magnesium being the limiting nutrient.

REFERENCES

Altinbas, M., Yangin, C. and Ozturk, I. (2002). Struvite precipitation from anaerobically treated municipal and landfill wastewaters. *Water Science and Technology*, 46, 271–278.

Burns, R.T., Moody, L.B., Walker, F.R. and Raman D.R. (2001). Laboratory and in-site reductions of soluble phosphorus in swine waste slurries. *Environ. Technol.*, 22, 1273–1278.

Battistoni, P., Pavan, P., Cecchi, F. and Mata-Alvarez, J. (1998). Phosphate removal in real anaerobic supernatants: modeling and performance of a fluidized bed reactor. *Water Science and Technology*, 38, 275–283.

Çelen, I., Buchanan J.R., Burns, R.T., Robinson, R.B. and Raman, D.R. (2007). Using a chemical equilibrium model to predict amendments required to precipitate phosphorus as struvite in liquid swine manure. *Water Research*, 41, 1689–1696.

Harada, H., Shimizu, Y., Miyagoshi, Y., Matsui, S., Matsuda, T. and Nagasaka, T. (2006). Predicting struvite formation for phosphorus recovery from human urine using an equilibrium model. *Water Science and Technology*, 54, 247–255.

Loewenthal, R.E., Kornmüller, U.R.C. and van Heerden E.P. (1994). Modeling struvite precipitation in anaerobic treatment systems. *Water Science and Technology*, 30, 107–116.

Moon, Y.H., Kim, J.G., Ahn, J.S., Lee, G.H. and Moon, H. (2007). Phosphate removal using sludge from fuller's earth production. *Journal of hazardous materials*, 143(1–2), p. 41–48.

Munch, E.V. and Barr, K. (2001). Controlled struvite crystallization for removing phosphorus from anaerobic digester sidestreams. *Water Research*, 35, 151–159.

Ohlinger, K.N., Young, T.M. and Schroeder, E.D. (1998). Predicting struvite formation in digestion. *Water Research*, 32, 3607–3614.

Sarkar, A.K., 1990. Phosphate cement-based fast setting binders. *J. Am. Ceram. Soc. Bull.*, 69, pp. 234–238.

Schuiling, R.D. and Andrade, A. (1999). Recovery of struvite from calf manure. *Environ. Technol.*, 20, pp. 765–768.

Scott, B. (2001). Maple for Environmental Science: Springer.

Scott, W.D., Wrigley, T.J. and Webb, K.M. (1991). A computer model of struvite solution chemistry. *Talanta*, 38, 889–895.

Stratful, I., Scrimshaw, M.D. and Lester, N.J. (2001). Conditions influencing the precipitation of magnesium ammonium phosphate. *Water Research*, 35, 4191–4199.

Tünay, O., Kabdasli, I., Orhon, D. and Kolçak, S. (1997). Ammonia removal by magnesium ammonium phosphate precipitation in industrial wastewaters. *Water Science and Technology*, 36, 225–228.

Von Münch, E. and Barr, K. (2001). Controlled crystallisation for removing phosphorus from anaerobic digester side stream. *Water Research*, 35, 151–159.

Wang, J.S., Song, Y.H., Yuan, P., Peng, J.F. and Fan, M.H. (2006). Modeling the crystallization of magnesium ammonium phosphate for phosphorus recovery. *Chemosphere*, 65, 1182–1187.

Wilsenach, J.A., Schuurbiers, C.A.H. and van Loosdrecht, M.C.M. (2007). Phosphate and potassium recovery from source separated urine through struvite precipitation. *Water Research*, 41, 458–466.

Yoshino, M., Yao, M., Tsuno, H. and Somiya, I. (2003). Removal and recovery of phosphate and ammonium as struvite from supernatant in anaerobic digestion. *Water Science and Technology*, 48, 171–178.

A thermochemical approach for struvite precipitation modelling from wastewater

Mary Hanhoun[a,b], Catherine Azzaro-Pantel[a,b,c], Béatrice Biscans[a,b], Michèle Frèche[b,c,d], Ludovic Montastruc[a,b,c], Luc Pibouleau[a,b,c] and Serge Domenech[b,c]

[a]Université de Toulouse; INP; UPS; LGC (Laboratoire de Génie Chimique), 5 rue Paulin Talabot, BP1301, F31106 Toulouse cedex 01
[b]CNRS, LGC, F-31106 Toulouse cedex 01
[c]ENSIACET INPT, 118, Route de Narbonne 31077 TOULOUSE Cedex 4
[d]Université de Toulouse; INP; UPS, CNRS – CIRIMAT, Route de Narbonne 31077 TOULOUSE Cedex 4

Abstract The cost of environmental protection and pollution prevention is increasing, above all because of stringent effluent quality standard. In that context, phosphate impact on water pollution plays a major role.

This work addresses the problem of phosphorus recovery from wastewater by precipitation of struvite which is chemically known as magnesium ammonium phosphate hexahydrate $MgNH_4PO_4.6H_2O$. A thermodynamic model for phosphate precipitation is proposed involving Davies activity coefficients.

The model aims to determine the phosphate conversion rate and is based on numerical equilibrium prediction of the study system $Mg-NH_4-PO_4-6H_2O$. The parameters include the solubility product of struvite.

The mathematical problem is represented by a set of nonlinear equations that turns out to be hard to solve, mainly due to the various order of magnitude of the involved variables. These equations have first been solved by a Genetic Algorithm strategy to perform a preliminary search in the solution space. The procedure helps to identify a good initialization point for the subsequent Newton-Raphson method.

The model developed in this study will be used for validation and determination of process operating conditions for struvite precipitation in an agitated reactor.

INTRODUCTION

Wastewater discharges of nitrogen and phosphorus to the environment in the recent years have produced an increase in water pollution because these nutrients accelerate eutrophication, producing grave consequences for aquatic life as well as for the water supply for industrial and domestic uses.

One proposed solution to this problem is the recovery of nutrients using crystallization. Two major crystallization processes have been developed for phosphorus recovery from wastewater, respectively the so-called calcium phosphate (CP) crystallization process and the magnesium ammonium phosphate

(MAP) or struvite crystallization. Struvite is a crystalline substance consisting of magnesium, ammonium and phosphorus in equal molar concentrations ($MgNH_4PO_4$-$6H_2O$).

Struvite crystallisation from wastewater effluents is seen as an alternative to traditional biological and chemical phosphorus removal processes used widely in the wastewater treatment industry. It presents the advantage of not only removing phosphorus but also generating a compound that could be reused as a commercial fertilizer (Durrant et al., 1999; Schipper et al., 2001; Nelson et al., 2003).

The formation of struvite in aqueous solutions takes place following the development of supersaturation, the driving force to all crystallization processes. Supersaturation is a measure of the deviation of a dissolved salt from its equilibrium value. The struvite formation occurs relatively quickly because of the presence of excess supersaturation in the liquid, as a result of the chemical reaction of magnesium with phosphate in the presence of ammonium. Supersaturation may be developed by increasing the aqueous medium content in ammonium, magnesium or orthophosphate and/or pH. Although H^+ concentration does not directly enter the solubility product equation for struvite, struvite precipitation is highly pH dependent.

This paper first presents the development of a simple thermochemical model, enough representative of struvite precipitation using a Davies activity coefficient modelling. The mathematical problem is represented by a set of 15 nonlinear equations. A two-stage solution strategy is proposed, combining a Genetic Algorithm for initialization purpose with a standard Newton-Raphson method implemented within MATLAB environment. Some typical numerical results are then compared with experimental and literature results.

CHEMICAL EQUILIBRIUM MODEL FOR STRUVITE PRECIPITATION

Model formulation

The objective is to propose a mathematical model for the calculation of the conversion for the system Mg-NH_4-PO_4-$6H_2O$ as a function of pH.

The struvite forms according to the following reaction (Bouropoulos and Koutsoukos, 2000):

$$Mg^{2+} + NH_4^+ + PO_4^{3-} + 6H_2O \rightarrow MgNH_4PO_4 \cdot 6H_2O \qquad (1)$$

The model involves the mass balances for magnesium, ammonium, and phosphate as a function of MAP conversion as well as the electroneutrality equation.

The sources of Mg, on the one hand, and of NH_4 and PO_4, on the other hand, consist respectively of $MgCl_2$ and of $NH_4H_2PO_4$. In addition, NaOH is used in order to increase pH. The formed ions and complexes include NH_4^+, $NH_3(aq)$, Mg^{2+}, $MgH_2PO_4^+$, $MgHPO_4(aq)$, $MgPO_4^-$, $MgOH^+$, $H_3PO_4(aq)$, $H_2PO_4^-$, HPO_4^{2-}, PO_4^{3-}, OH^-, Na^+, Cl^-, H^+.

The MAP supersaturation is defined by the S parameter set at equilibrium (i.e. taken equal to zero); which means that if supersaturation S is greater than zero, precipitation of MAP is likely to occur (Montastruc *et al.*, 2003).

$$S = \frac{1}{3}\ln\left(\frac{[Mg^{2+}]\lambda_{Mg^{2+}}[PO_4]\lambda_{PO_4^{3-}}[NH_4]\lambda_{NH_4^+}}{K_{SP}}\right) \tag{2}$$

In this expression, λ is the coefficient of K_{SP} represents struvite solubility product. The published values of pK_{sp} range from 9.4 to 13.26: 9.4 (Borgerding, 1972), 12.94 (Aage *et al.*, 1997), and 13.26 (Ohlinger *et al.*, 1998).

The model inputs are the concentrations of magnesium, ammonium, phosphate and NaOH, whereas the model outputs are the conversion of MAP, the concentration of the different ions and pH.

The different mass balances in the liquid phase include:

• a mass balance for magnesium:

$$[Mg^{2+}] + [MgH_2PO_4^+] + [MgHPO_4]$$
$$+ [MgPO_4^-] + [MgOH^+] = Mg_{tot} - P_{tot}X \tag{3}$$

where X is the conversion ratio relative to struvite form, defined as:

$$X = \frac{P_{tot} - P_{sol}}{P_{tot}} \tag{4}$$

Mg_{tot}, P_{tot} and P_{sol} refer respectively to the total quantity of magnesium and phosphorus as well as the quantity of phosphorus remaining in solution.

$$[H_3PO_4] + [H_2PO_4^-] + [HPO_4^{2-}] + [PO_4^{3-}]$$
$$+ [MgH_2PO_4^+] + [MgHPO_4] + [MgPO_4^{2+}] = P_{tot}(1 - X) \tag{5}$$

• a mass balance for ammonium:

$$[NH_3] + [NH_4^+] = NH_{4tot} - P_{tot}X \tag{6}$$

- the electroneutrality requirement gives:

$$[H^+] + [NH_4^+] + [MgH_2PO_4^+] + 2[Mg^{2+}] + [Na^+] + [MgOH^+]$$
$$= [OH^-] + [Cl^-] + 3[PO_4^{3-}] + 2[HPO_4^{2-}] + [H_2PO_4^-] + [MgPO_4^-] \quad (7)$$

The activity coefficients are calculated from the extended form of the Debye-Hückel equation proposed by Davies :

$$\text{Log}\lambda = A_{DH}Z_i^2 \left\{ \frac{\sqrt{\mu}}{1 + \sqrt{\mu}} \right\} - 0,3\mu \quad (8)$$

$$\mu = 0,5 \sum_{i=1}^{n} Z_i^2 C_i \quad (9)$$

where

A_{DH} = Debye-Hückel constant, has a value of 0.493, 0.499, and 0.509 at 5, 15, 25 and 35°C, respectively. (Mullin, 1993).
λ = Activity coefficients
μ = Ionic strength in molar
Z_i = Valency of the corresponding ions
C_i = Concentration of the corresponding ions in mol/l
    The concentrations of ions and complexes are determined from chemical equilibrium relations (K_i equilibrium constants of the specified ion complex are given for a temperature 25°C in molar units). (Table 1)

$$K_i = \frac{\alpha_A \cdot \alpha_B}{\alpha_{AB}}$$

α represents the activity based concentration of each ion and complex involved ($\alpha = \lambda C_i$).

The resulting mathematical formulation for MAP precipitation is based on a set of nonlinear equations, involving the concentrations of the aqueous species and the rate of conversion as variables.

Solution strategy

The low order of magnitude of the involved concentrations and equilibrium constants leads no surprisingly to a difficult numerical problem. This is why we

Table 1. Values of equilibrium constants for complexes presented in the equilibrium equations.

Equilibrium	Equilibrium constant K_i	Value	References
$H_2PO_4^- + H^+ \leftrightarrow H_3PO_4$ (10)	$K_{H_3PO_4} = \dfrac{\alpha_{H^+} \cdot \alpha_{H_2PO_4^-}}{\alpha_{H_3PO_4}}$	$10^{-2.15}$	Martell and Smith (1989)
$HPO_4^{2-} + H^+ \leftrightarrow H_2PO_4^-$ (11)	$K_{H_2PO_4^-} = \dfrac{\alpha_{H^+} \cdot \alpha_{HPO_4^{2-}}}{\alpha_{H_2PO_4^-}}$	$10^{-7.20}$	Morel and Hering (1993)
$PO_4^{3-} + H^+ \leftrightarrow HPO_4^{2-}$ (12)	$K_{HPO_4^{2-}} = \dfrac{\alpha_{H^+} \cdot \alpha_{PO_4^{3-}}}{\alpha_{HPO_4^{2-}}}$	$10^{-12.35}$	Morel and Hering (1993)
$H_2PO_4^- + Mg^{2+} \leftrightarrow MgH_2PO_4^+$ (13)	$K_{MgH_2PO_4^+} = \dfrac{\alpha_{Mg^{2+}} \cdot \alpha_{H_2PO_4^-}}{\alpha_{MgH_2PO_4^+}}$	$10^{-0.45}$	Martell and Smith (1989)
$HPO_4^{2-} + Mg^{2+} \leftrightarrow MgHPO_4$ (14)	$K_{MgHPO_4} = \dfrac{\alpha_{Mg^{2+}} \cdot \alpha_{H_2PO_4^{2-}}}{\alpha_{MgHPO_4}}$	$10^{-2.91}$	Martell and Smith (1989)
$PO_4^{3-} + Mg^{2+} \leftrightarrow MgPO_4^-$ (15)	$K_{MgPO_4^-} = \dfrac{\alpha_{Mg^{2+}} \cdot \alpha_{PO_4^{3-}}}{\alpha_{MgPO_4^-}}$	$10^{-4.80}$	Martell and Smith (1989)
$OH^- + Mg^{2+} \leftrightarrow MgOH^+$ (16)	$K_{MgOH^+} = \dfrac{\alpha_{Mg^{2+}} \cdot \alpha_{OH^-}}{\alpha_{MgOH^+}}$	$10^{-2.56}$	Childs (1970)
$H^+ + NH_3 \leftrightarrow NH_4$ (17)	$K_{NH_4^+} = \dfrac{\alpha_{H^+} \cdot \alpha_{NH_3}}{\alpha_{NH_4^+}}$	$10^{-9.25}$	Taylor *et al.* (1963)
$H_2O \leftrightarrow H^+ + OH^-$ (18)	$K_w = [H^+][OH^-]$	10^{-14}	Snoeyink (1980)

find it useful to propose the trial-and-error process followed in this study, to give some guidelines to tackle similar problem. A classical way to solve the set of nonlinear equations is to use a Newton-Raphson method. This was performed with the MATLAB environment (*fsolve* function). A preliminary step consists in considering the ionic strength and activity coefficient as unknowns, thus leading to a system of 16 nonlinear equations (the so-called set S_1). The variables were initialized with some experimental values, that will be presented in the next section. Unfortunately, this procedure always leads to a failure.

Another idea was to reduce the problem size by substituting some variables (set S_2) in order to identify optimization variables with process variables, so that the ionic strength and activity coefficients are no more considered as variables.

For this purpose, it was first necessary to initialize the ionic strength by considering the value relative to the initial conditions, that is a weighted sum reflecting the respective quantity of magnesium, phosphate, ammonium, chloride and sodium.

$$\mu = 0.5(4[Mg_{tot}] + [PO_{4tot}] + [NH_{4tot}] + [Cl] + [Na])$$

The activity coefficient can thus be computed explicitly from equation 6.

The final unknowns of the system are reduced to the following concentrations $[Mg^{2+}]$, $[PO_4^{3-}]$, $[NH_4^+]$, $[H^+]$ and to phosphate conversion. As previously, the initialization was carried out by the values corresponding to experimental runs. Once more, the solution strategy leads to a failure, with concentrations ranging in an unfeasible physical domain.

Another formulation for the substituted problem was thus initiated as an optimization one with the following inequality constraints: concentrations ≥ 0, $0 \leq X \leq 1$ (X is the conversion rate of phosphate). The optimization criterion is based on the minimization of a quadratic multivariate function (the sum of the square of the terms involved in equations [1] to [6]). This was solved by the *fmicon* function in MATLAB. Once more, the search was unsuccessful.

These numerical experiments showed us that the strong nonlinear feature of the problem along with the low values of the involved concentrations are mainly responsible for these successive failures. Due to the lack of place, only the philosophy of the final solution strategy is presented. Taking all these elements into account, the set of 13 equations is rearranged (set S_3) so that a linear feature is preserved as much as possible. Since the initialization phase still constitutes a tricky problem, we resort to a Multiobjective Genetic Algorithm previously developed (Gomez *et al.*, 2008) in which each equality constraint of the basic problem is transformed in a criterion to optimize. With this good initial point, it was easy to solve the so-called S_3 set with *fsolve* function.

This system equations imply the following unknowns: $[NH_4^+]$, $[Mg^{2+}]$, $[PO_4^{3-}]$, $[MgH_2PO_4^+]$, $[MgHPO_4]$, $[MgPO_4^-]$, $[MgOH^+]$, $[H_3PO_4]$, $[H_2PO_4^-]$, $[HPO_4^{2-}]$, $[NH_3]$, pH and the phosphate conversion rate for struvite precipitation. The system is solved for various concentrations in NaOH in order to analyze pH influence on conversion. Since magnesium exists in the form of magnesium chloride, this concentration has been taken to $2 [Mg^{2+}]$.

RESULTS AND DISCUSSION

Figure 1 shows the evolution of phosphate conversion as a function of pH, generated by the predictive model.

The initial concentration is set at 0.0058 mol/l for magnesium, phosphate and ammonium, which means that a molar ratio $Mg:NH_4:PO_4$ equal to 1:1:1 is adopted. At this concentration, the phosphate conversion is maximal, i.e., 88% for a pH equal to 9.5. When increasing magnesium concentration to 0.0116 mol/l, (i.e. molar ratio of $Mg:NH_4:PO_4$ equal to 2:1:1), the phosphate conversion is equal to 93% for a pH value of 9.8. For a molar ratio equal to 1:1:2, the phosphate converted as struvite reaches now 96.35% for a pH equal to 10. These

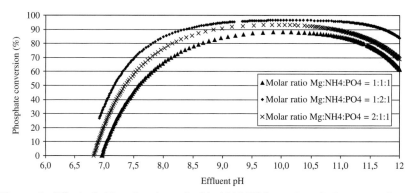

Figure 1. Effect of pH and molar ratio Mg:NH4:PO4 on phosphate conversion.

results show that the optimal pH is 9.5; with a $Mg:NH_4:PO_4$ molar ratio equal to unity. This is consistent with the results presented in the literature showing that an increase in $Mg:NH_4:PO_4$ molar ratio leads to a slight increase in optimal values for pH (around 10) for struvite precipitation (Wang *et al.*, 2006).

Some experimental points are used to validate the model proposed here (Figure 2). The experimental protocol and the characterization techniques used are not detailed here. Only some key points are presented to show the efficiency of the model. The runs correspond to an initial phosphorus concentration of 350 mg/l with a $Mg:NH_4:PO_4$ molar ratio equal to 1. Experiments were carried out at 25°C in a 500 ml vessel. Stock solutions of magnesium and dihydrogen ammonium phosphates were prepared from corresponding crystalline solids $MgCl_2.6H_2O$ and $NH_4H_2PO_4$. Distilled water was used to prepare the synthetic wastewater solution. The supersaturated solutions were prepared by rapidly mixing $NH_4H_2PO_4$. Next, the solution pH was adjusted, by the addition of the appropriate amount of a standard solution of sodium hydroxide, followed by the addition of the appropriate volume of stock magnesium chloride solution. Finally, the solution pH was readjusted as needed. The homogeneity of the solution was ensured by a magnetic stirrer. Experiments have been performed for several values of pH [7–12] for comparison purpose with the modelling results. In Figure 2, it must be pointed out that there is a good agreement between experimental and predicted values: optimal pH for struvite precipitation is 10 (respectively 9.8) with the model (respectively with the experiments).

The model was also used to predict struvite precipitation for the conditions of (Pastor *et al.*, 2008), corresponding to a PO_4 concentration of 80 mg/l. It can be seen that the predictive values are close to the experimental values (Figure 3) meaning that the proposed model can be extrapolated.

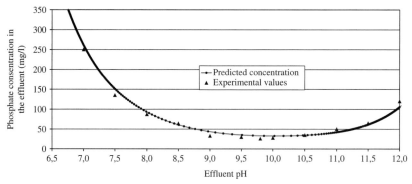

Figure 2. Comparison between the experimental values and the precipitation model.

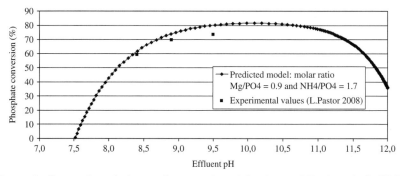

Figure 3. Comparison between the experimental values of Pastor *et al.* (2008) and the predicted model.

CONCLUSION

This research investigated the potential of struvite precipitation as a method to recover phosphorus from wastewater, while concurrently removing ammonium and magnesium. For this purpose, this study has presented a thermodynamic model for struvite precipitation. It gives for high pH values a domain in which struvite precipitation can occur. The results obtained have shown that an increase in pH and Mg:NH$_4$:PO$_4$ molar ratio increases the precipitation efficiency. The model developed in this study will be now used for validation and determination of process operating conditions for phosphate precipitation in a stirred reactor.

REFERENCES

Aage, H.K., Andersen, B.L., Blom, A. and Jensen, I. (1997). The solubility of struvite. *Journal of Radioanalytical and Nuclear Chemistry*, **223**(1–2) 213–215.

Borgerding, J. (1972). Phosphate deposits in digestion systems. *J Water Pollution Control Fed*, **44**, 813–9.

Bouropoulos, N.Ch. and Koutsoukos, P.G. (2000). Spontaneous precipitation of struvite from aqueous solutions. *Journal of Crystal Growth*, **213**(3–4) 381–388.

Childs, C.W. (1970). A potentiometric study of equilibria in Aqueous Divalent Metal orthophosphate solutions. *Journal of Inorganic Chemistry*, **9**(11), 2465–2469.

Durrant, A.E., Scrimshaw, M.D., Stratful, I. and Lester, J.N. (1999). Review of the feasibility of recovering phosphate from wastewater for use as a raw material by the phosphate industry. *Environ. Technol.*, **20**, 749–758.

Gomez, A., Domenech, S., Azzaro-Pantel, C. and Pibouleau, L., *MULTIGEN: procédures d'optimisation multicritères, 76ème Congrès de l'Association Francophone pour le Savoir (ACFAS)*, 5–9 mai 2008, Monréal (Quebec).

Martell, A.E. and Smith, J.C. (1989). Critically selected stability constants of metal complexes. *Texas University, College station, TX*.

Montastruc, L., Azzaro-Pantel, C., Biscans, B., Cabassud, M. and Domenech, S., (2003). A thermochemical approach for calcium phosphate precipitation modeling in a pellet reactor, *Chem. Eng. J.*, **94**, 41–50.

Morel, F.M.M. and Hering, J.G. (1993). Principles and Applications of Aquatic Chemistry. New York, John Wiley & Sons Inc.

Mullin, J.W. (1993). Crystallization. 3rd, *Butterworth-Heinemann Publications, Ipswich, UK*.

Nelson, N.O., Mikkelsen, R. and Hesterberg, D. (2003). Struvite precipitation in anaerobic swine liquid: effect oh pH and Mg:P ratio and determination of rateconstant. *Biosource Technol.*, **89**, 229–236.

Ohlinger, K.N., Young, T.M. and Schroeder, E.D. (1998). Predicting struvite formation in digestion. *Water Research*, **32**, 3607–14.

Pastor, L., Mangin, D., Barat, R. and Seco, A. (2008). A pilot-scale study of struvite precipitation in a stirred tank reactor: Condition influencing the process. *Biosource Technol*, **99**, 6285–6291.

Schipper, W.J., Klapwijk, A., Potjer, B., Rulkens, W.H., Temmink, B.G., Kiestra, F.D.G. and Lijmbach, A.C.M. (2001). Phosphate recycling in the phosphorus industry. *Environ. Technol.*, **22**, 1337–1345.

Snoeyink, V.L. and Jenkins, D. (1980). Water Chemistry. *John Wiley and Sons, USA*.

Taylor, A.W., Frazier, A.W. and Gurney, E.L., (1963). Solubility product of magnesium ammonium. *Transaction Faraday Society*, **59**, 1580–1589.

Wang, J., Burken, J.G. and Zhang, X. (2006). *Effect* of seeding materials and mixing strength on struvite precipitation. *Water Environment Research*, **78** (2) 125–132.

Numerical investigations of the hydrodynamics of the UBC MAP fluidized bed crystallizer

M.S. Rahaman[a], D.S. Mavinic[a], A.T. Briton[b], M. Zhang[b], K.P. Fattah[a] and F.A. Koch[a]

[a]Environmental Engineering Group, Department of Civil Engineering, University of British Columbia (UBC), 2010 – 6250 Applied Science Lane, Vancouver, BC, V6T 1Z4, Canada (Email: *mrahaman@interchange.ubc.ca*; *dsm@civil.ubc.ca*; *parvez@interchange.ubc.ca*; *koch@civil.ubc.ca*)
[b]Ostara Nutrient Recovery Technologies Inc. 690-1199 West Pender St, Vancouver, BC, V6E 2R1 (Email: *abritton@onrti.com*; *mzhang@ostara.com*)

Abstract This paper is an attempt to model the hydrodynamics of the UBC (University of British Columbia) MAP (Magnesium Ammonium Phosphate) fluidized bed crystallizer. In this study, a numerical investigation of hydrodynamics of the UBC MAP crystallizer was performed, using commercial Computational Fluid Dynamics (CFD) software, Fluent 6.3. One of the main findings of this modelling effort is the relative distribution of the different phases, in this case solids (struvite crystals) and liquid (water), within the reactor. This information is very important, in the sense that it helps to fix the size of the crystallizer; this can be a significant factor, affecting the total cost of the process. Another finding of this simulation is the local particle size distributions, which will help determine the position of the harvesting port for a desired size range of struvite crystals. The effects of aspect ratios (height/diameter) on crystallizer hydrodynamics were also investigated in this study.

INTRODUCTION

The cost associated with the disposal of additional phosphate sludge, the stringent regulations to limit phosphate discharge to the aquatic environments and resource shortage, due to limited phosphorus rock reserves, have diverted attention to phosphorus recovery from wastewater, leading to production of a reusable fertilizer called struvite (Yoshino *et al.*, 2003). Over the last eight years, the Environmental Engineering Group at UBC has developed a novel fluidized bed reactor configuration that converts 80–90% of soluble phosphate in wastewater into crystalline struvite (Adnan *et al.*, 2003). More recently, Ostara Nutrient Recovery Technologies Inc. (a spin-off company of the UBC P-recovery group) has installed a full scale demonstration nutrient recovery plant in Edmonton, AB, Canada. In order to optimize the UBC MAP crystallization process, a handful of research has been performed concerning

© 2009 The Authors, *International Conference on Nutrient Recovery from Wastewater Streams*. Edited by Ken Ashley, Don Mavinic and Fred Koch. ISBN: 9781843392323. Published by IWA Publishing, London, UK.

the chemistry of struvite formation and growth; however, very little initiative has been taken to look into the hydrodynamic behavior of the crystallizer. Hydrodynamics play a key role in the crystallization process, especially having a critical effect on mixing and mass transfer within the reactor; thus it affects the growth rate and quality of the crystals in terms of size and shape.

In a fluidized bed crystallizer, the simultaneous progress of two processes-fluidization and crystallization yield very complex phenomena, which requires comprehensive studies of the process hydrodynamics to help design an efficient reactor. As the supersaturated solution flows upward through a fluidized bed crystallizer, the liquor contacting the bed relives its supersaturation on the growing crystals and subsequently the supersaturation decreases along the upward direction. As a result, crystals near the bottom grow faster than those near the top of the crystallizer. Such behaviour results in the variation of particle size along the height of the reactor. When the bed is composed of particles of different sizes, the particle size distribution is influenced by two opposite phenomenon: classification and dispersion. Classification results from the movement of particles of different weights; larger particles tends to reach the bottom of the reactor, whereas smaller particles rises. At the same time, dispersion is induced by irregular motions of the solid particles. Perfect classification and complete mixing of the solids in fluidized bed are the two extreme situations. More often, a mixing zone is created between two layers of classification particles, allowing partial classification within the bed. For simplicity, design methods for fluidized bed crystallizers are generally based on the perfect size classification of the crystals (Mullin and Nyvlt, 1970). On the other hand, the liquid phase is considered as plug flow or near plug flow patterns, in most of the modeling effort for liquid fluidized bed crystallizer (Shiau and Lu, 2001). However, the majority of actual crystallizer systems exhibit macromixing behaviour somewhere between the two extremes of plug flow and ideally mixed flow. Hence accurate knowledge of mixing behaviour of the phases and is essential for modeling and optimizing the fluidized bed crystallizers. Another very important aspect of fluidized bed crystallizer is the expansion characteristics of crystal bed. A quantitative knowledge of the bed expansion, as a function of the liquid superficial velocity, is essential for efficient design of a fluidized bed.

Computational Fluid Dynamics (CFD) is becoming an important tool for studying the hydrodynamic behaviour of the conventional industrial crystal-lization processes. This technique allows the prediction of flow patterns, local solids concentration and local kinetic energy values, taking into account the reactor shape. However, to the authors' knowledge, none of the studies have dealt with the fluidized bed crystallizers. Therefore, in this current project, a numerical investigation of hydrodynamics of the UBC MAP fluidized bed crystallizer was performed, using the commercial CFD package, Fluent 6.3 (Fluent, 2006).

CRYSTALLIZATION PROCESS IN UBC MAP FLUIDIZED BED REACTOR

The basic design of the UBC MAP crystallizer follows the concept of a fluidized bed reactor. As depicted in Figure 1, the reactor has four distinct zones depending on the diameter of the column. The bottom part of the fluidized bed reactor is called the harvest zone (A); above that the active zone (B), while the top fluidized section is the 'fine zones' (C). There is a settling zone, also called 'seed hopper' (D) at the top. Both a lab-scale and the pilot scale UBC MAP crystallizer were used for the numerical investigations and the dimensions of the reactors are presented in Table 1. In the struvite crystallization process, synthetic anaerobic digester supernatant is fed into the bottom of the reactor along with the recycle stream. Magnesium chloride and sodium hydroxide are added to the reactor through the injection ports, just above the feed and recycle flows. Seed crystals are added into the crystallizer from the seed hopper and are allowed to grow in the supersaturated solution. The solution velocity is maintained in such a way that all the particles in the crystal bed are fluidized in the solution. Since the fresh influent is pumped into the bottom of the reactor, the reactive solution contains the maximum supersaturation at the bottom of the reactor and the crystals grow faster than those near the top of the reactor. As a result, the bigger crystals tend to settle in the bottom and the smaller crystals rise to the top of the crystallizer. The larger crystals at the bottom, once having achieved the desired size, are settled into the harvest zone and are withdrawn from the bottom of the reactor.

CFD MODELLING

Numerical models for particulate multiphase flows usually employ Eulerian continuum descriptions of the phases (Gidaspow, 1994), or a Lagrangian descriptions of the particulate phase, and an Eulerian continuum description of the fluid phase (Hu, 1996; Patankar and Joseph, 2001). The latter approach employs the lagrangian tracking method for calculating the behaviour of individual particles, and thus it is capable of simulating phase interaction with high spatial resolution. However, this technique is limited by the total number of

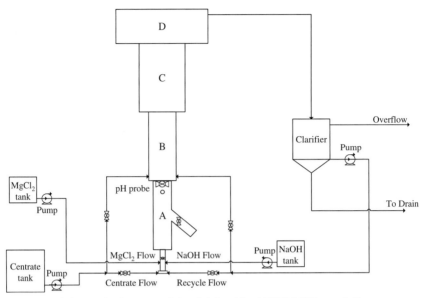

Figure 1. Schematic diagram of the fluidized bed UBC MAP crystallizer.

Table 1. Dimensions of the lab-scale and pilot scale reactor.

	Lab scale				Pilot scale			
	Harvest zone	Active zone	Fines zone	Seed hopper	Harvest zone	Active zone	Fines zone	Seed hopper
Dimensions	(A)	(B)	(C)	(D)	(A)	(B)	(C)	(D)
Diameter, D (mm)	76	102	152	381	76	102	152	381
Height, H (mm)	410	410	410	305	749	1549	1270	457
Aspect ratio (H/D)	5.4	4.02	2.7	0.8	9.86	15.19	8.36	1.2

particles employed in the multiphase systems. On the other hand, the Eulerian continuum approach considers the particulate phase to be a continuous 'fluid' interpenetrating and interacting with the fluid phase. Dense particulate systems are commonly modeled using this approach. Continuum models allow for the modeling of particle-particle stresses using spatial gradients of the volume fraction and velocities (Gidaspow, 1994). This modeling approach can be used to simulate the fluidized bed consisting of heterogeneous solids. Different solid phases, each with a characteristic size and density along with a continuous liquid phase can be simulated. The total of all phase volume fractions is still constrained to unity. However, the introduction of more than one solid phase significantly complicates the formulation and requires the introduction of additional equations, to represent the interaction among the different solid phases and the dynamics of the solid phases.

Both the lab and pilot scale configuration of UBC MAP fluidized-bed crystallizer were used for the numerical investigations. For numerical investigation, simulated two-dimensional (2-D) geometry of the UBC MAP crystallizer was created, using Gambit 2.3.16 and meshed with the grid sizes of 2 mm by 2 mm. In order to specify the initial conditions, the grid is divided into upper and lower zone. At time, t = 0, the lower part is filled with solids at an appropriate volume fractions, while the upper portion doesn't contain any solids initially. A multi-fluid Eulerian CFD model, with a granular flow extension, was used to investigate the hydrodynamics of the UBC MAP fluidized bed crystallizer. The solids and liquid phases are treated as fully interpenetrating continua. Conservation of mass and momentum relations, for each phase, provide the governing equations. A set of constitutive equations was then used, to close the governing equations. The dense, solid phase, containing inelastic spherical particles of struvite crystals, was modeled based on the kinetic theory of dense gases. A complete set of governing equations can be found elsewhere (Fluent, 2006). The model equations, along with the appropriate boundary and initial conditions, were solved by Fluent 6.3, using the turbulent flow model in double precision mode. Since the crystallization process is a multi-particle system, for both the lab and pilot scale configurations, simulations were performed for a mixture of different sizes of struvite crystals (2, 1.5, 1 and 0.5 mm; volume ratio, 1:1:1:1), with initial bed height of 0.254 m, with a packed bed solids volume fraction of 0.65. A solution velocity of 0.065 m/s was selected to fluidize the mixture of particles. The simulations were performed for 120 s, with a time step of 0.001 s. A summary of the model settings and the phase properties are presented in Table 2. The results were then analyzed for the solids volume fractions and the overall bed expansion behavior of struvite crystals.

RESULTS AND DISCUSSION

The snapshots of volume fraction of solid phases in pilot scale crystallizer at three different time intervals (30, 60 and 120 s) are plotted in Figures 2, 3 and 4, respectively. At time 30 s, the struvite crystals bed is found to be reasonably well-mixed, pretty much throughout the entire bed height. However, some portion of the solid particles 1 and 2 already started to settle down and to be

Table 2. Summary of phase properties and simulation settings.

Phase properties		Model settings	
Liquid properties at 20°C		Initial bed height	254 mm
Density	998.2 kg/m^3	Initial solids packing	0.65
Viscosity	0.001003 Pa.s	Outlet boundary condition	Pressure outlet
		Wall boundary condition	No slip condition
Solid properties		Inlet boundary condition	Uniform velocity inlet
Crystal group	*Diameters*	Liquid fluidization velocity	0.065 m/s
Group 1	2.0 mm	Mesh resolution	2 × 2
Group 2	1.5 mm	Convergence criteria	10^{-3}
Group 3	1.0 mm	Maximum iterations	20
Group 4	0.5 mm	Discretization method	First order upwind
Wet density	1450 (kg/m^3)	Time step	0.001 s

segregated. The particles of group 3 and 4 remained almost completely mixed in the bottom zone (A) of the reactor. However, there exists some localized areas, where the volume fraction of different particles are higher than the average, and similarly, localized pockets of low concentrations are also evident. The crystal bed height is found to be expanded and some portion of the particles of group 4 moved to the second zone (B) of the reactor. As depicted from Figure 3, at time 60 s, the particles 1 and 2 are found to be segregated completely according to their sizes and confined in the bottom portion of the reactor. However, the other two size ranges, solid 3 and 4, appear to be reasonably well mixed at their interface. Although the bed (in terms of particles 3 and 4) remains sparsely dispersed, pockets of low concentration of those particles as shown in Figure 2, indicate a modest degree of non-uniformity in terms of the struvite crystals distribution. After running for 120 s, all four different particle sizes clearly segregated according to their sizes and the largest particles (#1) are found to settle at the bottom of the reactor and the smallest ones (#4) are floating on the top of crystals bed. However, a reasonable mixed condition of particles 3 and 4 is prevailing in the transition between zone 'A' and 'B'. Accurate knowledge of mixing behaviour of the phases is essential in modeling and design optimizing the fluidized bed crystallizers. At steady state conditions, it is observed that a mixing zone is created between two layers of classified particles, allowing partial classification within the bed in UBC MAP fluidized bed crystallizer.

At any particular bed height, the average solid volume fractions can be calculated by integrating the volume fractions across the bed cross-sectional area. Figure 5 displays the area-averaged solid volume fractions of different sizes of struvite crystals along the bed height. It is observed that the bed, up to a height of 0.1 m, is entirely composed of the largest particles (solid#1) and the

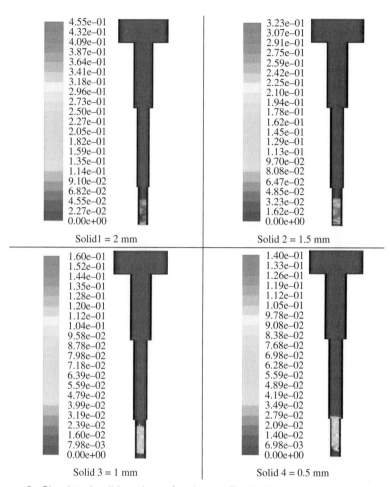

Figure 2. Simulated solids volume fraction profile in pilot scale reactor at 30 s.

average solid volume fraction is found to be 0.45. The second largest particle (solid#2) also appears to be segregated above the largest ones (solid1). However, in the remaining portion of the expanded bed height, only the solid particles #3 and #4 appear to be presented. Though the particles 3 and 4 are reasonably segregated, a considerable mixing is observed at their interface. Due to segregation and mixing of the struvite crystals, the overall solid volume fraction along the bed height is found to vary; a relatively dense bed is observed at the bottom of the reactor and it becomes very lean at the top of the crystals bed. It is also

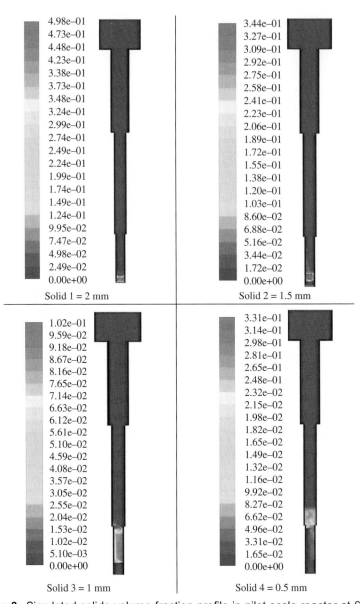

Figure 3. Simulated solids volume fraction profile in pilot scale reactor at 60 s.

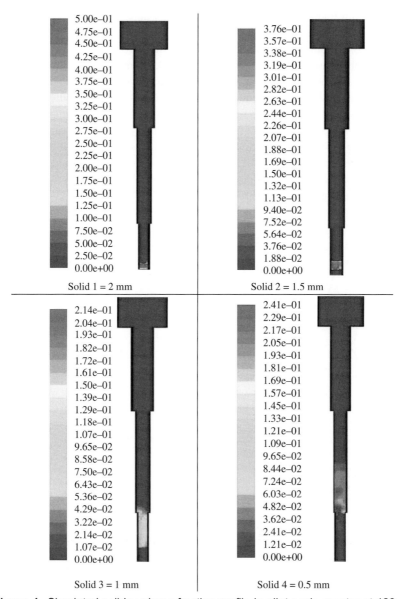

Figure 4. Simulated solids volume fraction profile in pilot scale reactor at 120 s.

obvious from Figure 5 that bed expansion greatly depends on the particles sizes. For the same up-flow liquid velocity, the smaller particles exhibit higher expansion, in comparison to their larger counterparts. A total expanded bed height is observed at 1.55 m, which is almost six times higher than the initial static bed height of 0.254 m. Since a quantitative knowledge of the bed expansion at any particular liquid superficial velocity is essential for efficient design of a fluidized bed, the results of this study generate an important data base for design engineers. Also, the knowledge of crystal size distribution along the bed height will help determining the positioning of the port for harvesting a desired size range of struvite crystals. Thus, the predicted bed expansion behavior and the crystal size distribution would provide useful insights for the crystallizer design and operation.

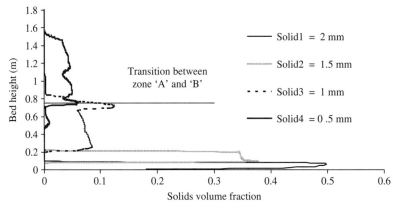

Figure 5. Simulated average solids volume fraction along the crystal bed height at time $=$ 120 s in pilot scale reactor.

In scaling up a reactor for plant scale operations, it is extremely important to investigate the effects of any changes in reactors' aspect ratios (height/diameter) on the hydrodynamics of the crystallizer. In order to identify those effects, the hydrodynamics of a scale-down version (lab-scale) of the pilot scale reactor was also investigated in this current study. The lab-scale reactor has the same diameters as of the pilot scale one, but with smaller heights. The reason for choosing this configuration is that we have a lab scale setup which has the similar dimensions and, can be used to compare the numerical simulation results with the experimental values. The snapshots of solid volume fractions for the lab scale reactor at time 60 s are displayed in Figure 6. It is obvious that, although the particles #1 and #2 segregated reasonably, some small fractions of the size ranges are observed to be mixed with the remaining sizes of particles (#3 and #4). At the same time, particles #3 and #4 are found to be considerably mixed and the height of the mixed portion is higher in this case. Since the lab scale reactor has lower aspect ratios, it exhibits both the mixing and segregation within the crystals bed. Therefore, from the operational point of view, a full scale reactor having the same aspect ratio may not be suitable, if it is designed to harvest crystals of different sizes from different zones of the reactor. That means it will be poorly designed, in the cases where the crystallizer will be used as a simultaneous particle classifier. On the other hand, crystals are found not to be confined only in the bottom zone, but also they make their way up to zone 3; this means a major portion of the reactor is used for this fluidization system. A very important point is that if this configuration provides significant reduction in the soluble phosphorus and the reactor is only being used for harvesting single size struvite crystals from the bottom of the reactor, this configuration may provide the optimal design for the case.

The solid volume fractions for different sizes of particles in the lab-scale reactor are presented in Figure 7. A maximum solid volume fraction (0.485) is observed at a height of 0.045 m from the bottom of the reactor and the bed is comprised only of particles of size group 1. A mixing interface between particle 1 and 2 is clear from the solid volume fraction profile. However, almost all the particles of group1 and 2 are found to be remained in zone 1 and a very small portion of it is moved to the second zone, which is known as the active zone for crystallization. Another interesting point is that the total expanded bed height for this particular case is 0.85 m which is only 3.35 times of the initial static bed height. So by lowering the aspect ratios, a considerable reduction in bed expansion can also be achieved. With the same amount of in-reactor struvite particles, the expanded bed height in the reactor, with lower aspect ratios, is found to be lowered by almost 50% from that of the higher aspect ratios' reactor. As the total height of the reactor depends on expansion characteristics of the crystal bed and the overall cost of the reactor significantly depends on the height of the reactor, an accurate knowledge of bed expansion can provide not only optimal design information, but also can reduce the construction cost considerably by selecting an appropriate height of the reactor.

CONCLUSIONS

The simulation of hydrodynamics of the UBC MAP crystallizer, using FLUENT 6.3, provides insights into the mixing behavior of the phases and the size distributions of struvite crystals in the reactor. In a pilot scale reactor, at an

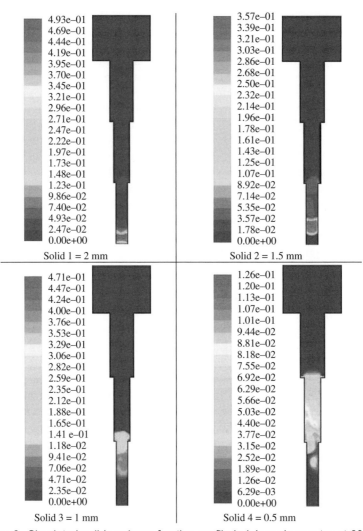

Figure 6. Simulated solids volume fraction profile in lab-scale reactor at 60 s.

operating upflow liquid velocity of 0.065 m/s, the particles were found to be completely segregated. However, the smaller particles (solid 3 and 4) appear to be well-mixed at their interface. The predicted expanded bed height was 1.55 m, which is 6 times higher than the initial static bed height. The lab-scale reactor, with smaller aspect ratios and, operated at the same upflow velocity, provides

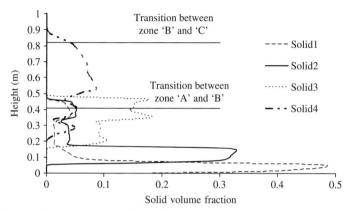

Figure 7. Simulated average solids volume fraction along the crystal bed height at time = 60 s in lab scale reactor.

considerable mixing of the stuvite particles. The bed expansion was significantly lower than that of the reactor with higher aspect ratio. All these simulation results provide useful insights for the UBC MAP crystallizer design and operation.

REFERENCES

Adnan, A., Mavinic, D.S. and Koch, F.A. (2003). Pilot-scale study of phosphorus recovery through struvite crystallization – examining the process feasibility. *J. Environ. Eng. Sci.*, **2**, 315–324.

Fluent (2006). *Fluent User's Guide*, Fluent Inc., Lebanon, NH, USA.

Gidaspow, D. (1994) *Multiphase Flow and Fluidization Continuum and Kinetic Theory Descriptions*, Academic Press, Boston, MA.

Hu, H.H. (1996) Direct simulation of flows of solid-liquid mixtures. *Int. J. Multiphase flow*, **22**, 335–352.

Mullin, J.W. and Nyvlt, J. (1970), Design of classifying crystallizers, *Trans. Instn. Chem. Engrs.*, **48**, T7–T14.

Patankar, N.A. and Joseph, D.D. (2001) Modeling and numerical simulation of particle flows by Eulerian-Lagrangian approach. *Int. J. Multiphase flow*, **27**, 1659–1684.

Shiau, L-D. and Lu, T-S (2001) Interactive effects of particle mixing and segregation on the performance characteristics of fluidized bed crystallizer, *Ind. Eng. Chem. Res.*, **40**, 707–713.

Yoshino, M., Tsuno H. and Somiya I. (2003). Removal and recovery of phosphate and ammonium as struvite from supernatant in anaerobic digestion. *Water Sci. Technol.*, **48**, 171–178.

About the economy of phosphorus recovery

T. Dockhorn

Institute of Sanitary and Environmental Engineering, Technische Universität Braunschweig, Braunschweig, Germany

Abstract Phosphorus is an essential and non-renewable resource. Up to now our economic system works as a dead-end system regarding the resource phosphorus. Thus, a significant part (approx. 50%) of the overall mass of used phosphorus can be detected in wastewater and sewage sludge but still remains unused for further production processes.

The price of phosphorus on world markets rose during the past few years from approx. 1 €/kg P to currently more than 3 €/kg P. This could give an increasing economic incentive regarding the full-scale implementation of phosphorus recovery technologies.

A systematic investigation of cost dependency of the struvite (MAP) precipitation process showed that costs for the Mg-source are a main cost factor and can contribute up to 75% of overall production costs. In general, increasing phosphate concentrations lead to a decrease in MAP-production costs, because the percentage of investment-related costs decreases at increasing concentrations so that a break-even point for the process was calculated at 200–300 mg P/L in the current study.

The remobilization of phosphate from sewage sludge ash was calculated at additional chemical costs of 1 €/kg P. While costs for consumption of chemicals linearly depend on ash quantities, specific process costs decreased with an increase in the phosphorus yield (amount of phosphorus that could be recovered from the ash).

If phosphorus were recovered from process streams at WWTPs, costs for P-removal within the WWTP theoretically would have to be added to respective recovery costs. Thus economically more favorable P-recovery approaches use highly concentrated wastewater streams prior to any treatment at a WWTP.

The lowest MAP-production costs in the current study were calculated at 160 €/t MAP for the MAP-precipitation from urine using seawater as Mg-source.

INRODUCTION

Regarding the increasing scarcity of global phosphorus resources, the recovery of phosphate from wastewater and sewage sludge will inevitably become more important in the future. Up to now our economic systems works as a dead-end system regarding the resource phosphorus. Thus a significant part of the overall mass of the phosphorus used can be retraced in wastewater and sewage sludge but still remains inaccessible for further production processes. In Germany almost 46% of the phosphorus net import enter the wastewater path (Dockhorn 2007a).

Mass flow balances show that approximately 28% of the worldwide phosphorus fertilizer consumption could be covered by the amount of phosphorus found in human excreta.

The costs for the removal of phosphate during wastewater treatment can be calculated to 2–3 €/kg P at minimum and may amount to more than 10 €/kg P under specific conditions (Dockhorn 2007a). Taking these figures into consideration, calculative costs for state-of-the-art wastewater treatment would sum up to approx. 9 billion €/a worldwide only for the removal of phosphate from municipal wastewater (estimated at approx. 4.3 million t P/a). On the other side, actual market value of this phosphorus would amount to approximately 13.5 billion €/a.

However, the recovery of phosphate (and also other nutrients) from wastewater and sewage sludge also requires technical and financial efforts. Schaum (2007) reported costs in a wide range of 2.2 – 8.8 €/kg P depending on the process. As long as calculated market prices of recycling products are below costs for recycling processes, there is no economic incentive for most stakeholders in the wastewater sector to implement and establish phosphorus recovery technologies.

The following study investigates which costs result from individual phosphorus recycling steps and which factors mainly influence these process costs. Because calculation of costs as well as a resource-economic evaluation of appropriate recycling processes requires a defined database, a population equivalent of 350,000 was assumed for this study.

THE RESOURCE ECONOMY OF PHOSPHORUS

The importance of a resource can usually be estimated by the fact whether the resource is essential or could be substituted by another resource and also whether it is renewable or non-renewable. Especially the combination of essential and non-renewable is sensitive, because the unlimited use of such a resource will, some day, definitely become a limiting factor for the whole system.

In our daily perception, the importance of a resource is mainly defined by its market price. Although, the price does not allow a differentiation between essential (e.g. phosphorus) or not essential (e.g. mineral oil) resources and mainly represents current market demand, it can be assumed that an increasing market price as scarcity indicator causes consumer adjustment.

Therefore it will be finally just a question of time until excessively used resources will lead to a further increase in market prices as it has already been observed on international markets for some years. Especially essential and non-renewable resources will be affected in such a way.

The price of phosphorus

At international commodities markets, a steep increase in prices for important industrial and agrarian resources could be observed since the year 2002. Since 2006, this has affected the prices for fertilizers to a growing extent. As an example, the price for rock phosphate increased by 800% (USDA 2008) between the middle of the year 2005 and the beginning of 2008 (on the basis of US$), which equals an increase from 0.78 €/kg P to 3.11 €/kg P during the same time.

A similar development of the phosphate price has already been observed in the years 1973/1974, where the price for phosphate more than tripled within one year (Bernhardt 1978). The relative development of the phosphate prices in percentage for the years 1969–1977 and 2002–2008 is shown comparatively in Figure 1.

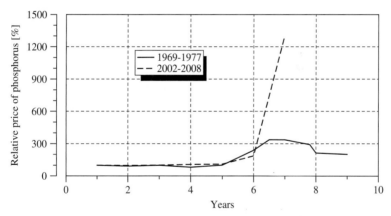

Figure 1. The relative development of the phosphate prices in percentage for the years 1969–1977 and 2002–2008.

Even if a certain speculative margin must be assumed for the current rapid price increase at international markets, a long-term increase in the phosphate prices is to be expected due to the decreasing availability of phosphate resources, which is accompanied by a decrease in the quality of rock phosphates and an increase in worldwide fertilizer consumption (FAO 2008). Therefore economic incentives for the recovery of phosphorus from wastewater and sewage sludge will be strengthened in future.

The value of nutrients

Because commercially available fertilizers are often multi-component products, it is not easy to determine the price of each individual nutrient component. On the other hand, it is obvious that the price of each individual compound in fertilizers is directly coupled to the prices for commodities at world markets and/ or the prices for energy as well. In other words: the price of a multi-component fertilizer should equal the sum of the prices for the individual components at equal product qualities.

For the current approach, first the specific nutrient prices of single component fertilizers were determined. Afterwards these values were used for the determination of individual nutrient prices within multi-component products.

Table 1 exemplarily shows selected commercially available fertilizers and their calculated component specific prices. To facilitate the transferability of the nutrient specific prices to the wastewater sector the values are given related to the individual components and not to the nutrients, as it is common for the agricultural sector.

Table 1. Exemplary selection of commercially available fertilizers, their specific components and prices (basis for the costs: July 2008; in italics: average values).

Product	Price [€/t]	Elementary content [%]					Specific price [€/kg element]				
		N	P	K	S	Mg	N	P	K	S	Mg
CAN	319.50	27					1.18				
Urea	450.00	46					0.98				
Ammonium-nitrate-urea	286.30	28					1.02				
Triple-phosphate	715.00		20					3.58			
Diammon-phosphate	775.00	18	20				*1.03*	2.95			
K (from KCl)	717.20			100					0.72		
NPK (15/15/15)	420.50	15	7	12			*1.03*	*3.13*	0.48		
K/Mg-fertiliser (40 K₂O + 6 MgO)	310.00			33		3.6			*0.60*		3.11
Mg (metal)	3312.00					100					3.31
Kalimagnesia (30/10)	405.50			25		6			*0.60*		4.26
Molten sulfur	282.90				100					0.28	
Ammoniumsulfate	323.80	26			13		*1.03*			0.44	
Calculated values							1.03	3.13	0.60	0.36	3.11

The probably most promising product for phosphorus recovery in the wastewater sector is magnesium ammonium phosphate (MAP or struvite), because it shows favorable reaction properties under wastewater treatment plant conditions and is furthermore a valuable agricultural fertilizer (Römer 2006). All three components of MAP Mg^{2+}, NH_4^+ and PO_4^{3-} have a fertilizing and valuable effect. Taking the calculated prices for commercial fertilizer from Table 1 into consideration, the calculative value of MAP sums up to 763 €/t MAP, where proportionately 51.8% are related to P, 7.7% to N and 40.4% to Mg^{2+} (Figure 2).

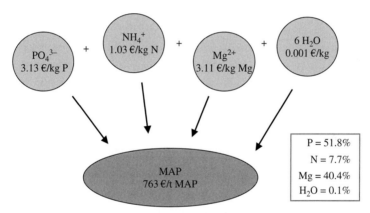

Figure 2. Calculative value of MAP and its individual components (based on prices from July 2008).

Such a resource economic determination of products from phosphate recycling is important for the economic assessment of phosphate recycling processes, because overall process costs have to be related to the whole value of the product. If process costs were only related to the mass of the recovered phosphorus, resulting treatment costs would be overestimated by approx. 100%. Additionally, the evaluation shows the significant influence of costs for the Mg-source since magnesium alone contributes to the product value of MAP by 40%. If MAP precipitation is not used primarily for the elimination of phosphate, but for the removal of nitrogen, besides Mg^{2+} also PO_4^{3-} must be added to the process. In this case costs incurred for supplies would sum up to more than 90% of the calculative product value.

THE COSTS OF PHOSPHORUS RECOVERY

Costs for phosphorus recycling processes largely depend on material flows, which are used and treated, since process and operation expenses depend on the properties of these flows. Normally, process streams at wastewater treatment plants, primarily centrate from digested sludge dewatering, sewage sludge or sewage sludge ashes, are investigated in regard to their suitability for phosphorus recycling (Maier et al., 2005, Pinnekamp et al., 2005). Besides that, especially highly concentrated, original, partial wastewater streams like urine (Lind et al., 2000) are appropriate for recycling, if they can be collected separately prior to the actual wastewater treatment. Ideally this could lead to a decrease in treatment costs at the WWTP and would thus raise general profitability of the process.

If, finally, phosphorus recycling (and not the public service wastewater treatment) would be the entrepreneurial objective of e.g. the wastewater treatment plant operator, the complete process chain would have to be considered for determining process and production costs. Then for the determination of costs all process steps before the actual recycling process would also have to be considered.

Thus, basically the following optional process steps with pertinent relevant cost factors can be determined.

1. Wastewater treatment (proportional costs from phosphorus removal).
2. Anaerobic sludge digestion: in dependency on ancillary conditions, a partial P remobilization occurs which is the prerequisite for recoverability of phosphorus from sludge liquor.
3. Acid hydrolysis of sewage sludge or sludge ashes: containing the highest P concentrations within the wastewater and sludge treatment process. Costs result mainly from acid hydrolysis, subsequent neutralization and if necessary incineration.
4. MAP precipitation: as the actual recovery step is necessary in any case. Costs largely depend on costs for supplies such as Mg^{2+} and lye.

Process components

In the following, costs for three different theoretical scenarios for phosphorus recycling will be discussed. Costs are considered for each process step with respective conditions and significantly influencing factors. For the examination of the individual scenarios, material flow data from a case study with 350,000 PE are used as input. For each scenario, necessary treatment processes were dimensioned, costs were determined and related to produced MAP quantities in order to calculate pertinent specific production costs.

The following scenarios were discussed:

1. MAP precipitation in the centrate from digested sludge dewatering.
2. Acid hydrolysis of sewage sludge ashes (additional costs for supplies such as Mg^{2+} and lye).
3. MAP-precipitation from yellow water (using seawater as Mg^{2+} source).

Input data for calculating above described scenarios were compiled using loads listed in Table 2, which were assessed on the basis of 350,000 PE.

Table 2. Concentrations and loads for three selected scenarios (*related to 100% NaOH and 100% H_2SO_4).

	Scenario 1 centrate liquor	Scenario 2 sludge ashes	Scenario 3 yellow water
Q $[m^3/d]$	500	2,630 kg TS ash/d (in 10–150 m^3 suspension)	480
NH_4-N [mg/L]	1,400	/	7,000
NH_4-N [kg/d]	700	/	3,360
PO_4-P [mg/L]	50–800	/	540
PO_4-P [kg/d]	25–400	460	259
PH	7.2	/	9.2
Mg-source	$MgCl_2$	$MgCl_2$	seawater
Mg-consumption (kg/d)	24–377	433	244
OH^- source	NaOH	NaOH	/
NaOH consumption* [kg/d]	445	340	/
Acid source	/	H_2SO_4	/
Acid consumption* [kg/d]	/	417	/
MAP produced [kg/d]	198–3,165	3,640	2,049

Costs for MAP precipitation

In order to be able to identify relevant cost drivers of the actual recovery step (MAP precipitation), only the process step MAP precipitation is investigated here. This step consists of precipitation and flocculation reactors, a decanter for separating the precipitation product and the necessary machinery and electro-technical equipment.

The wastewater stream to be treated is, according to scenario 1 (see Table 2), centrate from the dewatering of digested sludge. Variations in the phosphate concentration between 50 and 800 mg/L PO_4-P were assumed caused by varying operating conditions in digesters and due to variations in upstream wastewater treatment systems.

Total investment costs of 600,000 € for a precipitation step treating 500 m^3 centrate/d were determined. A uniform depreciation period of 10 years and an interest rate of 6% were defined. Besides annual costs resulting from the above consumptions, also, costs for maintenance and repair (3% of investment costs), energy costs, expenses for means of production as well as labor costs were considered.

In regard to magnesium required for precipitation a (slightly over-stoichiometric) ratio of Mg:P = 1.2:1 was assumed at specific costs of 3.11 €/kg Mg^{2+} (see Table 1). NaOH quantities required for raising the pH value of the centrate were accounted for with 0.89 kg NaOH (100%) per m^3 centrate.

Total annual costs for each investigated scenario were related to the respective produced annual amount of the MAP precipitation product, which resulted in specific production costs (€/t MAP). Possible profits, which could be realized by the sale of MAP, purposely were not considered. However, the calculative value of MAP at 763 €/t MAP (see Figure 2) was considered as a benchmark. In order to explicitly investigate the influence of required Mg^{2+} and NaOH supplies on production costs for MAP, other options were looked at besides the full costs variant. For one discussed option, a possible Mg^{2+} cost reduction to 30% of the original amount was assumed due to the employment of Mg^{2+} containing secondary raw materials. For another option it was assumed that no NaOH is required for raising the pH value. Figure 3 systematically shows production costs for all three above-mentioned options in dependency on phosphate concentrations in the wastewater streams to be treated.

Comparative investigations indicate that production costs highly depend on phosphate concentrations and decrease significantly with increasing concentration. The break-even point for all three investigated scenarios lies slightly below 200 or 300 mg/L PO$_4$-P respectively, which means that from these concentrations on specific production costs are below the calculated value of MAP. Altogether, resulting production costs for a full cost calculation range between just under 2,800 €/t MAP (at 50 mg/L PO$_4$-P) and 520 €/t MAP (at 800 mg/L PO$_4$-P). For the option with a Mg^{2+} cost reduction to 30%, calculated production costs for equal concentrations range between 2,500 and 260 €/t MAP.

For the full cost variant presented in Figure 3, the distribution of production costs by percentage (Figure 4) shows that fixed costs (related to investments, labor, maintenance and repair, NaOH and energy) decrease with increasing phosphate concentration, while variable costs for magnesium supply increase strongly and make up 75% of costs for the highest presented P concentrations.

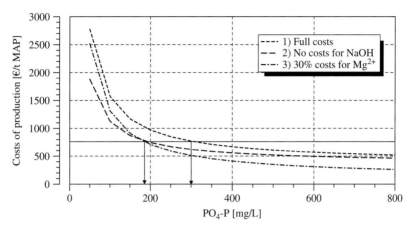

Figure 3. Costs for the production of MAP for three scenarios depending on phosphate concentration.

These findings indicate that on the one hand high phosphate concentrations in the wastewater to be treated are favorable for a reduction of fixed costs, but on the other hand at high phosphate concentrations a further significant reduction of process costs can only be realized, if a cost-efficient or cost-neutral magnesium source is used.

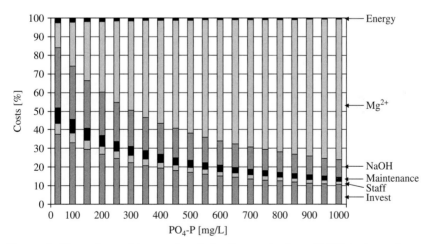

Figure 4. Proportions of types of costs for the production of MAP in dependency on the phosphate concentration.

Costs of phosphate remobilization

Remobilization of phosphate from sewage sludge or sewage sludge ashes normally is achieved by acid hydrolysis. The objective is to improve P yield by dissolution of phosphorus bound in the sludge. If resuspended sludge ash is used, phosphate concentrations in the supernatant can be significantly increased in comparison to the use of sewage sludge, since phosphate concentrations in sludge ashes are usually 2–3 times higher due to the nature of ashes. According to investigations by Schaum (2007), quantities of sulfuric acid required for the re-dissolution of P in sludge ashes lie at 0.6 t H_2SO_4/t ashes. Furthermore after liquid-solid-separation, the pH value must be raised by NaOH. Schaum (2007) has determined a specific NaOH consumption of 0.3 t/t ashes. Thus additional costs for a re-dissolution of P in this case would amount to 105 €/t ashes or 1,310 €/t P. According to current market prices, which have risen significantly also for sulfuric acid, costs would even sum up to 300 €/t ashes or 3,750 €/t P. This comparison indicates that costs just needed for chemicals for phosphate remobilization already reach the market value of phosphorus.

Own investigations of ashes originating from sewage sludge gasification showed that acid consumption for hydrolysis amounted to 0.18 t H_2SO_4/t ashes and additionally 0.062 t NaOH/t ashes were required for subsequent neutralization. For calculated market prices of 297 €/t for H_2SO_4 (100%) and 430 €/t for NaOH (100%), process costs for chemicals thus add up to 80 €/t ashes or 1000 €/t P (at a P concentration of 8% in the ashes). Although additional treatment costs still make up a large share of the calculated value of the product, apparently significant differences can be observed for the demand in supplies depending on the used ashes.

In the case under consideration, ashes re-suspended with volumes between 10 and 150 m^3 (see Table 2, scenario 2) were regarded. Calculated P concentrations in the supernatant (after dewatering) between approximately 2,900 mg P/L (at 150 m^3/d) and 43,000 mg P/L (at 10 m^3/d) could be determined.

A systematic investigation of costs for acid hydrolysis and neutralization showed that due to the proportional relation between solids content (ash quantities) and chemical consumption, in this case, this share of costs remained constant. The decisive factor was rather the amount of water used for re-suspension, since it ultimately determines the achievable product yield (see Figure 5). Small water quantities indeed led to high phosphate concentrations in the liquid phase, which according to the results described in section 3.1 should have a positive effect on production costs. However, a reduction of water quantities simultaneously leads to high phosphate loss (due to water content of the separated ash) when subsequently separating ashes from the suspension.

This results in a significantly decreased phosphate yield (percentage of dissolved P in the centrate after dewatering in relation to P_{total} in ashes).

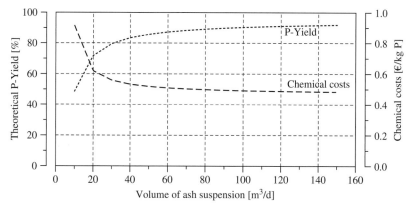

Figure 5. Additional chemical costs and theoretical phosphorus yield in dependency on the volume of the liquid phase for the resuspension of the ash.

Reducing supplies input

While the above described approach discussed costs for P precipitation from centrate as well as phosphate remobilization from sludge ashes in dependency on relevant conditions and under standard treatment procedures, in the following, possible production costs will be discussed which could be achieved, if no or very few cost-intensive supplies were necessary. This will be discussed on the basis of the data for phosphate precipitation from yellow water (scenario 3) shown in Table 2.

Within municipal wastewater, yellow water is the partial stream with the highest phosphate concentrations. Therefore it is especially suitable for a separate phosphorus recovery (Maurer *et al.*, 2003). Furthermore urine already exhibits a pH value of 9.2 after hydrolysis of urea, thus no lye must be added. Since costs for the addition of magnesium make up a significant share of total process costs, it will be investigated how production costs for MAP vary, if seawater, which contains about 1,200 mg/L Mg^{2+} (Terramare 2003), is used as a cost-neutral additive (Matsumiya *et al.*, 2000, Dockhorn 2007b). Identical investment costs, labor costs and costs for maintenance and repair were assumed as in scenario 1. The additional energy consumption for pumping of seawater is also regarded.

Resulting process costs for this calculation amounted to 160 €/t MAP. In comparison to this, the treatment of a wastewater stream with equal phosphate concentrations, but where lye and magnesium would have to be added, would cost 605 €/t MAP. In this last-mentioned case, costs for supplies alone amount to 75% of total process costs.

In addition the treatment of yellow water (or any other originally highly concentrated partial wastewater stream) has a significant cost advantage, since the wastewater stream is treated before it enters the general wastewater treatment system. This saves treatment costs for conventional wastewater treatment, which in fact should be added to process costs of P recovery in the scenarios 1 and 2.

CONCLUSIONS

As shown in the above systematic approach, expenses for phosphorus recycling, which are here considered as production costs for MAP, strongly depend on relevant ancillary conditions and the respective raw substrate.

Costs of MAP precipitation itself are highly concentration dependent. Increasing P concentration in the wastewater led to a disproportionately high decrease in production costs and simultaneously to an increase of the cost percentage that supplies (for Mg^{2+}) have in total costs.

A cost-effective realization of MAP-precipitation can be achieved by adjusting the following parameters:

1. Realization of a high phosphate concentration in the wastewater to be treated.
2. Reduction of the lye quantities used for raising the pH value.
3. Employment of secondary raw materials as cost-efficient Mg-source.
4. Cost-efficient plant construction for reduction of the investment cost fraction.

Calculated MAP production costs for the precipitation step alone amounted to 520–2,800 €/t MAP (or 2–11 €/kg P) depending on concentrations and were thus below the calculated break-even point of the process for concentrations above 200–300 mg P/L.

Required supplies costs for acid hydrolysis alone were determined in the case considered at 1 €/kg P. Process costs largely depended on achievable P yield. This fact must be taken into account regarding the amount of water used for the re-suspension of ashes.

The most cost-efficient MAP production is possible when using a highly concentrated wastewater stream, which is separated before the actual wastewater treatment. Thus conventional wastewater treatment costs of at least 2–3 €/kg P can be saved additionally. An especially cost-efficient production is possible, if as it is the case for yellow water no NaOH addition is necessary and a cost-neutral Mg-source (such as seawater) is available. In this case, where most favorable conditions were assumed, calculated production costs amounted to 160 €/t MAP or 630 €/t P respectively.

In summary, treatment costs for the most favorable case (scenario 3) add up to 0.63 €/kg P, while the treatment chain of conventional wastewater treatment and MAP precipitation from centrate leads to costs in the range of 4–14 €/kg P. Costs for a process combination of wastewater treatment and subsequent phosphorus recovery from sludge ashes, however, (not considering incineration costs) can be estimated at 5–8 €/kg P.

REFERENCES

Bernhardt, H. (1978). Phosphor, Wege und Verbleib in der Bundesrepublik Deutschland. Verlag Chemie, Weinheim–New York.

Dockhorn, T. (2007a). Stoffstrommanagement und Ressourcenökonomie in der kommunalen Abwasserwirtschaft. Habilitation treatise. Veröffentlichungen des Instituts für Siedlungswasserwirtschaft, TU Braunschweig 74, ISSN 0934-9731.

Dockhorn, T. (2007b). Rückgewinnung von Phosphat aus Abwasser und Klärschlamm mit dem Peco-Verfahren. *Müll und Abfall*, 8, 380–386.

FAO (2008). Current world fertilizer trends and outlook to 2012. Food and Agriculture Organization of the United Nations, Rome.

Lind, B.B., Ban, Z. and Bydén, S. (2000). Nutrient recovery from human urine by struvite crystallisation with ammonia adsorption on zeolite and wollastonite. *Bioresource Technology*, 73, 169–174.

Maier, W., Weidelener, A., Krampe, J. and Rott, U. (2005). Entwicklung eines Verfahrens zur Phosphat-Rückgewinnung aus ausgefaultem Nassschlamm oder entwässertem Faulschlamm als gut Pflanzenverfügbares Magnesium-Ammonium-Phosphat (MAP). Schlussbericht des durch die Deutsche Bundesstiftung Umwelt (Osnabrück) geförderten Forschungsvorhabens AZ 21042.

Matsumiya, Y., Yamasita, T. and Nawamura Y. (2000). Phosphorus removal from sidestreams by crystallisation of magnesium-ammonium-phosphate using seawater. J. CIWEM, 2000, 14.

Maurer, M., Schwegler, P. and Larsen, T.A. (2003). Nutrients in urine: energetical aspects of removal and recovery. *Wat. Sci. Tech.*, 48(1), 37–46.

Pinnekamp, J., Gethke, K. and Montag, D. (2005). Stand der Forschung zur Phosphorrückgewinnung. 38. Essener Tagung für Wasser- und Abfallwirtschaft, Aachen, 11.3.2005, Schriftenreihe Gewässerschutz-Wasser-Abwasser, Nr. 198, Aachen 2005.

Römer, W. (2006). Vergleichende Untersuchungen zur Pflanzenverfügbarkeit von Phosphat aus verschiedenen P-Recycling-Produkten im Keimpflanzenversuch. *J. Plant Nutr. Soil Sci.*, 169, 826–832.

Schaum, C.A. (2007). Verfahren für eine zukünftige Klärschlammbehandlung – Klärschlammkonditionierung und Rückgewinnung von Phosphor aus Klärschlammasche. *Schriftenreihe WAR*, TU Darmstadt **185**.

Terramare (2003). Forschungszentrum Wilhelmshaven: Vorlesungsscript Meereschemie I, WS 2003/2004. www.terramare.de

USGS (2008). Mineral commodity summaries. http://minerals. Usgs.gov/minerals/pubs/mcs/

Different strategies for recovering phosphorus: Technologies and costs

D. Montag, K. Gethke and J. Pinnekamp

Institute for Environmental Engineering of RWTH Aachen University, Aachen, Germany

Abstract In Germany, sewage sludge is being used less and less as a fertiliser in agriculture because the pollutants therein pose a significant risk to the environment. Therefore, in order to utilise the phosphorus and the nitrogen contained in the sludge, methods for phosphorus recovery are necessary. Investigations on laboratory scale and pilot scale were carried out in order to develop technical concepts and estimate preliminary costs. From the effluent, a recovery of up to 40% of the phosphorus load that enters the WWTP is possible. This option will cause specific net costs of 4.7 €/(I·a). For process water from sludge treatment, a potential of about 36% at net costs of 1.3 €/(I·a) was calculated. Approximately 80% of the phosphorus inflow load to the WWTP can be recovered from sewage sludge ash from mono-incineration plants at net costs of 2.3 €/(I·a). No long-term large scale operating experience, however, is available for any of these technologies. Hence, it is to be expected that these initial costs, which still contain several uncertainties, can be reduced once sufficient operating experience has been acquired at industrial scale plants.

INTRODUCTION

In Germany, sewage sludge is being utilised less and less as a nutrient source in agriculture because public and scientific discussions have highlighted the potential risks that pollutants in the sludge constitute for the environment. Nowadays, approximately 40% of the sludge is incinerated in coal-fired power plants, cement works or sludge incineration plants. Therefore, in order to utilise the phosphorus and the nitrogen contained in the wastewater and consequently in the sludge, technologies for nutrient recovery are required before the sludge or – in case of an incineration – the ash is finally disposed.

At the Institute for Environmental Engineering (ISA) of RWTH Aachen University, several approaches for phosphorus recovery within the framework of wastewater treatment have been investigated and developed. Basically, three points of implementation to the wastewater and sludge treatment process can be distinguished:

- Effluent of wastewater treatment plants (WWTP),
- Sludge/process water of sludge treatment and
- Sewage sludge ash.

Investigations into these possibilities for recovering phosphorus took place on a laboratory scale and pilot scale. Technical concepts were developed and costs were estimated.

PROCESS DESCRIPTIONS

Post precipitation in the effluent of a WWTP

The recovery of phosphorus from the wastewater treatment plant effluent by post precipitation (Figure 1) uses the same technology as the post precipitation for phosphorus removal. The only difference is the kind of precipitant. To recover phosphorus from the secondary clarifier effluent, the ideal precipitant is magnesium oxide (MgO).

The precipitant is dosed into the reaction vessel of which three should be designed as periodically fed reactors. The first step of the process is the filling combined with adjusting the pH value using caustic soda (NaOH) and dosing MgO. Hereby, phosphate crystallises (in the second step) as magnesium phosphate (MP). The third step is the emptying into a clarifier, where magnesium phosphate is separated by a lamella separator. The clear water effluent has to be neutralised, i.e. lowering its pH value, before conveying it into the receiving water. The product magnesium phosphate is dewatered by bag filtration, and the effluent of this dewatering is pumped back to the biological wastewater treatment process.

One of the three process steps mentioned above can take place simultaneously in each of the three reaction vessels. Thus, the process can run quasi-continuously.

Phosphate release and struvite crystallisation (PRISA process)

The PRISA process (Figure 2) can become integrated in the sludge treatment of a municipal WWTP. The first step of the process is the biological acidification of surplus sludge from enhanced biological phosphorus removal (EBPR) in order to release phosphates prior to anaerobic digestion. The release of phosphates from the sludge takes place in the pre-thickener and is basically achieved by extension of the hydraulic retention time and periodical, careful stirring of the sludge. Under these (anaerobic) conditions the EBPR bacteria take up fatty acid molecules and convert these to poly-ß-hydroxybutyrate (PHB). The energy required for the uptake comes from the hydrolysis of intracellular polyphosphate reserves which have been stored in the sludge during the EBPR. Thus, a considerable phosphorus load can be released into the supernatant liquor systematically.

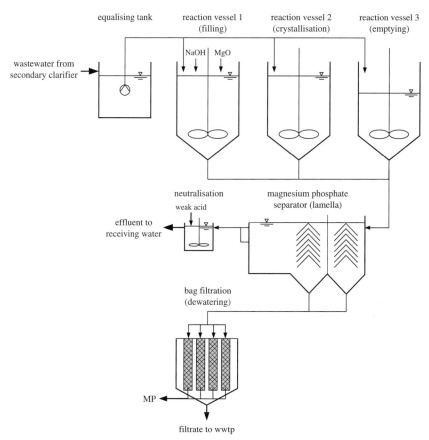

Figure 1. Phosphorus recovery from wwtp effluent (post-precipitation).

The hydraulic retention time of the pre-thickener has to increase from a conventional 1 or 1.5 days to 3 days. Since three pre-thickeners (instead of a single one with bigger volume) have been designed, each one can be used as periodically fed "release and thickening reactor". The procedure in the thickener can be divided into three steps: filling, phosphorus (P) release, and emptying. Using three pre-thickeners, each of these three procedures can take place simultaneously and the operation is quasi-continuous.

It is particularly important to release as much of the phosphorus prior to sludge digestion as possible, during which a certain amount of phosphate dissolved beforehand is fixed again to sludge particles.

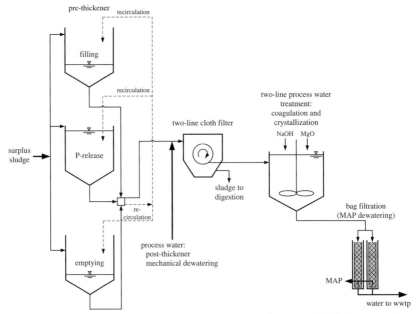

Figure 2. Phosphorus recovery from process water/sludge (PRISA process).

Solids are removed from the whole process water stream – i.e. not only supernatant liquor from the pre-thickening but also process water from post-thickening and mechanical dewatering – by a cloth filter and are conveyed to anaerobic digestion.

In an aqueous solution comprising magnesium, ammonium and phosphate, struvite (magnesium ammonium phosphate – MAP) precipitates if the pH value is approximately 9. In case of a molar ratio of 1:1:1 between the ions, a complete nutrient removal is theoretically possible. In the PRISA process, by dosing magnesium oxide (MgO) and after pH adjustment using caustic soda (NaOH), struvite crystallisation takes place. This stage should be designed with two parallel lines, one being redundant and guaranteeing a quasi-continuous operation.

Finally, MAP is dewatered by bag filtration. The liquid effluent of the dewatering is re-circulated to the biological wastewater treatment process, thus limit values for direct discharge need not be achieved. (Montag *et al.*, 2006, Montag *et al.*, 2007b)

Phosphorus extraction and recovery from ash (RPA process)

To recover phosphorus from sewage sludge ash (RPA process – Recovery of Phosphorus from Ash, Figure 3) the phosphorus has to become released from the solids first. This leaching is carried out by the means of hydrochloric acid for approximately 60 minutes. The temperature need not be controlled. But besides phosphorus, iron and most of the other metals contained in the ash are also dissolved at these settings. Thus, it is necessary to remove most of these elements from the solution.

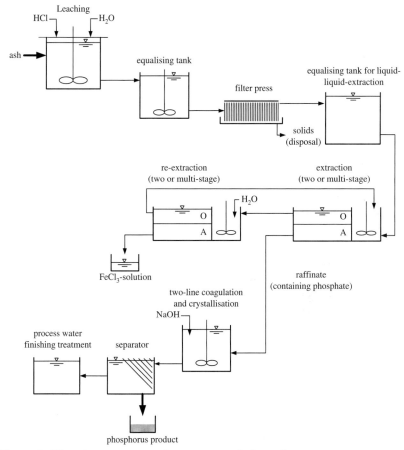

Figure 3. Phosphorus recovery from sewage sludge ash.

After the removal of solids (mainly sand), the filtrate produced enters the liquid-liquid extraction, which was operated with different organic extracting agents (tri-butyl-phosphate (TBP), 3-methyl-1-butanol (i-amyl) and tricaprylyl amine (Alamine®336)) in kerosene dilution in order to remove the iron ions and heavy metals from the filtrate. Conventional mixer-settlers are applicable for the liquid-liquid extraction. Extraction time was 15 minutes and the organic to aqueous ratio was 1:1. Using a two or multi-stage extraction unit, complete removal of the iron ions from the feed is possible. Additionally, most of the heavy metals can be extracted into the extract. The dissolved phosphate is not affected by the extraction and remains in the raffinate.

Finally, the pH value will be raised by adding caustic soda, and, hence, phosphate precipitates as aluminium phosphate (Al-P with a P_2O_5 content of up to 37%) and is separated. The effluent has to receive a finishing treatment due to some loads of heavy metals that are still contained in the liquid. (Montag et al., 2007a)

ECONOMICS

Investment (capital) costs and operation costs were considered to calculate the annual costs. Using an interest rate of 5% the depreciation periods were taken as follows:

- building and civil engineering works (e.g. tanks, structures): 30 years,
- mechanical equipment, wearing parts which normally require periodic replacement: 10 years, and
- electrical equipment, programmable logic control (PLC) equipment, telemetry: 10 years.

For the post-precipitation and the PRISA process a capacity of 100,000 inhabitants (I) was assumed. The plant calculated to recover phosphorus from ash is able to treat 15,000 tons (corresponding to approximately 1.5 million I) annually, which is a feasible size for an incineration plant. The average phosphorus inflow load to German wwtps is approximately 1.8 g $P/(I \cdot d)$. For all options the efficiency of the final phosphorus precipitation/crystallisation is set to 90%. The estimated costs for the three technologies presented before are summarised in Table 1.

The highest annual costs occur for the RPA process. Due to the relatively high phosphorus load recovered, the specific costs amount to 4.3 €/kg P. The highest specific costs are caused by the post-precipitation with 17.75 €/kg P,

Table 1. Estimated costs.

	Unit	Post precipitation	Process water (PRISA)	Ash (RPA)
Absolute costs				
investment costs	€	3,732,549	1,417,739	11,026,720
Capital costs	€/a	338,342	132,432	1,246,573
operation costs	€/a	213,917	65,372	3,604,834
annual costs	€/a	552,259	197,804	4,851,407
revenues for product	€/a	79,827	70,956	1,255,500
Net annual costs	€/a	**472,432**	**126,848**	**3,595,907**
Specific costs				
revenues for product	€/kg $P_{recovered}$	3.00	3.00	1.50
net annual costs	€/kg $P_{recovered}$	17.75	5.36	4.30
	€/(I · a)	4.72	1.27	2.29
Explanatory notes				
Annually recovered P	kg P/a	26,609	23,652	837,000
phosphorus product		magnesium phosphate	struvite	aluminium phosphate

which is more than three times the price of a phosphorus recovery by means of the PRISA process (5.36 €/kg P).

It is important to divide between the aluminium phosphate and the magnesium-based secondary phosphates. The struvite and the magnesium phosphate are appropriate products for direct agricultural use or can be mixed with ready-made fertilisers. The aluminium phosphate can only be used in the phosphate industry or as a raw-material in the fertiliser industry due to the poor fertilising effect. This is why different revenues have been calculated for the secondary phosphate products.

COMPARISON AND OUTLOOK

From the effluent, a recovery of up to 40% of the phosphorus load that enters the WWTP is possible at specific net costs of 4.7 €/(I · a). For process water from sludge treatment, a potential of about 36% at net costs of 1.3 €/(I · a) was calculated. Sewage sludge ash from mono-incineration plants allow a recovery of approximately 80% of the phosphorus inflow load at net costs of 2.3 €/(I · a).

No long-term large scale operating experience, however, is available for any of these technologies. Hence, it is to be expected that these initial costs, which still contain several uncertainties, can be reduced once sufficient operating experience has been acquired in industrial scale plants.

The future of phosphorus recycling is closely linked with the future of sludge management. Mono-incineration plants necessarily have an annual capacity of several 10,000 tons of sludge. If nearly all the sludge ends in these plants, its disposal will have almost become centralised. In this case installing a few centralised phosphorus recovery plants next to some of the biggest incineration plants seems to be a logical conclusion. Ashes from other incineration plants could be transported to these centralised facilities.

If large amounts of sewage sludge are co-incinerated, phosphorus content in the ashes is too low for an economically viable recycling process. Thus, recovering phosphorus prior to the sludge incineration must be carried out, e.g. via the PRISA process which utilises struvite crystallisation as a final recovery step. Basically, this option will also be the most promising method if ongoing research and development work proves that recovery from ash is generally not feasible.

Post precipitation causes the significantly highest specific costs and, therefore, does not seem to be a feasible way of recovering phosphorus at municipal wastewater treatment plants.

ACKNOWLEDGEMENT

The research and development projects in the field of phosphorus recovery carried out at the Institute for Environmental Engineering of RWTH Aachen University were funded by the German Federal Environment Agency and by the Ministry of the Environment and Nature Conservation, Agriculture and Consumer Protection of the Federal State of North Rhine-Westphalia. The current research work is funded within the framework of the German research cluster on phosphorus recycling which is supervised by the Project Management Agency Forschungszentrum Karlsruhe and funded by the Federal Ministry of Education and Research.

REFERENCES

Montag, D., Gethke, K. and Pinnekamp, J. (2006). Phosphate Recovery – An integral part of sustainable sludge management and resource protection. In: William G. Lyon, Steven K. Starrett, Jihua J. Hong. *Environmental Science and Technology*, Vol. (I), American Science Press, Houston, 2006. ISBN 0-9768853-6-0.

Montag, D., Gethke, K. and Pinnekamp, J. (2007a). Different Approaches for Prospective Sludge Management Incorporating Phosphorus Recovery. In: Filibeli, A., Sanin, F. D., Ayol, A., Sanin, S. L.: *Facing Sludge Diversities: Challenges, Risks and*

Opportunities, page 289-296, IWA-Congress, March 28th – 30th 2007, Antalya, Turkey, ISBN 978-975-441-238-3.

Montag, D., Gethke, K. and Pinnekamp, J. (2007b). A feasible approach of integrating phosphate recovery as struvite at waste water treatment plants. In: LeBlanc, R. J., Laughton, P. J., Tyagi, R. IWA-Conference Proceedings "Moving Forward: Wastewater Biosolids Sustainability: Technical, Managerial, and Public Synergy", page 551–558, June 24th – 27th 2007, Moncton, New Brunswick, Canada.

Social and economic feasibility of struvite recovery from Urine at the community level in Nepal

E. Tilley, B. Gantenbein, R. Khadka, C. Zurbrügg and K.M. Udert

Eawag, Swiss Federal Institute of Aquatic Science and Technology, 8600 Dübendorf, Switzerland

Abstract In the face of increasing phosphorus prices, Nepal, like all agrarian societies is in imminent need of a low-cost, sustainable source of phosphorus and nitrogen.

The town of Siddhipur in the Kathmandu valley was selected as a study site to determine the feasibility of establishing a community-scale struvite processing centre using source-separated urine as a feedstock. Focus groups, workshops and household questionnaires were used to determine cultural acceptability and willingness to pay. Urine quality and quantity measurements, along with fertilizer cost analysis were used to determine the economic feasibility. A regression analysis using the current, local value of the nutrients was used to calculate the theoretical value of struvite; the estimated nutrient value of struvite was found to be 24–41 NRp per kilogram, which is in the same range as locally available Diammonium Phosphate (DAP). The production of struvite would currently cost about 25 NRp/kg due to high prices of available magnesium sources. Nevertheless, struvite precipitation can be economically feasible for at least three reasons. First, Nepal possesses its own magnesite deposits, which might allow it to produce cheaper magnesium sources. Second, customers are likely to pay more for locally produced struvite than for imported synthetic fertilizers. Third, the effluent from struvite precipitation can also be used as a valuable and efficient nitrogen and potassium fertilizer. Despite shadow-pricing based on the constituent nutrients, the true value of struvite can not be known until it is produced and sold.

INTRODUCTION

After a long period of a state regulated fertilizer supply in Nepal, the fertilizer market was liberalized by the government in 1997. The Nepalese government, in an attempt to control long lines and shortages, removed import duties and value-added tax on fertilizer products and granted other tax reductions. However, these small gestures were not enough to compensate for the dramatic increase, and a new black market of imported fertilizer from India emerged. India continues to subsidize fertilizer nationally, and thus a lucrative business of smuggling Indian fertilizer into Nepal has emerged to take advantage of the significant price gap. As of 2006, roughly 60% of the total fertilizer sold in

Nepal came from informal sources and of that, the majority can be attributed to India (Thapa, 2006).

As fertilizer, especially phosphate, prices continue to rise at unprecedented rates (The Market, 2008), it is imperative for Nepal and other countries with low income to explore alternative sources for nutrient supply such as human urine (Lienert *et al.*, 2003).

Struvite

A great deal of research over the past 20 years has optimized phosphorus recovery through struvite precipitation from different side streams in WWTPs, especially from digester supernatant (Ueno and Fujii, 2001; Britton *et al.*, 2005; Forrest *et al.*, 2008). Until recently, however, relatively little work has been done on the recovery of struvite from urine, although the body of work is growing (Ronteltap *et al.*, 2007; Wilsenach *et al.*, 2007; Tilley *et al.*, 2008). Despite the fact that reactor design, dosing regimes and pH control can optimize the process and improve the product, the fundamentals of struvite recovery remain unchanged. A soluble magnesium source, a high pH value, a stirring mechanism and an effective method of solid separation are, essentially, the elements needed for struvite recovery with a high rate of phosphate recovery ($>90\%$ and usually higher). Due to the low technological requirements, phosphate recovery through struvite precipitation is an ideal process for low-tech applications, e.g. in decentralized setups or in the developing world.

Urine is often promoted as a liquid fertilizer although it has a strong smell, it is heavy to transport, it is voluminous, it requires a large amount of storage space and its nutrient content is often unknown. Application of urine as a fertilizer is further hampered by the volatilization of ammonia. Struvite, on the other hand, is odourless, dense, compact and efficient to transport, it can be stored during winter, or dry seasons and used when needed, and the nutrient quantity is consistent. Struvite from urine offers the simplicity and quality of chemical fertilizer without the high cost or technical requirements. However, struvite is essentially a phosphate fertilizer, with little nitrogen and no potassium except if struvite is precipitated from an ammonium free solution (Wilsenach *et al.*, 2007).

Siddhipur

As of 2006, 73% of people in Nepal still lacked access to improved sanitation, but between 1990 and 2004, intensive national and international projects increased the number of people with improved sanitation by 25% (WHO, 2008).

According to an ENPHO report (ENPHO, 2007) there are 481 Urine-Diverting Dry Toilets (UDDTs) in the Kathmandu valley, but informal information suggests it could be closer to 600.

Siddhipur is a farming village located about 10 km south east of Kathmandu. Of the 6000 residents, almost 90% of them work in agriculture. Although there is a sewer network in Kathmandu proper, Siddihipur is not connected. Currently there are 100, family-owned UDDTs in Siddhipur.

Because of the number of urine-collecting toilets, the proximity to agriculture, the tradition of nutrient reuse and the proven sensitivity to intervention programs, Siddhipur was selected as the study site for examining the possibility of community-scale struvite production.

METHODS

To determine the economic feasibility of collecting the urine generated at the household level and processing it into struvite at the community level, the quality, quantity and current uses of urine were determined.

The cultural acceptability and willingness to pay was determined in focus group disccusions, in workshops and with household.

Urine quality

Urine samples were taken from 14 household urine containers. To determine the average concentration of nutrients, 10 samples were mixed together and analyzed for nitrogen, phosphorus and potassium. The remaining 4 samples were analyzed individually for phosphorus. Additionally, struvite was prepared with the mixed sample and the process effluent was subsequently analysed for N, P and K as well. The results are summarized in Table 1.

Compared to literature data for stored urine (Udert *et al.*, 2006; Vinneras *et al.*, 2004; Jönsson *et al.*, 2005), the urine of Siddihipur is considerably less concentrated. Ek *et al.* (2006) monitored urine from UDDTs and found values that were more similar to those found in Siddhipur, although they attributed this to dilution of the urine with water. Citizens of Siddhipur also dilute the urine as a way to 'reduce smell', although the effectiveness of this technique, is not completely clear.

To assess the amount of urine that could be collected in Siddhipur, a questionnaire with 26 UDDT owners was conducted. People were asked how often they empty their urine tanks and based on that information, it was deduced that they collect, on average, 155 L of urine per person per year, which is one-third to one-half of the normally cited values (Ciba-Geigy 1977). Some families

Table 1. Urine analysis.

	pH	EC [mS/cm]	P [mg/L]	NH$_4$-N [mg/L]	K [mg/L]
Individual samples					
Sample 1	8.83	62.6	374	–	–
Sample 2	8.57	40.1	264	–	–
Sample 3	8.00	24.0	123	–	–
Sample 4	8.48	31.1	182	–	–
Mixed samples					
Untreated urine	8.67	38.2	259	2352	802
Mix struvite effluent	8.59	38.5	16.47	2240	744
Values from Ek et al. (2006)					
Stored urine	9.1	–	310	3600	900

claim as much as 355 L per person per year, while others reported as low as 51 L were collected per person per year.

To verify these data, the urine tanks at six households were monitored continuously for six days. On average about 0.5 L per person and day was recorded, although some families are collecting close to 1 L per person and day.

Interviews about how the families use the urine indicated that people either pour the urine on their compost pile (the Nepalese 'saaga'), use it on a local plot, dump it in the drain or give it to someone for agricultural use. So while there is not a strong tendency to use it directly on fields, there is little doubt that the use of urine is well accepted.

FINANCIAL FEASIBILITY

To determine the perceived value of urine, people were asked to answer questions about how much they would consider paying for urine as a liquid fertilizer. The results were too erratic to obtain a useful data set. Some people reported that they would not sell their urine for any price (despite the fact that they were currently dumping it down the drain) while still another reported that he would pay 2000 NRp for 100 L of urine (where 100 NRp is approximately 1 Euro). This clearly illustrates that estimating or valuing the nutrient worth of urine or struvite is something that has no precedent in this community and therefore, will likely be unpredictable until the point of sale.

The price of 16 different local fertilizers including NPK mixes, urea and diammonium phosphate (DAP) were collected in local fertilizer shops. The fertilizers and their prices are listed in Table 2.

Table 2. Currently available fertilizers.

Fertilizer	Prices (NRp/kg)
Urea ($NH_2)_2CO$	20, 29, 30, 20, 20
NP 20:20	29, 30, 27, 30
NPK 20:20:10	32
Diammonium Phosphate $(NH_4)_2HPO_4$	38, 44
Ammonium Sulfate $(NH_4)_2SO_4$	24, 25
Myriate of Potash KCl	22
Magnesium Sulfate $MgSO_4$	24

To determine the current value of each nutrient of interest, a regression model was used to disaggregate the price of each of the nutrients (N, P, K, S, Mg) from the total fertilizer price using the following:

$$\text{Price} = a_1[N] + a_2[P] + a_3[K] + a_4[Mg] + a_5[S] \tag{1}$$

The value for each element, as well as the standard error, for the determination of each, is shown in Table 3.

Table 3. Regression estimates for nutrient values.

	Value (NRp/ kg)	Std. error (%)
Nitrogen	60	9
Phosphorus	159	10
Potassium	54	20
Magnesium	173	34
Sulfur	61	27

Phosphorus has more than twice the value of nitrogen, and nearly three times the value of potassium. The low standard error for nitrogen and phosphorus reflects the fact that all, except for 4 of the fertilizers used in the regression include nitrogen and phosphorus, whereas the calculations for K, S and Mg were based on only two, two and one fertilizer(s), respectively.

Using the values shown in Table 4, the price of fertilizers were calculated. Figure 1 shows a comparison between the calculated values for each fertilizer as determined by the regression analysis, and the actual prices. Clearly, the regression was able to predict the price of the elements such that the fertilizers prices could be estimated closely using the regression results. The same regression results were used to estimate the value of each of the elements in struvite as well as the total nutrient value of struvite. Although magnesium is not typically sought out by farmers, it is important for healthy soils; the value of struvite both including and not including the value of magnesium is presented in Table 4.

Table 4. Regression estimates for nutrient value of struvite. Upper and lower values are estimates for the limits of the 95% confidence interval (average ± 2 standard deviations).

	N	P	Mg	Struvite (no Mg)	Struvite (incl. Mg)
Weight fraction (%)	5.7	13	10		
Upper value (NRp/kg)	4.0	24	29	28	57
Est. value (NRp/kg)	3.4	20	17	24	41
Lower value (NRp/kg)	2.8	16	5.5	19	25

The weight fraction for each element was based on the molecular weight of struvite being 245 g/mol; the upper and lower values were calculated using the percent standard deviation determined in the regression. The calculated value of 1 kg struvite is similar to locally available DAP (38 and 44 NRp/kg). The financial feasibility of struvite production strongly depends on the costs for the magnesium source. If the magnesium sulfate listed in Table 2 were used for struvite precipitation, the costs for 1 kg struvite would be at least 25 NRp/kg, which is close to the nutrient value of struvite. Nevertheless, we argue that there are at least three reasons why struvite precipitation is economically feasible. First, cheaper magnesium sources can be made accessible in Nepal, since the country has own magnesite deposits (USGS 2007). Second, it is very likely that customers are willing to pay more for locally produced struvite than for imported synthetic fertilizers. Third, the effluent from struvite precipitation is a

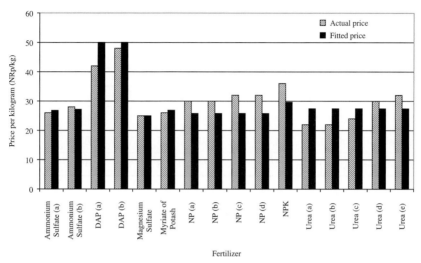

Figure 1. Comparison of modelled and actual fertilizer prices.

valuable fertilizer, which might be especially suited for micro-irrigation. When struvite is produced, a nitrogen-rich, high pH solution remains. As noted in Table 1 there is over 2000 mg/L of nitrogen remaining in the effluent after struvite has been precipitated. Table 5 presents the theoretical nutrient values of the struvite effluent. Non-potable freshwater can be bought in Siddhipur for 100 NRp per cubic metre. Since the effluent could potentially serve as a drought-proof irrigation-water source, as well as a year-round fertilizer, the value for effluent including and excluding the value of water is shown. The use of the effluent as a fertilizer can strongly increase the economic feasibility of urine processing. Assuming a struvite yield of 1.92 kg struvite/m^3 urine, the effluent values can be estimated as 149 NRp/kg struvite and 96 NRp/kg struvite with or without water, respectively (calculated with an ammonium to phosphate ratio in collected urine of 9 kgN/kgP).

Although untreated urine has a similar composition, effluent is arguably more suitable for modern fertilization methods such as micro-irrigation than untreated urine. Effluent, unlike untreated urine, does not have the ability to precipitate spontaneously the same way that full strength urine does (Tilley *et al.*, 2008). Urine precipitation is one of the major problems facing urine-separating technologies and components: pipes, values, and fittings routinely become blocked and must be cleaned with acid or by hand (Udert *et al.*, 2003). Although there is little published evidence, anecdotal evidence points to the same problem

Table 5. Regression estimates for value of struvite processing effluent. Upper and lower values are estimates for the limits of the 95% confidence interval (average $\pm$ 2 standard deviation). A ratio of 16 gNH_4-N/gSO_4-S was used to estimate the sulfur content of urine (Udert et al., 2006).

	N	P	K	S	Water	NPK	NPK, Water
Concentration (kg/m^3)	2.240	0.017	0.744	0.140	1000	–	–
Max value (NRp/m^3)	157	3.1	56	13	100	229	329
Est value (NRp/m^3)	134	2.6	40	8.5	100	185	285
Min value (NRp/m^3)	111	2.1	24	3.8	100	141	241

with using urine in micro irrigation. Micro irrigation is an attractive alternative to hand-application since it reduces lifting, carrying, smells, nitrogen losses and the fertilizer can be targeted to the ideal root zone. It is hypothesized that precipitates from urine would clog and disrupt the functioning of irrigation systems using urine: for this reason it is proposed to use the effluent generated as a precipitate/clog-proof source of irrigation water and nitrogen. Studies are currently underway to assess the technical feasibility of this solution, given that the economic feasibility has already been established.

CONCLUSIONS

Struvite recovery from urine in Siddhipur is both culturally and economically feasible.

The assessment of urine in Siddhipur indicates that most people are not using the UDDTs constantly, and therefore the current volume generated is less than initially thought. Public urinals and education campaigns about the hygienic, environmental and financial benefits of consistently using the UDDTs and collecting urine may be methods to improve the quality and quantity of urine, and therefore the amount of struvite that could be produced.

To reap increased value, future research will address the potential of using the nitrogen-rich processing effluent as a liquid fertilizer. If this is successful, the effluent could be used for free locally, while the struvite could be used, with any surplus sold for export, as a cash-earning product.

The methods used, i.e. shadow-pricing, do not, and can not, reflect the true 'value' of either the struvite or the effluent since there is inherent value which can not be numerically assessed. The fact that locally produced struvite will be made onsite, will be associated with a person or a group who is visible in the community, and that the product is consistently reliable (as opposed to foreign imports which are subject to natural disasters like in China, or foreign export rules) may add extra value. Alternatively, the reverse may hold true, if fears exist about the quality or if local fertilizer merchants see it as a threat to their business.

The price of magnesium has a large effect on the price of producing struvite, and it would be unrealistic to attempt producing struvite if a low cost source were not available. It is very likely that cheaper magnesium sources can be accessed in Nepal. Currently, a variety of local sources are being investigated.

Future work will examine the actual value given, in the current market, to both struvite and effluent which will become increasingly important if nutrient independence for Nepal is to be achieved.

ACKNOWLEDGEMENTS

The authors would like to thank the Swiss Development Corporation, the National Centres for Competence in Research North-South, UNHABITAT Nepal, ENPHO, and the people of Siddhipur for their generous support.

REFERENCES

Britton, A., Koch, F.A., Mavinic, D.S., Adnan, A., Oldham, W.K. and Udala, B. (2005). Pilot-scale struvite recovery from anaerobic digester supernatant at an enhanced biological phosphorus removal wastewater treatment plant. *J. Environ. Eng. Sci.*, **4**(4), 265–277.

Ciba-Geigy (1977). WissenschaftlicheTabellen Geigy, Teilband Körperflüussigkeiten (Scientific tables Geigy. Volume body fluids). 8 ed. Basel (in German).

Ek, M., Bergström, R., Bjurhem, J.-E., Björlenius, B. and Hellström, D. (2006). Concentration of nutrients from urine and reject water from anaerobically digested sludge. *Water Sci. Technol.*, **54**(11–12), 437–444.

ENPHO (2007). EcoSan Toilets in Nepal. Internal Report. Kathmandu, Nepal.

Forrest, A. Fattah, K. Mavinic, D. and Koch, F. (2008). Optimizing struvite production for phosphate recovery in WWTP. *J. Envir. Engrg.*, **134**(5), 395–402.

Jönsson, H., Baky, A., Jeppsson, U., Hellström, D. and Kärrman, E. (2005). Composition of urine, faeces, greywater and biowaste for utilisation in the URWARE model. Report 2005:6, Urban Water, Chalmers University of Technology, Gothenburg.

Lienert, J., Haller, M., Berner, A., Stauffacher, M. and Larsen T.A. (2003). How farmers in Switzerland perceive fertilizers from recycled anthropogenic nutrients (urine). *Water Sci. Technol.*, **48**(1), 47–56.

Thapa, Y.B. (2006). Constraints and approach for improving fertilizer supply for meeting domestic deman: Policy Paper 30. Economic Policy, Government of Nepal/ Ministry of Finance Network Singha Durbar, Kathmandu, Nepal.

The Market: Fertilizer News and Analysis. October 2008. London, England. http://fertilizerworks.com/html/market/TheMarket.pdf

Tilley, E., Atwater, J. and Mavinic, D. (2008). Recovery of struvite from stored human urine. *Environ. Technol.*, **29**(7), 797–806.

Ronteltap, M. , Maurer, M. and Gujer W. (2007). Struvite precipitation thermodynamics in source-separated urine. *Water Res.*, **41**(5), 977–984.

Ueno, Y. and Fujii, M. (2001). Three years experience of operating and selling recovered struvite from full-scale plant. *Environ. Technol.*, **22**(11), 1373–1381.

Udert K.M., Larsen T.A. and Gujer W. (2003). Biologically induced precipitation in NoMix systems and urinal traps. *Water Sci. Technol. Water Supply*, **3**(3), 71–78.

Udert, K.M., Larsen, T.A. and Gujer, W. (2006). Fate of major compounds in source-separated urine. *Water Sci. Technol.*, **54**(11), 413–420.

USGS (2007). Minerals Yearbook, Vol . I, Metals & Minerals, United States Geological Survey, United States Government Printing Office, Washington D.C.

Vinnerås, B., Palmquist, H., Balmér, P. and Jönsson, H. (2004). The characteristics of household wastewater and biodegradable solid waste – a proposal for new Swedish desing values. In *Proc. of the 2nd IWA Leading-Edge on Sustainability in Water-Limited Environments*, Sidney, 2006.

Wilsenach, J.A., Schuurbiers, C.A.H. and van Loosdrecht M.C.M. (2007). Phosphate and potassium recovery from source separated urine through struvite precipitation. *Water Res.*, **41**, 458–466.

World Health Organization (WHO) and United Nations Children's Fund Joint Monitoring Programme for Water Supply and Sanitation (JMP). (2008). Progress on Drinking Water and Sanitation: Special Focus on Sanitation. UNICEF, New York and WHO, Geneva, 2008.

Induced struvite precipitation in an airlift reactor for phosphorus recovery

D. Stumpf[a], B. Heinzmann[b], R.-J. Schwarz[b], R. Gnirss[b] and M. Kraume[a]

[a]Technische Universität Berlin (Berlin Institute of Technology), Chair of Chemical Engineering, ACK 7, Ackerstraße 71–76, 13355 Berlin, Germany,
Email: daniel.stumpf@tu-berlin.de
[b]Department of Research and Development, Berliner Wasserbetriebe (Berlin Water), Neue Jüdenstraße 1, 10179 Berlin, Germany

Abstract An attractive way of recovering phosphorus from digested sludge is the precipitation directly in the sludge as a step in the continuous process of wastewater treatment. The Berliner Wasserbetriebe (Berlin Water) implemented an induced magnesium ammonium phosphate (MAP or struvite) precipitation to avoid incrustations in their sludge treatment equipment. The required pH-values for precipitating MAP were reached by stripping CO_2 with air. To verify the process for phosphorus recovery and to optimize it in order to get a maximized yield of phosphorus, experiments in a 45-litre bench scale reactor were performed, which are described in this paper. This bench scale reactor was designed as an airlift reactor. The main objective was to find optimized operating parameters for the reactor which assured a maximum MAP-precipitation, particularly with regard to an efficient separation of the crystals to obtain a high P-recovery. Trials using a model solution were conducted to determine the yield of recovered phosphorus as precipitated MAP at different air flow rates. Furthermore, the crystal size plays a prominent role for a successful separation of the MAP-crystals. Thus, particle size distributions were measured. For all flow rates about 90% of the possible phosphorus was recovered as MAP in a batch and a continuous process. However, at lower aeration rates the particle size of the crystals shifted to larger sizes. Trials were carried out to clarify the influence of residence times on crystal sizes in the continuously operated airlift reactor. Increasing residence times led to larger crystal sizes.

INTRODUCTION

The sludge of a wastewater treatment plant with enhanced biological phosphorus removal is an effective source for phosphorus recovery (Pinnekamp *et al.*, 2007). There are different technical options to recover phosphorus (Corre *et al.*, 2007; UBA, 2007). One possible process is the precipitation and separation of magnesium ammonium phosphate (MAP or struvite) directly in digested sludge as part of the continuous treatment process. The sludge of a digester is under anaerobic conditions, so that the assimilated phosphorus of the microorganisms from the biological phosphorus removal process is released (Heinzmann and Engel, 2003, 2005). The Berliner Wasserbetriebe (Berlin Water) implemented an induced MAP precipitation by aerating digested sludge to avoid incrustations in the sludge chain, especially in the treatment equipment. As a result of aeration, carbon dioxide is stripped, raising the pH-value (Merkel and Krauth, 1999). Magnesium chloride is added and phosphorus is precipitated as MAP. After separating and cleaning, MAP could be reused as a fertilizer. This process can easily be included into an established wastewater treatment plant (wwtp) and can make the phosphorus recovery worthwhile. At Berlin Water a cylindrical storage tank between the digestion tank and the sludge treatment equipment is currently being used for the precipitation (Figure 1). Internal studies indicated that the current process is suitable for an induced MAP precipitation though it is not yet optimized. The current tank is not designed for a maximum separation of the precipitated MAP-crystals. Therefore, the yield of recovered phosphorus is too low. But an optimized precipitation tank is currently being planned and will be built in the near future to get a possible maximum yield of recovered phosphorus (as MAP).

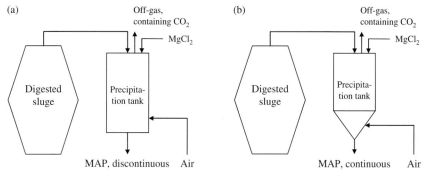

Figure 1. Scheme of MAP-precipitation at Berlin Water, a) current tank, b) optimized tank.

In order to verify the process and to optimize it, experiments in a 45-litre bench scale airlift reactor were performed. The main objective was to find optimized geometrical and operating parameters for a suitable precipitation reactor which assured a maximum MAP-precipitation, particularly with regard to an efficient separation of the crystals to obtain a successful P-recovery. In this case the results could be assigned for a planned large-scale precipitation reactor at the wwtp of Berlin Water.

For that reason the bench scale reactor was designed as an airlift reactor for an improved liquid phase mixing and stripping of the dissolved CO_2 and therewith an enhanced separation of the MAP crystals. The circulation flow of an airlift reactor keeps crystals suspended longer so that they can grow. The results of the trials in the batch mode and the results in continuous operation with different hydraulic residence times will be presented in this paper.

MATERIALS AND METHODS

A model system was used to examine the MAP-crystallization kinetics and the separation of the crystals. At first, the investigated bench scale airlift reactor was operated in batch mode, afterwards in continuous operation. Different parameters such as the aeration flow rate were varied. The resulting particle sizes distributions and the yields of MAP were investigated.

The investigated model system

For the bench scale experiments a model solution was used ($n_{PO4-P}:n_{Mg} = 1:1$, $c_{PO4-P} = 330$ mg L^{-1}, $c_{Mg} = 258,95$ mg L^{-1}, $c_{NH4-N} = 1250$ mg L^{-1}, buffer) which contained the respective ion concentrations typical for the real digested sludge of a wastewater treatment plant of Berlin Water. In the beginning of the experiments the model solution without magnesium was saturated with CO_2 to get the same pH-value as in the real system (pH 7). The CO_2 was stripped by supplying air. Furthermore, the third crystallization component was added by a liquid flow of $MgCl_2$ during the experiment. The pH was increased by CO_2 stripping and MAP precipitates as saturation is exceeded. To determine the amount of precipitated crystals, the residual dissolved concentrations of magnesium, ammonium and phosphate in the bench scale airlift reactor were measured by an ion chromatographic analyser. The pH-value was also measured at the run time of the experiment.

Airlift reactor in batch mode

For the precipitation experiments a bench scale airlift reactor was designed on the basis of the results of 1-L labscale batch tests (Stumpf et al., 2008). First experiments were conducted in batch mode with a liquid volume of 45 L. Therefore, a transferability of the previous 1-L tests results and a verification of the precipitation in the new reactor configuration were possible.

The investigated airlift reactor for batch operation is shown in Figure 2. The airlift reactor consists of two concentrical cylinders, one with an inner diameter of 290 mm and a draft tube with an inner diameter of 194 mm.

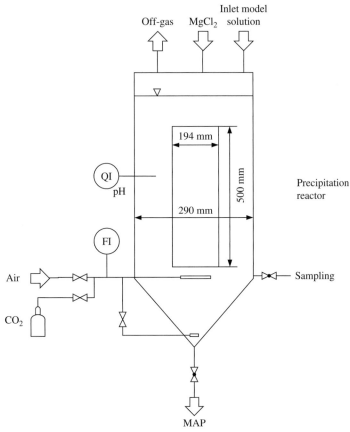

Figure 2. Flow sheet of the bench scale airlift reactor in batch mode (with added aeration for the cone-shaped bottom).

Underneath the internal draft tube the gas sparger is located. The bottom of the airlift reactor is cone-shaped with a valve for the outlet of the precipitated crystals. The aeration of the draft tube induces a loop circulation as it is known for airlift reactors with two sections: a riser and a downcomer (superficial gas velocities at conducted aeration rates: Table 1).

Table 1. Superficial gas velocities at conducted aeration rates.

Aeration rate (L h^{-1})	Superficial gas velocity (m s^{-1})
100	0.0028
300	0.0047
500	0.0090

In this airlift reactor the aeration has two functions: on the one hand to induce an internal recirculation where the MAP crystals precipitate in a well-mixed chemical solution and on the other hand to strip the solved carbon dioxide. The internal circulation is also necessary to keep the crystals suspended so that they grow by added salt solution (crystallization partners) to a size which makes the following separation step easier. The MAP crystals leave the internal circulation if the settling velocity of large crystals is higher than the recirculation velocity. Consequently, they settle down into the cone bottom (settling zone).

In several tests a gas sparger in the cone-shaped bottom was added. Thus a liquid recirculation and thereby reintroduction of the settled crystals into the circulation flow was induced. That way the crystals can grow more owing to the newly added salt solution.

Airlift reactor in continuous operation

In a second step the airlift reactor was extended by storage tanks and pumps for the continuous operation. Figure 3 shows a scheme of the continuous operation. Precipitation experiments with residence times between 2, 8 and 16 hours were conducted in the 45-L airlift reactor in continuous mode. Therefore, a storage tank was used for the feed of the model solution. The same amount of liquid volume was pumped out in the outlet storage tank. Different liquid flow rates were adjusted for the pumps for realizing the residence times.

Particle sizes of precipitated MAP

At the end of the experiments the whole volume of the airlift reactor in batch mode was filtered by a filter bag. In the case of the continuous operation the precipitated and settled crystals of the cone bottom and of the outlet liquid flow were collected in filter bags. In each case the filtrate was dried at 30°C until it reached a constant weight and the total mass of the precipitated crystals was measured (Sartorius BP221S). In addition, samples of the precipitated MAP were identified under a microscope (light-optical microscope, Zeiss) to gain information about the shape of the crystals. Typical shapes of MAP crystals such as the orthogonal and cross shape were identified, which were described by Pschyrembel et al. (1977).

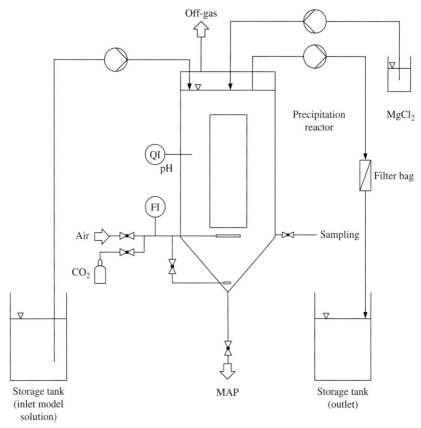

Figure 3. Flow sheet of the bench scale airlift reactor in continuous operation.

Furthermore, the particle size distribution of the dried crystals was measured by a particle size analyzer (Malvern Mastersizer). In certain cases analyses were conducted with sieves of different mesh sizes (300 μm, 200 μm, 150 μm, 90 μm, 60 μm, 33 μm) in order to control the results. Various research works on particle sizes of MAP are described by Corre *et al.* (2007).

RESULTS AND DISCUSSION

Airlift reactor in batch mode

As shown in Figure 4, MAP crystals precipitated after a few minutes by stripping CO_2 (indicated by the decreasing residual dissolved phosphate concentration and rising pH). Different air flow rates were tested to optimize the parameters for the precipitation reactor. When a high air flow rate $\dot{V}$ was adjusted ($\dot{V}_3 = 500\,L\,h^{-1}$), the precipitation took less time than at lower aeration rates ($\dot{V}_1 = 100\,L\,h^{-1}$ and $\dot{V}_2 = 300\,L\,h^{-1}$). This means that a fast precipitation of MAP requires a high air flow rate and a fast stripping of CO_2.

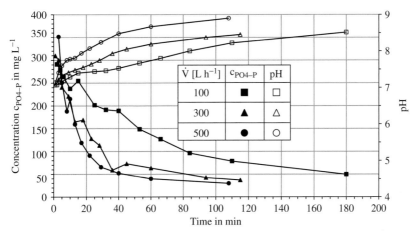

Figure 4. MAP precipitation in a model solution at different air flow rates ($\dot{V}_{1-3}$) in the airlift reactor in batch mode.

The main objective of the precipitation process is to recover a maximum amount of phosphorus (calculated as yield, equation 1).

$$y = m_{MAP,real}/m_{MAP,possible} \tag{1}$$

$$m_{MAP,possible} = c_{PO4-P} \cdot V_R \cdot M_{MAP}/M_{PO4-P} \tag{2}$$

Figure 5 shows the time dependent on the mass of precipitated MAP and the resulting yield of phosphorus. For a high air flow rate ($\dot{V}_3 = 500\,L\,h^{-1}$), this amount was reached after one hour. At the same time the yield was 58% at $\dot{V}_1 = 100\,L\,h^{-1}$ and 80% at $\dot{V}_2 = 300\,L\,h^{-1}$. However, at all tested air flow rates the result in terms of efficiency was the same: almost always approximately 85–90% of the phosphorus was recovered as MAP at the end of the experimental time. Therefore, it can be concluded that the final result depended on the amount of the stripped CO_2 and therewith on the increased pH-value.

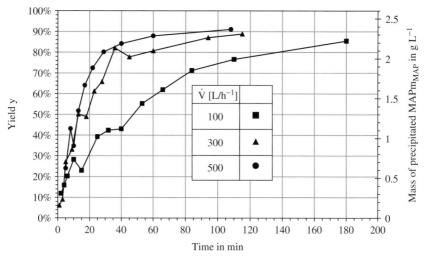

Figure 5. Yield of phosphorus and mass of precipitated MAP crystals at different air flow rates ($\dot{V}_{1-3}$) in the airlift reactor at batch mode.

Figure 6 shows the results of the particle size analyses. The identified average particle size was 90–150 µm for all air flow rates in the airlift reactor. However, crystal sizes shift to higher values at a low aeration rate. The fraction of sizes between 200–300 µm is higher at $\dot{V}_1$ than at higher aeration rates. At lower aeration rates the stripping of CO_2 is too slow, thus there are low supersaturation conditions. At low supersaturation conditions larger crystals can build up (Hofmann G., 2004).

For an efficient separation of the crystals, a certain size of the crystals is necessary. Large crystals with several 100 micrometers quickly settle down at the cone bottom and can be separated from the inner reaction zone through a valve. Moreover, in the real system (digested sludge) the crystals have to be cleaned by a washing process after the separation for a reuse as fertilizer. The cleaning of particles becomes easier the larger the particle size is. A special cone bottom aeration was implemented in the airlift reactor to resuspend the settled crystals back into the circulating flow where they can grow further owing to the newly added salt solution. Figure 7 shows the results of the particle size analyses with cone aeration and without it. The identified average particle size in the experiments with the cone aeration was also 90–150 µm with a small shift to larger crystals. Thus, it can be concluded that although a liquid recirculation and thereby reintroduction of the crystals into the

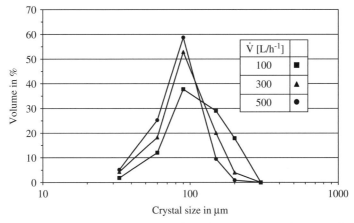

Figure 6. Final particle size distribution of MAP crystals for different air flow rates ($\dot{V}_{1-3}$) in the airlift reactor in batch mode.

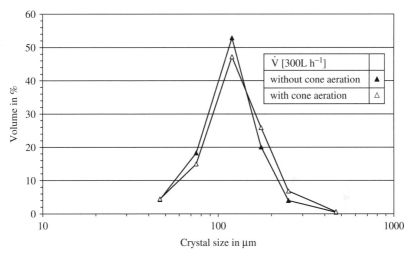

Figure 7. Final particle size distribution of MAP crystals with and without cone-aeration in the airlift reactor in batch mode ($\dot{V} = \dot{V}_2 = 300\,L\,h^{-1}$).

circulation flow was induced, there was no significant effect on crystal sizes at the end of the experiments.

In addition to the particle size distribution, MAP crystals were identified under a microscope. Figure 8 shows pictures of two samples of the batch mode.

The samples are taken after 5 minutes and after 75 minutes during the precipitation in the airlift reactor in batch mode. As can be seen, the crystals grow from about 30 μm to nearly 300 μm in about 70 minutes at an aeration rate of 300 L h^{-1}. This was the maximum crystal size for the whole experimental time at $\dot{V}_2$.

(a)   (b)

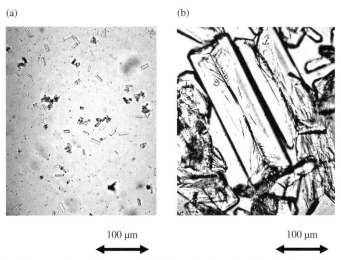

100 μm   100 μm

Figure 8. Microscopic pictures of MAP crystals, after 5 minutes (a) and after 75 minutes (b) of aeration (300 L h^{-1}) in the airlift reactor in batch mode.

Airlift reactor in continuous operation

As in to the experiments in batch mode (chapter 3.1) high yields of recovered phosphorus as precipitated MAP were also determined for continuous operation. Figure 9 shows the decreasing residual dissolved phosphate concentration during the start up of the continuous process for residence times of 3, 8 and 16 hours. In addition to the lower liquid flows at higher residence times the crystallization process needed more time than at lower residence times. However, as shown in Figure 10 yields of recovered phosphorus of 86–98% were achieved although steady conditions have not been reached. Some scattered points occurred due to an accidentally unsteady dosing of the inlet liquid flow of $MgCl_2$ caused by technical problems. However, the second main objective was to determine the influence of the residence time to the growth size of the MAP crystals. It was expected that at higher residence times the crystals would grow to

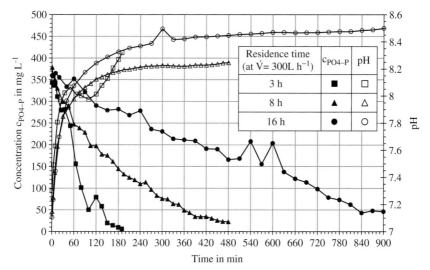

Figure 9. MAP precipitation of a model solution in the airlift reactor in continuous operation for different residence times ($\dot{V} = \dot{V}_2 = 300\,L\,h^{-1}$).

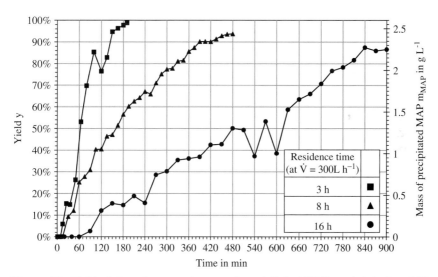

Figure 10. Yield of phosphorus and mass of precipitated MAP crystals in the airlift reactor at continuous operation for different residence times ($\dot{V} = \dot{V}_2 = 300\,L\,h^{-1}$).

larger sizes. MgCl$_2$ was also dosed by a pump. Previous experiments had shown that when the supersaturation is low, MAP precipitates as larger crystals (Stumpf *et al.*, 2008). For that reason the solution of MgCl$_2$ was added in a low supersaturation.

Figure 11 shows that with an increasing residence time the sizes of the MAP-crystals are actually larger. The average particle size shifts from 150 μm to 250 μm when the residence time is increased from 3 hours to 16 hours. Crystals with larger sizes leave the internal circulation of the airlift reactor more easily and settle down in the cone-shaped bottom, because the settling velocity of these crystals is higher than the circulation velocity. Small crystals with a low settling velocity could leave the reaction zone with the effluent leading to a reduced yield of MAP. Thus, a better separation by larger crystals is achieved when the residence time in the airlift reactor is higher. Hence, it is important to know with which maximum flow rate and therewith maximum residence time the precipitation reactor can be operated in a real continuous treatment process. Ueno and Fujii (2001) determined crystal sizes until 1 mm at residence times up to 10 days which are, however, not suitable for the process at Berlin Water (Ueno and Fujii, 2001).

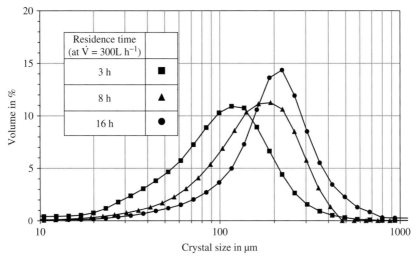

Figure 11. Final particle size distributions of MAP crystals after several residence times in the airlift reactor in continuous operation ($\dot{V} = \dot{V}_2 = 300\,L\,h^{-1}$).

CONCLUSION

Experiments in a 45-litre bench scale reactor were performed to find optimized conditions for a precipitation process which assured a maximum MAP-precipitation, particularly with regard to an efficient separation of the crystals to obtain a high P-recovery. The reactor was designed as an airlift reactor and the required pH-values for precipitating MAP were reached by stripping CO_2 of a model solution. At different air flow rates about 90% of the possible phosphorus was recovered as MAP in batch and continuous operation. These results are similar to other research results (Ohlinger *et al.*, 1998; Münch and Barr, 2001). Furthermore, trials were carried out to measure the crystal sizes in addition to the residence times in the airlift reactor trials. At a residence time of 3 hours the average size was 150 μm. At a residence time of 16 hours an average size of almost 250 μm was determined. Crystals with larger sizes left the loop circulation of the airlift reactor more easily and settled down in the cone-shaped bottom, whereas small crystals with a low settling velocity are entrained out of the reaction zone with the effluent and get lost for a separation. Thus, a better separation due to larger crystals is achieved when the residence time in the airlift reactor is higher. That way it can be concluded that a precipitation reactor in airlift configuration with optimized operating parameters is suitable for a feasible phosphorus recovery by MAP-precipitation. For further experiments digested sludge will be examined in the airlift reactor to verify the results.

Symbols

c	concentration	[mg L^{-1}]
CO_2	carbon dioxide	
MAP	magnesium ammonium phosphate (struvite)	
m	mass	[g]
M	molecular weight	[g mol^{-1}]
n	amount of substance	[mol]
pH	pH-value	
$\dot{V}_{1,2,3}$	air flow rates	[L h^{-1}]
V_R	reactor volume	[L]
y	yield of MAP (the possible amount of total mass of MAP)	[%]

REFERENCES

Corre, K., Valsami-Jones, E., Hobbes, P. and Jefferson, B. (2007). Agglomeration of struvite crystals. *Water Research*, **41**, 419–425.

Heinzmann, B. and Engel, G. (2003). Phosphor-Recycling bei Kläranlagen mit biologischer Phosphorelimination (Phosphorus recycling in WWTPs with biological phosphorus removal. English translation available from the authors). Tagungsband zum Berliner Symposium (Rückgewinnung von Phosphor in der Landwirtschaft und aus Abwasser und Abfall), Berlin.

Heinzmann, B. and Engel, G. (2005). Stand der Phosphorrückgewinnung bei Kläranlagen mit Biologischer Phosphorelimination der Berliner Wasserbetriebe (State of affairs of P-recovery at WWTPs with Bio-P at Berlin Water). Tagungsband zum 75. Darmstädter Seminar (Abwassertechnik), Darmstadt.

Hofmann, G. (2004). Kristallisation in der industriellen Praxis (crystallization in industrial practice). Wiley-VCH Verlag, ISBN 978-3-527-30995-5.

Merkel, W. and Krauth, K. (1999). Mass transfer of carbon dioxide in anaerobic reactors under dynamic substrate loading conditions. *Water Research*, **33**(9), 2011–2020.

Münch, E. and Barr, K. (2001). Controlled struvite crystallisation for removing phosphorus from anaerobic digester sidestreams. *Water Research*, **35**(1), 151–159.

Ohlinger, K., Young, T. and Schroeder, E. (1998). Predicting struvite formation in digestion. *Water Research*, **32**(12), 3607–3614.

Pinnekamp, J., Gethke, K. and Montag, D. (2007). Stand der Forschung über die Verfahren zur Phosphorrückgewinnung (State of research of P-Recovery processes). Tagungsband der 5. DWA-Klärschlammtage, Hildesheim, 21–23, May 2007.

Pschyrembel, W. (1977). Klinisches Wörterbuch (clinical encyclopedia), 253. Auflage, de Gruyter Verlag, Berlin. ISBN 3-11-007018-9, 1977.

Stumpf, D., Zhu, H., Heinzmann, B. and Kraume, M. (2008). Phosphorus recovery in aerated systems by MAP precipitation: optimizing operational conditions. *Water Science & Technology*, **58**(10), 1977–1983.

Umweltbundesamt (UBA-Texte) (2007). Rückgewinnung eines schadstofffreien, mineralischen Kombinationsdüngers "Magnesiumammoniumphosphat – MAP" aus Abwasser und Klärschlamm (Recovery of an environmentally compatible mineral fertilizer MAP from wastewater and digested sludge). Umweltforschungsplan des Bundesministeriums für Umwelt, Naturschutz und Reaktorsicherheit. Texte 25-07, ISS 1862–4804.

Ueno and Fujii (2001). Three years experience of operating and selling recovered struvite from full-scale plant. *Environmental Technology*, **22**, 1373–1381.

Pilot testing and economic evaluation of struvite recovery from dewatering centrate at HRSD's Nansemond WWTP

Ahren Britton[a], Ram Prasad[b], Bill Balzer[c] and Laurissa Cubbage[d]

[a] Ostara Nutrient Recovery Technologies Inc Vancouver, BC, Canada
[b] Old Dominion University Norfolk, VA, USA
[c] Hampton Roads Sanitation District, Suffolk, VA, USA
[d] Hazen and Sawyer, Raleigh, NC, USA

Abstract The Nansemond Treatment Plant (NTP), operated by the Hampton Road Sanitation District, services the Cities of Portsmouth, Chesapeake, Suffolk and Isle of Wight County and town of Holland in Virginia, and ultimately discharges treated effluent to the Chesapeake Bay watershed. It is a 30 MGD design capacity biological nutrient removal plant using the VIP® process and currently operates at an average flow of 18.3 MGD. The plant uses an anaerobic digestion process for solids stabilization and centrifugation for dewatering. Centrate, high in nutrient concentration, from the solids dewatering is returned to the head of the plant. Centrate liquor contributes significant phosphate and ammonia loading to the BNR process. The high loading of these nutrients one of the suspected causes biological instability in the treatment process leading to spikes in effluent phosphorus concentrations. These spikes require addition of metal salts ($FeCl_3$) to precipitate phosphorus to achieve compliance with a phosphorus discharge limit. While this method of precipitating phosphorus is effective, it is not considered a sustainable solution for phosphorus removal due to the high costs of chemicals, solids handling and disposal.

The formation of struvite has recently been commercialized as a treatment process for phosphorous and ammonia recovery from sludge dewatering side stream and is being considered for treating centrate at the Nansemond Treatment Plant. A full scale design for the Nansemond Treatment Plant has been prepared based on results from a 6 month pilot demonstration carried out in 2006/2007. The process would recover 1650 kg of struvite fertilizer per day from the Nansemond plant. Process implications and economic evaluation of the struvite recovery process have been carried out and peer reviewed. The evaluation compares struvite recovery to continued use of ferric chloride for phosphate precipitation and methanol addition for de-nitrification and demonstrates a net benefit for struvite recovery.

INTRODUCTION

The Commonwealth of Virginia is promulgating new nutrient limits (through the VPDES process) for effluents discharged from municipal wastewater treatment plants that discharge to tributaries of (or directly to) the Chesapeake Bay. While HRSD is implementing a multiple treatment plant strategy to address the lower limits, the elimination of phosphorus spikes and greater ammonia removal from secondary treatment is desirable from a plant operability, regulatory compliance, and economics perspective. In an effort to satisfy each of these goals a pilot struvite crystallization process provided by Ostara Nutrient Recovery Technologies Inc. was operated at NTP to determine the feasibility of the process for implementation at this site. Specifically, the ability of the pilot process to be operated with minimal operator intervention, to achieve the N and P removal goals, to run reliably for extended periods of time, and to generate a product of commercial value was studied closely. Subsequent to this demonstration, a full scale design and commercial proposal was prepared by Ostara Nutrient Recovery Technologies Inc. and evaluated by Hazen and Sawyer on behalf of the HRSD.

THE NANSEMOND TREATMENT PLANT

The Nansemond Treatment Plant (NTP), operated by the Hampton Roads Sanitation District (HRSD), services the Cities of Portsmouth, Chesapeake, Suffolk and Isle of Wight County and town of Holland in Virginia, and ultimately discharges treated effluent to the Chesapeake Bay watershed. It is a 30 MGD design capacity biological nutrient removal (BNR) plant using the VIP® process and currently operates at an average flow of 18.3 MGD. The plant uses an anaerobic digestion process for solids stabilization and centrifugation for dewatering. Centrate, high in nutrient concentration, from the solids dewatering is returned to the head of the plant where it is mixed with gravity belt thickener (GBT) filtrate before passing through the grit chamber and into the primary clarifier. Centrate liquor with high phosphate (up to 900 mg PO_4-P/l) and ammonia (500–800 mg NH_4-N/l) contributes significant loading to the BNR process. Figure 1 shows an aerial photo of the Nansemond treatment plant and Figure 2 shows how the phosphorus loading on the plant is affected by the centrate recycle.

The high nutrient loading in the centrate recycle is suspected of causing biological instability in the treatment process, leading to frequent spikes in effluent phosphorus concentrations (e.g. March, October and November 2005). During periods of biological process instability, metal salts ($FeCl_3$) have been used to precipitate phosphorus to achieve compliance with a phosphorus discharge limit.

Adding struvite recovery to the wastewater treatment process at the NTP has a dual purpose: reducing the phosphorus and ammonia load into the BNR process while producing a struvite fertilizer which can be sold to offset the operating costs of the process. Ostara markets the fertilizer under the Crystal Green™ brand name on behalf of its clients. By implementing struvite recovery a treatment plant benefits from a significant increase in the nutrient removal capacity for phosphorus, ammonia and total nitrogen while fostering more stable operating conditions for the biological processes due to significant reduction and equalizing of the phosphorus and ammonia load from sludge dewatering liquor.

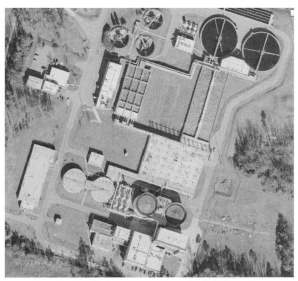

Figure 1. Aerial View of the Nansemond Treatment Plant.

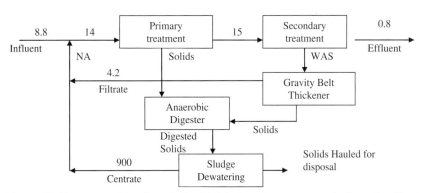

Figure 2. Nansemond treatment plant total phosphorus concentrations (mg/l) at various locations in the treatment process. (NA = not analyzed).

STRUVITE RECOVERY PROCESS DESCRIPTION

Struvite recovery from sludge dewatering side streams at wastewater treatment plants practicing biological nutrient removal is becoming an increasingly accepted way of recovering phosphorus and ammonia from wastewaters for re-use as a fertilizer resource. Over the past 10 years, such a process has been under development at the University of British Columbia (Adnan *et al.*, 2003) and more recently by Ostara Nutrient Recovery Technologies Inc. (Britton *et al.*, 2008). This process is based on an up-flow fluidized bed reactor with multiple reactive zones of increasing diameters, as shown in Figure 3. This process has the advantage of allowing large struvite pellets up to 8 mm in diameter to be kept in suspension in the bottom of the reactor without washing out fine crystal nuclei from the top of the reactor. It also provides better particle size classification than a typical single diameter fluid bed reactor, thus allowing selective harvesting of product particles based on size. The high fluid velocity in the bottom of the reactor also results in the washout of residual sludge solids, and therefore a more pure struvite product free of organic material and pathogens. Struvite crystallization is controlled by a combination of magnesium dose, pH control and by means of a treated effluent recycle.

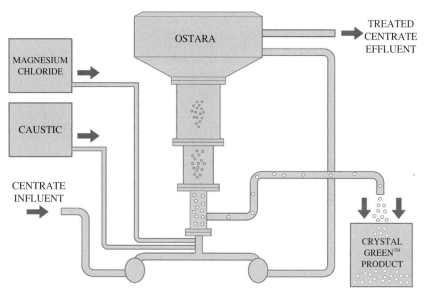

Figure 3. Ostara Process Schematic.

The process is typically integrated into a municipal wastewater treatment in the sludge thickening and dewatering reject water systems. This is where soluble nutrient concentrations are highest and solids levels are sufficiently low to operate a fluid bed reactor without interference. Figure 4 shows a typical treatment plant flow diagram integrating struvite recovery. The inclusion of struvite recovery into a plant using biological phosphorus removal has the advantage of allowing onsite sludge digestion while eliminating the recycle of nutrient loads from dewatering, and protecting downstream equipment from struvite scale. From Ostara's experience the inclusion of struvite recovery in a bio-P plant results in a reduction in phosphorus loading of 20–40% and a reduction in ammonia loading of 5–15% depending on wastewater character-istics and the processes employed at the plant. In contrast to biological processes, the struvite recovery process can be started and stopped nearly instantaneously without affecting the process.

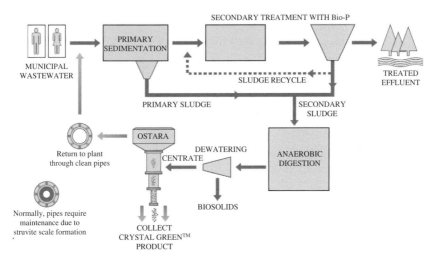

Figure 4. Wastewater treatment flow sheet integrating struvite recovery.

PILOT RESULTS

Figure 5 shows the pilot unit that was used for the testing. It consists of a process control and pump skid, a reactor tower and a product drying rack. Influent phosphorus concentration ranged from a high of 700 mg PO_4-P/L to a low of 140 mg PO_4-P/L during the pilot plant operation. Influent orthophosphorus concentrations decreased significantly, from 650 mg N/L to 140 mg N/L, when a process upset required the dosing of ferric salts to precipitate phosphorus in the secondary clarifiers to maintain regulatory compliance. Ferric dosing started on November 1 and continued till November 30, 2006. The orthophosphate concentration in the influent to the struvite recovery reactor, after decreasing to 140 mg/L, rose during the 3rd week of December after ferric dosing stopped (Figure 6). Excluding the month of ferric dosing and the following 2 weeks, the influent to the pilot plant averaged 550 mg P/L and effluent orthophosphate was 130 mg P/L.

Figure 5. Pilot System.

Ammonia did not reflect any fluctuation associated with the period of ferric dosing (Figure 7), and influent ammonia concentration averaged 536 mg/L while the average effluent ammonia concentration was 334 mg/L.

The pilot plant was operated at a pH range between 6.9–8.0 and at a variety of super-saturation ratios (SSR), to determine optimum operating conditions and process flexibility. Table 1 shows the operational results for each test phase. The overall average removal observed was 80% for phosphorus and 42% for ammonia.

Struvite formation and high nutrient removals have generally been reported at higher pH typically in the range of 8–10 (e.g. Battistoni *et al.*, 2001).

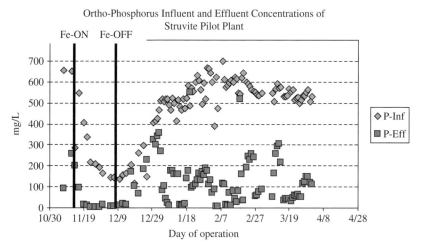

Figure 6. Pilot influent and effluent ortho-phosphorus concentrations as PO4-P.

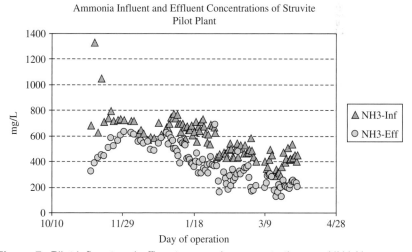

Figure 7. Pilot influent and effluent ammonia concentrations as NH4-N.

However, in this study it was demonstrated that struvite readily formed and achieved high levels of nutrient removal at lower pH values, using various supersaturation ratios.

While overall total phosphorus removal averaged 80% and ammonia at 42% during this pilot study, removals were potentially skewed with lower removal

Table 1. Pilot results and operating set points.

	SSR	P% removal	NH$_3$% removal
pH 7.2			
	3.25	85.4	42.0
	3.75	74.0	37.1
	4.5	88.2	38.4
pH 6.9			
	2.2	90.8	48.8
	2.75	62.1	34.3
	3.25	91.0	44.2
	3.75	71.7	43.5
pH 8			
	5.5	78.1	48.9
Average		**80.2**	**42.2**

rates during the pH 6.9 and SSR 2.75 operating conditions. During this period, pilot results were the lowest for nutrient removals. It should be noted that the pilot plant was shut down for 6 days before the start up of this run which might have caused the excursions in the performance of the pilot. Although it appears that magnesium was consumed at a higher ratio during this run compared to the rest of the study, phosphorus removal was lower, indicating potential error in the makeup of the magnesium solution concentration or process set points.

DISCUSSION

Pilot plant operation during this project explored the minimal use of caustic for pH adjustment. Due to the relatively high phosphorus concentrations in the centrate at the Nansemond plant and to avoid spontaneous nucleation, lower pH values were selected for pilot operation compared to higher pH (8–10) typically reported in literature. During each experimental run, a decrease in pH was observed to occur as struvite was formed and caustic was added to bring the pH up to the required set point irrespective of influent pH and super saturation ratio set points. Caustic requirements appeared correlated to the amount of struvite formation, hence caustic consumption was not significantly different during pilot operation at both pH 6.9 and 7.2.

Overall the variability in the centrate ammonia and phosphate concentrations demonstrated the need for a rapid sample analysis to be available in order to control magnesium dosing to the system in proportion to the influent composition. 24 hour or better turnaround on sample analysis appears to be optimal, indicating that either a field test such as a Hach Kit or an online instrument would be required for optimizing operation.

From Figure 1, the centrate phosphorus load to the treatment plant is around 37% of the total plant load. Therefore, by adopting the struvite crystallization process, the phosphorus load can be reduced by over 30% while simultaneously providing a significant reduction in the ammonia nitrogen load. These combined benefits result in significant increases in plant phosphorus removal capacity along with reduced demand for readily biodegradable COD (rbCOD) for PAO organisms. This in turn would leave more rbCOD available for denitrification, while the reduced ammonia load will help increase nitrification capacity and reduce effluent total nitrogen without modification to the liquid treatment train. This combination of effects is likely to reduce the upgrades required for compliance with stringent nitrogen and phosphorus limits as the plant capacity grows, while improving the stability of the VIP® process and reducing ongoing operating costs.

COST/BENEFIT ANALYSIS

Based on the results of the pilot project and further site investigations, Ostara presented HRSD with a proposal to build a full scale plant to recover struvite from the entire centrate flow at the Nansemond plant. Such a system was estimated to produce approximately 1650 kg/day of struvite product to be sold under Ostara's Crystal Green™ brand. In exchange for the right to distribute the product, Ostara would pay HRSD for the required magnesium chloride as well as a royalty intended to offset the costs of operating and maintaining the facility. In essence the facility operating and maintenance costs would be partially covered by the sale of the product to Ostara.

Independent analysis of the benefits to the plant by Hazen and Sawyer indicates that an annual operating cost savings of approximately $600,000 can be achieved through implementing the struvite recovery process, resulting in a payback of about 6 years (Cubbage *et al.*, 2007). These savings result from reduced ferric chloride use, reduced chemical sludge disposal, as well as reduced aeration and methanol requirements for nitrification and denitrification. Additional capital cost reductions can be expected during future capacity upgrades once reduced side stream nutrient loads are integrated into the design of the liquid treatment train.

CONCLUSIONS

The pilot struvite recovery project demonstrated that significant reductions in side stream ammonia and phosphate concentrations can be achieved through struvite recovery at the Nansemond treatment plant. Based on a full scale proposal and economic analysis, the struvite recovery system will pay for itself in approximately 6 years relative to current treatment methods, and has the potential to increase nutrient removal capacity in the liquid treatment train during future capacity upgrades.

REFERENCES

Adnan A., Mavinic, D.S and Koch, F.A. (2003). Pilot-scale study of phosphorus recovery through struvite crystallization-examining the process feasibility. *Journal of Environmental Engineering and Science*, **2**(5), 315–324.
Battistoni, P., De Angelis, A., Pavan, P., Prisciandaro, M. and Cecchi, F. (2001). Phosphorous removal from a real anaerobic supernatant by struvite crystallization. *Water Research*, **35**(9), 2167–2178.
Baur, R., Prasad, R. and Britton, A. (2008). Reducing ammonia and phosphorus recycle loads by struvite harvesting. Conference Proceedings WEFTEC 2008.
Cubbage, L., Stone, A. and Bilyk, K. (2007). Nansemond treatment plant nutrient reduction improvements technical memo. Internal Report – Hazen and Sawyer.

Standardizing the struvite solubility product for field trial optimization

A.L. Forrest[*], K.P. Fattah, D.S. Mavinic and F.A. Koch

Department of Civil Engineering, University of British Columbia, Vancouver, B.C., V6T1Z4, Canada (E-mail: forrest@civil.ubc.ca; parvez@interchange.ubc.ca; dsm@civil.ubc.ca; koch@civil.ubc.ca)

[*] Corresponding author

Abstract Phosphate (P) recovery technologies, in the form of struvite harvesting, are reaching the point of commercial viability. As the supersaturation condition of the bulk fluid, relative to struvite, is integral for process control, knowing the true value of K_{sp} (solubility product) of struvite is essential. This parameter normalizes combinations of the three constituent ions (magnesium, ammonium or phosphate) or the supersaturation ratio. Although extensive studies on the true value of K_{sp}, more typically referred to as pK_{sp} ($-\log K_{sp}$) of struvite have been conducted, significant variation exists between reported values. Widely varying experimental methodologies, in addition to experimental conditions, account for much of the observed discrepancy that exists between reported values for the struvite solubility product. This study conducted experiments with four working solutions; digester supernatant, centrate, distilled water and tap water, at various operating pH values and temperatures. A range of values for these physical parameters was used in order to determine a working value for pK_{sp} that reflects typical operating field conditions for struvite recovery. Three different solubility (pK_{sp}) prediction models were then examined to determine which provided the best estimate of the true value of pK_{sp}. It was found that the pK_{sp} had a linear relationship with pH below an inflection point at a pH value of 7.0, before becoming relatively constant in the working pH range of 7.0 to 9.0 (typical pH values used for struvite recovery). The model that provided the best estimate was then used to generate values for 3 different temperatures tested.

INTRODUCTION

Struvite solubility product

Struvite production for recovering phosphorus from domestic wastewater has gained substantial interest and progress in recent times. Discrepancies continue to exist between reported values of some of the most important operating parameters for struvite crystallization. One such parameter is the solubility product or K_{sp}, more commonly referred to as the $pK_{sp}(-\log K_{sp})$ value. This parameter is important in the determination of the supersaturation ratio (SSR), and the corresponding supersaturation condition, of the crystallization system.

SSR is defined in Equation (1) (Snoeyink and Jenkins 1980):

$$SSR = \frac{P_S}{P_S^{eq}} \tag{1}$$

where P_S is defined as the *conditional* solubility product and P_S^{eq} is defined as the *equilibrium* conditional solubility product. For struvite, the conditional solubility product (P_S) can be written as the product of the measured component species:

$$P_S = [Mg^{+2}] \cdot [NH_4 - N] \cdot [PO_4 - P] \tag{2}$$

The equilibrium conditional solubility product (P_S^{eq}), describing the equilibrium condition for any given pH level can then be expressed as a function of the solubility product, the ionization fraction (α_i) of each of the respective species and the activity coefficient for the respective ion species (γ_i):

$$P_S^{eq} = \frac{K_{sp}}{\left(\alpha_{Mg^{+2}} \cdot \gamma_{Mg^{+2}}\right)\left(\alpha_{NH_4^+} \cdot \gamma_{NH_4^+}\right)\left(\alpha_{PO_4^{-3}} \cdot \gamma_{PO_4^{-3}}\right)} \tag{3}$$

Although K_{sp} is theoretically constant, P_S^{eq} is highly correlated with pH, due to the changing component concentrations, each time a new equilibrium is reached. If SSR is to be used as a control parameter for struvite recovery, it is essential to know the true value of K_{sp} for the pH range that the systems are expected to operate in.

The SSR of the bulk fluid is the primary control variable used by the P-recovery team at the University of British Columbia (Fattah et al., 2008; Forrest et al., 2008; Britton et al., 2005). Although extensive studies on the value of K_{sp} of struvite have been conducted, there still exists a significant variation between reported values: $5.50 \times 10^{-14} - 3.89 \times 10^{-10}$ corresponding to pK_{sp} values of 9.41 to 13.27 (Rahaman et al., 2006). This variation may be related to the large range of experimental methodologies. The standard method for the experimental determination of a K_{sp} value of a particular reaction involves either the formation of precipitate or the dissolution of a previously formed salt in distilled water. In either approach, experiments are conducted under carefully controlled conditions in which constant mixing energy, constant pH, constant temperature or a set conductivity is maintained. Widely varying procedures in experimental methodologies account for much of the discrepancy that exist between the reported values for the struvite solubility product. In addition, some of the studies neglected the influence of ionic strength in the determination of the solubility product. Some other factors that may also influence the value of K_{sp} are given in Rahaman et al., 2006.

Struvite chemistry

Struvite is a sparingly soluble crystal that has a distinctive orthorhombic structure and is comprised of equimolar amounts of magnesium, ammonia, and phosphate. In addition, there are six waters of hydration attached. The structure of this compound is generally accepted to be $MgNH_4PO_4 \cdot 6H_2O$; however, there is dispute as to which reaction pathway is correct. The general reaction pathway that is most commonly seen is shown in equation (4) (Abbona, 1984).

$$Mg^{+2} + NH_4^+ + PO_4^{-3} + 6H_2O \leftrightarrow MgNH_4PO_4 \cdot 6H_2O \tag{4}$$

A different pathway has been suggested, using the reaction pathway as defined below in equation (5) (Shimamura 2003).

$$Mg^{+2} + NH_4^+ + HPO_4^{-2} + 5H_2O + OH^- \leftrightarrow MgNH_4PO_4 \cdot 6H_2O \tag{5}$$

This reaction pathway accounts for the fact that the dominant form of phosphate, in the working pH range for struvite formation, is primarily HPO_4^{-2} and not PO_4^{-3}.

In either case, many different side reactions also acts concurrently to struvite formation. These include the interactions of each of the various components and are summarized in Table 1. In wastewater systems, many other different species are present which may indirectly influence struvite equilibrium. Examples of these include carbonate, sulphate, and metals that precipitate phosphates such as calcium and aluminum (Babić-Ivančić et al., 2002; Nelson et al., 2003).

Table 1. Competing reactions in struvite formation in a distilled water system.

Equilibrium	pK (25°C)	ΔH° (kcal/mole)	Reference
$MgOH^+ \leftrightarrow Mg^{+2} + OH^-$	2.56	−2.60	Morel et al., 1983
$NH_4^+ \leftrightarrow H^+ + NH_3$	9.3	12.46	Snoeyink et al., 1980
$H_3PO_4 \leftrightarrow H_2PO_4^- + H^+$	2.15	−2.42	Martell et al., 1989
$H_2PO_4^- \leftrightarrow HPO_4^{-2} + H^+$	7.2	1.17	Martell et al., 1989
$HPO_4^{-2} \leftrightarrow PO_4^{-3} + H^+$	12.35	3.31	Martell et al., 1989
$MgH_2PO_4^+ \leftrightarrow H_2PO_4^- + Mg^{+2}$	0.45	−3.40	Morel et al., 1983
$MgHPO_4^+ \leftrightarrow HPO_4^{-2} + Mg^{+2}$	2.91	−3.30	Taylor et al., 1963
$MgPO_4^+ \leftrightarrow PO_4^{-3} + Mg^{+2}$	4.8	−3.10	Childs 1970
$H_2O \leftrightarrow H^+ + OH^-$	14	13.35	Martell et al., 1989

Methodology

Experimental setup

A six-station paddle stirrer (Phipps and Bird) was used with square jars in a temperature-controlled room. About 10 grams of struvite was added to 1.5 L of centrate in each jar. The idea was to insert enough struvite in the jars to attain saturation. The paddle stirrers were set to operate at 70 ± 2 RPM. Samples of 20 L of each working solution were brought to the testing temperature by storing them in a controlled temperature room. Since high levels of ammonia and phosphate were already present in the wastewater systems, only magnesium chloride was added in order to supersaturate the solutions with respect to each of the component ions. In the distilled and tap water systems, additional $(NH_4)_2PO_4$ had to be added in order to achieve the same concentration levels. The rationale for using struvite previously grown in a given matrix was to maintain the chemical uniqueness of a water matrix as much as possible. The final concentration ranges at a pH of 4.0 (i.e. no struvite formation possible) that were being targeted for Mg^{+2}, NH_4-N and PO_4-P were 180–200 mg/L, 150–180 mg/L and >300 mg/L respectively.

Experimental solutions

Operating fluids for typical struvite recovery installations range from advanced wastewater treatment plant (AWWTP) supernatant to wastewater centrate making it necessary to eliminate as much of the chemical variability from the operating control loop as possible. These experiments were conducted to determine a working value for K_{sp}, which reflect the operating conditions (in terms of pH and temperature) of the various field installations. These experiments were designed to establish equilibrium between struvite crystals in the solid phase and their dissolved component ions in an aqueous system. Ideally, this would be done under a closed system; however, due to experimental constraints, tests were conducted under open conditions. Real fluids from two local treatment plants, and artificial solutions formed from both distilled and tap water, were used as the bulk testing fluids. Multiple fluids were used in order to quantify appreciable differences, if any, in K_{sp} (due to inherent chemical composition). The conducted solubility experiments incorporated concepts of both dissolution and formation.

Operating conditions

Since the value of K_{sp} is highly temperature dependant, trials were conducted at 10, 15 and 20°C (chosen to reflect seasonal operational temperatures). As Ksp is also pH dependant, tests for each of the trial temperatures were conducted for multiple pH values, starting from 6.0 to 9.0 in 0.2 – 0.3 pH increments. Dilute hydrochloric acid or sodium hydroxide solutions were added to each of the jars to keep the pH constant. It was assumed that the equilibrium had been achieved after 24 hours, as shown by previous studies at UBC (Ping Liao, Department of Civil Engineering, UBC, pers. comm.), provided that the pH had remained constant within 3 hours of the final sampling. This assumption seemed valid since the NaOH needed, in order to achieve a given set point, would be large at first and then greatly diminish within 3 hours (and be low within 12 hours). This indicates that equilibrium was being achieved.

Analytical techniques

After equilibrium was established, the pH and conductivity in each jar were measured. Samples were taken from each jar to determine total magnesium, total ammonia-nitrogen and total ortho-phosphate concentrations. Mg^{+2} was measured using flame atomic absorption spectrophotometry (Varian Inc. SpectrAA220) whereas orthophosphate and ammonia were measured using a flow injection method (model Lachat QuikChem 8000). pH measurements were performed using an *in situ* Beckman Φ44 pH meter, equipped with a Oakton pH probe and the conductivity was measured using a Hanna Instruments HI9033 multi range conductivity meter.

Modeling techniques

Since a number of different techniques are available to estimate the K_{sp}, the measured results were used to test the accuracy of three different modelling techniques available. The importance of this lies in being able to accurately predict the value of the *SSR* of the system and is essential for the development of on-line control (Fattah *et al.*, 2009).

Formation model

The first model, the formation model, considered the reaction pathway of the formation of struvite without correcting for the ionic activity of each individual species. The determination of K_{sp} follows the alternate reaction pathway for struvite equilibrium generating reaction (Equation (6)) (Yoshino *et al.*, 2003):

$$K_{sp} = [Mg^{+2}][NH_4^+][HPO_4^{-2}][OH^-] \qquad (6)$$

For this model, it was assumed that the speciation of NH_4^+ into $NH_{3(g)}$ was minimal (i.e. $[NH_4^+]$ was equal to $[NH_4\text{-}N]$). It cannot be equivalently assumed that $[PO_4\text{-}P]$ is equivalent to $[HPO_4^{-3}]$ in the working range. The value of

$[HPO_4^{-2}]$ was calculated using the equilibrium constants for phosphate (re: equilibrium equations from Table 1). This is summarized in Equation (7).

$$\left[HPO_4^{-2}\right] = \frac{[PO_4 - P]}{\left[1 + \frac{K_3}{[H^+]} + \frac{[H^+]}{K_2} + \frac{[H^+]^2}{K_1 K_2}\right]} \tag{7}$$

PHREEQC model

Derived values for K_{sp} can be determined using a combination of experimental data and statistical methods. Measured concentrations of each component ion can be measured once an experiment has reached equilibrium. These values can then be compared, through a least squares analysis, with values that are derived theoretically using commercially available PHREEQC v.1.5.10. This program is commonly used in the calculation of chemical equilibria using an internal database for the thermodynamic constants of all possible reactions of a given scenario. Since struvite is not common enough to be included in the database, the reaction and the K_{sp} value must be user-defined. The program will then calculate the theoretical equilibrium concentrations by setting the SSR to unity.

Selecting a range of K_{sp} values and the initial concentrations of each ion, a range of theoretical concentrations were generated and compared against measured experimental values, using the weighted least squares method. Since it was impossible to know the initial concentrations explicitly, they were estimated using the concentrations measured at the initial pH value of 4.0. An objective function (OF) was derived by ignoring the variance of each species, known as a least squared analysis, as defined below in Equation (8) (Ohlinger 1999).

$$OF = \sum \left[\left([Mg^{+2}]_{theor} - [Mg^{+2}]_{meas}\right)^2 + \left([NH_4 - N]_{theor} - [NH_4 - N]_{meas}\right)^2 \right.$$
$$\left. + \left([PO_4 - P]_{theor} - [PO_4 - P]_{meas}\right)^2 \right] \tag{8}$$

Using this technique, the true value of K_{sp} for a given pH value was found by minimizing the OF over a range of K_{sp} values. It should be mentioned briefly that PHREEQC considers all of the complexes outlined in Table 1. As a further note, the reaction pathway that was defined in the program for struvite was the commonly accepted equation involving $[PO_4^{-3}]$ and not $[HPO_4^{-2}]$ that was used in the previous section. The justification is that this is the more frequently seen reaction pathway found in struvite literature.

Speciation model

The final model used an in-house program, entitled SimpleMAP v1.0, to perform the chemical speciation of the tested solutions. pK_{sp} was determined using the speciation model, by entering the measured values for each of the three species of interest, as well as the measured pH value. An iterative analysis was then conducted in order to calculate the value of pK_{sp} that would set the *SSR* value to 1.0. The assumption that is being made here is that each batch test has reached equilibrium. This model differs from the PHREEQC Model in only that it uses a simplified version of the alternate reactions.

RESULTS AND DISCUSSIONS

The ensuing discussion attempts to compare the results of the three different models detailed above. The purpose of this study was to determine which of the three models provides the closest prediction to the true value of pK_{sp} and then to use the best model to examine the influence of both temperature and water matrix on the results. Model comparison was done using the results from using the analyses conducted on the digester supernatant.

Solubility curve

Test data showed very similar trends for each of the three study temperatures (see Figure 1). As demonstrated, there are two distinct regions in the data with the inflection point at a pH value of ∼ 7.0. As the typical pH operating range of the reactors is 7.0 and 8.5, K_{sp} determination was restricted to this region. Essential to this determination were the generated conditional solubility curves as functions of pH. These acted as the in model input data for each test condition.

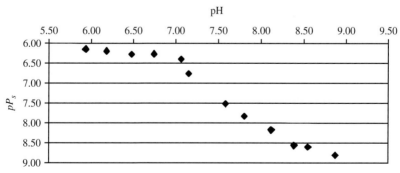

Figure 1. pP_s^{eq} curve of a typical batch experiment trial (Penticton AWWTP supernatant, 10°C).

Model comparison

At each of the three testing temperatures (10, 15 and 20°C), the dependence of pK_{sp} on pH produced very similar results. Typical results are shown in Figure 2.

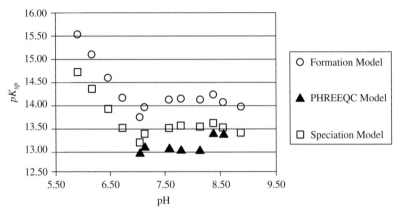

Figure 2. Typical pK_{sp} predictions of the three models (Penticton AWWTP supernatant, 10°C).

In this figure, it is shown that the Formation Model predicts the highest values for pK_{sp} and the Speciation Model predicts the lowest, a trend observed at each of the three testing temperatures. In the Formation Model, the OH^- concentration was considered in the formula, causing the model to be more sensitive to experimental error; this would lead to a higher than expected value. The one exception to the general pattern was the trial conducted at 10°C, in which the PHREEQC model predicted the lowest values for pK_{sp}. This is most likely reflective of the methodology used in developing the model itself, with inherent instability at lower temperatures. It is also interesting to note that, at each temperature, the results for the Speciation Model closely mirrored those of the Formation Model. This would seem to indicate a systemic offset. The most likely cause of this is the fact that the Speciation Model takes into account the formation of several additional complexes that the Formation Model does not.

In order to calculate a pK_{sp} constant for each of the models, the pK_{sp} values were averaged over the range of pH values where they appeared constant (from a pH value of 7.0 to 9.0). It should be noted that the sample population at 15°C was higher, since a repeat run was conducted during the course of the study. Although this was somewhat inadvertent, it did serve as a form of quality control, since the standard deviation was not severely impacted. The model predictions for at the three different temperatures are presented in Table 2.

Table 2. Model predictions of pK_{sp} values at 10, 15 and 20°C.

Model	10°C (n = 12)		15°C (n = 18)		20°C (n = 12)	
	Average	S.D.	Average	S.D.	Average	S.D.
Formation	14.01	0.15	14.24	0.20	14.11	0.15
PHREEQC	13.42	0.14	13.63	0.24	13.45	0.17
SimpleMAP	13.11	0.18	14.13	0.11	13.86	0.19

Recalling the relationship of pH and pP_s (as exemplified in Figure 1), it is a simple matter to calculate the value of pP_S for the 3 temperatures employed in the trial runs. Recalling that the *absolute SSR* (SSR_{abs}) was given as the ratio of measured solubility product (P_S) to the equilibrium value (P_S^{eq}), SSR_{abs} was then calculated for each of the measured samples that ranged in pH values from 7.1 to 7.9. Each of the values provided in Table 2 were then taken and inserted in Speciation, to calculate the *model SSR* value (SSR_{mo}). The difference of the model and the absolute *SSR* was then used to calculate the sum of squared error (*SSE*). The *SSE* values determined were 2214, 523 and 10 for the Formation, PHREEQC and Speciation models, respectively (n = 52). Assuming that a minimum value of *SSE* represents the best representation of the system, than it is concluded that the Speciation model provides the best estimation of the pK_{sp} value.

Once it had been determined that this model was the most robust in application, it was then applied to all the additional runs for each of the working fluids at each temperature. Table 3 provides a breakdown of each of the pK_{sp} values determined using the Speciation Model.

Table 3. $pKsp$ values for each water matrix as calculated using the speciation model.

Model	10°C (n = 6)		15°C (n = 6)		20°C (n = 6)	
	Average	S.D.	Average	S.D.	Average	S.D.
Distilled Water	13.63	0.19	n/a	n/a	13.39	0.20
Tap Water	13.65	0.20	13.44	0.16	13.26	0.06
Supernatant	13.42	0.14	13.63	0.24	13.30	0.03
Centrate	13.63	0.19	13.44	0.16	13.45	0.17

It should be noted that the previously discussed inflection point was observed, regardless of the water matrix that was being worked used. In addition, no value was calculated for the distilled water system at 15°C as two runs were accidentally conducted at 20°C. The predicted values compared well with values taken from literature and demonstrated that the value was independent (i.e. within the standard deviation) of the working fluid. The results are then 13.58 ± 0.11, 13.50 ± 0.11 and 13.35 ± 0.09 for 10, 15 and $20°C$ respectively. This clear temperature dependence can then be used to calculate the enthalpy change ($\Delta H°$) for struvite. Defined as the amount of heat absorbed or released during the course of the reaction, reported values for struvite display as much discrepancy as do reported pK_{sp} values. Determining this thermodynamic constant and the corresponding temperature correction factor is the source of ongoing work.

CONCLUSIONS

This investigation demonstrated that the in-house speciation model, SimpleMAP v1.0, of the three speciation models tested, provided the best estimate of pK_{sp}. The predicted values compared well with values taken from literature and demonstrated that the value was independent of the working fluid. This work also revealed how the pK_{sp} are pH independent in the typical working range of these reactors. Finally, the temperature relationship that was measured agreed closely with literature values. Future work is now looking at developing new temperature correction coefficients that are matrix independent. These conclusions have significant impact on potential control systems for both field studies and full-scale installations for struvite recovery.

ACKNOWLEDGEMENTS

The authors greatly appreciate the cooperation and assistance received from the staff at both the Lulu Island WWTP, Richmond, B.C. and the Penticton AWWTP. The authors would also like to thank the Natural Science and Engineering Research Council of Canada (NSERC), Metro Vancouver, Ostara Nutrient Recovery Technologies Inc. and Stantec Consulting Ltd., for their generous funding towards this study.

REFERENCES

Abbona, F., Calerri, M. and Ivaldi, G. (1984). Synthetic struvite, $MgNH_4PO_4 \cdot 6H_2O$; correct polarity and surface features of some complementary forms. *Acta. Cryst. (b) – structural science*, **40**, 223–227.

Babić-Ivančić, V., Kontrec, J., Kralj, D. and Brečeivć, L. (2002). Precipitation diagrams of struvite and dissolution kinetics of different struvite morphologies. *Croatica Chemica Acta,* **75**(1), 89–106.

Britton, A., Koch, F.A., Mavinic, D.S., Adnan, A., Oldham, W.K. and Udala, B. (2005). Pilot scale struvite recovery from anaerobic digester supernatant at an enhanced biological phosphorus removal wastewater treatment plant. *J. Environ. Eng. Sci.*, **4**(4), 265–277.

Childs, C.W. (1970). A potentiometric study of equilibria in aqueous divalent metal orthophosphate solutions. *Inorg. Chem.*, **9**(11), 2465–2469.

Fattah, K.P., Mavinic, D.S., Koch, F.A. and Jacob, C. (2008). Determining the feasibility of phosphorus recovery as struvite from filter press centrate in a secondary wastewater treatment plant. *Journal of Environmental Science and Health: Part A.*, **43**(7), 756–764.

Fattah, K.P., Rahaman, M.S., Mavinic, D.S. and Koch, F.A. (2009). Development of a Process Control System for Online Monitoring and Control of a Struvite Crystallization Process. In: *Proc. of the International conference on nutrient recovery from wastewater streams.* Vancouver, British Columbia, Canada. May 10–13, 2009.

Forrest, A., Fattah, K.P., Mavinic, D.S. and Koch, F.A. (2008). Optimizing struvite production for phosphate recovery In: WWTP. *Journal of Environmental Engineering – ASCE.*, **134**(5), 395–402.

Environmental Research Software. MINEQL+ – The First Choice in Chemical Equilibrium Modeling. Retrieved January 17, 2004, from http://www.mineql.com/mineql.html

Morel, F.M.M. (1983). Principles of aquatic chemistry. New York, NY: *John Wiley & Sons.*

Martell, A.E. and Smith, R.M. (1989). Critical stability constants. Plenum Press, New York

Nelson, N.O., Mikkelsen, R.L. and Hesterberg, D.L. (2003). Struvite precipitation in anaerobic swine lagoon liquid: effect of pH and Mg:P ratio and determination of rate constant. *Bioresour Technol,* **89**(3), 229–236.

Rahaman, M.S., Mavinic, D.S., Bhuiyan, M.I.H. and Koch, F.A. (2006). Exploring the determination of struvite solubility product from analytical results. *Environmental Technology*, **27**, 951–961.

Shimamura, K., Tanaka, T., Miura, Y. and Ishikawa, H. (2003). Development of a high-efficiency phosphorous recovery method using a fluidized-bed crystallized phosphorous removal system. *Wat. Sci. Technol.*, **48**(1), 163–170.

Snoeyink, V.L. and Jenkins, D. (1980). Water chemistry. New York, NY: *John Wiley & Sons.*

Taylor, A.W., Frazier, A.W. and Gurney, E.L. (1963). Solubility products of di- and trimagnesium phosphates and the dissociation of magnesium phosphate solutions. *Trans. Faraday Soc.*, **59**, 1580–1584.

Yoshino, M., Yao, M., Tsuno, H. and Somiya, I. (2003). Removal and recovery of phosphate and ammonium as struvite from supernatant in anaerobic digestion. *Water Sci. Tech.*, **48**(1), 171–178.

Plant availability of P fertilizers recycled from sewage sludge and meat-and-bone meal in field and pot experiments

R. Cabeza Pérez, B. Steingrobe, W. Römer, N. Claassen

Department of Crop Sciences-Plant Nutrition; Georg-August-University, Göttingen Carl-Sprengel-Weg 1, 37075 Göttingen rcabeza@gwdg.de, bsteing@gwdg.de

Abstract This project is part of the German Research Cluster on P-Recycling. The fertilizer effects of selected recycled P compounds were determined in field and pot experiments in comparison to commercial fertilizers (Triplesuperphosphat (TSP) and rock phosphate). For the field experiments, two magnesium-ammonium-phosphates (MAP) of different sewage treatment plants, a heavy metal depleted sewage sludge ash, and an alkali sinter phosphate made from meat-and-bone meal were used. Additionally in the pot experiments, a meat-and-bone meal ash, a cupola furnace slag made from sewage sludge, and a precipitated apatite of municipal waste water were tested. The field experiments were performed in three loamy soils poor in available P and differing mainly in pH. Phosphate fertilization was for a whole three-year crop rotation with a suboptimal amount of 60 kg of P/ha plus a well fertilized (100 kg P/ha) and unfertilized control. The pot experiment with maize was performed with 5.5 kg of mainly the same soils as in the field in 6L pots. Phosphate was given at a suboptimal rate of 60 mg/kg soil for each fertilizer, plus an unfertilized and well fertilized (200 mg P/kg) control.

In the field, the fertilization with the different materials had no significant effect on the CAL-extractable P or P concentration in soil solution in all three soils. Also grain yield and P uptake was not different. However, the two MAPs tended to perform slightly better than the other compounds. Differences in plant availability became more obvious in the pot experiments. The chemical characterisation of plant availability by CAL-extraction showed that the MAPs, apatite, cupola slag were similar to TSP. These results were partly confirmed by the relative fertilizer efficiency (P uptake from fertilizer in relation to TSP) which was in average highest for MAP, apatite and cupola slag, but not consistent in each soil. Hence, none of the tested compounds reached the plant availability of a water-soluble P fertilizer in the pot experiment in the first year. Long term effects will be studied in the next years.

INTRODUCTION

Phosphorus (P) is an essential plant nutrient necessary for several metabolic processes. The concentration of plant available P is low in most soils, hence a fertilization is needed. However, phosphate resources are limited. Assuming a constant rate of consumption, the known P resources will be depleted in about 100 years (Berg and Schaum, 2005; Stewart et al., 2005). On the other hand, anthropogenic waste material contains P that can be hazardous for the environment, e.g. like the discharge of municipal wastewaters can lead to eutrophication of waterbodies (Johnston and Richards, 2003). It is possible by several ways to recover P of different sources such as sewage sludge or animal bones and meal. The soluble P of wastewaters can be precipitated as struvite, i.e. Magnesium-ammonium-phosphate (MgNH4PO4, MAP), as well as in other forms such as calcium, aluminium or iron phosphate (Johnston and Richards, 2003). Furthermore, it is possible to recover P with thermo-chemical processes from sewage sludge, as well as from animal residues. It must be taken into account, that the different processes to obtain P from waste material result in fertilizers differing in P solubility and P availability to plants. In previous studies it was demonstrated that finely ground struvites can be as effective as water soluble P fertilizers (Ghosh *et al.*, 1996; Goto, 1998) although struvite is limited in water solubility. This shows that the water solubility of P compounds and their plant availability in the soil are not necessarily related. Hence, the agronomic effectiveness of each recovered-P product must be evaluated before it can be used as commercial fertilizer. In this study, the fertilizer effect of some MAPs and thermally recovered P was investigated in field and pot experiments.

MATERIALS AND METHODS

Phosphorus sources and soils. Seven different materials were tested with regard to their P fertilizer effect in field and pot experiments on four different soils. Additionally, two commercial fertilizers with different P availability, i.e. triple-superphosphate (TSP) and rock phosphate were taken as control. In the field experiments, two Magnesium-ammonium-phosphates of different sewage treatment plants (MAP-Sb, MAP-Gf), a heavy metal depleted sewage sludge ash (sl-ash), and an alkali sinter phosphate made from meat-and-bone meal (sinter-P) were used. Additionally in the pot experiments, a meat-and-bone meal ash (MB meal ash), a cupola furnace slag (Mephrec procedure, cupola slag) made from sewage sludge, and a precipitated apatite of municipal waste water were tested. These products were not available in sufficient amounts to test them also in the field. The P concentrations of the products are shown in Table 1. All products were finely ground before fertilization.

Field experiments. The field experiments were performed on farmland with different crops in three loamy soils low in available P. The soils differed in their pH (Table 2). The phosphate fertilization took place in March 2007 for a whole three-year crop rotation with an amount of 60 kg of P ha^{-1} for each fertilizer. This is a suboptimal rate in order to be able to differentiate the P available for the plants among the P fertilizers. Well fertilized (100 kg P ha^{-1}, TSP-100) and unfertilized controls were also established. The crop cultivation was performed

Table 1. Description of the P compounds from different P recycled works.

Material	Description	P concentration (%P)
MAP-Sb	Precipitated as $MgNH_4PO_4$	11.0
MAP-Gf	Precipitated as $MgNH_4PO_4$	9.6
sl-ash	Sewage sludge burned over 1000°C	7.8
sinter-P	Meat and Bone meal sintered over 1000°C	11.3
MB meal ash	Ashes from animal meal	16.4
cupola slag	Sewage sludge slagged over 1000°C	2.9
apatite	Calcium phosphate precipitated	11.1
*Triple superphosphate	–	20.1 (19.5 water soluble)
*Rock phosphate	–	11.8

*Were used as reference.

Table 2. Calcium-acetate-lactate extractable P (CAL-P) and pH of the soils used in field and pot experiments.

Soil	CAL-P (mg P kg^{-1})	pH CaCl$_2$ (1:2.5)	Crops
Sattenhausen	19.0	5.6	Oilseed rape
Lutterbeck	18.0	7.1	Winter barley
Gieboldehausen	21.0	6.8	Winter wheat
Düshorn	24.2	5.5	–

by the farmers as usual. Each treatment was replicated four times. Plot size was 24 m^2.

Pot experiments. The pot experiment was performed with 5.5 kg of soil in 6 L pots. The soils were taken from two sites of the field experiments (Sattenhausen and Gieboldehausen). Unfortunately, the farmer of Lutterbeck did not allow to remove soil from the farm. Additionally, an acid sandy soil (Düshorn) was also used (see Table 2) in the pot experiment, which is not in agricultural use. Phosphate was given at a suboptimal rate of 60 mg of P kg^{-1} soil for each fertilizer, plus unfertilized and well fertilized (200 mg P kg^{-1}, TSP-200) controls. Furthermore, the pots were fertilized with 1 g K (K_2SO_4), 0.2 g Mg ($MgSO_4 * 7H_2O$) and 0.5 g N (NH_4NO_3) per pot. Two further applications of N were realized as $Ca(NO_3)_2$ in doses of 0.3 g per pot each time. After 3 weeks of incubation without plants, maize (*Zea mais* L., cv. Atletico) was sown (2 plants per pot). Each pot was irrigated daily with demineralized water during the experiment. The plants were harvested at the 80th day after the sowing.

Measurement. Soil samples were collected from the upper 30 cm soil depth before fertilizing and at harvest in the field experiments and from the total soil volume in the pot experiments. The available P was determined by extraction with calcium-acetate-lactate (CAL-P; Schueller, 1969) as usual in Germany. Soil solution was obtained according to the displacement method described by Adams (1974). The P concentration in both methods was measured colorimetrically by the method of Murphy and Riley (1962).

In the field, plant samples were taken from 6 m^2 in the middle of the experimental plots. Unfortunately, the farmer of Gieboldehausen harvested the whole field before we could take the samples. Plants were dried at 65°C to constant weight before grain and shoot yield was determined. In the pot experiment, the whole above ground plants were taken as yield. A division in cobs and shoot was not done because the selected maize genotype is usually used as energy plant. Subsamples of plant dry matter were digested in conc. HNO_3 at 180°C and the P concentration herein was measured colorimetrically by the molybdenum-vanadate method (Scheffer and Pajenkamp, 1952). Total P uptake was calculated from shoot dry matter and P concentration.

Data analysis. Statistical analysis was performed using SigmaStat 2.0. The experimental design for field experiments was in blocks with 8 treatments and 4 replicates. The pot experiment was completely randomized with 4 replicates for each treatment. The effect of P fertilizers was analyzed with one-way analysis of variance between the treatments as the source of variation, followed with a Tukey test at the 0.05 level of significance to separate the means.

RESULTS

Field experiments. The fertilization with the different materials had no significant effect on the CAL-extractable P at harvest in all three soils (Table 3). At "Sattenhausen" and "Lutterbeck" the TSP fertilization increased CAL-P in tendency, whereas rock phosphate had not become available at all. The tested P compounds grouped between TSP-60 and rock phosphate. Similar results were obtained for P soil solution concentration (Table 3) on both soils. Only in "Sattenhausen", the concentration of TSP-100 was significantly higher than that of rock phosphate or sl-ash. The CAL-P of all fertilizers in "Gieboldehausen" did not differ from the unfertilized treatment and also not from the CAL-P before fertilization (not shown). Even the fertilization with water soluble TSP had not increased the CAL-P. This and the very low P soil solution concentration point to the fact that P was fixed after fertilization in non-CAL extractable fractions.

Table 3. Influence of recycled P fertilizer on CAL-extractable P and P concentration in soil solution at harvest in the field experiments.

	Sattenhausen (pH 5.6)		Gieboldehausen (pH 6.8)		Lutterbeck (pH 7.1)	
	CAL-P	Solution P	CAL-P	Solution P	CAL-P	Solution P
	mg kg^{-1}	µmol L^{-1}	mg kg^{-1}	µmol L^{-1}	mg kg^{-1}	µmol L^{-1}
Unfertilized	26.0	6.6 b	22.2	0.6	14.3	2.5
TSP-60	30.3	11.7 ab	21.1	2.7	23.8	5.0
TSP-100	34.1	17.8 a	22.3	2.7	20.5	5.1
Rock-P	23.0	6.2 b	21.3	0.7	11.1	1.3
sinter-P	26.2	10.0 ab	23.7	1.2	23.7	2.5
sl-ash	24.3	7.3 b	19.6	0.7	16.0	1.9
MAP-Sb	27.4	12.9 ab	25.1	1.4	14.7	2.3
MAP-Gf	27.3	9.8 ab	24.3	1.2	14.1	2.5

Different letters denote significant differences between treatments (Tukey, $p < 0.05$). Treatments without lettering are not significantly different.

Grain yield on both soils "Sattenhausen" and "Lutterbeck" did not differ between the treatments (Table 4). Unfortunately, there are no yield data for "Gieboldehausen" because the farmer harvested the whole field before we could harvest the experimental plots. On "Sattenhausen" total P uptake was also not different between treatments. However, on "Lutterbeck" a higher P uptake of MAP-Sb indicated a better plant availability of this material compared to the other recycled P fertilizers and rock phosphate. The small differences between the treatments might be due to unusually dry April in 2007. This reduced plant growth and concomitantly P demand. Already the soil-borne P in the unfertilized plots seemed to be sufficient. Furthermore, the fertilizers were not washed into the soil during this period and, hence, were not plant available at this time.

Pot experiments. Differences in plant availability between the compounds became more obvius in the pot experiments. The chemical characterizations of plant availability by CAL-extraction after harvest showed that MAP-Sb, apatite, cupola slag and partly MAP-Gf were similar to TSP-60 (Table 5). Rock phosphate and the two ashes of MB meal and sewage sludge were not different to the unfertilized control. Plants take up P only from soil solution, hence, P soil solution concentration may characterize the direct P availability. Only a fertilization with MAP-Sb resulted in a P concentration similar to TSP-60 in all three soils. In "Sattenhausen", sinter-P and MAP-Gf reached a higher P soil solution concentration than the unfertilized control, whereas all other compounds were not different to the unfertilized control regardless of the soil.

Table 4. Yield and P uptake of oilseed rape and winter barley after fertilization with recycled P compounds grown on different soils in the field experiments.

| | Sattenhausen (pH 5.6) Oilseed rape | | Lutterbeck (pH 6.8) Winter barley | |
	Grain yield	P uptake	Grain yield	P uptake
	t ha^{-1}	kg ha^{-1}	t ha^{-1}	kg ha^{-1}
unfertilized	2.5	22.9	6.0	24.3 ab
TSP-60	2.6	27.4	6.3	29.6 ab
TSP-100	2.5	28.6	5.6	26.6 ab
Rock-P	2.6	24.6	5.7	23.4 b
sinter-P	2.9	28.6	5.8	24.0 b
sl-ash	2.7	26.7	5.9	23.6 b
MAP-Sb	3.0	33.1	6.6	32.9 a
MAP-Gf	3.3	31.9	6.0	25.3 b

Different letters denote significant differences between treatments (Tukey, $p < 0.05$). Treatments without lettering are not significantly different.

Table 5. Influence of recycled P fertilizer on CAL-extractable P and P concentration in soil solution at harvest in the pot experiments.

| | Sattenhausen (pH 5.6) | | Gieboldehausen (pH 6.8) | | Düshorn (pH 5.5) | |
	CAL-P	Solution P	CAL-P	Solution P	CAL-P	Solution P
	mg kg^{-1}	μmol L^{-1}	mg kg^{-1}	μmol L^{-1}	mg kg^{-1}	μmol L^{-1}
unfertilized	25.0 e	0.6 d	10.0 d	0.3 de	19.0 e	0.3 cde
TSP-60	40.2 bcd	2.3 b	22.2 c	0.9 bc	40.4 b	0.6 bc
TSP-200	76.1 a	14.8 a	74.4 a	15.4 a	78.9 a	12.8 a
rock-P	27.4 e	0.5 d	10.9 d	0.2 de	27.5 d	0.3 cde
sinter-P	39.0 cd	1.2 c	17.9 c	0.3 de	33.1 cd	0.4 bcde
sl-slash	29.6 e	0.7 d	12.1 d	0.5 cde	28.9 d	0.1 de
MAP-Sb	44.1 bc	2.0 b	21.9 c	0.9 bc	43.5 b	0.8 b
MAP-Gf	39.5 cd	1.4 c	19.7 c	0.5 cde	38.3 bc	0.4 cde
MB meal ash	29.4 e	0.8 d	12.8 d	0.2 e	26.1 de	0.1 e
apatite	44.5 b	0.6 d	28.0 b	0.4 de	36.6 bc	0.5 bcd
cupola slag	Not tested		28.2 b	1.1 b	40.7 b	0.1 e

Different letters denote significant differences between treatments (Tukey, $p < 0.05$).

The chemical characterization of P availability does not necessarily reflect the 'real' P availability as can be deduced from P uptake and yield. This is due to the fact that plants are able to mobilize P fractions in the soil which are not taken into account by extraction methods. At least 90% of the yield of the TSP-60 treatment was achieved by sinter-P and MAP-Sb in all soils, as well as by apatite in the acid soils and sl-ash in "Sattenhausen" (Table 6). Phosphate uptake by plants was highest for TSP-200 in all three soils, indicating that a fertilization of 60 mg kg^{-1} was suboptimal as intended. The fertilizer effect of the compounds was different for each soil. In "Sattenhausen" only MAP-Sb had a similar uptake to TSP-60. In the second acid soil "Düshorn", MAP-Sb, apatite and sinter-P were similar to TSP-60, whereas in "Gieboldehausen" uptake of cupola slag and the both MAP's was even slightly higher than that of TSP-60.

Table 6. Yield and P uptake of maize after fertilization with recycled P compounds grown on different soils in pot experiments.

	Sattenhausen (pH 5.6)		Gieboldehausen (pH 6.8)		Düshorn (pH 5.5)	
	yield	P uptake	yield	P uptake	yield	P uptake
	g pot^{-1}	mg pot^{-1}	g pot^{-1}	mg pot^{-1}	g pot^{-1}	mg pot^{-1}
unfertilized	85.6 bcd	92.0 de	55.3 cd	55.6 e	48.2 d	70.0 f
TSP-60	103.3 ab	143.4 b	102.7 a	113.8 b	101.5 ab	120.3 bc
TSP-200	120.5 a	242.2 a	105.5 a	196.2 a	119.4 a	235.9 a
rock-P	71.0 cd	95.0 de	37.8 d	48.0 e	63.7 cd	84.2 def
sinter-P	98.1 abc	117.1 c	92.5 ab	96.0 bc	103.5 ab	108.6 cd
sl-slash	92.4 abcd	110.1 cd	72.6 bc	89.0 bcd	45.3 d	76.3 f
MAP-Sb	103.2 ab	125.0 bc	95.2 ab	116.2 b	104.1 ab	139.2 b
MAP-Gf	86.5 bcd	119.9 c	88.6 ab	116.0 b	87.0 bc	104.0 cde
MB meal ash	69.2 d	88.8 e	67.4 bc	75.8 cde	57.2 d	77.5 ef
apatite	113.5 ab	89.1 e	35.6 d	63.7 de	95.1 ab	111.6 bcd
cupola slag	Not tested		85.0 abc	123.8 b	44.0 d	75.8 f

Different letters denote significant differences between treatments (Tukey, p < 0.05).

DISCUSSION

In the field experiment, the fertilizer effect was very low regardless of the P compound. Even TSP increased CAL extractable P or solution P only in tendency. This indicates a fixation of P in fractions that are not CAL soluble. Hence, it can not clearly be deduced whether the recycled P compounds became soluble and were immediately fixed as soil-P or not. However, P uptake and grain yield points to a slightly better plant availability of TSP, MAP-Gf, MAP-Sb and sinter-P compared to the other compounds, although the yield effects were not significant. The similar P uptake of the TSP-100 treatment and the unfertilized control indicates that plants were able to use mainly soil-born P despite the low P supply level of the soils as indicated by the CAL-P.

The pot experiments confirmed the field results with a somewhat better differentiation between the treatments. The CAL extractable P and solution P of both MAPs, sinter-P and partly the apatite was comparable to the TSP-60 treatment whereas the other compounds differed not much from the unfertilized control. The plant availability can be assessed by deriving the relative fertilizer effect (RFE). The difference between P uptake of a specific fertilizer treatment and the unfertilized control reveals the amount of P that came from the fertilizer. This is compared to the fertilizer effect of a readily available fertilizer, i.e. the TSP-60 treatment, which is set to 100% (Table 7). Depending on soil and fertilizer this RFE ranged from 12% (cupola slag, "Düshorn") up to 138% (MAP-Sb, "Düshorn"). A RFE close to TSP, i.e. more than 70%, was only achieved by MAP-Sb, MAP-Gf and cupola slag in "Gieboldehausen" soil and MAP-Sb, apatite and sinter-P in "Düshorn" soil. In "Sattenhausen" none of the tested compounds reached a RFE higher than 70% (Closest was MAP-Sb with 64%).

Table 7. Relative fertilizer efficiency for P recycled products utilized in pot experiments.

	Sattenhausen (pH 5.6)	Gieboldehausen (6.8) %	Düshorn (5.5)
TSP-60	100.0	100.0	100.0
TSP-200	292.2	241.6	329.8
rock-P	5.8	−13.1	28.2
sinter-P	48.8	69.4	76.7
sl-slash	35.2	57.4	12.5
MAP-Sb	64.2	104.1	137.6
MAP-Gf	54.3	103.8	67.6
MB meal ash	−6.2	34.7	14.9
apatite	−5.6	13.9	82.7
cupola slag	−	117.2	11.5

At all, the MAPs performed best of all tested compounds. These results are consistent with other studies. Plaza *et al.* (2007) tested the effectiveness of MAP in comparison with sewage sludge and conventional P fertilizer (TSP) by calculating the relative agronomic effectiveness (RAE) in relation to TSP and found also that MAP reached 94% of effectiveness. The P uptake of ryegrass was equal in all levels of P, indicating that MAP under certain conditions can be totally available. Johnston and Richards (2003) noted that struvite was as effective as monocalcium phosphate (MCP). A comparison with rock phosphate on acid soils indicates that struvite had the same fertilizer potential as rock phosphate (Gonzalez Ponce and De Sa, 2007). Ghosh *et al.* (1996) found that struvite was equally efficient as diammonium phosphate which is also a water soluble fertilizer. These evidences that struvite can be as effective as water soluble P fertilizers are somewhat astonishing because the P in struvite is not water soluble (Johnston and Richards, 2003). There is also no clear indication that a low soil pH would be necessary to bring MAP in solution. The RFE of both MAPs on the acid soil "Sattenhausen" was comparably low (Table 7) and much higher on the neutral soil "Gieboldehausen", whereas on the second acid soil "Düshorn" they react differently. Gonzalez Ponce and De Sa (2007) attributed the good effectiveness of struvite to the fine particle size of the material. It is well documented that the particle size, i.e. the surface area, affects the solubility of fertilizers. In our experiment all P compounds were finely ground, which might explain the relatively good availability of the MAPs. Looking closer at the soil and yield data revealed that both used MAPs differ in there solubility and plant availability, i.e. MAP-Sb performs better than MAP-Gf. This is most probably due to the high Fe concentration of MAP-Gf (26.6 g Fe/kg) compared to MAP-Sb (5.9 g Fe/kg). The phosphate in the Gf sewage works is precipitated by Fe leading to these high concentrations. Römer and Samie (2002) and Römer et al. (2004) could show that P availability of sewage sludge is strongly reduced by a high Fe concentration in the sludge.

CONCLUSION

The results showed that none of the tested compounds reached in all soils the plant availability of a water-soluble P fertilizer in the pot experiment. Closest to this was the MAP-Sb in most soils. The ashes had a low availability, whereas the availability of sinter-P and cupola slag depended much on soil properties. However, these results so far are preliminary. Both, the field and pot experiments, were designed for a three-year period. They will be continued without a further P fertilization to examine the long-term fertilization effect of the tested compounds.

REFERENCES

Adams, F. (1974). Soil solution. In: *The plant root and its environment* (ed. E.W. Carson), pp. 441–481, University Press of Virginia, USA.

Berg, U. and Schaum, C. (2005). Recovery of phosphorus from sewage sludge and sludge ashes. Applications in Germany and Northern Europe. 1. Ulusal Aritma Çamurlari Sempozyumu, AÇS2005, pp. 87–98.

Ghosh, G.K., Mohan K.S. and Sarkar, A.K. (1996). Characterization of soil-fertilizer P reaction products and their evaluation as sources of P for gram (*Cicer arietinum L.*). *Nutrient Cycling in Agroecosystems*, **46**, 71–79.

Gonzales Ponce, R. and De Sa, M. (2007). Evaluation of struvite as a fertilizer: a comparison with traditional P sources. *Agrochimica*, **51**(6), 301–308.

Goto, I. (1998). Application of phosphorus recovered from sewage plants. *Environmental Conservation Engineering*, **27**, 418–422.

Johnston, A.E. and Richards, I.R. (2003). Effectiveness of different precipitated phosphates as phosphorus sources for plants. *Soil Use and Management*, **19**, 45–49.

Murphy, J. and Riley, J. (1962). A modified single solution method for determination of phosphate in natural waters. *Analytica Chimica Acta*, **26**, 31–36.

Plaza, C., Sanz, R., Clemente, C., Fernández, J., González, R., Polo, A. and Colmenarejo, M. (2007). Greenhouse evaluation of struvite and sludges from municipal wastewater treatment works as phosphorus sources for plants. *Journal of Agricultural and Food Chemistry*, **55**, 8206–8212.

Römer, W. and Samie, I.F. (2002). Phosphordüngewirkung eisenhaltiger Klärschlämme, *J. Plant Nutr. Soil Sci.*, **165**, 83–91 (English summary).

Römer, W., Samie, I.F., Neubert, M. and Merkel, D. (2004). The influence of iron content of sewage sludges on P uptake of rye grass (pot experiments). Proceedings of the Meeting of the German Society of Plant Nutrition, Goettingen, Germany, 1–3 September 2004, p. 86.

Scheffer, V.F. and Pajenkamp, H. (1952). Phosphatbestimmung in Pflanzenaschen nach der Molybdän-Vanadin-methode. Zeitschrift für Pflanzenernährung, Düngung, Bodenkunde **56**, 2–8.

Stewart, W. (2005). Phosphorus as a natural resource. In: *Phosphorus: Agriculture and the Environment. Agronomy monograph N 46* (eds. Sims, T. and Sharpley, A.), Madison, Wisconsin, USA.

Schueller, H. (1969). Eine neue Methode zur Bestimmung des pflanzenverfügbaren Phosphates in Böden. Z. Pflanzenernähr. und Bodenk. **123**, 48–63.

Ecological testing of products from phosphorus recovery processes – first results

K. Weinfurtner[a], S.A. Gäth[b], W. Kördel[a] and C. Waida[b]

[a]Fraunhofer Institute for Molecularbiology and Applied Ecology, 57392 Schmallenberg, Germany
[b]Institute of Landscape Ecology and Resources Management, University of Giessen, 35392 Giessen, Germany

Abstract Phosphorus is the essential nutrient with the lowest reserves and phosphorus can not be substituted by other elements. Therefore there is a great interest in using other available phosphate resources like waste water materials as source for phosphate fertilizers. Since 2004 the German Federal Ministry of Education and Research (BMBF) and the German Federal Ministry for the Environment, Nature Conservation and Nuclear Safety (BMU) launched the funding programme 'Recycling management of plant nutrients, especially phosphorus' which comprise the proving of novel and – up to now – not employed technologies and processes for recovery of phosphorus and if applicable further plant nutrients from waste materials – especially from municipal waste water and sewage sludge as well as from further applicable secondary raw materials.

Thus the aim of the project in question is to examine the products generated in the funding programme as for their effect as fertilizer, for possible harmful impacts on the environment and for the technical qualities of the observed products.

First results are presented which gives information about solubility of phosphorus and the concentrations of trace metals in the tested products. Further more the effect of the products on Phosphorus availability in soil and the uptake into plants was investigated.

INTRODUCTION

Phosphorus is together with nitrogen and potassium one of the essential nutrients and a limiting factor for plant growth. In its function as fertilizer phosphorus can not be substituted by other elements. In Western Europe about 80% of rock phosphate is used in the fertilizer industry. The world wide phosphate reserves are limited and considerably exhausted. The known phosphate reserves are sufficient for approx. 100 years (Pradt 2003), nethertheless phosphate is categorised as a lack resource. Another problem is the amount of contaminants in rock phosphates especially cadmium which can be a limiting factor for using phosphates as fertilizer.

In Germany about 115,000 tons of Phosphorus were used as fertilizer in the farm year 2006/2007 (IVA, 2008). Germany does not have its own phosphate reserves and is depending on phosphate imports. Therefore there is a great interest in using other available phosphate resources such as waste water materials as a source for phosphate fertilizers. The calculated potential of phosphorus from secondary raw materials (e.g. sewage sludge, bio wastes, feeding bone meal) is about 130,000 tons of phosphorus mainly in sewage sludge (about 57,000 tons) and meat and bone meal ashes (about 30,000 tons) (Fricke and Bidlingmaier, 2003). About 50% of the Phosphorus in sewage sludge was used as fertilizer in 2000 (Frede and Bach, 2003) but this amount is decreasing because of stronger regulations for the use of sewage sludge in some states in Germany.

Therefore in 2004 the German Federal Ministry of Education and Research (BMBF) and the German Federal Ministry for the Environment, Nature Conservation and Nuclear Safety (BMU) launched the funding programme "Recycling management of plant nutrients, especially phosphorus". The German Federal Ministry of Education and Research funds research- and development projects which comprise the proving of novel and – up to now – not employed technologies and processes for recovery of phosphorus and if applicable further plant nutrients from waste materials – especially from municipal waste water and sewage sludge as well as from further applicable secondary raw materials.

AIM OF THE STUDY

The recovery of phosphorus from waste materials takes central part of the shared funding programme. The produced materials have to fulfil some conditions for the use as fertilizer:

* the phosphorus contained in the materials have to be sufficient plant available to ensure the plant nutrition
* the amount of organic and inorganic pollutants should considerably fall below the legal requirements for fertilizers and should not lead to an accumulation of pollutants in the soil.

Thus the aim of the project in question is to examine the products generated in the funding programme as for their effect as fertilizer, for possible harmful impacts on the environment and for the technical qualities of the observed products.

MATERIAL AND METHODS

Test strategy

In the study about 15 interstage and final products generated in the funding program are analysed. The test strategy comprises four sub-goals divided in several subtasks (Figure 1):

Sub-goal 1: analysis of the effect as fertilizer
Sub-goal 2: analysis of possible harmful impacts
Sub-goal 3: test of technical quality
Sub-goal 4: study of ecological aspects of sustainability

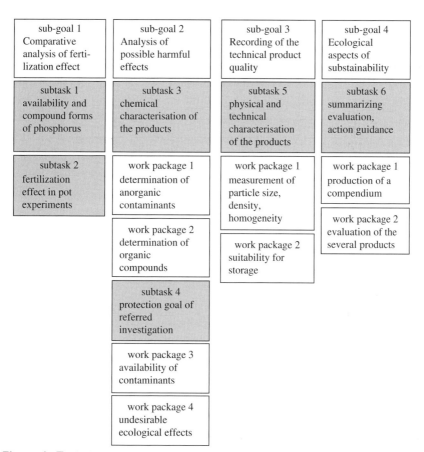

Figure 1. Test strategy.

In sub-goal 1 the solubility and availability of phosphorus measured in different extractants (e.g. water soluble P, H_2SO_4-soluble P) is determined as well as the fertilizer effect in pot experiments with maize over a period of two years.

In sub-goal 2 the amount of especially organic persistent compounds and trace metals is analysed. As the second step the availability of these compounds in batch and column experiments and the influence on the habitat function in soils will be examined.

The physical and technical characterisation (homogeneity of the products, density, particle size and suitability for storage) of the products is the task of sub-goal 3.

In sub-goal 4 a summerized assessment of all products is given and a compendium for the testing of a secondary raw fertilizer is established.

Material

In the experiments seven products from different sources and processing technologies were used and compared with two mineral fertilizers (Table 1).

Table 1. Tested products.

Product name	Type of material
ULO-Phos	Meat meal ash
P 18	Sewage sludge ash
Seaborne	Sewage sludge recyclate
MAP Gifhorn	Sewage sludge recyclate
MAP Stuttgart	Sewage sludge recyclate
PASCH	Sewage sludge ash
Prophos	Waste water recyclate
Dolophos	Rock phosphate fertilizer
TSP	Triple super phosphate

Two soils with different soil characteristics were used for the tests (Table 2).

Methods

The availability of P in the products was determined according to German fertilizer ordinance and was performed under the methods of VDLUFA (VDLUFA, 2004):

- Mineral acid soluble P: Extraction with concentrated H_2SO_4
- Formic acid soluble P: Extraction with formic acid with concentration of 2%
- Citric acid soluble P: Extraction with citric acid with concentration of 2%
- Water soluble P: Extraction of P with distilled water

Table 2. Characteristics of test soils.

	unit	Düshorn	Gieboldehausen
Sand	g/kg	890	440
Silt	g/kg	1	270
Clay	g/kg	10	290
pH (0,01 M $CaCl_2$)		4.86	6.32
C (total)	g/kg	26.8	16.8
N (total)	mg/kg	1389	2123
P (total)	mg/kg	330	580
Ca (total)	mg/kg	1040	3770
K (total)	mg/kg	80	3210
Mg (total)	mg/kg	150	5320
Al (total)	mg/kg	1660	19980
Fe (total)	mg/kg	570	12140
P (CAL[1])	mg/kg	57	49
CEC_{eff}	mmolc/kg	56.1	188.7
WHC[2]	g/kg	326	449

[1] calcium-acetate-lactate extract
[2] water holding capacity

The concentration of trace metals was determined in aqua regia extract according to DIN ISO 11466 with ICP-OES.

The effect of the products on parameters of P-supply in the soil was determined in an incubation experiment without plants, the effect on plant yield and P-uptake in a pot experiment.

In both cases 6 kg of dry soil was fertilized with 360 mg P in four replicates for every product and additionally for a 0-variant without P application and two mineral fertilizers (rock phosphate (Dolophos) and TSP). Additionally each pot is fertilized with 0,25g N and 0,25g K as well as with 1g MgSO except of those products (MAP-Gifhorn, MAP Stuttgart) that already contain Mg due to their process of production. After fertilization the soils were adjusted up to 70% of WHC.

The incubation experiment was performed in a hall with outdoor temperature but without natural precipitation. After 2 months of incubation the first soil sampling was performed. The samples were air dried and passed through a 2 mm mesh. The plant available phosphorus was determined with the calcium-acetate-lactate extract according to VDLUFA (VDLUFA, 1995), pH value according to DIN ISO 10390.

The pot experiment is performed with maize under greenhouse conditions. The temperature in the climate chamber is kept constantly at 25°C by day (16h with 10klx illumination-time) and is reduced to 20°C during night

(8h without illumination). Each pot was provided with seven seeds. After a month, in the juvenile growing stage and a length between 80 and 100cm, three of the seven plants per pot were cut, dried out for three days at 105°C, ground in a mill and analysed in the lab. At a final growth length of 200 cm the remaining four adult plans were also cut, dried out, ground and analysed in the lab. The ground plant material was digested in a microwave (HNO3- and H2O2-digestion) according to the instructions of the microwave producer (MLS GmbH, Germany). Afterwards the digestion was optically detected in a photo-meter to establish the absorbed phosphorous fraction in the plants (VDLUFA, 1991). To analyse the metal uptake from the used products, the plant digestion was also detected with ICP-MS subjected to DIN EN ISO 17294-2.

RESULTS

Solubility of phosphorus

In Table 3 the solubility of P of the tested products in different extractants is given as percentage of soluble P related to total P. The solubility in mineral acid of two products (PASCH and Prophos) is very high and close to the solubility of TSP. Two other products (P 18 and Seaborne) lie in between the solubility of TSP and Dolophos. The solubility of the other products is comparable with Dolophos.

Table 3. Solubility of tested products in different extractants in percentage related to total phosphorus.

Product	$P_{minacid}$	$P_{formiate}$	$P_{citrate}$	P_{H2O}
MAP Gifhorn	69.0	44.1	48.8	0.3
MAP Stuttgart	73.8	68.4	59.0	0.9
ULO-Phos	77.8	43.5	46.6	0.1
Seaborne	89.2	81.3	65.6	0.8
P 18	91.5	42.6	46.4	5.7
PASCH	94.2	98.3	85.2	0.1
Prophos	99.6	100.5	82.6	0.1
Dolophos	76.3	60	18.5	0.1
TSP	102.6	106.1	102.8	78.3

In formic acid PASCH and Prophos show again a solubility close to that of TSP. Seaborne and MAP Stuttgart lie in between the solubility of TSP and Dolophos. MAP Gifhorn, ULO-Phos and P 18 show a lower solubility as

Dolophos. In citric acid the solubility of all products lie in between Dolophos and TSP but PASCH and Prophos show again the highest solubility. The water solubility of all products is very low and ranged under 1% of total P except for P 18 with 5.7%.

Trace metals

The trace metal concentrations show a large variation. With exception of ULO-Phos the highest concentrations were always observed for Zinc (Figure 2). The highest concentrations for Cr, Cu, Ni and Zn were detected for P 18, a sewage sludge ash. Very low concentrations of these elements were observed for PASCH and Prophos. The cadmium concentrations were mostly low and in most cases less than the detection limit (Figure 3). Only TSP shows a high cadmium concentration of 20 mg/kg. The lead concentrations were mostly about 5 mg/kg, only P 18 and PASCH had concentrations over 10 mg/kg.

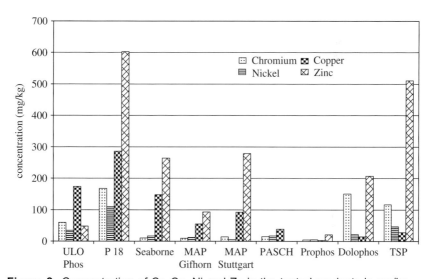

Figure 2. Concentration of Cr, Cu, Ni and Zn in the tested products in mg/kg.

The arsenic concentrations were low, too (Figure 3). Taking into consideration the German fertilizer ordinance ULO-Phos, P 18, Seaborne and MAP Stuttgart can not be used directly as fertilizer because they exceed the threshold values for Copper (70 mg/kg) and /or Nickel (80 mg/kg).

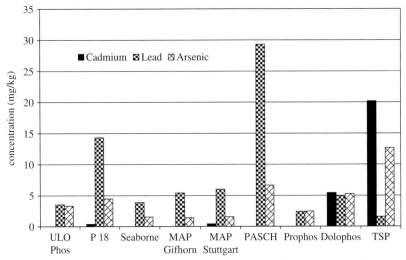

Figure 3. Concentration of Cd, Pb and As in the tested products in mg/kg.

Plant available phosphorus

The tested products affected different by the tested soils. In the sandy soil only the products Seaborne, PASCH and Prophos showed a significant increase of P_{CAL}. PASCH and Prophos were as efficient as TSP (Figure 4).

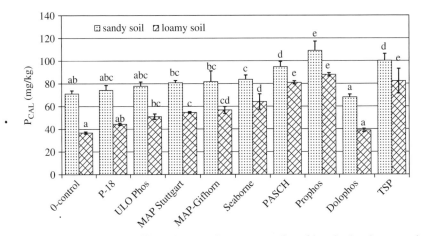

Figure 4. Plant available Phosphorus after two months of incubation in a sandy and loamy soil (bars represent the standard deviation, different characters represent difference in significance).

In the loamy soil only P 18 did not increase P_{CAL} significantly. Again PASCH and Prophos were as efficient as TSP. In both soils all products resulted in a higher increase of P_{CAL} as Dolophos. Between the soils the products did not differ in their efficiency. The effect of the products is not influenced by soil characteristics. The relatively better effect on the loamy soil is caused by the lower P concentration of this soil.

P-uptake

In the loamy soil the efficiency of all fertilizers including TSP was low. Only Seaborne caused a significant increase of P-uptake (Figure 5). In the sandy soil the P-uptake induced by fertilization was higher for all products compared to the P uptake in the loamy soil, but a significant increase could only be observed for MAP Stuttgart, Seaborne and TSP.

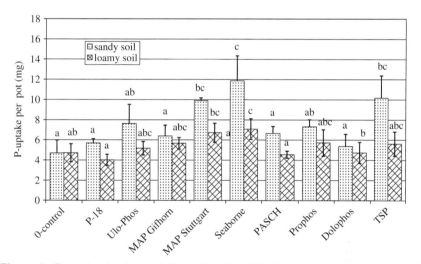

Figure 5. P-uptake 1st harvest (1 month after drilling).

CONCLUSIONS

The first results showed that some of the products can actually not directly be used as fertilizers because of legal restrictions. As a result of the small solubility of Phosphorus in water, the tested products should not be used as fertilizer for plants with a nutrient deficiency but for soils as base fertilization. A correlation between the efficiency of the products on plant available Phosphorus and the yield and P-uptake could not be observed.

REFERENCES

Frede, H.G. and Bach, M. (2003). Heutiger und künftiger Beitrag der Gülle bei der Phosphorversorgung der Böden. In "Rückgewinnung von Phosphor in der Landwirtschaft und aus Abwasser und Abfall", pp. 4/1–4/11. Umweltbundesamt, RWTH Aachen, Berlin.

Fricke, K. and Bidlingmaier, W. (2003). Phosphorpotenziale qualitative hochwertiger organischer Siedlungsabfälle und deren Nutzung. In "Rückgewinnung von Phosphor in der Landwirtschaft und aus Abwasser und Abfall", pp. 4/1–4/11. Umweltbundesamt, RWTH Aachen, Berlin.

IVA (2008). Jahresbericht 2007/2008. Report Industrieverband Agrar, Frankfurt, Germany

VDLUFA (1995). Methodenbuch Band I. Die Untersuchung von Böden, VDLUFA-Verlag, Darmstadt, Germany.

VDLUFA (2004). Methodenbuch Band II.1. Die Untersuchung von Düngemittel, VDLUFA-Verlag, Darmstadt, Germany.

Strategy for separation of manure P through flocculation

Maibritt Hjorth[a]*, Morten Lykkegaard Christensen[b]

[a]Department of Agricultural Engineering, Aarhus University, Blichers Allé 20, P.O. 50, DK-8830 Tjele, Denmark
[b]Department of Biotechnology, Chemistry and Environmental Engineering, Aalborg University, Sohngaardsholmsvej 57, DK-9000 Aalborg, Denmark

*Corresponding author. Tel.: +45-8999-3049; fax: +45-8999-1919,
E-mail address: Maibritt.Hjorth@agrsci.dk

Abstract To increase the application of solid-liquid separation of animal manure and thereby enjoy the environmental and practical benefits, procedure recommendations are necessary. Flocculation of manure prior to separation can improve phosphorus separation significantly; however the efficiency of the separation depends largely on how the of the flocculation was carried out. Two polymers and $FeCl_3$ were tested. After flocculation, the manure was gravity drained. At the optimum flocculation, the dewatering velocity was improved by a factor 150, the solid fractions dry matter content increased by a factor 3, the liquid fractions turbidity reduced by a factor 50, and the P removal into the solid fraction increased from 65% to 95% compared to the control experiment when no chemicals was added. The optimum P separation are carried out by adding $FeCl_3$ or $Al_2(SO_4)_3$) and a large, medium charged cationic, linear polymer. The added amount should be determined on the individual manure, and the stirring procedure must be carefully considered. A good solid-liquid separation of the flocculated manure was obtained by using the low cost technique drainage.

INTRODUCTION

Solid-liquid separation of animal manure is still not commonly used (Burton, 2006) despite large environmental and practical advantages. However, an increased application may be expected if recommendations were available, which were based on studies on how to perform the optimum separation to obtain specific manure products; a phosphorus (P) rich solid fraction, a dry solid fraction, a liquid fraction with low content of nutrients, a solid fraction with large content of volatile solids or merely an easily functioning separation. The focus in this study is the separation of P.

A separation can produce a solid fraction rich on P, which can be used e.g. as fertilizer on P deficient fields or, potentially, for production of a commercial fertilizer. The increased concentration of phosphorus in a manure fraction reduces the need for transportation. The solid-liquid separation additionally produces a liquid fraction with a nutrient content matching the plant require-ments better than the raw manure, i.e. with reduced nitrogen (N) and P contents and an increased N:P ratio (Vanotti et al., 2005). Thus, field application of the liquid fraction instead of the raw manure would result in reduced eutrophication of the surrounding aquatic systems.

Solid-liquid separation is performed mechanically by centrifugation, sedimen-tation, drainage or pressurized filtration. It has been shown that flocculation, coagulation and precipitation pre-treatment of the raw manure prior to the separation improve the manure separation, e.g. of P (Timby et al., 2004; Powers and Flatow, 2002). Flocculation can be carried out by a polymer, as the polymer charges induce particle bridging and to a less extend patch flocculation (Figure 1). Coagulation can be performed by multivalent ions, because these cause charge neutralisation of the particles (Figure 1). Precipitation of dissolved P may addi-tionally be performed by multivalent ions, e.g. by precipitation of $Fe_5(PO_4)_2(OH)_9$.

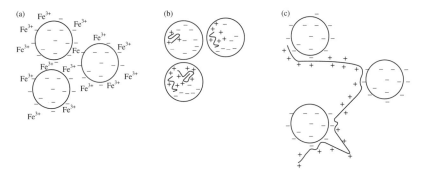

Figure 1. Principle of (a) coagulation, (b) patch flocculation and (c) polymer bridging (Hjorth et al., 2008).

The outcome of the separation depends on choice of polymer (structure, molecular weight, charge and charge density), polymer amount, choice of multivalent ion, amount of multivalent ion, mixing procedure, and separation technique. In addition, because manure varies between estates, the specific manure may require specific separation conditions and results in specific products. Further, as characteristics of the separation products are adjustable, the desired separation products may vary between applications. Hence, literature comparing the consequence of the many different separation options is needed, however little is available.

In this paper, comparisons of additions multivalent ion ($FeCl_3$) and polymer (linear and branched) to manure were performed in order to discuss how to perform the best flocculation and solid-liquid separation with regards to P separation.

MATERIALS AND METHODS

Swine manure was collected from an agitated tank at a commercial farm producing slaughtering pigs.

Sub samples were flocculated using two different polymers: the cationic, high molecular weight, linear polymer with medium charge density, Superfloc c-2260 (Cytec, West Paterson, NJ, USA), and the cationic, medium molecular weight, branched polymer with large charge density, Zetag 7878 FS40 (Ciba, Bradford, UK).

The added multivalent ion was ferric chloride; added to a final concentration at 10 mM in the samples. Two flocculation series with varying concentration of the linear polymer were produced with and without ferric chloride, as were two flocculation series with varying concentration of the branched polymer with and without ferric chloride. The amount of a charged polymer may be reported in terms of the amount of added charges (eq = mol charged group), and in this study the applied linear and branched polymer solution (0.5%) contained 16.5 eq/ml respectively 23.5 eq/ml. The polymer was added in small doses with 5 sec of rapid mixing (220 rpm) for distributing the polymers and 2 min of slow mixing (75 rpm) for reaction between polymers and sample carried out in-between every addition. Subsequently, a mechanical solid-liquid separation was performed by draining 200 ml of the manure through a 200 μm filter for 30 min.

Five characteristics of the samples were determined. The floc shape in the flocculated samples was ranked visually with respect to size and compactness. The dewatering velocity was determined by logging the drained volume continuously. Drainage velocity decreases with time because of increased clogging of filter; hence to obtain one constant value per sample, the velocity pr time unit (ml/s^2) was calculated. The turbidity (600 nm) and the P content of the obtained liquid fractions were measured, as was the dry matter content of the solid fractions.

RESULTS AND DISCUSSION

Effect of flocculation

Floc shape observations indicated the floc size to increase with polymer amount; a floc is a unit consisting of manure samples particles surrounded by multi-valent ions and entangled by polymer. However, the floc size decreased at the largest polymer amounts, as deflocculation occurred due to steric hindrance, i.e. the amount of charges on the polymers by far exceed the amount of available charges on the manure particles (bindings sites), hence the chance of polymer bridging are reduced as the possibility that a polymer attaches to bindings sites on more than one particle is low. Coagulation prior to the polymer addition may be expected to induce an increase in the compactness of the flocs, as the charge neutralisation, carried out by the multivalent ions, causes the smaller manure particles to be flocculated additionally. As could be expected from the polymer structures, the addition of the relatively small and branched polymer was observed visually to cause relatively small but compact flocs to be formed, as the polymer size causes relatively few manure particles to be caught, while the branched and thus rigid structure and additionally the relatively higher charge density, increases the chance of a polymer to adsorb onto multiple binding sites of one manure particles. On the contrary, the larger linear polymer caused larger but looser flocs to be formed.

Raw manure is usually very difficult to filtrate, however the dewatering improved considerably at flocculation (Figure 2), i.e. drainage of 200 ml raw manure resulted in 45 ml of the liquid fraction, while 200 ml flocculated manure resulted in approximately 140 ml of the liquid fraction. When observing the drainage velocity per time unit of the filtrations, the dewaterability was seen to increase from 0.3 ml/s^2 to 50 ml/s^2 (Figure 2), equalling an increase in dewaterability by a factor of 150. The smaller branched polymer was observed to induce the largest improvement of the dewatering, which was caused by the compact shape of the flocs formed by the small branched polymer, around which the liquid easily drains. The volume of branched polymer applied to obtain the maximum dewatering are smaller than the volume of linear polymer (Figure 2), however when considering the amount of charge added, the amount of linear and branched polymer required are equal, i.e. 2 meq/kg manure.

The dry matter content of the solid fraction after separation increased with polymer amount (Figure 3). The dry matter content of the solid fraction obtained by draining raw manure was 5% (w/w), while drainage of flocculated manure resulted in a dry matter content up to 13% (w/w) and hence a factor 3 increase in the dry matter content. The smaller branched polymer induced the largest dry matter content of the solid fraction as less water are bound in the smaller and more compact flocs.

Turbidity is a function of the particle concentration and particle sizes. Though the particle size distribution may change depending on the polymer addition, turbidity does provide a good estimate of the particle concentration. The turbidity measurements on the liquid fraction show flocculation to increase the particle removal from the liquid fraction (Figure 4), i.e. the turbidity in the liquid fraction decrease 30 arbitrary units (AU) and down to 0.6 AU if the manure

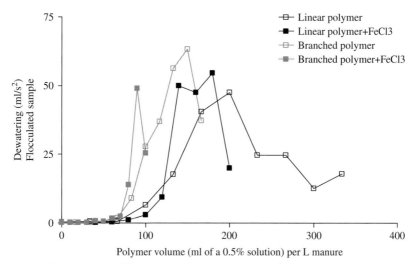

Figure 2. Dewatering of flocculated sample during drainage.

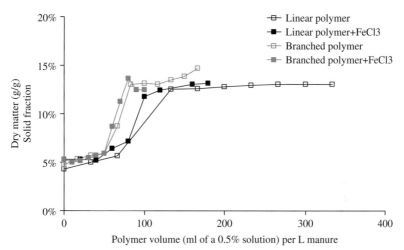

Figure 3. Dry matter content of solid fraction after separation.

is flocculated, which is a factor 50 reduction. Addition of the multivalent ion caused a reduction in the required polymer amount, as the multivalent ion causes coagulation and thereby eases the collection of the smallest particles by reducing the particle surface charge or increasing the particle size. This was also observed in the dewatering measurements (Figure 2). The lowest turbidity was generated

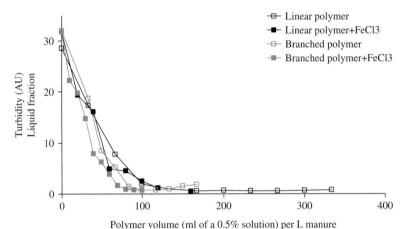

Figure 4. Turbidity of liquid fraction after separation.

by the larger linear polymer. This must be due to the produced large and loose flocs that catch the small particles more easily than the smaller and more compact flocs formed by the branched polymer.

The phosphorus content of the solid fraction obtained at the separation was increased at flocculation, with 65% of the phosphorus ending up in the solid fraction after drainage of untreated manure and up to 95% ending up in the solid fraction after drainage of flocculated manure (Figure 5). The manure phosphorus

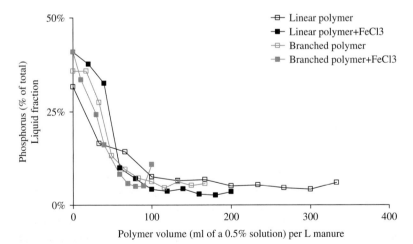

Figure 5. Total phosphorus content of liquid fraction after separation.

additionally being retained in the filtrate at flocculation was mainly the phosphorus bound in the smaller particles, as they were included in the flocs and removed from the liquid. An increasing phosphorus content of the solid fraction was obtained after addition of the multivalent ion, because of the initial precipitation of dissolved phosphorus e.g. as $Fe_5(PO_4)_2(OH)_9$, and afterwards, the precipitated phosphorus could be caught by the polymer and be integrated into the flocs. The small particles formed at precipitation were most easily caught by the large and loose floc shape created by the larger linear polymer compared to the smaller branched polymer.

Strategy for flocculation

The outcome of the separation, including the efficiency of the P separation, can be improved significantly by flocculating manure. Further, polymer consumption can be reduced slightly by adding ferric chloride. Flocculation with a large, cationic, linear polymer with medium charge density proved to provide the best separation of e.g. phosphorus. Previous studies comparing e.g. the polymer charge or charge density support this conclusion (Vanotti *et al.*, 2005; Timby *et al.*, 2004). The data indicate the polymer amount to be of large significance when performing separation, i.e. both too little and too much may be added. At continuous operation, the manure content may change and thus the required polymer amount. Hence, an apparatus for observing and controlling the operation would be beneficial. The liquid fractions turbidity correlated well with P content (Figures 4 and 5) and is simple to measure, and can therefore be used. Other possibilities exist such as measurements of the liquid fraction's viscosity (Hjorth and Christensen, 2008).

At addition of the polymer, the stirring procedure is of large importance for the floc formation. Specifically, after the initial rapid stirring to mix the polymer and manure, the velocity must be low enough to avoid significant floc break-up, but still sufficient to cause most small particles to be caught by the flocs. In this study, however, the mixing procedure was partly chosen coincidentally, as no thorough studies on this have been performed previously. Hence, especially the turbidity and the phosphorus separation observed in this study may be improved further by optimising the mixing procedure.

Considering addition of the multivalent ion, ferric chloride was in this study used to coagulate the manure and precipitate dissolved phosphorus. Comparing studies in the literature, this and aluminium sulphate should be the preferred multivalent ions when treating manure with respect to volume, dry matter and phosphorus separation (Powers and Flatow, 2002). The data indicate the addition of the multivalent ion to be advantageous if a large phosphorus removal from the liquid fraction is desired. It is, however, less important if e.g. merely a large dry matter content of the solid fraction is required. The required amount of multivalent ion is determined by the amount of dissolved P to be precipitated and, which is very difficult to asses, the amount of charges to be neutralized on the particles at coagulation.

Additionally, the flocculation strategy involves selection of the mechanical solid-liquid separation technique to be used after the flocculation. In this study drainage was performed, as a comparison of previous studies on floccula- tion and solid-liquid separation with respect to separation of dry matter and phosphorus indicate drainage to be superior to centrifugation and sedimentation (Powers and Flatow, 2002; Estevez Rodriguez et al., 2005; Møller et al., 2007). The specifications of the solid-liquid separations technique are adjustable, e.g. upon drainage the mesh size can be regulated; relatively large mesh was applied in this study, 200 μm, as the aim is mainly to retains the large flocs efficiently and allowing the liquid to drain through easily. In addition, it is necessary to note that the optimum flocculation strategy may depend on the specific separation technique (Hjorth et al., 2008).

CONCLUSION

When separating manure through flocculation, determining the flocculation strategy involves considering many aspects. A cationic polymer of large molecular weight is a good choice. Ferric chloride/aluminium sulphate addition is not necessary but does reduce the required polymer amount and improve the phosphorus separation slightly. Additionally, the stirring procedure needs care- ful attention. If the proper strategy is chosen, an efficient P separation may be achieved causing reduced eutrophication of the surrounding aquatic systems and creating the possibility of an efficient use of the manure phosphorus as fertilizer.

REFERENCES

Burton, C. (2006). Contributions of separation technologies to the management of live- stock manure. In: Dias Report No. 122, 12th RAMIRAN International Conference, pp. 43–48, Danish Institute of Agricultural Sciences, Foulum.

Estevez Rodríguez, M.D., Gomez del Puerto, A.M., Montealegre Meléndez, M.L., Adamsen, A.P.S., Gullov, P. and Sommer, S.G. (2005). Separation of phosphorus from pig slurry using chemical additives. Appl. Eng. Agric., 21, 739–742.

Hjorth, M. and Christensen M.L. (2008). Evaluation of methods to determine flocculation procedure for swine manure separation. Trans. ASABE 51, 2093–2103.

Hjorth, M., Christensen, M.L. and Christensen, P.V. (2008). Flocculation, coagulation, and precipitation of manure affecting three separation techniques. Biores. Technol., 99, 8598–8604.

Møller, H.B., Hansen, J.D. and Sørensen, C.A.G. (2007). Nutrient recovery by solid-liquid separation and methane productivity of solids. *Trans. ASABE*, **50**, 193–200.

Powers, W.J. and Flatow, L.A. (2002). Flocculation of swine manure: Influence of flocculant, rate of addition, and diet. *Appl. Eng. Agric.*, **18**, 609–614.

Timby, G.G., Daniel, T.C., McNew, R.W. and Moore, P.A. (2004). Polymer type and aluminium chloride affect screened solids and phosphorus removal from liquid dairy manure. *Appl. Eng. Agric.*, **20**, 57–64.

Vanotti, M.B., Rice, J.M., Ellison, A.Q., Hunt, P.G., Humenik, F.J. and Baird, C.L. (2005). Solid-liquid separation of swine manure with polymer treatment and sand filtration. *Trans. ASABE*, **48**, 1567–1574.

Phosphate removal in agro-industry: pilot and full-scale operational considerations of struvite crystallisation

W. Moerman[a], M. Carballa[b], A. Vandekerckhove[c], D. Derycke[d] and W. Verstraete[b]

[a]Akwadok Hoekstraat 3, 8540 Deerlijk Belgium
[b]LabMET Coupure Links 653, 9000 Gent Belgium
[c]Clarebout Potatoes nv Heirweg 26, 8950 Nieuwkerke Belgium
[d]Biotim Antwerpsesteenweg45, 2830 Willebroek Belgium

Abstract Pilot-scale struvite crystallisation tests using anaerobic effluent from potato processing industries were performed at three different plants. Two plants (P1 & P2) showed high phosphate removal efficiencies, $89 \pm 3\%$ and $75 \pm 8\%$, resulting in final effluent levels of 12 ± 3 mg PO_4^{3-}-P/L and 11 ± 3 mg PO_4^{3-}-P/L, respectively. In contrast, poor phosphate removal ($19 \pm 8\%$) was obtained at the third location (P3). A noticeable difference in the influent Ca^{2+}/PO_4^{3-}-P molar ratio was observed between the test sites, ranging from 0.27 ± 0.08 (P1), 0.62 ± 0.18 (P2) and 0.41 ± 0.04 (P3). A negative effect on struvite formation occurred when a Ca^{2+}/PO_4^{3-}-P molar ratio of 1.25 ± 0.11 was obtained after initial pH increase in the stripper at P3. A full-scale struvite plant treating 90–110 m^3/h of anaerobic effluent from a diary industry also showed Ca^{2+} interference. Initially in this plant, influent phosphate levels ranging from 40 to 45 mg PO_4^{3-}-P/L were decreased to below 10 mg PO_4^{3-}-P/L, but no struvite was produced. A shift in Ca^{2+}/PO_4^{3-}-P molar ratio from 2.69 to 1.36 by an increased phosphate concentration resulted in average total phosphorus removal of $78 \pm 7\%$, corresponding with effluent levels of 14 ± 4 mg P_{total}/L(9 ± 3 mg PO_4^{3-}-P/L). Under these conditions pure spherical struvite pellets of 2–6 mm were produced.

INTRODUCTION

Although the advantages of anaerobic treatment are obvious, subsequent nutrient removal still remains an important issue. Readily biodegradable organic matter needs to be bypassed towards aerobic post-treatment in order to achieve the final nutrient effluent standards, thus reducing the potential biogas yield and increasing the waste sludge production. Nutrient removal by struvite ($MgNH_4PO_4 \cdot 6H_2O$) or magnesium ammonium phosphate (MAP) precipitation is an interesting alternative approach to address phosphorus removal (von Munch and Barr, 2001; Gonzalez and De Sa, 2007). MAP crystallisation can be applied for several purposes, such as to prevent scaling problems

(Doyle and Parsons, 2004) and to remove phosphate (Battistoni *et al.*, 1997) or nitrogen (Altinbas *et al.*, 2002; Laridi *et al.*, 2005). Recent publications show an increasing interest in struvite precipitation as a technology for phosphorus recovery taking into account the economic impact of increasing energy costs and limited natural phosphorus resources (Durrant *et al.*, 1999; Shu *et al.*, 2006; Carballa *et al.*, 2008; Forrest *et al.*, 2008). Consequently, integration of struvite formation as specific treatment of side-stream wastewaters is becoming a common practise (Caffaz *et al.*, 2008).

Increasing the operational pH and adjustment of the molar ratios of magnesium, ammonium and phoshate are the most important process parameters in MAP crystallisation (Ohlinger *et al.*, 1998). In addition, presence of calcium has been shown to be determinative for both crystal size and purity (Le Corre *et al.*, 2005; Pastor *et al.*, 2007). When anaerobically processed wastewaters are treated, pH can be increased by simple air stripping (Williams, 1999). If air stripping does not suffice, additional alkaline reagents must be added. Alkaline addition can be combined with the required magnesium supplementation by using either MgO or Mg(OH)$_2$ (von Münch and Barr, 2001). Other magnesium sources include magnesium chloride, magnesium sulphate or seawater bittern waste (Li and Zhao, 2002). If ammonium removal is targeted, both magnesium and phosphate have to be supplied. Yet, to minimize reagent use, internal recycling of the magnesium and phosphate is possible by thermal decomposition of the recovered MAP (Stefanowicz *et al.*, 1992; He *et al.*, 2007).

Different pilot and full-scale units using fluidized bed (Ueno and Fujii, 2001; Forrest *et al.*, 2008) or continuously stirred tank reactors (Mangin and Klein, 2004; Laridi *et al.*, 2005) have been operated. Most full-scale struvite plants do treat anaerobic liquor originating from primary and secondary sludge digestion (Ueno and Fujii, 2001; Battistoni *et al.*, 2005; Forrest *et al.*, 2008). This paper describes the successful use of a fairly straightforward stirred tank crystallizer for full-scale phosphate recovery by struvite crystallization treating anaerobic effluent from a dairy processing industry at an average flow rate of 100 m^3/h. It also points out the need of preliminary feasibility tests before full-scale application to exclude excessive calcium interference, which renders this technology inefficient.

MATERIAL AND METHODS

Pilot plant description

Figure 1 shows a scheme of the pilot plant used. The anaerobic effluent was continuously fed into the stripper through peristaltic pump 1. The total volume of the stripper was 180 L and air was supplied as coarse bubble aeration at 8 L/s.

The 200 L-crystallizer was equipped with a 3 blade top entry impeller (impeller diameter/tank diameter = 0.5) and 120° interval baffles. The crystals were retained by a transient quiescent settling zone. Online control was used to adjust pH with 29% NaOH (pump 2). $MgCl_2$ was used as magnesium source and Mg^{2+} dosage was controlled by setting the flowrate (pump 3) of $MgCl_2$ solution (30%, v/v) according to the influent flowrate. Under steady state conditions, pH was controlled between 8.50 and 8.70 and an Mg^{2+}/PO_4^{3-}-P molar ratio of 1 to 1.2 was maintained. Struvite crystals were removed by intermittent purging.

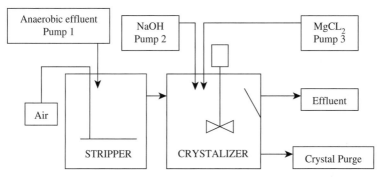

Figure 1. Pilot scale unit used in the struvite crystallisation tests.

Influent and effluent samples were taken 2–3 times a week for ammonium, phosphate and magnesium determinations. Recovered MAP was dried at ambient room temperature to avoid ammonia loss and then used for X-ray diffraction (XRD) analysis.

Anaerobic effluent characteristics

Pilot-scale tests were performed over a 2–3 month period at 3 different potato processing plants. On average, 5 m^3/day of anaerobic effluent coming from the full-scale upflow anaerobic sludge bed (UASB) reactor was treated in the struvite crystallisation. Table 1 shows the initial characteristics of the 3 different anaerobic effluents used.

Analytical methods

pH was measured with a pH meter (Consort C532). Ammonium was determined colorimetrically with Nessler reagent according to standard methods (APHA, 1992). Phosphate and magnesium were determined using a Metrohm 761 compact ion chromatograph equipped with a conductivity detector and flame atomic absorption spectrometry, respectively. Total phosphorus, dry matter content and its mineral fraction were determined according to standard methods (APHA, 1992). XRD analyses were done by using a Siemens D5000 unit equipped with Cu K_α radiation at 1.54 Å.

Table 1. Characteristics of the UASB effluent (influent of struvite pilot plant) of the 3 test sites.

	Plant 1	Plant 2	Plant 3
pH range	7.00–7.45	7.10–7.50	7.00–7.85
PO_4^{3-}-P (mg/L)	115 ± 13	43 ± 7	127 ± 8
NH_4^+-N (mg/L)	426 ± 45	208 ± 27	254 ± 36
Ca^{2+} (mg/L)	40 ± 8	36 ± 7	65 ± 7
Ca^{2+}/PO_4^{3-}-P molar ratio	0.27 ± 0.08[a]	0.62 ± 0.18[a]	0.41 ± 0.04[a]
			1.25 ± 0.11[b]
Average N/PO_4^{3-}-P molar ratio	8.22	10.76	4.54

[a]Influent of struvite plant (stripper inlet)
[b]Influent of crystallisation reactor (stripper outlet)

RESULTS

Pilot-scale tests

The anaerobic effluents from plants 1 and 3 were characterized by high phosphate levels (Table 1), with average PO_4^{3-}-P concentrations above 110 mg PO_4^{3-}-P/L. In contrast, plant 2 showed significantly lower phosphate levels, between 38 to 52 mg PO_4^{3-}-P/L. During the testing period, the operation of the full-scale UASB reactor was stable, resulting in effluent pH values of 7.00–7.85 and residual volatile fatty acid concentrations between 0.5 and 2.5 meq acetic acid/L. Therefore, besides normal full-scale fluctuations, the composition of the UASB effluent remained constant during the testing period (Table 1).

The results obtained in the struvite pilot plant are summarized in Table 2.

Table 2. Effluent phosphate levels and phosphate removal efficiencies at the 3 test sites.

	Plant 1	Plant 2	Plant 3
PO_4^{3-}-P (mg/L)	12 ± 3	11 ± 3	103 ± 11
Average PO_4^{3-}-P removal (%)	89 ± 3	75 ± 8	19 ± 8
Maximum PO_4^{3-}-P removal (%)	95	88	31
Minimum PO_4^{3-}-P removal (%)	79	49	5

High phosphate removal efficiencies were obtained in P1 and P2, with average values of 90% and 75%, respectively. Moreover, despite the initial phosphate levels differed by a factor of 2.67, similar concentrations were obtained in the effluent, around 10 mg PO_4^{3-}-P/L. However, while the elimination remained constant in P1, P2 showed occasionally a limited efficiency not exceeding 50%. Spherical self-retaining crystals of pure struvite, as confirmed by XRD analysis and with 98% of dry matter (after ambient temperature drying) were obtained in both plants. Heavy metal analysis only showed copper (21 mg Cu/kg dry matter) to be present in the struvite pellets of P1.

In contrast, low phosphate removal was noted in P3 (around 20%). In addition, no struvite formation was observed at any conditions tested. The reason was probably the formation of amorphous calcium and/or magnesium phosphate flocculent matter. Detailed analysis of Ca levels in the influent and effluent of P3 showed Ca elimination of 9–26%. A simple pH increase of the UASB effluent up to 8.30–8.50 was sufficient to initiate the amorphous sludge formation without $MgCl_2$ addition. Addition of $MgCl_2$ in the air stripping tank prior to pH adjustment in the crystallisation tank resulted in the formation of maiden orthorhombic pyramidal struvite crystals, which accumulated in the crystallisation tank. These smaller crystals, known as fines, could apparently not grow further, and thus not giving rise to spherical self-retaining crystals, such as those obtained in P1 and P2. Another difference was in the NaOH dose required to obtain the pH set-point. P3 needed a significantly lower amount (0.08–0.20 L/m^3) compared to the NaOH consumption in P1 (0.80–1.20 L/m^3) and P2 (0.50–1.00 L/m^3).

The clear difference in the potential to produce struvite at the different tested plants can be probably explained by the varying Ca^{2+}/PO_4^{3-}-P ratios (Table 1). However, it should be noted that not only the Ca^{2+}/PO_4^{3-}-P molar ratio affects struvite formation, but also the absolute Ca^{2+} concentration. The Ca^{2+}/PO_4^{3-}-P molar ratio differed significantly between P1 and P2, while high phosphate removal efficiencies were obtained in both plants as well as similar struvite crystals. The reason is probably that no distinction in Ca^{2+} concentrations in the UASB effluent was noted, 40 ± 8 mg Ca^{2+}/L in P1 and 36 ± 7 mg Ca^{2+}/L in P2. P3 had clearly higher Ca^{2+} influent levels of 65 ± 7 mg Ca^{2+}/L, which combined with the lower phosphate levels after the stripper (around 40 mg PO_4^{3-}-P/L), resulted in a Ca^{2+}/PO_4^{3-}-P molar ratio at the inlet of the crystallisation reactor of 1.25 (Table 1).

The NH_4^+-N concentrations were not limiting at any of the examined sites. The NH_4^+-N/PO_4^{3-}-P molar ratios in P1, P2 and P3 were 8.22, 10.76 and 4.54, respectively (Table 1). In terms of NH_4^+-N concentrations, no difference was observed between P2 (208 $\pm$ 27 mg NH_4^+-N/L) and P3 (254 $\pm$ 36 mg NH_4^+-N/L), while higher levels were present in P1 (426 $\pm$ 45 mg NH_4^+-N/L).

More detailed data at P1 revealed that most soluble phosphate removal occurred in the crystallizer (80%) and only a 20% decrease was noted in the stripper (Figure 2).

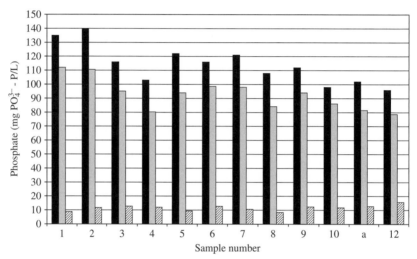

Figure 2. Orthophosphate levels in 12 samples taken during the pilot run in P1 Influent (black), air stripper (grey) and final effluent (striped).

The results shown in Figure 2 and Tables 1 and 2 refer to orthophosphate. After anaerobic treatment, around 90–95% of the total phosphorus was present as orthophosphate. Total phosphorus levels were analyzed once a week, approximately. In general, the total phosphorus effluent concentrations were 17 to 28% higher than the orthophosphate levels. Microscopic control of final effluent showed typical maiden orthorhombic fines in the effluent of P1 and P2.

Full scale unit

Based on the pilot tests, two full-scale units were designed and a process approach was developed, which is marketed as NuReSys® (Table 3). The main common features of both units are: air stripper, crystallisation reactor equipped with a top-entry mixer and a transient quiescent settling zone, pH control using NaOH and magnesium chloride as magnesium source.

The full-scale plant 1 contains a 200 m^3 air-stripper and 125 m^3 crystallisation reactor. The produced struvite crystals are retained in the tubular lamellae

Table 3. Main characteristics of the full-scale NuReSys plants.

	Average capacity	Operational characteristics	Status
1	100 m^3/h (192 kg P$_{total}$/d)	Dairy processing industry: HRT*: 3.25 h Footprint: 180 m^2	Operational
2	60 m^3/h (180 kg P$_{total}$/d)	Potato processing industry HRT*: 2.23 h Footprint: 92 m^2	Start-up in October 2008

*HRT: Hydraulic Retention Time.

settling zone located in the upper part of the crystallisation reactor and further purged towards a dewatering container by intermediate manual offset. These produced struvite crystals have been accredited for reuse in agriculture.

During the start-up of the plant, although a 70 to 75% decrease in phosphate was obtained, no struvite was produced due to similar Ca interference as observed during the pilot-scale tests at plant 3. The calcium and phosphate levels in the UASB effluent of the dairy plant averaged 140 mg Ca^{2+}/L and 40–45 mg PO$_4^{3-}$-P/L, resulting in an average Ca^{2+}/PO$_4^{3-}$-P molar ratio of 2.69. Moreover, no significant decrease of ammonia was observed during this period and a flocculent material was produced, as observed in the pilot-scale tests at P3. The produced amorphous matter contained mainly calcium, magnesium and phosphorus at a respective molar ratio of 2.50/0.75/1.80. Accordingly, Ca^{2+} concentrations decreased by 40–50%. The flocculent matter was also produced in the stripper unit suggesting its production to be driven by pH increase. Changing the position of the inlet ports of alkaline reagent NaOH and MgCl$_2$ to prevent the pH-driven formation of the amorphous matter was not successful.

A batch experiment was performed with 2 L of full-scale UASB effluent spiked with concentrated PO$_4^{3-}$-P solution to obtain a phosphate level of about 150 mg PO$_4^{3-}$-P/L. After phosphate spiking, pH was adjusted and Mg^{2+} was added. Phosphate was removed from 145 to 14.7 mg PO$_4^{3-}$-P/L. In addition, microscopic examination of the settled material clearly showed the formation of struvite crystals. Therefore, under these conditions corresponding to a Ca^{2+}/ PO$_4^{3-}$-P molar ratio of 0.77, struvite formation was possible, but amorphous calcium/magnesium flocculent matter was still the main mechanism responsible for the phosphate decrease.

A gradual shift in UASB effluent due to an increase in phosphate concentrations up to an average of 69 mg P$_{total}$/L resulted in a decrease of the Ca^{2+}/PO$_4^{3-}$-P molar ratio from 2.69 to 1.36. The full-scale plant was operated under these conditions over a 23-week period and the results obtained are shown in Table 4.

Table 4. Average, minimum and maximum phosphorus levels in the influent and effluent of the full-scale unit (n = 169).

	Influent	Effluent		
	P_{total} (mg/L)	P_{total} (mg/L)	PO_4^{3-}-P (mg/L)	P_{total} removal (%)
Average	64	14	9	78
Stand. Dev.	13	4	3	7
Maximum	108	33	26	91
Minimum	38	7	4	38

After the increase of phosphate influent levels, the formation of the flocculent amorphous matter has been largely suppressed and replaced by the growth of 2–6 mm in diameter spherical crystals struvite crystals (confirmed by XRD). The average ammonium influent levels during this period were 110 ± 18 mg NH_4^+-N/L, resulting in a molar NH_4^+-N/PO_4^{3-}-P ratio of 3.45. Contrary to the start-up period, the phosphate removal was accompanied by an ammonium decrease of around 22 mg NH_4^+-N/L, which would account for an average 46 mg PO_4^{3-}-P/L immobilized as struvite. Taking into account the overall phosphate removal over the examined period, ca. 55 mg PO_4^{3-}-P/L, the amorphous matter was responsible for only 16% of the total phosphate removal.

DISCUSSION

Results obtained at pilot-scale in plant 3 and during start-up of the full-scale plant 1 showed a negative interference of calcium. Pastor *et al.* (2007) observed a similar effect treating 2 anaerobic digestion liquors with Ca^{2+}/PO_4^{3-}-P molar ratios of 0.37 and 2.34, respectively. The high Ca^{2+}/PO_4^{3-}-P molar ratio liquor only removed 35% of the phosphorus as struvite, whilst the low Ca^{2+}/PO_4^{3-}-P molar ratio liquor achieved 73%. Overall phosphorus removal efficiencies were as high as 83% (ratio Ca^{2+}/PO_4^{3-}-P = 2.34) and 91% (ratio Ca^{2+}/PO_4^{3-}-P = 0.37), indicating that phosphorus removal was partially attributed to amorphous calcium phosphate. Similarly, Le Corre *et al.* (2005) reported that Ca^{2+}/Mg^{2+} molar ratios between 0.5 and 1.0 affected significantly struvite formation, and values exceeding 1 excluded nearly completely the formation of crystalline struvite. These findings are in accordance with the results obtained in this study, where Ca^{2+}/Mg^{2+} molar ratios are nearly equal to the Ca^{2+}/PO_4^{3-}-P ratios since the Mg^{2+}/PO_4^{3-}-P molar ratio used was 1–1.2. Full-scale results obtained at the average Ca^{2+}/PO_4^{3-}-P molar ratio of 1.36 are not in accordance with the limited struvite formation observed during the batch phosphate spiking test at Ca^{2+}/PO_4^{3-}-P of 0.77. However, in order to compare these results rigorously, the differences in concentration profiles and mixing, and the presence of secondary nucleation surfaces need to be considered.

The use of top-entry mixers has been shown to be an alternative to fluidized bed systems for struvite formation (Pastor *et al.*, 2008; Kim *et al.*, 2008). The results obtained in this work confirm that this operational approach can be successfully applied, even at considerable high flow rates of 100–125 m^3/h. Furthermore, adjustable mixing rotary speed and flexible selection of reagent injection points are major advantages of this system (Mangin and Klein, 2004).

From mass balance calculations, it becomes obvious that not all of the phosphate is removed as struvite. Moreover, not the entire quantity of immobilized orthophosphate, either as struvite or as amorphous phosphate, is retained within the crystallisation reactor. The washed-out immobilized phosphates will be determinative in achieving the final effluent standards in terms of total phosphorus. Sperandio *et al.* (2008) have shown that the pH of activated sludge systems, influenced by nitrification/denitrification and aeration-driven stripping effect, is a major factor affecting the formation and conservation of crystalline phosphate compounds, and thus contributing to the overall phosphate removal efficiency.

The produced full-scale crystals were shown to be mainly composed of pure struvite. This fact confirms earlier reports related to the production of a high quality end product suitable for agricultural reuse (Miles and Ellis, 2001; Shu *et al.*, 2006; Forrest *et al.*, 2008).

CONCLUSIONS

1. Phosphate recovery by struvite formation is a high value added technique, mainly after anaerobic treatment as typically applied in the agro-industrial sector.
2. The interference of calcium is an important factor to be addressed, which could exclude the use of this technology.
3. Since the final Ca^{2+}/Mg^{2+} molar ratios are a direct consequence of the initial Ca^{2+}/PO_4^{3-}-P molar ratios, the latter should be used as a determinative parameter instead of the Ca^{2+}/Mg^{2+} molar ratios. The Ca^{2+}/PO_4^{3-}-P molar ratio should be low, preferably below 1.0.
4. Formation of amorphous calcium and magnesium phosphates may contribute to overall phosphate removal, but as a non-recoverable product.

5. Feasibility of phosphate removal via struvite at high flowrates of 100–125 m^3/h has been demonstrated without affecting the high efficiency.
6. The high-quality of the obtained struvite crystals indicates that minimal processing is required prior to reuse.

REFERENCES

Altinbas, M., Oztruk, I. and Aydin, A.F. (2002). Ammonium recovery from high strength agro industry effluents. *Wat. Sci. & Technol.*, **45**, 189–196.

Battistoni, P., Fava, G., Pavan, P., Musacco, A. and Cecchi, F. (1997). Phosphate removal in anaerobic liquors by struvite crystallisation without addition of chemicals. Preliminary results. *Water Res.*, **31**, 2925–2929.

Battistoni, P., Boccadora, R., Fatone, F. and Pavan, P. (2005). Autonucleation and crystal growth of struvite in a demonstrative fluidized bed reactor (FBR). *Environ. Tech.*, **26**, 975–982.

Caffaz, S., Bettazzi, E., Scaglione, D. and Lubello, C. (2008). An integrated approach in a municipal WWTP: anaerobic codigestion of sludge with organic waste and nutrient removal from supernatant. *Wat. Sci. and Technol.*, **58**, 669–676.

Carballa, M., Moerman, W., De Windt, W., Grootaerd, H. and Verstraete, W. (2008). Strategies to optimize phosphate removal from industrial anaerobic effluents by magnesium ammonium phosphate (MAP) production. *J. Chem. Technol. Biotechnol.*, **doi**: 10.1002/jctb.2006.

Durrant, A.E., Scrimshaw, M.D., Stratful, I. and Lester, J.N. (1999). Review of the feasibility of recovering phosphate from wastewater for use as a raw material by the phosphate industry. *Environ. Tech.*, **20**, 749–758.

Doyle, J.D. and Parsons, S.A. (2004). Struvite scale formation and prevention. *Wat. Sci. Technol.*, **49**, 177–182.

Forrest, A.L., Fattah, K.P., Mavinic, D.S. and Koch, F.A. (2008). Optimizing struvite production for phosphate recovery in WWTP. *J. Environ. Eng.*, **134**, 395–402.

Gonzalez, P.R. and De Sa, M.E.G. (2007). Evaluation of struvite as a fertilizer: a comparison with traditional P sources. *Agrochimica*, **51**, 301–308.

Greenberg, A.E., Clesceri, L.S. and Eaton, A.D. (2002). *Standard Methods for the Examination of Water and Wastewater*. American Public Health Association (APHA), Washington, DC.

He, S., Zhang, Y., Yang, M., Du, W. and Harada, H. (2007). Repeated use of MAP decomposition for the removal of high ammonium concentration from landfill leachte. *Chemosphere*, **66**, 2233–2238.

Kim, D., Kim, J. Ryu, H.D. and Lee, S.I. (2008). Effect of mixing on spontaneous precipitation from semiconductor wastewater. *Bioresource Technology*, **100**, 74–78. **doi**: 10.1016/j.biortech.2008.05.024.

Laridi, R., Auclair, J.C. and Benmoussa, H. (2005). Laboratory and Pilot-scale phosphate and ammonium removal by controlled struvite precipitation following coagulation and flocculation of swine wastewater. *Environ. Tech.*, **26**, 525–536.

Le Corre, K.S., Valsami-Jones, E., Hobbs, P. and Parsons, S.A. (2005). Impact of calcium on struvite crystal size, shape and purity. *J. Cryst. Growth*, **283**, 514–522.

Li, X.Z. and Zhao, Q.L. (2002). MAP precipitation form landfill leachate and seawater bittern waste. *Environ. Tech.*, **23**, 989–1000.

Mangin, D. and Klein, J.P. (2004). Fluid dynamics concepts for a phosphate precipitation reactor design. In: *Phosphorus in Environmental Technologies, principles and applications*. Valsami-Jones, E., pp. 358–400, IWA Publishing.

Milles, A. and Ellis, T.G. (2001). Struvite precipitation potential for nutrient recovery from anaerobically treated wastes. *Wat. Sci. & Technol.*, **43**, 259–266.

Ohlinger, K.N., Young, T.M. and Schroeder, E.D. (1998). Predicting struvite formation in digestion. *Water Res.*, **32**, 3607–3614.

Pastor, L., Marti, N., Bouzas, A. and Seco, A. (2007). Calcium effect on struvite crystallization of liquors from an anaerobic digestion of prefermented and EBPR sludge. *Proceedings of Nutrient Removal 2007: state of the art*, 4–7 March 2007, Baltimore, USA, 136–144.

Pastor, L., Mangin, D., Barat, R. and Seco, A. (2008). A pilot-scale study of struvite precipitation in a stirred tank reactor: conditions influencing the process. *Bioresour. Technol.*, **doi**: 10.1016/j.biortech.2007.12.003.

Shu, L., Schneider, P., Jegatheesan, V. and Johnson, J. (2006). An economic evaluation of phosphorus recovery as struvite from digester supernatant. *Bioresour. Technol.*, **97**, 2211–2216.

Stefanowicz, T., Napieralska-Zagozda, S., Osińska, M. and Samsonowska, K. (1992). Ammonium removal from waste solutions by precipitation of $MgNH_4PO_4$ II. Ammonium removal and recovery with recycling regenerate. *Resour. Conservat. Recycl.*, **6**, 339–345.

Sperandio, M., Pambrun, V. and Paul, E. (2008). Simultaneous removal of N and P is SBR with production of valuable compounds: application to concentrated wastewaters. *Wat. Sci. & Technol.*, **58**, 859–864.

Ueno, Y. and Fujii, M. (2001). Three years experience of operating and selling recovered struvite from full-scale plant. *Environ. Tech.*, **22**, 1373–1381.

von Munch, E. and Barr, E. (2001). Controlled struvite crystallisation for removing phosphorus from anaerobic digester side streams. *Water Res.*, **35**(1), 151–159.

Williams, S. (1999). Struvite precipitation in the sludge stream at Slough wastewater treatment plant opportunities for phosphorus recovery. *Environ. Tech.*, **20**, 743–747.

Development of a process control system for online monitoring and control of a struvite crystallization process

K.P. Fattah*, D.S. Mavinic, M.S. Rahaman and F.A. Koch

Department of Civil Engineering, University of British Columbia, 2002-6250 Applied Science Lane, Vancouver, B.C., V6T 1Z4. Canada
(E-mail: parvez@interchange.ubc.ca; dsm@civil.ubc.ca; mrahaman@interchange.ubc.ca; koch@civil.ubc.ca)
*Corresponding author

Abstract The production of struvite in a crystallizer depends on a number of variables which bring about complex and non-linear changes in the process chemistry. This complex situation can be handled diligently by gaining an insight into the process using models that define the conditions for higher process efficiency. Among the technologies used to recover phosphorus from wastewater through struvite crystallization, manipulation of the systems' pH and/or supersaturation ratio (SSR) appears to be the major process control parameter. In this study, a new chemical model was developed, and currently being optimized, that will provide dynamic prediction, and subsequent control, of the struvite crystallization process. The model incorporates both chemistry as well as control programs aimed at providing smoother process control. It is expected that through this model, variations in the process parameters can be minimized and also allow for future automation of the process.

INTRODUCTION

Struvite crystallization is a complex process that involves a number of process variables, as well as complex process chemistry. In most cases, control of all the factors involved with the process is not possible, or, not feasible. However, there exists some control parameters that can effectively keep the process running smoothly and efficiently. At the University of British Columbia (UBC), different chemical-based models have been developed that have been successfully used in process control. However, all the models lack real-time control and the possibility of automation. During this study, a dynamic control model was developed that incorporates both chemistry and control software that can be used to increase the efficiency of the process. This process model is the basis of an automatic controller that will be able to manipulate flows and chemical additions, and thereby control the system at a desired set point. It is expected that, when applied to large scale installations, this controller will be able to increase efficiency of phosphorus removal/recovery through stable process conditions and ease of process operations through lower operation downtime. Although various chemical reaction models related to struvite formation and its kinetics have been developed previously (Ali and Schneider 2008; Rahaman *et al.*, 2008), none of the models actually involve process optimization or control.

Among the technologies used to recover phosphorus from wastewater through struvite crystallization, manipulation of the systems' pH and/or supersaturation ratio (SSR) appears to be the major process control parameter (Adnan *et al.*, 2003; Britton *et al.*, 2005; Munch and Barr 2001; van der Houwen and Valsami-Jones 2001). At the University of British Columbia, SSR in the struvite crystallizer has been used as the primary control variable with great success (Fattah *et al.*, 2008a; Britton *et al.*, 2005). Control of SSR can be achieved by individually controlling the reactor pH, magnesium flow into the reactor and the rates of feed and recycle flows. It is also possible that all four or any combinations can provide better control than individual changes. The most efficient mode of control is yet to be determined. The development of such a process control and automation system should help to reduce reactor operating costs and optimize conditions for maximizing revenues. It will also have a direct bearing on the ease of operation and influence product quality, when implemented in full-scale setups at future struvite recovery installations, in wastewater treatment plants.

Objective

The study attempts to optimize struvite crystal and pellet formation in a struvite recovery reactor, by developing a process control system that is capable of providing efficient and effective control of process parameters – through real time, online monitoring, development of computer based process model and the application of process automation.

METHODS AND MATERIALS

Process description

The struvite crystallization process utilized by the UBC research group consists of a fluidized bed reactor, a clarifier, a stripper, tanks for storing chemicals, sensors for monitoring and pumps for various liquid flows. Detailed process description can be found in Fattah *et al.* (2008a). The process utilizes filter press centrate to recover the phosphorus, as struvite.

Terminology

A few terminologies, used to describe the process and results, are explained in the following paragraphs. Details of the terms can be found elsewhere (Fattah *et al.*, 2008a).

Supersaturation ratio (SSR)

This is the saturation ratio in the crystallizer and is given by Equation 1. Ps-sample represents the conditional solubility product of the sample and Ps-equilibrium is the equilibrium conditional solubility product. This ratio determines if the conditions are right for struvite crystals to form. Studies (Fattah *et al.*, 2008a; Britton *et al.*, 2005) have shown that, although, theoretically a value of greater than one (SSR > 1) should precipitate struvite, a value in the range of SSR 3–5 is effective in producing good quality (in terms of size and shape) struvite.

$$SSR = Ps-sample/Ps-equilibrium \tag{1}$$

In order to determine the solubility product, the total soluble ortho-phosphate, ammonium and magnesium have to be determined. The total ion is a combination of species other than PO_4^{3-}, NH_4^+ and Mg^{2+}, as shown in Equations 2 to 4. The [ ] brackets indicate ion concentration in moles per liter, without correction for activity. However, in the calculation for the supersaturation ratio, these concentrations are corrected for activity. The Ps-equilibrium value was obtained from a series of laboratory experiments (Forrest *et al.*, 2009) conducted by the research group.

$$T-PO_4 = [H_3PO_4] + [H_2PO_4^-] + [HPO_4^{2-}] + [PO_4^{3-}] \tag{2}$$

$$T-NH_4-N = [NH_3] + [NH_4^+] \tag{3}$$

$$T-Mg = [Mg^{2+}] + [MgOH^+] \tag{4}$$

Struvite solubility product

$$K_{sp} = \{Mg^{2+}\}\{NH_4^+\}\{PO_4^{3-}\} \tag{5}$$

The struvite solubility product is calculated according to Equation 5. The {} brackets indicate ionic concentration in moles per liter, corrected for activity.

This involves the speciation of analytically determined concentrations using published acid and base dissociation constants, as well as an adjustment for activity. The activity is a function of the concentration of the ion and its activity coefficient, γ. The activity is given by the Güntelberg approximation of the Debye-Hückel equation shown in Equation 6 (Sawyer *et al.*, 1994).

$$\log \gamma = \frac{0.5z^2 \sqrt{\mu}}{1 + \sqrt{\mu}} \tag{6}$$

where, γ = the activity coefficient for the species of interest
z = the ionic charge of the species of interest
μ = ionic strength
The ionic strength of the solution can be determined based on conductivity measurements using the conversion factor described in Equation 7.

$$\mu = 1.6 \times 10^{-5} EC \tag{7}$$

The controlled variable

This is the variable that must be maintained, or controlled, at some desired value. Sometimes the term process variable is also used to refer to the controlled variable. As applied to struvite crystallization, this value is the supersaturation ratio.

Set point (SP)

The set point is the desired value of the controlled variable. Thus, the function of a control system is to maintain the controlled variable at its set point. In this study, the set point was taken in the range of SSR 3–5.

Manipulated variable

This is the variable that is used to maintain the controlled variable at its set point. The SSR was controlled by pH, once a relationship between SSR and pH was developed.

Instrumentation and process monitoring

Automated control can be described as carrying out a balancing act (Pitt and Preece 1990). In order to keep a process in a desired state, it is necessary to counteract any disturbing influences and balance the energy and mass flow streams that constitute the process. The role of the automatic controller is to restore the balance of flows and turn the system to the desired point. Thus, automatic control can be used to increase efficiency and economy of process operations.

In order to efficiently apply a control model to a process, a fully functional and complete process monitoring system should be in place. In this study, the important parameters that determine the condition of the process will be monitored. These include the pH, conductivity, temperature, flows, ammonia and phosphate concentrations. Figure 1 illustrates the basic process components and the locations of sensors and other control parameters for monitoring the process. The actual pilot-scale setup is located at the Lulu Island Wastewater Treatment Plant (LIWWTP) in Richmond, B.C. Canada.

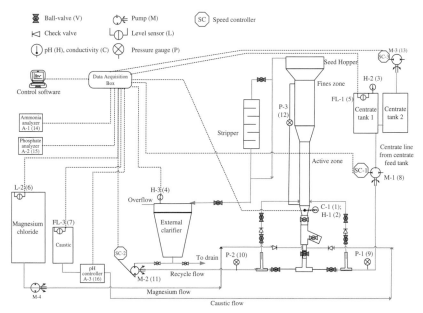

Figure 1. Schematic of phosphorus recovery process with instrumentation locations at the Lulu Island Wastewater Treatment Plant.

Process control

The dominant technique used in process control is the feedback control loop (Pitt and Preece 1990). A representation of a feedback control that can be used in the struvite crystallizer is given in Figure 2. After measuring the phosphate, ammonia, magnesium and pH in the crystallizer, the information is fed into the feedback control function which then determines the pH required in the crystallizer for a particular set SSR. This information is transmitted back to the pH controller where the set point is changed, as desired. In essence, the objective of an automatic process control system is to adjust the manipulated variable to maintain the controlled variable at its set point, in spite of disturbances.

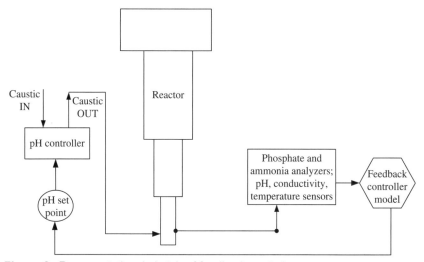

Figure 2. Representational sketch of feedback control.

Feedback controller model program development

The production of struvite in a crystallizer depends on a number of variables which bring about complex and non-linear changes in the process chemistry. This complex situation can be handled diligently by gaining an insight into the process using models. A model can be described as a simplified representation of those aspects of an actual process that are being investigated. A mathematical model of a chemical process is a mathematical description, which combines experimental facts and establishes relationships among the process variables. The primary objective of a mathematical model is to predict the behavior of a process and to work out ways to control its course (Babu 2004).

The struvite process control model is based on controlling the reactor supersaturation ratio by manipulating the pH in the reactor. This is achieved by calculating the effects of different chemical reactions that can take place in the reactor. These, well established chemical reactions, are given in Table 1. The effect of temperature on these reactions were also considered in deriving the equilibrium constants. The influence of these reactions, as well as pH, conductivity, temperature on the system, are taken into account. The model is based on two programs – one in which pH is the controlling parameter and the other, where reactor supersaturation ratio is the controlling parameter. The first program determines the supersaturation ratio for a particular pH. The second program determines the required operating pH for a particular, desired supersaturation ratio. These two programs are then used to control the crystallization process at the desired, optimum condition. The control model was developed using Matlab and Simulink software from MathWorks™. The programs were chosen because of their wide use in development of industrial process designs and the ability to process complicated chemical reactions, as well as provide tools for monitoring and control of the process. The program also has the ability to smoothly integrate with current industrial software and hardware.

Table 1. Reactions that have been used to develop the model.

$$H_3PO_4 \rightarrow H^+ + H_2PO_4^-$$
$$H_2PO_4^- \rightarrow H^+ + HPO_4^{2-}$$
$$HPO_4^{2-} \rightarrow H^+ + PO_4^{3-}$$
$$MgH_2PO_4^+ \rightarrow Mg^{2+} + H_2PO_4^-$$
$$MgHPO_4 \rightarrow Mg^{2+} + HPO_4^{2-}$$
$$MgPO_4^- \rightarrow Mg^{2+} + PO_4^{3-}$$
$$NH_4^+ \rightarrow H^+ + NH_3$$
$$H_2O \rightarrow H^+ + OH^-$$
$$Mg^{2+} + NH_4^+ + PO_4^{3-} + 6H_2O \rightarrow MgNH_4PO_4 \cdot 6H_2O$$

Figure 3 illustrates the basic schematic of the control model. The first level control block obtains the various process parameter signals. These signals are then used in the second level control block, where the two aforementioned program codes are contained. These control blocks are then used to send out signals to the pH controller.

Carbon dioxide stripping model

As illustrated in Figure 1, the process has an option for using a carbon dioxide stripper to strip off the gas, and thereby increasing the pH of the return flow. Previous studies (Fattah *et al.*, 2008b) at the treatment plant found that the cascade stripper was very effective in saving caustic usage, ranging from 35% to 86%, depending on the operating conditions. A carbon dioxide stripper model was also developed (Fattah *et al.*, 2008c), and is expected to be integrated into the process control model. This model takes into account various factors that influence the stripping of carbon dioxide, namely, the baffle number, the effluent

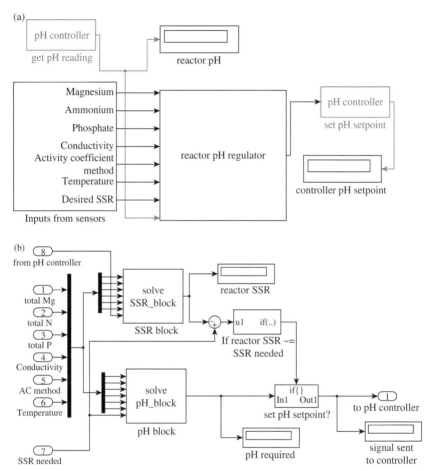

Figure 3. (a) First level of control block (b) Reactor pH regulator block (Second level control block).

recycle ratio, the influent flow rate, the air supply rate, the influent temperature, the influent dissolved CO_2 concentration and the influent buffering capacity. By using this model, it is possible to predict the change in pH within the stripper. Basic equation of the stripping model is given by Equation 8.

$$SE = SE_\Theta * A * B * C * D * E * F * G \qquad (8)$$

where, SE = predicted CO_2 stripping efficiency, %

SE_Θ = CO_2 stripping efficiency under reference conditions, 74%

A = coefficient for baffle number (BN)
B = coefficient for effluent recycle ratio (ERR)
C = coefficient for influent flow rate (IFR)
D = coefficient for air supply rate (ASR)
E = coefficient for influent temperature (IT)
F = coefficient for influent's dissolved CO_2 concentration ($[CO_2]$inf)
G = coefficient for influent's buffering capacity (IBC)

Detailed expressions for A~ G are given in Fattah *et al.* (2008c), based on a statistical evaluation of the results and cross correlations (Zhang 2006).

Graphical User Interface (GUI)

Operation and control of complicated programs and codes by non-technical users may pose challenges in smooth operation of the control model. Therefore, simple and user friendly graphical user interfaces were also developed using the software. Figures 4 and 5 illustrate the GUIs for model control.

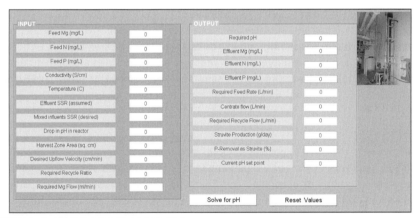

Figure 4. GUI for reactor operation.

Analytical methods

The measurement of parameters necessary for running the crystallizer will be measured in two locations – firstly, onsite at LIWWTP, with the help of online analyzers and probes, and secondly, from samples taken at LIWWTP, but analyzed in the Civil Engineering Department laboratory at University of British Columbia. Since no suitable online analyzer was found for determining the magnesium concentration, this chemical will be analyzed in the laboratory only.

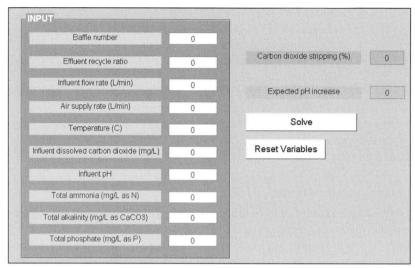

Figure 5. GUI for stripper model.

Given that magnesium samples are tested once a week (to reduce sample preparation and testing time and chemical wastage), further testing is being carried out to find a suitable, faster and less expensive method to determine the magnesium concentration in the samples. These experiments are based on preliminary experimental data obtained from Forrest *et al.* (2008).

Expectations from the model

The model is expected to provide smooth operation of the crystallization process, despite variations in the process parameters. Parameters such as pH, temperature and concentrations of phosphate and ammonia in the centrate are expected to vary from day to day, and occasionally within each day. By diminishing, or lessening, of the negative effects of these changes towards process operation, it is anticipated that the product quality and quantity will be improved.

CONCLUSIONS

This paper describes the process of developing a control model and strategy for efficient operation of a struvite crystallization process. At the time of writing this manuscript, no experimental validation was possible. However, it is expected that simulation and real operational validation of the model will be available at the time of the conference.

ACKNOWLEDGEMENT

The authors would like to thank the Natural Science and Engineering Research Council of Canada (NSERC), Metro Vancouver, Ostara Nutrient Recovery Technologies Inc. and Stantec Consulting Ltd., for their generous funding towards this study. Also, deeply appreciated is the helping hand provided by staff at the Lulu Island Wastewater Treatment Plant.

REFERENCES

Adnan, A., Mavinic, D.S. and Koch, F.A. (2003). Pilot-scale study of phosphorus recovery through struvite crystallization — examining the process feasibility. *J. Environ. Eng. Sci.*, **2**, 315–324.

Ali, M.I. and Schneider, P.A. (2008). An approach of estimating struvite growth kinetic incorporating thermodynamic and solution chemistry, kinetic and process description. *Chemical Engineering Science*, **63**(13), 3514–3525.

Babu, B.V. (2004). Process Plant Simulation. Oxford University Press, New Delhi, India.

Britton A., Koch F.A., Mavinic D.S., Adnan A., Oldham W.K. and Udala B. (2005). Pilot scale struvite recovery from anaerobic digester supernatant at an enhanced biological phosphorus removal wastewater treatment plant. *J. Environ. Eng. Sci.*, **4**(4), 265–277.

Fattah, K.P., Mavinic, D.S., Koch, F.A. and Jacob, C. (2008a). Determining the feasibility of phosphorus recovery as struvite from filter press centrate in a secondary wastewater treatment plant. *Journal of Environmental Science and Health: Part A.*, **43**(7), 756–764.

Fattah, K.P., Sabrina, N., Mavinic, D.S. and Koch, F.A. (2008b). Reducing operating costs for struvite formation with carbon dioxide stripper. *Wat. Sci. and Tech.*, **58**(4), 957–962.

Fattah, K.P., Zhang, Y., Mavinic, D.S. and Koch, F.A. (2008c). Application of carbon dioxide stripping for struvite crystallization – I: Development of a carbon dioxide stripper model to predict CO_2 removal and pH changes. *J. of Environ. Eng. and Sci.*, **7**(4), 345–356.

Forrest, A.L., Fattah, K.P., Mavinic, D.S. and Koch, F.A. (2009). Standardizing the struvite solubility product for field trial optimization. In: *Proc. of the International Conference on Nutrient Recovery from Wastewater Streams.* Vancouver, British Columbia, Canada. May 10–13, 2009.

Forrest, A.L., Mavinic, D.S. and Koch, F.A. (2008). The measurement of magnesium: A possible key to struvite production and process control. *Environmental Technology*, **29**(6), 603–612.

Münch, E. and Barr, K. (2001). Controlled struvite crystallization for removing phosphorus from anaerobic digester sidestreams. *Water Research*, **35**(1), 151–159.

Pitt, M.J. and Preece, P.E. (1990). Instrumentation and automation in process control. *Ellis Horwood Limited*, West Sussex, England.

Rahaman, M.S., Ellis, N. and Mavinic, D.S. (2008). Effects of various process parameters on struvite precipitation kinetics and subsequent determination of rate constants. *Wat. Sci. Tech.*, **57**(5), 647–654.

Sawyer, C., McCarty, P and Parkin, G. (1994). Chemistry for environmental engineering. McGraw-Hill Series in water Resources and Environmental Engineering, New York, U.S.A.

van Der Houwen, J.A.M. and Valsami-Jones, E. (2001). The application of calcium phosphate precipitation chemistry to phosphorus recovery: The influence of organic ligands. *Environmental Technology*, **22**(11), 1325–1335.

Zhang, Y. (2006). Struvite crystallization from digester supernatant – reducing caustic chemical addition by CO_2 stripping. *M.A.Sc. Thesis, Department of Civil Engineering, University of British Columbia*, Vancouver, BC, Canada.

Increasing cost efficiency of struvite precipitation by using alternative precipitants and P-remobilization from sewage sludge

T. Esemen, W. Rand, T. Dockhorn and N. Dichtl

Institute of Sanitary and Environmental Engineering, Technische Universität Braunschweig, Braunschweig, Germany

Abstract Precipitation of magnesium ammonium phosphate ($MgNH_4PO_4 \cdot 6H_2O$), better known as struvite, is a promising method to recover valuable nutrients from wastewater and to produce a fertilizer simultaneously, which enables an agricultural application of the precipitation product (Römer, 2006). For such a recovery by struvite precipitation, missing magnesium ions have to be compensated by adding magnesium compounds as precipitants, resulting in further consumption of natural resources, which decreases the sustainability as well as cost-efficiency of this process. Process costs are not only influenced by the need for external chemicals but also by nutrient concentration of the wastewater stream, since the application of highly concentrated streams enhances cost-effective recovery. The presented research work aimed to investigate different possibilities for decreasing the process costs of struvite precipitation. Experiments targeting a biological remobilization of phosphorus from waste activated sludge were conducted under anaerobic conditions. During the experiments remobilization degrees of up to 70% were achieved at mesophilic temperatures by adding easily degradable carbon sources, which served as substrate for iron(III) reducing bacteria. The microbial reduction of iron(III) enabled a re-dissolution of phosphate from iron(III)phosphate and a significant increase of soluble phosphate concentration. Various precipitation tests were performed in order to investigate the applicability of cost neutral precipitants containing high amounts of dissolved magnesium. During these batch experiments chemical effects on the struvite precipitation process caused by the application of seawater and salty industrial wastewater from potash production were investigated. Additionally, a pilot scale reactor was operated, to examine the applicability of seawater as precipitant. The theoretically required ratio for the struvite crystallization between P and Mg of 1:1 and a phosphorus recovery ratio of 98% were achieved during the tests with seawater at pH-values above 9. Similar stoichiometric ratios and phosphate recovery degrees were also achieved inside the pilot-scale upflow precipitation reactor. Using potash wastewater as precipitant, up to 96% of the initial phosphate content could be precipitated from digested sludge at pH 9.5 and at a stoichiometric ratio (Mg:P) of 1.89.

INTRODUCTION

Phosphorus and Nitrogen, both agriculturally essential nutrients, are commonly found in many wastewater streams and sewage sludge types. Especially the increasing scarcity of geological deposits of phosphorus will, without doubt, make large–scale recovery of phosphorus particularly from wastewater and sewage sludge absolutely necessary. The development of new methods for the simultaneous recovery of phosphorus and nitrogen in one single product therefore attracts more and more attention among scientists. The precipitation of phosphorus and nitrogen in the form of magnesium-ammonium-phosphate (MAP), better known as struvite, has proven to be technically suitable for the recovery of these nutrients. Several technologies for extracting nutrients from wastewater and sewage sludge by struvite precipitation have already been investigated and some of them have even been implemented on full-scale. But so far, the major challenge has been to make the phosphorus recovery processes economically competitive in comparison to conventional industrial manufac-turing of phosphates. This is due, amongst other things to the significant need in chemical supplies, which cause costs. The overall profitability of the struvite precipitation can significantly be improved by using cost neutral magnesium sources such as seawater or wastewater from the potash industry.

Another possibility to achieve a cost-effective recovery is the application of the precipitation process in wastewater streams with high phosphate concen-trations. Legally required phosphorus removal from wastewater is often achieved by iron(III) phosphate precipitation and/or by biological phosphorus elimination. During these processes, phosphorus is absorbed and removed together with waste-activated sludge. For the recovery of phosphorus from sewage sludge by struvite crystallization, phosphorus bound in the sludge has to be remobilized. The biological remobilization of phosphorus using iron(III) reducing bacteria can be a cost effective biological pretreatment for higher phosphate concentra-tions in the digested sludge dewatering liquor (centrate).

Considering resources contained in different partial wastewater streams, urine obviously dominates the resource potential but at the same time strongly influences specific treatment costs of a wastewater treatment plant. Although this partial stream makes up only 0.77% of the total wastewater quantity, the portion of its treatment costs amounts to 12.6% (Dockhorn, 2007). Thus, urine is the most suitable wastewater stream for separation with subsequent nutrient recovery. Therefore in the presented study the applicability of seawater as precipitant for struvite precipitation from urine was investigated.

MATERIALS AND METHODS

Biological P-remobilization from activated sludge

In order to investigate the possibilities for increasing phosphate concentration in digested sludge centrate, which consequently would increase the efficiency of the struvite precipitation process, the implementation of a previous extraction of phosphate from sewage sludge by microbial reduction was investigated. The main objective was to extract phosphate not only from bacteria cells but also from iron phosphate by biological reduction of iron(III) to iron(II). Anaerobic batch experiments were conducted at different (psychrophilic, mesophilic and thermophilic) temperatures and with addition of external carbon sources such as saccharose, ethanol and acetic acid as substrate for iron(III) reducing bacteria. The incubation period for all runs averaged 3.5 days. All tests were accomplished using waste activated sludge from municipal wastewater treatment plants operated with iron(III) phosphate precipitation as phosphorus removal step. During the batch experiments, phosphorus remobilization from bacteria cells as well as the redissolution of phosphorus and iron from iron(III)phosphate were observed in order to evaluate the additional extraction of phosphorus cauesd by the investigated biological remobilization method.

Seawater and wastewater from potash production as magnesium sources

Another focus of the presented study was on the applicability of cost-neutral precipitants containing high amounts of dissolved magnesium. Regarding struvite precipitation, experiments were conducted to investigate the chemical properties of seawater as well as salty industrial wastewater from potash production, as examples for cost neutral precipitants. By using the precipitants mentioned above, influences of different pH-values as well as stoichiometric ratios on the precipitation process were investigated in laboratory batch tests. Seawater containing 1250 mg Mg/l was added to digested sludge centrate in stirred 1 litre glass containers in order to enable struvite precipitation. The influent and effluent phosphate concentrations were determined for runs with pH values between 7.8 and 9.5, adjusted by adding sodium hydroxide NaOH.

During the beneficiation of potash ore waste streams, which predominantly consist of salt tailings and magnesium chloride ($MgCl_2$) brines, are produced. The salt tailings are generally discharged to a salt stack on the surface for disposal. Rainfall leads to significant quantities of additional brine being produced, which are disposed of together with magnesium chloride brines by either an injection into a suitable pervious dolomite formation or discharged to river systems. At potash beneficiation plants these wastewater streams containing high amounts of dissolved magnesium are usually collected in retention ponds before discharging, thus being available for reclamation in almost unlimited quantity. The applicability of this potash wastewater containing 28 g Mg/l (on the average) for struvite precipitation from digested sludge was examined in batch experiments. Potash wastewater was added to digested sludge in stirred 1 litre glass containers at different pH values and stoichiometric ratios. The composition of the wastewater from potash beneficiation, used for the precipitation experiments is shown in Table 1.

Table 1. Composition of the potash wastewater used for the precipitation experiments.

Water [ml/l]	KCl [g/l]	MgSO$_4$ [g/l]	MgCl$_2$ [g/l]	NaCl [g/l]
919	37	74	58	122

Similar precipitation experiments were performed, to investigate the recovery of phosphorus and nitrogen from urine. The influence of different reactants and pH-values was studied. The urine samples used for the struvite precipitation experiments were collected from a waterless urinal installed at the sanitary facilities of the Institute of Sanitary and Environmental Engineering. After collecting the urine, it was stored at room temperature for 48 h, in order to achieve a complete hydrolysis.

All batch experiments explained above were also performed by replacing seawater and/or salty wastewater with MgO, in order to compare the effects of these cost-neutral resources with commonly used reactants.

Pilot-scale precipitation reactor (seawater as precipitant)

To verify the applicability of the developed precipitation method, a pilot scale reactor was employed. The upflow precipitation reactor was operated with seawater as Mg-source to recover phosphate and nitrogen from the centrate of digested sludge. The influences of pH and the stoichometrical ratio between phosphorus, nitrogen and magnesium were examined. The performance of the system has also been explored under conditions of different supersaturation ratios. Seawater, used for the operation of the pilot scale reactor was produced synthetically at the laboratory according to the formula shown in Table 2. It contained 1250 mg Mg/l, which corresponds to the usual magnesium content of natural seawater. The influent to the MAP reactor was centrate from the centrifuge that dewaters anaerobically digested sludge at the Steinhof waste-water treatment plant in Braunschweig/Germany.

The reactor, as seen in Figure 1 was a (100 litre) cylinder with diameters of 24 cm in the lower and 56 cm in the upper section. Struvite crystals were held back in the lower part of the reactor by using a filter system, composed of styrofoam pellets, which also separated the two sections from each other. In order to assure an effluent free of particles, the upper section of the reactor

Table 2. Composition of the seawater used for the precipitation experiments.

Water	NaCl	MgSO$_4$ 7H$_2$O	MgCl$_2$ 6H$_2$O	CaCl$_2$ 6H$_2$O	NaHCO$_3$
985 ml	28 g	7 g	5 g	2.4 g	0.2 g

was enlarged (from 24 cm to a diameter 56 cm), thus decreasing upflow velocity and allowing the struvite crystals to sink in case of filter overstraining. The struvite precipitation product was withdrawn at the bottom of the cylinder regularly after reaching the desired amount. In order to control the precipitation product quality, struvite samples were collected and analyzed by X-ray Diffraction (XRD) and compared with synthetically produced pure struvite samples. The influent and effluent of the reactor were sampled regularly and phosphate as well as nitrogen concentrations were analyzed. The pH value was 9.5 on the average. It was adjusted by adding sodium hydroxide NaOH.

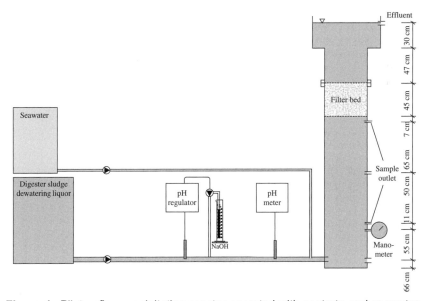

Figure 1. Pilot upflow precipitation reactor operated with centrate and seawater.

RESULTS AND DISCUSSION

Remobilization of phosphorus

During the experiments for biological extraction of phosphorus from waste activated sludge, remobilization degrees up to 70% were achieved under anaerobic conditions as shown in Figure 2. The remobilization of iron could also be explicitly detected, confirming that the achieved extraction of phosphate was not only a result of an extraction from biomass but also from re-dissolution of phasphate from iron(III)phosphate. Saccharose was identified as the optimal carbon source as substrate for the iron(III) reducing bacteria. It was also observed that the biological remobilization strongly depended on the process temperature and 37–38°C could be defined as the optimum process temperature range. A further temperature increase did not result in higher remobilization degrees. Although the addition of saccharose enhanced the extraction of phosphate significantly, remobilization degrees up to 48% could also be achieved without adding external carbon sources. This was still twice the remobilization degrees, which is usually achieved during conventional anaerobic sludge stabilization. In case of a full-scale implementation of the investigated method, process costs have to be analyzed individually for each alternative in order to determine the most efficient option by weighing the costs against desired phosphorus recovery degrees. The conducted experiments served as basic research in this field of biological P-remobilization by using iron-reducing bacteria and currently further experiments are conducted to optimize the investigated process. Focusing on process improvement, experiments for an extraction without an external addition of carbon sources are performed. Furthermore the addition of primary sludge as carbon source is also investigated.

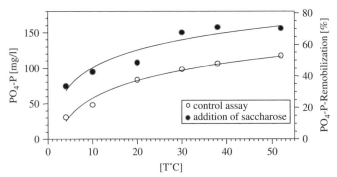

Figure 2. P-extraction from waste activated sludge by biological remobilization against process temperature with and without an addition of saccharose as external carbon source.

Seawater as precipitant

The theoretically required ratio for the struvite cristallization between P and Mg of 1:1 was achieved during the tests with seawater at pH-values above 9. It was not possible to achieve the same ratio in the tests with MgO. In both series phosphorus recovery degrees of 98% were reached. The results showed that the application of seawater leads to lower stoichometric relations between the reactants, compared to the precipitation by using MgO. Also at lower pH-values, the recovery degrees for both phosphorus and nitrogen were higher than those achieved in the tests with MgO. The amount of recovered ammonia was considerably higher than a 1:1 stoichiometry in all tests with pH values over 9. This is presumed to be a result of ammonia stripping. Figure 3 shows phosphorus recovery results achieved during the tests with MgO and seawater as precipitants at different pH values.

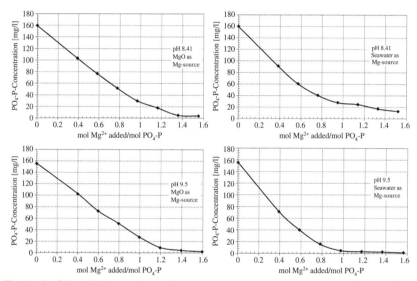

Figure 3. Phosphorus recovery degrees achieved during batch experiments from centrate at different pH values and stoichiometric relations.

During the precipitation experiments for the recovery of phosphorus and nitrogen from urine using seawater as precipitant, a recovery degree of 98% was achieved at a stoichiometric ratio of 1:1 (Mg:P) as shown in Figure 4. During the hydrolysis process (48h at room temperature), the pH of the collected urine rose from 7.32 to 9.22. The achieved pH value was absolutely sufficient for the struvite precipitation with seawater as precipitant. Subsequent experiments showed that a further increase of the pH value did not improve the precipitation performance significantly. The amount of recovered ammonia was also higher than a 1:1 (up to 1:3; Mg:N) stoichiometric ratio as observed during the precipitation experiments from digested sludge centrate at pH values over 9.

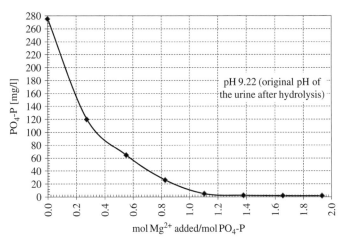

Figure 4. Phosphate concentrations achieved during precipitation experiments from urine at different stoichiometric ratios.

Wastewater from potash production as precipitant

During precipitation tests from digested sludge, up to 96% of the initial phosphate content could be precipitated at pH 9.5 and at a stoichiometric ratio (Mg:P) of 1.89. Experiments at lower pH values showed that struvite precipitation could also be conducted without changing the original pH of the digested sludge. Precipitation rates up to 80% were achieved at pH 7.9 by adjusting a stoichiometric ratio between Mg and P of 1.5. The Figures 5 and 6 show the results achieved during the precipitation experiments at different pH-Values using wastewater from potash production as an alternative magnesium source.

Upflow precipitation reactor

The seawater addition was adjusted during the operation of the upflow precipitation reactor at a 1:1 stoichiometric ratio. Thus addition of seawater at around 10% was calculated to achieve the desired P-recovery degrees > 90%. The influent PO_4-P concentration of the digested sludge centrate was on the average 161 mg/l.

P-recovery degrees up to 97%, reached during the batch experiments, were also achieved inside the reactor, which led to an effluent PO_4-P concentration of 5 mg/l on the average. This phosphorus recovery performance could be achieved reliably at pH 9.5 and at a hydraulic retention time inside the reactor of 0.2 h. 1.6 m^3/h were identified as the maximum achievable flow rate for a proper operation. At higher flow rates struvite particles were observed in the

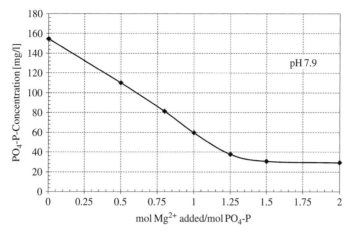

Figure 5. P-concentration plotted against Mg:P-Ratio at pH 7.9 during precipitation experiments using potash wastewater.

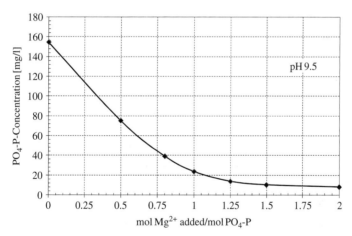

Figure 6. P-concentration plotted against Mg:P-Ratio at pH 9.5 during precipitation experiments using potash wastewater.

effluent. No influence of the hydraulic retention time, which was varied between 0.1 to 4 h, on the P-recovery performance could be detected. It was observed that the performance of the filter increased in conjunction with the amount of struvite precipitated inside the reactor. However, the influence of the amount of precipitation product on the effluent phosphate concentration was not significant, but the reactor could be operated at higher flow rates due to an increase of the struvite load. The reactor performance observed during an operational period of 30 days is shown in Figure 7.

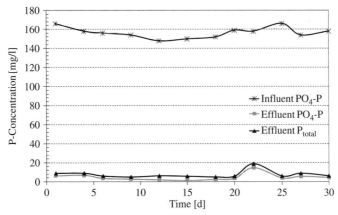

Figure 7. Reactor performance at pH 9.5 regarding influent and effluent phosphorus concentrations.

In order to examine the performance of the system at the original pH value of the digested sludge centrate, the pilot scale MAP reactor was also operated without pH-adjustment. During this period an effluent phosphate concentration of 20 mg/l was achieved at pH 8.3 (on the average), if the stoichiometric ratio between magnesium and phosphorus was increased from 1:1 to 1:2.

The XRD-Analysis of the samples showed that all struvite samples collected during the operation period had almost identical characteristics. The precipitation product consisted of fine crystals containing 6% N, 12% P, 10% Mg and 43% water on the average, close to theoretical values for struvite ($MgNH_4PO_4 \cdot 6H_2O$). The samples collected from the reactor showed no significant difference in comparison to synthetically produced pure struvite samples.

CONCLUSIONS

During the experiments for biological extraction of phosphorus from waste activated sludge, remobilization degrees up to 70% were achieved under anaerobic conditions by microbial reduction of iron(III). The precipitation experiments showed P-recovery degrees of up to 98% at pH 9.5 using seawater. Precipitation of struvite using salty industrial wastewater was also studied and phosphate precipitation degrees up to 96% were achieved at a stoichiometric ratio of 1.89 (Mg:P) and pH 9.5.

The pilot-scale MAP reactor operated with seawater as precipitant achieved a P-recovery ratio of 98%. An effluent phosphate concentration of 5 mg/l was reached on the average at a pH of around 9.5, from an average influent phosphate concentration of 161 mg/l. During the operation without pH-adjustment an average effluent phosphate concentration of 20 mg/l was achieved at pH 8.7, if the stoichiometric ratio of Mg:P was increased to 1:2. The batch experiments as well as the operation of the pilot scale MAP reactor showed that the investigated alternative magnesium sources are, without doubt, suitable for struvite precipitation and can replace commonly used reactants.

The experiments also showed that seawater can be used for struvite precipitation from highly concentrated partial streams such as urine. Besides its high resource potential, urine causes high treatment costs at conventional wastewater treatment plants despite its comparatively low quantity. Therefore the implementation of the investigated struvite precipitation from urine using seawater not only saves costs but represents a suitable approach regarding a cost effective resource recovery from wastewater in case of partial stream separation.

Of all methods already investigated for the recovery of nutrients from wastewater streams, those that require the lowest costs will establish themselves. The overall profitability of struvite precipitation processes at wastewater treatment plants can significantly be improved by using cost-neutral chemical supplies such as seawater or wastewater from the potash industry as sources of magnesium ions. The pretreatment of waste activated sludge by the investigated biological P-remobilization method could further improve cost efficiency.

REFERENCES

Dockhorn, T. (2007). Stoffstrommanagement und Ressourcenökonomie in der kommunalen Abwasserwirtschaft. Habilitation treatise. Veröffentlichungen des Instituts für Siedlungswasserwirtschaft, TU Braunschweig 74, ISSN 0934-9731.

Römer, W. (2006). Vergleichende Untersuchungen zur Pflanzenverfügbarkeit von Phosphat aus verschiedenen P-Recycling-Produkten im Keimpflanzenversuch. *J. Plant Nutr. Soil Sci.*, **169**, 826–832.

Temperature dependence of electrical conductivity and its relationship with ionic strength for struvite precipitation system

M. Iqbal, H. Bhuiyan[a]* and Donald S. Mavinic[b]

[a]Sperling Hansen Associates, #8-1225 East Keith Road, North Vancouver, BC, Canada, V7J 1J3
[b]Environmental Engineering Group, Department of Civil Engineering, 6250 Applied Science Lane, University of British Columbia, Vancouver, BC, Canada, V6T 1Z4

*Corresponding author. Tel.:1-604-225-0143; fax:1-604-822-6901;
e-mail address: iqbal@civil.ubc.ca

Abstract Depletion of naturally occurring phosphorus resources is gradually making the recovery and recycling of phosphorus from potentially rich waste streams an increasingly attractive option. A phosphorus compound of particular interest is magnesium ammonium phosphate hexahydrate, also called struvite, which under certain process conditions has been found to precipitate readily. Spontaneous struvite formation in wastewater treatment works, where high levels of soluble phosphorus and ammoniacal nitrogen exist, for example in flows associated with post anaerobic digestion stages of treatment, is well documented. Once formed, this struvite will encrust the walls of pipes and pipe fittings in the plant and eventually lead to costly repairs for the plant and downtime for operations. A novel solution to this problem is to recover phosphate as struvite before it forms and accumulates on wastewater treatment equipments. This provides an environmentally sound and renewable nutrient source to the agriculture industries. In addition, the ability of phosphates to have significant recovery potential gives full meaning to what should be a core principle of true sustainable development.

The supernatant from an anaerobic digester may contain numerous ions, in addition to those that constitute struvite. Electrical Conductivity (EC) measurements can predict struvite saturation in wastewater solutions. The measured EC value at any temperature needs to be corrected for a standard temperature since EC is dependent on temperature. A temperature compensation factor has been derived for conductivity correction from anaerobic digester supernatant/centrate samples of five different wastewater treatment plants in western Canada. For a temperature compensation factor of $0.0198°C^{-1}$, corresponding to the standard temperature at 25°C, the estimated electrical conductivity values were found to fairly accurately match the measured values. Considering the temperature dependence of EC, a relationship between EC and ionic strength (I) was developed as $I = 7.22 \times 10^{-6} EC_{25}$ in this study for anaerobic digester supernatant/centrate samples.

INTRODUCTION

Phosphate recovery through struvite formation before it spontaneously precipitate in the wastewater treatment equipments provides a novel solution to the wastewater treatment problems and ensures sustainable development through generation of an environmentally sound and renewable nutrient source.

EC measurements can be used to assess total ion concentration in a solution, as well as the solution ionic strength (I), an important thermodynamic property, which can be defined as $I = 1/2 \sum m_i \cdot z_i^2$ where, m_i is the molar concentration of ion i and z_i is the charge of ion i. The mobility of ions, and therefore the conductivity of the solution, also changes as a function of temperature. The relationship between EC and ionic strength and the effect of temperature on this relationship for waste water solutions that are typical of conditions under which struvite precipitation occurs was examined in this paper.

The relationship between EC and I is complex, depending on the chemical composition and ionic strength. However, there are many instances where the relative composition of the water is reasonably constant for a particular use and, therefore, the electrical conductivity-ionic strength (EC-I) relationship can be established with a reasonable degree of certainty over a wide concentration range (Ponnamperuma et al., 1966; Griffin and Jurinak, 1973; Russel, 1976). Due to the difficulty in the determination of ionic strength of complex wastewater solutions, an approximation of ionic strength from an EC-I relationship would be quicker and inexpensive.

EC is strongly dependent on temperature. Since EC measurements cannot always be at a standard temperature, it is imperative that the EC of water samples measured at various temperatures be corrected to values corresponding to a standard temperature. As the degree of nonlinearity in the electrical conductivity-temperature (EC-T) relation is relatively small for a practical temperature range (0–45°C), a linear equation (Equation 1) is commonly used to represent the relation (Sorensen and Glass, 1987; Hayashi, 2004):

$$EC_T = EC_{25}[1 + a(T - 25)] \tag{1}$$

where, EC_T is electrical conductivity at temperatures $T(°C)$, EC_{25} is electrical conductivity at a standard temperature of 25°C, and $a(°C^{-1})$ is a temperature compensation factor.

None of the relationships between ionic strength and conductivity found in the literature have been derived for wastewater or digester supernatant/centrate (Ponnamperuma et al., 1966; Griffin and Jurinak, 1973; Russell, 1976) which is the potential source of struvite formation.

This study aimed to develop a relationship between electrical conductivity and ionic strength with the determination of a representative temperature compensation factor for conductivity correction in a system associated with struvite formation.

MATERIALS AND METHODS

Temperature dependence of EC

Digester supernatant/centrate samples were collected from the Annacis Island Wastewater Treatment plant (secondary), Lion's Gate Wastewater Treatment Plant (primary), City of Penticton Wastewater Treatment Plant (BNR), and Lulu Island Wastewater Treatment Plant (secondary) in British Columbia, Canada, and the Edmonton Goldbar Wastewater Treatment Plant (secondary) in Alberta, Canada. The samples were transported in a cooler and stored at 40C in the laboratory, until the analyses were made. EC and temperature were recorded at approximately every 20C, while the samples slowly warmed to room temperature, using a Radiometer conductivity meter (CDM3) equipped with a Radiometer CDC 304 conductivity probe and a mercury thermometer.

EC-I relationship

Samples collected from the same treatment plants were used to develop an EC-I relationship. To minimize the effects of ion-ion interactions and to appropriately account for the contribution of each separate ion to the measured EC, it is recommended that the solutions be diluted (APHA et al., 1998). After the required dilution with deionized water, conductivity and temperature were both recorded, using the same conductivity meter and thermometer as mentioned in the above section. pH was measured using an Orion 420A bench top pH meter, equipped with a VWR Symphony temperature compensated probe, which was calibrated using pH 7 and 10 buffers.

Analyses

Analyses for ortho-phosphate ammonia, nitrate and chloride were made, using the flow injection method on a LaChat QuickChem 8000 instrument, as described in the method number 4500-P G, 4500-NH_3 H, 4500-NO_3- I, and 4500-Cl^- of the Standard Methods for the Examination of Water and Wastewater respectively (APHA et al., 1998). Calcium, magnesium, potassium, sodium, aluminum, and iron analyses determined by flame atomic absorption spectrophotometry, using a Varian Inc. SpectrAA220 Fast Sequential Atomic

Absorption Spectrophotometer. Sulphate was quantified by the turbidimetric method. TC and IC of the filtered (0.45 µm) samples were analyzed by Shimadzu Total Carbon Analyzer TOC-500. Filtered TOC or dissolved organic carbon (DOC) was calculated from their difference. Knowing the pH of the solutions, concentrations of HCO_3^- and CO_3^{2-} were calculated from IC ($IC = H_2CO_3^* + HCO_3^- + CO_3^{2-}$), as described in the method number 4500 CO_2. D of the Standard Methods for the Examination of Water and Wastewater (APHA *et al.*, 1998).Volatile fatty acids (acetic, propionic and butyric) were analyzed using a gas chromatograph (Hewlett Packard 5890 A), equipped with a FID detector and an HPFFAP column.

As the correction of the analytical concentrations used to compute ionic strength is necessary to provide an accurate measure of the *EC-I* relation (Griffin and Jurinak, 1973), PHREEQC version 2.12 (Parkhurst and Appello, 1999) was used for speciation and ionic strength calculation.

RESULTS AND DISCUSSION

Temperature dependence of EC

Figure 1 shows the *EC-T* relations of the anaerobic digester supernatant/centrate samples of Annacis Island WWTP, Goldbar WWTP, Lion's Gate WWTP, City of Penticton WWTP, and Lulu Island WWTP. Within the temperature range tested, the relationships, in all five cases, were well represented by a linear equation ($R^2 > 0.99$).

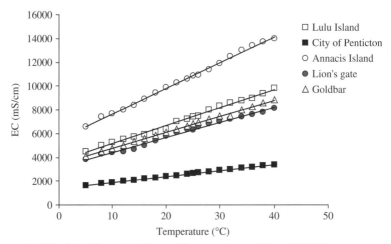

Figure 1. Electrical Conductivity vs. Temperature of Five WWTPs.

The solid lines in Figure 1 indicate Equation 1 with a value of a ranging from 0.0197–$0.0205°C^{-1}$ with an average of $0.0198°C^{-1}$ (Table 1), determined by the least-squares method. The root-mean squared (RMS) percentage error of the equation is defined by (Hayashi, 2004)

$$e = \frac{100}{EC_{25}} \sqrt{\frac{1}{m} \sum_{i=1}^{m} (EC_{meas} - EC_{eqn})^2} \qquad (2)$$

where e is the RMS percentage error, m is the number of data points, EC_{meas} is measured EC, and EC_{eqn} is the predicted EC. The value of e of Equation 1 was found to range from 0.38–2.01% (Table 1). The estimated EC_{25} values were found to be reasonably accurate, with a maximum error of 2.01% (Table 1).

Table 1. Temperature compensation factor (a) and RMS percentage (e).

	Annacis Island	Goldbar	Lion's Gate	City of Penticton	Lulu Island
A ($°C^{-1}$)	0.0198	0.0197	0.0199	0.0191	0.0205
E	1.02	0.99	2.01	0.38	1.71
$E_{0.0198}$	1.02	0.99	2.01	0.87	1.82

Equation 1 is useful for calculating EC_{25}, but it may be necessary to calculate EC at a standard temperature different from 25°C. As modified from Equation 1,

$$EC_{T_0} = \frac{EC_T}{[1 + c(T - T_0)]} \qquad (3)$$

where EC_{T_0} is the electrical conductivity at a standard temperature T_0, and c is a constant defined by

$$c = \frac{a}{[1 + a(T_0 - 25)]} \qquad (4)$$

Equation 4 indicates that c varies with T_0. For example, $c = a = 0.0198$ at 25°C, and $c = 0.0229$ when $a = 0.0198$ and $T_0 = 18°C$. Therefore, the temperature compensation factor is dependent on the standard temperature used, even though electrical conductivity and temperature maintain a linear relationship, with a certain a value for a specific standard temperature. However, the EC value can be reasonably estimated based on any standard temperature. Figure 2 shows estimated EC values at 10°C for $T_0 = 25°C$ ($EC_{10est@25}$) and for

$T_0 = 18°C$ ($EC_{10est@18}$), and they were found to be statistically similar to each other ($p = 0.51$) by the paired-t test.

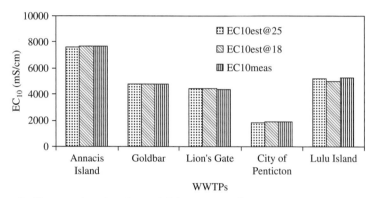

Figure 2. Measured and estimated EC_{10} using different standard temperatures.

EC-I relationship

Table 2 shows compositions of 20 samples, of different dilutions, from the 5 WWTPs with 4 from each. PHREEQC version 2.12 with the Lawrence Livermore National Laboratory Database (llnl.dat), was used in this study for speciation and other thermodynamic equilibria. The charge balance of the selected samples was found be below 10%, using PHREEQC.

Since the DOC level of the samples varied from 103 to 1458 mg l^{-1}, low molecular fatty acids were suspected to make a contribution to the charge balance. However, volatile fatty acid analysis indicated that the contribution from acetic, propionic and buteric acids was insignificant.

Table 3 shows pH, electrical conductivity and corresponding temperature readings of the samples. All EC_T values have been converted to EC_{25} (Table 3), using a temperature compensation factor of 0.0198 in Equation 1. The ionic strength of the samples was calculated, using PHREEQC where most of the ion-pairs were included in the calculation. Table 3 also shows the calculated I values.

From a regression analysis using least squares method, the following linear EC-I relation was found:

$$I = 7.22 \times 10^{-6} EC_{25} \tag{5}$$

where the ionic strength, I, is in mol l^{-1} and EC_{25} is in μS cm^{-1} at 25°C.

Table 2. Chemical composition of the anaerobic digester supernatant/centrate samples. All values in mg l^{-1}.

Sample	PO_4-P	NH_4-N	Cl^-	SO_4^{2-}	NO_3-N	Mg^{2+}	Na^+	K^+	Al^{3+}	Fe^{2+}	Ca^{2+}	HCO_3^-	CO_3^{2-}	DOC
Annacis I.	108	1290	38	10	0.7	1.0	24	150	1.0	1.4	11.5	4606	16.4	570
	55	578	17	5	0.5	0.8	12	75	0.5	0.7	6.0	2534	6.6	900
	30	266	11	3	0.3	1.1	6	37	0.3	0.2	3.9	1136	2.7	541
	23	289	8	2	0.0	0.8	5	26	0.2	0.2	3.0	1269	3.4	299
Goldbar	118	744	101	12	0.2	2.3	93	259	0.5	0.7	11.6	3408	8.0	589
	58	285	46	6	0.2	1.5	46	129	0.3	0.3	6.2	1263	3.1	385
	28	148	21	3	0.1	0.9	23	65	0.1	0.2	3.5	627	1.3	214
	23	118	20	2	0.1	0.9	19	52	0.1	0.1	3.2	500	1.0	154
L/Gate	6	286	78	11	0.5	6.1	45	128	1.0	1.2	20.6	1675	34.4	910
	3	156	33	6	0.3	2.5	22	64	0.5	1.0	13.4	903	17.3	937
	3	117	32	3	0.2	11.5	11	33	0.3	0.7	6.8	646	10.3	257
	7	181	36	2	0.1	1.8	9	27	0.2	0.6	5.4	775	12.4	274
C/Penticton	3	175	30	7	0.3	12.5	14	34	4.7	0.5	28.6	1035	7.5	653
	3	110	15	4	0.3	5.4	7	16	2.4	0.2	14.5	517	3.5	240
	2	56	8	2	0.3	3.2	4	9	1.2	0.2	7.7	257	1.1	163
	2	43	7	1	0.1	2.7	3	7	1.0	0.2	6.3	193	1.0	103
Lulu I.	67	699	48	7	0.3	3.5	25	123	2.4	0.4	19.1	3190	10.4	1458
	37	422	25	3	0.4	2.2	12	62	1.2	0.3	9.7	1914	6.2	600
	18	194	12	2	0.1	1.3	6	31	0.6	0.1	5.0	890	2.5	291
	14	181	10	1	0.3	1.2	5	24	0.5	0.1	3.9	761	2.1	274

Table 3. pH, Temperature, Electrical conductivity and calculated ionic strength.

Sample source	pH	EC_T (mS cm^{-1})	T (°C)	EC_{25} (µS cm^{-1})	I (mol l^{-1})
Annacis Island	7.89	10790	21.1	11693	0.094
	7.75	5790	21.4	6234	0.046
	7.71	3230	21.5	3471	0.021
	7.77	2770	21.4	2983	0.023
Goldbar	7.71	6810	18.4	7834	0.069
	7.72	3610	20.7	3946	0.027
	7.65	1710	21.2	1849	0.014
	7.62	1430	21.2	1546	0.011
Lion's Gate	8.65	6800	19.3	7665	0.030
	8.62	3670	20.6	4020	0.016
	8.54	1700	21.0	1846	0.012
	8.54	1430	21.1	1550	0.014
City of Penticton	8.2	2520	20.3	2779	0.018
	8.17	1260	20.7	1377	0.010
	7.98	690	20.9	751	0.005
	8.04	590	20.9	642	0.004
Lulu Island	7.85	8120	20.1	8992	0.059
	7.85	4380	20.8	4777	0.035
	7.79	2540	21.1	2753	0.016
	7.77	1860	21.0	2020	0.014

Relationships between EC_{25} and I for different types of water samples, have been developed by others. A relationship of I (mol l^{-1}) $= 16 \times 10^{-6} EC_{25}$ (µS cm^{-1}) was developed for 13 waters of varying composition (Russell, 1976), while for extracts of flood soils and electrolyte solutions of I less than 0.06 mol l^{-1}, the same relationship of I (mol l^{-1}) $= 16 \times 10^{-6} EC_{25}$ (µS cm^{-1}) (Punnamperuma et al., 1966), and for arid-zone soil extracts and river waters, a relationship of I (mol l^{-1}) $= 13 \times 10^{-6} EC_{25}$ (µS cm^{-1}) (Griffin and Jurinak, 1973) were developed. In all three cases, the relationship was developed for a temperature of 25°C.

The proposed relationship between the conductivity data obtained during the struvite solubility tests in different kinds of solutions, including supernatant/centrate, and the ionic strength manually calculated using major equilibia equations (Rahaman et al., 2006), was found to be I (mol l^{-1}) $= 5 \times 10^{-6} EC_{25}$ (µS cm^{-1}). Temperature dependence of conductivity was not considered in developing that relationship. The relationship developed in this study (Equation 5) was developed exclusively from anaerobic digester supernatant/centrate, using the computer model PHREEQC. It involved most of the equilibrium equations and appropriately considered the temperature dependence of EC.

Figure 3 shows the observed EC_{25}-I relationship in this study, as described in Equation 5. It can be seen from Figure 3 that the EC_{25} values correlate better with I in the lower range of EC_{25}. The higher variability of I, at a higher range of EC_{25} values, is due to the higher ion-ion interaction (Hall and Northcote, 1986). However, to minimize this effect, diluted samples of varying dilution factors were used to develop the relationship. Also, correction for ion-pairing was done by using the computer model PHREEQC.

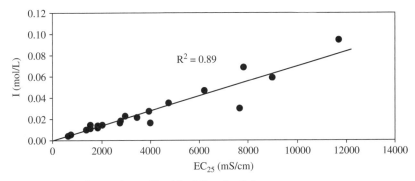

Figure 3. Ionic Strength vs. Electrical Conductivity.

CONCLUSIONS

A representative temperature compensation factor for conductivity has been derived in a system associated with struvite formation from anaerobic digester supernatant/centrate samples of five different wastewater treatment plants. Using a temperature compensation factor, $a = 0.0198°C^{-1}$ for all samples, the estimated electrical conductivity values were found to fairly accurately match the measured values ($R^2 \sim 1$). Although the value of the temperature compensation factor varies with the standard temperature used, the estimated EC values, based on any standard temperature, were found to be statistically similar to each other by paired t-test.

Considering the temperature dependence of EC, a relationship, including correction for ion-pairing, was developed from anaerobic digester supernatant/centrate. This relationship can be used to estimate the ionic strength of the solution in a system associated with struvite formation, from anaerobic digester supernatant/centrate. An *in situ* estimation of ionic strength, from electrical conductivity, would help predict struvite precipitation potential faster and effectively monitor the system performance.

REFERENCES

American Public Health Association (APHA), American Water Works Association, and Water Pollution Control Federation (1998). *Standard Methods for the Examination of Water and Wastewater*, 20th ed., Washington, DC.

Griffin, R.A. and Jurinak, J.J. (1973). Estimation of activity coefficients from the electrical conductivity of natural aquatic systems and soil extracts. *Soil Science*, **115**(1), 26–30.

Hall, K.J. and Northcote, T.G. (1986). Conductivity-temperature standardization and dissolved solids estimation in a meromictic lake. *Canadian Journal of Fish and Aquatic Science*, **43**, 2450–2454.

Hayashi, M. (2004). Temperature-electrical conductivity relation of water for environmental monitoring and geophysical data inversion. *Environmental Monitoring and Assessment*, **96**(1–3), 119–128.

Lawrence Livermore National Laboratory database (llnl.dat) derived from thermo.com. V8.r6t.dat

Parkhurst, D.L. and Appelo, C.A.J. (1999). *User's guide to PHREEQC (Version 2) – a computer program for speciation, reaction- path, advective-transport, and inverse geochemical calculations*. USGS water-Resources Investigation Report 99–4259.

Ponnamperuma, F.N., Tianco, E.M. and Loy, T.A. (1966). Ionic strengths of the solutions of flooded soils and other natural aqueous solutions from specific conductance. *Soil Science*, **102**, 408–413.

Rahman, M.S., Mavinic, D.S., Bhuiyan, M.I.H. and Koch, F.A. (2006). Exploring the determination of struvite solubility product from analytical results. *Environmental Technology*, **27**, 951–961 (2006).

Russell, L.L. (1976). *Chemical Aspects of Groundwater Recharge with Wastewaters*, PhD Thesis, University of California, Barkley, USA.

Sorensen, J.A. and Glass, J.E. (1987). Ion and temperature dependence of electrical conductance for natural waters, *Analytical Chemistry*, **59**, 1594–1597.

Study on phosphorus recovery by calcium phosphate precipitation from wastewater treatment plants

Wang Hui-Zhen, Zhang Ya-jun, Feng Cui-Min, Xu Ping and Wang Shao-Gui

Department of Municipal Engineering, Beijing University of Civil Engineering and Architecture, Beijing 100044, China

Abstract Phosphorus (P) is a non-renewable resource and is also the key element of water Eutrophication. The research on P recovery by calcium phosphate precipitation from supernatant at the end of the anaerobic region was carried out at a Beijing wastewater treatment plant. The impact of pH, temperature and hydraulic retention time on P recovery was investigated.

The research proved that pH has a major impact on calcium phosphate precipitation (HAP). Addition alkalinity to adjust the pH of the wastewater to 9.3 or 9.5 is able to recover 80% of P within 5 minutes of reaction time. $Ca(OH)_2$ addition will not only adjust pH but also increase Ca ion in raw wastewater. But it requires large amounts of $Ca(OH)_2$. Adjusting pH by NaOH will significantly reduce the addition amount of the reagent but needs higher chemical costs.

Aeration will not only increase the pH but also mix the reactor as well. However it can only increase pH to a certain level and requires relatively longer reaction times. Aeration is suitable for raw wastewater with high SP concentration. If the SP is low, a combination of aeration and alkalinity addition is required to adjust the pH to optimize P recovery.

The raw wastewater temperature will not impact the P recovery rate.

The recovery product is suitable for acidy phosphorus fertilizer manufacture process and has great potential around Beijing area.

INTRODUCTION

Phosphorus is the key element of Eutrophication. If the concentration of total phosphorus reaches to 0.02 mg/L in waters, the ecological balance will be broken. The survey results of 23 lakes in China in 1996 indicated that total phosphorus of 92% lakes was more than 0.02 mg/L, and of 50% lakes up to 0.2~1.0mg/L (Yang Li-Hua and Zhuo Fen, 1996). So National standards of China "Wastewater discharge standard" (GB 8978–1996) and "The discharge standard for urban wastewater treatment plants" (GB18918–2002) both limit the effluent phosphorus of wastewater treatment plants. Biological nutrient removal process should be applied in new wastewater treatment plants and existing wastewater treatment plants should be upgraded to meet the national discharge standard.

On the other hand phosphorus is a non-renewable resource. Phosphorus mine reserve in China occupies the third place in the world. But rich ore in which the amount of P_2O_5 is more than 30% is only 8.4% in the total P reserve and will be used up in 10~15 years. So National Resource Board of China has listed phosphorus mine resource as one of twenty mines which would not meet the economic development requirement in 2010 (Wu Chu-Guo, 2002; Wu Xi-Yan, 2001).

There are several sub-streams containing concentrated P in a BPR waste-water treatment plant such as the supernatant of anaerobic region, sludge thickening, sludge digester and sludge dewatering streams. Normally those streams will be recycled to the inlet screen and increased P loading of the plant by 10–15%. Recovering P from those streams will not only sustainably utilize P, but also improve the plant operation (Bernd Heinzmann and Berliner Wasser Betriebe, 2001).

The P can be recovered by struvite precipitation (MAP) or Calcium phosphate. Considering the production process of P fertilizer plants near Beijing, a pilot trial on calcium phosphate recovery was conducted (Wang Hui-Zhen and Wang Shao-Gui., 2004).

EXPERIMENT METHODS

Principle

Calcium phosphate precipitation can be encouraged by increasing the calcium concentration and promoting pH. Multiple forms of calcium phosphate exist such as $CaHPO_4$, $CaHPO_4 \cdot 2H_2O$, $Ca_4H(PO_4)_3 \cdot 2.5H_2O$, $Ca_3(PO_4)_2$, $Ca_5(PO_4)_3OH$ etc. HAP with formula $Ca_5(PO_4)_3OH$ and Ksp $= 10^{-55.9}$ at 25°C (equation 1) can been recrystallized from other kinds of calcium phosphate and is stable.

$$Ca_5(PO_4)_3OH \Leftrightarrow 5Ca^{2+} + 3PO_4^{3-} + OH^- \qquad (1)$$

Raw wastewater

Raw water of the experiment was taken from a wastewater treatment plant with A^2O process in Beijing (Table 1). The inlet soluble P (SP) was around 4 mg/L. It increased to 10-20 mg/L at the end of the anaerobic region because of P

Table 1. Water quality of each unit in the WWTP.

Index	pH	$NH_4^+-N/$ $(mg\ L^{-1})$	$TP/$ $(mg\ L^{-1})$	$SP/$ $(mg\ L^{-1})$	$Mg^{2+}/$ $(mg\ L^{-1})$	$Ca^{2+}/$ $(mg\ L^{-1})+$	$S\ S/$ $(mg\ L^{-1})$	$COD/$ $(mg\ L^{-1})$
WWTP influent	7.73	64.71	8.51	4.38	29.1	69.7	278	292.1
Anaerobic tank terminal	7.21	35.73	40.80	10.64	42.9	66.9	2862	125.7
Final clarifier influent	7.00	12.74	63.60	0.83	47.8	63.3	1963	99.1
Final clarifier effluent	7.02	14.36	1.10	0.35	51.1	66.1	12.1	45.7

releasing. Therefore the supernatant at the end of the anaerobic region was used as one kind of raw wastewater for the experiment.

Supernatant of sludge dewatering filtrate and sludge digester contained SP as high as 168 mg/L in the plant, so it was taken as another kind of raw wastewater.

Experiment focus

From the molecular structure of HAP, the ratio of Calcium to P is 1.67. The calculated ratio of the raw wastewater was around 4.9 (Table 1). Since there was enough calcium in the wastewater for HAP production, the focus of the experiment was to investigate the impact of pH on the P precipitation.

The experiment was divided into 3 stages: adding $Ca(OH)_2$, adding NaOH and adjusting pH by aeration.

TEST RESULTS

Adjust pH by adding Ca(OH)₂

The effect of pH

Under the condition of adding $Ca(OH)_2$,the P recovery rate was increasing with rising pH and reached 80% when pH was over 9.5 (Figure 1). The reaction was so quick that it only took about 5 minutes to reach the equivalent (Figure 2). Longer reaction time (over 10 minutes) will reduce the P recovery rate. This was because some unstable materials such as $CaHPO_4 \cdot 2H_2O$; $Ca_3(PO_4)_2$ etc. in HAP were dissolved into the solution again.

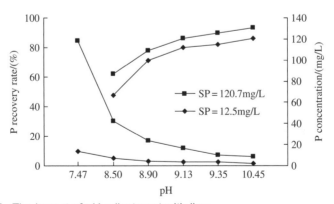

Figure 1. The impact of pH adjustment with lime.

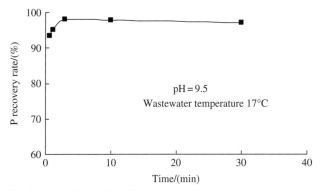

Figure 2. The impact of reaction time.

The short reaction time means that the hydraulic retention time for the P recovery device will be shorter and volume smaller.

Using Ca(OH)$_2$ to adjust pH required large amounts of reagent. For example, increasing pH of 1 liter raw water from 7.5 to 9.5 required 220 ml of saturated Ca(OH)$_2$. It is difficult to use lime from an engineering point of view because lime is hard to dissolve into solution, hard to store and hard to be slake.

The effect of temperature

Wastewater temperature varies between 10°C and 30°C in most of treatment plants. Water temperature affects the solubility and ionic activity of calcium orthophosphate.

The P recovery rates under different temperature were identified when adjusting pH to 9.3 by adding lime with 1 minute reaction time. From Figure 3 it was noticed that the effect of water temperature was not significant.

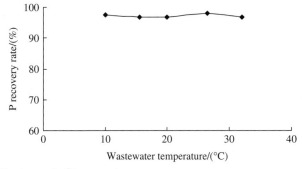

Figure 3. The impact of temperature.

Adjusting pH by adding NaOH

From Figure 4 it can be seen that increasing pH will gradually decrease SP as the P recovery rate increasing. The P recovery rate reached 80% when pH was over 9.5. Comparing with adding $Ca(OH)_2$ to adjust pH, using NaOH had slightly lower P recovery rate. This was because the addition of $Ca(OH)_2$ increased Ca ion concentration at the same time.

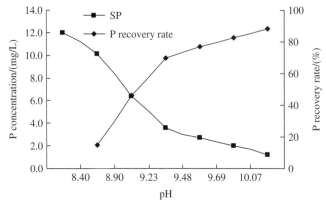

Figure 4. The impact of pH adjust with NaOH.

In engineering, NaOH addition has several advantages because it is easily dissolved in water and requiring less reagent. For example to increase pH of 1 liter raw wastewater from 7.5 to 9.5, only 10 ml NaOH of 1 mmol/L would be required.

Adjusting pH by aeration

Aeration can blow off ammonia and CO_2 from the raw wastewater and result in increasing pH. The rising pH is only controlled by the amount of CO_2 and is not related with aeration amount or intention.

The effect of aeration time

Within first 10 minutes of aeration pH quickly raised to 8.4 linearly and then gradually reached to 8.6 within 30 minutes of aeration. Further aeration would only increase pH very slowly (Figure 5).

The effect of SP concentration

pH of the raw wastewater could be adjusted to 8.6 by aeration with P recovery rate about 50%. However if the raw wastewater contained high SP such as

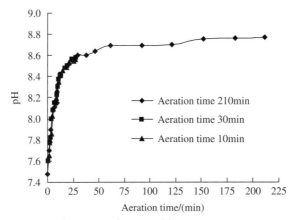

Figure 5. The impact of aeration time on pH.

supernatant of sludge dewatering filtrate or sludge digester, the P recovery rate would be over 80% at pH 9.5 (Figure 6). Therefore aeration to adjust pH was more feasible for wastewater with high SP. Otherwise the external alkalinity would be required to increase P recovery rate.

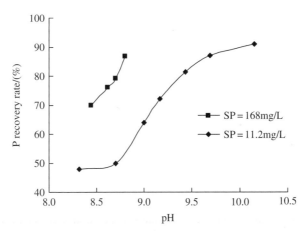

Figure 6. The impact of SP concentration.

Adjusting pH by aeration can save certain amounts of alkalinity addition and also mix the solution. So it is recommended to combine the aeration and NaOH addition to increase the P recovery rate.

THE REUSE FEASIBILITY OF RECOVERED HAP

Content analysis of the product

The recovery product from the experiments by alkalinity addition and aeration looks like grey powder. Based on analysis with the electron microscope and X-ray, the components of the product are in Table 2. The major compound is HAP.

Table 2. Main components of the recovered product.

Components	P	Ca	Mg	Al	Fe	Si	Cl	K
%	20.0	71.6	3.3	0.2	0.8	1.1	1.6	0.3

Reuse potential of recovered product

Most fertilizer manufacturers around Beijing area use acidy phosphorus fertilizer production process. They are interested in the recovered product, because it meets the quality requirement for acidy or thermo phosphorus fertilizer manufacture process and furthermore it doesn't contain toxins and hazardous content such as F etc. as in the raw ore. It means that using the recovered HAP not only can save cost for ore transportation from South China but also avoid air pollution caused by hydrogen fluoride gas discharged during the production.

Some experiment results show that struvite can be recovered from wastewater either but it contains 14.3% Mg and MgO/P_2O_5 is as high as 26.7%. For the acidy phosphorus fertilizer manufacture process MgO/P_2O_5 in raw material must be below 8% so struvite is not suitable for the manufacturers. Compared with struvite recovery, HAP recovery has great potential near Beijing area.

CONCLUSIONS

- pH has a major impact on calcium phosphate precipitation (HAP). Addition alkalinity to adjust the pH of the wastewater to 9.3 or 9.5 is able to recover 80% of P within 5 minutes of reaction time. The raw wastewater temperature will not impact the P recovery rate.
- $Ca(OH)_2$ addition will not only adjust pH but also increase Ca ion in raw wastewater. But it requires large amounts of $Ca(OH)_2$. Adjusting pH by NaOH will significantly reduce the addition amount of the reagent but needs higher chemical costs.
- Aeration will not only increase the pH but also mix the reactor as well. However it can only increase pH to a certain level and requires relatively longer reaction times. Aeration is suitable for raw wastewater with high SP

concentration. If the SP is low, a combination of aeration and alkalinity addition is required to adjust the pH to optimize P recovery.
- The recovery product is suitable for acidy phosphorus fertilizer manufacture process and has great potential around Beijing area.

REFERENCES

Bernd Heinzmann and Berliner Wasser Betriebe (2001). Phosphorus recovery in waste-water treatment plants. *Paper of the 2nd International Conference on Phosphate Recovery for Recycling From Sewage and Animal Wastes.* Noordwilkerhout, The Netherlands.

Wang Hui-Zhen and Wang Shao-Gui (2004). The influence of pH on phosphorus recovery in wastewater treatment plant. *Learned Journal of Beijing University of Civil Engineering and Architecture*, **20**(4), 5–8.

Wu Chu-Guo (2002). The sustainable development of phosphorus mine resource and phosphorus fertilizer industry in China. *Fertilizer Industry*, **29**(4), 19–21.

Wu Xi-Yan (2001). Existing state and prospect of phosphorus fertilizer industry in China. *Phosphorus Fertilizer and Compound Fertilizer*, **17**(4), 1–5.

Yang Li-Hua and Zhuo Fen (1996). Phosphorus pollution and preventing countermeasure in lake waters. *The Preventing and Curing Technology of Pollution*, **9**(2), 47.

Phosphorus removal and recovery from sewage sludge as calcium phosphate by addition of calcium silicate hydrate compounds (CSH)

Sebastian Petzet and Peter Cornel

TU-Darmstadt, Institut WAR, Petersenstrasse 13, 64287 Darmstadt, Germany (Email: s.petzet@iwar.tu-darmstadt.de)

Abstract Phosphorus removal and recovery from sewage sludge with calcium silicate hydrate compounds (CSH) was investigated. The material is known to trigger the kinetically inhibited calciumphosphate precipitation onto its surface. CSH was added to continuously operated laboratory scale mesophilic reactors for sewage sludge digestion followed by separation of the P-loaded material. It is demonstrated that the addition of CSH did not impair the anaerobic digestion process. Phosphorus recovery and removal mainly depended on the PO_4-P concentration in the sludge liquor and the concentration of the added CSH. Therefore the highest phosphorus load, phosphorus removal and recovery rates were found using waste activated sludge from enhanced biological phosphorus removal (EBPR) with a high PO_4-P release during anaerobic digestion. At CSH concentrations of 5 g/L, 35% of the total phosphorus could be recovered from the digested sludge as calciumphosphate. Total phosphorus loads of CSH of about 8 wt% P could be achieved. The CSH proofed to be inefficient when it was applied to waste activated sludge from chemical phosphorus removal. The investigation showed that CSH can be used for direct phosphorus removal and recovery from sewage sludge. It might also provide a solution where struvite scaling occurs and for high phosphorus return loads, two problems frequently encountered in relation with the EBPR process.

INTRODUCTION

Phosphorus (P) is essential for all life forms on earth and cannot be substituted. It is used as raw material for fertilizer- and other industries. Without P-fertilizer the growing world population cannot be nourished. At the current rate of extraction and consumption, the known economically exploitable reserves will be depleted, well within the next 60 to 130 years (Steen 1998).

In this context research to recover P from different sources such as wastewater, where it is mainly present because of human faeces and urine, was initiated.

Phosphorus in waste water treatment

During waste water treatment P is incorporated into the sewage sludge by biological incorporation into the biomass and if so by chemical precipitation with iron, aluminium or calcium salts in addition. As direct recycling of P by land application of sewage sludge is becoming more difficult due to concerns about its content of heavy metals (mostly zinc and copper) and organic pollutants, incineration is increasingly seen as a solution. This is not a sustainable practise in respect to P-recovery since the P containing ashes are disposed of to landfill with the P past recovery.

During the sewage sludge treatment process (anaerobic digestion or aerobic treatment) the biologically bound P is released into the liquid phase (sludge liquor) and then partly refixated into the sludge; mainly by precipitation with ferrous/ferric, aluminium, calcium and magnesium present in the sludge (Jardin and Pöpel 2001, Frossard et al., 1997 and Wild et al., 1996). Wild et al. (1997) have shown that the amount of PO_4-P refixation and the resulting PO_4-P concentration in the sludge liquor mainly depend on the concentration and availability of these cations in the sewage sludge and, consequently, on the applied PO_4-P removal process.

During anaerobic digestion sewage sludge from wastewater treatment plants (WWTP) with EBPR (with no precipitants added) releases PO_4-P at high concentrations into the sludge liquor. This often leads to operational problems due to the scaling of struvite ($MgNH_3PO_4 \times 3\ H_2O$) (Borgerding, 1972) and high PO_4-P return loads to the head of the WWTP. High PO_4-P concentrations in the sludge water also provide an excellent opportunity for P-recovery by precipitation as struvite or calciumphosphate (Doyle and Parsons, 2002). In practice, the dissolved PO_4-P is often fixated into the digested sludge by adding ferric chloride (Mamais et al., 1994), which has the effect of binding sulphur as well. Adding iron brings along the disadvantage of additional sludge formation and further decreases the quality of the biosolids since iron salts often contain impurities such as heavy metals.

Heinzmann (2001) investigated the addition of magnesiumhydroxide/chloride to digested sludge from EBPR, followed by aeration in order to trigger the struvite precipitation – a process that has been put into practise on several large WWTP in Germany. With regard to P-recovery, the precipitation and fixation of PO_4-P into the digested sludge is disadvantageous as the precipitate is embedded into a complex organic sludge matrix hampering subsequent separation and recovery. The formation and separation of MAP crystals within the sludge matrix is currently investigated by Stumpf et al. (2008).

In this context a novel approach to fixate and recover PO_4-P in sewage sludge by adding and subsequently separating CSH containing material was investigated in this study. The aim was to influence the phosphorus fixation process within the digester. CSH is a by-product from the building industry which is known to trigger the kinetically inhibited calciumphosphate precipitation onto its surface due to a lowered activation energy (Berg *et al.*, 2006). The material has proven to be suitable for P-removal and recovery from the liquid phase. However, its applicability within the digester at very high solid concentrations of 30–50 g/L and temperatures of 37°C has not been investigated. The experiments of this study were conducted with waste activated sludge (WAS) from a WWTP with phosphorus removal by a combination of chemical precipitation with iron dosing and EBPR, as well as with WAS from a WWTP with EBPR only.

MATERIAL AND METHODS

Analytical methods

Dissolved components in the sludge liquor were analysed after centrifugation at 30,000 rpm (Sigma 3K30) and filtration through 0.45 μm membrane filters (Schleicher & Schuell ME 25). Total concentrations of P, Mg, Ca, K, Al, Fe, S, Cu, and Zn in the filtered sludge liquor were measured with ICP-OES (Perkin Elmer, Optima 3200 DV) according to DIN EN ISO 11885 (1998). PO_4-P was measured photometrically according to DIN EN 1189 (1996) (Zeiss PMQ 3). Dissolved chemical oxygen demand (COD) was determined with Hach Lange Cuvette Tests (Dr. Lange LCK 514; DIN ISO 15705, 2003) and Ammonium (NH_4-N) with Hach Lange Cuvette Tests (Dr. Lange LCK 302 und 303). The total solid concentration (TS) in the sewage sludge was determined according to DIN EN 12880 (2000) and loss on ignition (LOI) according to DIN EN 12879 (2000).

The overall composition of the sewage sludge and the CSH was determined by aqua regia dissolution, according to DIN EN 13346 followed by determination of P, Mg, Ca, Al, Fe, K, S, Cu and Zn by ICP-OES. Total COD of the sewage sludge was determined according to DIN 38414-S9. The pH in the sludge was determined with a WTM pH 197-S and Mettler Toledo InLab 1003 electrode.

Long-term experiments

After preliminary experiments confirmed the applicability of CSH in sewage sludge in principle, a series of long term experiments was conducted in order to investigate if the P-refixation process in the digester can be influenced by CSH addition. The aim was to determine the optimal CSH-concentration for P-removal and recovery and to test whether the additive inhibits the digestion process.

To that end in a first step CSH was added at different concentrations and two different grain sizes (1.5 mm and 3 mm) to three mesophilic stirred 15 L digesters (R-1, R-2, and R-3) that where placed in a climatic chamber (37°C), with a reference digester (R-0) without CSH addition (Figure 1). All reactors were inoculated with digested sludge from a 150 L pilot plant. The WAS was derived from a WWTP using a combination of chemical P-elimination by iron dosing and EBPR. The sludge was characterised by high total solid concentrations of 50–60 g/L (dewatering by centrifugation with addition of polymers). The overall composition of the sludge is given in Table 1.

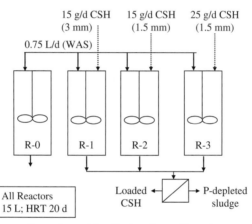

Figure 1. Reactor configuration of the first series of long term experiments.

Table 1. Composition of the sieved sewage sludge and feed WAS from the first series of experiments.

	TS [%]	LOI [%]	COD$_{tot}$ [mg/L]	Fe [mg/g]	Al [mg/g]	Mg [mg/g]	Ca [mg/g]	P [mg/g]
WAS	6.2	69	67683	12.6	32.8	5.0	15.0	24.9
R-0	4.4	57	40621	24.3	32.8	7.9	25.3	41.4
R-1	4.7	59	40215	21.5	31.0	7.3	39.0	36.0
R-2	4.8	57	39593	21.1	29.0	7.0	50.1	35.6
R-3	5.2	54	43012	22.0	30.8	7.5	40.6	37.8

The CSH was added daily together with 0.75 L of WAS, thus increasing the CSH concentration stepwise in order to determine the optimal CSH concentration. The same amount of digested sludge was discharged daily, resulting in a hydraulic retention time (HRT) of 20 days. From day 24 on the CSH concentrations were kept constant at 10 g/L (R-1 and R-2) and 16 g/L (R-3), respectively. Later the CSH concentration in reactor R-3 was reduced to 5 g/L on day 48 for the rest of the experiment. Samples for the analysis of PO_4-P, NH_4-N, COD, pH, total COD, TS content and LOI were taken 2–3 times a week. After 3, 5, 8, 9 and 12 weeks samples of CSH were taken from the reactor and analysed for chemical composition.

In a second series of long term continuous experiments, WAS from a WWTP with EBPR process was used that had a lower total solid concentration of 30 g/L and a higher phosphate content (Table 4).

Two digesters A-0 and A-1 (15 L) were fed daily with 0.75 L of WAS and the same amount was discharged. Digester A-0 served as a reference without CSH addition while A-1 was fed with 3.75 g of CSH daily resulting in a CSH concentration of 5 g/L (Figure 2). The reactors B-0 and B-1 were fed with a mixture of 0.4 L of WAS and 0.35 L of primary sludge. Digester B-0 served as a reference and digester B-1 received 3.75 g of CSH daily mixed with the feed sludge, resulting in a CSH concentration of 5 g/L.

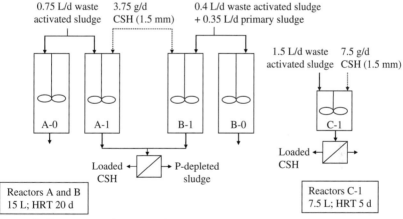

Figure 2. Reactor configuration for the second series of long term experiments with WAS from a WWTP with EBPR.

Another stirred mesophilic 7.5 L reactor (C-1) was operated at a HRT of 5 days and received 1.5 L of WAS and 7.5 g of CSH daily, resulting in a CSH concentration of 5 g/L (Figure 2). The CSH was separated from the sludge by sieving (0.625 mm mesh size). The CSH was then washed, dried and analysed for overall composition. All digesters were inoculated with digested sludge from the same WWTP.

RESULTS AND DISCUSSION

Long-term experiments

Figure 3 shows the PO_4-P and CSH concentration in the first series of long term experiments with continuous CSH application to the digesters R-1, R-2 and R-3. The reference digester R-0 without CSH addition showed a relatively high PO_4-P concentration of around 230 mg/L in the first 50 days of the experiment. This was because of very high solid concentrations in the feed WAS and combined phosphorus removal with iron precipitation and EBPR (Table 1). From day 50 on the PO_4-P concentration in the reference digester R-0 decreased from 240 to 98 mg/L (Figure 4), which was caused by an increase in aluminium content of the WAS (use of Na-aluminat at the WWPT).

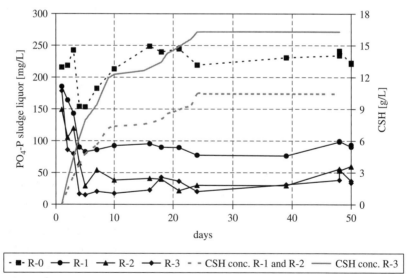

Figure 3. Phosphorus and CSH concentrations in digester R-0, R-1, R-2 and R-3 with rising CSH concentrations (until day 25) and at constant CSH concentrations (10 g/L in R-1 and R-2; 5 g/L in R-3) (from day 25).

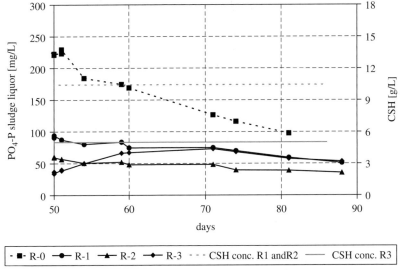

Figure 4. Phosphorus and CSH concentrations in digester R-0, R-1, R-2 and R-3 at constant CSH concentrations (10 g/L in R-1 and R-2; 5 g/L in R-3).

Figure 3 shows that the PO_4-P concentrations in the sludge water of the three digesters with CSH addition decreased immediately as the CSH concentrations increased. The minimum PO_4-P concentration in the sludge liquor of the reactors was reached after 5 days at CSH concentrations of about 5 g/L (R-1 and R-2) and 8 g/L (R-3), respectively. The CSH concentrations were kept constant from day 25 at 10 g/L (R-1 and R-2) and at 16 g/L (R-3) since the results suggested that this range of concentrations was sufficient for P-removal. The lowest PO_4-P concentrations could be found in R-3 due to higher CSH concentrations of 16 g/L. The PO_4-P concentration in R-1 was higher than in R-2, showing that the material with 3 mm grain size was less effective than the material with 1.5 mm grain size. The CSH concentration in R-3 was then reduced to 5 g/L on day 48 using new sewage sludge from the pilot plant while the concentrations in R-2 and R-3 were kept constant.

Analysis of dissolved COD, total COD, dried solid content, volatile solids and NH4-N showed that the anaerobic degradation was not impaired by the addition of CSH at these concentrations (results not shown).

The achieved P-load of the CSH, the corresponding PO_4-P concentrations in the sludge liquor, and the pH values are depicted in Table 2.

Table 2. pH values, PO$_4$-P concentration in the sludge liquor, P-load of the CSH material and CSH concentrations for the first series of experiments.

Day			25	39	60	69	84
R-0	pH		7.37	7.43	7.36	7.50	7.45
	PO$_4$–P sludge liquor	[mg/L]	219	230	169	126	98
R-1	pH		7.55	7.75	7.64	7.86	7.60
	PO$_4$–P sludge liquor	[mg/L]	78	77	75	75	59
	P–load of CSH	[wt%]	2.2	1.8	2.1	2.2	2.2
	CSH$_{conc.}$	[g/L]			10 g/L		
R-2	pH		7.57	7.87	7.70	7.75	7.55
	PO$_4$–P sludge liquor	[mg/L]	31	30	48	48	39
	P–load of CSH	[wt%]	1.4	1.7	2.0	1.9	1.7
	CSH$_{conc.}$	[g/L]			10 g/L		
R-3	pH		7,63	7,59	7,5	7,74	7,55
	PO$_4$–P sludge liquor	[mg/L]	20	31	67	73	55
	P–load of CSH	[wt%]	0.7	1	4.4	3.1	2.9
	CSH$_{conc.}$	[g/L]	16 g/L			5 g/L	

During the whole experiment the P-load of the CSH in R-1 and R-2 remained around 2.1 and 1.7%, respectively. However, the P-load of the CSH was highest in reactor R-3 at low CSH concentrations of 5 g/L (day 60 to 84). These results show that the P-load of the CSH increases when the ratio of CSH-concentration and PO$_4$-P concentration in the sludge water decreases.

In order to determine the overall P-extraction from the digested sludge, the sludges were sieved for CSH separation and then analysed for total P. The P-content of the sieved and P-depleted sludge was then compared with the P-content of the reference sludge (Table 3). It can be seen that the total P-removal was higher at a CSH-concentration of 10 g/L (R-1 and R-2) and lower at a CSH-concentration of 5 g/L (R-3). Thus a high P-removal can only be achieved at high CSH concentrations, which in turn reduce the P-load of the CSH.

The total composition of the sewage sludge (reference sludge R-0 and sieved sludge from R-1, R-2 and R-3), show that calcium is released from the CSH into the sludge (Table 1). Accordingly, the calcium content of the used CSH decreases, compared to the new material (results not shown). The pH increases in all reactors due to the addition of CSH (Table 2). This could trigger further precipitation of calciumphosphate or MAP into the sludge, thus decreasing the overall efficiency.

Table 3. Calculation of the P-removal from the treated sludges on day 69.

Feed WAS, V = 15 L; HRT = 20 d				
Day 69		R-1	R-2	R-3
		10 g/L	10 g/L	5 g/L
		3 mm	1.5 mm	1.5 mm
Total P–sieved sludge [mg/L]	(1)	1578	1559	1656
P–load of CSH [wt%]	(2)	2.2	1.9	3.1
P–removal by CSH [mg/L]				
(2)*conc.$_{CSH}$[g/L]		220	190	155
Sum: (1)+(2) [mg/L]		1798	1749	1811
Reference sludge (3)		1811	1811	1811
P–removal [%] 1-((1)/(3))		13	14	9

Long-term experiments with EBPR sludges

Even though the first experiments clearly confirmed that P could effectively be removed and recovered from the sewage sludge by addition of CSH, the overall P-recovery rate of 14% and the P-load of the CSH of only 4.4 wt% were not yet satisfactory.

Thus, a second series of continuous experiments was conducted with WAS from a German EBPR-WWTP without iron or aluminium addition. It was expected that higher total P-removal rates and higher P-loads of the CSH could be achieved at higher PO_4-P concentrations in the sludge liquor and lower total solid concentrations of around 30 g/L.

Two digesters (A-0 and A-1) were operated with pure WAS (one reference and one with CSH). Two additional reactors (B-0 and B-1) were fed with a mixture of WAS and primary sludge (one reference and one with CSH) in order to test how the material works with a realistic sewage sludge composition.

Another reactor C-1 was operated with pure WAS and CSH addition at a short hydraulic retention time (HRT) of 5 days in order to investigate if the material could be effective during a pre-treatment step. This configuration would allow to separate the CSH before the WAS enters the digester for final treatment thus avoiding possible problems of settleable CSH particles in the digester. Furthermore the separation of CSH from the sludge by sieving is easier for pure WAS than for mixed sludge, which would be another advantage of such a configuration.

The composition of the EBPR-WAS is given in Table 4. The physiological P-content of this WAS was estimated by Beier (2008) to 3.3% MLVSS conducting several P re-dissolution tests (unpublished data). The PO_4-P concentration in the sludge liquor of the reference reactor A-0 remained around 400 mg/L (Figure 5) while the reference digester with mixed sludge (B-0) showed lower PO_4-P concentration of around 150 mg/L (Figure 6). This is in line with the low P-content of primary sludge (Table 4) and its high P-refixation capacity due to a high aluminium and calcium content (Jardin, 1995).

Table 4. Composition of the EBPR-WAS, primary sludge (PS) and the sieved digested sludges from all reactors for the second series of experiments.

	TS [%]	LOI [%]	COD$_{tot}$ [mg/g]	Fe [mg/g]	Al [mg/g]	K [mg/g]	Mg [mg/g]	Ca [mg/g]	P [mg/g]
WAS	3.3	75	1065	7.3	12.8	14.0	9.1	18.9	39.5
PS	3.3	75	1055	11.2	9.1	2.7	3.3	19.5	5.2
A-0	2.4	67	836	9.9	16.9	16.8	10.7	25.4	45.7
A-1	2.4	62	895	9.2	16.8	17.1	9.6	38.5	34.2
B-0	2.5	56	866	16.0	16.0	11.0	8.3	28.7	28.7
B-1	2.6	55	792	13.8	15.2	10.8	7.9	43.7	20.8
C-1	2.6	71	941	7.6	14.7	16.0	8.6	30.7	31.9

Table 5. Composition of the sludge liquors for the second series of experiments.

	Ca [mg/L]	Mg [mg/L]	K [mg/L]	NH$_4$ [mg/L]	pH [-]
A-0	29	24	366	707	7.00
A-1	51	28	361	805	7.45
B-0	44	14	238	637	7.01
B-1	195	91	250	687	7.36
C-1	68	44	366	559	7.23

In the reactor A-1 with a CSH concentration of 5 g/L the PO_4-P concentration decreased to 80–150 mg/L. The achieved P-load of the CSH in this reactor was between 8.0 to 8.7 wt% (Table 6). These values are considerably higher than in the first row of experiments when WAS from chemical P-removal with iron was used. This is also true for the overall P-recovery (sieved sludge/reference sludge) which increased to 25–32% (Table 6).

In Figure 5 it can be seen that the PO_4-P concentration in the sludge water of reactor C-1 is almost as low as in reactor A-1 despite the lower HRT of 5 days

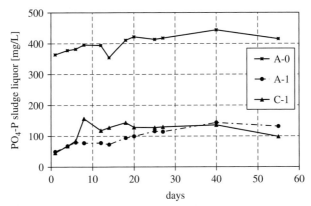

Figure 5. Phosphorus concentrations in the second series of experiments with pure EBPR-WAS.

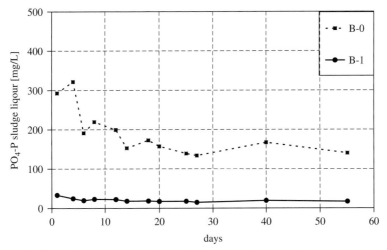

Figure 6. Phosphorus concentrations in the second series of experiments with a mixture of primary sludge and EBPR-WAS.

compared to 20 days in A-1. This indicates that the material efficiently reduces PO_4-P during the shorter pre-treatment. The P-load of the CSH from this reactor reached 6.3–8.1 wt% (Table 6) which was lower than the CSH from reactor A-1. The overall P-extraction from C-1 was in the range of 17–36% (Table 6).

In the reactor B-1 the PO_4-P concentrations decreased to less than 20 mg/L (Figure 6) and the P-load of the CSH reached around 5 wt% (Table 7) which is

Table 6. P-content of the sieved sewage sludge in comparison to the reference sludge and the P-load of the CSH for the reactors that were operated with WAS only.

Reactor A-1	$CSH_{conc.}$ = 5 g/L	1.5 mm			
WAS; V = 15 L; HRT = 20 d					
Day		11	19	26	35
Total P–sieved sludge [mg/L]	(1)	783	806	906	753
P–load CSH [wt%]	(2)	8.0	8.1	8.7	8.0
P–removal by CSH [mg/L] (2)* 5 g_{CSH}/L		399	407	435	398
Sum: (1)+(2) [mg/L]		1183	1213	1341	1151
Reference sludge (3)		1160	1089	1208	1004
P–removal [%] 1-((1)(3))		32	26	25	25

Reactor C-1	$CSH_{conc.}$ = 5 g/L	1.5 mm			
WAS; V = 7.5 L; HRT = 5 d					
Day		11	19	26	35
Total P–sieved sludge [mg/L]	(1)	964	888	858	642
P–load CSH [wt%]	(2)	6.3	6.9	7.6	8.1
P–removal by CSH [mg/L] (2)* 5 g_{CSH}/L		313	347	382	403
Sum: (1)+(2) [mg/L]		1277	1235	1240	1044
Reference sludge (3)		1160	1089	1208	1004
P–removal [%] 1-((1)(3))		17	18	29	36

lower than reactor A-1 and C-1 that were operated with WAS only. This confirms that the P-load decreases at lower PO_4-P concentrations in the sludge liquor. Consequently it should be possible to reach higher P-loads of the CSH by reducing the CSH concentration in this reactor. Table 7 shows that the overall P-extraction in reactor B-1 was between 17% and 34% which is in the same range as the reactors A-1 and C-1 that were fed solely with WAS.

Table 7. P-content of the sieved sewage sludge in comparison to the reference sludge and the P-load of the CSH for the reactor B-1 with mixed sludge.

Reactor B-1		$CSH_{conc.}$ = 5 g/L		1.5 mm	
WAS + PS (0.4 : 0.35); V = 15 L; HRT = 20 d					
Day		11	19	26	35
Total P–sieved sludge [mg/L]	(1)	463	555	658	455
P–load CSH [wt%]	(2)	5.5	5.1	3.8	5.2
P–removal by CSH [mg/L]					
(2)∗ 5g$_{CSH}$/L		276	256	190	260
Sum: (1)+(2) [mg/L]		739	811	847	715
Reference sludge (3)		706	716	794	622
P–removal [%] 1-((1)/(3))		34	22	17	27

Table 6 and 7 depicts that the sum of the theoretical amount of P on the CSH (P-load × CSH concentration) and the P-content of the depleted sludge always exceed the P-content of the reference sludge. This can be explained by the fact that the CSH retention was not 100% when the material was sieved since a fraction of fine CSH particles passed the mesh and remained in the sieved sludge.

The second series of experiments demonstrated that high P-removal and recovery rates could be achieved, when CSH was added to digesters with WAS as well as to digesters with mixed sludge from EBPR.

Higher P-loads of the CSH could be achieved when pure WAS from EBPR-process without primary sludge was used. The experiments also demonstrated that a 20 days contact time during the whole digestion was not necessary for the reaction. A shorter contact time of only 5 days already resulted in satisfactory P-loads and recovery rates. This can be attributed to the fact that the poly–P from EBPR sludge is completely released during the 5 day pre-treatment.

As the separation of CSH is more complicated for mixed sludge than for WAS and given that the addition of CSH to full scale digesters might lead to operational problems due to sedimentation of the additive, the option of a pre-treatment of WAS with CSH addition appears to be the most promising approach.

However, it needs to be investigated further if higher P-removal and recovery rates can be achieved by optimising the pre-treatment step. Furthermore it should be tested if the use of CSH helps to avoid operational problems due to struvite scaling since most soluble phosphorus is removed during the pre-treatment.

Economical aspects

The P-load of the CSH is the most important cost factor of the process as it influences both, the CSH-demand and the value of the recovered product. Phosphate rock, the material CSH is supposed to substitute, has a P-content of around 13 wt%.

The price for CSH is approximately 0.4 €/kg. According to Cornel *et al.* (2004) about 78% of the daily P-load per capita (cap) of approx. 1.8 g P/(cap · d) is incorporated into the waste activated sludge during wastewater treatment with EBPR. If 30–40% would be removed and recovered with CSH addition (153–204 g $P_{recovered}$/(cap · a)), the annual cost for the material would add up to 0.77 €/(cap · a) to 1.02 €/(cap · a). This equals around 5 €/kg $P_{recovered}$ which compares to today's world market price of 2.18 €/kg P for phosphate rock. However, possible savings due to reduced demand of chemicals for P-precipitation, reduced sludge volume, revenues for recovered P and prevention of operational problems have not been considered.

CONCLUSIONS

- Up to 36% of total phosphorus can be recovered from EBPR sewage sludge as calciumphosphate by the addition and subsequent separation of CSH to the digester
- CSH effectively reduces the PO_4-P concentration in the sludge liquor and might provide a solution for problems such as struvite scaling and high phosphorus return loads in WWTP with EBPR
- The P-load of the CSH increases with a decreasing ratio of CSH and available PO_4-P in the sludge water while the overall P-recovery increases with the CSH concentration
- High P-loads of more than 8 wt% P can be achieved when sewage sludge from EBPR process is used
- The best configuration seems to be a pre-treatment of waste activated sludge from EBPR followed by CSH separation and final treatment.

ACKNOWLEDGEMENTS

This work is part of the "German Research Clusters on P-Recycling" and was founded by the German Federal Ministry of Education and Research.

REFERENCES

Beier, M. (2008). Personal communication. Leibniz Universität Hannover, ISAH.

Berg, U. and Schwotzer, M. and Weidler, P.G. and Nüesch, R. (2006). Calcium Silicate Hydrate Triggered Phosphorus Recovery – An Efficient Way to Tap the Potential of Waste-and Process Waters as Key Resource. Water Environment Federation WEFTEC®.06.

Borgerding, J. (1972). Phosphate deposits in digestion systems. *J. Wat. Pollut. Control Fed.*, **44**(5), 813–819.

Cornel, P., Jardin, N. and Schaum, C. (2004). Rückgewinnung von Phosphor aus Klärschlammasche Teil 1: Ergebnisse von Laborversuchen zur Extraktion von Phosphor. GWF-Wasser/Abwasser **145**(9), 627–632.

Doyle, J.D. and Parsons, S.A. (2002). Struvite formation, control and recovery. *Water Res.*, **36**(16), 3925–3940.

Frossard, E., Bauer, J.P. and Lothe, F. (1997). Evidence of vivianite in FeSO4-flocculated sludges. *Wat. Res.*, **31**(10), 2449–2454.

Heinzmann, B. (2001). Options for P-recovery from Waßmannsdorf Bio-P wwtp, Berlin. Implications for wwtp operation and phosphorus recovery potential at different locations in the Bio-P and sludge treatment process. 2nd international Conference on Recovery of phosphates, Noordwijkerhout, NL, 12–13th March, 2001.

Jardin, N. (1995). Untersuchungen zum Einfluss der erhöhten biologischen Phosphor-elimination auf die Phosphordynamik bei der Schlammbehandlung. Dissertation TH Darmstadt, Schriftenreihe WAR, Band 87.

Jardin, N., and Pöpel, H.J. (2001). Refixation of Phosphates during bio-p sludge handling as struvite or aluminium phosphate. *Environ. Technol.*, **22**(11), 1253–1262.

Mamais, D., Pitt, P.A., Yao Wen Cheung, Loicano J. and Jenkins D. (1994). Determination of ferric chloride dose to control struvite precipitation in anaerobic digesters. *Water Environ. Res.*, **66**, 912–918.

Steen, I. (1998). Management of a non-renewable resource. *Phosphorus and potassium* **217**, 25–31.

Stumpf, D., Zhu, H., Heinzmann, B. and Kraume, M. (2008). Phosphorus Recovery in Aerated Systems by MAP Precipitation: Optimizing Operational Conditions. 5th IWA Leading-Edge Technology, Zürich, 1–4. Juni 2008.

Wild, D., Kisliakova, A. and Siegrist, H. (1996). P-fixation by Mg, Ca and zeolite a during stabilization of excess sludge from enhanced biological P-removal. *Wat. Sci. Tech.*, **34**(1), 391–398.

Wild, D., Kisliakova, A. and Siegrist, H. (1997). Prediction of recycle phosphorus loads from anaerobic digestion. *Water Res.*, **31**(9), 2300–2308.

Field application methods for the liquid fraction of separated animal slurry in growing cereal crops

T. Nyord

Tavs Nyord, Ph.D. student, Institute of Agricultural Engineering, University of Aarhus, Blichers Allé 20, DK-8830 Tjele. Email: tavs.nyord@agrsci.dk

Abstract High concentrations of farm animals can cause high impact on non farmed land and vulnerable ecosystems. By separating animal slurry in nutrient rich fractions where the liquid nitrogen (N) rich fraction can be used locally on the farm, can potential reduce the environmental impact. To do so, it is important to avoid ammonia (NH_3) emission from land spreading of the liquid fraction. Soil injection can be used as a technique to eliminate NH_3 emissions. However, existing soil injection technologies is highly power consuming and create unacceptable crop damage due to low working width of the injector which brings extra passing's in the crop by the tractor and slurry tanker wheels. The aim of this study was to design an injection tine operating shallow and combining soil opening and high-pressure injection in preparations for reducing the power requirements and thereby increase the potential working width of slurry injectors. Three high-pressure injection tines with nozzles pointing forward, downward and backward were made. Horizontal and vertical force requirements for each of the nozzle configurations were measured and tested using water in stead of liquid separated slurry, in an indoor sand and soil bin facility. Using high-pressure injection where jets pointing backward compared to the travel direction reduced the horizontal (draught) force significantly compared to operating the tine without water jets.

INTRODUCTION

High concentrations of livestock in areas with relatively little agricultural land can lead to high impacts on the environment due to sub optimal use of nutrients in manure and waste waters. Advanced technologies for separation of animal slurry are now available. By using this technology it is possible to produce nitrogen (N) rich liquid fraction and a phosphorus (P) rich solid residue. The liquid part is used locally on the farm and the solid part is transported to either; 1) areas with a low density of farm animals 2) centralized biogas plants, or 3) incineration plants.

Land spreading of the liquid fraction can lead to environmental problems, such as odour nuisance and ammonia (NH_3) emission. To avoid this, slurry can be injected into soil. Equipment is available for injecting slurry into un-cropped soil and grassland. However, due to unacceptably high levels power requirements and of crop damage it is not practical to use equipment made for un-cropped soil injection in growing crops (Birkmose, 2007). A major problem with using existing equipment for slurry injection, into growing crops, is that injection requires a significant additional draught force compared to surface application (Huijsmans, Hendriks and Vermeulen, 1998), this implies that the working width has to be lower than the distance between existing tramlines and thereby increases the risk of crop damage by the wheels dramatically (Arvidsson, Keller and Gustafsson, 2004). To reduce draught fore requirements the injection tines can be operated very shallow. However, injection of the liquid in the very top soil including a risk of not covering the slurry by soil which increases the NH_3 emission significantly.

Araya (1985) introduced a new technology for soil injection of sewage sludge. The sludge was pressurized from the tip of a subsoiler chisel. The liquid saturated the soil zone in front of the subsoiler and the liquid also lubricated the soil-metal interface. The liquid broke down the soil structure in front of the tip of the injector tine, which permitted the subsoiler to operate in looser soil conditions which reduced the draught force.

The aim of this study was 1) to develop a high-pressure injection tine with reduced horizontal force requirements which could lead to application machinery with a working width like common surface applicators and 2) examine the effect on NH_3 emission depending on injection depth.

MATERIALS AND METHODS

Measuring draught force

The trial including high-pressure injection tines, see Figure 1, with three different placements of injection nozzles (forward, downward and backward), operating at three different depths was conducted in an instrumented sand bin and a soil bin, located at Cranfield University at Silsoe, Bedfordshire, UK in 2007.

Sand bin

The sand bin is 6 m long, 1 m wide and 1 m deep. For both sand and soil bin a reservoir tank supplied a Grundfoss centrifugal pump (CR16) with water. The pump was used supplying water under pressure to the injection tines. A Danfoss Magflo® (MAG 5000) flow meter mounted on the pump outlet was used for measuring water flow. The carriage was driven at 0.54 m s^{-1} and the velocity was checked several times during the experiment. The two bins were equipped with instrumented carriages fitted with an Extended Octagonal Ring Transducer (EORT) (Godwin 1975) to which the tines were mounted. A pressure transducer

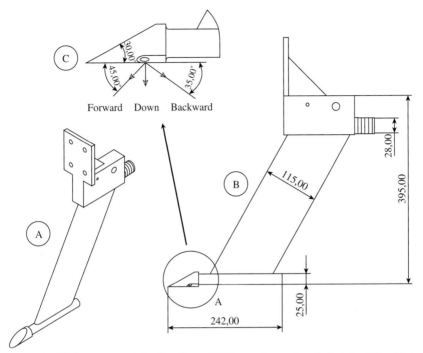

Figure 1. High-pressure injection tine. A) Three dimensional sketch of the high-pressure tine B) Measurements of the tine in mm. C) Close up of the tip of the tine, with the three different angles of the water jets.

from Applied Measurements LTD (P805C9 – 20 bar) was used to measure the water pressure just before the inlet to the tine.

The sand bin facility is equipped with a grid of perforated water tubes in the bottom of the bin. These allow water jets to submerge and liquefy the sand. When the water is pumped out and the sand settles down, a flat surface with a repeatable dry bulk density profile is created. To ensure uniformity of dry bulk density and water content between each preparation the water flow through the perforated tubes needs to be carefully regulated the same. Knight (2005) found that a flow of 1.3 l s^{-1} created the most uniform sand preparation.

In a field situation typically slurry injection would be carried out at 2.2 m/s, however, the maximum speed attainable in the sand bin was 0.55 m/s, and therefore these experiments were limited to 0.55 m/s.

To enables the analysis of draught force with depth to be considered, the tines were tested for draught force requirements at three depths, 50, 70 and 90 mm.

Based on a pilot study in Denmark, 70 mm was found to be the depth where the tines could incorporate up to 50 m^3 ha^{-1}. The application rate was 30 tons of water/ha with 30 cm spacing between the individual injector tines, this is the most common application rate in Denmark (Birkmose, 2007). This application rate corresponds to approximately 1 litre m^{-1} in the injection slit.

Soil bin

The soil bin is 25 m long, 1.7 m wide and 1 m deep. The factors tested in the soil bin experiment were; placement of nozzle and high pressure or not. The same three tines used in the sand bin experiment were tested for draught and vertical forces in the soil bin, with and without high-pressure, at one depth, replicated 3 times. The soil bin was prepared to give a dry bulk density of approximately 1450 kg/m^3 at a moisture content of 8 to 9% (vol/vol). Due to practical limitations all three replicates for each individual tine were tested in the same preparation of the soil bin; one replicate in the left side of the bin, one in the middle and one in the right side of the bin. This means that in the statistical analysis each preparation of the bin is considered as a block and dry-bulk density and water content were measured for each preparation and used as covariate in the statistical analysis to explain some of the variation between the replicates.

Pressure and flow were the same as used in the sand bin the water supply was generated by the same pump fitted with an extended supply hose.

NH_3

Slurry from fattening pigs (30–100 kg) was separated by using a centrifuge which produced a liquid fraction that contained less than 1.5% of dry matter with a maximum particle size < 1 mm. In laboratory experiment conducted in autumn 2008, NH_3 emission will be measured by using dynamic chamber technique like described by (Sommer *et al.*, 2006).

RESULTS AND DISCUSSION

Pressure and flow

Pressure and flow were measured for all treatments (see Table 1).

The flow was equal for all treatments in both the sand and the soil bin experiment, but the pressure changed with the treatments and was significantly higher when nozzles pointing forward compared to the case when the jets pointed down or back. The reason for this difference was clearly a minor difference in nozzle size. Even though all nozzles were made with a 6 mm drill

Table 1. Pressure and flow for the high-pressure tines used in the sand and soil bin experiments.

Treatment	Press (bar)	Flow (l s^{-1})
Back	1.1	0.48
Down	0.9	0.48
Forward	1.3	0.48

there was a minor difference in the shape of the nozzles, due to the different angles of the drillings. These small differences affected the pressure at the inlet of the tine. It was decided to keep the flow stable between the treatments and therefore the flow was adjusted before each run and this created the different pressures for each tine. The reason for keeping the flow constant and adjusting the pressure was that it was assumed that the flow was more important for the draught force, than the pressure, because an increased water flow may create a bigger saturated soil zone and thereby affect the soil structure and the draught force. Pressure was not included in the statistical analyses, because it would have been impossible due to the design of the experiment to separate the pressure effect from the treatment effect because the pressure was different from treatment to treatment.

Draught force

Jet pointing forward

Figure 2 shows a tendency for higher draught requirement when the nozzles are pointing forwards, irrespective of high-pressure or not, compared with the two other tines. This raise two questions: 1) why are there observed higher draught requirement for the tine where nozzles are pointing forward compared to the two other tines and 2) why are there not observed any decrease in draught when high-pressure injection is compared with no pressure?

An explanation of the first question could be a small difference in design of the tine compared to the two other tines, as earlier described. This difference is specially seen in the soil-bin experiment, see Figure 3, but there is also a tendency to higher draught force for this tine in the sand bin experiment.

Regarding question 2, the answer seems much more complex. Logically, kinetic energy in the water jet should increase the draught force by creating a forward movement with the force from the jets. The reason to make a tine with the nozzles pointing forward was because of the theory of degrading the soil structure in front of the tine allowing the tine to run in soil with lower shear stress and, therefore, reduced draught force. Araya (1985), Araya (1994a) and

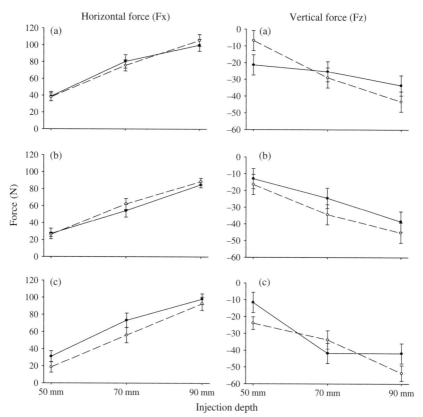

Figure 2. Sand bin experiment. Draught force and vertical force measured for three different high-pressure tines; (a) nozzles pointing forward (b) nozzles pointing down and (c) nozzles pointing back. Broken lines represent high-pressure injection and solid lines no injection. Error bars are equal to the standard deviation (n = 3).

Knight (2005) showed reduced draught force by injecting sewage sludge and water under pressure in front of a tine. The high-pressure injection changed the soil failure and soil structure in front to the tine and thereby reduced the energy needed for soil disturbance (Araya, 1994b).

The explanation for not achieving the same results in this experiment could be that due to the placement of the nozzle the water jet did not reach the tip of the tine. It would have been useful to test if the water had reached the tip of the tine, however, this was not possible in the available time.

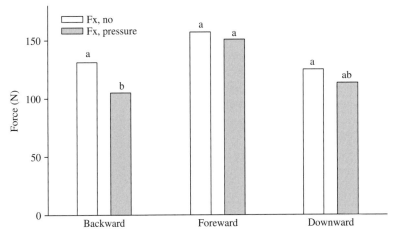

Figure 3. Soil bin experiment. Draught force measured for three different high-pressure tines; (a) nozzles pointing forward (b) nozzles pointing down and (c) nozzles pointing back, in the soil bin. Letters indicate significant differences on a 5 % level (n = 3).

Another effect of injected water in front of the tine was to lubricate the tip and the leg (or the shank) of the tine which has been reported to reduce draught force (Araya 1994a; Schafer *et al.*, 1975; Schafer *et al.*, 1979). If the water reached the tip of the tine the effect may have been neutralized by the wider design of the tip. If the water was not pressurised in front of the tine no lubrication of the tip and the leg took place.

This experiment was carried out at very shallow injection depths which could explain why there was no effect on draught force by pointing the jet forward. Because the total draught force required for pulling a tine in these upper soil layers is relatively small compared to the experiments reported in the literature. Shafer *et al.* (1979) found that lubrication of soil-metal interfaces could reduce the draught force by up to 40% for mouldboard ploughs. But it is also concluded that there need to be a relatively high pressure on the soil-metal interface before lubrication has an effect. Also Araya (1994b) conclude that the reducing effect of pointing the jet forward was greatest when the draught force used for pulling the tine itself was high.

A back pressure is created when water is pressurised in front of the tine (Araya 1994b). Araya (1994b) found that the back pressure at the nozzles port was up to 0.1 MPa. The measured gauge pressure at the inlet of the tine was approximately 0.1 MPa, data not shown, and it is therefore assumed that the back pressure will be the same. But if the energy gain from lubricating the tine and saturate the soil in front of the tine is approximately the same as the back pressure from the water, then this may explain why there was no effect of high-pressure in this experiment when the water jet was facing forward. The neutral effect on draught force could be due to some of the effects of lubrication and degradation of soil structure which is mentioned above and could lead to further investigations of these potential effects.

Jet pointing down

There is no significant difference between high-pressure or not for the tine with the downward facing nozzles, which is logically.

Jet pointing back

This arrangement experienced a significant reduction in draught force with pressurised injection as well as an increase in up thrust (data not shown). The reduction was almost 20% compared to non pressurised injection. One possible explanation could be that the water jets is pointing slightly outward and will create a back pressure on the undisturbed. The reduction is almost 25 N. This pressure will create a forward movement of the tine and thereby reducing draught force.

One key point when talking about reducing draught force by lubricating or changing soil failure by high-pressure injection is that the power required for the fluid injection should not exceed the power gain resulting from reduced draught force.

Araya (1994b) used the following equation to calculate the total power requirement for high-pressure injection:

$$W_f = W_{f1} + W_{f2} \tag{1}$$

W_f is the power supplied to the injection tine and is defined as the sum of draught power, W_{f1}, and the power required for fluid injection, W_{f2}.

$$W_{f2} = \frac{p_{\text{inf}} \times G_t}{p_l} + \frac{G_t^3}{2 \times a^2 \times p_l^2} \tag{2}$$

Where p_{inf} (Pa) is gauge pressure at the nozzle port (in this case it is assumed that the gauge pressure at nozzle port and at the inlet to the tine is equal), G_t (Kg s^{-1}) is mass flow rate, p_l is density of liquid, a (m^2) is area of nozzle port.

For each tine is required power of 0.118 kW (118 W) to build up the pressure to 1 bar in the tine. This means that 72 tines (equal to a boom of 18 m width) need at least 8.5 kW. This power requirement has to be compared to the measured reduction in draught force of approximately of 25 N (~27 W) from the tine where the nozzles were pointing backward. 25 N * 72 tines gives 1.8 kN which has to be compare to the 72 kN in horizontal force needed to pull an 18 m wide injector, which corresponds to a reduction of 2.5%, which is a minor reduction. This means that the tractor has to deliver 8.5 kW extra power through the power take off, but at the same time the energy use on the driving wheels will be reduced by 2.5%. As the extra power would be used through the power take off and not through the driving wheels, some of the tractor power could be transferred to the power take off from the tractor wheels.

It could be a problem constructing an 18 m wide injector boom having sufficient strength to pull 72 injector tines. Therefore it may be considered acceptable to use a system where the overall power requirement for high pressure injection was greater than for low pressure injection if the high pressure injection reduced draught force to a level that allowed more tines to be used and, therefore, a wider boom that would fit in with the tramline system.

Ammonia emission

No results are yet achieved about NH_3 emission, but will be presented at the conference.

REFERENCES

Araya, K. (1985). Soil failure by introducing sewage-sludge under pressure. *Transactions of the Asae*, **28**(2), 397–401.

Araya, K. (1994a). Optimization of the design of subsoiling and pressurized fluid injection equipment. *Journal of Agricultural Engineering Research*, **57**(1), 39–52.

Araya, K. (1994b). Soil failure caused by subsoilers with pressurized water injection. *Journal of Agricultural Engineering Research*, **58**(4), 279–287.

Arvidsson, J., Keller, T. and Gustafsson, K. (2004). Specific draught for mouldboard plough, chisel plough and disc harrow at different water contents. *Soil and Tillage Research*, **79**(2), 221–231.

Birkmose, T. (2007). Application of slurry in Denmark. *Personal Communication*.

Godwin, R.J. (1975). An extended octagonal ring transducer for use in tillage studies. *Journal of Agricultural Engineering Research*, **20**(4), 347–352.

Huijsmans, J.F.M., Hendriks, J.G.L. and Vermeulen, G.D. (1998). Draught requirement of trailing-foot and shallow injection equipment for applying slurry to grassland. *Journal of Agricultural Engineering Research*, **71**(4), 347–356.

Knight, C.S. (2005). An invastigation into jet assisted submarine cable burial ploughs, *PhD Thesis*, Cranfield University at Silsoe, National Soils Resources Institute, Engineering Group.

Schafer, R.L., Gill, W.R. and Reaves, C.A. (1975). Lubrication of soil-metal interfaces. *Transactions of the Asae*, **18**(5), 848–851.

Schafer, R.L., Gill, W.R. and Reaves, C.A. (1979). Experiences with lubricated plows. *Transactions of the Asae*, **22**(1), 7–12.

Sommer, S.G., Jensen, L.S., Clausen, S.B. and Sogaard, H.T. (2006). Ammonia volatilization from surface-applied livestock slurry as affected by slurry composition and slurry infiltration depth. *Journal of Agricultural Sciences*, **144**, 229–235.

Research on nutrient removal and recovery from swine wastewater in China

Yong-hui Song, Peng Yuan, Guang-lei Qiu, Jian-feng Peng, Xiao-yu Cui, Ping Zeng

Chinese Research Academy of Environmental Sciences, Dayangfang 8, Anwai Beiyuan, Beijing 100012, China. songyh@craes.org.cn

Abstract Heavy pollution of livestock and poultry wastewater has been a problem in China. It is already noticed that nutrient removal is necessary for water environment protection, but nutrient recovery is still a rather fresh topic. In the present study, research on nutrient removal and recovery by magnesium ammonium phosphate (MAP) from swine wastewater was carried out at bench and pilot scale. It showed that for MAP crystallization the optimum pH value was at 9.0–10.0, the coexistence of Ca in solution would increase the P removal efficiency, but would affect the morphology of MAP; the coexistence of carbonate would slightly decrease the crystallization efficiency of MAP. The pilot scale experiments of 10 m^3/d in the piggery showed that both discontinuous and continuous flow MAP crystallization reactors were efficient for phosphorus (P) removal and recovery from anaerobically treated swine wastewater, while the discontinuous reactor was more efficient; the reactors were resistant to temperature variation in the range of 12 to 30°C, and to influent phosphorus concentration variation in the range of 30 to 80 mg/L; the effluent P concentration could meet the national discharge standard of livestock and poultry wastewater. Ammonia could be also removed with MAP crystallization, but more ammonia was removed by carbonate stripping for pH increase purpose instead of the MAP crystallization reaction.

INTRODUCTION

In China, with urbanization and the improvement of people's living level, the livestock and poultry industry has been developed quickly since 1990s, and the discharge of livestock and poultry wastewater has caused severe water pollution in the rural areas and the rural-urban transient areas (SEPA, 2002a). Anaerobic treatment of livestock and poultry wastewater has been taken as the main treatment process to recover methane gas and to reduce the pollution load, but the anaerobically treated wastewater cannot meet the discharge standards, and this is still a difficult problem to overcome for many piggeries. Eutrophication has become one of the main problems of Chinese water bodies, and this is closely related to nutrients discharge to the environment, among which livestock and poultry industry takes an important part.

Nutrient removal and recovery, especially phosphorus (P) recovery from wastewater has been paid much attention worldwide since end of 1990s (Doyle and Parsons, 2002; Suzuki *et al.*, 2004). The MAP crystallization process has been proved to be a feasible process for removing and recovering N and P from swine wastewater, simultaneously. Our previous theoretical research (Wang *et al.*, 2006) developed models with PHREEQC Program for thermodynamic simulation of MAP crystallization from wastewater. The results show that the saturation index (SI) value of MAP is the logarithmic functions of the concentrations of P, ammonium–N and Mg, and increases with the concentration increase of each element. The SI value of MAP is a polynomial function of pH value of the solution, and the optimum pH value for the crystallization of MAP is 9.0 but increases slightly with the increase of the N/P. The adjustment of Mg concentration and the control of solution pH are two effective methods for controlling MAP crystallization. For MAP crystallization process there are still some knowledge gaps, such as the influences of carbonate and calcium concentrations on the process, and in China there is still few research and practice for P recovery from wastewater. In our research both bench scale and pilot scale experiments were carried out, and essential and practical results on the process of MAP crystallization have been obtained. This would provide technological supports to nutrient pollution control of livestock and poultry wastewater in China.

MATERIALS AND METHODS

Bench scale experiments

Phosphate removal and recovery by MAP crystallization was studied at bench scale by using simulated wastewater. Referring to the components of swine wastewater (SEPA, 2002a), P concentration was kept at 80 mg L^{-1}, and an N/P molar ratio of 8 was adopted in all the experiments. The phosphate removal rate and efficiency were traced temporally, and the effects of solution conditions were evaluated.

Before the batch experiments 5 stock solutions containing N, P, Mg, Ca and CO_3^{2-} were prepared by dissolving NH_4Cl, $NaH_2PO_4 \cdot 2H_2O$, $MgCl_2 \cdot 6H_2O$, $CaCl_2 \cdot 2H_2O$ and $NaHCO_3$ into deionized water, respectively. All the reagents used were analytically pure reagents and deionized water was used in the bench scale experiments. Batch experiments were performed on a magnetic stirrer with a stirring rate of 1000 rpm at room temperature of 22 to 25°C. The N and P stock solutions were mixed in a beaker of 2 L with an N/P molar ratio of 8 and diluted to 950 mL, and its pH value was adjusted by addition of 10 M NaOH to the designed value. According to the Mg/P (or Ca/Mg, CO_3^{2-}/Mg) molar ratios designed, the Mg stock solution (or both Mg and Ca stock solutions, both Mg and CO_3^{2-} stock solutions) was added to the above solution under rapid stirring within 10 s to make the total volume of the mixed solution into 1000 mL. Throughout the experiment the pH value of the mixture was kept at the defined constant value ($\pm$0.02) by NaOH supplement. The experiment lasted for 2 h, and water samples of 10 mL were taken at frequent intervals and filtrated rapidly with 0.45 μm-membranes. 2 μL of 6 M HCl was added to the filtrate rapidly to prepare samples for components analyses. The crystals formed were dried naturally at room temperature (Le Correa et al., 2005) for instrumental analysis.

Pilot scale experiments

For technology demonstration of nutrient removal and recovery from swine wastewater, pilot scale experiment was carried out with a treatment flow rate of 10 m^3/d. The experiments were carried out at a piggery of Beijing suburb. The swine wastewater was firstly treated with anaerobic process, which resulted in a wastewater with average concentrations of COD 1728 mg/L, ammonia 582 mg/L, total phosphorus (TP) 59.0 mg/L, phosphate (as P) 40.4 mg/L, and the wastewater pH value was around 7.2, the highest water temperature in summer was 30°C and the lowest temperature was 12°C in late autumn so far.

The MAP crystallization reactor was a part of the treatment system consisting of anaerobic treatment, nutrient removal and recovery, and moving bed biofilm reactor unit for wastewater treatment (Figure 1).

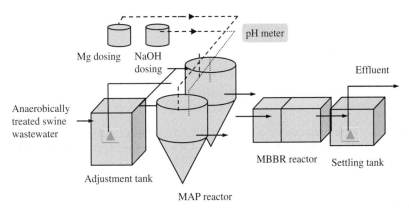

Figure 1. Schematic diagram of swine wastewater treatment system with MAP crystallization.

Two MAP crystallization reactors were built up, a sequencing batch MAP reactor and a continuous flow MAP reactor (Figure 2). The batch MAP reactor was a barrel with a conical bottom, with a height of 2.5 m, a diameter of 1.0 m, and a reacting area volume of 1.0 m³. A stainless steel crystal collector consisting of four concentric cylindrical meshes was set in the reactor, so that the MAP crystal could grow on it. The reactor was operated in the sequencing batch mode with four phases: loading, aeration, sedimentation, drainage. The influence of aeration time on MAP crystallization and P removal was studied, and four aeration time of 1 h, 2 h, 3 h and 4 h were adopted in the tests.

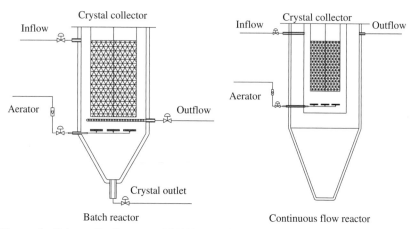

Figure 2. Schematic diagram of MAP crystallization reactors.

The continuous flow MAP reactor was a cubic column with a pyramid bottom, with a total height of 4.0 m, a cross-section side length of 1.6 m, and a volume of 4.0 m³. The reactor was separated into a reacting zone of 1.3 m³ and a settling zone of 2.6 m³ by a square iron barrel. A same crystal collector as used in the batch MAP reactor was set in the reacting zone. The reactor was operated in the continuous flow mode. The influence of hydraulic retention time (HRT) on MAP crystallization and P removal was studied, and four HRT of 6 h, 9 h, 12 h and 15 h were adopted in the tests.

Analytical Methods and Instrumentation

Water samples were analyzed according to Monitoring and Analytical Methods of Water and Wastewater (SEPA, 2002b). Ammonia nitrogen and phosphate were analyzed with a spectrophotometer (752N, China), Ca and Mg were measured with an atomic adsorption photometer (Shimadzu AA-6800, Japan).

The morphology of the crystals obtained was observed by using scanning electron spectroscopy with energy dispersive x-ray (SEM-EDX) analysis (Cambridge S-360, UK), and the morphology of the crystals was analyzed with x-ray diffraction (XRD) (Rigaku D_{MAX}-RB, Japan).

RESULTS AND DISCUSSION

Bench scale experiments

Influence of pH value

Solution pH value is one of the most important factors controlling the crystallization of MAP (Hoffmann *et al.*, 2004). A series of experiments were performed under the conditions of the pH range of 8.0–12.0 and N:P:Mg molar ratio of 8:1:1. The removal efficiencies of N and P at the reaction time of 2 h are shown in Figure 3. The P removal efficiency increases obviously at pH 8.0–9.0 and keeps at 90% stably at pH 9.0–11.0, then decreases rapidly from 90% to 40% at pH 11.0–12.0. The N removal efficiency increases obviously at pH 8.0–10.5, and then rapidly decreases, but the highest removal efficiency of only 13% is reached at pH 10.5.

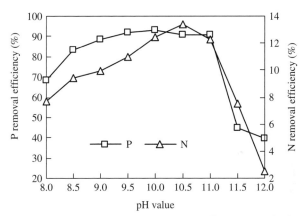

Figure 3. Effect of pH value on the removal efficiencies of phosphate and ammonium nitrogen.

At pH 8.0–10.0, a large amount of small white powder appeared and settled rapidly. At pH above 10.0 the crystals kept on increasing, but their settling ability was getting worse; the crystals were loose and lots of small particles suspended in the supernatant. So the optimum pH value for MAP crystallization was in the range of 9.0–10.0.

Influence of Mg^{2+}

To study the effect of Mg^{2+} on MAP crystallization, Mg/P molar ratios of 1.0, 1.2, 1.4, 1.6, and 2.0 were adopted. The P removal efficiencies at the reaction time of 2 h are shown in Figure 4. It shows that the P removal efficiency increases with the increase of Mg/P molar ratio at the tested pH values. At a relatively low pH value of 8.0 the P removal efficiency increases more significantly; at higher pH range of 9.5–10.5 the P removal efficiencies all reach 97% at the Mg/P molar ratio of 1.4, and does not increase significantly any more with further increase of Mg/P molar ratio.

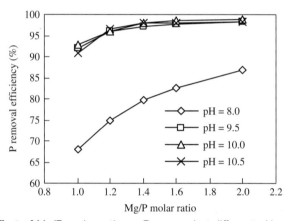

Figure 4. Effect of Mg/P molar ratio on P removal at different pH values.

Influence of Ca^{2+} and carbonate

Both Ca^{2+} and CO$_3^{2-}$ are common components in wastewater, and the anaerobically treated swine wastewater contains abundant carbonate. In the MAP crystallization system, the presence of Ca^{2+} and CO$_3^{2-}$ may influence the MAP crystallization due to the formation of calcium phosphate, magnesium carbonate or calcium carbonate precipitate. To understand the effect of Ca^{2+} concentration, the following solution conditions were designed: the pH value was 9.5, Mg^{2+} molar concentration (C$_{Mg}$) was 3.61 mM, N:P:Mg molar ratio was 8:1:1.4, Ca/Mg molar ratios were 0.5, 1.0, and 2.0, respectively. The removal amounts of P, Mg and Ca are shown in Table 1. With the increase of Ca/Mg molar ratio the Mg removal decreases significantly while the removal of Ca increases. The amount of removed P increases only slightly because of the limited initial P concentration. Figure 5 shows the SEM and EDX analyses of the crystals obtained with different Ca/Mg molar ratios.

Table 1. The removed P, Mg and Ca from solution with different initial Ca/Mg molar ratios.

Initial Ca/Mg molar ratio	P removal (mM)	Mg removal (mM)	Ca removal (mM)
0.5:1	2.46	1.52	1.09
1:1	2.52	0.53	2.22
2:1	2.56	0.11	2.26

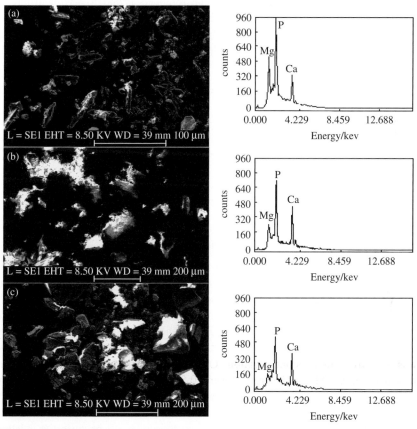

Figure 5. SEM pictures and their respective EDX of the crystals obtained with different Ca/Mg molar ratios. $C_{Mg} = 3.61$ mM. (a) Ca/Mg $= 0.5$; (b) Ca/Mg $= 1.0$; (c) Ca/Mg $= 2.0$.

The SEM pictures show that the morphology of the crystals can be distinguished at a low Ca/Mg molar ratio of 0.5, although the surface of the crystals is not smooth and covered by other precipitates. However the irregular crystals appeared with the increase of Ca^{2+} concentration. The EDX spectra reveal that the peak of Ca strengthened while the peak of Mg weakened with the increase of Ca/Mg molar ratio.

To understand the effect of carbonate concentration, the following solution conditions were designed: the pH value was 9.5, C_{Mg} was 3.61 mM, N:P:Mg molar ratio was 8:1:1.4, CO_3^{2-}/Mg molar ratios were 0.5, 1.0, and 2.0, respectively. Although carbonate and bicarbonate coexist at the pH of 9.5, here we use carbonate to represent both of them. The removed P and Mg with different CO_3^{2-}/Mg molar ratios are shown in Table 2. The molar ratios of the removed P and Mg are close to the theoretical P/Mg molar ratio of MAP, i.e. 1.0. However, the amounts of the removed P and Mg decrease slightly with the increase of CO_3^{2-}/Mg molar ratio.

Table 2. The removed P and Mg with different initial CO_3^{2-}/Mg molar ratios.

Initial CO_3^{2-}/Mg molar ratio	P removal (mM)	Mg removal (mM)	Removed P/Mg
0.5:1	2.51	2.48	1.01:1
1:1	2.50	2.44	1.02:1
2:1	2.46	2.27	1.08:1

Pliot scale experiments

Nutrient removal and recovery in the sequencing batch MAP reactor

The influence of the aeration time on MAP crystallization and P removal in the sequencing batch MAP reactor was shown in Figure 6 and Figure 7. The aeration rate used throughout the operation period was 15 m^3/h.

It can be seen from Figure 6 and Figure 7 that as the aeration time shortened form 4 h to 1 h, the removal rate of TP and phosphate still kept high, because under these four aeration time conditions, the pH value in the reactor was almost the same, all about 8.5 because of the carbonate stripping; and the similar pH conditions resulted the similar MAP crystallization efficiencies. It was also observed that at the beginning of operation, there was little MAP crystal attached on the crystal collector, but after the reactor worked for a period of time, there were MAP crystals attached on the crystal collector, because the existed crystal could induce the formation new crystals (Wu and Bishop, 2004; Adnan et al., 2004).

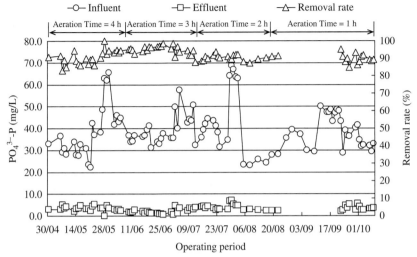

Figure 6. The phosphate removal in the sequencing batch MAP reactor.

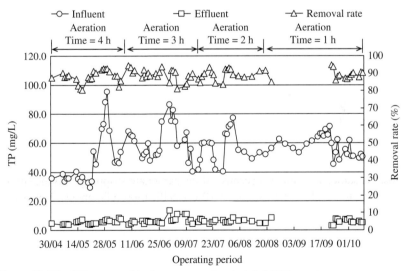

Figure 7. The TP removal in the sequencing batch MAP reactor.

Ammonium ions can be also removed in the process of MAP crystallization. However, according to the stoichiometric relationship, for the crystallization of 1.0 mg PO_4^{3-}-P, there will be 0.45 mg NH_4^+-N removal. And, during the operation period, the average removal of PO_4^{3-}-P was 36.9 mg/L, so the NH_4^+-N combined in the MAP crystal will be 16.6 mg/L. However, during the operation period, the average removal of NH_4^+-N was 108.5 mg/L, this is probably due to ammonia stripping because of the relatively high pH value of 8.5 in the reactor.

Under the above crystallization conditions, the average TP of the effluent was lower than 8mg/L, which met the Discharge Standard of Pollutants for Livestock and Poultry Breeding (SEPA, 2001).

Nutrient removal in the continuous flow MAP reactor

The influence of HRT on MAP crystallization and P removal in the continuous flow MAP reactor was shown in Figure 8 and Figure 9. The aeration rate throughout the whole operation period was 25 m^3/h.

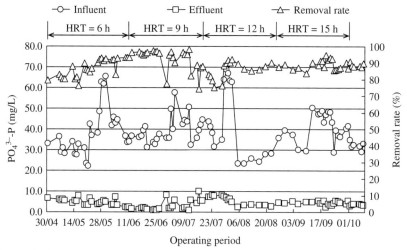

Figure 8. Phosphate removal in the continuous flow MAP reactor.

The removal rates of TP and phosphate in the continuous flow MAP reactor were a little low than that in the sequencing batch MAP reactor, because under the continuous flow condition, the aeration effect on the pH increase was weaker; the average pH in the continuous flow reactor was 0.1 pH unit lower than that in the sequencing batch MAP, and lower pH value and weaker air

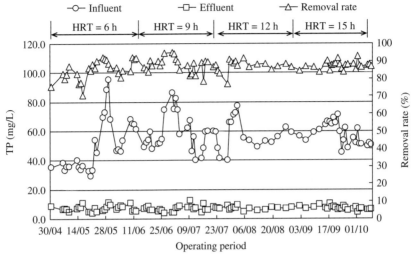

Figure 9. TP removal in the continuous flow MAP reactor.

agitation resulted in lower P removal efficiency (Ohlinger *et al.*, 1999). It showed that during the pilot scale experiment, the nutrient removal efficiency of the reactor was stable, which was almost not affected by the water temperature (12–30°C) and the influent P concentration (30–80 mg/L).

CONCLUSIONS

(1) The optimum pH value for MAP crystallization is at 9.0–10.0. The increase of Mg^{2+} concentration can enhance the P removal efficiency; at the phosphate concentration of 80 mg/L the P removal efficiency reaches 97% under the conditions of Mg/P molar ratio of 1.4 and pH 9.5.

(2) At Ca/Mg molar ratio > 0.5, calcium ion can inhibit the crystallization reaction of MAP and significantly affect the purity of MAP crystal; the coexistence of carbonate in solution affects the phosphate removal efficiency slightly

(3) For the actual swine wastewater, both of the sequencing batch reactor and the continuous flow reactor can remove and recover phosphate by MAP crystallization efficiently, and the effluent P concentration can meet the national discharge standard of China. The process has good resistance to temperature and influent P concentration variations.

ACKNOWLEDGEMENTS

The supports of the National Natural Science Fund of China (50678162) and the National Science & Technology Pillar Program during the 11th Five-Year Plan Period of China (2006BAD14B09) are acknowledged.

REFERENCES

Adnan, A., Dastur, M., Mavinic, D.S. and Koch, F.A. (2004). Preliminary investigation into factors affecting controlled struvite crystallization at the bench scale. *J. Environ. Eng. Sci.*, **3**, 195–202.

Doyle, J.D. and Parsons, S.A. (2002). Struvite formation, control and recovery. *Water Res.*, **36**, 3925–3940.

Le Correa, K.S., Valsami-Jones, E., Hobbs, P. and Parsons, S.A. (2005). Impact of calcium on struvite crystal size, shape and purity. *J. Cryst. Growth*, **18**, 1020–1024.

Ohlinger, K.N., Young, T.M. and Schroeder, E.D. (1999). Kinetics effects on preferential struvite accumulation in wastewater. *J. Environ. Eng.*, **125**, 730–737.

Suzuki, K., Tanaka, Y., Kuroda, K., Hanajima, D. and Fukumoto, Y. (2004). Phosphorous in swine wastewater and its recovery as struvite in Japan. In: *International Conference for Struvite: Its Role in Phosphorus Recovery and Reuse.* Cranfield University, UK.

The State Environmental Protection Administration of China (SEPA) (2001). *Discharge Standard of Pollutants for Livestock and Poultry Breeding.* GB18596-2001.

The State Environmental Protection Administration of China (SEPA) (2002a). *National Situation Investigation and Pollution Prevention Policy of Concentrated Livestock and Poultry Industries.* China Environmental Science Press, Beijing, 35–38.

The State Environmental Protection Administration of China (SEPA) (2002b). *Monitoring and Analytical Methods of Water and Wastewater*, 4th edn (Editorial Committee of "Monitoring and Analytical Methods of Water and Wastewater"), China Environmental Science Press, Beijing.

Wang, J.S., Song, Y.H., Yuan, P., Peng, J.F. and Fan, M.H. (2006). Modeling the crystallization of magnesium ammonium phosphate for phosphorus recovery. *Chemosphere*, **65**, 1182–1187.

Wu, Q.Z. and Bishop, P.L. (2004). Enhancing struvite crystallization from anaerobic supernatant. *J. Environ. Eng. Sci.*, **3**, 21–29.

Chemical recycling of phosphorus from piggery wastewater

M-L Daumer[a], F. Béline[a], S.A. Parsons[b]

[a]Environmental management and biological treatment research unit, Cemagref, 17 avenue de Cucillé, CS64427, 35044 Rennes Cedex, France.
Email: marie-line.daumer@cemagref.fr
[b]Centre of Water Sciences, Building 52, Cranfield campus, U.K.

Abstract The technical feasibility of a dissolution/precipitation chemical phosphorus recycling process from piggery wastewater was assessed. Several combinations of acidifying/precipitating reactants were evaluated for their impact on liquid effluent quality and solids formed. Chloride and sodium concentration in the liquid effluent, which could contribute to soil salinisation, were reduced by a third by using acetic acid/magnesium oxide when compared to hydrochloric acid/sodium carbonate.

Finally more than 95% of the initial dissolved phosphorus from the acidified supernatant was recovered. Struvite crystals and amorphous calcium phosphate were the main components in the solid, identified by X-ray diffraction, optical and SEM-EDS microscopy. The size and the shape of struvite crystals were increased by increasing the magnesium or ammonium/phosphate ratio which made them more suitable for the filtration and drying steps required to export phosphorus as a dry mineral product.

INTRODUCTION

In intensive animal production areas, phosphorus (P) brought to the soils by livestock wastes is higher than crop requirements. P is fixed by the soils but rainfall, runoff and erosion lead to the transfer of soil particles in surface water. After solubilisation in water, eutrophication can occur. The best way to regulate the overloading on soils is to export the P in excess as mineral fertiliser (struvite or calcium phosphate) to intensive cropping areas. Moreover, this is in compliance with the fact that P is a limited resource which has to be saved.

Some successful attempts to recycle P as struvite (Ammonium magnesium phosphate) from the liquid phase of piggery wastewater, have been performed at laboratory and pilot scales (Burns *et al.*, 2003; Kim *et al.*, 2004; Nelson *et al.*, 2003; Suzuki *et al.*, 2005). Depending on the precipitation/crystallisation route, solid phosphate is recovered by filtration or by scrapping solid from the suspended crystallisation support (Suzuki *et al.*, 2007). Another way to recycle P as calcium phosphate was developed by Vanotti *et al.* (2003). The amount of reactant needed to precipitate the mineral P was reduced by treating the wastewater biologically to decrease ammoniacal nitrogen concentration and consequently the buffer capacity. Filtration bags and natural drying were used to recover the solid phosphate.

All these processes were developed to recycle dissolved P. However, the recycling potential could be enhanced by recycling also the mineral solid fraction, which is the prevailing fraction of P (up to 80%) in piggery wastewater from Western Europe. Dissolution of this fraction before decantation combined with a side-stream recycling process from the enriched supernatant had been proposed by Greaves and Haygarth (2001). The idea was to use the enhanced biological phosphorus removal metabolism to act as a "phosphate pump". These first experiments were undertaken using a diluted piggery wastewater from which about 80% of P had been previously removed by centrifugation. So the recycling rate was low. More recent works have shown the limited effect of biological treatment on dissolving P using a raw piggery wastewater (Daumer *et al.*, 2007 a and b). Therefore, chemical solubilisation appears as a possible option to increase the dissolved P concentration needed to improve the recycling rate. This option has been previously tested on wastewater sludge (Weidelener *et al.*, 2004) and on poultry litter (Szögi *et al.*, 2008). Depending on (i) the solubility of calcium or magnesium salts in the effluent, (ii) the compounds to be formed and (iii) the pH required for the P precipitation, different combination of dissolution/precipitation reactants could be used. This choice of reactants will determine the particles size, density and the hygroscopic properties of the precipitate and, consequently, its filterability but also the phosphate concentration and availability and hence its agronomic properties. The chemicals used will also determine the quality of the liquid effluent which is typically spread locally.

The aim of this study was to assess the technical feasibility of a chemical recycling process from non centrifuged piggery wastewaters and identify the best reactants to use.

MATERIAL AND METHODS

Piggery wastewater supernatants

Two different piggery wastewaters were used. The first one was sampled in the biological reactor of farm treatment plant located near Lamballe in France while the second one was sampled in the biological reactor of a pilot treatment plant located in Cemagref (Rennes, France). In the first case, wastewater was acidified by hydrochloric addition (0.14 $M \cdot kg^{-1}$) (AS1). In order to decrease the chloride concentration in final liquid effluent, the second wastewater was acidified with acetic acid (0.26 $M \cdot kg^{-1}$) (AS2). In both cases, the supernatant was pumped gently after acidification followed by 72 hours of decantation. Composition of the supernatants finally obtained is given in Table 1.

Table 1. Composition of the acidified supernatants. AS1: biologically treated piggery wastewater acidified with hydrochloric acid; AS2: biologically treated piggery wastewater acidified with acetic acid.

	pH	N-NH$_4^+$	P-PO$_4^{3-}$	Ca^{2+}	Mg^{2+}	Na$^+$	K$^+$	Cl$^-$
				mg.kg^{-1} $\pm$ s.d. (mmoles.kg^{-1})				
AS1	5.41	630 $\pm$ 32	620 $\pm$ 18	1720 $\pm$ 53	460 $\pm$ 21	1100 $\pm$ 45	3200 $\pm$ 145	7300 $\pm$ 260
		(45)	(20)	(43)	(19)	(48)	(82)	(209)
AS2	4.57	370 $\pm$ 15	800 $\pm$ 17	1300 $\pm$ 42	400 $\pm$ 16	800 $\pm$ 38	2400 $\pm$ 115	2600 $\pm$ 125
		(26)	(25)	(33)	(17)	(35)	(62)	(74)

Biochemical analyses

Total solids (TS) and volatile solids (VS) were measured by the APHA method (2540B and E). The suspended solid were measured by centrifugation (18000G, 20 min, 4°C). Total phosphorus was measured by a flow injection analyser (Lachat Instruments, Milwaukee, Wisc., USA) with a blue molybdate method after mineralisation (ashes digestion with peroxodisulfate and sulfuric + nitric acid at 120°C under pressure 1 bar). Dissolved ortho-phosphate were analysed by ionic chromatography after centrifugation and filtration. All the samples were analysed for cations by ionic chromatography (DIONEX, Sunnyvale, Cal.).

Results for each sample are the mean of 3 analyses.

Description of the precipitation runs

Run 1: The supernatant AS1 was divided in 100 ml fractions in Erlenmeyer flasks. Increasing amounts of reactants (NaOH, MgO and Na$_2$CO$_3$) were added in each flask to reach pH values between 5.4 and 9. The flasks were slowly mixed for 3 hours and centrifuged.

Run 2: The supernatant AS2 was divided in 11 fractions placed in 2 litres reactors equipped with a continuous mixer and a pH probe. Increasing amount of MgO was added and the pH was measured every 15 minutes during 3 hours and then, time to time, up to a pH plateau was reached.

In the both cases, the supernatant from the decantation was analysed for its ionic concentrations and the phase containing solid was kept for analysis.

Chemical reactants were laboratory pure reactant obtained from Sigma-Aldrich Chemical (Saint-Louis, Missouri, USA).

Solid analyses

Phases containing solid were centrifuged (10510G, 20 min, 20°C). The mineral crystallised solid phase was at the bottom of the pellet. The pellet was put upside down on a plate and carefully scratched. Optical microscopic analyses were performed, at once, on the wet pellet. Scanning electronic spectroscopy coupled with energy dispersive X-ray analysis (SEM-EDS) (Philips XL 30 SFEG) and X-ray diffraction analysis (XRD) (Power X-ray diffract meter D 5005 Siemens) were performed after drying of the scratched particles at room temperature.

RESULTS AND DISCUSSION

Run 1:

The pH required to precipitate all the dissolved P in AS1 was around pH 8 whatever the reactant used (Figure 1). This required about 45–50 mmoles $\cdot$ kg^{-1} of reactant to reach such a pH and precipitate all the P from AS1 supernatant.

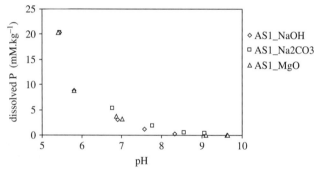

Figure 1. Dissolved P as function of pH during piggery wastewaters precipitation tests.

For all the reactant used, the crystals observed by optical microscopy between pH 5 and pH 8, were very translucent and had no regular shape. The main components determined by SEM-EDS and expressed as atomic composition were O (59%), Ca (17%) et P 11(%). These crystals were clearly identified as brushite (CaPO$_4$(OH), 2H$_2$O) by X-ray diffraction (Figure 2). At higher pH (>8), no brushite was observed. A very few crystals were observed in NaOH samples but, they could not be identified. When Na$_2$CO$_3$ was used as the reagent, some small crystals predominated. They were needle shape and were typically 20–30 μm length and width <10 μm. When MgO was used a large amount of orthorhombic medium size crystals was observed (length ≈ 30–50 μm, width ≈ 20 μm). The atomic composition of the needle and orthorhombic crystal surface was in the same range O (60–67%), Mg (9–13%) and P (7–19%). The presence of was due to the coating agent used for sample preparation. Even if the peak intensity was low, struvite ($MgNH_4PO_4$, $6H_2O$) was identified in MgO samples by XRD analysis (Figure 3). No clear spectrum was obtained with Na_2CO_3.

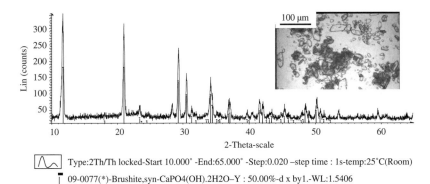

Figure 2. X-ray spectra and optical microscopy picture of brushite crystrals obtained at pH 5.6.

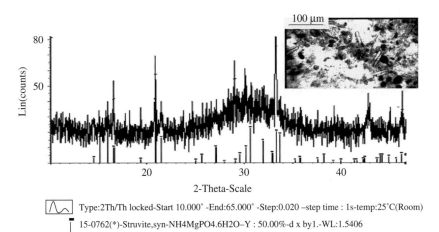

Figure 3. X-ray spectra and optical microscopy picture of struvite crystals obtained at pH 8.

The observation of brushite precipitation followed by struvite with Na_2CO_3 and MgO addition was in agreement to dissolved calcium and magnesium curves (Figure 4). Calcium started to precipitate between pH 5 and 7 while precipitation of magnesium occurred only from pH 7. Decrease of dissolved magnesium due to precipitation was not observed using MgO because it was brought in excess.

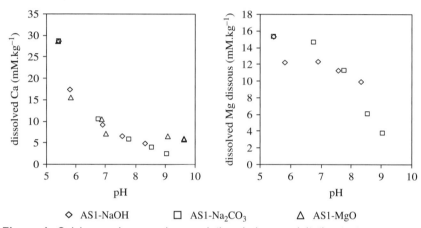

Figure 4. Calcium and magnesium evolution during precipitation tests.

An amorphous calcium phosphate was probably the main mineral phosphate in NaOH samples where very few crystals were observed. In contrast to MgO or Na_2CO_3, NaOH was added on a liquid form. As crystallisation is increased by suspended particles (Valsami-Jones, 2004), nucleation without crystallisation could occur, when NaoH was added.

In MgO and Na_2CO_3 samples, dissolved calcium did not increase after brushite dissolution (pH > 7). Formation of calcite ($CaCO_3$) is possible as is the formation of an amorphous calcium phosphate even in the samples where struvite was formed. In fact, brushite is the most stable form of calcium phosphate in acidic conditions (Valsami-Jones, 2001) but amorphous forms of octocalcium phosphate, tricalcium phosphate and monetite ($CaHPO_4$) have been identified at neutral or basic pH in sewage sludge (Frossard *et al.*, 1994; Giesen *et al.*, 1999; Angel, 1999). This was confirmed by the precipitated Mg/P ratio in Na_2CO_3 samples, which was 0.58 while Mg/P ratio in struvite is 1. Between one third and half of phosphorus could be precipitated as calcium phosphate. The same succession of solid phases, brushite, struvite and amorphous calcium phosphate, with pH evolution, had been described by Abbona *et al.* (1986, 1988) in synthetic solution containing calcium magnesium and phosphate.

Usually, because of its higher solubility, $MgCl_2$, is used instead of MgO, to supply magnesium. However $MgCl_2$ has an acidifying effect and NaOH addition is required to increase the pH for P precipitation (Türker and Celen, 2007; Le Corre *et al.*, 2007; Suzuki *et al.*, 2007). In our study, the supernatant was already acid, increasing the solubility of the MgO which could supply magnesium while increasing the pH without NaOH addition.

Struvite cristallisation depends on several factors and the size and the shape of crystals can be modified by ionic strength, foreign ions among which calcium and Ca/Mg ratio or organic matter (Le Corre *et al.*, 2005; Valsami-Jones, 2004; Antakyali *et al.*, 2004). This could explain the difference of the crystal shape between Na_2CO_3 and MgO precipitation.

Finally, precipitation with MgO increased the crystal size which could help to further filtration of the solid without increasing the amount of reactant. Moreover only one solid reactant easy to handle and use was needed and the sodium concentration of the final liquid effluent was decreased compare to sodium reactants which was 2 and 3 $g \cdot kg^{-1}$ using NaOH or Na_2CO_3 respectively. So MgO was chosen as precipitant for further experiments.

Run 2:

Chloride concentration in AS2 was decreased from 7 to 2 $g \cdot kg^{-1}$ with acetic acid compare to hydrochloric acid.

Firstly, increasing amounts of MgO were added to AS2. After precipitation, 50–100 μm needle shape crystals were observed, comparable to those obtained with Na_2CO_3 in the run with AS1. Kabdasli *et al.* (2004) described the same struvite crystal needle shape from a model solution containing an equimolar ratio of ammonium, phosphate and magnesium. In our study, the main difference between AS1 and AS2 was the N/P molar ratio which was 2.25 and 1, respectively. Thus secondly, some complementary trials with ammonium additiond before precipitation, either as raw slurry supernatant or as ammonium salt, were performed increasing the N/P molar ratio from 1 to 1.4. In the both cases larger crystals were formed when NH_4^+ was added previously (Figure 5).

The pH increased from 4.65 to 6 in less than 30 minutes after MgO addition. Then the MgO dissolution was reduced and the maximum was reached in about 3–4 hours (Figure 6). In this run, 100 mM of MgO were required to reach pH 8 in 3 hours. It was significantly more than in run 1. The lower pH of the acidified supernatant and the higher buffer effect of acetic acid compared to hydrochloric acid could explain this difference.

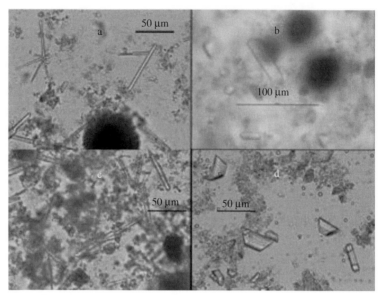

Figure 5. Optical microscopic pictures of crystals formed by MgO addition in acidified supernatant of biologically treated pig slurry (AS2). a: AS2+MgO-pH 8.6; b: AS2+MgO+NH$_4$-pH 8.5; c: AS2 + MgO-pH 9.6;d: AS2 + MgO + RS-pH 9.2.

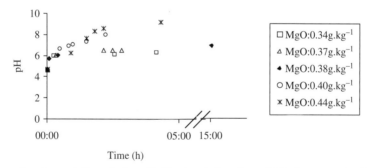

Figure 6. pH evolution during MgO dissolution in acidified supernatant from biologically treated piggery wastewater.

The decantation of the solid formed took less than 1 hour and the supernatant represented about 90% of the initial volume, the solid phase around 10%. Characteristics of the liquid phase are presented in Table 2.

Table 2. Characteristics of the liquid phase obtained by MgO precipitation of P from the acidified supernatant of biologically treated piggery wastewater.

	AS2+MgO			AS2+RS+MgO		AS2+NH$_4$+MgO
	8	8.6	9.6	8	9.2	8.5
TS (g.kg^{-1})	27 ± 1	28 ±	30 ± 1	28 ± 1	27 ± 1	27 ± 1
MM (g.kg^{-1})	14 ± 1	14 ±	17 ± 1	14 ± 1	14 ± 1	14 ± 1
VS (g.kg^{-1})	13 ± 1	14 ±	14 ± 1	14 ± 1	13 ± 1	13 ± 1
TKN (mg.kg^{-1})	500 ± 20	400 ± 8	—	—	600 ± 17	500 ± 13
N−NH$_4^+$ (mg.kg^{-1})	350 ± 16	270 ± 10	230 ± 17	470 ± 23	440 ± 21	470 ± 15
P total (mg.kg^{-1})	37 ± 2	39 ± 2	24 ± 1	27 ± 1	33 ± 2	33 ± 3
Ca (mg.kg^{-1})	595 ± 27	655 ± 13	551 ± 11	679 ± 11	618 ± 9	725 ± 12
Mg (mg.kg^{-1})	1881 ± 93	1724 ± 85	1677 ± 75	1577 ± 73	1415 ± 63	1495 ± 71
Na (mg.kg^{-1})	805 ± 18	865 ± 40	863 ± 39	885 ± 27	680 ± 30	715 ± 25
K (mg.kg^{-1})	2525 ± 112	2495 ± 98	2527 ± 118	2640 ± 120	2320 ± 95	2340 ± 105
Cl (mg.kg^{-1})	2550 ± 110	2563 ± 113	2600 ± 125	2588 ± 118	2525 ± 105	3175 ± 134

Total solids in the solid phase containing were 6–8%. The distribution between particles and interstitial liquid in this phase was calculated from the suspended solid content and the composition of the liquid phase (Table 3).

Table 3. Composition of the solid phase ($mg \cdot gDM^{-1}$)

	AS2	AS2	AS2	AS2 + RS	AS2 + RS	AS2 + NH4+
PH	8	8.6	9.6	8	9.2	8.5
NH4+	3 ± 1	12 ±	13 ±	1 ±	16 ±	12 ±
P total	86 ± 5	86 ± 4	69 ± 4	87 ± 7	75 ± 6	75 ± 5
Cl	11 ± 2	6 ± 1	9 ± 1	–	4 ± 1	3 ± 1
Ca	96 ± 5	86 ± 4	73 ± 4	88 ± 5	85 ± 4	79 ± 3
Mg	127 ± 5	147 ± 7	156 ± 5	190 ± 6	213 ± 8	192 ± 6

Ammonium in solid was increasing when pH was about 9 which is close to the *optima* for struvite precipitation in literature (Doyle and Parsons, 2002). In our experiments, Mg supply and pH could not be dissociated due to the use of MgO; the Mg/Ca ratio was also more favourable to struvite precipitation at higher pH.

If we consider that all the ammonium in the solid phase was struvite, 7 to 47% of phosphorus was struvite form which was increased with pH. Unfortunately, SEM-EDS microscopy and X-ray diffraction were not performed on these samples.

Finally, the run showed that acidification with acetic acid decreased the chloride content in liquid phase but increased the amount of MgO required to precipitate P. In this test, a 3–4 hours mixing period was also required to reach pH 8 and a 1 hour decantation period was sufficient to separate the liquid phase from the phase containing solid. The collected solid was only 6–7% DM. A filtration step could be useful to decrease the duration of the drying step needed to export the product as a dry mineral fertilizer. More struvite was formed at higher pH. The shape and the size of the crystals were more appropriate for filtration when the N/P molar ratio was higher than 1.

CONCLUSIONS

Struvite crystals and amorphous calcium phosphate were the main components identified in the solid phase.

The struvite crystals were larger when the magnesium or ammonium/phosphate ratio was high. This could improve the solid filterability and reduce the drying duration.

Three hours mixing and one hour decantation were sufficient to reach pH 8 which was needed to precipitate and separate more than 90% of the phosphorus from the acidified supernatant of biologically treated piggery wastewater.

Using acetic acid and magnesium oxide instead of hydrochloric acid and sodium hydroxide decreased the chloride and sodium concentration in the liquid phase, limiting the risk of soil salinisation by spreading of the effluent.

Overall this work has shown that 95% of the initial total phosphorus content of acidified supernatant from biologically treated piggery wastewater could be recycled as mineral fertiliser without increasing the salinity of the liquid effluent, locally spread. However an economic study and an optimisation of the process at a larger scale are needed to conclude on the economic feasibility of the dissolution/precipitation recycling process.

REFERENCES

Abbona, F., Lundager, Madsen H.E. and Boistelle, R. (1986). Initial phases of calcium and magnesium phosphates precipitated from solutions of high to medium concentrations. *Journal of Crystal Growth*, **74**(3), 581–590.

Abbona, F., Lundager Madsen, H.E. and Boistelle, R. (1988). The final phases of calcium and magnesium phosphates precipitated from solutions of high to medium concentration. *Journal of Crystal Growth*, **89**(4), 592–602.

Angel, R. (1999). Removal of phosphate from sewage as amorphous calcium phosphate. *Environmental Technology*, **20**(7), 709–720.

Antakyali, D., Ketibuah, E., Schmitz, S., Krampe, J. and Rott, U. (2004). Struvite precipitation from municipal sludge liquor focusing on the development of a mobile unit. International conference on Struvite: its role in Phosphorus recovery and reuse, Cranfield.

APHA (1998). Standard methods for the examination of water and waste, 18th edn. American Public Health Association, Washington, DC, USA.

Burns, R.T., Moody, L.B., Celen, I. and Buchanan, J.R. (2003). Optimization of phosphorus precipitation from swine manure slurries to enhance recovery. *Water Sciences and Technology*, **48**(1), 139–146.

Daumer, M.L, Béline, F., Guiziou, F. and Sperandio, M. (2007a). Influence of pH and biological metabolism on dissolved phosphorus during biological treatment of piggery wastewater. *Biosystems Engineering*, **96**(3), 379–386.

Daumer, M.L., Béline, F., Guiziou, F. and Sperandio, M. (2007b). Effect of nitrification on phosphorus dissolving in a piggery effluent treated by a sequencing batch reactor. *Biosystems Engineering*, **96**(4), 551–557.

Doyle, J.D. and Parsons, S.A.U. (2002). Struvite formation, control and recovery. *Water Research*, **36**(16), 3925–3940.

Frossard, E., Tekely, P. and Grimal, J.Y. (1994). Characterization of phosphate species in urban sewage sludges by high-resolution solid-state 31P NMR. *European Journal of Soil Science*, **45**(4), 403–408.

Giesen, A. (1999). Crystallization process enables low-cost fluoride removal. *Ultrapure Water*, **16**(3), 56–60.

Greaves, J. and Haygarth, P. (2001). A novel biological phosphate pump. Second international conference on Recovery of Phosphates from sewage and animal wastes, Noordwijkerhout, Holland.

Kabdasli, I., Parsons, S.A. and Tünay, O. (2004). Affect of major ions on struvite crystallization. International conference on struvite: its role in phosphorus recovery and reuse, Cranfield.

Kim, B.U., Lee, W.H., Lee, H.J. and Rim, J.M. (2004). Ammonium nitrogen removal from slurry-type swine wastewater by pretreatment using struvite crystallization for nitrogen control of anaerobic digestion. *Water Science and Technology*, **49**(5–6), 215–222.

Le Corre, K.S., Valsami-Jones, E., Hobbs, P. and Parsons, S.A. (2005). Impact of calcium on struvite crystal size, shape and purity. *Journal of Crystal Growth*, **283**(3–4), 514–522.

Le Corre, K.S., Valsami-Jones, E., Hobbs, P., Jefferson, B. and Parsons, S.A. (2007). Agglomeration of struvite crystals. *Water Research*, **41**(2), 419–425.

Nelson, N.O., Mikkelsen, R.L. and Hesterberg, D.L. (2003). Struvite precipitation in anaerobic swine lagoon liquid: effect of pH and Mg : P ratio and determination of rate constant. *Bioresource Technology*, **89**(3), 229–236.

Suzuki, K., Tanaka, Y., Kuroda, K., Hanajima, D. and Fukumoto, Y. (2005). Recovery of phosphorous from swine wastewater through crystallization. *Bioresource Technology*, **96**(14), 1544–1550.

Suzuki, K., Tanaka, Y., Kuroda, K., Hanajima, D., Fukumoto, Y., Yasuda, T. and Waki, M. (2007). Removal and recovery of phosphorous from swine wastewater by demonstration crystallization reactor and struvite accumulation device. *Bioresource Technology*, **98**(8), 1573–1578.

Szögi, A.A., Bauer, P.J. and Vanotti, M.B. (2008). Agronomic effectiveness of phosphorus material recovered from manure. 13th RAMIRAN conference, Albena, Bulgaria, 52–56

Türker, M. and Celen, I. (2007). Removal of ammonia as struvite from anaerobic digester effluents and recycling of magnesium and phosphate. *Bioresource Technology*, **98**(8), 1529–1534.

Valsami-Jones, E. (2001). Mineralogical controls on phosphorus recovery from wastewaters. *Mineralogical Magazine*, **65**(5), 611–620.

Valsami-Jones, E. (2004). Phosphorus in environmental technology, IWA publishing, London, UK, 656 p.

Vanotti, M.B., Szogi, A.A. and Hunt, P.G. (2003). Extraction of soluble phosphorus from swine wastewater. *Transactions of the ASAE*, **46**(6), 1665–1674

Weidelener, A. (2004). Recovery of phosphorus of sewage sludge as MAP. International conference on Struvite: its role in phosphorus recovery and reuse, Cranfield.

Struvite harvesting to reduce ammonia and phosphorus recycle

R. Baur[a], R. Prasad[b] and A.Britton[c]

[a]Clean Water Services, Tigard, OR, USA
[b]Old Dominion University Norfolk, VA, USA
[c]Ostara Nutrient Recovery Inc Vancouver, BC, Canada

Abstract Anaerobically digested sludge dewatering centrate recycle is a significant source of phosphorus and ammonia load to the aeration basins of the Clean Water Services Durham Advanced Wastewater Treatment plant. A small scale demonstration reactor from www.ostara.com was piloted on the centrate stream to see if it would be a practical method to reduce phosphorus and ammonia recycle to both reduce treatment costs and gain treatment capacity. The reactor is an up flow fluidized bed that adds magnesium to the centrate to create 1 to 3 mm spherical struvite particles. The particles are harvested, dewatered and dried to produce a clean product that looks just like commercial fertilizer. It is not considered a biosolid, but regulated by the Dept. of Agriculture as a fertilizer rated as 5-27-0 10% Mg (5% as N, 27% as P_2O_5, and zero potassium). The fertilizer will be marketed locally by Ostara to the \$USD 1billion container nursery market in Oregon.

The reactor reduced the 600 mg/l l T-PO_4-P centrate by 90% and the 1,200 mg/l NH_3-N by 20%. The aeration basin influent phosphorus will be reduced by 24%. Clean Water Services expects to increase the reliability of biological phosphorus removal by reducing the influent phosphorus and the amount of volatile fatty acids required for anaerobic phosphorus release. Treatment costs should be reduced and the struvite produced will be an increasing revenue source. Start up of the first plant scale Ostara process will be Spring 2009.

INTRODUCTION

Clean Water Services operates the 76,000 m^3/d (20 mgd) Durham Advanced Wastewater Treatment Plant serving suburban communities south and west of Portland Oregon. The plant's phosphorus discharge permit is a monthly median of 0.10 mg/l l T-PO_4-P from May to November, dependent on river flow. The Tualatin River receiving stream is a slow moving, warm, effluent dominated river. To meet the stringent permit limits initially a primary and tertiary dose of alum was used to remove phosphorus. To reduce the environmental impact of chemical treatment and the chemical sludge disposal costs, the plant converted to Enhanced Biological Phosphorus Removal (EBPR).

On line ammonia analyzers showed that secondary effluent ammonia bleed through was occurring at peak diurnal flow and ammonia loading. 30% of the ammonia load was from dewatering centrate recycle. The initial approach was to shift dewatering ammonia load to off peak times where there was adequate plant capacity and lower power rates. An additional approach was to convert the complete mix portion of the aeration basins to plug flow, that project increased nitrification capacity 18% in each basin which resulted in delaying the need for expansion for many years. The final approach was side stream treatment of the centrate to remove ammonia.

Sidestream treatment options

Side stream nitrification was investigated and rejected due to lack of existing tankage and the potential for struvite formation. We had problems with struvite in the initial pilot centrate storage tank. Pipes and drains quickly plugged with struvite. We tested pipe lining materials and found struvite would not stick to KYNAR™ (PVDF) plastic. The ultra smooth material does not allow struvite to adhere to the walls.

We learned a lot about struvite and solved many of the issues. Baur (2008). We were very interested when Ostara (www.ostara.com), developed a process to make struvite on purpose from centrate and market the struvite as a slow release fertilizer. The benefits of struvite precipitation over other side stream process are nutrient recovery, rather than nutrient removal, and removing both ammonia and phosphorus in a low energy process with a marketable end product, a slow release fertilizer.

PILOTING THE OSTARA PROCESS

Reactor operation

Ostara's 1 liter per minute pilot reactor (Figure 1) was installed to treat the dewatering centrate that had approximately 600 mg/l T-PO_4-P and 1,200 NH_3-N. To provide centrate to the reactor when dewatering was off line, two tanks in series were used to store the centrate. The centrate also off gassed CO_2 in these tanks since the pressure was reduced to atmospheric pressure from the pressure in the digesters. The release of CO_2 raised the pH prior to the reactor.

The Ostara reactor is an up flow fluidized bed reactor with the diameters increasing in several steps from the bottom to the top. Initially the recirculation flow was 20 times the feed flow. The recirculation flow fluidizes the bed and dilutes the reactants so the struvite precipitates on existing particles rather than spontaneously forming tiny crystals.

Figure 1. Ostara pilot reactor.

After seeding the reactor with struvite 1 to 3 millimeter struvite prills (spherical fertilizer particles are called prills) from previous pilots, reactor operation began. After a short while a new coating of struvite was visible on the prills. The recirculation rate and different diameters maximize uplift and turbulence in the lower sections so the largest, heaviest prills are located there where the concentration of the reactants are the highest. As particles move up the column, up flow velocity and turbulence decrease until at the top there is a clarifier with weir overflow. Particle size decreases as you move up the column since the heaviest prills sink and the lightest rise due to the up flow velocity. Harvesting removes the largest prills from the bottom. Slightly smaller prills are now at the bottom where the fresh centrate and magnesium is introduced causing them to grow in diameter. The increasingly smaller particles above have lots of surface area to scavenge the reactants and accumulate struvite. Fine particles are swept to the top of the column. The prills accumulate concentric layers of struvite and work their way to the bottom as the get larger where they would be harvested at the desired size. The accumulation of material is like a pearl growing in an oyster where layer after layer is precipitated on the surface resulting in a dense, hard, round prill. The sections of the reactor act similar to a thundercloud making hail. As the prills circulate due to the up flow velocity, they gain weight until the can no longer be supported by the up flow velocity. They fall to the next lower section where the struvite reactants are at a higher concentration and further deposition occurs causing them to sink lower in the reactor to where they are finally harvested. The particle size is determined by the harvesting rate, a longer retention time will produce larger prills.

Ostara supplied a graduate student to run the pilot and analyze operating data while CWS did the laboratory analysis. The precipitation rate is controlled by the concentration of the reactants, feed flow, recirculation rate and pH, all controlled by the Ostara PLC. In our case with high ammonia and phosphorus concentrations and CO_2 off gassing raising the pH in the holding tank, no additional pH adjustment was needed. The operating variables were MgCl2 dose and recirculation flow to generate up flow velocity (Table 1).

The reactor typically achieved a 95% removal of phosphorus and a 20% removal of ammonia. Struvite is a 1:1:1 molar ratio of $Mg:NH_3:PO_4$ with 6 waters of hydration, so we removed an equal number of moles of ammonia and phosphorus, but their atomic weight are much different along with their molar concentration in the centrate. The operation of the reactor was very stable. We had several power interruptions that stopped the reactor for several hours until it was restarted. The prills in the reactor did not stick together and when recirculation flow was re-established, the bed expanded and the prills were again sorted by size with no large clumps generated. When fresh

Table 1. Overall nutrient removal and molar ratio during the pilot study.

	Up flow velocity	Average % removal		Average molar ratios		
		P % removal	NH3 % removal	P	NH3	Mg
	400 cm/min	97	19	1.00	1.05	0.89
	500 cm/min	95	18	1.00	0.99	0.89
	450 cm/min	92	18	1.00	1.05	0.91
	350 cm/min	94	21	1.00	1.18	0.94
Overall Average		**95**	**19**	**1.00**	**1.07**	**0.91**

centrate and MgCl2 flow was re-established the process resumed as if nothing had happened.

Running the pilot at different set points to determine the optimum operating envelope resulted in some non-optimum performance in phosphorus capture or prill quality (Figure 2). In one case the particles became soft and crumbly. The process recovered when the set points were returned to known good values and the prill hardness and phosphorus removal returned to normal.

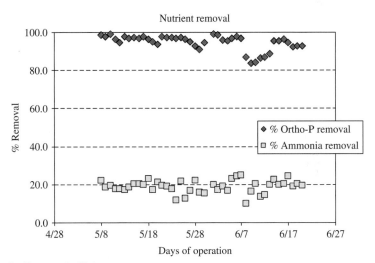

Figure 2. Removal efficiency.

The pilot demonstrated the process stability, the power interruptions or process upsets could be recovered from without difficulty.

Reactor products

The struvite prills produced look like typical fertilizer prills (Figure 3). Ostara markets it as Crystal Green™. The fertilizer rating is 5% nitrogen as N, 27% phosphorus as P_2O_5, zero potassium and 10% magnesium as Mg. (5-27-0 10% Mg). Since struvite is relatively insoluble, it acts as a slow release fertilizer over a 9 month period. In the fertilizer market it will compete with slow release fertilizers that are made from polymer coated soluble fertilizers. Testing showed the absence of bacteria and viruses and very low heavy metals, less than commercially mined phosphorus fertilizers.

Figure 3. Struvite prills from pilot.

Reactor effluent

The treated centrate showed no tendency to continue to deposit struvite since the phosphorus levels were reduced to around 25 mg/l $T-PO_4-P$.

FULL SCALE STRUVITE RECOVERY ANALYSIS

Ostara has one full size reactor in Edmonton, Alberta, Canada. We visited the site and were impressed with the equipment and Crystal Green™ product. The full scale reactor made a prill that was even better than the pilot. It was harder and completely dustless. We investigated the cost and benefit to operate a full scale Ostara reactor system. An engineering study was commissioned to develop a basic design and rough cost numbers. To get the cost down, we were able to place the three reactors in an old influent pump station that was being decommissioned, that avoided $USD 1 million to design and build a building for the equipment (Figure 4). Since we have a large centrate storage tank, we eliminated caustic dosing equipment planning on off gassing of CO_2 to raise the pH a bit.

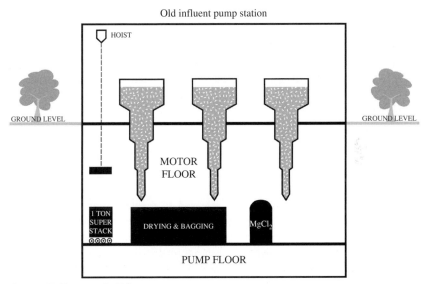

Figure 4. Reactor building.

We felt that the concentration of ammonia and phosphorus in the centrate was stable enough that on line analyzers were not needed, the lab turnaround would be adequate. If needed, those systems could be added later.

We investigated the market for the struvite product. The Durham Plant is located in the Willamette Valley which has a $USD 1 billion container nursery industry. Oregon State University Extension Service and local fertilizer blenders were very interested in the product. The Crystal Green™ prill looks and handles just like normal fertilizer and can be applied or handled with existing equipment. The product is not a biosolid; it is a sterile and dry chemical precipitant. The fact that it was generated locally from a sustainable process was a plus.

Phosphate ore is being rapidly depleted and phosphate fertilizer prices have quadrupled or more in a relatively short time.

Combining projected cost savings from reduced chemical and energy use with savings from increased plant capacity and incentives from Ostara to be the first full scale plant resulted in a contract that meets our 5 year payback requirement. CWS will pay for a design build turnkey system and operate it with Ostara remotely monitoring and controlling the process. Ostara will buy 100% of all Crystal Green™ produced and market it to the local container nursery industry and other users. The estimated production is 36 tonnes (40 US tons) per month. There is capacity for increased production as flows and loads go up.

The benefit to the plant will be significant, a State wide phosphate detergent ban reduced plant influent phosphorus 22% in the 1980's, the Ostara reactors will reduce it a further 24% along with a 6% ammonia reduction while producing revenue from a marketable product.

As of this writing, construction is ongoing with startup scheduled for spring 2009. The Oregon State Dept. of Agriculture has licensed the facility as a fertilizer producer.

CONCLUSIONS

EBPR simply moves the phosphorus from the liquid stream, concentrating it in the waste activated sludge which is then transferred to the solids stream. In the anaerobic digester the phosphorus is released to create nuisance struvite and a dewatering centrate high in phosphorus and ammonia. We were originally looking at ammonia side stream treatment but quickly realized the importance of removing phosphorus and converting it into a revenue stream. Treating the side stream by adding magnesium to the centrate will precipitate both phosphorus and ammonia as struvite using very little energy. The Ostara process makes the struvite as a value added, slow release, 5-27-0 10% Mg fertilizer in a compact, dense, spherical dustless prill that is readily acceptable to the end user. The magnesium added is an important plant nutrient and the process increases the value of the magnesium by converting it to a slow release nutrient. Every chlorophyll molecule has magnesium in the center and is important in growing a vibrant green plant. Dramatic increases in the cost of transportation, phosphate raw materials and slow release coating materials have increased interest in a locally produced, sustainable product. The product is not a biosolid; it is a sterile and dry chemical precipitant. The Oregon State Dept. of Agriculture has licensed the facility as a fertilizer producer.

The reduced phosphorus recycle load should result in more stable operation of EBPR with less VFA needed for EBPR while delaying future expansion needs. The system has low energy requirements and the revenue will repay the initial investment and become a profit center. Carbon credits may be available because the energy and transportation demands are much less when recovering nutrients compared to mining, processing and transporting fertilizer. In some locations nutrient trading credits may be available for the nitrogen and phosphorus removed.

REFERENCE

Baur, R., Struvite control and prevention in centrate. *WEFTEC 2008. Proceedings Session* 67.

The application of process systems engineering to the development of struvite recovery systems

P.A. Schneider* and Md. I. Ali

School of Engineering, James Cook University, Townsville, Queensland, 4811, Australia

*Corresponding author

Abstract A model for struvite suspension crystallisation from seed, incorporating solution thermodynamics and crystal growth kinetics, is developed. The set of differential algebraic equations is implemented within the gPROMS simulation environment. The aqueous phase chemical equilibria employed in the model are validated against PHREEQc. Unknown kinetic model parameters related to crystal growth rates were successfully estimated, by combining this model with an ensemble of data generated in three separate experiments carried out in a 44-l suspension crystalliser. In principle this model can be used for more confident process design of suspension crystallisation of struvite. The approach described can potentially be applied to precipitation-based struvite recovery systems.

INTRODUCTION

Increased levels of research have been undertaken in the past decade to elucidate the behaviour of struvite in relation to solution thermodynamics and its precipitation from solution. In order to confidently design production-scale struvite recovery units, it is now necessary to apply more process-relevant research methodologies.

Solution thermodynamic behaviour has been well characterised by Ohlinger *et al.* (1998) through their regression of the activity solubility product for struvite. Of course, this does not imply that solubility predictions can be made for *real* wastewater solutions. In that case, extra constituent elements and their ion speciations must be known and can only be taken into account with accurate knowledge of the relevant equilibrium constants.

A number of workers have also evaluated struvite precipitation kinetics, based on one of two operational modes; fluidised bed or suspension crystallisation. This paper will focus on modelling suspension crystallisation based recovery systems, although the concepts developed could well be applied to either mode of operation.

The proposed model combines solution thermodynamics with a description of crystal growth, incorporating mass and simplified population balance relations.

The model is implemented within gPROMS (i.e. general process Modelling System). Estimation of unknown model parameters via regression analysis, using a single model with an ensemble of data from one, or more, experiments is described. This leads to a set of valid kinetic model parameters describing struvite crystallisation.

PROCESS MODELLING

The following sections outline the process chemistry and thermodynamics, crystallisation kinetics and process topography for a fed-batch, seeded struvite suspension crystalliser.

Process chemistry and thermodynamics

This section gives a short review of struvite thermodynamic modelling, a brief introduction to relevant crystal growth modelling and finally the overall process model employed to describe the fed-batch system under investigation.

Solid phase struvite is created from the following ionic precipitation reaction.

$$Mg^{2+} + NH_4^+ + PO_4^{3-} + 6H_2O \rightleftarrows MgNH_4PO_4 \cdot 6H_2O$$

Struvite precipitation is governed by the saturation of solution with respect to the activities of the reactive ions. This is conveniently captured in the supersaturation ratio, Ω, which is the logarithm of the ratio of the ion activity product, IAP, to the solubility product of struvite, K_{sp}.

$$\Omega = Log\left(\frac{IAP}{K_{sp}}\right)$$

where

$$IAP = \{Mg^{+2}\}\{NH_4^{+1}\}\{PO_4^{-3}\}$$

Determination of ion activities is accomplished by solving the set of nonlinear algebraic relations related to equilibrium ion speciation relations, mole balances on magnesium, nitrogen and phosphorus, an appropriate formulation for activity coefficients (using the Davies equation) and a charge balance based on electroneutrality conditions. The charge balance can be replaced by setting the system pH, which was the case in the present work. Ion speciations and their equilibrium constants are presented in Table 1. The struvite K_{sp} employed was $10^{-13.26}$ and it was assumed that no other solid phases would form.

Table 1. Ion speciation considered in this study.

Equilibrium	Value	Reference
K_{MgOH^+}	$10^{-2.56}$	(Childs, 1970)
$K_{NH_4^+}$	$10^{-9.25}$	(Taylor et al., 1963)
$K_{HPO_4^{2-}}$	$10^{-12.35}$	(Morel and Hering, 1993)
$K_{H_2PO_4^-}$	$10^{-7.20}$	(Morel and Hering, 1993)
$K_{H_3PO_4}$	$10^{-2.15}$	(Martell and Smith, 1989)
$K_{MgH_2PO_4^+}$	$10^{-0.45}$	(Martell and Smith, 1989)
K_{MgHPO_4}	$10^{-2.91}$	(Martell and Smith, 1989)
$K_{MgPO_4^-}$	$10^{-4.80}$	(Martell and Smith, 1989)
K_{H_2O}	10^{-14}	(Snoeyink and Jenkins, 1980)

Process kinetics

In order to characterise the growth of struvite crystals, experimental conditions were designed so that secondary nucleation was avoided, by keeping the solution in the metastable zone. As such, seed crystals introduced to the system should grow, owing to the relative supersaturation in the system, which is defined as follows.

$$\sigma = \Omega^{1/2} - 1$$

In general, the growth rate of crystal from aqueous solution depends on the supersaturation, temperature, fluid hydrodynamics (agitation and mixing), impurities concentration and the past history of crystals including imperfections, cracks and size dispersions (White 1971). Crystal growth is characterised by the rate of change of its linear dimension, L. The rate of change of L depends on the relative supersaturation, σ, according to

$$G = \frac{dL}{dt} = k\sigma^n$$

where G is the linear growth rate and k and n are the growth rate constant and the order of the dependency of growth on relative supersaturation, respectively.

Based on the assumption of a volume equivalent spherical diameter, the mass, m_{MAP}, of a single crystal thus changes according to

$$\frac{dm_{MAP}}{dt} = \frac{1}{2}\pi\rho_c L^2 G$$

where ρ_c is the crystal density. The total mass of the crystal population can only be determined if the size distribution is known at any point in time. In a batch or semi-batch system, if we assume that the crystal population has a point size distribution (i.e. all crystals are the same size) and that the number of crystals, n, is preserved throughout the batch time (i.e. no nucleation, agglomeration, breakage or dissolution), then the rate of change of the crystal population mass, M_{MAP}, becomes

$$\frac{dM_{MAP}}{dt} = \frac{1}{2}n\pi\rho_c L^2 G$$

Process description

The experimental system under consideration is a fed-batch suspension crystalliser. A fed-batch system is employed in an effort to maintain a constant relative saturation as long as possible. Continuous mode operation was not chosen, since this would require significant amounts of feedstock and the system would eventually be polluted by struvite nuclei, which was counter to this study's goals.

Crystalliser volume changes, subject to inlet flow, $F_{in}(t)$, according to

$$\frac{dV}{dt} = F_{in}(t)$$

The general dynamic balance on the mass of struvite constituent ions, M_i, is

$$\frac{dM_i}{dt} = F_{in}C_{i,in} - \frac{dM_{MAP}}{dt}\left(\frac{MW_i}{MW_{MAP}}\right)$$

where the subscript, i, represents either Mg^{2+}, NH_4^+ or PO_4^{3-}, $C_{i,in}$ is the concentration of the feed in the ith ion and MW is molecular weight.

Model simulation and thermodynamic validation

The thermodynamic, kinetic and process relations lead to a set of nonlinear differential algebraic equations (DAEs) that is coded into gPROMS and solved over a batch time that depends on when the system reaches its final volume, simulating the behaviour of the fed-batch suspension crystalliser.

The thermodynamic relations employed in the model of struvite crystallisation were validated against predictions made by MINTEQ (Allison *et al.*, 1991). Figure 1 shows good agreement between the two, apart from the speciation of MgH_2PO_4, across a range of pH. This difference makes only a small impact on the system mass balance ($< 1\%$) and, as a result, can be ignored.

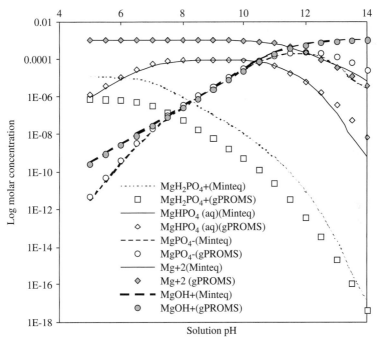

Figure 1. Thermodynamic comparison between gPROMS and MINTEQ.

To simulate this system it was necessary to assume values for the kinetic parameters, k and n. The next section describes the experimental and computational approach to estimating the values of these hitherto unknown parameters.

Solving this system of nonlinear DAEs is a nontrivial exercise and care must be taken to ensure appropriate scaling of variables and appropriate formulation of model equations. While this was the most challenging aspect of the model development, it is necessary to develop a dynamic simulation of the system that is robust. Otherwise the regression of the unknown kinetic parameters – which requires a large number of model simulations across the parameter space – may not have been feasible.

PARAMETER ESTIMATION

This section details the experimental and computational steps taken to estimate the unknown growth rate parameters, k and n.

Experimental design

Experiments were carried out in a 44-l acrylic vessel in fed-batch mode, shown in schematic in Figure 2. The system is essentially a scaled-up version of the apparatus used by Bouropoulos and Koutsoukos (2000) in which the two stock solutions are fed to the reactor in order to maintain supersaturation at a constant value. Details of the reactor design and operation may be found in Ali and Schneider (2006).

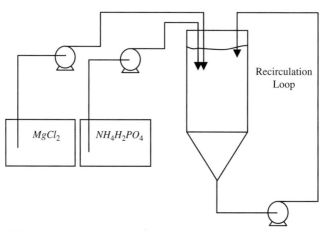

Figure 2. Schematic of pilot scale struvite crystalliser.

Figure 3 shows a typical time course throughout the fed-batch crystallisation. Solution pH control was affected using an industrial controller, which signalled both dosing pumps to add reagents subject to a proportional-integral control algorithm. It is clear that total magnesium and phosphorus concentrations are reasonably close to their desired concentrations; in this case 0.0045 M.

Samples of crystal suspension were extracted from the system and filtered so that crystals could be sized by laser scattering and ICP-AES analysis done to quantify magnesium and phosphorus concentration in the solvent.

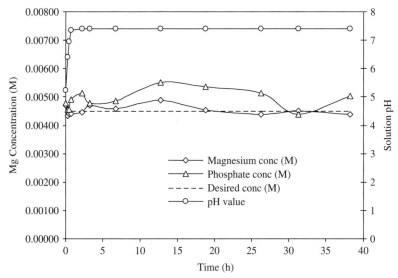

Figure 3. Solution pH and ion concentrations in typical fed-batch experiment.

Regression of growth rate parameters *k* and *n*

The numeric values chosen for k and n will affect the rate at which struvite formation is predicted by the differential-algebraic model. This can be quantified experimentally by measuring the mean size of the crystal size distribution (CSD), since the model predicts this quantity. This in turn affects the amount of feed solution required the reactor, due to the feedback control system in place on the crystalliser. For example, higher rates of growth would demand greater feed flow rates into the system.

The parameter space (i.e. $[k, n]$) was searched in order to find the minimum of the sum of the square of the residuals between predicted and measured process outputs. gPROMS is ideally suited to this task, since it incorporates the previously implemented DAE simulation in its parameter estimation block and can employ any number of experimental trials (three in this case). See Ali and Schneider (2008) for further details.

RESULTS AND DISCUSSION

Figure 4 shows the relationship between the model-predicted crystal size and solution concentrations to those of one of the three experiments used in the regression of the unknown parameters. Good agreement is seen between the two, which is encouraging at pilot scale.

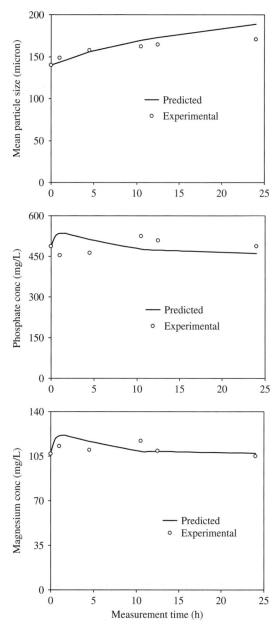

Figure 4. Model-predicted versus measured process outputs.

The proposed growth rate for struvite grown in a suspension crystalliser is

$$\frac{dL}{dt} = (46.64 \pm 8.026)\sigma^{1.48 \pm 0.162}$$

where the units for k are $\mu m/h$. The value for n is encouraging, since it lies between 1 and 2, which is expected from theory and observed in other inorganic systems (Mullin 1993). The confidence intervals for each parameter value indicate one standard deviation of its estimated value, which gives reasonable confidence in these estimates.

The two kinetic parameters are somewhat strongly correlated, based on the confidence ellipsoid shown in Figure 5. This may improve if supplementary data, possibly spanning wider operating conditions, were added to the ensemble data set used in the regression analysis.

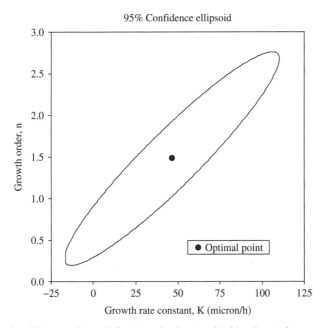

Figure 5. Confidence ellipsoid for k and n in struvite kinetic study.

CONCLUSIONS AND RECOMENDATIONS

An important result from this work is that, for the first time, a mathematical model for struvite crystallisation has been developed and implemented. This model incorporates validated chemical equilibria and proposed a kinetic model describing the growth of struvite crystals from seed. This DAE model was implemented using the gPROMS simulation environment using assumed values for the unknown growth rate parameters.

The unknown kinetic parameters that characterise struvite crystal growth were then estimated by combining the model with an ensemble of pilot-scale experimental data across three experiments.

In principle, this growth model can now be incorporated into process configurations, such as batch- or continuous-mode. Thus more rigorous process designs can be affected, leading to greater confidence for process engineers responsible for their design.

Similar studies should be undertaken to describe the intentional formation of struvite by precipitation, rather than crystal growth. While the mechanisms, and therefore the kinetic structure of the model, will differ for precipitation – with the possible inclusion of nucleation, growth and agglomeration – the approach described herein could well be applied.

REFERENCES

Ali, M.I. and Schneider, P.A. (2006). A fed-batch design approach of struvite system in controlled supersaturation. *Chemical Engineering Science*, **61**(12), 3951–3961.

Ali, M.I. and Schneider, P.A. (2008). An approach of estimating struvite growth kinetic incorporating thermodynamic and solution chemistry, kinetic and process description. *Chemical Engineering Science*, **63**, 3514–3525.

Allison, J.D., Brown, D.S. and Novo-Gradac, K.J. (1991). MINTEQA2/PRODEFA2: A Geochemical Assessment Model for Environmental Systems (Version 3.0 User Guide). Athens: Georgia, U.S Environmental Protection Agency (EPA).

Bouropoulos, C.C. and Koutsoukos, P.G. (2000). Spontaneous precipitation of struvite from aqueous solutions. *Journal of Crystal Growth*, **213**, 381–388.

Childs, C.W. (1970). A potentiometric study of equilibria in aqueous divalent metal orthophosphate solutions. *Inorganic Chemistry*, **11**, 2465–2469.

Martell, A.E. and Smith, R. M. (1989). *Critical Stability Constants*, New York, Plenum Press.

Morel, F.M.M. and Hering, J.G. (1993). *Principles and Applications of Aquatic Chemistry*, New York, John Wiley and Sons.

Mullin, J.W. (1993). *Crystallization*. 3rd, Butterworth-Heinemann Publications, Ipswich, UK.

Ohlinger, K.N., Young, T.M. and Schroeder, E.D. (1998). Predicting struvite formation in digestion. *Wat. Res.*, **32**(12), 3607–3614.

Snoeyink, V.L. and Jenkins, D. (1980). *Water Chemistry*. John Wiley and Sons, USA.

Taylor, A.W., Frazier, A.W. and Gurney, E.L. (1963). Solubility product of magnesium ammonium. *Transaction Faraday Society*, **59**, 1580–1589.

White, E.T. (1971). *Industrial Crystallization: A Short Course*. University of Queensland.

Membrane EBPR for phosphorus removal and recovery using a sidestream flow system: preliminary assessment

H. Srinivas[a], F.A. Koch[b], A. Monti[c], D.S. Mavinic[b] and E. Hall[b]

[a]Levelton Consultants Ltd., Richmond, BC, Canada
[b]Department of Civil Engineering, University of British Columbia, Vancouver, BC, Canada
[c]GE Water and Process Technologies, Oakville, ON, Canada

Abstract Phosphate removal and recovery was combined in a Membrane Enhanced Biological Phosphorus Removal (MEBPR) process. This was achieved by struvite precipitation, using a magnesium ammonium phosphate (MAP) crystallizer, from the supernatant of the sludge drawn from the anoxic zone in an MEBPR process. A modeling technique was employed in order to answer several questions regarding the behaviour of an MEBPR system for the removal of sludge from different compartments, to obtain a supernatant. Simulations, using the ASM2-TUD Bio-P model, revealed that the sidestream from an anoxic zone yielded better conditions for struvite precipitation, when compared to sidestream from aerobic and anaerobic zones. Up to 78% of the incoming phosphorus was estimated to be recovered by implementing a sidestream MEBPR process. This configuration was successfully implemented at pilot scale, as part of a feasibility study.

INTRODUCTION

Processes which allow for simultaneous phosphorus recovery and removal have been discussed at length in recent years. The prime reasons for this are (1) a dwindling presence of high-quality phosphate ore as a natural resource, which is expected to exhaust globally in next 50–60 years (Britton *et al.*, 2005), (2) stringent discharge standards regarding phosphorus and nitrogen compounds worldwide (Britton *et al.*, 2005), and (3) the formation of struvite (PO_4MgNH_4 6 H_2O) in the piping of sludge treatment processes, thereby increasing the cost of pumping and maintenance (Britton *et al.*, 2005; Doyle *et al.*, 2002). This paper describes the process of removal and recovery of phosphorus by integrating a Membrane Enhanced Biological Phosphorus Removal (MEBPR) process and a MAP (magnesium ammonium phosphate) crystallizer, through a sidestream originating from one of the zones of the MEBPR process itself.

In the past, several variants of sidestream processes have been discussed or implemented for the recovery of phosphorus (Saktaywin *et al.*, 2005; Drnevich 1979; Woods *et al.*, 1999; Smolders *et al.*, 1996). In these processes, a sidestream usually originates from the secondary clarifier. For example, there is the Phostrip process (Drnevich, 1979), in which phosphorus is recovered from the return sludge. In this study, a pilot-scale MAP crystallizer, developed at The University of British Columbia (UBC), was used for the recovery of phosphorus. Traditionally, MAP crystallizers are used to recover phosphate (in the form of struvite) from the secondary sludge, anaerobic digester supernatant stream (Munch and Barr, 2001; Britton *et al.*, 2005). In the present study, however, a phosphorus rich supernatant was not obtained from the secondary sludge digestion, but directly from one of the MEBPR process zones, as part of a preliminary assessment of its feasibility. The sidestream process proposed in this study has the potential to integrate the best available processes for the removal (MEBPR) and recovery (MAP crystallizer) of phosphorus. The sidestream configuration was first simulated using the ASM2–TUD Bio-P model on the AQUASIM platform, to gain more understanding of the process. The experimental investigation, itself, lasted more than a year. Based on the simulation results, a MAP crystallizer was fed with the supernatant obtained from the anoxic zone of the MEBPR process and it was operated in a batch mode, for short period of time. The aim of this research was to test the sidestream MEBPR process, at a pilot scale, to recover phosphate in the form of struvite.

METHODOLOGY

Model and process design

The ASM2-Delft metabolic Bio-P model was implemented and calibrated on an AQUASIM platform (simulation software) for the MEBPR process of the UBC pilot plant (Figure 1) by Al-Atar (2007). The calibration protocol used and the calibrated parameter values, are fully described by Al-Atar (2007).

The simulation software used was the second version of AQUASIM (Richert, 1994). The MEBPR model, in AQUASIM, was extended to include a sidestream process which consisted of a phosphorus releasing unit (PRU) and a clarifier (Figure 1). The PRU was simulated in a similar fashion as for the anaerobic zone of the main MEPBR process. The clarifier was assumed to be an ideal separator, except that a residual X_{eff} component was added to simulate include the loss of sludge through overflow. All simulations were calculated over a period of 600 days, to attain steady state conditions. Further, all zones were constructed as completely-mixed reactors. However, in the sidestream clarifier, particulate matter

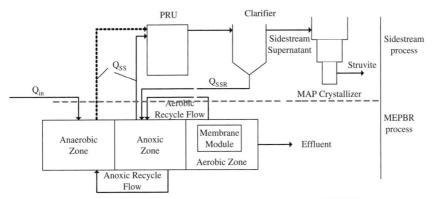

Figure 1. Schematic of sidestream process integrated with MEBPR process.

was separated from the liquid. Under steady state conditions, the portion of the flow going to the sidestream (Q_{SS}) was the only variable parameter in this study.

The sidestream flow was calculated as the percentage of the inflow (Q_{in}).

$$Q_{SS} = f_1 Q_{in} \qquad (1)$$

where f_1 is the fraction of influent to be taken as sidestream and which varied from 0.01 to 1.

Q_{SSR} is the flow going back to the anoxic zone from the sidestream clarifier and was given as:

$$Q_{SSR} = f_2 Q_{SS} \qquad (2)$$

where f_2 is the fraction of Q_{SS} to be taken as underflow and was set to 0.2 throughout this study.

MEBPR process

The MEBPR process at the UBC pilot plant is a modified version of the University of Cape Town (UCT) process that consists of three zones; an anaerobic zone, followed by an anoxic zone and then an aerobic zone. The general characteristics of the influent wastewater are presented in Table 1. All recycle flows were set to the influent flow rate, which was 5.4 m³/d. A hydraulic retention time (HRT) of 10 hours and solids retention time (SRT) of 20 days was maintained throughout this study. Two coupled membrane modules were used for this study and were directly immersed in the aeration zone. The surface area of each module was 12 m² and the membranes had a nominal pore size of 0.04 μm. MEPBR process control and membrane operations at the pilot plant are fully described by Monti (2006) and Hemanth (2007).

Table 1. Influent wastewater characteristics.

Parameter	Mean	Min-Max
TSS (mg/L)	90.1	16–180
COD_{tot} (mg/L)	291	122–590
COD_{sol} (mg/L)	186	80–467
Acetate (mg/L)	18.2	0.0–43.4
Propionate (mg/L)	5.2	0.0–24.4
Tot VFA's (mg COD/L)	19.0	0.0–76.57
TKN (mg N/L)	35.9	25.8–47.3
NH_4-N (mg N/L)	26.7	9.1–39.2
NO_3-N (mg N/L)	Not detec.	
TP (mg/L)	4.2	2.1–7.9
PO_4-P (mg/L)	3.0	1.1–6.7
Mg	1.2	
T°C	20.2	14.0–24.0
pH	7.2	6.4–7.8

Sidestream configuration

The sidestream (Q_{SS}) from the MEBPR process entered the phosphorus release unit (PRU), the volume of which was 120 liters. The PRU was well mixed, with the aid of a mechanical mixer, so that the acetic acid added was distributed throughout the unit. The outlet was connected to a clarifier with a volume of 57 liters. The strength of the acetic acid was the same as that added to the anaerobic zone. The acetate flow was maintained at 0.015 mL/min, thus ensuring that a sufficient amount of volatile fatty acid (VFA) remained (in the range 40–50 mg/L) in the supernatant produced. This was required to confirm that the VFA was not a limiting factor and that maximum phosphorus release was obtained. Supernatant from the sidestream clarifier was collected in a separate tank, and which formed the feed to run the crystallizer. Underflow from the clarifier was returned to the anoxic zone (Q_{SSR}), at a rate of 20% of the incoming sidestream flow.

Struvite crystallizer

A pilot scale reactor (MAP Crystallizer), developed previously at The University of British Columbia (UBC), was used in this study (Figure 2). The reactor design is based on the fluidized bed reactor concept. More information on the reactor design and operations are provided elsewhere (Srinivas, 2007; Britton, 2005).

The sidestream supernatant was stored in a collection tank (~3000 L) for batch wise processing. The crystallizer, once started, was maintained in continuous operation for three to four days at a time. The feed to the crystallizer consisted of

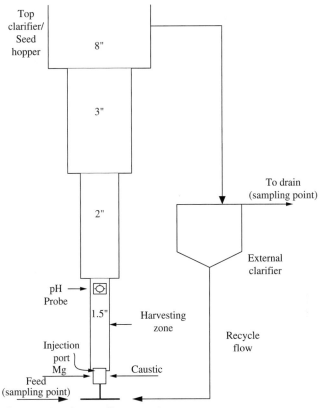

Figure 2. Schematic of crystallizer reactor.

sidestream supernatant, a mix of magnesium chloride and ammonium chloride solution and caustic soda, all of which were fed to the injection port of the crystallizer. In the injection port, the solutions were mixed to achieve the required super-saturation ratio (SSR). Details on SSR and method of controlling the reactor operations are provided elsewhere (Britton *et al.*, 2005; Adnan *et al.*, 2003a, 2003b, Srinivas 2007).

Sampling and analyses

A systematic sampling and analytical program was designed for monitoring the MEBPR process performance, sidestream performance and crystallizer performance. Details on sampling and analyses are given by Srinivas (2007).

Sidestream wasting

After the sidestream approach was implemented into the MEBPR process, there was poor settling of sludge in the sidestream clarifier. In order to obtain a clear supernatant, it was decided that the clarifier itself would be drained once daily. Initially, it was drained into the anoxic zone, but later it was removed from the system as waste sludge (referred to as sidestream wasting), so as to help control the SRT (Srinivas, 2007).

RESULTS AND DISCUSSION

Simulation results

The ASM2 bio-P model was useful in predicting the most suitable conditions for operations of sidestream process. Two ways of taking sidestream from the MEBPR process were considered; sidestream from a single zone and sidestream from two zones (combined). Simulation conditions were identical to the UBC pilot plant MEBPR process (see Methodology section). Some of the key observations are discussed below.

When sidestream was taken from only one zone, the anoxic zone produced a supernatant which was most suitable for struvite production as it exhibited the optimum NH_4-N and PO_4-P release (Figure 3). The molar ratio of NH_4-$N_{(tot)}$: PO_4-$P_{(tot)}$ was 1:1 (Figure 4) which favored struvite formation, when compared with the anaerobic and aerobic zones. The anaerobic zone supernatant also exhibited favorable characteristics for struvite formation; however, PO_4-P release from the anoxic zone sidestream was higher than from that of the anaerobic zone, which indicates more phosphate was diverted for recovery. The aerobic and anoxic zones were the primary sources of harvestable phosphorus, as a result of the phosphate uptake phenomenon occurring in these two zones. Ammonia was not present in the aerobic zone due to the nitrification process occurring there. Moreover, the anaerobic zone has both phosphorus and ammonia supplied by the influent and the anoxic recycle line.

From Figure 5, it is clear that phosphate and ammonia release are not optimum (ammonia deficient) when sidestream originated from anaerobic/aerobic zones. Theoretically, an aerobic/anoxic combination would produce a super-rich phosphorus supernatant, but with very little ammonia; this means the addition of ammonia would be required to produce struvite. A second option would be to mix this particular supernatant with the ammonia rich supernatant, usually obtained from the anaerobic digestion of return sludge. Overall, the advantage of the anaerobic/anoxic combination was that more NH_4-N and PO_4-P were already available for recovery, when compared with the other

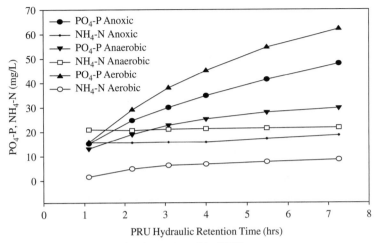

Figure 3. PO$_4$-P and NH$_4$-N release trend in PRU.

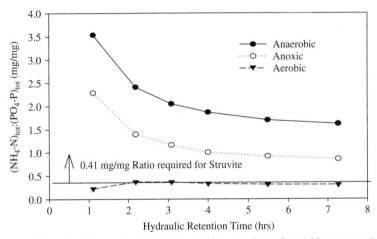

Figure 4. NH$_4$-N: PO$_4$-P ratio in the anaerobic, anoxic and aerobic supernatants.

combinations. In other words, the required molar ratio for struvite formation was best achieved by the anaerobic/anoxic combination.

For both individual and combined sidestreams, the maximum PO$_4$-P release was reached within one hour of retention time in the phosphate release unit (Figure 6). Subsequently, there was no significant additional PO$_4$-P release, with time. In Figure 6, the retention time was constant for each f_1 value

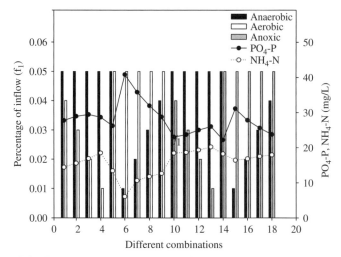

Figure 5. PO$_4$-P and NH$_4$-N release trend for combined sidestream flow.

(i.e. 0.05, 0.1 and 0.2) by varying the volume of the simulated PRU. The initial concentration of PO$_4$-P in the anoxic zone (4.5 mg/L) of the main MEBPR process is also plotted on the y axis (at zero HRT in PRU). The initial concentration of PO$_4$-P in the anaerobic zone was 12 mg/L, while the initial concentration of PO$_4$-P for the combined sidestream (anoxic/anaerobic), was calculated to be approximately 8.2 mg/L. It can be seen from Figure 6 that, for conditions $f_1 = 0.05$ (anoxic/anaerobic) and $f_1 = 0.1$ (anoxic), the phosphate release was similar. The difference observed was in the ammonia release; for the combined sidestream process, it was around 20 mg N/L (Figure 5) and for the individual sidestream, it was around 12 mg N/L (see Figure 3). As discussed earlier, (if not for this) the anoxic/anaerobic combination would be preferred, since the required molar ratio of PO$_4$-P to NH$_4$-N for struvite formation could be achieved. Further, from Figure 6 it appears that, at higher sidestream flows ($f_1 = 0.2$), there is either no or little release of PO$_4$-P in the PRU.

At lower f_1 values (<0.2), effluent PO$_4$-P and NH$_4$-N values of the main MEBPR process remained unaffected. When f_1 was increased with f_2 (fraction of sidestream returning to main process) kept at zero, there was an increase in effluent PO$_4$-P and NH$_4$-N values of the MEBPR process. When f_2 was set at 0.1, the process regained its stability. Thus, the MEBPR process required a minimum f_2 to be set at 0.1, otherwise there was noticeable deterioration in the effluent quality (Figure 7a, b and c). This was somewhat expected since, at higher values of f_1, more phosphorus accumulating organisms (PAOs) containing

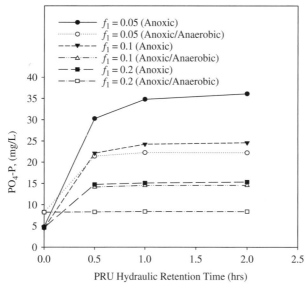

Figure 6. PO₄-P release with time for different sidestream flow.

polyphosphate (along with the bulk liquid) are removed from the main MEBPR system (Figure 7d); this causes a significant amount of phosphorus to be removed. The leftover phosphorus in the main MEBPR process would thus be insufficient for the physiological requirements of cell metabolism and the growth of biomass. As well, at higher f_1 (at $f_1 = 0.2$, Q_{SS} is 1080 L) values, the SRT of the main MEBPR process would fall below the recommended SRT (2–4 days) for the biological phosphorus removal process itself, to work (Tchobanoglous *et al.*, 2003).

A total phosphorus mass balance was used to evaluate potential phosphate recovery. The phosphorus contents in the effluent, activated sludge, and side stream supernatants were calculated as a percentage of the influent total phosphate load. When f_2 was kept constant at 0.2 and f_1 was varied between 0.01 and 1, 30% to 78% of the incoming phosphorus was able to be diverted for recovery (Figure 8). One additional factor to be considered here is the efficiency at which the MAP crystallizer is operated. For calculation purposes, the MAP crystallizer process was assumed to operate with 100% efficiency, although practically, this would never be the case. Typically efficiencies reported in much of the literature range from 65–85%, with some of the previous work at UBC achieving as much as 90% (Fattah, 2004). This "field" situation would have to be recognized in any final process design calculation.

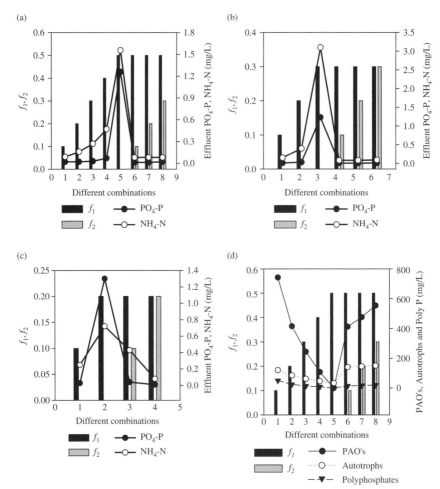

Figure 7. (a) EBPR process failure due to anaerobic sidestream; (b) EBPR process failure due to anoxic sidestream; (c) Failure due to combined sidestream (anoxic/anaerobic); (d) Response of biomass to changes in f_1 and f_2.

Experimental results

The entire experimental investigation lasted for more than a year and was divided into two periods. Period I occurred before implementing the sidestream process, and lasted for five months. During this period, the MEBPR process ran efficiently, at the desired SRT and HRT. Period II took place after implementing

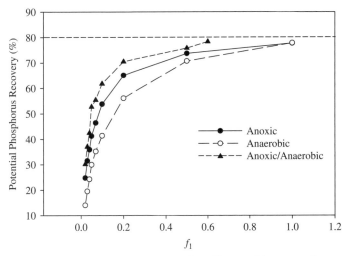

Figure 8. Estimated phosphorus recovery for different sidestream flows.

the sidestream, and remained in place for the rest of the study. Period II was subdivided into two parts, the first without sidestream wasting and the second with sidestream wasting (as discussed earlier). The time series of the influent TP and effluent phosphate of the MEBPR process are shown in Figure 9. Phosphorus removal performance was characterized by consistent removal, both in Period I and II[1]. The phosphorus removal efficiencies were comparable before and after the implementation of the sidestream system.

Based on simulation results, the sidestream was taken from the anoxic zone. The required HRT in the phosphate release unit was a minimum of one hour, with f_2 set to 0.2. For simulation purposes, the volume of the PRU used was 50 liters. At the pilot plant, the volume of the PRU was increased to 120 liters. The f_1 value was set at 0.025 (0.0925 m^3/d); thus, the HRT was nearly 30 hrs. These conditions were adopted, considering the practical difficulties at the plant in setting up a very small PRU, with a clarifier, for the chosen f_1. Furthermore, a larger PRU volume would allow for an increase of the f_1 value, while maintaining the required hydraulic retention time, if needed in the future.

The purpose of the sidestream clarifier was to produce a suspended solids-free supernatant. As expected, for the MEBPR sidestream process, the settling in the clarifier was very poor. The sludge volume index of the anoxic zone was

---

[1]There was an increase in the effluent phosphorus during Period III, as shown in Figure 9, which was the Mg addition period and is not discussed in this paper.

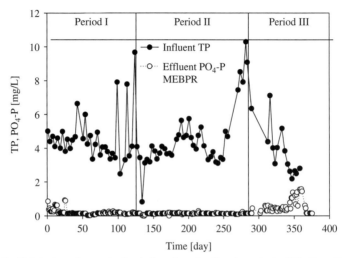

Figure 9. Total phosphorus in the influent and orthophosphate-P in the effluent for MEBPR process.

around 350 mL/g, whereas a value of 100 mL/g or less is considered representative of a good settling sludge (Tchobanoglous *et al.*, 2003). A high SVI is to be expected in an membrane-based EBPR process, since membrane-bioreactor, mixed liquors are subjected to coarse bubble aeration in the aerobic tank; this can destroy the flocculating nature of the sludge, resulting in poor settling (Monti, 2006). Poor settling subsequently resulted in an increased total suspended solids (TSS) content of the PRU clarifier supernatant, thus decreasing the efficiency of phosphate recovery. To overcome this problem, the settled sludge was drained once daily from the sidestream clarifier.

PO₄-P and NH₄-N release in PRU

The initial, low phosphate release was mainly due to the lack of proper mixing in the PRU and the limiting VFA (Figure 10). One other possible explanation was the higher levels of nitrate in the anoxic zone of the MEBPR process, such that the HAc's were utilized, first, by denitrifying microorganisms. If the HAc added was the limiting factor, then PO_4-P release will be decreased further. The unusual peak of phosphate in the initial stages (Figure 10) is likely due to release in the clarifier under poor mixing conditions.

Ammonia measured in the PRU was approximately equal to the ammonia present in the anoxic zone (Figure 11) in the initial stage (approximately 100 days). This indicated that there was no actual "release" of additional

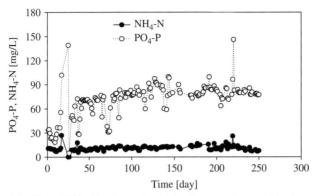

Figure 10. PO$_4$-P and NH$_4$-N release trend for the membrane sidestream process.

ammonia nitrogen in the sidestream process, beyond the concentration present in the anoxic zone. In Figure 11, it can be seen that there was a gradual increase in ammonia nitrogen concentration after fifty days of PRU operation and the increase was at a high level, after eighty days. This additional release can be attributed to biomass decay, and the subsequent conversion of organic nitrogen to ammonia nitrogen. The observed increase happened gradually, since biomass decay under anaerobic conditions is a slow process, when compared to aerobic and anoxic conditions (Siegrist *et al.*, 1999). The mean value of anoxic PO$_4$-P was around 7.2 mg/L and the mean value of sidestream PO$_4$-P release was 72 mg/L. Therefore, it can be concluded that there was a considerable amount (almost ten fold) of PO$_4$-P release in the PRU.

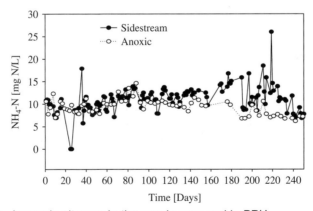

Figure 11. Ammonia nitrogen in the anoxic zone and in PRU.

The sidestream supernatant produced, contained soluble COD in the range of 20 to 50 mg/L, while the pH ranged from 6.89 to 7.34. Initially in the supernatant, TSS were very high due to poor sludge settling in the clarifier. After implementing sidestream wasting, clear supernatant was obtained. The average values of TP and TKN measured were 21 and 29 mg/L, respectively. Residual VFA from the PRU was also present in the supernatant. The average concentrations of PO_4-P and NH_4-N were 72 mg P/L and 11 mg N/L, respectively.

P recovery study

The MAP crystallizer was operated in a batch mode, in this study. This was a preliminary study only, to assess its feasibility, in conjuction with an MEBPR process. The sidestream supernatant was stored in a tank ($\sim$3000 L), after which it was fed to the crystallizer. The storage tank was stirred thoroughly, to determine the concentrations of PO_4-P, NH_4-N and Mg before starting the crystallizer. In the formation of pure struvite, equimolar amounts of ammonia, phosphate and magnesium are removed (Fattah, 2004). Thus, to attain good P removal, phosphate ions should be limiting. Since the ammonium nitrogen and magnesium concentrations in the sidestream supernatant obtained were low, compared to that of the PO_4-P, it was decided to supplement the supernatant with ammonium chloride and magnesium chloride.

P removal and recovery efficiency

For the first run, the PO_4-P removal efficiency varied from 39 to 61%. This removal efficiency was calculated by comparing the theoretical mass of struvite grown, based on the amount of phosphate removed, against the amount harvested (Table 2). In this first run, the recovered struvite efficiency was very low (around 30%). This was due to a higher supersaturation ratio (SSR) inside the reactor than required. It was observed that the effluent in the clarifier was turning cloudy, which was an indication that nucleation was occurring, but that there was no agglomeration. Most of the struvite was deposited as fines in the clarifier and in the hopper; these were difficult to recover. Some fine struvite escaped the reactor and settled at the bottom of the reactor; these losses were not taken into account. Also, some losses invariantly occurred during the recovery process, from harvesting to sieving. Thus, in reality, the actual mass of struvite produced was higher than that harvested, when these losses are taken into consideration.

In the second run, the removal efficiency of PO_4-P varied from 46 to 57%. The pH was decreased from 8.4 to 8.2, when the effluent in the clarifier was turning cloudy, to avoid the situation of greater SSR than required. Thereafter, the struvite

Table 2. Struvite recovery.

	Run I	Run II
Initial seed weight	612 g	1091 g
Final weight	915 g	1752 g
Struvite production (model prediction)	1059 g @ 80% removal rate	1335 g @ 76% removal rate
Struvite production obtained	303 g	661 g
Recovery rate	30%	50%

recovery efficiency was around 50% only, whereas a recovery efficiency of over 80 % had been reported during previous studies (Fattah, 2004; Adnan, 2002).

The low removal efficiency in this study may be attributed to a number of factors, including difficulty in maintaining a steady flow and the desired SSR. Operating the MAP crystallizer continuously, instead of batchwise, would serve to avoid flow variation by maintaining constant water head in the feed tank. This would help in maintaining the desired SSR at the inlet. Since the sidestream flow was very low, the batch processing lasted only around 3 days per run, in this study. Prolonged runs would yield much better performance, as experienced in previous studies (Fattah, 2004.). Further studies are underway to improve on this aspect of the process train.

N removal

In this study, up to 30% ammonia removal was achieved in both runs. As noted earlier, ammonium chloride was added to provide ammonium ions to optimize P removal. In principle, the addition of ammonium chloride should be avoided. The best alternative is to mix this supernatant with the centrate obtained from the anaerobic digestion of return sludge, which is usually high in ammonia. This would enhance the recovery/removal of ammonia. In addition, if desired, the process could also be engineered for better ammonia removal by keeping ammonium ions as the limiting factor.

CONCLUSIONS

Based on the results obtained from the simulation exercises and preliminary pilot scale study of MEBPR sidestream process for struvite recovery, the following conclusions can be drawn.

- The anoxic zone was more suitable for taking a sidestream flow when alternative zones were considered individually. When considered in combination, an anaerobic/anoxic zone combination yielded better conditions for struvite formation. One hour of hydraulic retention time was enough for the maximum phosphate release in the PRU.

- The ASM2 TUD Bio-P model predicted that 78% of the incoming phosphorus could be diverted for recovery as struvite, using a sidestream process. The model predicted the failure of the MEBPR process when the percentage of sidestream flow (f_1) was more than 0.2.
- The sidestream operation, to recover phosphorus, was successfully implemented at a pilot scale. An average P removal efficiency of 50% was achieved with a sidestream supernatant, using a MAP crystallizer, and 50% of the phosphorus removed was recovered as struvite pellets in this initial study. The refined conditions as suggested, should yield higher recovery efficiency.

REFERENCES

Adnan, A. (2002). *Pilot-scale Study of Phosphorus Recovery Through Struvite Crystallization*. M.A.Sc Thesis, Department of Civil Engineering, University of British Columbia, Vancouver, B.C. Canada.

Adnan, A., Koch, F.A. and Mavinic, D.S. (2003a). Pilot-scale study of phosphorus recovery through struvite crystallization – examining the process feasibility. *J. Environ. Eng Sci.*, **2**, 315–324.

Adnan, A., Koch, F.A. and Mavinic, D.S. (2003b). Pilot-scale study of phosphorus recovery through struvite crystallization – II: Applying in-reactor supersaturation ratio as a process control parameter. *J. Environ. Eng Sci.*, **2**, 315–324.

Al-Atar, E. (2007). *Dynamic Modeling and Optimization Membrane Enhanced Biological Phosphorus Removal Process for Municipal Wastewater Treatment*. Ph.D. Thesis, University of British Columbia, Vancouver, B.C. Canada.

Britton, A., Koch, F.A., Mavinic, D.S., Adnan, A., Oldham, W.K. and Udala, B (2005). Pilot scale struvite recovery from anaerobic digester supernatant at an enhanced biological phosphrous removal wastewater treatment plant. *J. Environ. Eng. Sci.*, **4**, 265–277.

Doyle, J.D. and Parsons, S.A. (2002). Struvite formation, control and recovery. *Water Res.*, **36**(16), 3925–3940.

Drnevich R.F. (1979). *Biological-chemical process for removing phosphorus at Reno/Sparks*. NV. **USEPA – 600/2-79-007**. U.S. Environmental Protection Agency, Washington, D.C.

Fattah, K.P. (2004). *Pilot-scale Struvite Recovery Potential from Centrate at Lulu Island Wastewater Treatment Plant*. M.A.Sc Thesis, Department of Civil Engineering, University of British Columbia, Vancouver, B.C. Canada.

Monti, A. (2006). *A Comparative Study of Biological Nutrient Removal Processes With Gravity and Membrane Solids-Liquid Separation*. Ph.D. Thesis, Department of Civil Engineering, The University of British Columbia, Vancouver, B.C., Canada.

Munch, E. and Barr, K. (2001). Controlled struvite crystallization for removing phosphorus and anaerobic digester sidestreams. *Water Research*, **35**, 151–159.

Reichert, P. (1994). Aquasim- A tool for simulation and data analysis of aquatic systems. *Wat. Sci. Tech.*, **30**(2), 21–30.

Saktaywin, W., Tsuno, H., Nagare, H., Soyama, T. and Weerapakkaroon, J. (2005). Advanced sewage treatment process with excess sludge reduction and phosphorus recovery. *Water Research*, **39**(5), 902–910.

Siegrist, H., Brunner, I., Koch, G., Phan, L.C. and Le, V.C. (1999). Reduction of biomass decay rate under anoxic and anaerobic conditions. *Water Science and Technology*, **39**(1), 129–137.

Smolders, G.J.F., Van Loosdrecht, M.C.M. and Heijnen, J.J. (1996). Steady-state analysis to evaluate the phosphate removal capacity and acetate requirement of biological phosphorus removing mainstream and sidestream process configurations. *Water Research*, **30**(11), 2748–2760.

Srinivas, K.H. (2007). *Coupling of Phosphorus Recovery to an Enhanced Biological Phosphorus Removal Process Through a Sidestream: A Pilot Scale Study*. M.A.Sc Thesis, Department of Civil Engineering, University of British Columbia, Vancouver, B.C. Canada.

Tchobanoglous, G., Burton, F.L. and Stensel, H.D. (2003). Wastewater Engineering: Treatment and Reuse, 4th ed. McGraw-Hill Higher Education, New York.

Woods, N.C., Sock, S.M. and Daigger, G.T (1999). Phosphorus recovery technology modeling and feasibility evaluation for municipal wastewater treatment plants. *Environmental Technology*, **20**(7), 663–679.

Phosphorus recovery from eluated sewage sludge ashes by nanofiltration

Claudia Niewersch[a], Sebastian Petzet[b], Jochen Henkel[b], Thomas Wintgens[c], Thomas Melin[a] and Peter Cornel[b]

[a]Department of Chemical Engineering, RWTH-Aachen, Turmstrasse 46, 52056 Aachen, Germany
[b]TU-Darmstadt, Institut WAR, Petersenstrasse 13, 64287 Darmstadt, Germany
[c]University of Applied Sciences Northwestern Switzerland, Gründenstrasse 40, 4132 Muttenz, Switzerland

Abstract A combined process for a phosphorus recovery from sewage sludge ash was studied. The process consists of an acidic elution, ultrafiltration for the separation of particles, nanofiltration for the separation of phosphorus and neutralisation of the nanofiltration permeate. Filtration experiments were conducted with the nanofiltration polyamide membrane DL (GE Osmonics) with eluates of three different sewage sludge ashes and two different dilution rates as well as ultrafiltration experiments with polymeric membranes (UP150 by Microdyn Nadir) and ceramic membranes (UF50n by Atech). It was shown that the selectivity of the nanofiltration membrane depended significantly on the dilution rate of the ash elution. With the help of model solution experiments, a dependency of the phosphorus retention on the concentration of multivalent cations in the feed could be demonstrated. The performance of the ultrafiltration turned out to be relatively high (220 L/m^2∗h) and stable, also with an increasing concentration. In a preliminary economical analysis, the main cost factors were estimated for two different dilution rates and the three different ashes.

INTRODUCTION

Global phosphate reserves are limited and at the current rate of extraction, the world will run out of phosphate reserves within the next 60 to 240 years (Steen, 1998). Phosphorus is an essential and irreplaceable nutrient for all forms of life on earth. Hence, phosphorus recovery from waste streams is essential to guarantee the long-term fertiliser supply for a growing world population. Wastewater treatment plants are important sinks for phosphorus, as phosphorus is transferred during the wastewater treatment process from the liquid phase into solids waste (sewage sludge) with a high phosphorus content (2–5%).

Recycled phosphorus could be used as agricultural fertiliser, so that the product should not contain organic or inorganic pollutants, be hygienically safe and guarantee a high bioavailability of phosphorus. For other fields of application, such as raw material for detergents, animal feeding or food industry, the required quality standards are even higher. In any phosphorus recovery process from sewage sludge, a separation of phosphorus from harmful substances also present in the waste (e.g. heavy metals, organic pollutants) is required in order to create a high quality phosphorus containing product stream (Driver *et al.*, 1999).

This paper reports about the feasibility and process efficiency of nanofiltration as a separation step for phosphorus recovery processes. In principle, nanofiltration can be used as a separation step in phosphorus recovery processes using sewage sludge, sewage sludge ashes or sludge liquors as phosphorus recycling source. However, this study focuses on sewage sludge ashes due to the growing trend towards sewage sludge incineration. In particular, the separation of phosphorus and metals in acidic eluates from sewage sludge ashes was investigated with the aim to create a high quality phosphate stream. In the case of sewage sludge ashes, the separation is restricted to the elimination of inorganic contaminants since organic pollutants are destroyed during incineration.

THEORETICAL BACKGROUND OF THE PROCESS

Nanofiltration is a pressure driven membrane filtration process that is placed between ultrafiltration and reverse osmosis in terms of molecular weight cut-off (MWCO, 90% separation limit in unit of molecular weight) for uncharged molecules and according to the required transmembrane pressure. The MWCO of nanofiltration membranes covers a range from 200 to 1000 Da, whereas the MWCO of 200 Da (corresponding to about 1 nm) is responsible for the notation nanofiltration. In contrast to reverse osmosis, this relatively young technology offers the option to realise a salt retention far below 100% and a charge-dependent ion selectivity. The membranes have fixed charges on the surface and inside the pores, leading to the above mentioned characteristics (Melin and Rautenbach (2007); Schäfer *et al.*, 2005).

In line with the ion selectivity of nanofiltration membranes, dissolved phosphate is expected to pass the membrane in the case of acidic conditions (pH < 2) since phosphoric acid is expected to be existent mainly as the neutral H_3PO_4 and monovalent anion $H_2PO_4^-$. The pollutants (positively charged heavy metals) are retained by the membrane and remain in the retentate while a phosphorus-containing product stream (permeate) is created.

These special properties of nanofiltration membranes can be utilized to gain a high quality phosphorus recycle product from sewage sludge. Due to the fact that acidic conditions are essential for the separation efficiency (Niewersch *et al.*, 2007), nanofiltration requires a sludge pretreatment process with acid.

The separation principle has already been applied under different process conditions for the purification of sulphuric and phosphoric acid (Soldenhoff *et al.*, 2005; González *et al.*, 2002; Tanninen *et al.*, 2003; Skidmore and Hutter, 1999).

In this study, three sewage sludge ashes have been used originating from different sources. Two of them (Al-ash and Fe-ash) originate from incinerated sewage sludge of waste water treatment plants with phosphorus elimination using aluminium and iron precipitation, respectively. The third ash had special characteristics containing a relatively high amount of calcium (Ca-ash).

To ensure the dissolution of phosphorus from the sludge into an aqueous phase, deionised water and sulphuric acid has been used. The use of sulfuric acid is expected to be advantageous in comparison to other acids. This is due to the fact that sulfuric acid is a relatively strong acid and existent mainly as divalent anions at acidic conditions. Therefore it is retained to a higher extent than the phosphoric acid molecules and ions. The use of e.g. hydrochloric acid would lead to a formation of Cl^--ions which can permeate relatively freely through the nanofiltration membrane.

The pH has an influence on the performance of the nanofiltration process since the dissociation of phosphoric acid, the membrane charge and the amount of easily permeating counter cations in form of H^+-ions are pH-dependent. The influence of pH on the phosphorus permeability has been studied with the help of relatively simple model solutions verifying that a low pH increases the efficiency of the separation (Niewersch et al., 2007). On the other hand, a low pH decreases the durability of the membrane material and leads to a higher consumption of sulfuric acid and consequently higher chemical costs. A pH of 1.5 has been identified as the optimum and adjusted in the ash eluates by titration with sulfuric acid. The ratio of ash mass and elution volume, which is indicating the degree of dilution, has been varied to study its influence on the process efficiency.

To ensure stability of the nanofiltration membranes, it is necessary to realise a particle-free feed. Hence, particles were removed from the ash eluates by decantation and ultrafiltration. The efficiency of the ultrafiltration process step was studied in detail with the help of a cross-flow ultrafiltration experiment which was conducted with the test facility shown in Figure 2 and a ceramic membrane manufactured out of TiO_2. With this experiment, the operation in recirculation mode and with an increasing concentration has been investigated during the application of a periodical backflushing.

MATERIAL AND METHODS

Preparation of ash eluates

In a first experimental series, 400 g of each ash type has been mixed with 3.8 L of deionised water, followed by titration with sulfuric acid to achieve a pH of 1.5. During the titration, the suspension was stirred with a magnetic stirrer.

The titration took place for 2 h to ensure a stable pH, before deionised water was added until reaching a total volume of 4 L. Additional elutions were conducted analogously with two higher dilution rates, namely 40 g ash with a final elution volume of 4 L and 80 g with a final volume of 16 L.

Filtration experiments

After elution and 10 min of sedimentation, the sewage sludge ash suspensions of the concentration of 100 g_{ash}/L and 10 g_{ash}/L were filtered with ultrafiltration polymeric membranes of the type UP150 (Microdyn Nadir).

Both ultrafiltration and nanofiltration with polymeric membranes were performed in a stirred membrane test cell containing a flat-sheet membrane with an active membrane area of 121 cm^2. The temperature was adjusted using an external thermostat, the pressure was adjusted using nitrogen gas from a gas bottle and the permeate flux was measured with a balance. Figure 1 gives an overview of the experimental set up for the filtration experiments.

Figure 1. Experimental set up for the filtration tests.

The nanofiltration was performed using the polyamid membrane DL (GE Osmonics) at 25°C and at a varying transmembrane pressure between 45 and 70 bar for the ash eluates with the lower dilution rate and at 25 bar transmembrane pressure for the higher diluted eluates.

Before and after each membrane filtration test, the pure water permeability and the retention of magnesium sulfate were measured to ensure membrane integrity.

In the nanofiltration experiments, the retention was measured as a parameter for the separation ability of the membrane. The retention is defined as follows:

$$R_i = \frac{c_{Fi} - c_{Pi}}{c_{Pi}} \tag{1}$$

[c_{Fi} = concentration of the component i in the feed at one point in time during the filtration; c_{Pi} = concentration of the component i in the permeate at one point in time during the filtration]

Therefore, the retention characterises the filtration at one point in time and is not the rate between the concentrations in the feed and the permeate for the whole process. The retention for the entire process would be the ratio of the concentrations measured in the feed before starting the filtration and the concentrations measured in the cumulative permeate after the whole filtration.

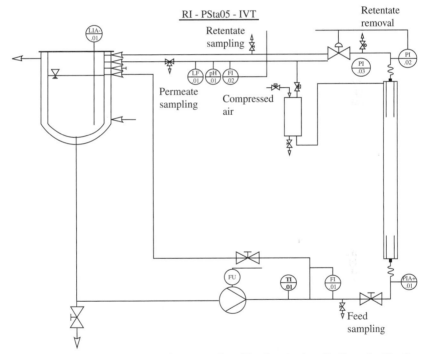

Figure 2. Experimental set up for cross-flow filtration tests with the ultrafiltration membrane UF50n.

Figure 2 shows the experimental set up used for the cross-flow ultrafiltration experiment. It was performed according to the timetable shown in Table 1 with the ceramic ultrafiltration membrane UF50n supplied by Atech (nominal pore diameter = 0.05 μm) at a transmembrane pressure of 2.1 bar and a backflush each 10 minutes. The membrane was used as a one-channel module with an inner diameter of 6 mm and a cross-flow velocity of 4 m/s. As feed solution, the ash eluate of the highest dilution rate (5 g/L) was used.

Table 1. Timetable of the cross-flow filtration experiment with the ultrafiltration membrane UF50n.

Exp phase	Description	Duration
1	Pure water permeability measurement	26 h
2	Filtration of ash eluate, recirculation	0:45 h
3	Removal of permeate → increasing concentration	24 h
4	Filtration of ash eluate, recirculation	3:40 h
5	Removal of permeate → increasing concentration	24 h
6	Pure water permability measurement	2:50 h

Analytics

The concentrations of P, Mg, Ca, Al, Fe, S, Cu and Zn in the eluates, retentates and permeates of the nanofiltration were measured with ICP-OES (inductively coupled plasma with optical emission spectometry) while the composition of the applied sewage sludge ashes was determined by aqua regia according to DIN EN ISO 11885 (E22) (see Table 2).

Table 2. Chemical composition of the sewage sludge ashes [g/kg] determined by Schaum (2007).

	P	Al	Fe	Ca	Mg	Zn	Cu
Al-ash	85	114	22	90	12	2.1	1.1
Fe-ash	78	49	129	97	12	2.8	0.8
Ca-ash	32	21	66	232	10	3.2	0.3

Preparation of model solutions

The model solutions for the investigation of the nanofiltration mass transport mechanisms were prepared according to the composition given in Table 3 by using the corresponding metal sulfates, sulfuric acid, phosphoric acid and deionised water.

Table 3. Composition of model solutions used for the nanofiltration.

[g/L]	M1	M3	M7	M9	M10	M12	M16	M18
Al	3	0.1	3	0.1	0.02	0.02	0.02	0.02
P	0.1	0.1	8	8	0.1	0.1	8	8
Fe	0.02	0.02	0.02	0.02	0.1	1.2	0.1	1.2
Ca	0.02	0.02	0.02	0.02	0.02	0.02	0.02	0.02
Cu	0.004							
Mg	0.02							
S(acid)	1.15	1.44	0	0	1.73	2.88	0	0
S(salts)	5.4	0.23	5.4	0.23	0.14	0.77	0.14	0.77
S(sum)	6.56	1.68	5.4	0.23	1.87	3.65	0.14	0.77
Na	0	0	0	0	0	0	0	0
	M19	M21	M25	M27	M28	M30	M34	M36
Al	0.02	0.02	0.02	0.02	0.02	0.02	0.02	0.02
P	0.1	0.1	8	8	0.1	0.1	8	8
Fe	0.02	0.02	0.02	0.02	0.02	0.02	0.02	0.02
Ca	0.02	0.5	0.02	0.5	0.02	0.02	0.02	0.02
Cu	0.004							
Mg	0.02							
S(acid)	1.44	1.44	0	0	1.7	10	1.7	10
S(salts)	0.09	0.48	0.09	0.48	0.09	0.09	0.09	0.09
S(sum)	1.53	1.92	0.09	0.48	1.79	10.09	1.79	10.09
Na	0	0	0	0	869	11178	2782	18505

RESULTS AND DISCUSSION

Nanofiltration

Figure 3 shows the retention of phosphorus vs. the permeate recovery at different pressures for the sewage sludge eluates with a low dilution rate (100 g/L). In contrast to studies for the purification of acids and the performed pre-tests with model solutions (Niewersch *et al.*, 2007), the membrane showed a significantly higher retention for phosphorus resulting in a lower selectivity and process efficiency. The retention further increased with permeate recovery and no significant dependency on the pressure was observed. However, the results cannot be compared directly as in the case of the purification of phosphoric and sulfuric acids, the acid concentrations were significantly higher and in the case of the model solutions used in the pre-testing, concentrations of cations were significantly lower than in the sewage sludge ash eluates. The retention for phosphorus was highest in the case of Al-ash which contained high concentrations of dissolved multivalent cations, followed by the iron and calcium ash (Figure 3).

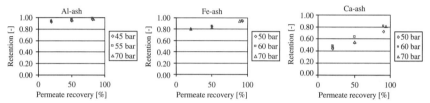

Figure 3. Phosphorous retention measured for the nanofiltration of the three ash eluates with low dilution (100 g/L).

In dead end test cell experiments, an increase in permeate recovery leads to higher concentrations of the retained components in the retentate and, consequently, to an increase of ion concentration and conductivity. Figure 4 depicts the retention of several elements vs. permeate recovery for all three ashes. It can be clearly seen that the retention of phosphorus and sulphur increases consistently with permeate recovery while the retention for the metals slightly decreases. Nevertheless, the differences between the retentions measured for one permeate recovery point is relatively high. This indicates that the retention depends on the permeate recovery as well as on the sewage sludge ash composition. According to these results, it was concluded that the concentrations of the dissolved components play an important role for phosphorus retention.

In the literature (Melin and Rautenbach, 2007), the pressure dependency is ascribed to the fact that both diffusive and convective transports play a role in nanofiltration. A higher pressure leads to a higher solvent flow through the membrane. Therefore, the water flux increases while the diffusive salt transport is not affected. The convective salt transport increases accordingly to the water flux. Hence, the salt retention increases with increasing pressure but not as much as observable for reverse osmosis membranes where only diffusive salt transport is relevant (Melin and Rautenbach, 2007). Additionally to this theory about transport mechanisms in nanofiltration membranes, an effect called critical flux is described in the literature (Nyström, 2007). This means that for solvent fluxes exceeding a certain level, the pressure dependency of the flux is decreasing until the flux reaches a constant level (Nyström, 2007). The exceedance of the critical flux can explain the retention-pressure independency for the experimental results of this study.

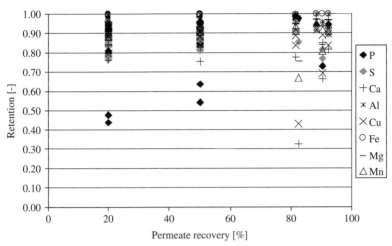

Figure 4. Retention vs. conductivity, measured in experiments with sewage sludge eluates with a low dilution (100 g/L).

In order to further investigate the influence of varying concentrations of different cations and the ionic strength, additional experiments with eluates of a higher dilution rate (10 g/L) and with model solutions simulating the eluates were carried out. The model solutions contained differing concentrations of Al, Fe, Ca, P and S in the range of the original ash eluates. The compositions are shown in Table 3 and the corresponding retention results are depicted in Figure 5. It can be seen that the total concentration of ions correlates with the selectivity for phosphorus.

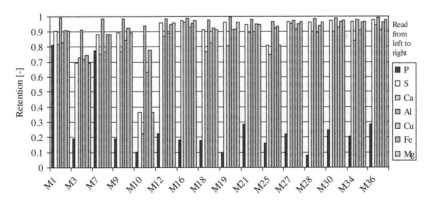

Figure 5. Retention of all components of the model solutions (composition see Table 3).

The results show that the retention for phosphorus and sulphur increased with the concentration of multivalent cations in the feed. Especially the Al^{3+} concentration – the only triple-charged cation in the model solutions – proved to have a significant effect on phosphorus and sulphur retention, confirming the results of the real ash experiments. At Al concentrations of 3000 mg/L (that means model solutions M1 and M7), which is in the range of the Al-ash eluate, phosphorus retention was above 78% and therefore comparable to the results from the ash experiments. The significant influence of the ionic strength can also be seen in Figure 6–8 where the retention of the dissolved compounds versus the cumulated ionic strength of two- and three-valent cations is depicted. The graphs indicate a dependency of phosphorus and sulphur retention on the concentrations of multivalent cations that can be fitted reasonably well by a log-trend curve.

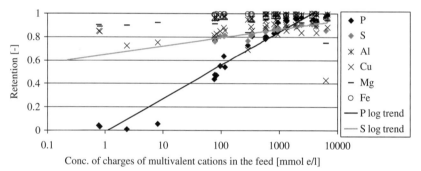

Figure 6. Dependency of the retention on the sum of charges of two- or three-valent cations in the feed (experimental series with ash eluates of lower dilution rates (10 g/L)).

According to the characteristic ion selectivity of nanofiltration membranes, multivalent cations are retained to a large extent and can only act to a minor degree as counter-ions for permeating anions such as HPO_4^-, SO_4^{2-} or HSO_4^-, thus forcing them to remain on the feed side for electric neutrality purposes. However, SO_4^{2-} and HSO_4^- are present at high concentrations in sewage sludge ash eluates that contain Al since larger amounts of sulphuric acid are required to reach a pH value of < 2. Given that the amount of positively charged counter-ions that can freely pass the membrane, such as Na^+ and H^+, is limited, high concentrations of sulphur could lead to a competition for the charged HPO_4^-, providing an explanation for the high phosphorus retentions.

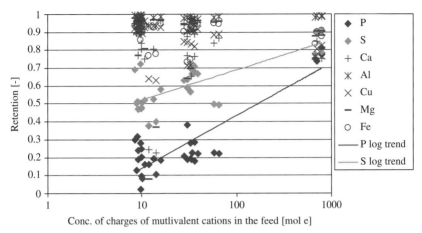

Figure 7. Dependency of the retention on the concentration of charges of two- or three-valent cations in the feed (experimental series with model solutions).

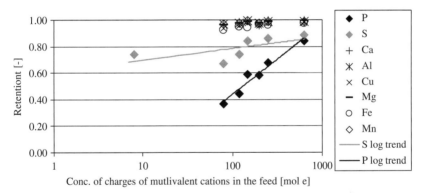

Figure 8. Dependency of the retention on the concentration of charges of two- or three-valent cations in the feed based on experimental series with ash eluates of higher dilution rates (10 g/L).

Ultrafiltration

The results from the experimental series with cross-flow ultrafiltration can lead to conclusions about the efficiency of the whole process since the particle elimination is an important pre-treatment for the nanofiltration. The results depicted in Figure 9 show that a stable filtration is possible at a cross-flow velocity of 4 m/s and a transmembrane pressure of 2.1 bar. The usage of an ultrafiltration membrane with the relatively low average pore size of 0.05 μm leads to a high permeate quality.

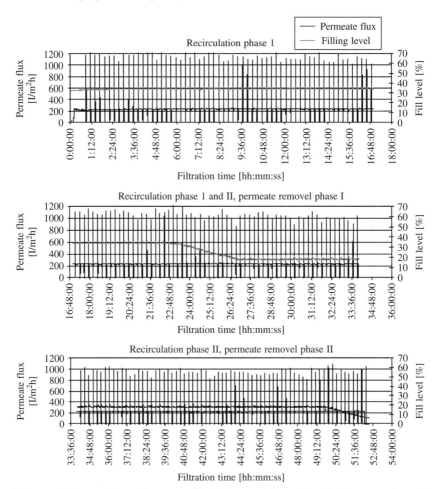

Figure 9. Permeate performance for the cross-flow ultrafiltration experimental series.

The permeate flux could be kept at a constant high level of 220 L/m^2∗h for the recirculation mode as well as for the permeate removal mode with increasing feed concentration and a final permeate recovery of 65%. The observed pure water flux of 280 L/m^2∗h measured in the pre-experiment is only slightly higher than the permeate flux measured for the filtration of the eluates.

Economical aspects

The most important aspects for the estimation of the cost-effectiveness have been assessed based on the previous experimental investigations. This assessment is still on a preliminary level but should give an idea about the potential of this process combination.

The following most important cost factors have been identified:

- Amount of sulphuric acid needed for the elution of the ashes.
- Energy consumption for ultrafiltration respectively nanofiltration.
- Membrane area of ultrafiltration respectively nanofiltration.
- Amount of caustic soda needed for the neutralisation of the nanofiltration permeate and the retentate of nanofiltration and ultrafiltration.

These factors have been calculated for the two dilution rates 1:10 (100 g/L) and 1:100 (10 g/L) on the basis of the above-described experimental results for an amount of 1 kg/h of sewage sludge ash. The energy consumption has been calculated as a first approximation according to the following equation and taking only the membrane filtration into account:

$$W_b = \mu \frac{\Delta P}{\rho \eta_p \eta_m} \qquad (2) \qquad \text{(Biegler } et\ al., 1997)$$

[μ = mass flow, ρ = density, η_p = pump efficiency = 0.5, η_m = motor efficiency = 0.9]

Furthermore, the amount of recovered phosphorus has been calculated which will generate revenues and therefore reduce the costs. Figure 10 shows an overview of all cost factors based on 1 kg sewage sludge ash and 1 kg sewage sludge ash throughput per hour for the calculation of the necessary membrane areas. For the Fe-ash eluate with the highest dilution (5 g/L) 484 g_{NaOH}/kg_{ash} was required for the neutralisation of all acidic process streams (the nanofiltration permeate and the retentate of both membrane filtration processes). Figure 11 additionally reflects the amount of energy necessary for the recovery of 1 kg phosphorus.

While the higher selectivity for the diluted sewage sludge ash eluates (1:100) significantly increases the amount of recovered phosphorus, all the other cost influencing factors increase at higher dilutions. The energy consumption and the membrane area increase with the volumetric flow despite the fact that a lower pressure is required to realise the permeate flux.

From the economical point of view, it is therefore important to compare the benefit from the phosphorus product and the costs due to energy, membrane material as well as consumption of chemicals to decide which dilution rate is the most advantageous.

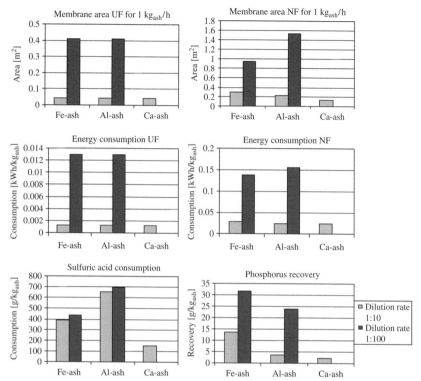

Figure 10. Cost factors for two different dilution rates 1:10 (according to 100 g/L) and 1:100 (according to 10 g/L).

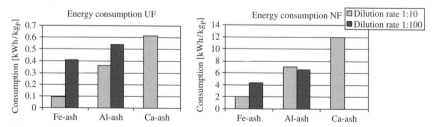

Figure 11. Energy consumption referring to the recovered phosphorus mass.

CONLCUSION AND OUTLOOK

The experiments confirmed the selectivity of nanofiltration membranes for phosphorus and the retention of dissolved multivalent cations, showing the feasibility of this process for phosphorus recovery. However, the increasing retention of phosphorus at high concentrations of dissolved compounds is still a problem. This effect appeared to be especially important for ash eluates that contained high concentrations of Al^{3+}. Hence, further investigations are to be undertaken to identify the influencing factors for the phosphorus selectivity in order to develop an efficient separation step for phosphorus recovery.

A first assessment of the main economic aspects identified the energy consumption, membrane area and consumption of chemicals for elution and neutralisation as the most important factors. While the amount of phosphorus that could be recovered increased with higher dilution rates, the costs for energy consumption, membrane area and chemicals were increasing consistently. In conclusion, there is still a need for further studies on a larger scale to gain a reliable economical analysis.

ACKNOWLEDGEMENT

This work is part of the "German Research Clusters on P-Recycling" and was founded by the German Federal Ministry of Education and Research. The authors gratefully acknowledge the supervision by the Project Management Agency Forschungszentrum Karlsruhe and the funding by the German Federal Ministry of Education and Research.

REFERENCES

Biegler, L.T., Grossmann, I.E. and Westerberg, A.W. (1997). Systematic Methods of Chemical Process Design. ISBN 0-13-492422-3, Prentice Hall PTR, New Jersey.

Driver, J., Lijmbach, D. and Steen, I. (1999). Why recover phosphorus, and how? *Environmental Technology*, **20**(7), 651–662.

González, M.P., Navarro, R., Saucedo, I., Avila, M., Revilla, J. and Bouchard, Ch. (2002). Purification of phosphoric acid solutions by reverse osmosis and nanofiltration. *Desalination*, **147**, 315–320.

Melin, T. and Rautenbach, R. (2007). Membranverfahren. Grundlagen der Modul-und Anlagenauslegung, 3. aktualis. u. erw. A.

Niewersch, C., Koh, C.N., Wintgens, T. and Melin, T. (2007). Separation of phosphorus from metal ions with nanofiltration – an examination of potentials using model solutions. Proceedings of 11th Aachener Membran Kolloquium March 28/29. pp. 433–440.

Niewersch, C., Koh, C.N., Wintgens, T. and Melin, T. (2007). Experimentelle Untersuchung des Rückgewinnungspotentials von Phosphor mit Nanofiltration. Poster und Proceedings auf der ProcessNet-Jahrestagung, 16–18. Oktober 2007, Aachen.

Nyström, M. (2007). Functional characterisation of Membranes for liquid filtration. Vortrag in: NanoMemCourse EF1: Nanostructured materials and membranes, synthesis and characterisation 7–16, November 2007, Saragossa.

Schäfer, A.I., Fane, A.G. and Waite, T.D. (2005). Nanofiltration Principles and Applications. Elsevier Advanced Technology, Oxford.

Schaum, C.A., Verfahren für eine zukünftige Klärschlammbehandlung – Klärschlamm-konditionierung und Rückgewinnung von Phosphor aus Klärschlammasche. Dissertation. Darmstadt 2007. Erschienen in: Schriftenreihe WAR 185.

Skidmore, H.J. and Hutter, K.J. (1999). Methods of purifying phosphoric acid. United States Patent Nb. 5,945,000.

Soldenhoff, K., McCulloch, J., Manis, A. and Macintosch, P. (2005). Nanofiltration in metal and acid recovery. In: A.I. Schäfer, A.G. Fane and T.D. Waite (eds), Nanofiltration Principles and Applications, Elsevier Advanced Technology, Oxford.

Steen, I. (1998). Phosphorus availability in the 21st century: Management of a non-renewable resource, Phosphorus & Potassium. British Sulphur Publishing, No. 217, 25–31.

Tanninen, J., Platt, S. and Nyström, M. (2003). Nanofiltration of sulphuric acid from metal sulphate solutions. Proceedings Imstec, Sydney.

P-recovery from sewage sludge ash – technology transfer from prototype to industrial manufacturing facilities

L. Hermann

ASH DEC Umwelt AG, A-1210 Wien, Donaufelderstrasse 101/4/5

Abstract Phosphorus (P) is required by every cell of living organisms. Serious prognoses anticipate that phosphate rock may be the first low running natural resource in the world. As a response, ASH DEC and BAM have developed a technology to produce P-fertilizers from sewage sludge and meat and bone meal ash. Both are secondary P-resources available in abundance and currently being disposed of in landfills and construction materials. After complete industrial implementation of the technology, P-fertilizers from sewage sludge and meat and bone meal ash in the EU27 could substitute up to 25% of the necessary rock phosphate imports.

The technology is based on a thermo-chemical process that removes heavy metals and increases the plant availability of phosphorus. It has been transferred from the laboratory scale to a continuous pilot plant operation in 2008 and will be further transferred to the industrial scale in 2009.

The development of a product that can fully replace customary mineral fertilizers was a key achievement for the technology upscale. The fully licensed complex fertilizer manufactured from sewage sludge ash was a prerequisite for obtaining the necessary financial funds to engineer, build and operate the pilot plant. At present, after a few months of successful operation of the pilot plant, the first industrial manufacturing plant is conceived, engineered and financed.

The following article deals with the procedures and benefits of the – so far successful – step-by-step upscale from laboratory to industrial scale manufacturing plants for new and innovative products that are based on nutrient rich incineration residues as a raw material source.

INTRODUCTION

Phosphorus (P) is required by every cell of living organisms. It is one of the major elements for plant nutrition and therefore essential for crop production. P supply in plant nutrition depends entirely on external inputs into soils. This is achieved by P-fertilisers deriving from rock phosphate deposits that are going into depletion (Jasinsky, 2008). Serious prognoses anticipate that rock phosphate will be the first low running resource in the world. The EU27, 2006, imported 2.813.000 (as P_2O_5) tons of phosphate rock and mined only 313.000 (as P_2O_5) tons (IFA Production and Trade Statistics 2008).

For these reasons P recycling is becoming an increasingly important issue. Suitable secondary raw materials are sewage sludge and meat and bone meal (Werner, 2003). In the EU close to 1 million tons of P_2O_5 are accumulated annually in theses materials, of which almost 65% are permanently lost to landfills and construction materials (Hermann, 2005). Of particular interest is P-recovery from ash after the incineration of municipal sewage sludge because of its increasing abundance and high P-concentration. Only from sludge ash, more than 200.000 tons P_2O_5 could be recovered annually in the EU.

THE PROCESS

As a response, ASH DEC and BAM with the participation of scientific and industrial partners have successfully developed a P-recovery technology for sewage sludge ashes, that removes heavy metals and organic pollutants and produces ready-to-use fertilizer granules (Adam *et al.*, 2007). A major part of research was carried out within the framework of the EU-FP6-project SUSAN. The technology has been tested at lab-, medium- and technical scales before ASH DEC 2008 has set-up a pilot plant. Its core process is a thermo-chemical treatment (Adam *et al.*, 2007).

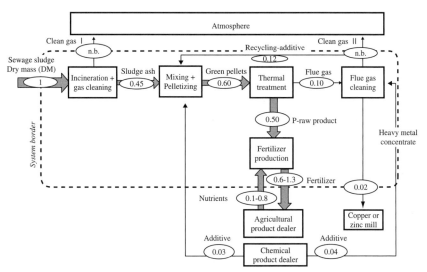

Figure 1. Process and mass flow diagram (ASH DEC Umwelt AG, 2008).

Sewage sludge ash is mixed with a measured quantity of solid chlorine donors and treated for 20–30 min. at 1,000°C in a thermal reactor. The amount of chlorine depends on the concentration of heavy metals in the ash and the target removal rates as required by national fertilizer legislation in European countries. In the process, organic pollutants are completely destroyed and heavy metals are removed as volatile heavy metal chlorides via the off-gas that is cleaned in a multi stage gas cleaning system. Additionally, the bioavailability of P is improved due to the formation of new P-bearing mineral phases in the thermo-chemical process (Peplinski *et al.*, 2008). The fertilizers produced from the thermo-chemically treated ashes have been licensed in Austria for unlimited use on crop- and woodland. The German parliament is currently discussing a revision of its fertilizer ordinance to legalize the utilization of ash as a fertilizer raw material, if in compliance with certain heavy metal concentration limits. These limits are fully achieved with the novel technology.

THE PRODUCT

By having conceived and tested the process, one has only half paved the pathway to the successful industrial integration of a new technology. Design, marketing and distribution of the recycled product are equally important and frequently neglected. A process that replaces 1/3 of rock phosphate imports and avoids landfill disposal of hundreds of thousands tons of ash interferes with vested interests of powerful stakeholders in the waste management and fertilizer industry. Consequently, one has to select carefully potential sources of information and know-how.

For this purpose, ASH DEC has identified stakeholders that have been adversely affected by either the technical or the market development. Among the first group are those distributors that used to sell secondary P-fertilizers produced from P-containing by-products of the iron- and steel industry – the so-called Thomas-fertilizers. The product disappeared years ago but the know-how still exists within specialized companies that are ready to help because the new P-fertilizer offers a future market potential to them. The second group is represented by many established agricultural product manufacturers and distributors that suffer from the increasing concentration of supply sources, particularly pronounced in the phosphate fertilizer market. Representatives of both groups have been involved in product development from an early stage and they have largely contributed to develop a marketable product.

If the goal is to make use of the enormous technology potential and replace a significant share of phosphate imports, a niche product strategy alone cannot work. As a consequence, a product has to be designed that finds acceptance from the mass market – mainstream cropland farmers as massive fertilizer consumers. French studies show that farmers are the most conservative of all consumer groups and require huge incentives to change their settled habits. As these incentives would quickly outbalance the possibilities of a small or medium enterprise, ASH DEC decided to design products that are directly comparable to traditional mineral fertilizers. As a consequence, they have to precisely meet market requirements in terms of physical and chemical properties. This requirement presented another challenge to the researchers of the company – how to produce compatible products in much smaller scale and with different raw materials.

In particular the following targets had to be met: i) comparable performance in terms of yield and P-uptake, ii) Similar appearance as mineral P-fertilizers, i.e. round, dust and abrasion free granules of about 3.25 mm diameter that can be stored and handled in exactly the same way as granules from the leading fertilizer brands.

To learn how to achieve and to judge fertilizer performance criteria, ASH DEC contracted wpa Advising Engineers, Vienna as permanent advisors on plant nutrition and started cooperation with specialized scientific institutes such as the German Federal Research Centre for Cultivated Plants, Braunschweig, The Institute for Plant Nutrition at the Swiss Federal Institute of Technology, Zürich and the Austrian Agency for Nutrition Safety, Vienna. Specialized scientists advised how to achieve and test the performance of the new P-and complex fertilizers.

As a quick reference test for the plant availability of P in the fertilizer, the P-solubility in 2% citric acid was determined. This test is one of the standardized P-solubility tests with reference to the solubility of rock-phosphates and other non water soluble phosphate fertilizers, in particular Thomas phosphates. According to the European Fertilizer Ordinance, a solubility of $>75\%$ should be achieved. ASH DEC products were tested accordingly and treatment conditions were optimized until a citric acid solubility of more than 80% was achieved regularly at treatment temperatures of 900–1,000°C and residence times of 10–30 minutes.

High citric acid solubility should result in a good yield and P-uptake of crops. As a first reference plant, ryegrass was taken because of its poor P-content in germs and its high response to P-fertilizers. Because of the non-acceptance of phosphate-mono fertilizers in the Austrian market, PK 12–20 (12% phosphorus, 20% potassium) fertilizers were produced and tested in pots with 3 P-concentrations and 4 repetitions. The tests demonstrated that fertilizers from thermo-chemically treated sludge ash produced a similar yield increase and P-uptake as traditional mineral fertilizers of the best selling brand in Austria. The pot tests were repeated with different soils and crops such as corn, wheat, rape and potatoes with similar results.

In parallel to the chemical composition and performance of the new product, its physical appearance and parameters had to be optimized. For this purpose, small and technical scale granulation tests had to be performed. Manufacturers of fertilizer granulation equipment were asked for their assistance and for the availability of test units. However, fertilizer granulation was more difficult than expected. Equipment manufacturers are used to design and produce equipment for plants with throughputs of several hundred thousand tons per year, whereas the new technology needs equipment for several ten thousand tons. Most of the equipment envisaged was either not suitable or much too expensive for the planned plant size. As a consequence, a research for appropriate granulation equipment was undertaken und pellet presses identified as most suitable machines with an acceptable price-to-throughput ratio. These machines have only one shortcoming: they do not produce round granules but cylindrical pellets of irregular lengths. Moreover, these machines are not designed to press inorganic and abrasive materials such as the ash with a silicon concentration of roughly 20%. Nonetheless, after 3 test runs at the site of one of the pellet-press producers, a machine was bought and installed at the test and demonstration site. It took about 6 months of intensive tests to achieve regularly solid pellets with the required diameter of 3.2 mm, the point load of >25 N and a good appearance. The machine was equipped with a steam generator to provide additional lubrication to the ash-nutrient mix and with additional knives to cut pellets to a length of 3–4 mm, in order to get small cylindrical pellets of similar diameter and length. This product, because of its nice shape, smooth and dust free surface and easy handling is well accepted – at least by the wholesalers and distributors. To a certain extent, pellets are superior to compacted fertilizer granules – they are more regular in shape and less dusty.

Until now, different PK and NPK fertilizers have been developed, tested and sold either in cooperation with distributors or in cooperation with small manufacturers that disposed of free capacity.

At this point the reader may ask why all these efforts have been necessary when the fertilizer industry could have integrated the recovered phosphate to its standard processes? The answer is simple. The fertilizer industry would compare the price of the recycled phosphate to phosphate rock. After thermo-chemical treatment, the recycled P-fertilizer has achieved a degree of purity and plant availability that phosphate rock does not have. It has become a product with its own beneficial properties that does not need further chemical treatment as provided by the standard fertilizer manufacturing processes. Consequently, production costs of the recycled P-fertilizers should not and cannot be compared to the mining costs of rock phosphates. The development of marketable products is the key factor in making P-recycling from sewage sludge ash a commercially viable process!

Figure 2. PhosKraft® NPK fertilizer granules (ASH DEC Umwelt AG, 2008).

TECHNOLOGY TRANSFER

The prototype manufacturing plant

A number of technical and commercial challenges had to be resolved before the process could be implemented in the pilot scale. Already the first investigation led to the conclusion that a straightforward upscale of the laboratory technology by the use of larger machinery and equipment of the same type was not possible. In the lab process, indirectly (through the wall) heated quartz, silicon and corundum pipes were used as reactors. These materials are extremely sensitive to mechanical strain and therefore not scaleable to the dimensions needed for industrial scale production. Strain resistant metal pipes cannot be used because of non-withstanding the corrosive gases produced during the process. It was determined that only a refractory lined furnace would be appropriate for the thermal process but it was not clear, if a rotary kiln would be the best choice.

Consequently, the type of reactor had to be chosen among the potentially available shaft, single or multiple hearth, fluidized bed or rotary furnaces. Laboratory and small industrial scale tests in shaft, multiple hearth and rotary furnaces were conducted at facilities operated by plant engineering companies. Fluidized bed test furnaces were not available. The tests lead to the selection of a gas-fired rotary kiln as the most appropriate reactor. However, this decision did not solve all reactor related problems and special attention had to be paid to the adequate drum and burner design, the effective separation of the solid from the gaseous mass flows and the corrosion resistant, metal free refractory lining. Fortunately, manufacturing of refractory materials that resist the process specific atmosphere with a high concentration of chemical chlorine compounds is an ongoing duty of specialized companies because of similar requirements from established industries that process selected metal ores. With the assistance of one of these refractory manufacturers, an adequate lining material could be selected. The furnace wall insulation by 300 mm refractory bricks made it impossible to heat the furnace indirectly. As a consequence, a natural gas-burner that fires into the reaction chamber had to be chosen as a source of heating. Those burners need relatively high volumes of air and/or oxygen and increase the speed of air and off-gas in the kiln by a factor 10 in comparison to the laboratory kilns. Consequently, the technology had to be adapted to the new process conditions.

The first challenge was how to avoid excessive fly-dust and potential particle clogging of the ash-additive mixture in the furnace. Apparently the mixture needed compaction, but which format and size would support the process and which would have adverse effects? Which compaction technology would be most appropriate for the raw materials – highly abrasive ash and hygroscopic chlorine donors? Again, tests were conducted with disk granulators, compaction, briquetting and pelleting presses at different manufacturers' test facilities until a pelleting press was selected because of its relatively low cost and satisfactory performance after 3 test series. The machine produces pressed pellets of acceptable hardness that remain stable during the reaction process and thus prevent large amounts of fly dust. Low dust evaporation reduces particle adhesion on furnace and flue-gas duct walls and helps to design an effective flue-gas cleaning system for almost waste-free operations.

Efficient gas cleaning in accordance with the European and national air pollution control regulations provided the next challenge. The process requires chlorine admixture in significant molar excess and produces a very high chlorine concentration in the flue-gas. If a state-of-the-art dry gas cleaning system is applied, comparatively large amounts of adsorbents have to be fed to the dust filter that would adsorb the metal chloride aerosols. Consequently, chlorines, heavy metals and absorbents would end up as a mixed, non-recoverable secondary waste material. The goal was instead to recycle the chlorines and – as a second step – recover the metals to avoid solid process residues. In cooperation with one supplier of air pollution control systems, a modified dry gas cleaning system was developed and installed. The system consists of a large spray-cooler that allows selective cooling between 180°C and 240°C. Into the cooler a recycled calcium-chloride solution is sprayed that is produced from the effluents of the acidic scrubber neutralized with calcium hydroxide. The intention was to produce a dry calcium chloride at the bottom of the spray-cooler. This target is partly reached – the problem is the selection of nozzles and the correct and fine dosing of the calcium chlorine solution. From time to time the calcium chloride becomes wet (because of inconsistent off-gas flow or overdosing the solution) and needs to be manually removed from the spray cooler. However, if working appropriately, the dry calcium chloride powder is almost heavy metal free and can be recycled as a process additive. After the spray cooler, the off-gas gets into a baghouse filter, where the heavy metals and fly dusts are removed and stored in a bin. About 40% of chlorines are passing the spray cooler and being removed in the baghouse filter. The remaining chlorines are filtered in the acidic scrubber. Its effluent, a 10% HCl solution, is neutralized with calcium-hydroxide. The flow passes a filter press, where a small quantity of gypsum is withheld that is mainly produced from the sulphur in the ash. The effluent from the filter press is recycled to the spray-cooler. Exhaust air is in accordance with the strict limits of waste incineration plants. ASH DEC plants, despite being legally considered as a manufacturing and not as a waste treatment facility, have nonetheless adhere to the off-gas limits being stipulated by the European Waste Incineration Directive and the corresponding national laws.

At present, the modified and up-scaled process is on-line in the pilot plant scale with a capacity of 3–7 tons of sewage sludge ash treated per day.

The industrial manufacturing plant

Induced by the concerted P-recovery Initiatives of the German Federal Ministry for the Environment, Nature Conservation and Nuclear Safety, the German Federal Ministry of Food Conservation, Agriculture and Consumer Protection and the German Federal Ministry of Education and Research, ASH DEC has selected Germany and in particular Altenstadt in Bavaria as the location of its first industrial P-recovery and fertilizer manufacturing plant. The plant will be located next to a new private municipal sludge incineration plant, from which it will receive 33 tons of ash per day. Because of the different grate incineration technology, the ash is coarser than the usual fly ashes from fluidized bed incinerators. The P-containing bottom ash contains in average 20–23% of P_2O_5. It will be extensively tested in the pilot plant facility during the next months. ASH DEC's first industrial manufacturing plant shall be on-line before the end of the year 2009. This manufacturing plant will finally demonstrate the industrial and economic feasibility of a process that can substitute up to 25% of the rock phosphate imports to Europe.

At this point of time, the conceptual design of the plant is almost completed. The value of the non-commercial pilot plant becomes evident now, because it has already revealed the reasons for potential downtimes and the areas in need of improvements. Conceptual mistakes will not be repeated and the experience from the pilot plant operation will avoid the downtimes and necessary retro-fittings that new industrial facilities are – today almost usually – facing.

Let's first start with the positive conclusions from the pilot plant operations: the technology works as expected and experienced in the small scale. The P-rich product as the raw material for the complex fertilizers being sold meets all requirements of current and potential future fertilizer ordinances. In the kiln and in the refractory lined exhaust gas pipes no build-up of material deposits is being observed, the refractory lining material proves to be corrosion resistant and appropriate. The core process of the technology – thermal treatment and cooling of the product – works as smoothly as one could wish and did not cause one single downtime in 5 months of operations. In less than 3 months from starting the pilot plant operations, the planned quantity of NPK fertilizers – 230 tons – could be delivered to a trading house that distributed the product among farmers in Hungary, without a single claim. However, certain problem areas have been identified that need some retrofitting in the pilot plant and a modified conceptual design of the industrial plants.

In the first place, we will tackle a fundamental problem of the technology. The corrosive atmosphere in the kiln prevents the hot off-gas from being used as a secondary energy source. The hot gas needs sudden cooling in a quench made from corrosion resistant materials, such as graphite or ceramics. Heat exchangers cannot be used because of the corrosive particles that would condensate on the heat exchanger surfaces. The rotary kiln being chosen as the most appropriate furnace contains only 8–12% of material to be heated – the rest is air. The material cannot be heated without heating the air and consequently, most of the energy – about 2/3 – is needed to heat the significant air-volume. If the process gas could be separated from the combustion gas, the energy from the combustion gas can be re-used and its energy does not need to be destroyed in the quench. For this reason, the kiln selection was again scrutinized and finally a system could be conceived that combines direct and indirect heating, whereas the direct gas-volume in the reactor is reduced to only 15% of the original amount of gas to be heated, cooled and cleaned. The chosen furnace is a rotary kiln that does not rotate completely. Because of working like a pendulum and swinging 120° around the centre axle, it is called a pendulum kiln. As a consequence of choosing this kiln, the energy consumption in the industrial plant can be reduced by 45% in comparison to the pilot plant. Furthermore, it will significantly contribute to avoid the main problem that has been identified in the pilot plant.

As single most frequent source of unplanned downtimes, the conditioning of the ash-additive mixture has been identified. Because of the hygroscopic additives, the mixture and – later on – the pellets tend to absorb the environmental humidity. This leads to humid residues in all parts of the conveying system and occasional jamming in conveyors, particularly in the elevators. As a consequence, the conditioning must be simplified to one step of mixing and pelleting in one single machine, from which the pellets are fed to the furnace in a closed and dry system. In combination with the pendulum kiln, the conditioning can be completely skipped. Due to the low and slow air-flow in the kiln, ash can be treated without prior compaction and ash and additives will be fed directly to the kiln. Following the principle that machines which are not there cannot cause troubles, no ash conditioning will be installed in the first industrial plant.

A second source of unplanned downtimes is the air-pollution control system. Because of its sensitivity to variations in the off-gas temperature and volume in combination with the cooling liquid flow, the dry system of the pilot plant will be replaced by a 4-stage wet cleaning system consisting of quench, acidic scrubber, alkaline scrubber and venturi scrubber in industrial units. The resulting liquid solution undergoes a chemical treatment by which heavy metals are precipitated as hydroxides whereas the chlorines are recovered to become new process additives. Some of the metal hydroxides could potentially be recycled and sold to metal processing industries – but this is only the next R&D task. The new system is much more expensive but expected to reduce downtimes dramatically. Design and engineering of the necessary equipment is provided by companies that have extensive experience in cleaning off-gases from chemical and thermal processes.

CONCLUSION

As a result of 5 years of intensive R&D with dedicated scientific support of institutions such as *German Federal Institute of Material Research and Testing (BAM), German Federal Research Centre for Cultivated Plants – Julius Kuehn Institute, Swiss Federal Institute of Technology and Vienna University of Technology* ASH DEC proudly presents the new type of phosphate fertilizers from renewable resources that has been perfectly accepted by fertilizer traders and farmers. Due to the effective engineering assistance from TREVIS AG, Basel, about 1.500 tons per month will be available from the industrial manufacturing plant in Altenstadt, Southern-Germany, from the beginning of 2010.

REFERENCES

Adam, C., Kley, G. and Simon, F.G. (2007). Thermal treatment of municipal sewage sludge aiming at marketable P-fertilisers. *Materials Transactions*, **48**(12), 3056–3061.

Adam, C., Peplinski, B., Kley, G. and Simon, F.G. (2007). Thermo-chemical treatment of sewage sludge ashes aiming at marketable P-fertiliser products, (paper presented at the R'07 World Congress – Recovery of Materials and Energy for Resource Efficiency, Davos, Hilty, L.M., Edelmann, X. and Ruf, A. (ed.), EMPA, St. Gallen, 14/1-7).

Hermann, L. (2005). Market and Economic Feasibility – P-fertilisers to be developed during the SUSAN-project (Report to the European Commission (2006); FP6-project SUSAN; Contract No. 016079).

International Fertilizer Industry Association IFA (2008). http://www.fertilizer.org/ifa/statistics/IFADATA/dataline.asp

Jasinski, S.M. (2008). PHOSPHATE ROCK. U.S. Geological Survey. http://minerals.usgs.gov/minerals/pubs/commodity/phosphate_rock/mcs-2008-phosp.pdf

Peplinski, B., Adam, C., Michaelis, M., Kley, G., Emmerling, F. and Simon, F.G. (2008). Reaction sequences in the thermo-chemical treatment of sewage sludge ashes revealed by X-ray powder diffraction – A contribution to the European project SUSAN.

Werner, W. (2003). Complementary Nutrient Sources (paper presented at the IFA-FAO Conference "Global Food Security and the Role of Sustainable Fertilization", Rome, March 26–28, 2003).

Phosphorus recovery by thermochemical treatment of sewage sludge ash – Results of the European FP6-project SUSAN

C. Adam[a], C. Vogel[a], S. Wellendorf[b], J. Schick[b], S. Kratz[b] and E. Schnug[b]

[a]Federal Institute for Materials Research and Testing, Berlin, Germany
[b]Julius Kuehn Institute, Braunschweig, Germany

Abstract Phosphorus (P) is an element essential to all living organisms and cannot be replaced by any other element in terms of its functions (DNA, RNA, ATP etc.). It is one of the major elements for plant nutrition and therefore essential for crop production. Currently P-fertilisers derive from rock phosphate – a non-renewable resource. Alternatively, secondary P-bearing resources such as sewage sludge can be used for fertilisation. However, direct land application of sewage sludge became a controversial issue as it contains organic pollutants and heavy metals.

The SUSAN-project (Sustainable and Safe Re-use of Municipal Sewage Sludge for Nutrient Recovery, www.susan.bam.de) is aimed to develop a sustainable and safe strategy for nutrient recovery from sewage sludges using thermal treatment. Mono-incineration of the sludges completely destructs the organic pollutants in a first step. The incineration residues are ashes with high phosphorus content but still contain heavy metal compounds above the limits for agricultural use. Phosphorus in the ashes exhibits low bioavailability – a disadvantage in farming. Therefore, in a subsequent thermochemical treatment step heavy metals are removed from the sewage sludge ashes and the containing P is transferred into mineral phases available for plants. The thermochemical treatment was investigated in lab-, medium- and technical-scale rotary furnaces. The results showed that volatile heavy metal chlorides are formed by adding a chlorine donor (e.g., magnesium chloride) at temperatures between 800 and 1000°C and are effectively separated from the P-bearing ash via the gaseous phase. The separated heavy metals can be post treated for recycling purposes. The thermochemically treated ashes are characterised by low pollutant concentrations and are suitable raw materials for the production of P-fertilisers. Greenhouse pot experiments showed that the ash based fertilisers reached comparable results to conventional fertilisers.

INTRODUCTION

Phosphorus is an essential element of all animal life forms. It is necessary for the metabolism process (ADP/ATP) and part of the DNS. For this reason phosphorus is a main nutrient and is used in form of phosphates in the fertilizer industry. Generally, P-fertilizers are produced from rock phosphates – a non-renewable resource. Due to the finite nature of the rock phosphate deposits and increasing prices for mineral phosphate on the world market, recycling options for phosphates from waste streams have to be developed and applied. One secondary source of phosphates is sewage sludge. Municipal sewage sludge is a carrier of nutrients but is often polluted with organics, such as hormones, antibiotics, endocrine disruptors and persistent organic pollutants (POP's), and inorganics such as heavy metal compounds. Thus, agricultural application of sewage sludge became a controversial issue. In the last couple of years, the agricultural application of sewage sludge has decreased, while the interest in alternative sludge disposal routes has increased e.g. thermal treatment. Pollutants should be removed or destroyed before agricultural application of sewage sludge to protect farmland and human health.

The European project SUSAN (Sustainable and Safe Re-use of Municipal Sewage Sludge for Nutrient Recovery) (SUSAN 2008) bundles the research efforts of seven partners from four countries, aiming at the development of a sustainable and safe strategy for phosphorus recovery from sewage sludge for agricultural utilisation using a two-step thermal treatment including mono-incineration of sewage sludge and subsequent thermochemical treatment of the resulting ash.

A schematic of the strategy is presented in Figure 1. Mono-incineration of the sludges completely destructs the organic pollutants in a first step. The incineration residues are ashes with high phosphorus content, but still contain heavy metal compounds above the legal limits for agricultural use. The fate of elements during incineration has been intensively studied (Chandler 1997). Matrix or lithophilic elements are silicon, calcium, aluminium and iron, which form stable oxides or anions (i.e. silicates or alumosilicates). The fraction of volatile elements consists of elements forming acid gases like halogens or sulphur, volatile trace metals (e.g. mercury), and trace metal compounds (e.g. heavy metal chlorides).

In the SUSAN-process, volatile heavy metal chlorides are formed by adding a chlorine donor (magnesium or calcium chloride) at temperatures around $1000\,°C$ to the sewage sludge ash. The volatile heavy metal chlorides are separated from the gaseous phase in a gas treatment system. For metal recycling purposes the different heavy metals must be separated from each other. This might become profitable in the future due to increasing prices for heavy metals (especially copper). For separation of the heavy metals a liquid-liquid extraction with the mining chemical LIX® reagent was investigated. The organic reagent forms a metal–hydroxyoxime complex (Kongolo et al., 2003; de San Miguel et al. 1997; Alguacil et al., 1999). Heavy metals were quantitatively removed from the watery phase by LIX® extraction and subsequent precipitation.

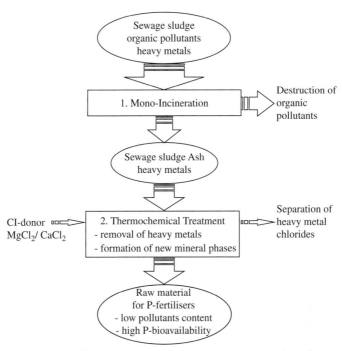

Figure 1. Schematic of the pathways of organic and inorganic pollutants during the proposed treatment steps.

METHODOLOGY

During the SUSAN-project a thermochemical process was developed for the treatment of sewage sludge ash (SSA) for the production of phosphorus fertilizers (Adam *et al.*, 2008a; Adam *et al.*, 2008b; Adam *et al.*, 2007). SSA was mixed with the chlorine donor ($MgCl_2$ or $CaCl_2$) and water before thermochemical treatment. The amount of water was calculated to receive a water content of 20–30% in the ash mixture, a practicable water content for pelleting of the ashes, which is applied in large-scale.

The raw ashes were intensively mixed with the solution consisting of the chlorine donor and water. For lab-scale experiments an amount of 400 g of the mixture was filled in a quartz glass reactor of the gas tight lab-scale rotary furnace. The total masses were determined before and after thermal treatment. The batch was treated at a certain temperature (750–1050°C) for a certain retention time (10–120 min) depending on the trials programme. The quartz glass reactor was continuously moved during thermal treatment and was continuously flushed with an air-flow rate of 3 L/min. The gas-outlet of the quartz glass reactor was discharged in a PTFE-tube for cooling the off-gas and condensation of heavy metals. The cooled off-gas from the outlet of the PTFE-tube passed 4 washing flasks (filled with 1 L water each) with integrated perforated plates for fine gas distribution and two wet scrubbers operated with NaOH-solution (Figure 2). The reactor was cooled down after thermal treatment. Characterisation of the mineral phases was done by XRD; chemical analysis by XRF and after chemical degradation also by ICP-OES. Furthermore, the P-solubility of the thermochemical treated SSA in citric acid (2%) was analysed.

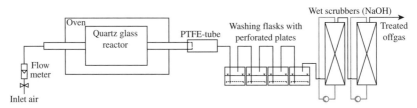

Figure 2. Schematic of the experimental set-up for thermochemical trials including lab-scale rotary furnace, PTFE-tube for off-gas cooling and condensation of heavy metals, washing flasks with perforated plates for heavy metals rejection and wet scrubbers (NaOH) for removal of HCl and SO_2.

For separation of the heavy metals from the acidic washing solutions (Figure 2, washing flasks) a liquid-liquid extraction with the LIX® 84-I reagent was applied at a ratio water/organic phase of 1:1 at room temperature. Complexes with copper and zinc where formed at pH > 0.5 and pH > 5.5, respectively. The (heavy metal) loaded organic phase was stripped with sulfuric acid. The concentrations of the heavy metals where analyzed by ICP-OES.

Plant performance (growth and development, health and yield) as well as availability and uptake of the different products' compounds were investigated by means of two greenhouse experiments in 2006 and 2007. In 2006, five ash-based PK-fertilisers that differed regarding their raw ash (SSA-SNB$_a$ – high iron content or SSA-SNB$_b$ – low iron content), the chlorine donor used (CaCl$_2$ or MgCl$_2$), heating technique of the rotary furnace (directly with gas or indirectly electrically heated) as well as the chemical form of potassium added (KCl or K$_2$SO$_4$) were tested with maize as test crop. For control, a current PK-fertiliser (Thomaskali 7-21-4-3) as well as a treatment without any P were set up. In the greenhouse experiment conducted in 2007, 8 pure ash-based P-fertilisers that also differed regarding their raw ash (SSA-SINDL – low iron content and high aluminium content or SSA-SNB$_a$ – high iron content), the chlorine donor used (CaCl$_2$ or MgCl$_2$), heating technique (direct or indirect) as well as their additional treatment (two fertiliser types were partly digested with H_2SO_4) were used. For control a treatment with a current SSP and one without any P were set up.

RESULTS

Thermochemical process

The ranges of the chemical compositions of 7 different sewage sludge ashes from different mono-incineration facilities (fluidised bed incinerators) from Germany and the Netherlands are presented in Table 1 together with legal limit values for heavy metal mass fractions of the German Fertilizer Ordinance.

Table 1. Ranges of the composition of 7 different sewage sludge ashes measured by XRF and after 'aqua regia' digests [microwave], measured by ICP-OES and limit values of the German Fertilizer Ordinance (Adam *et al.* 2007).

XRF	%	ICP-OES	mg/kg	Limit (DüMV Germany)
SiO_2	21.2–43.4	As	4.25–40	40
Al_2O3	5.7–15.5	Cd	0.20–4.71	4
Fe_2O_3	2.8–22.6	Cr	70–173	–
MnO	0.08–0.35	Cu	470–1267	70
CaO	11.7–19.4	Hg	<0.1–0.23	1
Na_2O	0.21–1.12	Mo	11.6–79.5	–
K_2O	0.61–2.81	Ni	39.5–98.0	80
P_2O_5	13.7–25.0	Pb	80.5–264	150
		Sn	36.1–60.0	–
		Zn	1540–2181	1000

The average P_2O_5-content of the 7 SSAs tested was 21.4%. The P_2O_5-content of the ashes ranged from 14% to 25%. Conventional P-fertilisers like Single super phosphate SSP have a P_2O_5-content of 18% and multi-nutrient fertilisers like PK- and NPK-fertilisers contain approx. 5–12% P_2O_5. Therefore, sewage sludge ashes are suitable secondary raw-materials for the production of P-fertilisers and could be used to save rock phosphates and thus contribute to resource conservation.

Sewage sludge ashes are carriers of valuable nutrients like phosphorus, calcium, magnesium and potassium but also contain considerable amounts of heavy metals (Table 1). The mass fractions of As, Cd, Cu, Ni, Pb and Zn in the untreated sewage sludge ashes exceeded the limit values of the German Fertilizer Ordinance in many cases.

In Table 2 the mass fractions of heavy metals are presented for the different SSAs that were thermochemically treated according to the SUSAN-process

(Cl-donor: $MgCl_2$, chloride-concentration: 150 g Cl/kg ash; retention time: 30 min.; temperature: 1000°C). All limit values of the German Fertilizer Ordinance for heavy metals were met by the thermochemically treated ashes except for copper. Two ashes (out of 7) exceeded the copper limit. However, the German limit value of 70 mg/kg for copper is relatively strict compared to other European countries and is currently under discussion. At the moment the German Fertiliser Ordinance is revised. An obligation to label of 500 mg/kg copper is currently discussed in the amendment process. To further decrease the copper concentration below the current German limit of 70 mg/kg a higher concentration of chloride must be used in the SUSAN-process. Table 3 shows the concentrations of copper for a thermochemically treated ash ($CaCl_2$; 1000°C, 60 min.) depending on the amount of chloride added. With increase of the chloride concentration from 150 g Cl/kg ash to 200 g Cl/kg ash the copper concentration was decreased below the limit value of the German Fertiliser Ordinance. However, this is achieved at higher treatment costs due to a higher demand of the Cl-donor.

Table 2. Heavy metal concentrations of different thermo-chemically treated sewage sludge ashes ('aqua regia' digests [microwave], ICP-OES) and limit values of the German Fertiliser Ordinance (Adam *et al.*, 2007).

ICP-OES	mg/kg	Limit
As	1.8–27.6	40
Cd	<0.1–0.1	4
Cr	58.1–169	–
Cu	10.5–249	70
Hg	<0.1	1
Mo	<0.1–21.5	–
Ni	32.1–72.4	80
Pb	<0.1–4.2	150
Sn	<0.1–32.5	–
Zn	18.3–172	1000

Table 3. Copper concentration for thermal treated ash ash with different concentration on $CaCl_2$ ('aqua regia' digests [microwave], ICP-OES).

g Cl/kg Ash	Cu (mg/kg)
50	564
100	275
150	128
200	55

To gain insight into the chemical processes accompanying the thermo-chemical treatment of sewage sludge ashes, XRD was applied. The results of the XRD analysis of a systematic series of samples of an iron-bearing ash thermochemical treated using identical conditions (Cl-donor $MgCl_2$, Cl-conc. 150 g Cl/kg, retention time 60 min), and varying only a single process parameter, namely the temperature (from 350°C to 1050°C), are displayed in Figure 3. Among the numerous crystalline phases present in the raw ash, only quartz and hematite continue to exist up to 1050°C. All other components undergo at least one decomposition–recrystallisation cycle, some of the components recrystallise even several times.

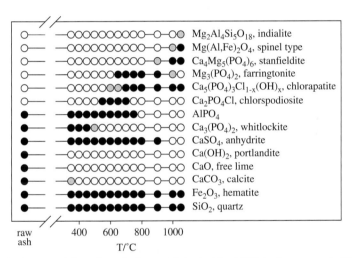

Figure 3. Some of the mineral phases detected by XRD in the raw ash SSA and the same ash after being thermochemical treated with $MgCl_2$ at temperatures between 350°C and 1050°C; ● = phase detected by XRD, ○ = phase not detected by XRD, ◐ = phase just (dis)appearing.

Free lime, most probably the decomposition product of limestone which is added to the sewage sludge during the incineration process for flue gas desulfurisation, as well as portlandite, $Ca(OH)_2$, disappear during the thermochemical treatment at temperatures below 400°C. Between 450°C and 600°C whitlockite reacts with the $CaCl_2$ and forms a mixture of chlorspodiosite and chlorapatite. While chlorspodiosite is only an intermediate species which disappears again with decreasing concentration of chlorine in the reactor, chlorapatite is stable and becomes the dominant phosphate-bearing phase up to the maximum reaction temperature of 1050°C. Magnesium phosphate is first detected at a reaction temperature of 750°C, whereas the mixed (Mg, Ca)-phosphate forms only above 850°C, possibly on the expense of the former which gradually disappears at about 1000°C. Another major step towards the understanding of the phase formation processes in sewage sludge ashes resulted from the proof of existence of crystalline $AlPO_4$ in raw sewage sludge ashes and the finding, that this phase is thermally considerably more stable than whitlockite. The aluminium oxide originating from the thermal decomposition of the aluminium phosphate reacts with other components of the ash and finally forms two aluminium- and magnesium-bearing compounds: the spinel type $Mg(Al,Fe)_2O_4$ as well as $Mg_2Al_4SiO_{18}$ (indialite or cordierite), which were detected in samples treated at 1000°C or 1050°C, respectively.

In Figure 4 the solubility of phosphorus compounds in citric acid (2%) is presented. The raw ashes showed P-solubilities of 25–40%, which is low compared to 87% that was determined for a conventional Single Super Phosphate (SSP). The P-bioavailability of the ashes was significantly increased by thermo-chemical treatment. The P-bioavailability was improved with increasing temperature, resulting in 97% P-solubility in citric acid after thermochemical treatment with $MgCl_2$ at 1000°C. In contrast to SSP with a P-water solubility of 78%, P in the treated ashes was almost insoluble in water (<1%).

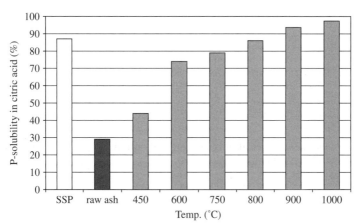

Figure 4. P-solubility in 2% citric acid of SSP and SSA-SNB$_a$ untreated (raw ash) and thermochemical treated at different process temperatures. [Constant parameters: Cl-donor $MgCl_2$, Cl-conc. 150 g Cl/kg, retention time 60 min.].

Heavy metal recycling

Figure 5a shows the pH-dependant (pH 0.5–2.0) transfer rates of copper in the organic phase after 5 minutes of extraction. Below a pH-value of 1.5 not all copper was transferred from the water phase into the organic phase of the LIX$^{®}$ reagent. At a pH-value of 1.5 and higher, copper was quantitatively transferred into the organic phase.

With increase of the pH-value also the extraction of other elements slightly increased (not shown). Besides the pH-value the contact time of organic and water phase is relevant. In Figure 5b the transfer rates of copper, iron and zinc are presented for the extraction of the heavy metal concentrate (pH-value 1.5) with different contact times. For all contact times (1, 2, 5 and 10 min.) copper was quantitatively extracted from the water phase. With increase of contact time the transfer rate of iron slightly increased. The transfer rate of zinc is constant. During the stripping process the extracted metal (mainly copper) in the organic phase is changed by protons of sulfuric acid. This process depends on the concentration of the sulfuric acid. Figure 5c shows the transfer rates of copper of the stripping process for different concentrations of sulphuric acid and pure water.

For a continuous extraction (mixer-settler-extraction) of copper from the heavy metal concentrate a pH-value of 1.5 for the extracted solution and 2 mol/L sulfuric acid for the stripping process in combination with short contact times should be adjusted. After extraction the obtained copper- and zinc/iron-bearing solutions can be precipitated with NaOH and Ca(OH)$_2$, respectively, followed by filtration of the metal hydroxides. The concentration of copper in the extracted and precipitated mass is 38 wt.%. The precipitated mass of the zinc/iron-bearing concentrate contains around 23 wt.% zinc and 37 wt.% iron, respectively. These materials are suited secondary raw materials for recycling.

Pot trials

Both, the ash-based PK-fertilisers and the control fertiliser used in 2006 showed a P-solubility in water which was below 1% of P$_{total}$. In contrast, the control fertiliser as well as the ash-based PK-fertilisers produced in the indirectly heated rotary furnace showed an almost complete P-solubility in citric acid, however, the P-solubility of the PK-fertilisers produced in the directly heated rotary furnace tended to be lower (73% of P$_{total}$ soluble in citric acid). In 2006, only the results of the variants fertilised at the lowest P-level (50 mg P per pot) were analysed because higher PK-applications led either to a delayed germination of the seeds or in some cases to no germination at all. This might have been caused by high K- resp. Cl-levels in the soil due to the high PK- application which was obviously not adequate for the small amount of substrate used.

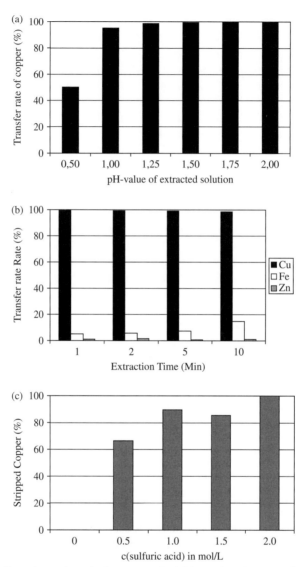

Figure 5. Transfer rates (ratio water (acid)/organic phase 1:1) for the a) pH-dependent extraction of copper b) time-dependent extraction of Cu, Fe and Zn at pH 1.5 and c) stripped copper from the organic phase with different concentration on sulfuric acid.

All fertilised variants showed a dry matter yield that differed significantly from the zero control (Figure 6). Furthermore, all ash-based PK-fertilisers achieved dry matter yields at the same level as the commercial control PK-fertiliser. The ash-based fertilisers produced with $MgCl_2$ as Cl-donor and the control PK-fertiliser showed a slightly higher dry matter yield than the fertilisers produced with the Cl-donor $CaCl_2$. However, these differences were not statistically significant.

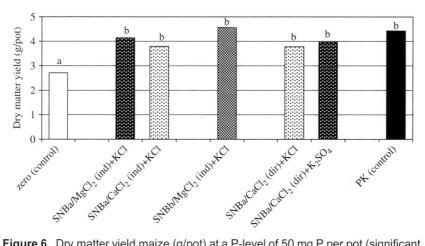

Figure 6. Dry matter yield maize (g/pot) at a P-level of 50 mg P per pot (significant differences between groups were determined by Tukey post-hoc test, $p < 0.05$ and are denoted by different letters).

The SSP, used as the control fertiliser in the greenhouse experiment in 2007, showed a good P-solubility in water (78% of P_{total}), while the water-solubility of the ash-based fertilisers was, analogous to the products tested in 2006, below 1% of P_{total}. However, the P-solubility in water of ash-based fertilisers that were partly digested with H_2SO_4 (+S) was increased by this treatment to 25% of P_{total} (fertiliser produced in directly heated rotary furnace) and 34% of P_{total} (fertiliser produced in indirectly heated rotary furnace). P-fertilisers produced in the indirectly heated rotary furnace tended to have a higher P-solubility in citric acid (81–100% of P_{total}) than the control fertiliser SSP (87% of P_{total}).

In contrast, thermochemical treatment in the directly heated rotary furnace resulted in a lower P-solubility in citric acid (67–72% of P_{total}). The yield curve of the experiment conducted in 2007 showed higher yields for all ash-based fertilisers than for the zero control at all P-levels. For the ash-based fertilisers,

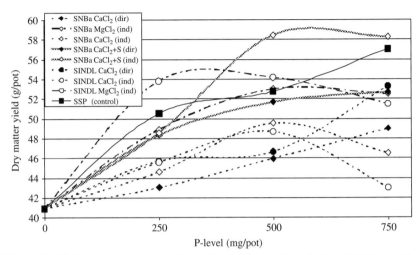

Figure 7. Yield response curves for maize for the test products of 2007 and the SSP-control. [Legend: type of SSA, type of Cl-donor (MgCl$_2$ or CaCl$_2$), +S = partly digested with H$_2$SO$_4$, (dir) = directly heated furnace, (ind) = indirectly heated furnace].

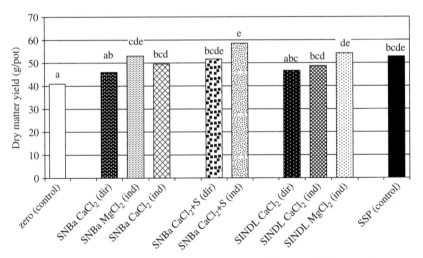

Figure 8. Dry matter yield (g/pot) (maize) at a P-level of 500 mg P per pot (significant differences between groups were determined by Tukey post-hoc test, p < 0.05 and denoted by different letters) [Legend: type of SSA, type of Cl-donor (MgCl$_2$ or CaCl$_2$), +S = partly digested with H$_2$SO$_4$ (dir) = directly heated furnace, (ind) = indirectly heated furnace].

except for the fertilisers produced in the directly heated rotary furnace, the strongest increase in dry matter yield could be observed from P-level 250 to P-level 500, further rising of the P-level led to a stagnation or decrease of dry matter yield. The yield curve for the products made in the directly heated rotary furnace as well as for the control fertiliser showed a continuous increase in dry matter yield with rising P-level (Figure 7).

At P-level 500, which was assumed to be the optimal level, maize obtained a dry matter yield comparable to the SSP-variant and significantly higher than the zero control when SSA-SNB$_a$/MgCl$_2$ (indirect), SSA-SNB$_a$/CaCl$_2$ (indirect) + S and SSA-SINDL/MgCl$_2$ (indirect) were applied (Figure 8). The remaining 5 variants showed a dry matter yield which ranged between that of the control fertiliser and the zero control, not being significantly different from either of them. The composition of the raw ashes (SSA-SNB$_a$ and SSA-SINDL), which differed concerning their aluminium and iron content, did not have an effect on the dry matter yield.

CONCLUSION

Sewage sludge ashes exhibit high phosphorus contents of approx. 20% P$_2$O$_5$ and are therefore suitable raw materials for P-fertilizer production. It was shown in a screening that heavy metals can be effectively removed from different types of sewage sludge ashes by a thermochemical treatment at operating temperatures of about 1000°C. The P-bioavailability was significantly increased during thermochemical treatment resulting in P-solubility in citric acid of up to 100%. XRD analyses showed that the thermochemical treatment is accompanied by a sequence of chemical reactions and transformations of the phosphate-bearing mineral phases.

The removed heavy metals (mainly copper, zinc and iron) could be separated by liquid-liquid-extraction in a copper- and zinc-bearing solution which can be precipitated with NaOH and Ca(OH)$_2$, respectively. The high concentration of copper and zinc in the precipitated masses (38 wt.% and 23 wt.%, respectively) are suitable secondary raw materials for the recycling industry.

The results of the pot experiments conducted in 2006 with ash-based PK-fertilisers and in 2007 with ash-based pure P-fertilisers showed that especially products made of MgCl$_2$-treated ashes have the potential to achieve dry matter yields comparable to current mineral PK- or P-fertilisers. A pure ash-based P-fertiliser that was partly digested with H$_2$SO$_4$, was only used in the second pot experiment (2007) and showed promising results, as well.

OUTLOOK

A pilot plant according to the SUSAN-process was built up by the Austrian company ASH DEC Umwelt AG and was taken into operation in June 2008. The capacity of this plant is 7 Mg per day. Approx. 200 Mg of an ash based NPK-fertiliser were already produced and sold. A demonstration facility is currently in the planning phase and will be build in 2009/2010 in Bavaria, Germany (by ASH DEC Umwelt AG).

REFERENCES

Adam, C., Peplinski, B., Kley, G., Kratz, S., Schick, J. and Schnug, E. (2008a). Phosphorrückgewinnung aus Klärschlammaschen – Ergebnisse aus dem Eu Projekt SUSAN. *Österr. Wasser- und Abfallwirtschaft*, **60**(3–4), 55–64.
Adam, C., Kley, G., Simon, F.-G. and Lehmann, A. (2008b). Recovery of nutrients from sewage sludge – Results of the European research-project SUSAN. *Water Practice and Technology*, **3**(1).
Adam, C., Kley, G. and Simon, F.-G. (2007). Thermal treatment of municipal sewage sludge aiming at marketable P-fertilisers. *Mater. Trans.*, **48**, 3056.
Alguacil, F.J. and Cobo, A. (1999). Solvent extraction with LIX 973N for the selective separation of copper and nickel. *J. Chem. Technol. Biotechnol.*, **74**, 467.
Chandler, A.J., Eighmy, T.T., Hartlen, J., Hjelmar, O., Kosson, D.S., Sawell, S.E., van der Sloot, H.A. and Vehlow, J. (1997). Municipal solid waste incineration residues: An international perspective on characterisation and management of residues from municipal solid waste Incineration. Studies in Environmental Science, Vol. 67, International Ash Working Group, Elsevier, Amsterdam.
Kongolo, K., Mwema, M.D., Banza, A.N. and Gock, E. (2003). Cobalt and zinc recovery from copper sulphate solution by solvent extraction. *Miner. Eng.*, **16**, 1371.
de San Miguel, E.R., Aguilar, J.C., Bernal, J.P., Ballinas, M.L., Rodriguez, M.T.J., de Gyves, J. and Schimmel, K. (1997). Extraction of Cu (II), Fe(III), Ga(III), Ni(II), In(III), Co(II), Zn(II) and Pb(II) with LIX® 984 dissolved in n-heptane. *Hydrometallurgy*, **47**, 19.
SUSAN (2008). http://www.susan.bam.de, Accessed 14-10-2008.

Remediation of phosphorus from animal slurry

A.M. Thygesen, E. Skou, O. Wernberg and S.G. Sommer

University of Southern Denmark (SDU), Faculty of Engineering, Institute of Chemical Engineering, Biotechnology and Environmental Engineering, Postal address: Campusvej 55, Denmark
E-mail: sgs@kbm.sdu.dk, Tel. +45 6550 7359

Abstract In the near future phosphorus (P) may be a limited resource that will be high in demand. Deficiency of P in mineral fertilizers for crop production may cause a worldwide shortage of food and feed. This will increase the incentives for recycling P in animal manure which is a huge source of raw material for green P fertilizer production. In this abstract we are presenting results from a study of remediating P from ash originating from incineration of the solid fraction of separated animal slurry. The intention of the study is to link incineration technology with recycling of P using the ash as a fertilizer.

INTRODUCTION

Phosphorus (P) is a plant nutrient which is of paramount importance for a sustainable crop production. Large amounts of phosphor is excavated from mines, processed and used in mineral fertilizers and in the industry, i.e. approximately 80% of phosphates used worldwide with the balance divided between detergents (12%), animal feeds (5%) and speciality applications (3%), for example, food grade and metal treatment (Steen, 1998; Diskowski and Hofmann, 2005).

It is acknowledged that phosphor is a very limited resource that will be exhausted within the next 60–130 years (Steen, 1998), and that a reduced supply may be a risk to the global feed and food supply. The incentives for recycling plant nutrients and especially P are therefore great and with depletion of the rock phosphor reserves the value of the recovered P will increase.

A significant source of P is animal manure which has not been used as a raw material for P fertilizer production (EFMA, 2000). Today livestock production is becoming increasingly specialised and is considered an industry, where huge amounts of manure is produced on the farms. In consequence the traditional environmental friendly application of manure to fields will become increasingly costly and a problem to the management. On the other hand the large amounts of manure produced on each unit will facilitate treatment and processing of the manure, which instead of being a waste product is a source of raw material for P production.

Thus, there is a global marked for developing sustainable and environmental friendly management of animal manure in processes producing energy and high value fertilizer products. There is a need of reducing cost and improving efficiency of the processes. In this research and development one must consider that manure management consist of several interrelated operations from slurry is removed in the animal house till it has been used for bio energy production and to fertilize crops (Petersen *et al.*, 2007).

The project presented in this article intend in a whole system approach to contribute to development of technology for optimising recycling of P in livestock production. The goal is to combine separation of manure in a solid phosphor rich fraction and incinerate the solid fraction, so that P in the ash fraction is available for plants and also for remediation. The barrier for using the ash as a fertilizer is that P availability of ash from pig slurry is 25–50% of the P availability from field applied P fertilizers and investigations indicate that the availability decrease with the incineration temperature, but it has not been possible to find systematic studies confirming this hypothesis (Sørensen and Rubæk, 2006; Petersen *et al.*, 2005).

The focus in this project is on assessing which form of incineration will be the best considering that the ash shall be used as raw material for ash-fertilizers or bio fertilizer production. Thus, the experiments include measurements of composition in the ash as affected by the temperature during incineration using thermal gravimetric analysis (TGA) and X-ray powder diffraction spectroscopy (XRD). In addition the concentrations of plant available and total P in the ashes are quantified.

MATERIALS AND METHODS

The bioavailability of P in ash from manure incineration was examined by determining the solubility of P in the bio ash. The intention was to study the differences in the quality of the ash by heating the dry matter (DM) rich fraction at four specific temperatures (450°C, 700°C, 900°C and 1050°C) in an oven in the presence of oxygen. The content of crystalline components and the availability of phosphorus in the bio ash was examined.

Three different solid fractions produced when separating animal slurries was used for the studies: KO, HA and FA (Table 1). These were chosen with the purpose of examining if the presence of additives has any influence on the plant availability of phosphorus in the ash. Characteristics of the manures are presented in Table 1. All manures were stored in a refrigerator at 10°C until they were treated.

Table 1. Description of the manures included in the study.

Manure	Additives	Description
KO	Yes	Pelleted, degassed manure produced by Peter Stoholm for Kommunekemi A/S (Nyborg, Denmark).
HA	No	Untreated mixture of faeces and straw from boars at Hatting A/S (Odense, Denmark).
FA	?	Degassed fiber fraction from separation from Fangel Bioengergy (Odense, Denmark). The manure is a mixture of fiber fraction from different separation system. In consequence the addition of additives is unknown. The consistens of the manure is like humus.

Dry matter (DM), ash fraction (AF) and volatile solids (VS)

Dry matter (DM) of the manures was determined after a 24-h drying period at 105°C (Sommer *et al.*, 2007; Møller *et al.*, 2007; Møller *et al.*, 2004, DS/EN 14346). Ash fractions (AF) of DM were determined after incineration at different temperatures. Concentration of DM and AF are shown in Table 2. The difference between DM and AF constitute the content of volatile solids (VS).

Table 2. Characteristics of the analysed manures and their ashes from different incineration temperatures. DM dry matter, AF ash fraction of dry matter.

Manure	DM	AF 400°C	AF 700°C	AF 900°C	AF 1050°C
KO	867	485	360	349	345
HA	180	332	265	255	254
FA	343	183	176	168	172

Thermal gravimetric analysis (TGA)

Information for assessing suitable incineration temperatures for the experiments was examined by determining the pattern of loss of VS using thermal gravimetric analysis (TGA, Setaram TG-DTA 92). The analysis gives data for weight loss against temperature and heating time. The TG analysis was carried out after a 24-h drying period of the used manure at 105°C.

X-ray diffraction spectroscopy (XRD)

Content of crystalline components in the ash was determined using x-ray diffraction spectroscopy (XRD, Siemens Diffraktometer D5000 Kristalloflex, Germany).

Extraction

The content of plant available P was extracted with neutralised ammonium citrate to dissolve the plant available P as orthophosphate, PO_4^{3-}. The ammonium citrate solution contains 185 mg citrate acid per litre and had pH 7.00. 500 mg of bio ash was extracted in 100 mL ammonium citrate at 65°C for one hour. Afterwards the solution was cooled rapidly to room temperature and the solution was diluted with demineralised water to 500 mL.

The content of total P was determined to facilitate the comparison content of plant available P in the different ashes. 500 mg ash was extracted as PO_4^{3-} in 32.5 mL aqua regia with 31 vol% sulfuric acid and 46 vol% nitric acid. The solution was boiled in 30 minutes and then diluted with 75 mL demineralised water and boiled for additional 15 minutes. Then the solution was cooled to room temperature and diluted to 250 mL.

Phosphorus analysis

The concentration of PO_4^{3-} in the extracts was analyzed with optical ion coupled plasma spectroscopy at Analytech A/S.

RESULTS AND DISCUSSION

Thermal gravimetric analysis

One representative weight loss studies is presented in Figure 1. The temperature is kept constant in intervals at 100, 200, 300, 400 and 500°C, and the weight loss is in consequence happening stepwise.

Up to a temperature at about 120°C the weight loss is caused by water evaporation. At higher temperatures the VS evaporates and only inorganic components remains in the ash from high temperature incineration. Most VS is lost at temperature up till 400°C (Figure 1). Further, 400°C is the lowest temperature at which the VS can be removed at a high rate.

The incineration temperature at power plants was chosen for treatment of the organic material to ensure that near all VS is lost or used and that the process is fast. The incineration temperatures at power plants are above 400°C, which was decided to be the lowest treatment temperature in this study. The option of removing a similar amount of VS at 300 or 350°C was studied, but the necessary incineration time to get the optimal weight loss was found to be high (Figure 2), e.g. when the temperature is only kept for 8.9 hours at 300 and 350°C respectively then the weight loss caused by afterwards increasing the temperature to 400°C is remarkable. In the TGA plot on Figure 2 where the

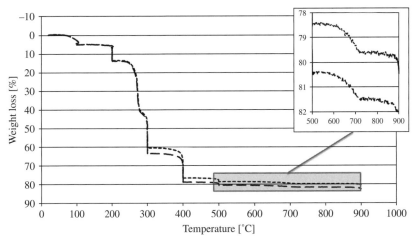

Figure 1. Data for weight loss of biomass against incineration temperature in the TGA study. The temperature is kept constant in 3 hours at 100, 200, 300, 400 and 500°C respectively. The incineration has been carried out in an atmosphere of mixed oxygen and argon.

temperature has been kept constant for 16.7 hours at 300 and 350°C the weight loss caused by afterwards increasing the temperature to 400°C is less significant. The weight loss sequence in this experiment is different from the two experiments pictured in Figure 2, because biomass fraction had a higher content of VS at very low temperatures. It can therefore not be said if there actually is a less significant temperature loss when increasing the temperature from 350 to 400°C in this experiment in relation to the other experiments. But still 400°C is chosen as the lowest incineration temperature, because an incineration time at 8.9 hours are assessed to be too high.

From the zoomed box at Figure 1 it appears that the weight loss has a bend at 700 to 900°C, indicating a phase change. Therefore it was chosen to carry out the experiments at 700 and 900°C.

The fourth incineration temperature chosen was 1050°C because this is in the range of normal incineration temperatures at power plants in the industry (Graversen 2006, Sørensen and Petersen 2006). In addition 1200°C was used as incineration temperature in some of the studies, but the high temperature resulted in overheating of the oven, and therefore only few determinations of these incinerations were carried out.

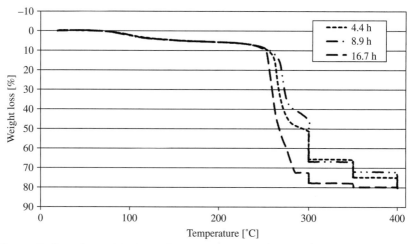

Figure 2. Data for weight loss of biomass against incineration temperature. The temperature is kept constant at 300, 350 and 400°C in respectively 4.4, 8.9 and 16.7 h. The incineration has been carried out in an atmosphere of mixed oxygen and argon.

X-ray diffraction

The identified components in the XRD spectres are listed in Table 3. For comparison d-spacing values from other literature are listed in Table 4.

The XRD spectres of KO and FA shows that hydroxapatite, $Ca_5(PO_4)_3(OH)$, is formed when the manure is incinerated at 700°C or at higher temperatures.

In the spectres of KO it is seen that calcium carbonate, $CaCO_3$, is converted to calcium oxide, CaO, at temperatures above 400°C (Figure 3a,b). Calcium carbonate is also abundant in the ash from FA incinerated at 400°C, but is not found in FA ashes incinerated at higher temperatures. It is though remarkable that calcium oxide is not determined in the ashes from high incineration temperatures. The decomposition of calcum carbonate takes place at 900°C at a pressure at 1 atmosphere of carbon dioxide (Housecroft and Sharpe, 2005), and it is therefore expected to take place at a lower temperature, when the pressure of carbon dioxide is lower than one atmosphere as it is under normal atmospheric conditions.

A set of d-spacing values at about 2.87, 2.60 and 3.19 are found in both HA and FA ashes incinerated at 700°C or higher temperatures telling that at least one crystalline component is formed, which is not permitted at lower temperatures. But we have not succeeded in identifying these components.

Table 3. Identified components in XRD spectres of the three manures incinerated at different temperatures and in the raw manure. The component A is an unknown component. The numbers listed are the characteristic d-spacing values read from the spectres and their relative intensities are listed in the brackets.

	KO	HA	FA
Raw	$CaCO_3$ 3.03 (1) 2.28 (0.4) 2.09 (0.3)	Noise 3.34 (1)	Noise
400°C	$CaCO_3$ 3.03 (1) 2.28 (0.2) 2.09 (0.2)	Noise 3.34 (1)	$CaCO_3$ 3.05 (1) 2.29 (0.2) 2.10 (0.2)
700°C	CaO 2.41 (1) 1.70 (0.7) 2.79 (0.7)	A 2.87 (1) 2.60 (0.8) 3.19 (0.6)	
	$Ca_5(PO_4)_3(OH)$ 2.83 (1) 2.73 (0.8) 2.79 (1.8)		$Ca_5(PO_4)_3(OH)$ 2.83 (1) 2.73 (0.9) 2.79 (0.9)
900°C	CaO 2.41 (1) 1.70 (0.5) 2.79 (0.5)	A 2.87 (1) 2.60 (0.7) 3.19 (0.6)	
	$Ca_5(PO_4)_3(OH)$ 2.83 (1) 2.72 (1) 2.79 (1.7)		$Ca_5(PO_4)_3(OH)$ 2.83 (1) 2.73 (0.8) 2.79 (0.8)
1050°C	CaO 2.40 (1) 1.70 (0.5) 2.78 (0.5)	A 2.86 (1) 2.59 (0.5) 3.19 (0.4)	A 2.87 (1) 2.60 (0.7) 3.19 (0.7)
	$Ca_5(PO_4)_3(OH)$ 2.83 (1) 2.72 (0.3) 2.78 (1.2)		$Ca_5(PO_4)_3(OH)$ 2.83 (1) 2.73 (0.4) 2.79 (0.5)
1200°C	CaO 2.40 (1) 1.70 (0.5) 2.78 (0.4)	A 2.86 (1) 2.59 (0.4) 3.18 (0.4)	A 2.86 (1) 2.60 (0.2) 3.19 (0.3)
	$Ca_5(PO_4)_3(OH)$ 2.83 (1) 2.72 (0.5) 2.78 (0.8)		

Table 4. d-spacing values and relative intensities from McClune 1979.

		d-spacing
Hydroxe apatite	$Ca_5(PO_4)_3(OH)$	2.82 (1) 2.72 (0.6) 2.78 (0.4)
Calciumcarbonate, calcite syn	$CaCO_3$	3.04 (1) 2.29 (0.2) 2.10 (0.2)
Calciumoxide	CaO	2.40 (1) 1.70 (0.5) 2.78 (0.3)

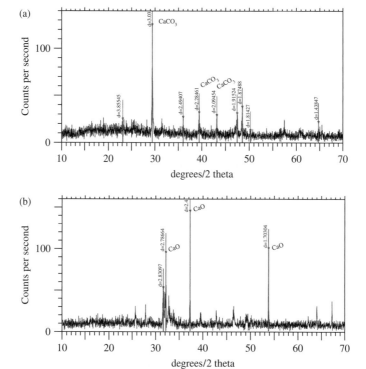

Figure 3. (a) XRD spectrum for the KO manure incinerated at 400°C. The d-spacings 3.03, 2.82 and 2.09 tells that the sample contains calcium carbonate; (b) XRD spectrum for the KO manure incinerated at 700°C. The d-spacings at 2.41, 1.70 and 2.79 tells that the sample contains calcium oxide and the d-spacings at 2.83, 2.73 and 2.79 tells that the sample contains hydroxapatite.

Generally more crystalline components can be identified at incineration temperatures above 400°C, but there is no significant change in the composition of the ash when raising the temperature from 700 to 1200°C as seen from Table 3.

Phosphorus content

The contents of total P in the ashes are shown in Table 5. Except from the concentrations in the raw material, the concentrations are nearly constant and independent of the incineration temperature. This tells that the extractions and concentration analysis are reliable.

Table 5. Concentration of total P in the ashes after different incineration temperatures and in the raw biomass. The concentrations are g PO_4^{3-}-P per kg TS.

	Raw	400°C	700°C	900°C	1050°C
KO	18.5	21.0	21.7	21.1	22.8
HA	6.14	21.8	25.4	25.8	26.8
FA	8.73	31.4	29.6	30.5	32.0

The concentrations of plant available P are about 1 g PO_4^{3-}-P per kg TS in ashes from biomass incinerated at 400°C and about 0.5 g PO_4^{3-}-P per kg TS in ashes from higher incineration temperatures as shown in Table 6. This reduction coincides with formation of hydroxapatite. We also extracted the P with water but the content of P in the water extracts was below the detection limit of the analysis. The dissolution in neutral ammonium citrate show that the temperature has to be kept below 700°C to give an ash that is a useful source of P. This temperature will not be an economical optimal temperature for energy production at traditional power plant and optimal energy production therefore conflicts with the demands for producing an ash with a high content of plant available P.

Table 6. Concentration of plant available P in the ashes after different incineration temperatures and in the raw biomass. The concentrations are g PO_4^{3-}-P per kg TS. The contents are measured as the content dissolvable in neutral ammonium citrate.

	Raw	400°C	700°C	900°C	1050°C
KO	0.6	1.1	0.5	0.4	0.6
HA	1.0	1.0	0.5	0.4	0.7
FA	1.2	1.2	0.6	0.4	0.4

CONCLUSIONS

Decrease in plant availability of P in the ash when the incineration temperature is increased from 400 to 700°C. At the same temperature increase the number of crystalline components in the ash increases. Hydroxapatite which is a crystalline phosphor component is made in the KO and FA biomasses. At the temperatures optimal for energy production i.e. above 700°C, the ash produced is not an optimal P fertilizer.

REFERENCES

Diskowski, H. and Hofmann, T. (2005). Phosphorus. I: *Ullmann's encyclopædia of industrial chemistry*, Wiley-VCH Verlag GmbH and Co.

DS/EN 14346:2007: Karakterisering af affald – Beregning af tørstofindhold ved bestemmelse af tørstof eller vandindhold. *Dansk Standard*, 1st ed.

EFMA (2000). Phosphorus essential element for food production. *European Fertilizer Manufacturers Association (EFMA)*. Downloaded in 2008 from http://www.efma.org/publications/phosphorus/understanding%20phosphorus/Final%20phosphorus.pdf

Graversen, P. (2006). *Odense Kraftvarmeværk – Grønt Regnskab 2006*, DONG Energy.

Housecroft, C.E. and Sharpe, A.G. (2005). *Inorganic Chemistry*. Pearson Education Limited, 2nd Ed.

McClune, W.F. (red.) (1979). *Inorganic Materials Alphabetical Index*. Powder diffraction file search manual, Hanawalt Method Inorganic, International centre for diffraction data (JCPDS).

Møller, H.B., Sommer, S.G. and Ahring, B.K. (2004). Methane productivity of manure, straw and solid fractions of manure. *Biomass and Bioenergy*, **26**, 485–495.

Møller, H.B., Nielsen, A.M., Nakakubo, R. and Olsen, H.J. (2007). Process performance of biogas digesters incorporating pre-separated manure. *Livestock Science*, **112**, 217–223.

Petersen, J., Sørensen, P., Rubæk, G. and Hansen, J.F. (2005). Konsekvenser for gødningsværdien ved afbrænding af husdyrgødning. I: Iversen P.A., *Rapport fra arbejdsgruppen om afbrænding af fraktioner af husdyrgødning*, Danish Ministry of Food, Agriculture and Fisheries, pp. 91–96.

Petersen, S.O., Sommer, S.G., Béline, F., Burton, C., Dach, J., Dourmad, J.Y., Leip, A., Misselbrook, T., Nicholson, F., Poulsen, H.D., Provolo, G., Sørensen, P., Vinnerås, B., Weiske, A., Bernal, M.-P., Böhm, R., Juhász, C. and Mihelic, R. (2007). Recycling of livestock wastes and other residues – a whole-farm perspective. *Livestock Science*, **112**, 180–191.

Sommer, S.G., Petersen, S.O., Sørensen, P., Poulsen, H.D. and Møller, H.B. (2007). Methane and carbon dioxide emissions and nitrogen turnover during liquid manure storage. *Nutrient Cycling in Agroecosystems*, **78**, 27–36.

Steen, I. (1998). Phosphate Recovery – Phosphorus availability in the 21st century – Management of a non-renewable resource. *Phosphorus and Potassium*, Issue No: 217

(September-October, 1998), p. 25–31. Downloaded in 2008 from http://www.nhm.ac. uk/research-curation/projects/phosphate-recovery/p&k217/steen.htm

Sørensen, L. and Petersen, U. (2006). *Frederikshavn affaldskraftvarmeværk A/S – Grønt Regnskab 2006*, DONG Energy.

Sørensen, P. and Rubæk, G.H. (2006). Asken skal syrebehandles for at bevare gødningsværdien af P og K. *Mark* (November 2006), p. 28.

Affecting corn processing nutrients using membrane separation and biological extraction and conversion

K.D. Rausch[a], R.L. Belyea[b], L.M. Raskin[c], V. Singh[a], D.B. Johnston[d], T.E. Clevenger[b], M.E. Tumbleson[a] and E. Morgenroth[a]

[a]University of Illinois at Urbana-Champaign, Urbana, IL, USA
[b]University of Missouri, Columbia, MO, USA
[c]University of Michigan, Ann Arbor, MI, USA
[d]Eastern Regional Research Center, ARS, US Department of Agriculture, Wyndmoor, PA, USA

Abstract During processing, coproducts are produced in parallel with a primary product. Corn coproducts typically have low value and form one fourth to one third of material output from modern processing facilities. Corn processing streams contain nutrients that have value not reflected by markets because they are mixtures of protein, fat and fiber following extraction or conversion of starch. Research has been undertaken to characterize nutrient composition, develop membrane separation processes and convert nutrients such as phosphorus in an effort to more efficiently use valuable nutrients. Characterization studies found that process streams presenting greatest opportunity for nutrient recovery were steepwater and process water from wet milling and thin stillage from dry grind processes. Characterization studies also identified that process streams had compositions with large variations; these fluctuations in nutrient contents reduce market value of coproducts. Process streams that appear the most readily adaptable to further processing are steepwater (wet milling) and thin stillage (dry grind). Membrane filtration research showed potential to remove water and phosphorus (ash) from corn gluten meal and distillers dried grains with solubles.

INTRODUCTION

During processing, coproducts are produced in parallel with a primary product. Corn coproducts typically have low value and form one fourth to one third of material output from modern processing facilities. Corn processing streams contain nutrients that have value not reflected by markets because they are mixtures of protein, fat and fiber following extraction or conversion of starch (Rausch and Belyea, 2006). The wet milling process produces modified starch products, corn sweeteners and fuel ethanol as well as coproducts corn gluten meal (CGM) and corn gluten feed (CGF) (Figure 1). In the US, the corn dry grind process is used to produce more than 80% of fuel ethanol and results in a single coproduct, distillers dried grains with solubles (DDGS, Figure 2). Coproducts are not designed specifically for an end use diet (human or animal)

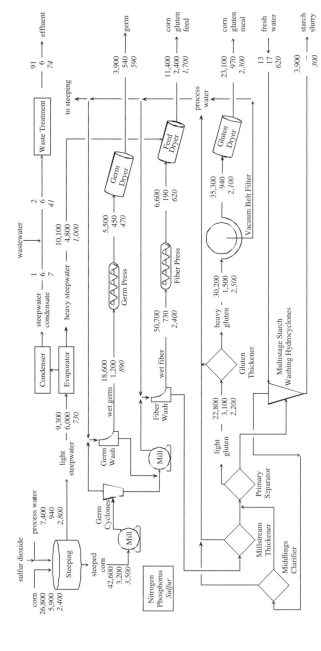

Figure 1. Nutrient flows of nitrogen, phosphorus and sulphur (kg/day) in a wet mill (2,700 tonne corn/day capacity).

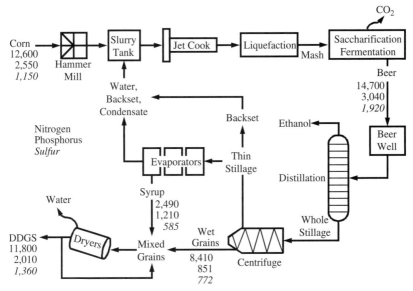

Figure 2. Nitrogen, phosphorus and sulphur (kg/day) at selected points for a 150 L/year (1,000 tonne corn/day) dry grind facility.

and are given a relatively low priority within the facility. This results in reduced value for most coproducts due to variable composition and/or undesired composition (Belyea *et al.*, 2004; Rausch *et al.*, 2003a; Rausch *et al.*, 2003b). These issues are of increasing importance since production of renewable fuels and biobased products will increase coproduct supply. This has been apparent, especially in the values of DDGS and corn gluten feed, during the past several years as production of fuel ethanol from corn has expanded (Figure 3).

Due to regulations for wastewater discharge, corn processors often are more concerned with nutrient compositions of wastewater discharged from their processes and entering waste treatment facilities than the compositions of coproducts used in animal diets. Processors routinely will collect and analyze information on nitrogen, phosphorus and sulphur (N, P and S, respectively) compounds within process streams so that waste treatment facilities will perform optimally and result in effluents that comply with regulations. In contrast, total protein for coproduct marketing is calculated simply as N × 6.25. Most specifications for nutrients in coproducts have open ended limits, either maximum or minimum levels, thereby increasing variation. Coproducts contain valuable and useful nutrients that are needed for an array of plant, animal and

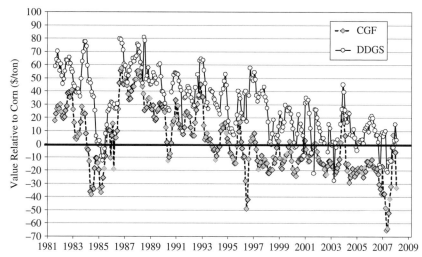

Figure 3. Coproduct value relative to corn grain (US dollars). CGF: corn gluten feed; DDGS: distillers dried grains with solubles (data: ERS 2008).

human needs; currently, coproducts fed to animals contain mixtures of nutrients not designed for the animal consuming the coproducts. Technologies are needed to separate, convert and use nutrients more fully.

CHARACTERIZATION OF CORN PROCESSING NUTRIENTS

An important initial step for nutrient recovery in corn processing was characterization of process streams and final coproducts from commercial processes. Three wet milling and nine dry grind facilities were included in characterization studies. Nutrient contents of process streams and coproducts were determined. This identified process streams that had high concentrations and large total flows of compounds of interest as well as identified streams representing the greatest opportunities for nutrient recovery. Nutrients reported included total phosphorus, nitrogen and sulphur in each process stream.

Three commercial wet milling facilities located in the Midwest US were sampled at 21 locations within the process. Each sample set consisted of three samples collected within one week. In most cases, four sample sets were collected during a period of one year. A simulation model was used to estimate the flows (kg/day) of each nutrient at each location. Nine dry grind facilities participated in a similar study. Samples at these facilities were collected so that material could be traced from original corn through fermentation to the production of coproducts. Four sample sets, consisting of three replicates each, were collected at five locations during a period of one year. A mass balance of the dry grind process was used to estimate the flows of N, P and S at a 40 million gal/yr (150 million L/yr) facility.

Wet milling

For wet milling, P concentrations in corn did not vary among processing plants (Rausch et al., 2005). Among plants, P concentrations in wet milling process streams varied due to a variety of factors. There were differences in P concentrations in coproducts (CGF, CGM) among the three wet milling plants (Table 1). Simulations estimated P flows ranged from < 10 to 6000 kg P/day for a wet milling plant (2700 tonne corn/day capacity; Figure 1). Several process streams, light steepwater, light gluten and process water carried large flows of P (5960, 3080 and 970 kg P/day, respectively). The wet milling process stream most likely to benefit from P reduction was light steepwater. Reduction in the P concentration of this stream could improve the market value of CGF.

Process water, process water + SO_2, light steepwater and light gluten had high orthophosphate concentrations (Table 2). Light steepwater had the highest concentrations at 560, 1300 and 630 mg/L for Plants A, B and C, respectively. High orthophosphate levels in the light steepwater stream may be attributed to inositol phosphates (phytate) as well as inorganic phosphates. Orthophosphate has been reported as a product of dephosphorylation of inositol phosphates as facilitated by phosphatases present in steepwater (Hull and Montgomery, 1995).

Streams carrying most of the N, P and S included steepwater and gluten; additional processing of these streams could improve coproduct quality (Table 3; Rausch et al., 2007). Processing of recycled water from the germ press and fiber press also could improve processing efficiency and coproduct market value. Two streams that carried large quantities of N were gluten and steepwater streams; in each case, the quantity of N carried across different processing steps within each of these streams was relatively consistent. The process water stream carried large quantities of nutrients and represented another opportunity for improving process efficiency and coproduct value.

A large variability in final coproduct composition was observed. Standard deviations relative to the means had ranges of 6 to 42, 28 to 61 and 10 to 83% for N, P, and S concentrations, respectively, in germ, CGF and CGM (Table 3). Incoming corn had lower variabilities of 28, 27 and 18% for N, P and S, respectively.

Table 1. Total phosphorus concentrations among three commercial wet milling facilities (Rausch et al., 2005).[*]

Process Stream	Total phosphorus (mg/kg sample)		
	Plant A	Plant B	Plant C
Corn	$2,190 \pm 619^a$	$2,260 \pm 589^a$	$2,030 \pm 557^a$
Process water	589 ± 162^a	302 ± 115^b	154 ± 59^b
Process water + SO_2	389 ± 130^a	325 ± 108^a	235 ± 66^a
Steeped corn	660 ± 181^b	913 ± 241^a	654 ± 181^b
Light steepwater	$3,830 \pm 1,090^b$	$5,240 \pm 1,250^a$	$3,340 \pm 921^b$
Steepwater condensate	bdl	3.8 ± 4.2^b	4.9 ± 1.8^a
Heavy steepwater	$15,600 \pm 4,270^a$	$14,700 \pm 4,040^{ab}$	$13,600 \pm 3.703^b$
Wet germ	915 ± 249^{ab}	$1,160 \pm 335^a$	710 ± 219^b
Pressed germ	1.320 ± 395^a	$1,390 \pm 407^a$	$1,200 \pm 330^a$
Dry germ	$3,140 \pm 916^a$	$2,690 \pm 795^{ab}$	$2,060 \pm 580^b$
Wet fiber	120 ± 38^a	314 ± 88^a	175 ± 51^a
Pressed fiber	302 ± 91^{ab}	407 ± 72^a	254 ± 73^b
Corn gluten feed (CGF)	$6,070 \pm 1,770^a$	$5,170 \pm 1,410^b$	$4,960 \pm 1,350^b$
Light gluten	458 ± 136^a	731 ± 427^a	513 ± 140^a
Heavy gluten	740 ± 184^c	$1,370 \pm 376^a$	$1,030 \pm 282^b$
Gluten cake	$1,740 \pm 230^{ab}$	$2,160 \pm 619^a$	$1,580 \pm 434$
Corn gluten meal (CGM)	$2,750 \pm 804^b$	$7,190 \pm 2,020^a$	$2,500 \pm 686^b$
Fresh water	bdl	bdl	bdl
Starch slurry	bdl	bdl	bdl
Wastewater	5.0 ± 2.7^a	5.0 ± 3.17^b	9.1 ± 7.2^b
Final effluent	6.4 ± 3.9^a	4.1 ± 1.5^b	3.1 ± 2.9

[*]Mean $\pm$ standard deviation (6 observations). Means in the same row with the same letter are not different ($p \leq 0.05$).

Based on concentrations and simulated flow rates, total N entering the wet milling process was 26,800 kg N/day (Figure 1). Total amount of N leaving the process in coproducts and various waste streams was estimated to be 42,400 kg N/day. About one third of P could not be accounted for, possibly due to sampling errors or simplifying assumptions of the process simulation. The amount of S entering the wet milling process was estimated to be 4,200 kg S/day, which included 2,400 kg S/day originating from corn and 1,800 kg S/day added as SO_2; total S leaving in coproduct and other streams was estimated to be 4,820 kg S/day. The difference between input and output S, 620 kg S/day (15% of S input), can be attributed to addition of S to control microbial growth in process tanks prior to dewatering and drying. Some loss of N and

Table 2. Orthophosphate concentrations of selected process streams among three wet milling facilities.[*]

Process Stream	Orthophosphate (mg/liter PO$_4$-P)		
	Plant A	Plant B	Plant C
Corn			
Process water	78 ± 19^b	150 ± 39^a	150 ± 25^a
Process water + SO$_2$	68 ± 28^c	200 ± 57^b	270 ± 120^a
Steeped corn			
Light steepwater	580 ± 170^b	$1,100 \pm 250^a$	520 ± 230^b
Steepwater condensate	3.4 ± 4.9^a	4.6 ± 3.9^a	4.1 ± 2.1^a
Heavy steepwater			
Wet germ			
Pressed germ			
Dry germ			
Wet fiber			
Pressed fiber			
Corn gluten feed			
Light gluten	84 ± 24^b	180 ± 52^b	160 ± 31^{ab}
Heavy gluten			
Gluten cake			
Corn gluten meal			
Fresh water	0.21 ± 0.19^b	0.17 ± 0.33^b	3.5 ± 2.0^a
Starch slurry			
Wastewater	4.1 ± 2.4^b	4.7 ± 3.1^b	7.2 ± 7.0^a
Final effluent	3.8 ± 4.4^a	4.1 ± 1.5^a	1.1 ± 2.1^b

[*]Mean $\pm$ standard deviation (12 observations). Means in the same row with the same letter were not different (p $\leq$ 0.05).

S compounds due to evaporation was anticipated, but not measured. The most important coproduct streams carrying S were gluten streams (2,300 kg S/day), wet fiber (2,400 kg S/day) and heavy steepwater (1,000 kg S/day). These three streams carried 75% of total S flow in the wet milling process. Low amounts of S were found in the starch and waste streams (<300 kg S/day).

For all plants, process water and process water + SO$_2$ had high levels of sulphite. Light steepwater had highest concentrations of sulphate (180 to 300 mg SO$_4$-S/litre). Sulphide concentrations for all samples and plants were below 1.2 mg/litre. Wastewater samples had low sulphate (6 to 72 mg SO$_4$-S/litre) and sulphite (0.8 to 15 mg SO$_3$ S/litre) concentrations.

Table 3. Overall means of total nitrogen, phosphorus and sulfur concentrations (mg/kg sample) in wet milling process streams (Raush et al., 2007).

Process Stream	Nitrogen		Phosphorus		Sulfur	
	Mean	Std Dev	Mean	Std Dev	Mean	Std Dev
Corn	10,400	2,900	2,160	574	873	157
Process water	1,650	634	344	209	338	69
Process water+SO_2	2,510	1,760	317	118	936	480
Steeped corn	9,800	2,940	749	234	836	240
Light steepwater	6,660	1,660	4,250	1,370	520	257
Heavy steepwater	29,400	3,190	14,600	3,950	2,980	712
Steepwater condensate	0.90	0.65	4.5	4.1	7.8	6.6
Wet germ	14,100	10,200	926	320	675	419
Pressed germ	16,200	5,380	1,300	372	1,430	680
Dry germ	18,900	3,120	2,600	862	2,850	1,010
Wet fiber	17,100	14,900	210	104	710	551
Pressed fiber	10,200	8,780	318	110	1,010	408
Corn gluten feed	24,100	10,200	5,380	1,530	3,720	3,090
Light gluten	4,370	1,500	588	312	418	230
Heavy gluten	23,900	33,500	1,060	392	1,860	387
Gluten cake	72,500	33,100	1,830	566	4,180	790
Corn gluten meal	100,000	6,230	4,200	2,560	10,200	1,030
Starch slurry	1,460	1,280	bdl		82	75
Fresh water	4.0	2.4	bdl		206	242
Wastewater	2.1	2.2	5.9	5.0	28	33
Final effluent	218	93	4.7	2.9	66	37

Dry grind

For dry grind, concentrations of most elements in corn did not vary among processing plants (Belyea et al., 2006). However, for other processing streams there were differences in element concentrations among plants. Sources of variation were not identified, but variations in processing conditions could have been a primary cause. Although syrup had the highest concentrations of total N, P and S, it would be difficult to process (Table 4). To reduce the element content of DDGS, thin stillage (the parent stream for syrup) appeared to be the most logical stream for processing. Phosphorus concentration and mass flow were highest in thin stillage.

Table 4. Total nitrogen, phosphorus and sulfur concentrations in dry grind processing streams.

	Nitrogen (% db)		Phosphorus (mg/kg db)		Sulfur (mg/kg db)	
	Mean	Std Dev	Mean	Std Dev	Mean	Std Dev
Corn	1.44	0.04	2,910	184	1,310	245
Beer	4.76	0.15	9,830	996	6,230	1,700
Wet Grains	5.34	0.27	5,410	356	4,900	523
Syrup	3.15	0.81	15,200	1,540	7,400	3,580
DDGS	5.00	0.17	8,500	714	5,770	1,450

MEMBRANE SEPARATIONS

Many process operations in corn processing remove water to make coproduct solids safe for storage and handling. While mechanical pressing is used as an initial stage to remove water, evaporation and drying are thermally driven process steps that reduce moisture content to levels that make transport and storage of coproducts economical as well as make the coproducts microbially stable. Membrane separations are nonthermal methods to remove water. Use of membranes is attractive due to their potential to reduce energy input, reduce heat damage to nutrients and provide more selective separations relative to evaporators and dryers. Membrane processing is not used widely in commercial corn processing due economic reasons and resistance by the industry (Rausch, 2002). However, membrane filtration has been accepted in other industries in which streams are difficult to process from a fouling standpoint (Table 5).

Gluten filtration and nutrient separations

During the corn wet milling process, endosperm proteins are separated from the germ, fiber and starch components of the kernel. This fraction is called corn

Table 5. Energy requirements for water removal (Rausch and Belyea, 2006).

Water Removal Method	Energy Input per Unit Water Removed (Btu/lb$_{H2O}$)
Evaporation	300 to 600
Drying	1100 to 1600
Centrifugation	2 to 3
Germ or fiber pressing	0.5 to 3
Vacuum belt filtration	10
Membrane filtration	3 to 50

gluten, although no gluten protein is contained in corn. Following germ and fiber separation, centrifuges are used to separate starch from the gluten (light gluten), further concentrated using a second centrifuge (heavy gluten), dewatered using vacuum belt filtration (gluten cake) and dried to form corn gluten meal (Figure 1). These processing steps represent significant capital expenditures and release small gluten particles that are recycled in the wet milling process, create difficulties elsewhere in the process and eventually end up in low valued coproducts (corn gluten feed). Because corn gluten meal is relatively valuable, even small amounts of lost gluten represent significant economic potential for the wet mill. Membrane filtration has potential to reduce the processing steps in gluten meal production by eliminating centrifuges and vacuum belt filters (VBF).

Table 6. Separation of light gluten using microfiltration:[1] comparison of methods and streams (Thompson et al., 2006).[*]

Comparison	DM[*]	TN	SN	Ash	TSS
Separation Method					
Gluten Thickener Centrifuge[2]					
Light gluten	4.94[b]	10.50[a]	5.53[b]	3.83	3.97[b]
Heavy gluten	14.24[a]	11.40[a]	1.40[c]	2.54	10.32[a]
Overflow	2.44[c]	8.87 [b]	7.51[a]	6.97	1.12[b]
SE[**]	1.02	0.40	0.71	1.59	1.75
Microfiltration (0.1 μm pore size; 65.3 LMH mean flux rate)					
Light gluten	3.65[b]	9.99	5.45[b]	5.34[b]	2.43[b]
Retentate	14.70[a]	11.32	1.35[b]	2.21[b]	12.78[a]
Permeate	1.98[b]	10.88	8.46[a]	16.27[a]	0.54[b]
SE	0.79	1.55	1.03	1.05	0.74
Comparison of Streams					
Heavy Gluten vs Concentrate					
Heavy gluten	13.14	11.41	1.52	1.44	11.92
Retentate	14.70	11.32	1.35	2.21	12.78
SE	2.30	0.14	0.40	0.32	2.13
Overflow vs Permeate					
Overflow	2.08	9.15	6.80	11.03	1.12[a]
Permeate	1.98	10.88	8.46	16.27	0.54[b]
SE	0.63	2.19	1.57	1.72	0.05

[abc]Means within same column and effect having unlike letters differ ($P < 0.01$).
[*]DM (dry matter) and TSS (total suspended solids): g/100 g sample; all others: g/100 g dry basis.
[**]SE: standard error.
[1]Overflow: supernatant stream from gluten thickener.
[2]Gluten thickener centrifuge located at a wet milling facility.

Batches of light and heavy gluten were obtained from a wet mill plant and processed by microfiltration. Samples of permeate and concentrate from microfiltration were analyzed and compared to corresponding streams from wet milling. Microfiltration of light gluten resulted in concentrate and permeate streams similar in total solids contents to conventionally processed light gluten using a centrifuge, suggesting that microfiltration is as effective as centrifugation in partitioning solids and water in light gluten (Table 6; Thompson *et al.*, 2006). When heavy gluten was dewatered, we found conventional VBF caused dry matter concentrations in gluten cake to be higher than concentrate from micro-filtration (Table 7). Permeate from microfiltration of heavy gluten had higher concentrations of ash and lower soluble nitrogen than filtrate from VBF.

Table 7. Separation of heavy gluten using microfiltration:[1] comparison of methods and streams (Thompson *et al.*, 2006).*

Comparison	DM*	TN	SN	Ash	TSS
Separation Method					
Vacuum Belt Filtration[2]					
Heavy gluten	14.24^b	11.40^a	1.40^b	4.29^a	10.32^a
Filtrate	3.61^c	9.74^b	6.91^a	6.97^a	1.50^b
Gluten cake	40.79^a	11.53^a	–	0.80^b	–
SE**	1.05	0.29	0.28	1.41	2.17
Microfiltration of heavy gluten (0.1 μm pore size; 51.3 LMH mean flux rate)					
Concentrate vs Permeate					
Heavy gluten (initial)	12.34^b	11.09	2.10^a	2.84^a	11.17^a
Retentate	20.81^a	11.66	1.33^a	2.26^a	15.19^b
Permeate	1.94^c	10.75	12.56^b	16.24^b	0.62^c
SE	0.77	0.26	0.87	0.43	1.15
Comparison of Streams					
Gluten Cake vs Concentrate					
Gluten cake	41.68^a	11.61	–	2.21	–
Retentate	21.22^b	11.64	–	2.14	–
SE	2.88	0.18	–	0.80	–
Filtrate vs Permeate					
Filtrate	3.03	10.38	7.25^a	9.74^b	1.73
Permeate	1.78	11.22	0.54^b	16.24^a	0.29
SE	0.90	0.44	0.34	0.86	0.56

[abc]Means within same column and effect having unlike letters differ (P < 0.01).
*DM (dry matter) and TSS (total suspended solids): g/100 g sample; all others: g/100 g dry basis.
**SE: standard error.
[1]HG: heavy gluten; Filtrate: vacuum belt filtrate.
[2]Vacuum belt filtration located at a wet milling facility.

We evaluated the effectiveness of laboratory scale microfiltration systems to remove water from wet mill processing streams (Templin *et al.*, 2006; Thompson *et al.*, 2006). We showed the potential of microfiltration to remove considerable amounts of water (and elements as well). Templin *et al.* (2006) found that microfiltration of light gluten recovered 67% of total ash in the permeate stream and nearly 80% of protein in the retentate stream, while solids were concentrated nearly sixfold in simple batch filtration experiments. In larger-scale membrane filtration work (Thompson *et al.*, 2006), light gluten separation achieved a nearly fivefold increase in total solids in membrane concentrate while permeate concentrations of total ash were five times higher than in the original light gluten stream. The ability to remove ash (and, presumably, phosphorus) from coproduct streams as well as the requirement of small amounts of energy for dewatering illustrate the potential for broader use of this technology in corn processes (Table 6).

Thin stillage filtration and nutrient separations

Relatively few researchers have evaluated the effectiveness of membrane filtration in modern fuel ethanol plants. Some have studied removal of water and reduction of wastewater strength (Wu, 1988a, 1988b; Wu and Sexson, 1985, Wu *et al.*, 1983), but this work was done before modern ethanol processing facilities came into production and before the latest advances in membrane materials were available.

In the dry grind process, 6 to 7 L thin stillage is produced concomitantly with 1 L ethanol (Arora *et al.*, 2008). Concentration of thin stillage requires evaporation of large amounts of water as well as evaporator maintenance. Evaporator operation and maintenance requires excess evaporator capacity at the facility, increasing capital expenses and requiring plant slow downs or shut downs. We used ultrafiltration (UF) to evaluate membranes as an alternative to evaporation. To obtain thin stillage, corn was fermented using laboratory methods. UF experiments were conducted in batch mode under constant temperature and flow rate conditions. Two regenerated cellulose membranes with molecular weight cut offs of 10 and 100 kDa were evaluated with the objective of retaining solids as well as maximizing permeate flux. Retentate obtained from thin stillage fractionation had mean total solids contents of 28% (Arora *et al.*, 2009). Total solids in retentate streams were similar to those from commercial evaporators used in industry (25 to 35% total solids). Fat contents in retentate streams ranged from 16.3 to 21.0% (db). Thin stillage ash content was reduced 60% in retentate streams. Retentate produced after fractionation had higher protein, fat and NDF contents and lower ash contents.

Retentates could be used as animal food and permeate could be further processed through membranes such as nanofiltration or reverse osmosis and recycled in the dry grind plant, increasing the amount of recycle and reducing water requirements.

We conducted experiments to investigate fouling of ultrafiltration membranes during processing of commercial thin stillage (Arora et al., 2008). Reversible resistance was the primary component in fouling of membranes which indicated that altering operating conditions could help in minimizing fouling and appropriate membrane materials and cleaning protocols could be used to make membrane filtration an attractive alternative to thermal processes currently in use.

EXTRACTION, CONVERSION AND USE OF NUTRIENTS

Another important step will be to recover, concentrate and use nutrients in coproducts suited for animal and human nutrition. For example, P concentrations in CGF and DDGS were higher than needed for ruminant diets (Belyea et al., 2006; Belyea et al., 2004; Rausch et al., 2005). Methods to recover P in dry grind and wet milling process streams and reduce P concentrations in CGF and DDGS would improve their composition for the intended animal (ruminant) diet and potential market value of these coproducts.

When characterizing thin stillage and syrup streams, we found the majority of P to be in these streams in the dry grind process. Based upon calculations, if two thirds of P were removed from the thin stillage stream, DDGS would contain approximately 4,000 mg P/kg and approach the levels found in whole corn (3,000 mg P/kg) and nearer to the levels needed in ruminant animal diets. This would facilitate increased inclusion rates of DDGS in ruminant diets and improve market demand for this coproduct. P recovered from thin stillage could be used where needed, such as in nonruminant animal diets or in crop production.

Membrane filtration has been effective in separating the ash/P component from process streams (Thompson et al., 2006; Templin et al., 2006). This resulted in concentration of nutrients desired for animal diets (protein) while creating a permeate stream high in ash/P. Historically, membrane filtration research has focused on membrane performance and durability, recovery of total solids and protein or removal of BOD/COD in permeate streams. For the corn processing industry, additional focus would be needed to investigate benefits of nutrient recovery in permeate streams in addition to changes in nutrient bioavailability in coproducts as a result of membrane processing.

Researchers (Kim *et al.*, 2008; Martinez Amezcua and Parsons, 2007; Martinez Amezcua *et al.*, 2004 and Martinez Amezcua *et al.*, 2007) found that processing affected bioavailability of nutrients such as P. While this has long been presumed, their quantified findings were indicative of P bioavailability from corn coproducts.

In the conventional dry grind process, whole stillage is converted to DDGS through energy intensive water removal steps (Figure 2). However, bacterial hydrolysis can be used to convert phytic acid into orthophosphate which subsequently can be stored in bacterial cells as polyphosphate (Seviour *et al.*, 2003). Using membrane separation methods, permeate streams could provide an input stream to the biological phosphorus treatment. This step, using bacterial hydrolysis, would convert phytic acid originating in the corn kernel into more bioavailable forms using biodegradable organic matter available in the membrane permeate. Effluent from the biological treatment step could be recycled in the dry grind process.

CONCLUSIONS

Corn processes have developed during a period of many years to produce efficiently starch based products. However, understanding of nutrient routing through these highly developed processes is in the early stages of understanding and quantification. Characterization studies identified that process streams had compositions with large variations; these fluctuations in nutrient contents reduce market value of coproducts.

Researchers have reported opportunities to improve coproduct composition, identify where nutrients are in high concentrations and have large mass flows and initiated work to develop processes that may recover nutrients for valuable use. Process streams that appear the most readily adaptable to further processing are steepwater (wet milling) and thin stillage (dry grind).

Adoption and implementation of new processes, such as membrane filtration and biological conversion, will require a shift in philosophy by the corn processing industry. The success of the industry to date has focused on a single component: starch. Continued success of the industry will depend on using the other components available from their feedstock: protein, fiber, oil and minerals.

REFERENCES

Arora, A., Wang, P., Belyea, R.L. Tumbleson, M.E., Singh, V. and Rausch, K.D. (2008). Thin stillage fractionation using ultrafiltration: resistance in series model. *Bioproc. Biosys. Eng.* (in press).

Arora, A., Wang, P., Belyea, R.L., Tumbleson, M.E., Singh, V. and Rausch, K.D. (2009). Ultrafiltration of thin stillage from conventional and E-Mill processes. *Cereal Chem.* (in press).

Belyea, R.L., Clevenger, T.E., Singh, V., Tumbleson, M.E. and Rausch, K.D. (2006). Element concentrations of dry grind corn processing streams. *Appl. Biochem. Biotechnol.*, **134**, 113–128.

Belyea, R.L., Rausch, K.D. and Tumbleson, M.E. (2004). Composition of corn and distillers dried grains with solubles in dry grind ethanol processing. *Biores. Technol.*, **94**, 293–298.

ERS. (2008). Feed grains database. www.ers.usda.gov/Data/feedgrains/. USDA Economic Research Service. Washington, DC.

Hull, S.R. and Montgomery, R. (1995). myo-Inositol phosphates in corn steep water. *J. Agric. Food Chem.*, **43**, 1526–1523.

Kim, E.J., Martinez Amezcua, C., Utterback, P.L. and Parsons, C.M. (2008). Phosphorus bioavailability, true metabolizable energy, and amino acid digestibilities of high protein corn distillers dried grains and dehydrated corn germ. *Poult. Sci.*, **87**, 700–705.

Martinez Amezcua, C., and Parsons, C.M. (2007). Effect of increased heat processing and particle size on phosphorus bioavailability in corn distillers dried grains with solubles. *Poult. Sci.*, **86**, 331–337.

Martinez Amezcua, C., Parsons, C.M, and Noll, S.L. (2004). Content and relative bioavailability of phosphorus in distillers dried grains with solubles in chicks. *Poult. Sci.*, **83**, 971–976.

Martinez Amezcua, C., Parsons, C.M., Singh, V., Srinivasan, R., and Murthy, G.S. (2007). Nutritional characteristics of corn distillers dried grains with solubles as affected by the amounts of grains versus solubles and different processing techniques. *Poult. Sci.*, **86**, 2624–2630.

Rausch, K.D. (2002). Front end to backpipe: membrane technology in the starch processing industry. *Starch/Staerke*, **54**, 273–284.

Rausch, K.D. and Belyea, R.L. (2006). The future of coproducts from corn processing. *Appl. Biochem. Biotechnol.*, **128**, 47–86.

Rausch, K.D., Raskin, L.M., Belyea, R.L., Agbisit, R.M., Daugherty, B.J., Clevenger, T.E. and Tumbleson, M.E. (2005). Phosphorus concentrations and flow in maize wet milling streams. *Cereal Chem.*, **82**, 431–435.

Rausch, K.D., Raskin, L.M., Belyea, R.L., Clevenger, T.E. and Tumbleson, M.E. (2007). Nitrogen and sulfur concentrations and flow rates of corn wet milling streams. *Cereal Chem.*, **84**, 260–264.

Rausch, K.D., Thompson, C.I., Belyea, R.L., Clevenger, T.E. and Tumbleson, M.E. (2003a). Characterization of gluten processing streams. *Biores. Technol.*, **89**, 163–167.

Rausch, K.D., Thompson, C.I., Belyea, R.L. and Tumbleson, M.E. (2003b). Character-ization of light gluten and light steep water from a wet milling plant. *Biores. Technol.*, **90**, 49–54.

Seviour, R.J., Mino, T. and Onuki, M. (2003). The microbiology of biological phosphorus removal in activated sludge systems. *FEMS Microbiol. Rev.*, **27**, 99–127.

Templin, T.L., Johnston, D.B., Singh, V., Tumbleson, M.E., Belyea, R.L. and Rausch, K.D. (2006). Membrane separation of solids from corn processing streams. *Biores. Technol.*, **97**, 1536–1545.

Thompson, C.I., Rausch, K.D., Belyea, R.L. and Tumbleson, M.E. (2006). Microfiltration of gluten processing streams from corn wet milling. *Biores. Technol.*, **97**, 348–354.

Wu, Y.V. (1988a). Recovery of stillage soluble solids from corn and dry-milled corn fractions by high-pressure reverse osmosis and ultrafiltration. *Cereal Chem.*, **65**, 345–348.

Wu, Y.V. (1988b). Reverse osmosis and ultrafiltration of corn light steepwater solubles. *Cereal Chem.*, **65**, 105–109.

Wu, Y.V. and Sexson, K.R. (1985). Reverse osmosis and ultrafiltration of stillage solubles from dry milled corn fractions. *JAOCS*, **62**, 92–96.

Wu, Y.V., Sexson, K.R. and Wall, J.S. (1983). Reverse osmosis of soluble fraction of corn stillage. *Cereal Chem.*, **60**, 248–251.

Technology for recovery of phosphorus from animal wastewater through calcium phosphate precipitation

M. Vanotti and A. Szogi

USDA-ARS, Florence, SC, USA

Abstract A wastewater treatment process was developed for removal of phosphorus from livestock wastewater. The phosphorus is recovered as calcium phosphate with addition of only small quantities of liquid lime. The process is based on the distinct chemical equilibrium between phosphorus and calcium ions when natural buffers are substantially eliminated. It was discovered that reduction of carbonate and ammonium buffers during nitrification substantially reduces the $Ca(OH)_2$ demand needed for optimum P precipitation and removal at high pH. The technology produced consistent results in pilot tests on ten swine farms and successfully demonstrated full-scale on two swine farms in North Carolina, USA. It can be used to retrofit animal lagoons or in new systems without lagoons. The recovered calcium phosphate can be recycled into a marketable fertilizer without further processing due to its high content (>90%) of plant available phosphorus. The concentration grade obtained during full-scale demonstration was $24.4 \pm 4.5\%$ P_2O_5. A second generation version of the technology is available for municipal and agricultural wastewater and includes the simultaneous separation of solids and phosphorus from wastewater and industrial effluents.

INTRODUCTION

The aspect of P reuse is important for crop producers because of increasing demand and cost of inorganic fertilizers. The merging of food and fuel economies has increased the demand of mineral P fertilizer, and its price increased over 200% in 2007 (Trostle, 2008). The increased P demand may stimulate new technologies and economic opportunities for P recovery from manure, specially using technologies that produce concentrated byproducts with nutrient values competitive with mineral fertilizers.

Reindl (2007) classified techniques for P removal from wastewater by precipitation of calcium phosphate into three groups: crystallizers, fluid bed reactors, and the new process used in this chapter. The technology was developed to remove P from animal wastewater and other high-ammonia strength effluents and has the advantage over previous art of requiring minimal chemical addition and producing a valuable by-product (Vanotti *et al.,* 2003; Vanotti *et al.,* 2005). It is based on the distinct chemical equilibrium between phosphorus and calcium ions when natural buffers (NH_4-N and alkalinity) are substantially eliminated.

Basic process configuration (Figure 1)

The processes involved in the new technology include (i) biological nitrification of liquid manure to oxidize ammonium (NH_4^+) to nitrate (NO_3^-), (ii) reduction of natural buffers, and (iii) increasing the pH of the nitrified wastewater through addition of Ca or Mg hydroxide to precipitate P (Vanotti *et al.*, 2003; Vanotti *et al.*, 2005; Szogi and Vanotti, 2009). Since NH_4^+ is mostly converted to NO_3^-, increased pH during P precipitation does not result in significant gaseous N loss in ammonia gas form. The final product is a Ca phosphate-rich sludge that can be used as P fertilizer (Szogi *et al.*, 2006; Bauer *et al.*, 2007).

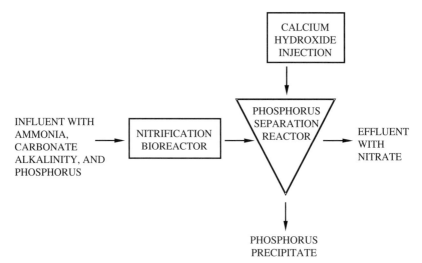

Figure 1. Schematic showing the basic configuration of the P removal process (Vanotti *et al.*, 2005).

Process chemistry

Animal wastewater is a mixture of urine, water, and feces. Livestock urine usually contains more than 55% of the excreted N of which more than 70% is in the form of urea (Sommer and Husted, 1995). Hydrolysis of urea by the enzyme urease produces NH_4^+ and carbonate according to the following reaction:

$$CO(NH_2)_2 + 2H_2O \rightarrow 2NH_4^+ + CO_3^{2-} \tag{1}$$

Therefore, a substantial part of the inorganic carbon in liquid manure is produced during decomposition of organic compounds. Carbonate and NH_4^+ alkalinity are the most important chemical components in liquid manure contributing to the buffering capacity in the alkaline pH range (Fordham and Schwertmann, 1977; Sommer and Husted, 1995). Alkaline pH is necessary to form a P precipitate with Ca and Mg compounds (House, 1999). When a Ca or Mg hydroxide is added to liquid manure, the hydroxide reacts with the existing bicarbonate to form carbonate, with NH_4^+ to form ammonia (NH_3), and with phosphate to form phosphate precipitate compounds (Loehr *et al.*, 1976; Tchobanoglous and Burton, 1991). For instance, using Ca hydroxide as an example, the following equations define the reactions:

$$Ca(OH)_2 + Ca(HCO_3)_2 \rightarrow 2CaCO_3 \downarrow +2H_2O \qquad (2)$$

$$5Ca^{++} + 4OH^- + 3HPO_4^= \rightarrow Ca_5OH(PO_4)_3 \downarrow +3H_2O \qquad (3)$$

The reaction in Equation 2 is complete at pH $\leq$ 9.5, while that of Equation 3 starts at pH $>$ 7.0, but the reaction is very slow below at pH $\leq$ 9.0. As the pH value of the wastewater increases beyond 9.0, excess Ca ions will then react with the phosphate, to precipitate as Ca phosphate (Equation 3). Not expressed in Equation 2 is the fact that in wastewater containing high NH_4^+ concentration, large amounts of lime are required to elevate the pH to required values since NH_4^+ reaction tends to neutralize the hydroxyl ions according to Equation 4:

$$Ca(OH)_2 + 2NH_4^+ \rightarrow 2NH_3 \uparrow + Ca^{++} + 2H_2O \qquad (4)$$

Consequently, precipitation of phosphate in animal wastewater using an alkaline compound such as lime is very difficult due to the inherent high buffering capacity of liquid manure (NH_4-N $\geq$ 200 mg L^{-1} and alkalinity $\geq$ 1200 mg L^{-1}). The buffer effect prevents rapid changes in pH. However, this problem is solved using a pre-nitrification step that reduces the concentration of both NH_4^+ (Equation 5) and bicarbonate alkalinity (Equation 6) (Vanotti *et al.*, 2005):

$$NH_4^+ + 2O_2 \rightarrow NO_3^- + 2H^+ + H_2O \qquad (5)$$

$$HCO_3^- + H^+ \rightarrow CO_2 \uparrow + H_2O \qquad (6)$$

The buffering effect of NH_4^+ (Equation 4) is reduced by biological nitrification of the NH_4^+ (Equation 5). Simultaneously, the buffering effect of bicarbonate (Equation 2) is greatly reduced with the acid produced during nitrification (Equation 6). These two simultaneous reactions leave a less buffered liquid in optimum pH conditions for phosphate removal with the addition of small amounts of lime (Equation 3).

PROCESS APPLICATIONS TO LIVESTOCK WASTEWATER

The new P removal technology was conceived to remove P in systems with lagoons (Vanotti *et al.*, 2003; Szogi and Vanotti, 2009) and systems without lagoons (Vanotti *et al.*, 2007; Vanotti and Szogi, 2008). In the livestock systems with anaerobic lagoons (or other anaerobic digesters), the anaerobically digested supernatant liquid, rich in NH_4-N and alkalinity, is nitrified and P is subsequently removed by adding hydrated lime. The effectiveness of the technology was tested in a pilot field study at ten swine farm in North Carolina, where 95–98% of the P was precipitated from the anaerobic lagoon effluent. In the systems without lagoon, raw liquid manure is first treated through an enhanced solid-liquid separation process with polymers to remove most of the carbonaceous material from the wastewater. The separated water is then treated with the nitrification and soluble P removal sequence. A denitrification tank can also be incorporated into the treatment system to provide total N removal in addition to the P removal. This configuration was tested full-scale in two finishing swine farms in North Carolina with removal efficiencies of 94% for soluble P (Vanotti *et al.*, 2007).

Phosphorus extraction from digested swine lagoon effluents (Figure 2)

Phosphorus was efficiently removed from lagoon liquid from ten diverse North Carolina swine production farms. The study consisted of ten consecutive experiments using the basic process configuration shown in Figures 1 & 2 to treat ten lagoon liquids (Szogi and Vanotti, 2009). In each experiment, swine wastewater first received biological treatment in a nitrification bioreactor, followed by chemical treatment with $Ca(OH)_2$ in a P separation reactor to precipitate phosphate. This configuration was compared with a control representing a control method that also received chemical treatment with $Ca(OH)_2$ but without the nitrification pre-treatment. All control and nitrified lagoon liquids were treated with $Ca(OH)_2$ applied at seven rates of 0, 2, 4, 6, 8, 10, and 12 mmol Ca L^{-1} of lagoon liquid.

Even though a pH $\geq$ 9.0 is needed to optimize precipitation of phosphate using Ca-based compounds, the pH of the liquid was initially lowered with the

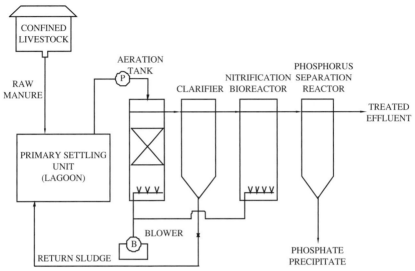

Figure 2. Schematic drawing of the unit used to remove phosphorus from swine lagoon liquid using a nitrification-lime treatment sequence (Vanotti *et al.*, 2003).

acid produced by nitrifying bacteria. In this way, the resulting nitrified liquid was low in NH_4^+ and bicarbonate buffers, and the total amount of alkali needed to increase the pH above 9.0 was significantly reduced. Without exception, the natural carbonate alkalinity of the lagoon liquid was significantly reduced in the process of biological NH_4^+ oxidation during the nitrification pre-treatment (average alkalinity changed from 2279 ± 300 mg L^{-1} to 91 ± 14 mg L^{-1}). On average, the rate of change of effluent pH (0.42 pH units/mmol Ca L^{-1}) for the pre-nitrified lagoon liquid was significantly different (P < 0.0001) and higher than the rate for the control without nitrification (0.05 pH units/mmol Ca L^{-1}). These results illustrate that precipitation of phosphate using lime in untreated livestock effluents is very inefficient because of the high natural buffer capacity of these effluents, which prevents rapid changes in pH.

Total P removal efficiencies with the non-nitrified (control) treatment never exceeded 50%, even at the highest Ca rate of 12 mmol L^{-1} used in the tests (Figure 3). Instead, high TP removal efficiencies of $> 90\%$ were achieved at Ca rates between 8 and 10 mmol L^{-1} (0.3 and 0.4 g Ca L^{-1}) in the nitrified samples. Nitrification pre-treatment significantly reduced the amount of $Ca(OH)_2$ needed for optimum P precipitation and removal while preventing N losses via NH_3 volatilization. The two final products of this wastewater treatment process were a liquid effluent for on-farm use and a solid calcium phosphate material.

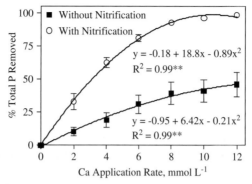

Figure 3. Phosphorus removal enhancement from swine lagoon wastewater using hydrated lime and nitrification pre-treatment. Results are means (s.e.) of the treatment of ten lagoon effluents in North Carolina (Szogi and Vanotti, 2009).

Manure treatment systems without lagoon (Figure 4)

The on-farm technology used liquid-solid separation, nitrification-denitrification, and soluble phosphorus removal processes linked together into a practical

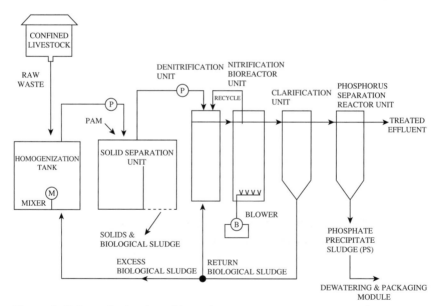

Figure 4. Schematic drawing of the swine waste treatment system without lagoon. The soluble P is removed after liquid-solid separation and ammonia removal (Vanotti *et al.*, 2007).

system. It was developed to replace anaerobic lagoon technology commonly used in the USA to treat swine waste (Vanotti *et al.*, 2007). Hydrated lime precipitated the phosphates and the phosphate precipitate was dewatered using polymer and filter bags (Szogi *et al.*, 2006) (Figure 5).

Figure 5. Phosphorus separation module at Goshen Ridge Farm. Pictures show reaction chamber and settling tank (above) and P precipitate dewatering and bagging (below).

Results obtained at full-scale in a 4360-swine farm showed removal efficiencies of the soluble P of 94.9% for wastewater containing 76 to 197 mg L^{-1} soluble P (Table 1). The recovered P precipitate solid had a concentration grade

Table 1. Wastewater treatment plant performance and system efficiency at Goshen Ridge Farm, North Carolina. Data are means (± s.d) for 18-month period (Vanotti and Szogi, 2008).

Water Quality Parameter	Raw Flushed Swine Manure	Treatment Step			System Efficiency
		After Solid-Liquid Separation	After Biological N Treatment	After Phosphorus Treatment	
		mg L^{-1} ‡			%
TSS	11,612 ± 6746	811 ± 674	134 ± 75	232 ± 152	98.0
BOD$_5$	3046 ± 2341	923 ± 984	40 ± 44	10 ± 16	99.7
TKN	1501 ± 567	895 ± 298	43 ± 34	26 ± 25	98.3
NH$_4$-N	838 ± 311	796 ± 297	31 ± 34	14 ± 19	98.3
NO$_2$+NO$_3$-N	1.5 ± 4.5	0.4 ± 2.6	228 ± 110	235 ± 116	–
TP	566 ± 237	168 ± 53	149 ± 33	26 ± 16	95.4
Soluble P	131 ± 39	116 ± 33	138 ± 28	7 ± 7	94.7
Alkalinity	5001 ± 1695	4154 ± 1463	624 ± 470	763 ± 353	84.7
pH	7.64 ± 0.22	7.93 ± 0.26	7.29 ± 0.70	10.53 ± 0.63	–

‡Except for pH.

of $24.4 \pm 4.5\%$ P_2O_5 and was $> 99\%$ plant available based on standard citrate P analysis used by the fertilizer industry (Bauer *et al.*, 2007). Although a high pH (10.5) in the phosphorus removal process is necessary to produce calcium phosphate and kill pathogens, the treated effluent is poorly buffered, and the high pH decreases readily once in contact with the CO_2 in the air. For example, Vanotti *et al.*, (2003) showed that short-term (2.5-h) aeration treatment of the effluent could create enough acidity to lower the pH from 10.5 to 8.5. However, natural aeration during storage may be equally effective to lower pH as seen in the converted lagoon described in the following section.

On average, the advanced treatment system reduced total N (TKN + NO_3-N) concentration from 1503 to 261 mg L^{-1} (83% reduction) and TP from 566 to 26 mg L^{-1} (95% reduction) (Table 1). In addition to substantial reductions in land requirement due to the reduced N and P loads after advanced treatment, the N:P ratio of the liquid was improved from 2.65 to 10.04. This higher N:P ratio resulted in a more balanced effluent from the point of view of crop utilization. A second generation version of the technology is available for municipal and agricultural wastewater and includes the simultaneous separation of solids and phosphorus from wastewater and industrial effluents (Garcia *et al.*, 2007). The combined separation process is more efficient in terms of equipment needs and chemical use. Thus, it reduces installation and operational cost of manure treatment.

CONCLUSIONS

Manure phosphorus (P) in excess of the assimilative capacity of land available on farms is an environmental concern often associated with confined livestock production. A wastewater treatment process was developed for removal of phosphorus from livestock wastewater. It includes nitrification of wastewater to remove ammonia and carbonate buffers and increasing the pH of the nitrified wastewater by adding an alkaline earth metal-containing compound to precipitate phosphorus. Since ammonia nitrogen has been mostly converted to nitrate, increased pH does not result in significant gaseous nitrogen loss. The amount of phosphorus removed, and consequently the N:P ratio of the effluent, can be adjusted in this process to match specific crop needs or remediate sprayfields. In addition to the phosphorus removal aspect, the high pH used in the process destroys pathogens in liquid swine manure. The final product is calcium phosphate that has the potential to be reused as fertilizer or processed to produce phosphate concentrates.

REFERENCES

Bauer, P., Szogi A.A. and Vanotti, M.B. (2007). Agronomic effectiveness of calcium phosphate recovered from liquid swine manure. *Agron. J.*, **99**(5), 1352–1356.

Fordham, A.W. and U. Schwertmann (1977). Composition and reactions of liquid manure (gulle), with particular reference to phosphate: III. pH-buffering capacity and organic components. *J. Environ. Qual.*, **6**(1), 140–144.

Garcia, M.C., Vanotti, M.B. and Szogi, A.A. (2007). Simultaneous separation of phosphorus sludge and manure solids with polymers. *Trans. ASABE*, **50**(6), 2205–2215.

House, W.A. (1999). The physico-chemical conditions for the precipitation of phosphate with calcium. *Environ. Technol.*, **20**(7), 727–733.

Loehr, R.C., Prakasam, T.B.S., Srinath, E.G. and Yoo Y.D. (1976). Development and demonstration of nutrient removal from animal wastes. U.S. Environmental Protection Agency, Washington, DC.

Reindl, J. (2007). Phosphorus removal from wastewater and manure through hydroxy-lapatite. An annotated bibliography. Available at: http://danedocs.countyofdane.com/webdocs/PDF/lwrd/lakes/hydroxylapatite.pdf

Sommer, S.G. and. Husted. S (1995). The chemical buffer system in raw and digested animal slurry. *J. Agric. Sci. Cambridge*, **124**(1), 45–53.

Szogi, A.A., Vanotti, M.B. and Hunt, P.G. (2006). Dewatering of phosphorus extracted from liquid swine waste. *Bioresour. Technol.*, **97**(1), 183–190.

Szogi, A.A. and Vanotti, M.B. (2009). Removal of phosphorus from livestock effluents. *J. Environ. Qual.*, **38**(2): (In press).

Tchobanoglous, G. and Burton, F.L. (1991). Wastewater engineering: Treatment, disposal, and reuse. Irwin/McGraw-Hill, Boston, MA.

Trostle, R. (2008). Global agricultural supply and demand: Factors contributing to the recent increase in food commodity prices. USDA – Economic Research Service, Outlook WRS-0801.

Vanotti, M. B., Szogi, A.A. and Hunt, P.G. (2003). Extraction of soluble phosphorus from swine wastewater. *Trans. ASAE*, **46**(6), 1665–1674.

Vanotti, M.B. and Szogi, A.A. (2008). Water quality improvements of wastewater from confined animal feeding operations after advanced treatment. *J. Environ. Qual.*, **37**, S86–S96.

Vanotti, M.B., Szogi, A.A. and Hunt, P.G. (2005). Wastewater treatment system. US Patent No. 6,893,567, Issued May 17, 2005. U.S. Patent & Trademark Office, Washington, D.C., USA.

Vanotti, M.B., Szogi, A.A., Hunt, P.G., Millner, P.D. and Humenik, F.J. (2007). Development of environmentally superior treatment system to replace anaerobic swine lagoons in the USA. *Bioresour. Technol.*, **98**(17), 3184–3194.

Determining the operational conditions required for homogeneous struvite precipitation from belt press supernatant

B. Lew[a], M. Kummel[b], C. Sheindorf[c], S. Phalah[d], M. Rebhum[d] and O. Lahav[d]

[a]Agriculture Research Organization, Bet Dagan-Israel
[b]Mekorot Israel Water Co., Tel-Aviv-Israel
[c]Shenkar College, Ramat Gan-Israel
[d]Technion, Haifa-Israel

Abstract The objective of this work was to define the best reactor operation conditions to attain homogeneous deliberate struvite precipitation, in order to relieve the ammonia and P flux on the WWTP and that could be reused as a fertilizer or as a fertilization additive. To this end, a laboratory-scale continuous completely-mixed reactor fed with the supernatant of a sludge belt filter press system was operated at different hydraulic retention times (one hour, 30 minutes and 15 minutes) and pH values (7.4, 8.0 and 8.6). 10 mM $MgCl_2$ was dosed into the influent reactor to improve struvite quantity and homogeneity.

The results indicate that short retention times are favorable for the attainment of more homogeneous struvite precipitant, even if it means a small decrease in the amount of precipitant formed. Moreover, an increase in pH does improve reactor performance for struvite precipitation however the improvement seems too small to justify the effort. Moreover, calcium showed a low and approximately constant removal percentage at all the HRTs and pH values studied, indicating that the precipitate was composed of some calcium minerals but at a low concentration.

SVI results (settling performance) of precipitated collected at different reactor operation conditions indicates that the best precipitate was obtained when the reactor was operated at a HRT of 0.25 h. No strong pH effect was found on the precipitate composition at this HRT. SRF (dewatering capacity) results showed best results for precipitate collected at 0.25 hours HRT and at pH higher than pH 8.0.

INTRODUCTION

Struvite, a white crystalline substance consisting of magnesium, ammonium and phosphorus at equal molar concentrations ($MgNH_4PO_4 \cdot 6H_2O$) is known to precipitate and clog pipes and pumps, causing operational difficulties and costs in wastewater treatment plants around the world. Struvite formation typically occurs in wastewater treatment plants downstream of the sludge anaerobic digester, where pH is increased due to CO_2 stripping to the atmosphere. The occurrence is particularly pronounced in activated sludge plants that include bio-phosphate removal by including an anaerobic section before the anaerobic zone.

While struvite is a recognized operational problem in wastewater treatment plants, it has been shown that more than 90% of the dissolved phosphate can be recovered from anaerobic digester supernatant through struvite crystallization, if controlled precipitation is applied (Battistoni et al., 2001; Munch and Barr, 2001; Yoshino et al., 2003). Recovery of struvite has been reported to reduce sludge volumes under specific conditions by up to 49% when compared to chemical phosphorus removal (Woods et al., 1999) and to reduce the phosphorus and nitrogen load of side-stream and sludge liquors recirculation to the head of wastewater treatment works. Also, owing to its low solubility in neutral pH solutions, struvite is an excellent slow-release fertilizer and can be used as part of other industrial products (cleaning products, chemicals, fire retardants) (Battistoni et al., 2002).

Phosphorus (P), which is the eleventh most common element on Earth, is of fundamental importance to living organisms, including humans, and as an essential nutrient for crop production, alongside nitrogen. Furthermore, there is no substitute for phosphorus in nature (USGS, 2005). It is estimated that there are 7000 million tons of phosphate rocks as P_2O_5 remaining in reserves that could be economically mined and that around 40–140 million tons of P as P_2O_5 are extracted each year around the world (Steen, 1998; Jasinski et al., 1999). It has been predicted that world P demand will increase by 1.5% each year (Steen, 1998), leading to an estimate resource exhausted in as little as 100–250 years.

Moreover, almost half of the phosphorus consumed every year find its way to wastewaters. Therefore, recovering at least some of the P and reusing it in agriculture has long ranging implications. Furthermore, removal of phosphorus compounds from wastewaters has important environmental implications to reduce and prevent eutrophication of sensitive inland and coastal waters.

Several authors reported on the effect of operational parameters for struvite recovery: pH value, the molar ratio of the three constituent elements and the potential risks of competitive reactions. It seems that a pH value higher than pH7 is essential for struvite precipitation under supersaturation conditions. According to Munch and Barr (2001), the efficiency of struvite recovery results directly from the pH value of the reaction and a pH value of 9 seems to be a technical optimum for the precipitation (Nelson et al., 2003).

In municipal wastewater magnesium is the limiting element for struvite precipitation. For this reason, if large amount of P is to be recovered addition of magnesium as a chemical becomes necessary. Different Mg/N/P ratios have been studied and in order to optimize this reaction. An increase in the Mg:P ratio has been reported to result in an increase in phosphate removal for any given pH (Nelson *et al.*, 2003).

The research main objective is to determine reactor conditions to obtain deliberate precipitation of struvite from the supernatant of the dewatering facility that is pumped back to the WWTP. The aim was to precipitate and separate struvite from this stream in order to decrease the phosphate and ammonia load on the plant and to study the possibility of struvite reuse in agriculture as fertilizer.

MATERIAL AND METHODS

A 32.1 liter completely stirred reactor (CSTR), followed by a 5.7 liter settling compartment was built to precipitate struvite in the domestic wastewater treatment plant located in Karmiel-Israel (Figure 1). The reactor was continuous fed with supernatant from the sludge dewatering system and operated at three different hydraulic retention times (HRT): one hour, 30 minutes and 15 minutes.

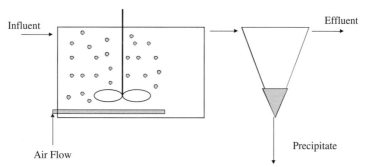

Figure 1. Reactor and settling compartment for struvite precipitation located in Karmiel.

Magnesium (Mg^{2+}) and calcium (Ca^{2+}) concentrations were measured in the reactor influent, effluent and precipitate using ICP. Ammonia (TAN-total ammonia nitrogen) and phosphate (P_T-total phosphate) were measured in the reactor influent, effluent and precipitate using a colorimetric analysis, according to Standard Methods. Reactor influent and effluent were regularly monitored for pH, EC and temperature.

The precipitate was at times studied for SVI (sludge volume index) according to Standard Methods and also by the SRF (sludge resistance to filtration) test. SRF was calculated based on the precipitate filtrate volume (V) measured at time interval (t) at a constant pressure of 49 kN/m^2. The slope (b) of the t/V by V line is applied to the formula: SRF $= (2*P*A^2*b)/(\mu*TSS)$, where, P is the pressure applied; A the filtration area in m^2 units; μ is the sludge viscosity (assumed similar to the water).

RESULTS AND DISCUSSION

The concentrations of the relevant components in the supernatant in the Karmiel WWTP were analyzed and appear in Table 1.

Table 1. Average composition of the supernatant of the dewatering system at the Karmiel WWTP.

Parameter	Units	Value (n = 20)
Total ammonia nitrogen	Mg/l as N	267 ± 40
Phosphate (P_T)	mg/l as P	330 ± 7.5
Magnesium (Mg^{2+})	mg/l	33 ± 11
Calcium (Ca^{2+})	mg/l	70 ± 8
pH		7.61–7.79
Alkalinity (Ref species: H_2CO_3*)	mg/l as $CaCO_3$	1379 ± 224
Total COD	mg/l	248 ± 26
TSS	mg/l	157 ± 39
VSS	mg/l	53 ± 36
EC	ms/cm	3.55 ± 0.38
Nitrate (NO_3^-)	Mg/l as N	11 ± 4

Ammonia, magnesium and calcium concentrations were high and in accordance with equivalent systems in Israel. Based on the magnesium, ammonia and phosphate concentrations, precipitation potential (PP) was calculated to be 136 mg/l.

Based on struvite species concentratin in the supernatant and in the calculated PP, a low quantity of struvite was expected to be obtained. Moreover, the precipitate was expected to be poor in struvite quality and homogeneity due to the low magnesium concentration in the supernatant.

To obtain higher precipitant mass reach in struvite (that is more appropriate for agricultural reuse) and with a higher P_T and TAN removal from the supernatant, experiments were conducted with the addition of $MgCl_2$ to reactor influent. The reasons for this were (1) Mg^{2+} was identified as the limiting factor for struvite precipitation and (2) there is clearly no logic in dosing ammonia or phosphate to the water. The aim was to precipitate higher concentrations of struvite that would lead to reduction in ammonia and phosphate in the supernatant and in the flux of these species back to the wastewater treatment plant. In this respect it is noted that the potential amount of P and N that can be removed amounts approximately between 15% and 20% of the total load of these species on the treatment plant.

To determine the necessary $MgCl_2$ concentration to be added a theoretical calculation was effected: the initial (i.e. prior to precipitation) PP of struvite was calculated for different $MgCl_2$ dosages at two pH values, 8.0 and 9.0. The theoretical results indicate that a ~10.0 mmol/l magnesium addition would result in a close to maximal struvite precipitation with the lowest chemical addition. This concentration addition was therefore used in all further experiments and according to the model an approximately 8.0 mmol/l struvite should be expected to precipitate, almost independently of reactor effluent pH.

Reactor operation at different HRTs and different pH values for each HRT

The continuous reactor was operated at three different HRTs (one hour, 30 minutes and 15 minutes). At each HRT the reactor was operated with three different pH values, 7.4, 8.0 and 8.6. pH increase was attained through stripping of supersaturated CO_2 by air bubbling and NaOH addition.

Species removal in the aqueous phase at each HRT (based on influent and effluent characteristics and ammonia stripping) are shown in Table 2. Ca^{2+} showed the lowest removal at all the HRT and pH studied, indicating that the precipitate is composed of calcium minerals at a low concentration. Between the three species that form the struvite crystal, ammonia showed the lowest removal at all the HRT and pH studied and struvite formation can be assumed to be equal to the ammonia removed, around 8.0 mmol/l according to the aqueous phase. This value is similar to the one observed in the modeled struvite precipitation. At 0.25 hour HRT and pH 7.4 a very similar ammonia, phosphate and magnesium removal was observed, around 8.3 mmol/l (aqueous phase).

In Table 2, the species concentration ratio between each other, in the dissolved precipitate and in the aqueous removed phase is very similar at each HRT studied, confirming the results.

Based on the removal rates observed in the aqueous phase and the knowledge that TAN can only precipitate as struvite, the conclusion is that around 8.0 mmol/l struvite can precipitate at 0.25 hours HRT, around 9.0 mmol/l struvite can precipitate at 0.5 hours HRT and around 7.0 mmol/l struvite can precipitate at 1.0 hours HRT.

Table 2. TAN, Mg^{2+}, P_T and Ca^{2+} removal at different HRT and pH values.

HRT (hour)	pH		TAN	Mg^{+2}	P_T	Ca^{+2}
0.25	7.4	Aqueous (mmol/l)	8.1	8.7	8.6	0.4
		Precipitate (mmol/g)	7.5	8.1	9.2	1.7
	8.0	Aqueous (mmol/l)	8.7	8.9	10.8	0.4
		Precipitate (mmol/g)	7.8	8.0	9.0	1.5
	8.6	Aqueous (mmol/l)	8.9	8.7	11.3	0.5
		Precipitate (mmol/g)	8.2	8.2	9.1	1.4
0.50	7.4	Aqueous (mmol/l)	9.0	10.3	12.3	1.1
		Precipitate (mmol/g)	6.7	7.7	9.0	2.1
	8.0	Aqueous (mmol/l)	10.0	11.0	14.4	1.4
		Precipitate (mmol/g)	6.9	7.7	9.3	2.4
	8.6	Aqueous (mmol/l)	10.2	11.1	15.9	1.5
		Precipitate (mmol/g)	7.0	7.6	9.2	2.5
1.0	7.4	Aqueous (mmol/l)	6.6	9.4	12.1	0.7
		Precipitate (mmol/g)	5.7	8.1	9.5	2.0
	8.0	Aqueous (mmol/l)	7.8	10.7	13.2	1.0
		Precipitate (mmol/g)	5.7	7.8	9.4	2.4
	8.6	Aqueous (mmol/l)	8.8	10.9	13.9	1.2
		Precipitate (mmol/g)	6.2	7.6	9.5	1.4

Species concentration measurement in the precipitate can be more accurate than mass balance in the aqueous phase. Based on the results observed in Table 2 (for the precipitate only), struvite concentration in the precipitate at each HRT and pH can be assumed based on the TAN concentration, Table 3. The excess phosphate and Mg^{2+} precipitation and Ca^{2+} concentration are also shown in Table 3.

Table 3. Calculated struvite, excess phosphate and magnesium and calcium concentration in the precipitate at different HRT and pH.

HRT (hour)	pH	Struvite (mM/g)	Excess P_T (mM/g, %)	Excess Mg^{+2} (mM/g, %)	Ca^{+2} (mM/g)
0.25	7.4	7.5	1.7, 18%	0.6, 8%	1.7
	8.0	7.8	1.2, 13%	0.2, 2%	1.5
	8.6	8.2	0.9, 10%	0.0, 0%	1.4
0.50	7.4	6.7	2.4, 27%	1.0, 13%	2.1
	8.0	6.9	2.4, 25%	0.8, 10%	2.4
	8.6	7.0	2.2, 24%	0.6, 8%	2.5
1.0	7.4	5.7	3.8, 41%	2.5, 30%	2.0
	8.0	5.7	3.7, 40%	2.1, 27%	2.4
	8.6	5.2	3.4, 35%	1.5, 20%	1.4

At each HRT, struvite concentration in the precipitate increase with the increase in pH, however, the increase was not very pronounced and the amount of NaOH necessary to increase the pH is not justified by the increase in struvite precipitant mass.

Excess phosphate and Mg^{2+} percentage in the precipitate decreased with the increase in pH for each HRT. Again, the decrease was not very pronounced and the amount of NaOH necessary to increase the pH is not justified based on the decrease in these percentages. An increase in these percentages was observed with the increase in HRT.

The increase in the percentage of excess phosphate and magnesium precipitation with the increase in HRT can be explained by the fact that struvite precipitation kinetics are much faster than other minerals with phosphate and magnesium and a decrease in HRT promotes a sludge richer in struvite.

According to Wentzel et al. (2001) struvite would be the first mineral to be formed under these conditions, excess Mg^{2+} would form newberyite and afterwards magnesite, excess phosphate would form newberyite and afterwards calcium phosphate and excess calcium would form calcium phosphate and afterwards calcite. In our study, based on the concentrations observed in Table 3, it appears that all excess Mg^{2+} precipitated formed newberyite ($MgHPO_4$); the excess phosphate that did not precipitate as newberyite, precipitated with calcium as calcium phosphate ($Ca_3(PO_4)_2.xH_2O$) and no calcite was observed (Table 4).

Table 4. Theoretical calculated struvite, newberyite and calcium phosphate concentrations in the precipitate at different HRT and pH.

HRT (hour)	pH	Struvite (mM/g)	Newberyite (mM/g)	Calcium Phosphate (mM/g)
	7.4	7.5	0.6	0.6
0.25	8.0	7.8	0.2	0.3
	8.6	8.2	0.0	0.3
	7.4	6.7	1.0	0.7
0.50	8.0	6.9	0.8	0.8
	8.6	7.0	0.6	0.8
	7.4	5.7	2.5	0.7
1.0	8.0	5.7	2.1	0.8
	8.6	5.2	1.5	0.5

At each HRT, newberyite and calcium phosphate concentrations in the precipitate decreased with the increase in pH, however, again the increase was not very pronounced and the amount of NaOH necessary to increase the pH is not justified.

These findings are important because they tend to indicate that short retention times are favorable for the attainment of more homogeneous struvite precipitant, even if it means a small drop in the amount of precipitant formed. Moreover, an increase in pH does improve reactor performance for struvite precipitation, however, the improvement is too low to justify the effort.

Based on previous results a high precipitate purity (high struvite content) seems to be achieved at short HRTs (around 0.25 hours) with only little pH effect. However, quantification of the cost of re-using the precipitate as fertilizer in agriculture cannot be done only by the precipitate purity, but dewatering, handling and storage costs must be taken into consideration as well.

To quantify precipitate storage volume, the precipitate collected at each HRT and pH were tested for SVI (sludge volume index). The results are shown in Figure 2a. SVI test shows the volume occupied by one gram of precipitate after 30 min of settling. Low SVI values are an indication of high settling efficiency.

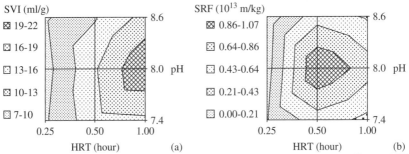

Figure 2. SVI (a) and SRF (b) values for the precipitate collected at different HRT and pH.

An increase in SVI from around 8 to around 17 ml/g, along with the increase in HRT from 0.25 to 1.0 h was observed for a constant pH value. This increase shows that the precipitate from 0.25 hour HRT operation occupies half the volume than the precipitate from the 1.0 h operation. Moreover, at 0.25 hours HRT, precipitate SVI values were constant for all the pH values studied (around 8 ml/g), however, at 0.5 and 1.0 h HRT a maximum SVI value was observed for pH8 with a decrease in SVI values for lower and higher pH values (7.4 and 8.6).

According to SVI values, the best (best settling properties) precipitate is obtained when the reactor is operated at a HRT of 0.25 h. No strong pH effect was found at this HRT.

To quantify the dewatering capacity of the precipitate, the solids collected at each HRT and pH were tested for SRF (sludge resistance to filtration). Results are shown in Figure 2b. SRF test simulates the resistance of the sludge/ precipitate to dewatering. Low SRF values are advantageous.

SRF change as function of HRT and pH showed a "hill" form, with the highest value of $1.07*10^{13}$ m/kg at 0.5 hour HRT and 8.0 pH. Increase and decrease in HRT and pH values led to a decrease in SRF values. Although the lowest SRF values were observed for HRT of 0.25 hours (around $0.1*10^{13}$ m/kg), in all the HRT and pH studied, SRF values are much lower than the ones observed for typical activated sludge solids, around $50*10^{13}$ m/kg.

Moreover, it has been reported by Swanwick et al. (1962) that a sludge/ precipitate with a SRF of $0.13*10^{13}$ m/kg could be dewatered on a pilot scale rotary vacuum filter without chemical conditioning. In our study, only the precipitate collected at 0.25 hours HRT and at pH higher than pH8.0 showed a SRF value lower than the $0.13*10^{13}$ m/kg.

CONCLUSIONS

A continuous reactor was operated at different HRT (two, one, half and 0.25 h) and different pH values (7.4, 8.0 and 8.6) to determine reactor operation conditions to reduce the ammonia and P flux on the WWTP and to generate struvite that could be reused as a fertilization additive.

For this aim, around 10.0 mmol/l Mg salt concentration was added to the reactor influent to attain maximal struvite precipitation with the lowest chemical addition. This value was based on theoretical calculation.

Three parameters were studied to determine precipitate characteristics for reuse in agriculture: precipitate purity, dewatering properties (SRF) and storage volume (SVI). Based on purity, SVI and SRF parameters, the reactor should be operated with 10 mM $MgCl_2$ addition, at 0.25 hours HRT and at pH value of 8.0 to obtain a precipitate with good characteristics (purity, dewatering and storage volume).

ACKNOWLEDGEMENTS

This research was supported by grants from the Mekorot, Israel's water company.

REFERENCES

Battistoni, P., Boccadoro, R., Pavan, P. and Cecchi, F. (2001). Struvite crystallization in sludge dewatering supernatant using air stripping: the new full scale plant at Treviso (Italy) sewage works. In: Proceedings of the 2nd International Conference on phosphorus Recovery for Recycling from Sewage and Animal Wastes, Noordwijkerhout, Holland, March 12–14.

Battistoni, P., de Angelis, A., Prisciandaro, M., Boccador, R. and Bolzonella, D. (2002). P removal from anaerobic supernatants by struvite crystallization: long term validation and process modeling. *Wat. Res.*, **36**, 1927–1938.

Jasinski, S.M., Kramer, D.A., Ober, J.A. and Searls, J.P. (1999). Fertilizers-Sustaining Global Food Supplies, USGS Fact Sheet FS-155-99.

Munch, E. and Barr, K. (2001). Controlled struvite crystallisatin for removing phosphorus from anaerobic digester sidestreams. *Wat. Res.*, **35**, 151–159.

Nelson, N.O., Mikkelsen, R.L. and Hesterberg, D.L. (2003). Struvite precipitation in anaerobic swine lagoon liquid: effect of pH and Mg:P ratio and determination of rate constant. *Bior. Tec.*, **89**, 229–236.

Steen, I. (1998). Phosphorus availability in the 21st century management of a non-renewable resource. *Phosphorus and Potassium*, **217**, 25–31.

Swanwick, J.D., White, K.J. and Davidson, M.F. (1962). Some recent investigations concerning the dewatering of sewage sludges. *Journal of Prog. Ins. Sewage Purif.*, **5**, 394–402.

US Geological Survey (2005). Phosphate rock. Available from: http://mineral.er.usgs.gov/minerals/pubs/commodity/phosphate_rocks/phospmcs05.pdf

Wentzel, M.C., Musvoto, E.V. and Ekama, G.A. (2001). Application of integrated chemical – physical process modelling to aeration treatment of anaerobic digester liquors. *Env. Tech.*, **22**, 1287–1293.

Woods, N.C., Sock, S.M. and Daigger, G.T. (1999). Phosphorus recovery technology modeling and feasibility evaluation for municipal wastewater treatment plants. *Env. Tec.*, **20**, 653–680.

Yoshino, M., Yao, M., Tsuno, H. and Somiya, I. (2003). Removal and recovery of phosphate and ammonium as truvite from supernatant in anaerobic digestion. *Wat. Sc. Tec.*, **48**(1), 171–178.

Involvement of filamentous bacteria in the phosphorus recovery cycle

J. Suschka[a], E. Kowalski[a] and K Grübel[b]

[a]Polish Academy of Sciences – Institute of Environmental Engineering, M. Sklodowskiej-Curie 34 str., 41-819 Zabrze, Poland
[b]University of Bielsko-Biala – Institute of Environmental Protection and Engineering, Willowa 2 str., 43-309 Bielsko-Biala, Poland

Abstract Filamentous bacteria like *Microthricx parvicella* or *Nocardia amarae* are responsible for scum or foam formation floating over the surface of bioreactors as well as secondary sedimentation tanks at advanced nutrients removal sewage treatment plants. At 'simple' biological treatment plants not aiming at phosphorus or nitrogen removal that phenomenon has not known. Evidently the alternating conditions – anaerobic, anoxic and aerobic – create favorable condition for the growth of the above filamentous. The filamentous bacteria like *Microthricx parvicella* or *Nocardia amarae* have the ability of phosphorus accumulation and release, respectively in aerobic and anaerobic conditions. In consequence the phosphorus accumulated by the scum forming organisms is similar or even higher than accumulated by the specific phosphorus microorganisms, like *Acinetobacter calcoaceticus* present in the bulk of activated sludge. In this paper examples of the presence of phosphorous in the scum and activated sludge were given, as well as the advantageous possibility of phosphorus recovery from the scum.

INTRODUCTION

Filamentous bacteria like *Microthricx parvicella* or *Nocardia amarae* are responsible for activated sludge bulking and scum (foam) formation at wastewater treatment plants aiming at nutrients removal. From the very beginning of enhanced nutrients removal processes introduced in the last century scum floating over the surface of bioreactors and secondary settling tanks raised a lot of attention. The reasons of the scum appearance and possible techniques of destruction or avoidance have been intensively investigated. Despite numerous investigations made, many published papers, discussion at meetings and conferences a consistent or satisfactory explanation of the mechanism of filamentous growth scum-forming has so fare not been given.

Surprisingly, the capacity of filamentous bacteria to accumulate phosphorus has found very little attention. It is well documented that specific micro-organisms have the ability of excess phosphorus assimilation, which enters into composition of several macromolecules in the cell. Some bacteria including *Acinetobacter* have the ability to store phosphorus as polyphosphates in special granules (volutin granules). It has been demonstrated that the mentioned before filamentous bacteria (but not only) also accumulate phosphorus as polyphosphates in volutin granules (Machnicka et al., 2004). The amount of stored phosphates is at least similar, or even higher than that of specialized phosphorus accumulating bacteria like *Acinetobacter*. The most often present in scum filamentous bacteria *Microthricx parvicella* and *Nocardia amarae* can accumulate about 20% more phosphorus than *Acinetobacter calcoaceticus*. (Machnicka, 2007).

Actinomycetes are synthesizing polyphosphates and poli-β-hydroxybutyrate as reserve material (Lemmer, 1985) and are able to use many nitrogen compounds, and thus have the potential to tolerate both nitrogen and phosphorus deficiencies. Phosphorus and nitrogen was found to be present in filamentous organisms in excess (Lemmer and Kroppenstedt, 1984). Marshall (1979) showed that nutrients can accumulate at interfaces, thus providing a relatively nutrient reach environment (haven). Naturally present surface active agents can render nutrients sufficiently hydrophobic to allow their adhesion to the gas/water interface to which acitnomycetes are bound (Lemmer, 1986). This process would certainly make nutrients more available to these microbes and explain increased growth of *N. amarae* on the addition of hydrophobic substrates like hexadecane, and industrial wastewater containing nonpolare hydrocarbons (Lemmer and Baumann, 1988).

The principle of phosphorus accumulation by filamentous bacteria follows the principle of the micro-organisms present in the activated sludge flocks. Under anaerobic condition phosphates are released, and accumulated in aerobic conditions.

The foam or scum floating over the surface of bioreactors, very often over all of the three sectors – anaerobic, anoxic and aerobic – does not only cause a visual negative effect but also cause serious operational troubles.

Various attempts and experiences to control foam-causing microorganisms have not resulted in an optimum method. Although the biological control utilizing mean cell residence time, food to microorganism ratio and sludge age, showed high efficiency for typical filamentous bulking, a reduction in *M. parvicella* was not accomplished in many cases.

In some cases, successful control seemed to have been achieved by instantaneous chlorination of return activated sludge (RAS) or the foam surface (Hong et al., 1984). At a dose of 100 g Cl/kg MLSS the microbial flocs in the mixed liquor were completely destroyed. A dose of 500 g Cl/kg MLSS was required to destroy almost completely *M. parvicella* (Hwang and Tanaka, 1998) Lemmer and Kroppenstedt (1984) have reported that addition of $FeCl_3$ at a concentration of 10 mg Fe/l inhibited the growth of *N. amarae*. In fact addition of iron or aluminum salts is relatively widely accepted as an effective foam controlling agent. The use of ferrous sulfate, ferric chlorine or aluminum sulfate, commercially modified by addition of specific polymers, and/or antifoaming chemicals, is very popular in Poland. The addition of ferrous or aluminum salts allows also removal of phosphorus to the required level of 1 mg/l P_{total} in the effluent. Most probable the low concentration of phosphorus is than limiting the growth of filamentous. A side effect of ferrous sulfate addition is increase of sewage salinity and concentration of Fe by about 22% (Muller, 1997).

Another approach is the use of a specific anti-filament polymer (AFP) which suppressed sufficiently the foaming and the growth of *M. parvicella* without effluent quality change (Hwang and Tanaka, 1998).

MATERIALS AND METHODS

Samples of activated sludge and scum were collected from, six different, existing municipal sewage treatment plants, designed and operated as biological nutrients removal plants. All of physical and chemical determinations were performed according to the Standard Methods (1998). The procedure of sludge or scum disintegration was given in a previous publication (Suschka *et al.*, 2007). Microscopic investigation were carried out using a microscope of bright field and contrast phase coupled with a camera. The microscope used – Nikon Alphaphot – 2 YS coupled with camera Panasonic GP – KR 222 allowed also for size measurements by a programme *Lucia* – ScMeas Version 4.51. Samples for microscopic investigations of the presence of polyphosphates were stained according to the Neisser method.

RESULTS AND DISCUSSION

The content of phosphorus in the scum biomass expressed as g P/kg MLSS most often were found to be somewhat higher (exceptionally much higher) than that in the activated sludge. The measured in our experiments average amount of accumulated phosphorus in the activated sludge biomass (from several large EBNR plants) was 26 g P/kg MLSS (2.6%), while in the scum biomass floating over the surface the accumulated phosphorus could be as high as 40.0 g P/kg MLSS, (4.0%). Four examples of the measured amounts of accumulated phosphorus in activated sludge and scum, have been presented in Figure 1.

Actually the possible amounts of phosphorus to be accumulated by e.g. *Acinetobacter calcoaceticus*, or filamentous organisms like *Nocardia amarae*, or *Microthrics parvicelal* are very similar (Machnicka, 2007). The higher content of phosphorus in the scum biomass in comparison to activated sludge flocks is the result of higher content of filamentous micro-organisms in the scum.

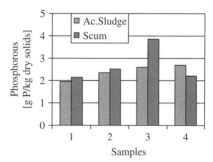

Figure 1. Examples of phosphorus concentration in activated sludge and scum.

Microscopic photo (Figure 2) is showing clearly the polyphosphorus volutins in filamentous organisms.

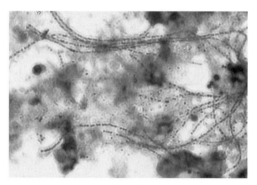

Figure 2. Micrograph of scum, showing phosphorus volutins in filamentous organisms (black dots).

Sludge or scum disintegration results in micro-organisms and associated polymeric matter solubilisation, as well as release of phosphates. The photos below (Figure 3) are showing clearly partially destructed filaments and phosphorus granules thrown out.

Although the contend of phosphorus in the scum is usually not drastically higher then in the activated sludge flocs, (Figure 1) the amount of phosphates which can be redissolved in the process of disintegration (Suschka *et al.*, 2007) are definitely much higher in the case of the scum. Two, in a sense extreme cases, of different number of filamentous organisms present (microscopic evaluation) in the scum are showing (Figure 4) the range of phosphates redissolved. The sample, both activated sludge and scum were collected from two different treatment plants.

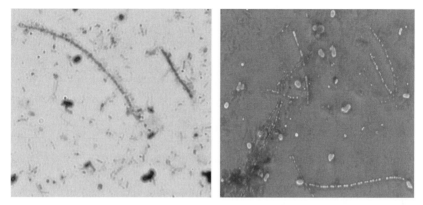

Figure 3. Partially disintegrated scum showing the shot out phosphorus volutins.

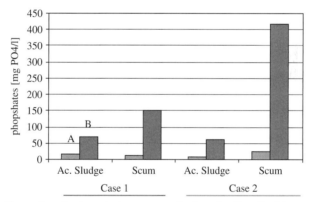

Figure 4. Example of phosphates concentration in the liquid associated to activate sludge and scum, before (A) and after (B) disintegration.

Although the contend of phosphorus in the scum is usually not drastically higher then in the activated sludge flocs, (Figure 1) the amount of phosphates which can be redissolved in the process of disintegration (Suschka *et al.*, 2007) are definitely much higher in the case of the scum. Two, in a sense extreme cases, of different number of filamentous organisms present (microscopic evaluation) in the scum are showing (Figure 4) the range of phosphates redissolved. The sample, both activated sludge and scum were collected from two different treatment plants.

The rates of phosphates released from activated sludge in the process of disintegration do not vary very much, independently from the source (different plants). Rarely the amount of phosphates dissolved attend a level of

100 mg PO$_4$/L. In contrast, the amount of phosphates dissolved in the process of scum disintegration reaches values over 400 mg PO$_4$/l (approximately 130 mg P/L). As shown in the two cases (Figure 4) the concentration of redissolved phosphates from the scum can be 2 or even 8 times larger that that from the activated sludge. Keeping in mind, that in order to achieve a similar degree of disintegration the time and consequently the power consumption is at least 50% lower in the case of scum disintegration, it becomes evident that the scum promise to be an effective substrate for phosphorus recovery.

Definitively the release of phosphorus is accompanied with the release of organic matter. Expressing the organic matter as dissolved COD (SCOD) a correlation between the released organic matter and phosphates was found. In order to present consistent results, further (Figures 5 and 6) only that results of organic matter and phosphates release are presented, which where obtained in the process of hydrodynamic disintegration of 90 minutes.

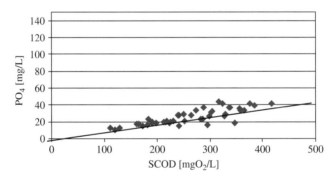

Figure 5. Correlation of release phosphates and soluble chemical oxygen demand for various samples of disintegrated activated sludge.

The ratio of SCOD/P for activated sludge is approximately 30,8. That value depends very much on the concentration of disintegrated sludge. The SCOD/P ratio for RAS (recirculated activated sludge) of concentration in the range of 8 to 10 kg/m^3 was most often about 18.6, while the respective ratio for scum with a concentrations of up to 20 kg/m^3 was about 14,6. The correlation in the full range of investigated samples biomass content is presented in Figure 6.

The given values do not have a universal meaning. Although the given values have been of a similar order for 4 different large municipal treatment plants, they can vary for other cases. Determined here, the SCOD/P ratios are corresponding well to results given e.g. by Kampas et al. (2007). They have given an approximate value for SAS ~13 g/g.

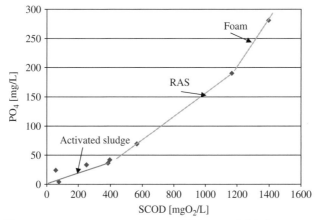

Figure 6. The relationship between dissolved PO_4 and COD for activated sludge, RAS (Recirculated Activated Sludge), and scum.

What is most interesting is the reverse ratio P/SCOD [g P/g SCOD]. With the concentration of disintegrated biomass increase there is an increase of P/SCOD ratio

Approximate values of P/SCOD;

Activated sludge	0.0325	or	3.25%
RAS	0.0538		5.35%
Foam	0.0685		6.85%

The message from the results is, that the amount of phosphorus released form the scum could be twice as high as from activated sludge taken from a bioreactor in relation to the amount of released organic matter (SCOD). That phenomenon could be explained by more effective destruction of filamentous micro-organisms. Beside the mechanical destruction of micro-organisms also the polyphosphates volutins are more easily "shot" out from the filaments (Figure 3).

The presented and discussed above results as mentioned before, refer to the disintegration time of 90 minutes. Obviously the effects of COD and phosphorus release, depends also on the disintegration conditions, or time for a given procedure. In our experiments the investigated time of disintegration was 15, 30, 60 and 90 minutes. In the case of scum usually the highest values of P/SCOD were obtained after 30 min of hydrodynamic disintegration (Figure 7).

Substantial release of phosphates by filamentous microorganisms occurs if the foam is left under anaerobic conditions. The measured concentration of phosphates in the liquid associated to the scum biomass floating over the anaerobic sectors is always much higher in comparison to the aerobic sectors.

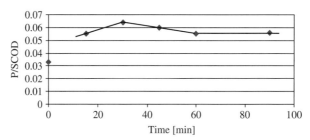

Figure 7. The effects of disintegration time on the P/SCOD ratio.

In one case a thick scum layer was not removed from the surface of the secondary settling tank for many weeks and was septic. The concentration of phosphates was surprisingly high in the order of 140 mg PO_4/l.

Under prolonged anaerobic conditions and higher temperature of 33°C, say for 15 days the determined concentration of released phosphates varied in the range from 210 up to 286 mg PO_4/l. That amount of released phosphates was further increased to a level 385–419 mg PO_4/l when a part (40%) disintegrated scum was added.

The liquor associated to the scum contains relatively low concentrations of ammonia – up to 18 mg NH_4/l. However, if left under anaerobic conditions the amount of ammonia can increase to as much as 110 mg $N-NH_4$/l. That means that in order to fulfill the struvit formula, (MAP – magnesium ammonia phosphate – $MgNH_4PO_4$) only magnesium has to added. It applies also to the classical process of anaerobic sludge digestion, eventually with the addition of disintegrated scum.

Recover of phosphorous in the form of struvit is already a well established process, therefore it seems to be the most appropriate technique.

PRACTICAL IMPLICATIONS

Micro-organisms responsible for the presence of scum (foam) in EBNRP are participating actively in the nutrients removal processes, primarily in phosphorus removal.

Collection of the scum from the bioreactors surface is a simple operation. Collection of the scum results in several positive effects to the overall process of wastewater treatment and sludge handling.

Two additional operations can substantially upgrade the phosphorus removal and recovery process. They are scum disintegration and/or anaerobic digestion.

Disintegration of the scum is an easy, low power consuming process, in which the concentration, in the associated liquor can be increased by a factor of 20, or even more. Precipitation of phosphates as MAP (magnesium ammonium phosphate) is already a well established procedure.

In order to enhance the precipitation of MAP, liquor storage under anaerobic conditions for few days in order to increase the concentration of ammonia could be applied. That means that only addition of limited amounts of magnesium would be required.

Disintegration of scum affects also positively the operation of anaerobic sludge digesters. By addition of disintegrated scum definitively the foaming in digesters is avoided.

Removal (recovery) of phosphorus from disintegrated scum liquor, prevent struvit precipitation in digesters and the peripheral equipment an installations.

No addition of chemicals to combat (control) scum will be required, lowering substantially the cost of treatment and avoiding side effects, like salinity or iron content increase in the effluent.

CONCLUSIONS

1. Micro-organisms responsible for scum forming are following the same pattern as phosphorus accumulating micro-organisms present in activated sludge flocs.
2. Formed scum is perceived as a burden to be combated, while it can be a valuable "byproduct" effectively assisting the process of phosphorus removal and recovery.
3. Foam subjected to disintegration and/or anaerobic conditions is releasing phosphates and ammonia to the liquor in sufficient concentration to make the process of recovery in the form of struvit feasible.
4. The negative aspects of chemicals including iron or aluminum salts addition to control scum forming, like the increase of iron concentration and salinity in the treated wastewaters can be avoided by a different approach to the problem. Also it has not to be forgotten that all of anti foaming chemicals, including specific ones add costs to the process of treatment.

REFERENCES

APHA, (1998). Standard methods for the Examination of water and wastewater, 20th ed. American Public Health Association, American Water Works Association and Water Environment Federation, Washington, USA.

Lemmer, H. and Baumann, M. (1988). Scum Actinomycetes In Sewage Treatment Plants – Part 2. The Effect of Hydrophobic Substrate. *Wat. Res.*, **22**, 761–763.

Lemmer, H. (1986). The ecology of scum causing Actinomycetes in sewage treatment plants. *Water Research*, **20**, 531–525.

Lemmer H. (1985). Mikrobiologische Untersuchungen zur Bildung von Schwimschlamm auf Klaranlagen. Thesis, Technischen Universität, Munchen, Germany.

Lemmer, H. and Kroppenstedt. R.M. (1984). Chemotaxonomy and physiology of some actinomycetes isolated from summing activated sludge. *Systematic a. Applied Microbiology*, **5**, 124–135.

Machnicka, A. Suschka, J. and Grübel, K. (2004). "Phosphorus uptake by filamentous bacteria" World Water Congress and Exhibition, Morocco, Marrakech.

Machnicka, A. (2007). Assimilation and release of phosphorus by filamentous micro-organisms in process of wastewater treatment. – PhD dissertation. (in polish) Silesian University, Gliwice, Poland.

Marshall, K.C. (1979). Growth at interfaces. In Strategies of Microbial Life in Extreme Environments ed. Shilo, M. 281–290, Berlin, Dahelm Konferenzen.

Muller, N. (1997). Eisen in Ablauf Kommunaler Klaranlagen – eine Gefahr für die Aquatische Fauna? Korrespondenz Abwasser 1, 80–87.

Hong, S. Krichten, D. and Rachwal, A. (1984). Biological phosphorus and nitrogen removal via the A/O™ process: recent experience in the United States and United Kingdom. *Wat. Sci. Tech.*, **16**, 151–172.

Hwang, Y. and Tanaka T. (1998). Control of Microthrix parvicella foaming in activated sludge. *Wat. Res.*, **32**(5), 1678–1686.

Kampas, P. Parsons, S.A. Pearce, P. Ledoux, S. Vale, P. Churchley, J. and Cartmell, E. (2007). Mechanical sludge disintegration for the production of carbon source for biological nutrient removal. *Wat. Res.* **41**(8), 1734–1742.

Su Jiang, Yinguang Chen, Qi Zhou, and Guowei Gu (2007). Biological short-chain fatty acids (SCFAs) production from waste-activated sludge affected by surfactant. *Wat. Res.*, **41**, 3112–3120.

Suschka, J. Machnicka, A. and Grubel, K. (2007). Surplus activated sludge disintegration for additional nutrients removal. *Archives of Environmental Protection, – Polish Academy of Sciences*, **33**(2), 55–65.

Carbon and struvite recovery from centrate at a biological nutrient removal plant

B.G. Dirk[a], A. Gibb[a], H. Kelly[a], F. Koch[b] and D.S. Mavinic[b]
[a]Dayton and Knight Ltd., North Vancouver, Canada
[b]University of British Columbia, Vancouver, Canada

Abstract Wastewater contains many valuable resources, however, recovering these resources is often a challenge. Nutrients such as phosphorus and nitrogen are valuable resources that should not be wasted. Struvite, a compound containing magnesium, ammonium and phosphate often forms naturally in wastewater treatment plants; sometimes accumulating in equipment and piping causing operational issues. However, struvite can also be recovered beneficially for sale which can result in improvements to other plant processes.

A pilot-scale demonstration study of volatile fatty acid (VFA) and phosphorus recovery from the centrate of an Autothermal Thermophilic Aerobic Digestion (ATAD) system was carried out at the City of Salmon Arm Water Pollution Control Centre in British Columbia, Canada. Removal of phosphorus from the ATAD centrate, which also has a high VFA concentration, using a struvite crystallizer, would allow the centrate to be returned to the biological nutrient removal process without the use of metal salts.

The ATAD process operates at high temperatures (40°C to 70°C), solubilising large amounts of phosphates and other nutrients. This results in higher concentrations in the centrate than would be expected from other solids digestion processes at conventional secondary treatment plants. In assessing the performance of the crystallizer, magnesium, ammonium, phosphate, pH, conductivity, temperature, solids, and VFA concentration were measured daily. Bench-scale batch tests were routinely carried out to demonstrate the impact of returning the VFA-rich centrate to the phosphorus removal process.

The study was successful in consistently producing good quality fertilizer struvite pellets (avg. 2 mm in diameter), removing approximately 80% of the orthophosphate and 35% of the ammonium-nitrogen from the centrate. This combination of fermenter/ATAD and struvite crystallizer improved and stabilized the performance of the nutrient removal process, eliminated the need for ferric or alum for phosphorus removal from the centrate, and provided consistent nutrient recovery in the form of a saleable end product.

INTRODUCTION

The City of Salmon Arm has a unique treatment process at their Water Pollution Control Centre (WPCC). The process includes a combination Fixed Growth Reactor (FGR) and suspended growth reactor (SGR) for biological nutrient removal (BNR) and an Autothermal Thermophilic Aerobic Digestion (ATAD) System for sludge digestion (Kelly *et al.*, 2005 and Kelly *et al.*, 2005). This combination provides a special opportunity for high amounts of nutrient

© 2009 The Authors, *International Conference on Nutrient Recovery from Wastewater Streams.* Edited by Ken Ashley, Don Mavinic and Fred Koch. ISBN: 9781843392323. Published by IWA Publishing, London, UK.

recovery. The combination of these technologies is unique and uses an integrative approach with a valuable end product. The struvite crystallizer uses dissolved orthophosphate and ammonium in wastewater to produce a high quality, slow-release fertilizer product that is particularly valuable in locations where fertilization is critical (e.g. golf courses). Nutrient removal processes, such as the one in use in Salmon Arm, move the nutrients from the liquid effluent to the biosolids and therefore, there are high concentrations of phosphorus in the digestors. The ATAD System also has high levels of volatile fatty acids (VFAs) due to the nature of the digestion process. The ATAD functions as an acid digester in the initial stages as described by Chu *et al.* (1996), which is common to the first phase of a two phase acid-gas digestion process.

By using centrate from the dewatering of the digested biosolids leaving the ATAD process, high levels of phosphorus can be captured in the crystallizer, and the treated centrate, containing high levels of VFAs, can be returned to the BNR process. This would also eliminate the need to use alum to precipitate phosphorus salts from the centrate.

Description of plant process

The sludge digestion (ATAD) process involves treating both the solids from the primary sedimentation tanks and the waste biological sludge from the BNR process. In the digestion process, as the bacterial cells are broken down the accumulated phosphate is released into solution as dissolved ortho-phosphate.

Because phosphorus is released into solution during digestion, the plant operators are normally required to remove the phosphate from the centrate by adding alum to the digested sludge prior to centrifuging. If this is not done, phosphorus will be returned via the centrate to the liquid treatment process, thereby increasing the load on the biological phosphorus removal process. For this project, in order to keep the phosphate in solution (in the centrate used for the study) alum was not added to the sludge before centrifuging. Instead, an increased amount of polymer was added, which clarifies the centrate without affecting the dissolved orthophosphate level.

Background

The key issues related to the need for the study include the following:

(1) Based on known global reserves, it is understood that the supply of phosphate will be exhausted by about 2050 and all means of phosphorus recovery and reuse will be of growing importance.

(2) It is well established in the literature that VFAs, such as acetic acid, are an important carbon source in the biological phosphorus removal process, as well as in denitrification, as they provide a readily accessible form of carbon (Tchobanoglous *et al.*, 2003).

Although several pilot studies have been undertaken with this struvite crystallizer using various methods for thickening and separating solids after digestion, none of the studies to date have used ATAD for solids digestion. Because this plant includes biological nutrient removal (BNR) and ATAD, with its high temperatures, this results in much higher nutrient concentrations. This study is even more unique because of the focus on recovering VFAs for return to the BNR process.

Objectives

This nutrient recovery pilot test work was designed to fulfil the following main objectives:

(1) To capture VFA from the crystallizer effluent for reuse in the biological nutrient removal process.
(2) To remove excess phosphorus from the digester centrate so that the centrate can be returned to the biological nutrient removal process.
(3) To recover dissolved phosphorus in the form of a saleable fertilizer using the UBC Crystallizer and associated equipment.

Acknowledgements

Funding for this pilot work was provided by:

(1) the Federation of Canadian Municipalities;
(2) the National Research Council under the Industrial Research Assistance Program, grant number 649456;
(3) the City of Salmon Arm, engineering and WPCC operations staff;
(4) Ostara Nutrient Recovery Technologies Inc.; and
(5) Dayton and Knight Ltd. Consulting Engineers.

In-kind support was provided by the University of British Columbia, Civil Engineering, Environmental Engineering Group including Mr. D.S. Mavinic, P.Eng., Ph.D.; Mr. Fred Koch, M.A.Sc.; Mr. Parvez Fattah, M.A.Sc.; and Ms. Paula Parkinson.

METHODS

To test the performance of this process, a pilot reactor was operated and several tests were carried out to assess its feasibility and efficiency. The struvite crystallizer was used to capture the phosphorus, while "batch tests" simulated the effect of returning the effluent to the treatment process.

During the study, the following parameters were monitored daily: dissolved ortho-phosphate, dissolved magnesium, dissolved ammonium, volatile fatty acid (VFA) content, pH, temperature, conductivity, and total suspended solids.

Crystallizer operation

The crystalizer is a fluidized bed reactor, with sections of increasing diameter and a settling zone on top. The changes in diameter result in turbulent eddies at each transition and ensure sufficient mixing in the reactor. These different sized sections also help to classify the fluidized particles by size, so that only the largest ones will be harvested (Britton et al., 2005). This pilot reactor treated 1.7 L/min of centrate which is approximately 10% of the overall centrate generated at the WPCC (18 L/min).

The key ingredients for forming struvite crystals are magnesium, ammonium, and phosphate and, therefore, these were the key parameters that were monitored. These chemicals combine according to the following chemical equation:

$$Mg^{2+} + NH_4^+ + HPO_4^{2-} + OH + 5H_2O \rightarrow MgNH_4PO_4 \cdot 6H_2O$$

Because of the interconnectedness of many of the operating parameters, a computer model was developed by Ostara Nutrient Recovery Technologies Inc. to determine the optimal values of some parameters based on key inputs.

A diagram showing the pilot plant set-up and all included tanks and pumps is shown on Figure 1.

Feed

During the pilot test, the ATAD centrate was stored and then pumped up through the bottom of the crystallizer. The overflow from the crystallizer went into the crystallizer clarifier, and the overflow from the clarifier was recycled back into the crystallizer at multiples of the influent flow rate. This recycling is done to dilute the feed and achieve the desired super-saturation ratio (SSR). The recycle ratio used for this study was ten.

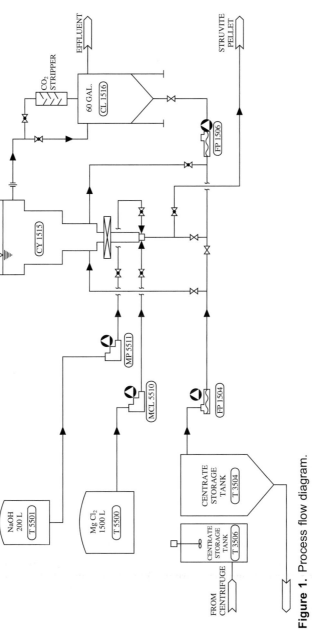

Figure 1. Process flow diagram.

Magnesium

Upon entering the crystallizer, the centrate was blended with magnesium chloride ($MgCl_2$) and caustic in a stainless steel injector block. The flow rates of these two chemicals were dictated by the computer spreadsheet model. The flow rate of $MgCl_2$ is determined based on the molar ratio of Mg:P that is desired, the concentration of magnesium in the centrate, and the concentration of magnesium in the magnesium chloride solution. For optimal struvite formation, a ratio of Mg:P of 1.2:1 or greater is needed (Fred Koch, pers. Comm.).

pH control

The caustic flow rate was determined by the pH controller, which adjusted the flow rate automatically in order to reach the pH set point. The pH set point on the controller was one of the outputs of the computer model and is based on the desired super-saturation rates (SSR), the recycle ratio, temperature of the influent, and the influent concentration of magnesium, ammonium, and phosphate. The pH was affected by all these things and was set so that all nutrients stayed in solution and were available for struvite formation.

The key parameter for optimal operation of the crystallizer is the super saturation ratio (SSR) (Rahaman et al., 2008). pH is the main parameter that can be manipulated in order to achieve the desired SSR.

Monitoring

Samples for VFA analysis were taken five days a week. The VFA samples were analyzed by a Gas Chromatograph using an HP-FFAP column. The Separation of Unsaturated Acids and Fatty Acid Methyl Esters was done using HP-FFAP and HP-INNOWax Columns.

Dissolved orthophosphate was measured as phosphorus, both in the influent and the effluent of the crystallizer, approximately five days a week. The samples were filtered through 0.45 μm filter paper and analyzed using a spectrophotometer according to Standard Methods Method 4500-P D (Greenber et al., 1992).

In order to measure the magnesium concentration in the centrate on a daily basis, the Calculation Method was used from Standard Methods, Method 3500-Mg E except that polyaluminum chloride (PAC) was added to remove soluble phosphate, which is known to interfere with the test. According to Forrest et al. (2007), this is the most reliable method of measuring magnesium on-site.

Because of the importance of pH, an in-line pH probe was used to regulate the supply of caustic to the reactor, so that the desired pH was achieved.

Dissolved ammonium concentration was determined using the Ammonia-Selective Electrode Method (Method 4500-NH₃ F from *Standard Methods*). This test was done five times per week and the results were entered into the model.

Temperature is important because of its effect on several other parameters, including conductivity and pH. The conductivity of a solution is dependent on the ions present within it. Therefore, as struvite crystals are formed in the crystallizer, the conductivity should increase. Temperature and conductivity were measured using the Oakton Instruments/Eutech Instruments PC 300 probe.

TSS is an important parameter because of the influence solids can have on the operation of the crystallizer. Not only do solids provide extra nucleation sites for struvite to form but often high solids levels are anticipated to cause clogging in the tubing and the reactor which causes occasional shut-down for clean-up. The TSS test was done approximately five days per week according to Standard Methods Method 2540D.

The bench-scale experiments were typically carried out once a week and were designed to mimic the flow rates and hydraulic retention times (HRTs) in the three phases of biological phosphorus removal: anaerobic, anoxic, and aerobic; in order to determine the effect of returning the crystallizer effluent to the process. Two experiments were run simultaneously each time, one without crystallizer effluent, as a control, and one with crystallizer effluent.

For the test which included crystallizer effluent, the appropriate amount was included based on the amount of centrate that would be produced at full scale. Samples were withdrawn from the batch reactor every five to ten minutes and analyzed for ortho-phosphate concentration (PO₄-P).

RESULTS AND DISCUSSION

Reactor operation

Although the reactor consistently produced struvite pellets, the size and quality of the pellets varied. Initially, pellets of up to 3.6 mm in diameter with high strength were produced, but over the length of the study the average size was approximately 2 mm in diameter. The quality and size of pellets can be improved by optimizing the reactor operation. Figure 2 shows struvite pellets formed during each of the first five weeks of the study.

Monitoring

Volatile fatty acids

The average volatile fatty acid (VFA) content of the crystallizer influent and effluent are shown in Table 1 (See also Figure 3). There was a decrease in VFAs

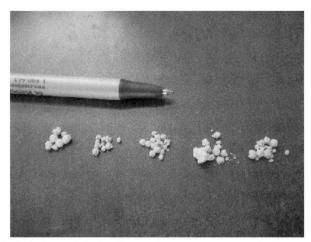

Figure 2. Pellets formed during the study.

through the crystallizer of 24%. The difference was significant and could be partially accounted for by measurement error. However, the decrease could also be due to entrained air from the recycle line which might lead to bacterial oxidation of VFA. Further research and analysis are required to confirm this hypothesis. However, the concentration of VFA in the effluent was still high enough to be beneficial to the biological nutrient removal process at the plant.

Table 1. Volatile fatty acid content.

	Influent (mg/L as acetic acid)	Effluent (mg/L as acetic acid)	% drop through crystallizer
Overall avg.	1259	946	24%
Overall St. dev.	620	518	20%

Phosphate

The average orthophosphate concentration of the crystallizer influent and effluent is shown in Table 2. The average phosphorus removal observed was 77% (see Figure 4). This is similar to what has been found in other pilot studies with similar influent phosphorus concentrations (Prasad *et al.*, 2007). However, with lower influent phosphorus concentrations, higher removal percentages can sometimes be anticipated (pers. Comm. Ahren Britton). It is suspected that higher removals could also be achieved using this process given further optimization.

The typical influent flow rate to the plant is 4600 m³/d with an influent phosphorus concentration of 7.5 mg/L. This gives a daily phosphorus load of

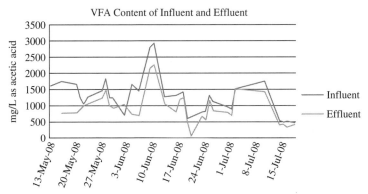

Figure 3. Variation in VFA in crystallizer influent and effluent.

Table 2. Phosphate concentrations and removal.

PO$_4$ -P (mg/L)	Influent	Effluent	% P removal
Overall average	433	92	77%
Overall Standard deviation	142	68	19%

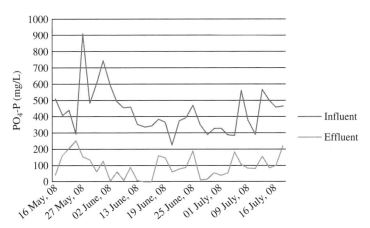

Figure 4. Variation in influent and effluent orthophosphate.

approximately 34.5 kg/d. The pilot crystallizer would remove 2.4% of this load if operated continuously and a full-scale reactor would remove approximately 26% of the total phosphorus load to the plant in the form of crystals. This phosphorus would be removed in the form of a saleable fertilizer and would decrease the amount of sludge produced through the current practice of adding alum to remove phosphorus.

Ammonium nitrogen

On average 34% removal of ammonium nitrogen (NH_4^+-N) was achieved (see Table 3). This removal was higher than is normally achieved in struvite reactors and was partially due to the high influent concentration. Normally, the molar ratio of N:P removal is approximately 1.1:1 (Ahren Britton, pers. Comm.). However, in this study, a molar ratio of 1.7:1 N: P removal from the liquid was achieved. Several samples of struvite crystals were tested for composition by weight, of the three key elements (magnesium, nitrogen, and phosphorus). All samples had approximately a 1:1:1 molar ratio of P:N:Mg. The pellets were very close to pure struvite with about 1.1% by weight impurities. This shows that there was a loss of nitrogen that is unaccounted for. It is unclear why this extra removal was occurring.

Table 3. Ammonium nitrogen concentration and removal.

NH_4-N (mg/L)	INFLUENT	EFFLUENT	% REMOVAL
Overall avg.	871	566	34%
Overall st. dev.	199	157	13%

Magnesium

The on-site magnesium test was difficult to perfect and several different methods were tested in order to remove all of the PO_4 in the samples, so that there would be no interference with the hardness test. Previously, for other struvite pilot studies, magnesium had to be measured in a lab off-site, which meant a delay before getting the results back.

pH

Since the influent centrate was mixed with magnesium chloride and caustic to encourage formation of the struvite pellets, the pH of the effluent was always higher than that of the influent. The set point is the pH that was entered into the pH controller in order to regulate the flow of caustic.

Table 4 shows the overall average influent, effluent and set point pH and their standard deviations.

Temperature

The average temperature of the influent to the crystallizer was 23.9°C. Although the temperature in the ATAD was always higher the centrate cooled in the

Table 4. pH results.

Date	Crystallizer Influent	Crystallizer Effluent	Set point
Overall Avg.	6.8	7.2	7.3
Overall St. Dev.	0.4	0.3	0.3

centrifuge and during storage. The difference between the maximum and minimum temperatures measured was 9°C.

Total suspended solids

Table 5 shows the average total suspended solids (TSS) concentration of the crystallizer influent and effluent. The solids level in the effluent was higher than the influent because of fines formed in the reactor. Solids were anticipated to be a major operational difficulty, as struvite reactors have not previously been run with such high levels of solids. However, because a higher polymer dosage was used, and sufficient time was allowed for settling, none of these problems were encountered. High solids however, were likely the source of many of the fines in the reactor as they can act as nucleation sites. This can hinder the formation of larger struvite pellets, since there are more nucleation sites available.

Table 5. Total suspended solids.

TSS (mg/L)	Influent	Effluent
Overall avg.	351	716
Overall standard deviation	235	489

Batch test results

Batch tests were typically done weekly, to determine the effect of returning the VFA-rich crystallizer effluent to the beginning of the nutrient removal process. Several different configurations were used in order to better understand the effect the effluent would have.

Since the centrifuge was installed in 1998, the centrate from the ATADs has been returned to the primary tank at the head of the plant through the use of alum. This has provided, in combination with the VFA in the influent sewage, sufficient fatty acids to achieve biological phosphorus (bio-P) removal. The average total P concentration in the plant effluent between Jan 2005 and June 2008 was 0.44 mg/L. Therefore, using the centrate return flow has consistently allowed the plant to meet its permit requirements regarding phosphorus removal.

The goal of adding extra VFA to the process is to encourage greater uptake of phosphorus by the phosphate accumulating organisms (PAO) and to stabilize the process. During the anaerobic stage of the nutrient removal process, the PAOs release phosphate and uptake VFAs such as acetate. The more acetate they assimilate in this stage (and the more phosphorus released), the greater the amount of phosphate they will be able to uptake in subsequent phases. Therefore, the amount of PO_4-P released during the anaerobic stage is a measure of the performance of the process. For all the batch tests, the phosphorus that was released in the anaerobic phase was assimilated by the PAOs during the subsequent aerated phase or more was assimilated with the centrate than without.

Overall, for all configurations, the average phosphorus release was higher in the experiment than in the control and had a lower standard deviation, indicating that addition of the crystallizer effluent led to more consistent phosphorus release in the anaerobic phase. This indicates that the VFAs from the crystallizer effluent would improve the performance of the phosphorus removal process and provide more consistent operation. Some results are shown in Table 6.

Table 6. Phosphate release during anaerobic phase.

Date	Control	Experiment
12-May	1.4	11.6
21-May	0	14
27-May	22.1	16.2
12-Jun	16.8	13
20-Jun	18	0
27-Jun	−1	16
11-Jul	0	9
15-Jul	12	8
17-Jul	0	7
Average	7.7	10.5
St. Dev.	8.9	4.9

CONCLUSIONS

Overall, this study demonstrated that a struvite crystallizer can be effectively used to capture phosphates and ammonium from ATAD centrate, and that the treated crystallizer effluent can improve and stabilize the biological phosphorus removal process. The following further conclusions are based on the results of this study.

(1) Biological phosphorus removal sludge can be combined with primary sludge, digested, and used for recovery of VFA and phosphate.

(2) Using the struvite crystallizer removed sufficient phosphorus from the centrate to allow the effluent to be beneficially used in the biological phosphorus removal process.

(3) Returning the VFA-rich stream to the BNR anaerobic phase increases the amount of phosphorus release and uptake, and provides consistent operation.

(4) An average of 77% phosphorus removal from the centrate was achieved consistently over the course of the short study.

(5) An average of 34% ammonium-nitrogen removal from the centrate was achieved throughout the course of the study.

(6) The average total suspended solids concentration in the influent to the crystallizer was 354 mg/L. The reactor operated well, even at this high concentration and no problems with clogging were encountered.

REFERENCES

Britton, A. (2008). Personal communication.

Britton, A., Koch, F.A., Mavinic, D.S., Adnan, A., Oldham, W.K. and Udala, B. (2005). Pilot-scale struvite recovery from anaerobic digester supernatant at an enhanced biological phosphorus removal wastewater treatment plant. *J. Environ. Eng. Sci.*, **4**, 265–77.

Chu, A., Mavinic., D.S., Ramey, W.D. and Kelly, H.G. (1996). A biochemical model describing volatile fatty acid metabolism in thermophilic aerobic digestion of wastewater sludge. *Wat. Res.*, **30**(8), 1759–1770.

Forrest, A.L., Mavinic, D.S. and Koch, F. (2007). The measurement of magnesium: A possible key to struvite production and process control. *Env. Tech.* (in press).

Greenber, A.E., Clesceri, L.S. and Eaton, A.D., Eds. (1992). *Standard Methods For the Examination of Water and Wastewater*, 18th ed. American Public Health Association, Washington, DC.

Kelly, H., Frese, H., Gibb, A. and Koch, F. (2005). Ten years of BNR operating experience for a trickling filter-activated sludge combination, Proceeding IWA Specialty Conference, Nutrient Management in Wastewater Treatment Processes and Recycle Streams, September 19–21, 2005, Krakow, Poland, pp. 845–854.

Kelly, Harlan G., Warren, R. and Urban, W. (2005). Design considerations for auto-thermal thermophilic aerobic digestion, ASCE Conferences, Anchorage, AK, May.

Koch, F. (2008). Personal communication.

Neethling, J.B., Bakke, B., Benisch, M., Gu, A., Stephens, H., Stensel, H.D. and Moore, R. (2005). *Factors Influencing the Reliability of Enhanced Biological Phosphorus Removal*, Final Report. Water Environment Research Foundation, IWA Publishing, London.

Prasad, R., Britton, A., Balzer, B. and Schafran, G. (2007). *WEFTEC 07.* Water Environment Federation, 344–58.

Rahaman, M.S., Maviniv, D.S. and Ellis, N. (2008). Effects of various process parameters on struvite precipitation kinetics and subsequent determination of rate constants. *Water Science and Technology*, **57**(5), 647–54.

Tchobanoglous, G., Burton, F.L. and Stensel, H.D. (2003). *Wastewater Engineering: Treatment and Reuse*, 4th ed. Metcalf and Eddy, McGraw-Hill, Boston.

Recovery of phosphorus from sewage sludge incineration ash by combined bioleaching and bioaccumulation

J. Zimmermann and W. Dott

Institute of Hygiene and Environmental Health, RWTH Aachen, Pauwelsstr. 30, D-52074 Aachen, Germany
(Email: jennifer.zimmermann@rwth-aachen.de; wdott@ukaachen.de)

Abstract The recovery of phosphorus from sewage sludge incineration ash as well as the separation of heavy metals from ash was investigated by using the biotechnological process of bioleaching and bioaccumulation of released phosphorus by newly developed syntrophic population of bioleaching bacteria, *Acidithiobacillus spec.* strains, and polyphosphate (poly-P) accumulating bacteria, the AEDS-population (*Acidithiobacillus spec.* enriched digested sludge). The biologically performed solubilization of phosphorus from sewage sludge incineration ash is accompanied by the release of toxic metals. Therefore a combined process to separate phosphorus from heavy metals by achieving a plant available phophorus-enriched product and a metal depleted ash was designed. Leaching experiments were conducted in laboratory scaled leaching reactor containing a bacterial stock culture of *Acidithiobacillus spec.* Next step was the enhancement of P-recovery in combining bioleaching with simultaneous bio-P-accumulation by AEDS-population. Incineration ash was treated for 11 days on the one hand with Acidithiobacillus spec. strains and on the other hand with AEDS-population (pH-range 2.0–2.3). The leaching efficiency with *Acidithiobacillus spec.* culture for phosphorus was up to 93%, with AEDS-population up to 66%. The leaching efficiencies with *Acidithiobacillus spec.* for metals were for Al up to 61%, Zn, Cu and Co 20–58%, Mn 67–75% and Cr 11–13%. The bioaccumulation of released phosphorus by the AEDS-population can be observed after 3 days, even in low pH-range (pH 2.3). The uptake of phosphorus in biomass reaches up to 66% of the mobilized phosphorus by bioleaching. The combined biologically performed technology of phosphorus leaching and bioaccumulation developed in this study is a promising process for economical and ecological recovery of phosphorus from waste solids.

INTRODUCTION

The increasing global demand of plant available phosphorus leads to a scarcity of natural phosphorus (P) resources. Investigating new P-resources, waste water treatment and sewage sludge can be regarded as promising P-sink. If sewage sludge is not suitable as fertilizer, due to high amounts of toxic metals or organic pollutants, it must be incinerated. The resulting ash is characterized as a sink for phosphorus and a concentrate of heavy metals. Therefore, technologies that allow the recovery of phosphorus from sewage sludge ashes have gained more and more interest. Leaching with inorganic acids allows recovering phosphorus but a costly separation from heavy metals is necessary. The combined process of bioleaching from phosphorus and metals by *Acidithiobacillus spec.* strains and the simultaneous separation of released phosphorus by P-accumulating bacteria by newly developed syntrophic bacteria population named AEDS-population, an environmentally compatible procedure for waste water treatment and nutrient recovery is possible. The principal microorganism involved in bioleaching process is *A. ferrooxidans*, a chemolithotrophic organism. This organism mainly receives its energy from oxidation of ferrous iron to ferric iron and metal sulphides to sulphates. Another bacteria involved is *A. thiooxidans*, which can use elemental sulphur as substrate for growth producing sulphuric acid. This leads to a reduction in pH-value and an increase of solved metals. Bioleaching is a common treatment in resolving metals from sewage sludge (Couillard and Mercier, 1990; Couillard and Chartier, 1991) and fly ash (Brombacher *et al.*, 1998; Yang *et al.*, 2008). In this project sewage sludge incineration ash was used as substrate for bioleaching. Simultaneously to the dissolution of metals, the amount of phosphorus increases as well and can be recovered by biological P-accumulation. This process is based on poly-P-accumulating organisms, named PAOs (Wu, 2006). PAOs were characterized as mixed consortium of different sludge indigenous bacteria in enhanced biological phosphorus removal (Hollender *et al.*, 2002).

Bio-P-accumulation by PAOs is common in the phosphate elimination as enhanced biological phosphorus removal (EPBR) in waste water treatment plants. Aerated tanks with sludge are subdivided in aerobic and following anaerobic zones. In this configuration, bacteria accumulate phosphate in anaerobic zones to store it as poly-phosphate.

In order to recover P and separate heavy metals from sewage sludge incineration ash, a mixture of *A. ferrooxidans* and *A. thiooxidans* strains was used as leaching media in comparison to common chemical treatment. To get the syntrophic bacteria population of *Acidithiobacillus spec.* strains and Poly-P bacteria, anaerobic stabilized digested sludge was used as substrate. The presence of *Acidithiobacillus spec.* species in digested sludge is well known (Blais, 1993). They can be enriched in sewage sludge by the donation of reduced sulphur compounds under aerobic conditions (Shanableh and Omar, 2003). The occurrence of poly-P bacteria, in general term PAOs (Lopez-Vazquez *et al.*, 2007), is also given in digested sludge, but was never investigated under low pH conditions. Hence, the syntrophic AEDS-population is the first attempt to combine a bioleaching process with a simultaneous P-recovery from sewage sludge incineration ash catalysed by microorganisms (MOs).

MATERIALS AND METHODS

Experiments were conducted in a lab-scale bioleaching reactor (Figure 1) using the following conditions: constant flow (25 ml/min) of media, medium volume of 500 ml, aeration (20 l/min ambient air) and 2 g of sewage sludge incineration ash of sludge combustion in wastewater treatment plant Bonn, Germany. Temperature was maintained between 21 and 23°C. Sewage sludge incineration ash was leached with *Acidithiobacillus spec.* species, AEDS-population and sulphuric acid (2M) for 11 days.

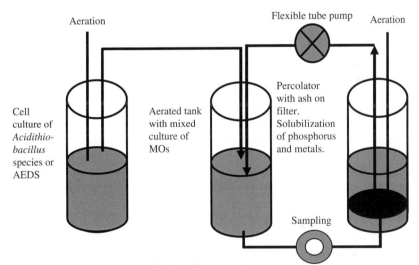

Figure 1. Schematic view of bioleaching reactor.

Acidithiobacillus spec. strains and AEDS- population

A. ferrooxidans and *A. thiooxidans* (DSM 583, DSM 14887) were cultivated in media for lithotrophic organisms (per liter: 3,0 g KH_2PO_4, 0,5 g $MgSO_4$, 3,0 g $(NH_4)SO_4$, 0,2 g $CaCl_2$, 2,0 g Na_2CO_3, 5,0 g $Na_2S_3O_3$, pH 4.4–4.7 using 0.1 N sulphuric acid) with additional elemental sulphur. For AEDS-population enrichment, digested sewage sludge from waste water treatment plant Aachen, Germany, was incubated with added sulphur for 15 days by 22°C. When reaching pH 2.3–2.0, sludge was centrifuged with 13.000 rpm. Supernatant was used as leaching medium in bioleaching reactor.

Sampling and analysis

PH was measured once a day with appropriate electrodes. Sampling of leaching medium was daily for phosphorus and metal content. A 6 ml sample was filtered with 0.45 μm membrane filter and divided for metal and phosphorus analysis. After adding 7M HNO_3, samples for metal analysis were stored at 4°C until analysis. Phosphorus samples were stored at 4°C until analysis. Metal analysis was performed with ICP-MS (Perkin Elmer). Ion chromatography (ICS 3000, Dionex) was used to determine phophorus as orthophosphate according to DIN 38405 method.

RESULTS AND DISCUSSION

Bioleaching of phosphorus and metals

The behaviour of phosphorus release from sewage sludge incineration ash by bioleaching process was carried out in the initial series of experiments. Therefore, leaching experiments with (1) mixed culture of *A. ferrooxidans* and *thiooxidans*, (2) AEDS- population and (3) 2M sulphuric acid were performed and P-release was compared (Figure 2). From the ash, containing 80 mg/g P, maximal releasing rates for P with bioleaching with *Acidithiobacillus spec.* species are 74–75 mg/g (93%) P, 53 mg/g (66%) with AEDS-population and 41 mg/g (51%) with chemical leaching. The recovered P is available as ortho-phosphate. A decreasing P-release (~30% less) was measured for leaching with AEDS-population in contrast to leaching of P by *Acidithiobacillus spec*. The P-content is influenced by the activity of the PAOs. They start P-accumulation already in lag-phase of growth. Significant P-accumulation can be seen after 3 days, when cell count increases exponentially.

Metal extraction experiments were performed for *Acidithiobacillus spec.-* strains, AEDS-population and 2M sulphuric acid, due to the promising results in P-leaching. The aim in these bioleaching experiments was to extract metals from ash into solution as efficient as possible to get a metal depleted ash. Alike, a high metal leaching efficiency even with AEDS-population had to be verified. Biological treatment was compared with chemical extraction (2M sulphuric acid). By achieving nearly the double contents for P-recovery in bioleaching process, the release of heavy metals also shows different rates (Figure 3). The results after 11 days indicated that the metal extraction yields in mixed culture of

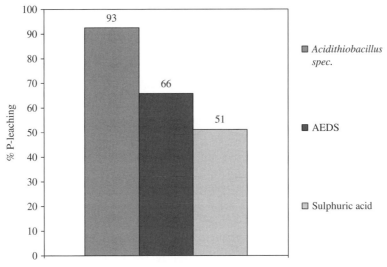

Figure 2. Rate of phosphate-leaching of sewage sludge ash using AEDS-population, medium with *acidithiobacillus spec.* and sulphuric acid 2M.

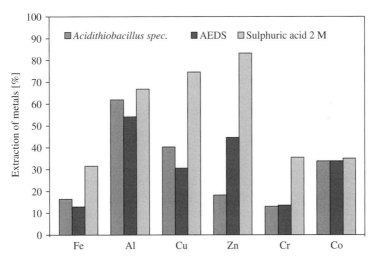

Figure 3. Metals extraction yields in bioleaching using *acidithiobacillus spec.*, AEDS-population and sulphuric acid 2M.

A. ferrooxidans and *thiooxidans* achieved nearly the same amounts like AEDS-population. Chemical leaching shows for all examined metals higher amounts. The maximum metal extraction yields with mixed culture for Fe, Al, Cu, Zn, Cr and Co were 16, 61, 41, 20, 13 and 34%, respectively. With AEDS-population metal recovery for Fe, Al, Cu, Zn, Cr and Co was 13, 55, 31, 45, 13 and 34%, for chemical leaching 32, 67, 75, 83, 36 and 35%, respectively. Hence, the process of resolving phosphorus from incineration ash as well as the extraction of metals is given by using the biological process of bioleaching. The remaining ash is stripped from metals and can be used devoid of mobilization of toxic metals. Even if metal amounts of chemical leaching achieve higher yields, a costly separation of P from metal solution is necessary. The bioleaching process takes advantage by the simultaneous separation of metals and the recovery of P by AEDS-population.

Bioaccumulation of phosphorus and separation of heavy metals

To get a plant available phophorus-product, released P must be separated from extracted metals. Therefore, the newly developed syntrophic population of AEDS-population was applied. In AEDS-population, bioleaching organisms live syntrophically together with P-accumulating bacteria (PAOs), even in low pH-values. The PAOs accumulate under aerobic conditions the through bioleaching released phosphate. For examining the ideal pH-conditions for syntrophic growth and activity, different acidities of the AEDS- population were tested (pH 4.0, 2.5, 2.3, 2.0 and 1.9). The ideal pH-value to get a sufficient metal dissolution combined with good amounts in phosphate accumulation is pH 2.3. Beneath this pH, no accumulation of phosphate by AEDS- population can be observed. Above pH 2.3, the dissolution of metals decreases. The experiments for observing the phosphate disposition for 11 days were conducted with mixed culture and AEDS- population. They were performed in lab-scale bioleaching reactor to ensure a separation of leaching material (incineration ash) and media. After 3 days the AEDS-population start to accumulate the released phophorus as polyphosphate and biomass increases (Figure 4). With mixed culture of *Acidithiobacillus spec.* and sulphuric acid, no significant accumulation of phophorus is apparent. AEDS-population can accumulate nearly 66% of released phosphorus in biomass. The accumulated phophorus is hereby shifted into biomass and can be departed from dissolved toxic metals as plant available phosphate product. The fact of an existing syntrophic population of leaching bacteria and PAOs in low pH-ranges is the first attempt to recover phosphate in a single biologically performed process.

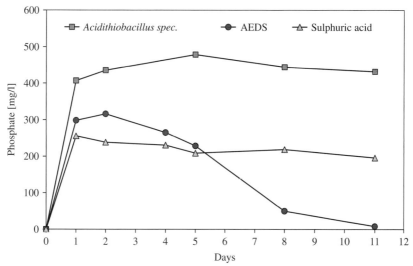

Figure 4. Kinetic of phosphate disposition in bioleaching and bio-P-accumulating process.

CONCLUSIONS

This research investigated the effect of bioleaching and bioaccumulation on the recovery of phosphorus. Aside the well known chemical leaching and precipitation of phophorus as plant available fertilizer, a simple, economical and ecological alternative was developed. The bioleaching process to release phosphorus and the contemporaneous biological phosphorus accumulation is a new way to recycle phophorus from waste waster treatment. As leaching material sewage sludge incineration ash from sludge combustion was used. Incineration ash is a sink for phophorus, but also for heavy metals. The treatment with *Acidithiobacillus spec.* solves phosphorus in higher yields than conventional chemical leaching. Equally, the yields in metal removal are adequate to perform an ecologically unobjectable ash. For separation of released phosphorus from metals, the newly developed syntrophic AEDS- population is able to leach metals and phosphorus and simultaneously accumulate it as polyphosphate. Even in low pH-ranges, the syntrophic association of *Acidithiobacillus spec.* and PAOs is possible and efficient. Hence, biologically resolved phophorus is separated from toxic metal solution and phosphate enriched biomass can be used as plant available product.

ACKNOWLEDGEMENT

Bioleaching studies are part of the project ÇRecycling of plant nutrients – in particular phosphorus from sewage sludge ashÇ (PASCH) supported by budget funds from the German Federal Ministry of Education and Research (BMBF project number AZ 02WA0793). The PASCH project is part of the "German Research Cluster on P-Recycling". For supplying AEDS-POPULATION and funding bioaccumulation experiments AIR UMWELT GmbH (Aachen, Germany) is gratefully acknowledged.

REFERENCES

Blais, J.F., Tyagi, R.D. and Auclais, J.C. (1993). Bioleaching of metals from sewage sludge: microorganisma and growth kinetics. *Water Research*, **27**(1), 101–110.
Brombacher, C., Bachofen, R. and Brandl, H. (1998). Development of a laboratory-scale leaching plant for metal extraction from fly ash by *Thiobacillus* strains. *Applied and Environmental Microbiology*, **64**(4), 1237–1241.
Couillard, D. and Chartier, M. (1991). Removal of metals from aerobic sludges by biological solubilization in batch reactors. *Journal of Biotechnology*, **20**, 163–180.
Couillard, D. and Mercier G. (1990). Bacterial leaching of heavy metals from sewage sludge – bioreactors comparison. *Environmental Pollution*, **66**, 237–252.
Hollender, J., Dreyer, U., Kronberger, L., Kämpfer, P. and Dott, W. (2002). Selective enrichment and characterization of a phosphorus-removing bacterial consortium from activated sludge. *Applied Microbiology and Biotechnology*, **58**, 106–111.
Lopez-Vasquez, C.M., Hooijmans, C.M., Brdjanovic, D., Gijzen, H.J. and van Loosdrecht, M.C.M. (2007). A practical method for quantification of phosphorus- and glycogen-accumulating organism population in activated sludge systems. *Water Environment Research*, **79**(13), 2487–2498.
Shanableh A. and Omar M. (2003). Bio-acidification of metals, nitrogen and phosphorus from soil and sludge mixtures. *Soil and Sediment Contamination*, **12**(4), 565–589.
Wu, Q., Bishop, P.L. and Keener, T.C. (2006). Biological phosphate uptake and release: effect of pH and magnesium ions. *Water Environment Research*, **78**(2), 196–201.
Yang, J., Wang, Q. and Wu, T. (2008). Comparisons of one-step and two-step bioleaching for heavy metals removed from municipal solid waste incineration fly ash. *Environmental Engineering Science*, **25**(5), 783–789.

Energy efficient nutrient recovery from household wastewater using struvite precipitation and zeolite adsorption techniques
A pilot plant study in Sweden

Zsófia Ganrot[a]*, Jan Broberg[b] and Stefan Bydén[a]

[a]Melica Environmental Consulting, Fjällgatan 3 E, 41317 Göteborg, Sweden
[b]Split Vision Development AB, Framtidsgatan 3, 26273 Ängelholm, Sweden
*Corresponding author: zsofia.ganrot@melica.se

Abstract The most N and P abundant part among the domestic waste components is the toilet wastewater; especially urine. Urine separation has been proposed to achieve maximum recovery and recirculation of nutrients in Sweden. However, storage, transportation of large amounts of urine, as well as spreading and hygienic aspects connected to handling of urine and faeces are still main obstacles in achieving full system efficiency. A decade of academic research on struvite precipitation and zeolite adsorption techniques aimed to solve some of these problems and 3 years of collaboration with Split Vision Development AB resulted in a commercially available, energy efficient system developed and tested in laboratory scale and in a pilot plant.

This paper presents the results from one-year testing of a pilot plant installed and integrated in a family house in northern Sweden. The pilot plant is based on the combined struvite precipitation and zeolite adsorption processes. The nutrient rich end-product obtained in the unit called SplitBox is vacuum-dried (fully hygienised) and this drying process connects the SplitBox to the energy system of the house. Moreover, spill-energy from wastewater and ventilation is efficiently reused. The pilot study shows 90–98% P and 95–98% N recovery from toilet wastes using urine separating toilet and SplitBox technique and a total energy saving of ca 50% for the household after one year operation.

INTRODUCTION

Two decades of research and pilot studies in Sweden have successfully pointed out the benefits of nutrient recovery from household wastewater from urban and rural areas and their big potential for an ecologically sustainable resource management and energy consumption. The most N and P abundant part among the domestic waste components is the toilet wastewater; especially urine. Urine separation has been proposed to achieve maximum recovery and recirculation of nutrients in Sweden (Jönsson, 1994; Kirchmann and Pettersson,

1995; Åsblad, 1998; Hellström and Johansson, 1999; Kärrman, 2000; Ganrot, 2005). Several types of toilets and small-scale separation systems have been installed tested and evaluated (Hellström and Johansson, 1999; Palm *et al.*, 2002; Adamsson *et al.*, 2003; Jönsson *et al.*, 2004). Even faeces is nutrient rich, mostly in P. Different solutions and systems were tested and are commercially available for faeces reuse as soil conditioner (Vinnerås *et al.*, 2003; Schönning and Stenström, 2004). All this projects are examples for decentralized, small scale systems from one-household to eco-village size. Storage, transportation of large amounts of urine, as well as spreading and hygienic aspects connected to handling of urine and faeces are still main obstacles in achieving full system efficiency. The biggest problem today in Sweden is connected to the scarce reuse of anthropogenic nutrients on arable land; the nutrient loop is far from closed. In addition to above mentioned obstacles, EU legislation and of course our own perception, behaviour and lack of acceptance of toilet waste as resource are two more serious obstacles.

A decade of academic research (see Ganrot, 2005) on struvite precipitation and zeolite adsorption techniques and 3 years of collaboration with Split Vision Development AB resulted in a commercially available, energy efficient household wastewater system developed and tested in laboratory scale and in pilot plant.

This paper will present the results from one-year testing of a pilot plant installed and integrated in a family house in northern Sweden. The pilot plant is based on the combined struvite precipitation and zeolite adsorption processes. The nutrient rich end-product obtained in the unit called SplitBox is vacuum-dried (fully hygienised) and this drying process connects the SplitBox to the energy system of the house. This unite and the drying unite together is called 'The SplitBox technique'. This technique, besides nutrient recovery, recycles the spill-energy from the wastewater (from kitchen, wash and bath) and from the ventilation system of the house. Moreover, the isolation system in the house is based on a unique, patented method called Koljern isolation technique.

METHOD AND MATERIALS

The pilot plant

During 2005 a lab-scale SplitBox unit was designed, constructed and tested with collected and stored human urine at the research laboratory at Göteborg University, Dept. of Applied Environmental Science. The test results confirmed earlier experiments made by Ganrot on nutrient recovery from human urine through struvite precipitation and zeolite adsorption, when 98–100% of P and 64–80% of N was recovered with this combined method (Ganrot, 2005).

During 2006 a full scale SplitBox unit was constructed and integrated in the energy system of a house (120 m^2 dwelling area, single plan, no cellar) in northern Sweden (Kvissleby) for testing. Figure 1 shows the concept for the energy and wastewater flows in the house. Figure 2 is a schedule over the wastewater flows in the SplitBox treatment units and Figure 3 shows the SplitBox under installation.

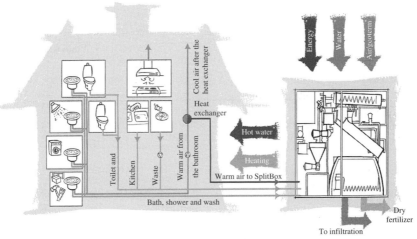

Figure 1. Two unique techniques working together for energy efficiency and nutrient recovery from different wastewater flows in the household.

Wastewater flows in the house

There are two adults living in the household, both working during daytime (9 to 17 h). 2007 their summer vacation weeks were between 1st of July and 15th of August, when the wastewater flow was very low. Table 1 presents the average wastewater flow calculated for the household.

During 2007 only the toilet wastewater and the dish- and bathroom wastewater flows were connected to the SplitBox. The urine processing flow was tested and samples analyzed. For the faeces fraction calculations were made based on previous experiences (Vinnerås, 2002) and applied to the present situation.

The separating toilet installed in this house is Aquatron 400. The separating technique is described on the seller's website www.aquatron.se

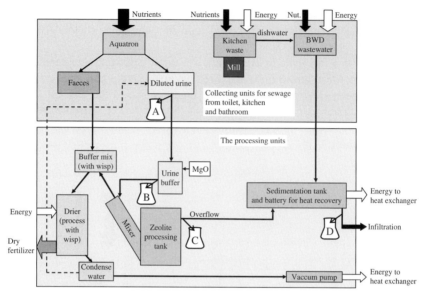

Figure 2. The wastewater sources and the treatment units in the SplitBox. The sampling locations are marked with A, B, C and D.

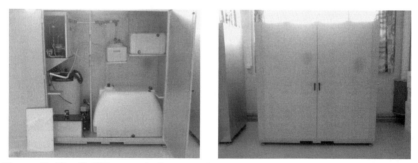

Figure 3. The SplitBox ready for operation.

Table 1. The average wastewater flow calculated for the household.

Wastewater type	Amount (kg/day)
Toilet (Aquatron, with 12 L flushing volume)	58 (urine 4.86; faeces 0.51; paper 0.08)
Bathroom (bath, shower, wash)	350
Kitchen (org. material after milling + water)	140

The MgO of puriss quality used for P precipitation was added as slurry with careful monitoring to reach pH 9 in the urine buffer tank, the best pH level for struvite precipitation (Ganrot, 2005).

The natural zeolite clinoptilolite used in the study was a single batch of 22 kg in the zeolite processing tank originating from USA (Cortaro, Arizona), with the commercial name CABSORB ZK406H, grain size 2 mm and CEC of 1.65 meq/g.

The energy system of the house

The energy system of the house is based on several aspects working together:

a) A unique insulating method, known as Koljern™-technique (using Foamglass as material in the basements, walls and roofs, www.koljern.se, international patent nr PCT/CH02/00711).

b) The SplitBox is integrated in the energy system of the house through energy reuse from the drying process, BWD-wastewater and the ventilation of the house (here the connection to the Koljern™-technique) through a heat exchanger. Figure 1 and Figure 2 are presenting this concept.

The energy consumption of the house was recorded during 2007 and is presented in Figure 5.

Sampling

During 2007 samples were taken from the SplitBox unit at 6 occasions from the sampling sites illustrated in Figure 2 and marked with A, B, C and D:

A – after the urine tank where even the condensate from the drier is recycled,

B – after the urine buffer tank, after the MgO addition and struvite precipitation,

C – after the zeolite processing tank,

D – after the sedimentation tank for BWD water, the outlet to an infiltration bad (if this is necessary).

Totally 24 samples were taken and analyzed. The pH of the samples were carefully measured each time with a field pH meter (Checker) and the samples were frozen ($-18°C$) and sent for analysis to the Göteborg University, Dept. of Applied Environmental Sciences.

Chemical analysis

The total-P and total-N concentrations were analyzed using a DR4000 spectrophotometer with relevant Hach reagents according to the Hach manual

(Hach Company, USA, 1997). The pH was measured by a calibrated Metrohm 605 pH-meter at the beginning of analysis. All necessary dilutions, washing, etc. was made by Millipore water (double ionized water).

RESULTS AND DISCUSSION

The results from total-P and total-N analysis are presented in Figure 4. The black columns represent the input values for each sampling, and in this case this was collected from the sampling place A (diluted urine after the urine tank). The grey columns are the corresponding output for each sample and were collected from sampling place D (outflow from the system). Sampling place D also includes a further wastewater flow, water from bath, wash and dish (BWD), not included in the processing of urine. Only sedimentation was used as treatment for the BWD water.

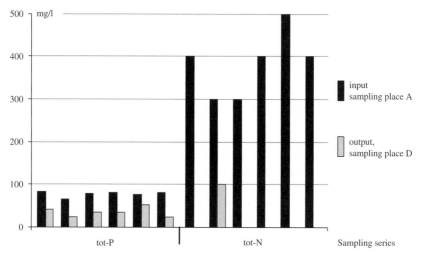

Figure 4. The result from the sample analysis, collected during 2007 (at 6 occasions). Sampling place A = diluted urine after the urine tank, sampling place D = outflow from the house, incl. the treated urine flow together with the bathroom wastewater (not treated). For the tot-N, where no grey columns are visible the tot-N content of outflow was not detectable in the sample.

Results from the sampling places B and C are not presented here, they served mainly for monitoring and intern control of the SplitBox process (pH and MgO addition rates). The results showed that the total-P and total-N recovery after precipitation with MgO and adsorption in the zeolite tank was high; however, the final result of the tot-P in the outflow (grey columns in Figure 4) shows a reduction of only 50–75%. This was caused by two sources in our tests:

a) The P content in the BWD water (Phosphorus containing detergents, hygienic products, etc.). 2007 the P-containing detergents were not banned by law in Sweden and at the testing time our test household was not avoiding such products. The P content of BWD water in Sweden is 0.5 g P/pd (Vinnerås, 2002).

b) Fine crystalline particles from P precipitation may follow the water stream over the zeolite tank ending up in the sedimentation tank.

Both sources of 'problem' for the P reduction in our outflow have been solved since 2007. From July 2008 it is prohibited by law to use P-containing detergents in Sweden. And for the SplitBox process, a micro filtering step was added between the zeolite tank and sedimentation tank.

The total-N reduction was around 100%, because a batch of 22 kg of zeolite is an excess amount for the N uptake of this system during the testing period. The BWD water may content small amounts of N (around 1 g/pd according to the Swedish EPA), however, being extremely diluted in this wastewater stream, this was under the detection level in our analysis, excepting one sampling occasion (Figure 4).

The total-P and total-N from the faeces fraction was not analysed. According to earlier experiments (Vinnerås, 2002) about 85% of nutrients from faeces can be recovered using the Aquatron toilet. However, the system Vinnerås described earlier does not use vacuum-dryer with wisp or does not recycle the condense water after drying (Figure 2), both methods being good to prohibit small nutrient losses during processing. Therefore, our assumption was that a 90–98% recovery for both P and N is achievable from faeces in the SplitBox unit.

The result from the energy consumption during 2007 is presented in the right column in Figure 5, and it was 67 kWh/m^2. The same figure also presents the average energy consumption for an average house in Sweden (139 kWh/m^2 year), the Boverkets (The Swedish Board of Housing, Building and Planning) norm for new-built houses in the Climate Zone North (130 kWh/m^2 year) and the average energy consumption for the most energy efficient new houses built in Sweden during the past 5 years (60 kWh/m^2 year). In all cases the electricity consumption for freeze, home entertainment,, lighting, etc. is not included.

During 2007 the energy system was adjusted because the original heat exchanger was over dimensioned and was changed to one with lower effect. Being a testing and evaluation period, adjustments were made with several occasions in the energy system. The 67 kWh/m^2 value can be further lowered when the system runs continuously in its final form.

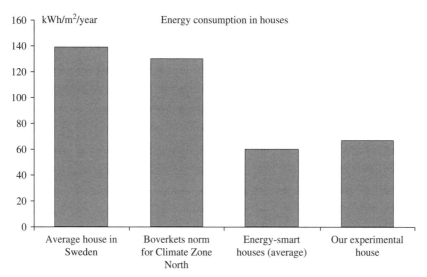

Figure 5. The energy consumption in our experimental house (left column) in proportion to Energy-smart new houses built recently in Sweden, Boverkets norm for new-built houses in climate zone North in Sweden and an average house's average consumption in Sweden.

During 2007–2008 the SplitBox unit was installed in another family house in Umeå (northern Sweden). Combined with Koljern™-technique and a solar panel the energy consumption of the house was extreme low, 20 kWh/m^2 year (Backman, *et al.*, 2008) Moreover, SplitBox units will be dimensioned and installed in an eco-building with 32 dwelling apartments during 2009, with testing and evaluation 2009–2010.

CONCLUSIONS

After one year of operation and testing of the SplitBox unit in a family house in northern Sweden the following conclusions can be drown:

Nutrient recovery

90–98% P and 95–98% N recovery from toilet wastes using urine separating toilet. From the urine the P is precipitated as MAP (mainly struvite) and the N is adsorbed by the zeolite. The final material is a mixture of MAP and ammonium-loaded zeolite. Both are well-known slow release fertilizers. This solid, nutrient rich material is mixed and dried together with the faeces. From faeces the P and N are recovered during mixing, drying and recycling of condense water. In this way the nutrients in faeces are completely recovered together with the organic matter and other macro- and micronutrients.

The thermally treated (fully hygienised) end-product is ready for use as a solid fertilizer.

In our case the BWD water was not treated for P and N recovery with the SplitBox technique, therefore, the analysis shows some lower level of the P recovery from the system (75%). However, this was solved later and the P recovery was much improved, during 2008.

The SplitBox technique may be able to solve some of the problems related to storage, transportation of large amounts of urine, spreading and hygienic aspects connected to handling of urine and faeces. Moreover, ongoing international research (Ronteltap *et al.*, 2007) shows promising results for treatment of hormone- and pharmaceutical residues from urine through MAP precipitation processes. The MAP does not contain these unwanted residues.

The SplitBox is technically simple, easily maintained and is flexible. Can be designed for many requirements from single houses to official buildings or summer residences with or without connection to sewerage.

Recycling and reusing of nutrients from domestic wastewater streams is itself one of the most promising way to minimize greenhouse gas emissions due to fossil fuel consumption in commercial fertilizer production industry today.

Energy saving and reduction of greenhouse gases

The energy need for the system is low due to the SplitBox technique, based on maximal recovery and reuse of the spill-energy from the wastewater and ventilation system in the house. Using the SplitBox technique in combination with the Koljern isolation technique of the house the energy consumption was 67 kWh/m^2 during 2007 including the energy usage for the operation and fertilizer drying in the SplitBox. Our pilot study shows a total energy saving of ca 50% for the household after one year operation with adjustments. Consequently, this energy saving contributes to the same percentage of greenhouse gas reduction if the energy sources are based on fossil fuel.

The energy system is robust and flexible, fully applicable to any type of buildings and energy sources available. However, the most sustainable and climate-friendly way is when renewable energy sources are applied.

ACKNOWLEDGEMENTS

Many thanks to Örjan Andersson and Per Nylander both from the Sollefteå office of Split Vision Development AB for their precious work in this project.

REFERENCES

Adamsson, M., Ban, Zs. and Dave, G. (2003). Sustainable utilization of human urine in urban areas – practical experiences. Peer reviewed paper in Conf. Proceedings: 2nd International Symposium on Ecological Sanitation, Lübeck, Germany, 643–650.

Backman, J., Nilsson L. and Nyqvist A. (2008). Energismart och kretsloppsanpassat byggande i kallt klimat. (In Swedish). Västerbotten County Government Board (www.ac.lst.se/miljomal), p. 35.

Ganrot, Zs. (2005). Urine processing for efficient nutrient recovery and reuse in agriculture, PhD thesis, ISBN 91 88376 29X, Dept. of Environmental Science and Conservation, Göteborg University, Sweden.

Hellström, D. and Johansson, E. (1999). Swedish experiences with urine-separating systems. *Wasser and Boden*, **51**(11), 26–29, Blackwell, Berlin.

Jönsson, H. (1994). Source-separation of human urine – towards a sustainable society. (In Swedish). Swedish University of Agricultural Sciences, Uppsala, Sweden, *Teknik*, **3**, 1–4.

Jönsson, H., Richert Stinzing, A., Vinnerås, B. and Salomon, E. (2004). Guidelines on the use of urine and faeces in crop production. EcoSanRes Publications, Report 2004–2, p. 35.

Kärrman, E. (2000). Environmental system analysis of wastewater management. PhD Thesis. Chalmers Technical University, Göteborg, Sweden.

Kirchmann, H. and Pettersson, S. (1995). Human urine – chemical composition and fertilizer use efficiency. *Fertilizer Research*, **40**, 149–154.

Palm, O., Malmén, L. and Jönsson, H. (2002). Sustainable small-scale wastewater treatment systems. (In Swedish). Rapport 5224, p. 119, Naturvårdsverket (Swedish EPA).

Ronteltap, M., Maurer, M. and Gujer, W. (2007). The behaviour of pharmaceuticals and heavy metals during struvite precipitation in urine. *Water Research*, **41**(9), 1859–1868.

Schönning, C. and Stenström, T.-A. (2004). Guidelines for the safe use of urine and faeces in ecological sanitation systems. EcoSanRes Publications, Report 2004-1, p. 38.

Vinnerås, B. (2002). Possibilities for sustainable nutrient recycling by faecal separation combined with urine diversion. PhD Thesis. Agraria 353, Swedish University of Agricultural Sciences, Uppsala, Sweden.

Vinnerås, B., Jönsson, H., Salomon, E. and Richert Stinzing, A. (2003). Tentative guidelines for agricultural use of urine and faeces. Peer reviewed paper in Conf. Proceedings: 2nd International Symposium on Ecological Sanitation, Lübeck, Germany, 23–30.

Crystallisation of calcium phosphate from sewage: efficiency of batch mode technology and quality of the generated products

Ehbrecht, A.[a]*, Patzig, D.[a], Schönauer, S.[b] Schwotzer, M.[b] and Schuhmann, R. [a,b]

[a]University of Karlsruhe, Competence Center for Material Moisture (CMM)
[b]Forschungszentrum Karlsruhe, Institute for Technical Chemistry, Division of Water Technology and Geotechnology (ITC-WGT)

*corresponding author: anke.ehbrecht@kit.edu

Abstract P-Recovery from sewage by crystallisation of calcium-phosphates triggered by a reactive substrate was applied in bench-scale and semi-technical-scale units with wastewater spiked with 50 and 300 mg/L P, respectively. It had been shown that hydrochemical parameters like P-concentration and pH-value control the reaction mechanisms with regard to the quality of the products. Furthermore grain size and Ca-content of the reactive substrate affect the P-elimination efficiency.

INTRODUCTION

The increasing use of fertilizers and the finite nature and contamination of natural phosphorus resources (WAFD and GFST, 2006–2007) necessitates a sustainable technology for P-recovery. As anthropogenic sources like sewage are an important nutrient sink, crystallisation of calcium-phosphate-phases (Ca-P-phases) triggered by a reactive substrate represents a promising method of P-Recovery from wastewater. In P-RoC-technology (Phosphorus Recovery from Waste- and Process Water by Crystallisation) P-elimination and -recovery occurs in one single step without the addition of chemicals except for the reactive substrate (Berg *et al.*, 2007). P-RoC wastewater treatment was applied to sewage with an average P-concentration of 10 mg/L and industrial effluents with P-concentrations up to 400 mg/L. These mentioned P-concentrations can be attained under certain operational management on wastewater treatment plants like sidestream-processes or biosolids disintegration.

OBJECTIVE

Goal of this study was to characterise the reaction mechanisms of Calcium-Silicate-Hydrate (CSH) substrates by using sewage added with the P-concentrations mentioned above in bench-scale and semi-technical units. Detailed knowledge of the chemical and mineralogical coherences between pH-value, P-concentration and the nature of the generated crystallisation product is of utmost importance, particularly with regard to the recycling of the generated P-containing products. The generated Ca-P-phases can be used in the P-industry (Schipper *et al.*, 2001) which demands a P-content of ≥ 10 wt.-% and as an alternative to mineral fertiliser manufactured by P-rock in agriculture because the background pollution of heavy metals (Berg *et al.*, 2006) is only marginal. Furthermore the plant availability of P could be assumed as being well due to the formation of poorly crystalline Hydroxylapatite-like Ca-P-phases showing higher water solubility compared to well crystallised products. Higher water solubility applies for brushite as generated product, too. Another focus was set on the efficiency of four different CSH substrates regarding to the P-elimination (Ehbrecht *et al.*, 2008).

MATERIAL AND METHODS

In order to characterise the reaction mechanisms subject to P-concentration and grain size of the applied CSH substrate crystallisation experiments were carried out both in bench-scale (5 L) and semi-scale units (80 L) and a solids content of 5 wt.-% CSH substrate. The reactor influent consisted of clarifier effluent of a municipal sewage plant ($SI_{Calcite} > 0$) spiked with KH_2PO_4 stock solution up to the P-concentrations of 50 and 300 mg/L P, respectively. In bench-scale experiments different grain sizes (0.063 mm to 0.7 mm) of the CSH substrate (several sieve fractions of substrate D, see Table 1) were used and the semi-scale experiments were carried out with the CSH substrate of an average grit size of 0.5 mm. The bench-scale (5 L) experiments were therefore filled once with wastewater and sampled in time intervals from 1 min to 24 h whereas the semi-scale units were constructed as Sequence Batch Reactor (SBR) with a hydraulic retention time (HRT) of 2 h and P-concentrations of the influent of 50 mg/L P and 300 mg/L P.

The experimental investigation of the different CSH substrates with regard to the efficiency of P-elimination was also conducted in bench-scale experiments (5 L) and semi-technical units (40 L) both with a solids content of 5 wt.-% and clarifier effluent spiked with KH_2PO_4 stock solution to an average P-concentration of 50 mg/L and 300 mg/L P. The applied CSH substrates differ in grain size, mineralogical and chemical composition such as Ca:Si-ratio and Ca-content, respectively (Table 1).

Table 1. Specification of the reactive CSH substrates.

	Substrate A	Substrate B	Substrate C	Substrate D
Grain size [mm]	1.5–2.0	1.5	1.5	0.5
Mineralogical composition	tobermorite	grained xonolite	tobermorite, quartz	tobermorite, quartz
Empirical formula	$Ca_5Si_6O_{17} \cdot$ $5\ H_2O$	$Ca_6Si_6O_{17} \cdot$ H_2O	$Ca_5Si_6O_{17} \cdot$ $5\ H_2O$	$Ca_5Si_6O_{17} \cdot$ $5\ H_2O$
Ca:Si-ratio	1	1	0.83	0.83
Ca-content [%]	28	34	22	22

Important hydrochemical parameters as the pH-value and conductivity were measured and the determination of P was carried out according to DIN EN ISO 6878. Solid samples were taken in intervals and one part of solid analysis was used in order to evaluate the content of total phosphorus (TP) via acid digestion by microwave according to DIN EN 13346. The identification of the Ca-P-compounds was possible due to the X-ray Diffraction (XRD) with the X-Ray Diffractometer D 5000 (Siemens). Furthermore an element linescan through a product particle was performed by means of EDX analysis.

RESULTS AND DISCUSSION

The laboratory experiments provided an overview for the differences in the kinetically driven crystallisation reaction. Using the finest grain size P-eliminations starts at a higher level and reaches a higher level within the first 30 min compared to an application of the coarser grain size [Figure 1]. So the feasibility of the crystallisation technology triggered by CSH substrates might be pre-estimated on the basis of bench-scale experiments with regard to the efficiency of P-elimination.

Figure 2 shows the temporal development of the P-elimination applying both P-concentrations of the influent in semi-technical scale experiments. While experiments with an influent of 50 mg/L P showed a P-elimination of 85 to 95% until it declined to 40% within approximately 70 h, the temporal sequence of P-elimination of the experiments with an influent of 300 mg/L P followed a nearly parabolic decline. This is caused by a faster Ca-depletion of the reactive substrate: the more P available for crystallisation the higher the diffusion potential causing greater Ca release. The transferability of the results from bench-scale to semi-technical experiments with regard to P-elimination proved the potential feasibility of P-RoC-technology with view to a large-scale application.

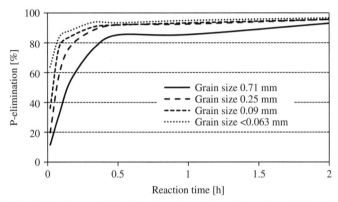

Figure 1. P-elimination obtained in bench-scale units using CSH-substrate of grain sizes from <0.063 mm to 0.71 mm and P-concentrations in the influent of 50 mg/L P.

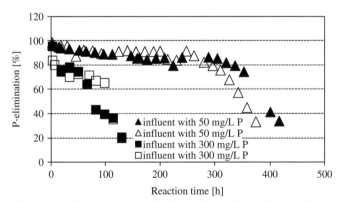

Figure 2. Temporal development of P-elimination obtained in batch experiments (80 L) with CSH substrate of an average grain size of 0.5 mm and P-concentrations in the influent of 50 mg/L P (▲/△) and 300 mg/L P (□/■). Experiments of both P-concentrations were carried out twice (graphed in filled and unfilled symbols).

The following hypothesis was proposed at first: A high P-concentration could presumably lead to the formation of a Ca-P-coating avoiding the further release of Ca. In view of Ca being the most important reactant for P in the crystallisation of Ca-P-phases, no further release of Ca disrupted the crystallisation. Consistent with hypothesis the measured P-content of the product was 7 wt.-% compared to the theoretical calculated P-content of about 12 wt.-%. To prove this hypothesis an EDX-linescan was performed through a product particle whether Ca and P were higher concentrated in the boundary area. But the hypothesis had to be disproved because Ca and P were evenly distributed (Figure 3). Slightly elevated Ca- and P-contents in the boundary area are due to initiating of crystallisation at the surface of the particle.

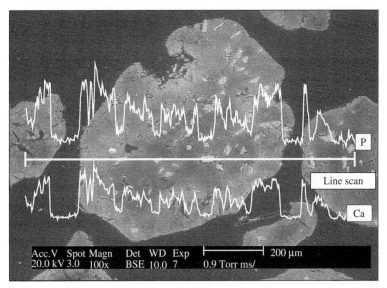

Figure 3. ESEM micrograph of product particles cast in epoxide resin (cross-section) after a reaction time of 130 h with effluent spiked to 300 mg/L P. The element scan along the white line through a particle shows the element distribution.

Furthermore the results imply that the hydrochemical parameters like pH-value and P-concentration affect the characteristics of the generated Ca-P-compounds regarding their chemical and mineralogical composition. In Figure 4 and Figure 5 the XRD pattern from solid samples of the experiments with 50 an 300 mg/L P, respectively are shown.

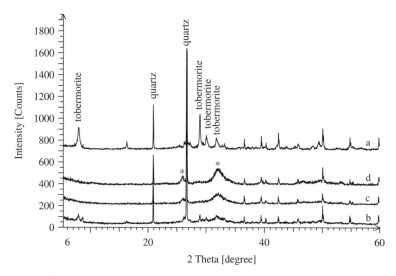

Figure 4. XRD patterns of CSH substrate and products generated in process after different reaction times with a P-concentration of 50 mg/L P in the influent: a. CSH substrate b. – d. products after 243 h (b), 354 h (c) and 420 h (d). ∗ = Hydroxylapatite-like Ca-P-phase.

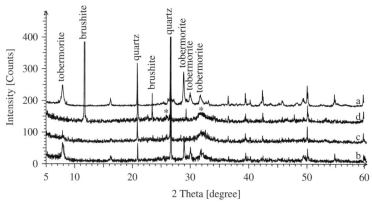

Figure 5. XRD patterns of CSH substrate and products generated in process after different reaction times with a P-concentration of 300 mg/L P in the influent: a. CSH substrate b. – d. products after 34 h (b), 84 h (c) and 130 h (d). ∗ = Hydroxylapatite-like Ca-P-phase.

In the experiment with wastewater spiked up to 50 mg/L P a degradation of CSH substrate occurred after a reaction time of 243 h during the crystallisation of poorly crystallised Hydroxylapatite-like (HAp-like) Ca-P-phase (Berg *et al.*, 2005) with a P-content of up to of 12 wt.-% P. Concomitantly residuals of quartz remained. The same phenomenon was observed at solid samples of the experiments spiked up to 300 mg/L P but towards completion of the experiment after 130 h, a crystallisation of brushite was detected – presumably at the expense of HAp [Figure 5]. This could be explained by a change of the pH-value during crystallisation, as brushite preferential crystallizes at a pH-value <7 compared to HAp (Abbona *et al.*, 1993; Boistelle and Lopez-Velero, 1990).

The development of the pH-value during crystallisation proceeds as follows: First there can be observed an increase in pH-value from a start pH-value of ca. pH 7 to a pH-value of approximately pH 9 due to release of Hydroxide ions (OH⁻). With depletion of the CSH substrate there can be observed a decrease of pH-value to the initial pH-value. Furthermore the crystallisation of HAp led to a shift of the equilibrium and composition of the residual liquid which had been additionally confirmed by a geochemical modelling (The Geochemist's Workbench 6.0). A further explanation can be the higher nucleation rate of brushite compared to HAp (Abbona *et al.*, 1988). Additionally, HAp requires a higher critical supersaturation for its formation (Abbona *et al.*, 1988).

Results of laboratory experiments to investigate the efficiency of the different CSH substrates are shown in Figure 6.

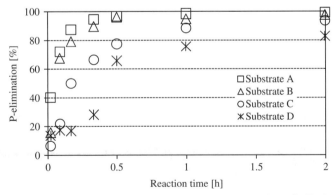

Figure 6. P-elimination from clarifier effluent spiked with 50 mg/L P obtained by addition of 5 wt.-% of the reactive substrates in bench-scale units.

The reaction time for the substrates A and B reaching a P-elimination over 90% was 30 min whereas the substrates C and D reached less P-elimination after 2 h. This can be explained by the lower Ca-content of the substrates C and D compared to the substrates A and B [Table 1]. Furthermore it was observed a more decelerated increase in P-elimination using substrate C compared to substrate D although it is the same mineral phase. This can be attributed to the different graining [Table 1]. The results of the semi-scale experiments are shown in Figure 7. All substrates exhibit an almost constant P-elimination of over 80% for different durations. So P-elimination using the substrates A and B decreased after 350 h whereas P-elimination with the substrates C and D decreased after 200 and 280 h, respectively.

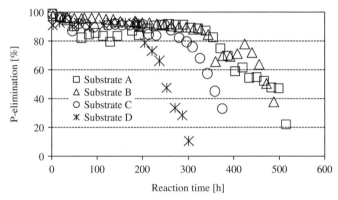

Figure 7. P-elimination from clarifier effluent spiked with 50 mg/L P attained by addition of 5 wt.-% of the several substrates depending on the reaction time carried out in SBR experiments with a hydraulic retention time (HRT) of 2 h.

The decrease in P-elimination can be assigned to a depletion of the substrates. The efficiency concerning the P-elimination depends on the Ca-content of the applied substrate, what additionally can be corroborated by the content of total Phosphorus (TP) of the generated Ca-P. Using the tobermorite-like CSH substrates C and D in Sequence Batch Reactor (SBR) technology TP-contents of about 11 wt.-% can be obtained whereas the application of the more Ca-rich xonolite or tobermorite without addition of quartz led to TP-contents of about 14 wt.-%.

CONCLUSION

Crystallisation of Ca-P-phases triggered by a reactive CSH substrate represents a promising technology to eliminate P from definite wastewater with P-concentrations of 50 mg/L and 300 mg/L. An essential finding of this work was the transferability of results concerning P-elimination obtained in short-time bench-scale experiments to long-term experiments in semi-technical scale. Also it had been shown that the particle size affects the efficiency of the crystallisation because of the higher surface of the finer grained substrates. Furthermore P-elimination efficiency is related to the Ca-content of the substrate. With increasing Ca-content of the CSH substrate a higher P-elimination efficiency was obtained resulting in a higher P-content of the generated products.

Generally, a recirculation of the effluent should be possible at wastewater treatment plants using this technology. According to current knowledge, this fact is crucial regarding the treatment of concentrates from biosolids disintegration following limit values of the effluent in spite of the high efficiency of P-elimination. The Ca-P-compounds generated by P-RoC-technology – composed of poorly crystalline silicon dioxide and HAp-like calcium phosphates, respectively brushite – could be used in P-industry as a surrogate for natural P-rocks (P-content up to 12 wt.-%). Another application could be its use as so-called "green" fertilizer in agricultural areas as the heavy metal content is only marginal and its water solubility makes P accessible to plants.

ACKNOWLEDGEMENT

We thank the Federal Ministry of Education and Research funding this project (No. 02WT0783). Our thanks go to Marita Heinle and Gudrun Hefner for the measurements. We thank Julia Scheiber for modelling with Geochemist's Workbench and Ute Schwotzer (former Berg) for the valuable discussion.

REFERENCES

Abbona, F., Christiansson, A. and Lundager-Madsen, H.E. (1993). *J. Crystal Growth*, **131**, 331–346.
Abbona, F., Lundager-Madsen, H.E. and Boistelle, R. (1988). *J. Crystal Growth*, **89**, 92–602.
Berg, U., Donnert, D., Ehbrecht, A., Weidler, P.G., Kusche, I., Bumiller, W. and Nüesch, R. (2005). *Colloids and Surfaces A: Physicochem. Eng. Aspects*, **265**, 141–148.
Berg, U., Knoll, G., Kaschka, E., Kreutzer, V., Weidler, P.G. and Nüesch, R. (2007). *J. Res. Sci. and Techn.*, **4**(3).
Berg, U., Schwotzer, M., Weidler, P.G. and Nüesch, R. (2006). WEFTEC, 21–25 Oct. 2006, Dallas.
Boistelle, R. and Lopez-Velero, I. (1990). *J. Crystal Growth*, **102**, 609–617.
DIN EN 13346.
DIN EN ISO 6878.

Ehbrecht, A., Patzig, D., Schönauer, S. and Schuhmann, R. (2008). Poster presentation at IWA World Water Congress, 7–12 Sept. 2008, Vienna.

Schipper, W.J., Klapwijk, A., Potjer, B., Rulkens, W.H., Temmink, B.G., Kiestra, F.D.G. and Lijmbach. A.C.M. (2001). *Env. Techn.*, **22**(11), 133.

World Agriculture and Fertilizer Demand, Global Fertilizer Supply and Trade 2006–2007. Summary Report of the 32nd IFA Enlarged Council Meeting, Buenos Aires, Argentina, 5–7 Dec. 2006.

Effect of osmotic pressure and substrate resistance on transmembrane flux during the concentration of pretreated swine manure with reverse osmosis membranes

L. Masse[a], D.I. Massé[a] and Y. Pellerin[b]

[a]Agriculture and Agri-Food Canada, Sherbrooke, Qc, Canada
[b]Technologies Osmosys, St-Adrien de Ham, Qc, Canada

Abstract In many livestock-producing countries, animal manure could supply a large part of the nitrogen required by the cultivated land base. The objective of the membrane technology presented here is the production of nitrogen concentrates from swine manure, that can be economically transported and used as fertilizers by feed producers. This paper specifically examines the effect of osmotic pressure and membrane resistance on transmembrane flux during the concentration of prefiltered swine (PFS) manure by reverse osmosis (RO) membranes. PFS manure at various concentration stages was filtered by four highly selective polyamide RO membranes, with maximum allowable pressures ranging from 4.1 to 8.3 MPa. The osmotic pressure created by the PFS manure on the RO membranes fitted a second-order equation with respect to manure conductivity, indicating that the rate of increase in osmotic pressure accelerated as manure was being concentrated. Average osmotic pressure increased by a factor of 7.4, from 0.47 to 3.54 MPa, as total ammonia-nitrogen was increased 6.1 times, from 1.5 g to 9.1 g/l. Substrate-related resistance (R_s), which has been attributed to specific membrane-solute interactions even in the absence of flow, also tended to increase slightly as PFS manure was concentrated. The R_s value was on average 67% higher with the two membranes built to sustain pressures up to 6.9 and 8.3 MPa than with the two membranes with maximum allowable pressure of 4.1 MPa. However, if the objective of the technology is to concentrate manure in small volumes with high nitrogen concentrations, RO systems have to be equipped with membranes that are able to sustain high applied pressures, because the decrease in flow due to increased osmotic pressure will be substantial.

INTRODUCTION

In the past century, human activities have doubled the natural rate of nitrogen entering the N cycle (Vitousek *et al.*, 1997). The fixation of atmospheric, non reactive nitrogen (N_2) to produce fertilizers accounts for over half the new available nitrogen. Large increases in nitrogen application on agricultural lands have contributed to air and water pollution. Additionally, the production of synthetic nitrogen fertilizer is a high energy-consuming process, that requires between 1.1 and 1.3 m^3 of natural gas per kg of anhydrous ammonia (Gelling and Parmenter, 2004). Thus, the production of N fertilizer is an important source of greenhouse gas emissions and fertilizer price increases with the rising cost of energy.

Steinfeld *et al.* (2006) estimated that between 20% and 25% of all chemical N fertilizers are used for livestock feed production. A large quantity of this nitrogen is still available in the manure, because animals have a low nitrogen assimilation efficiency of about 10% globally and 20% for pigs (Van der Hoek, 1998). It is thus imperative to conserve and reuse the nitrogen excreted with animal manure. However, the transport of manure to feed-producing farms is often not profitable because manure contains between 90% and 99% of water and is thus a highly diluted fertiliser. Manure is also an unbalanced N-P-K fertilizer and application rates are increasingly limited by the maximum allowable phosphorous load for a particular field. Thus, feed producers still rely heavily on chemical nitrogen fertilizers, even though in many livestock-producing regions, manure could supply most of the plant N requirements (MENVQ, 2003).

One solution is to isolate and concentrate manure nutrients in volumes that could be economically transported to other farms as fertilizers. In recent years, physico-chemical technologies have been developed to concentrate up to 85% of the phosphorus in a solid phase representing between 10% and 30% of the raw manure (Forbes *et al.*, 2005). Reverse osmosis (RO) can then be used with the liquid fraction to produce nitrogen concentrates, the by-product being relatively clean water that could be used to wash barns or irrigate nearby cultures. A few experiments with laboratory and farm-scale membrane systems have been described in the literature, but there is little information on long-term performance of the technology, optimum operating parameters, and effluent quality with respect to initial manure characteristics and volumetric concentration (Masse *et al.*, 2007). The effect of osmotic pressure, substrate-, pressure-, or time-related fouling, temperature and concentration polarization on flux decrease during manure concentration has not been reported (Bilstad *et al.*, 1992; Thorneby *et al.*, 1999; Zhang *et al.*, 2004). Manure has a high salt content, composed of bicarbonates, volatile fatty acids, ammonia, potassium, etc., and can thus exert a high osmotic pressure on RO membranes. A 62% decrease in flux was observed as permeate recovery rate was increased from 0 to 53% during the filtration of a wastewater containing 8.7 g/l of ammonia (Minhalma and Pinho, 2002). The decrease in flow was essentially attributed to increased osmotic pressure.

The objective of the experiment presented here was to quantify osmotic pressure and fouling resistance during the concentration of pretreated swine manure. This information is necessary to design full-scale systems for farm application.

MATERIALS AND METHOD

The raw manure was collected from the transfer storage tanks on a typical farrow-to-finish swine operation in Québec, Canada. The raw manure contained 47.8 g/l of dry matter (DM). It was filtered under vacuum through diatomaceous earth (DE), an inert material that forms a filtering cake with a mean pore size of 7 μm (Meeroff and Englehardt, 2001). Filtration through DE decreased suspended solids (SS), phosphorous and total nitrogen (TKN) by 97%, 78% and 24%, respectively. The liquid fraction, representing 77% of initial manure volume, corresponds to the prefiltered swine (PFS) manure #3 in Table 1. It contained 16.46 g/l of DM and had a conductivity of 33.4 mS. To simulate manure at various stages of the concentration process, the liquid fraction was concentrated with RO membranes or diluted with water to produce 7 PFS manures with conductivity ranging from 14.3 to 61.3 mS. Their characteristics are presented in Table 1.

Table 1. Characteristics of the prefiltered swine (PFS) manures fed to the four RO membranes.

Parameter	PFS manure						
	1	2	3	4	5	6	7
EC (mS)	14.3	23.8	33.4	41.3	51.6	54.2	61.3
DM (g/)	4.99	8.86	16.46	27.70	31.01	33.76	41.13
DM vol. (g/l)	1.89	3.80	7.63	14.88	13.68	16.92	21.18
TDS (g/l)	4.49	7.49	13.91	22.88	26.96	29.86	36.29
SS (mg/l)	503	1374	2550	4821	4050	3895	4840
VSS (mg/l)	388	1055	2070	3561	2945	2452	2930
TAN (mg/l)	1646	2982	4393	5734	7254	8174	9169
TKN (mg/l)	1669	3155	4929	6433	7957	8543	10317
K (mg/l)	1000	2004	2670	4300	5734	6477	7181
P (mg/l)	83	195	242	395	394	738	979
pH	8.23	7.98	7.98	8.43	8.61	8.41	8.42
Alk. (gCaCO$_3$/l)	5.00	11.10	16.25	21.83	27.38	30.00	33.50
VFA (g/l)	2.91	2.07	16.14	23.75	26.76	29.18	29.82

EC: electrical conductivity; DM: dry matter; DM vol: volatile DM; TDS: total dissolved solids; SS: suspended solids; VSS: volatile SS; TAN: total ammonia-nitrogen; TKN: total Kjeldahl Nitrogen; Alk: alkalinity; VFA: volatile fatty acid

The four polyamide RO membranes used in this experiment are presented in Table 2. Their maximum allowable pressure ranged from 4.1 to 8.3 MPa. Membrane RO1 is a low fouling composite membrane especially marketed for municipal wastewater treatment. Membrane RO2 and RO3 are designed for brackish water treatment. They have a slightly lower nominal rejection capacity than the other two membranes. Membrane RO4 is a high pressure seawater membrane, with a high rejection capacity.

Table 2. Membranes used in the study.

Membrane	Salt rejection[a] (%)	Max. allowable pressure (MPa)[a]
RO1	99.7	4.1
RO2	99.4	4.1
RO3	99.4	6.9
RO4	99.75	8.3

[a]Data provided by the manufacturer

The filtration runs were conducted with the laboratory-scale system presented in Figure 1. It consisted in four cells operated in parallel. The 11.6-cm^2 membrane coupons were placed between two detachable sections of the cells, with the membrane support side lined against a stainless steel porous plate, through which the permeate flowed at atmospheric pressure. The feed entered at the bottom right end of the cell cavity (7.1 cm^3) at 0.9 l/min and exited near the membrane surface at the far left. Initial PFS manure volume was approximately 9 l and feed temperature was maintained at $21\,°C \pm 2\,°C$ throughout the experiments.

Because of their similarities in allowable pressure range, membranes RO1 and RO2 were tested concurrently, with two replicate cells per membrane type, while RO3 and RO4 were tested together. With each of the 7 PFS manures, the system was operated at various pressures, up to 4.1 MPa for RO1/RO2 and 6.3 MPa for RO3/RO4, for about 1.5 h at each pressure. The permeate collected during a test at one pressure represented less 2% of the feed and was returned to the feed vessel to prevent concentration effect. The pressures were chosen at random to avoid fouling effect due to increasing pressure on the membrane, and tests were repeated at some of the pressures.

Distilled water flux was measured with all new coupons. Results were used to calculate intrinsic membrane resistance (R_m in m^{-1}), as followed:

$$J_w = \frac{\Delta P_t}{\mu R_m} \tag{1}$$

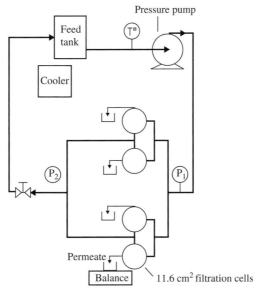

Figure 1. Schematic diagram of the laboratory scale filtration pilot used in the study. P and T stand for pressure and temperature gauges, respectively.

where J_w is distilled water flux (m/s), ΔP_t is applied pressure (Pa), and μ is the viscosity of the permeate (Pa-sec). Intrinsic resistance was measured at pressures of 2.7 and 4.1 MPa for membranes RO1/RO2 and RO3/RO4, respectively. Previous tests in our laboratory showed that resistance was similar at all pressures within the range used in this experiment. Distilled water flux was also measured after 3 to 4 test runs to monitor changes in membrane permeability. New coupons were installed if permeability had dropped by more than 10% and each time a new PFS manure was tested. Average R_m values for new coupons and after test runs are presented in Figure 2. Results indicated that there was not significant increase in membrane resistance during tests.

RESULTS AND DISCUSSION

Transmembrane flux and osmotic pressure

Figure 3 presents PFS manure flux with respect to applied pressure (ΔP_t) through the four membranes. Osmotic pressure ($\Delta\pi$) was calculated by extending the linear portion of the flux curves to the x-axis. With RO1 and RO2, the linearity between flux and applied pressure broke down at an effective pressure

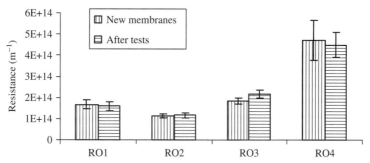

Figure 2. Membrane resistance (R_m) before and after test runs.

($\Delta P_t - \Delta \pi$) of about 2 MPa. At conductivities of 14.3, 23.8 and 33.4 mS, flux through RO1 and RO2 reached a maximum value, at which it became independent of applied pressure. Increasing concentration polarization at the membrane surface probably increased osmotic pressure at the same rate as applied pressure. At conductivities > 50 mS, flux through RO1 and RO2 increased linearly with pressure throughout the applied pressure range, because the effective pressure remained below 2 MPa due to high osmotic pressure.

With RO3, flux tended to deviate from linearity when effective pressure became higher than approximately 2.5 MPa. However an absolute maximum flux was never reached within the applied pressure range. Membrane RO3 produced the highest fluxes registered during this experiment. With RO4, flux tended to increase linearly with applied pressure up to 6.2 MPa at all PFS manure conductivities. However, membrane RO4 was not tested at its full pressure capacity (8.3 MPa).

Figure 4 presents osmotic pressure with respect to manure conductivity. Osmotic pressure was similar for all membranes except at 61.3 mS with RO1 and RO2. A that conductivity level, an average osmotic pressure of 3.5 MPa was measured with RO3 and RO4. Osmotic pressure was thus close to the pressure limit of RO1 and RO2 (4.1 MPa) and the two membranes presented very low flux at all applied pressures at that conductivity (Figure 3).

The osmotic pressure data (except values obtained with RO1/RO2 at 61.3 mS) were fitted to the empirical osmotic pressure equation:

$$\Delta \pi = aC + bC^2 + cC^3 + \ldots \tag{2}$$

where a, b, c, ... are the virial coefficients and C is solute concentration (Nikolova and Islam, 1998). The value of C was replaced by PFS manure conductivity (mS), but other parameters such as DM, TDS, TAN, potassium and

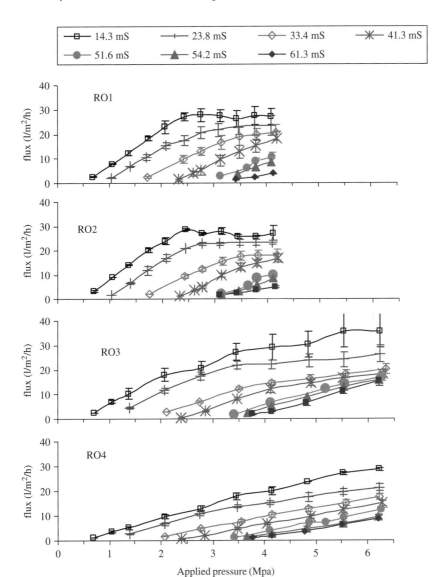

Figure 3. Transmembrane flux through the four reverse osmosis membranes during the filtration of prefiltered swine manure at various concentration level. The error bars present one standard deviation on either side of the average value of two to four replicates.

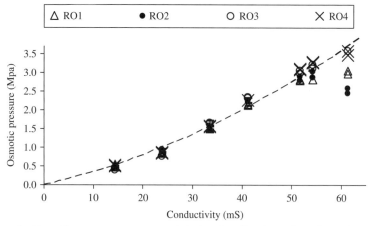

Figure 4. Osmotic pressure measured during the filtration of prefiltered manure at various concentration levels. Results from the two duplicate cells are presented for each membrane.

alkalinity concentrations, which were highly correlated ($p \geq 0.99$) to conductivity, could have been used. Equation 2 states that osmotic pressure reaches 0 at a concentration of 0, and if all coefficients are positive, the rate of increase in osmotic pressure accelerates as concentration increases. Nikolova and Islam (1998) suggested that the first virial coefficient (a) was sufficient to characterized osmotic pressure at dextran (polysaccharide of glucose) concentration up to 100 g/l in water. Yamada *et al.* (1991), on the other hand, reported that a third order equation was necessary to model osmotic pressure at protein (albumin) concentrations ranging from 0 to 180 g/l. With a complex medium such as manure and within the concentration range of this experiment, the first and second order coefficients of Equation 2 were necessary to describe osmotic pressure, with a = 0.0005 MPa/mS^2, b = 0.0297 MPa/mS and the correlation coefficient (R^2) = 0.98.

In full-scale RO systems, osmotic pressure increases along the length of membrane elements as permeate is removed and feed is being concentrated (Song *et al.*, 2003). Results of this experiment suggested that the osmotic pressure would reach the maximum pressure that can be applied on membranes RO1/RO2 (4.1 MPa), RO3 (6.9 MPa) and RO4 (8.3 MPa) at PFS manure conductivities of 66, 92 and 103 mS, respectively, corresponding to TAN concentrations of 9.8, 14.0 and 15.8 g/l, respectively. These values represent the maximum levels that should be allowed in the concentrate collected at the end of the RO system, in order to maintain flow all along the membrane elements, operated at maximum allowable pressure.

Resistance and concentration polarization

The resistance-in-series model is widely used to describe transmembrane flux through membranes:

$$J_s = \frac{\Delta P_t - \Delta \pi}{\mu(R_m + R_s + R_f)} \tag{3}$$

where J_s is transmembrane flux (m/s), $\Delta \pi$ is the osmotic pressure at the membrane surface (Pa), R_s is the resistance associated to the particular substrate being processed, and R_f represents the resistance due to pressure- and time-related fouling (m^{-1}). The substrate-associated resistance has been attributed to specific membrane-solute interactions, such as macromolecule adsorbtion on the membrane surface, even in the absence of flow (Nikolova and Islam, 1998). Chiang and Cheryan (1986) found that the value of R_s for ultrafiltration membranes filtering skim milk was a constant over a range of crossflow velocities (0.34 to 1.11 m/s), temperatures (40°C to 60°C), and protein concentrations (3.1% to 11.5%). Nikolova and Islam (1998), on the other hand, observed that the value of R_s was proportional to dextran concentrations ranging from 3 to 50 kg/m^3.

Figure 5 present the resistances $R_s + R_f$ for the four membranes at three conductivities. Results showed high variabilities, but within each conductivity, resistance values tended to be independent of pressure throughout the tested range for RO4, below an effective pressure of 2 MPa for membranes RO1/RO2 and below an effective pressure of 3 MPa for RO3. It was assumed these values represented the resistance related to the substrate being filtered (R_s) as opposed to fouling-related resistance (R_f). The increased in resistance values above 2 and 3 MPa for RO1/RO2 and RO3, respectively, was probably mostly due to increased concentration polarization, and could be represented as an increase in osmotic pressure if concentration at the membrane surface could be measured.

Figure 6 presents average R_s values with respect to conductivity. The R_s values were similar for RO1 and RO2. At conductivity $\leq$ 54.2 mS, R_s tended to increase with conductivity, at low rates of 7.1×10^{12} and 7.5×10^{12} m^{-1}/mS for RO1 and RO2, respectively. In absence of time-induced resistance, such as cake build up, the decrease in flow through RO1 and RO2 during the concentration of PFS manure would be mostly attributable to increased osmotic pressure, as opposed to increased substrate-related resistance. The large increase in R_s at a conductivity of 61.3 mS was mostly due to the low osmotic pressure calculated for RO1 and RO2 at that conductivity (Figure 4). Other phenomenon must have contributed to this increase.

Membranes RO3 and RO4 also had similar R_s values (Figure 6). For both membranes, there was an abrupt increase in R_s when PFS manure conductivity

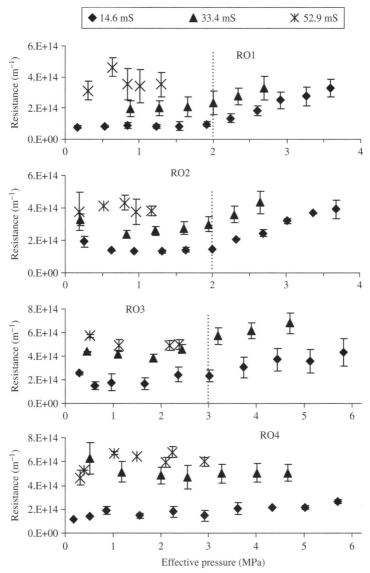

Figure 5. Membrane resistances $R_s + R_f$ calculated using Equation 3 with respect to effective pressure ($\Delta P_t - \Delta \pi$) at three conductivities of prefiltered swine manure. The error bars present one standard deviation on either side of the average value of two to four replicates.

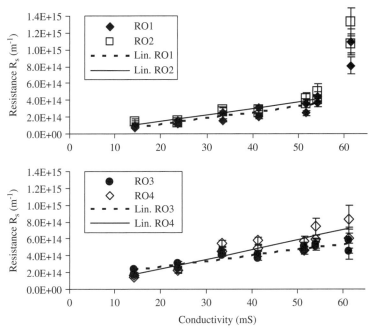

Figure 6. Substrate-related resistance (R_s) as a function of prefiltered swine manure conductivity. The error bars present one standard deviation on either side of the average value of the two cells and all pressures below effective pressures of 2 and 3 MPa for RO1/RO2 and RO3, respectively, all pressures for RO4.

was increased from 23.8 to 33.4 mS. Thereafter, the value of R_s tended to stabilize or increase very slowly with conductivity. At conductivities ≤54.2 mS, R_s was 67% ± 18% higher for RO3/RO4 than RO1/RO2. However, if the objective is to concentrate manure in small volumes with high TAN concentrations, RO systems have to equipped with membranes that are able to sustain high applied pressures, because the decrease in flow due to increased osmotic pressure will be substantial.

REFERENCES

Bilstad, T., Madland, M., Espedal, E. and Hanssen, P.H. (1992). Membrane separation of raw and anaerobically digested pig manure. *Water Sci. Technol.*, **25**, 19–26.
Chiang, B.H. and Cheryan, M. (1986). Ultrafiltration of skim milk in hollow fiber. *J. Food Sci.*, **51**(2), 340–344.

Forbes, E.G.A. Easson, D.L., Woods, V.B. and McKervey, Z. (2008). An evaluation of manure treatment systems designed to improve nutrient management, Agri-Food and Biosciences Institute, Occasional publication no. 5, Hillsborough, Northern Ireland, BT26 6DR, 2005. Available at: http://www.afbini.gov.uk/gru-report5-manure-treatment-systems.pdf, accessed November 3, 2008.

Gellings, C.W. and Parmenter, K.E. (2004). Energy efficiency in fertilizer production and use. in Gellings, C.W. and Blok, K. (eds). *Efficient Use and Conservation of Energy*, Eolss Oxford, UK.

Masse, L., Masse, D.I. and Pellerin, Y. (2007). The use of membranes for the treatment of manure: a critical literature review. *Biosystems Eng.*, **98**(4), 371–380.

MENVQ, 2003. Synthèse des informations environnementales disponibles en matière agricole au Québec, Direction des politiques du secteur agricole, Ministère de l'Environnement, Québec, Envirodoq ENV/2003/0025.

Meeroff, D.E. and Englehardt, J.D. (2001). Precoat filtration and ultrafiltration of emulsified bitumen from water. *J. Environ. Eng. ASCE*, **127**, 46–53.

Minhalma, M. and De Pinho, M.N. (2002). Development of nanofiltration/steam stripping sequence for coke plant wastewater treatment. *Desalination*, **149**(1–3), 95–100.

Nikolova, J.D. and Islam, M.A. (1998). Contribution of adsorbed layer resistance to the flux-decline in an ultrafiltration process. *J. Membr. Sci.*, **146**(1), 105–111.

Song, L.F., Hu, J.Y., Ong, S.L., Ng, W.J., Elimelech, M. and Wilf, M. (2003). Emergence of thermodynamic restriction and its implications for full-scale reverse osmosis Processes. *Desalination*, **155**(3), 213–228.

Steinfeld, H., Gerber, P., Wassenaar, T., Castel, V., Rosales, M. and de Haan, C. (2006). Livestock long shadow – Environmental issues and options, FAO, Rome.

Thorneby, L., Persson, K. and Tragardh, G. (1999). Treatment of liquid effluents from dairy cattle and pigs using reverse osmosis. *J. Agric. Eng. Res.*, **73**(2), 159–170.

Van Der Hoek, K.W. (1998). Nitrogen efficiency in global animal production. *Environ. Pollut.*, **102**(Suppl. 1), 127–132.

Vitousek, P.M., Aber, J.D., Howarth, R.W., Likens, G.E., Matson, P.A., Schindler, D.W., Schlesinger, W.H. and Tilman, D.G. (1997). Human alteration of the global nitrogen cycle: Sources and consequences. *Ecol. Appl.*, **7**(3), 737–750.

Yamada, S., Grady, M.K., Licko, V. and Staub, N.C. (1991). Plasma protein osmotic pressure equations and nomogram for sheep. *J. Appl. Physiol.*, **71**(2), 481–487.

Zhang, R.H., Yang, P., Pan, Z., Wolf, T.D. and Turnbull, J.H. (2004). Treatment of swine wastewater with biological conversion, filtration, and reverse osmosis: a laboratory study. *Trans. ASAE*, **47**(1), 243–250.

P and N in solids from manure separation: separation efficiency and particle size distribution

Karin Jorgensen[a]*, Maibritt Hjorth[b] and Jakob Magid[a]

[a]Karin Jorgensen and Jakob Magid, Department of Agriculture and Ecology, Faculty of Life Sciences, University of Copenhagen, Denmark
[b]Maibritt Hjorth, Department of Agricultural Engineering, Faculty of Agricultural Sciences, University of Aarhus, Denmark

*Corresponding author. Tel.: +45-3533-3412; E-mail address: karinj@life.ku.dk

Abstract Implementation of animal manure solid separation gives farmers in areas with high intensive livestock production an opportunity to reduce their overall environmental impact. Separation of slurry into a nutrient-rich solid fraction and a nutrient-poor liquid fraction is often evaluated on its efficiency of removing nutrients into the solid fraction. However, there are significant differences in the nutrient removal efficiency between different technologies. Hence, the produced solid fractions may vary in nutrient content and nutrient distribution dependent on the applied separation technology.

Swine and dairy cow slurries were separated with five different technologies (sedimentation, centrifugation, filtration, flocculation + drainage and coagulation + flocculation + drainage) in laboratory scale to simulate separation technologies in commercial operation. Each of the slurries and the solid fractions were after separation sieved into four different size categories (>1000 µm, 1000–250 µm, 250–25 µm and <25 µm) in which the P concentration was measured in order to determine the P distribution in the different particle sizes.

The DM-percent in the solid fraction decreased in the following order: filtration, centrifugation, coagulation + flocculation + drainage = flocculation + drainage and sedimentation. The removal efficiencies of N and P into the solid fraction by the five technologies decreased in the following order coagulation + flocculation + drainage, flocculation + drainage, sedimentation = centrifugation and filtration. The P distribution in the solid fractions was dependent on the efficiency of the separation technologies to separate the smallest particles from the liquid into the solid fraction. The low technology separations (sedimentation, centrifugation and filtration) produced solid fractions with a P distribution particle sizes similar to the untreated manures and the chemical separations (flocculation + drainage and coagulation + flocculation + drainage) produced solid fractions with an increased association of P to particles >25 µm.

INTRODUCTION

Intensification of livestock production into larger farms brings along challenges with handling of the large amount of produced slurry. Often the most limiting factor for increasing the livestock production in Denmark and other intensive livestock producing countries is the lack of farmland on which to spread the manure which may cause potential environmental problems. Separation of slurry into a nutrient-rich solid fraction and a nutrient-poor liquid fraction gains footing in the agricultural sector as a relatively inexpensive environmental friendly technology that alleviates the nutrient surplus in the livestock production, through facilitating transportation of nutrients in the solid fraction over longer distances due to reduced weight and volume.

Slurry separation is evaluated by its separation efficiency that expresses the removal efficiency of a certain compound (dry matter, nitrogen and phosphorus) into the solid fraction. Several slurry separation technologies are available, e.g. screwing presses, decanting centrifuges, chemical pre-treatment followed by filtering, sieving or screwing presses all working with different separation efficiencies of the nutrients N and P (Moller *et al.*, 2002). Removal of N and P into the solid fraction is therefore dependent on the applied separation technology and the efficiency of the single technologies on the individual compounds.

Phosphorus in slurry is associated with the smaller sized particles (<50–125 μm) (Masse *et al.*, 2005; Meyer *et al.*, 2007) which may produce solid fractions with varying amounts of P due to the efficiency of the separation technology to separate the smaller particles into the solid fraction. By separation of slurry using chemical treatment with polyacrylamide a large amount of smaller P rich particles is agglomerated into larger particles producing a solid fraction rich in P (Hjorth *et al.*, 2008). Knowledge about P distribution in various particle sizes of the solid fractions is interesting with an agronomic point of view because the fertilizer efficiency of P may be affected by the partitioning of P in different particle sizes. This may furthermore be important knowledge prior to industrial purification and utilization of the valuable P in the solid fractions.

Thus, the aim of this study was to obtain a better understanding of 1) the removal efficiencies of dry matter, nitrogen and phosphorus into the solid fraction by 5 different separation technologies on swine and dairy cow slurry and 2) the distribution of phosphorous in various particles sizes of the raw slurries and the solid fractions produced after separation.

MATERIALS AND METHODS

Untreated slurry was collected at a commercial slaughtering pig farm and at a commercial dairy cow farm. Five separation strategies were applied on the slurries at laboratory scale to simulate commercial available separation technologies: 1) sedimentation for 24 hours (possible only for swine manure) simulating gravity settling during storage, 2) centrifugation at 3.500 *g* for 1 min simulating a decanting centrifuge (Moller *et al.*, 2007), 3) filtration through a 1 mm filter at 6 bars pressure for 2 min simulating a screw press (Moller *et al.*, 2002), 4) flocculation and drainage (floc + drain) by adding polyacrylamide (linear very high molecular weight 40% charge density and cationic, superfloc C-2260, Kemira Kemwater, Finland) as a flocculating agent followed by drainage through a 0.2 mm filter for 30 min, and 5) coagulation, flocculation and drainage (coa + floc + drain) was carried out by adding $FeCl_3$ to the manure in addition to the flocculation + drainage method described above in no. 4, simulating two available commercial manure separators (Kemira Kemwater, Finland). All separations were carried out in triplicates. Removal efficiencies into the solid fraction can be compared by calculation of a simple separation index (Equation 1) (Svarovsky, 1985)

$$E_t(x) = \frac{m(x)_{solid}}{m(x)_{slurry}} \tag{1}$$

where m = the mass of the compound x and E_t = the removal efficiency index of the compound x into the solid fraction. This however gives no indication of the transfer of nutrients or DM into the solid fraction. The increase in concentration of nutrients in the solid fraction was considered in the reduced efficiency index (Equation 2) (Svarovsky, 1985).

$$E_t'(x) = \frac{E_t(x) - R_f}{1 - R_f} \tag{2}$$

where E_t' = the reduced separation efficiency index of the compound x, E_t = the simple separation index of x and Rf = m(solid)/m(slurry) (mass of solid fraction to mass of slurry ratio). This index was used to express the removal efficiency of DM, total N and total P into the solid fraction. E_t' becomes 0 when no separation of x takes place and 1 when the separation is complete.

Untreated slurries and the solid fractions from separation were subjected to size distribution into the following 4 size categories: >1000 µm, $1000-250$ µm, $250-25$ µm and <25 µm. The size distribution was done by sieving the manures in a water-jet sieving device, in which a spraying arm with 34 nozzles was forced to rotate over each sieve by the water pressure of the jet (Moller *et al.*, 2002). Total Kjeldahl N was measured in the untreated slurries and in the solid and liquid fractions after separation. Total P was measured in the untreated slurries, the liquid and the solid fraction after separation, and the 4 size categories of the solid fraction. A dried sub-sample of each size fraction and of the untreated solid fractions was digested in concentrated HNO_3 and H_2O_2 at

120°C on a MOD block digestion system (CPI International, Netherlands). Total P was determined in the digest at a spectrophotometer (Hitachi U-2000 Spectrophotometer, USA) with ammonium molybdate and ascorbic acid as reagents (Murphy and Riley, 1962).

RESULTS AND DISCUSSION

Separation efficiencies

One of the purposes of this study was to investigate the separation efficiencies of the 5 different separation technologies applied on swine and dairy cow slurry. The aim of applying slurry separation was to obtain the highest possible removal of nutrients into a solid fraction with the highest possible DM content (Table 1).

Table 1. DM contents in % of wet weight of the untreated slurries and the solid fractions after 5 different separations. The results are presented as means of three replicates and standard deviations in parenthesis. Note the high DM% obtained in the solid fraction at filtration and the low DM% at sedimentation.

Manure type	Untreated slurry	Sedimen- tation	Centrifu- gation	Filtration	Floc + drain	Coa + floc + drain
Swine	5.6	9.0	15.8	20.4	13.5	14.3
DM (%)	(0.6)	(0.1)	(0.8)	(1.0)	(0.6)	(1.9)
Dairy cow	9.2		15.1	23.0	11.3	10.0
DM (%)	(0.1)		(0.1)	(0.1)	(1.0)	(0.7)

The removal efficiencies of DM, total N and total P as reflected by the reduced separation indexes are presented in Table 2. Generally there were no large differences between the reduced separation efficiencies of swine and dairy cow slurry. The DM separation efficiency was lowest for the three mechanical separations, sedimentation, centrifugation and filtration whereas the efficiencies by the chemical separations were very much higher. E.g. by coagulation, flocculation and drainage of swine slurry 57% of the DM (Table 2) were in the solid fraction with a DM content of 14.3% (Table 1). In contrast only 14% of the DM was in the solid fraction with a DM content of 20.4% after filtration.

The reduced separations indexes for N followed the same pattern as for DM removal (Table 2) which may be due to separation of the organically bound N in the slurry. Only a minor part of the total N from the slurry is retained in the solid fraction, because much of the nitrogen in the slurry is in the form of dissolved ammonium-N (Sommer and Husted, 1995). Between 10–13% of the

Table 2. Reduced separation indexes of DM, total N and total P into the solid fraction by 5 different separations of swine and dairy cow manure. The results are presented as means of 3 replicates and standard deviations in parenthesis. A reduced separation index equals 0 when no separation of the compound occurs and 1 when the separation is complete. E.g. reduced separation index of e.g. 0.75 of a compound means that 80% of the total amount of the compound is located in the solid fraction that constitutes of 20% of the total volume slurry.

Manure type	Sedimen- tation	Centrifu- gation	Filtration	Floc + drain	Coa + floc + drain
			--DM--		
Swine	0.13 (0.05)	0.17 (0.02)	0.14 (0.03)	0.46 (0.07)	0.57 (0.04)
Dairy cow	–	0.10 (0.01)	0.14 (0.03)	0.50 (0.06)	0.51 (0.02)
			total N		
Swine	0.10 (0.02)	0.13 (0.02)	0.03 (0.005)	0.40 (0.04)	0.37 (0.06)
Dairy cow	–	0.12 (0.04)	0.07 (0.01)	0.39 (0.02)	0.40 (0.02)
			total N		
Swine	0.77 (0.02)	0.70 (0.02)	0.15 (0.02)	0.91 (0.02)	0.98 (0.004)
Dairy cow		0.70 (0.03)	0.37 (0.02)	0.91 (0.01)	0.93 (0.01)

N was transferred into the solid fraction by sedimentation or centrifugation (noticing that the volume of the solid fraction from sedimentation was 3 times higher than after centrifugation) of swine and dairy cow slurry whereas the filtration was not very efficient in removal of N, which was the same level as found by Moller et al. (2000). This indicates that centrifugation the most efficient of the mechanical separation technologies (sedimentation, centrifugation, filtration) in other to remove N into a solid fraction with the smallest possible volume and a relatively high DM content.

The highest removal efficiencies of N into the solid fraction were found for the chemical separations flocculation + drainage and coagulation + flocculation + drainage of both swine and dairy cow slurry ($E_t'(P) = 0.37$–0.40). The removal efficiencies of P into the solid fraction were generally much higher than for N due to the direct association between P and the particles in slurry.

P in slurry is especially associated smaller particles (<50 μm) and therefore is the separation efficiencies of P significantly improved by the flocculation + drainage and the coagulation + coagulation + drainage separations. Addition of polyacrylamide to the slurry increases the effective particle size by agglomeration of small particles into larger particles ('flocs') that separates from the liquid and dewaters more readily. These larger sized particles can significantly enhance manure solids retention by screens and separation of colloidal particles by settling (Hjorth *et al.*, 2008; Vanotti *et al.*, 2002). The P removal efficiency by the centrifuges was also relatively high which is also due to its approved affinity to remove smaller P containing particles (Moller *et al.*, 2002). Centrifugation removed P with the same efficiency as sedimentation but the volume of the solid fraction was 3 times higher for the sedimentation than for centrifugation, making sedimentation less interesting with a farmers point of view in order to remove P into a the smallest possible volume solid fraction. Filtration was the least efficient separation technology for P removal which may be explained by the reduced retention of smaller particles by the used 1 mm filter.

P distribution in particle sizes

The distribution of P in various particle sizes of the solid fractions was investigated in this study to obtain a better understanding of the mechanisms of P removal by different separation technologies. The distribution of P among particle sizes both in the untreated and the solid fractions of dairy cow and pig slurry was very identical (Figure 1). The proportion of P associated with particles larger than 25 μm was generally higher for untreated cow slurry compared to untreated pig slurry, which may be due to the difference in feed composition and digestion of feed by dairy cow vs. pigs, which will not be discussed further in the current study.

The distribution of P associated to particles in the solid fractions from sedimentation, centrifugation and filtration was very similar to the P distribution in the untreated manures because the original properties of the manures were not changed during separation. Centrifugation of both swine and dairy cow slurry produced a solid fraction with an increased amount of P associated to particles <25 μm compared to filtration, because of the increased 'position' of the smaller particles in the solid fraction. A significant increase of the P associated to the particles larger than 25 μm was observed in the chemically separated solids, due to the agglomeration of the particles <25 μm into the larger flocs by addition of polyacrylamide. There was a slight increase in P bound to particles between 25 and 250 μm in the solids from coagulation + flocculation + drainage separation compared to the solids from flocculation + drainage due the

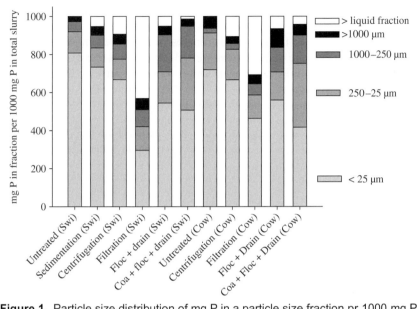

Figure 1. Particle size distribution of mg P in a particle size fraction pr 1000 mg P in total slurry, which was determined for the slurries and the solid fractions of separated swine and dairy cow slurry. The results are normalized to 1000 mg P in the slurry prior to separation and the P separation efficiency of the 5 different separation technologies are taken into account.

improved P removal efficiency by addition of coagulating agent ($FeCl_3$). Addition of $FeCl_3$ causes both coagulation, i.e. charge neutralization of the negatively surface charged slurry by the Fe^{3+} ions, and precipitation of phosphate that by addition of polyacrylamide may be absorbed in larger particles ('flocs') (Hjorth *et al.*, 2008). The formation of stable flocs may furthermore improve the functioning of the flocs as a filter for smaller particles during the drainage.

CONCLUSION

The five different technologies used in the current study were applied to simulate solid-liquid separation technologies in commercial operation. The most effective removal of N and P from slurry into the solid fraction was obtained using a chemical separation with flocculation, coagulation and drainage. The P distribution in the solid fractions was dependent on the efficiency of the separation technology to separate the smallest particles from the liquid into the solid fraction. The non-chemically enhanced technology separations (sedimentation, centrifugation and filtration) produced a solid fraction with a P distribution particle sizes similar to the untreated manures, while the chemical separations (flocculation + drainage and coagulation + flocculation + drainage) produced a solid fraction with an increased association of P to particles >25 μm.

REFERENCES

Hjorth, M., Christensen, M.L. and Christensen, P.V. (2008). Flocculation, coagulation, and precipitaion of manure affecting three separation techniques. *Bioresource Technology*, **99**, 8598–8604.

Masse, L., Masse, D.I., Beaudette, V. and Muir, M. (2005). Size distribution and composition of particles in raw and anaerobically digested swine manure. *Transactions of the Asae*, **48**, 1943–1949.

Meyer, D., Ristow, P.L. and Lie, M. (2007). Particle size and nutrient distribution in fresh dairy manure. *Applied Engineering in Agriculture*, **23**, 113–117.

Moller, H.B., Hansen, J.D. and Sorensen, C.A.G. (2007). Nutrient recovery by solid-liquid separation and methane productivity of solids. *Transactions of the Asabe*, **50**, 193–200.

Moller, H.B., Lund, I. and Sommer, S.G. (2000). Solid-liquid separation of livestock slurry: efficiency and cost. *Bioresource Technology*, **74**, 223–229.

Moller, H.B., Sommer, S.G. and Ahring, B.K. (2002). Separation efficiency and particle size distribution in relation to manure type and storage conditions. *Bioresource Technology*, **85**, 189–196.

Murphy, J. and Riley, J.P. (1962) A modified single solution method for determination of phosphate in natural waters. *Analytica Chimica Acta*, **27**, 31–36.

Sommer, S.G. and Husted, S. (1995). The chemical buffer system in raw and digested animal slurry. *Journal of Agricultural Science*, **124**, 45–53.

Svarovsky, L. (1985). Solid-liquid separation processes and technology. In: Handbook of Powder Technology. Eds. J C Williams and T Allen, pp. 18–22. Bradford, UK.

Vanotti, M.B., Rashash, D.M.C. and Hunt, P.G. (2002). Solid-liquid separation of flushed swine manure with PAM: Effect of wastewater strength. *Transactions of the Asae*, **45**, 1959–1969.

Treating solid dairy manure by using the microwave-enhanced advanced oxidation process

Anju A. Kenge, P.H. Liao and K.V. Lo*

*Department of Civil Engineering, University of British Columbia, Vancouver, B.C. Canada

Abstract The microwave enhanced advanced oxidation process was demonstrated in this paper that it could be used to treat separated solid dairy manure for nutrient release and solids reduction. It was conducted at a microwave temperature of 120°C for 10 minutes, and a hydrogen peroxide dosage approximately 2 mL per 1% TS for a 30 mL sample. Three pH conditions of 3.5, 7.3 and 12 were examined to find the favorable conditions for the nutrient release. The results indicated that substantial quantities of nutrients could be released into the solution at pH of 3.5. However, at neutral and basic conditions only volatile fatty acids and soluble chemical oxygen demand could be released. The analyses on orthophosphate, soluble chemical oxygen demands and volatile fatty acids were re-examined for dairy manure. It was found that the orthophosphate concentration for untreated samples at a higher %TS was suppressed and lesser than actual. To overcome this difficulty, the initial orthophosphate concentration had to be measured at 0.5% TS.

INTRODUCTION

The main challenge for dairy industry is to sustain and increase its production without degrading the environment. The management of dairy manure becomes an issue of considerable concern for the industry. The conventional management practice of dairy manure is for land application. It can pose problems like loss of nutrients from the soil to ground water and streams. These nutrients could enter water bodies, causing eutrophication, thus, resulting in water pollution. Furthermore, it is coming under environmental and regulatory scrutiny due to the limited amount of land available for manure disposal. Dairy manure contains a large quantity of labile carbon, such as fats, proteins, carbohydrates, and lignin, as well as nitrogen and phosphorus. It can be considered as a valuable bioresouce for recovery, instead of waste materials (Rico *et al.*, 2007). However, most of those nutrients and organic matters are contained in solid dairy manure, which are either not biodegradable or degrade very slowly.

For example, up to 65% of phosphorus is in an organic form, which is tied up in the solid (Barnett 1994). With proper pretreatments the solid dairy manure can be used as a feedstock for producing value-added products, such as volatile fatty acids, sugar and others.

The measurement of water extractable phosphorus (WEP) in manure has not been well established. To assess the WEP content in dairy manure, various extraction methods were used (Kleinman *et al.*, 2002; Chapuis-lardy *et al.*, 2003; Gungor *et al.*, 2005; Wolf *et al.*, 2005). Varying dry matter/distilled water ratio (1 to 20:200) revealed that the greater dilution of manure dry matter increased WEP (Chapuis-lardy *et al.*, 2003; Gungor *et al.*, 2005). Using repeated water extraction method, it was reported that there was a sharp initial increase in phosphorus extractability, and up to 70% of total phosphorus was extracted after third extraction out of seven (Gungor *et al.*, 2005). The experiments conducted in this laboratory showed that the same sample under different dilutions gave uncertain orthophosphate results (Pan, 2005). Four sets of manure and distilled water (1:0, 1:1, 1:4 and 1:9) were used to extract orthophosphate from diluted dairy manure. It revealed that a greater dilution of liquid dairy manure increased orthophosphate concentration. It was, however, found that it was due to a positive interference, a higher ratio of liquid/manure gave a higher interference. It was also reported that significant errors in phosphorus analysis by the colori-metric method were observed in soil and plant materials, the resulting values were not accurate (Kowalenko and Babuin 2008). The manure WEP method was proposed that it should be measured at the solution of 0.5% TS using either inductively coupled plasma (ICP) or colorimetric method (Wolf *et al.*, 2005). These two methods were highly correlated; the colorimetric procedure was approximately 7% higher than phosphorus measured by ICP. Pan also reported that it had less interference for orthophosphate determination at near 0.5% TS of liquid dairy manure using the colorimetric method (Pan, 2005).

The microwave-enhanced advanced oxidation process (MW/H_2O_2-AOP) uses microwave radiations to heat the waste and a strong oxidant like hydrogen peroxide to produce hydroxyl free radical. This process has first been developed for the treatment of sewage sludge. It is a very dynamic and efficient means to covert a significant portion of suspended solids into soluble organic compounds, leaving reduced amounts of sludge suspended solids in solution. Nutrient, such as nitrogen, phosphorus and metals are also solubilized in the process (Chan *et al.*, 2007). This study was aimed at exploring the use of the MW/H_2O_2-AOP to extract valuable products from dairy manure.

The objectives of this preliminary study were 1) to examine the effectiveness of the MW/H_2O_2-AOP on separated solid dairy manure, and 2) to re-examine the results obtained from chemical measurements for dairy manure, particularly for orthophosphate analysis.

MATERIALS AND METHODS

Apparatus and substrate

The Milestone Ethos TC microwave oven digestion system (Milestone Inc., USA) was used in this study. The solid dairy manure used in this study was obtained from the UBC Dairy Education and Research Centre in Agassiz, British Columbia, Canada. The solid portion of dairy manure obtained after solid-liquid separation was used in this study. Once collected from the farm the manure was stored at 4°C. The concentrations of the dairy manure used in samples were altered by hydrogen peroxide addition. The initial values have been adjusted using a dilution factor, which is calculated as the ratio of dairy manure volume by total sample volume in the vessel during the run. The initial concentrations of the dairy manure used in this part of the study are listed in Table 1.

Table 1. Initial characteristics of dairy manure used in part 1.

Substrate	Solid dairy manure[*]
Total solids (%)	1.6 ± 0.09
pH	7.3 ± 0.6
TCOD (gL^{-1})	32.2 ± 3.8
SCOD (mgL^{-1})	2272 ± 416
TP (mgL^{-1})	84.3 ± 1.8
PO$_4$-P (mgL^{-1})	38.3 ± 4.75
TKN (mgL^{-1})	570.2 ± 19.8
NH$_3$-N (mgL^{-1})	92.4 ± 7.6
NO$_x$ (mgL^{-1})	0.15 ± 0.05
VFA (mgL^{-1})	4.97 ± 1.8

[*]$\pm$ represents standard deviation.

Experimental design

The experiment was divided into two parts. The first part was to examine the effectiveness of the MW/H$_2$O$_2$-AOP on separated solid dairy manure. The second part was to evaluate the reliability of orthophosphate analysis, as well as others for dairy manure.

Part 1

The experimental design for this part is shown in Table 2.

Table 2. Experimental design for part 1.

Set No.	TS (%)	pH	Temp (°C)	H_2O_2 dosage (mL)	Sample volume (mL)	Heating time (mins)
1	1.6	3.5 ± 0.28	120	4	34	10
2	1.6	7.3 ± 0.6	120	4	34	10
3	1.6	12 ± 0.5	120	4	34	10

To examine the effect of pH on release of nutrients from dairy manure using the MW/H_2O_2-AOP, three sets of pH of 3.5, 7.3 and 12 were used in this study. The pH was adjusted to 3.5 by adding a few drops of 30% conc. sulfuric acid, while pH 12 was obtained by using 0.25 mL of 1M NaOH. In order to break down a high level of fibers in solid dairy manure, it was assumed that a high microwave heating temperature and long heating time are required. Therefore, a microwave tempera-ture of 120°C and heating time of 10 minutes was chosen. The ramp time, time taken by the microwave system to reach the set temperature, was 5 minutes, i.e. 20°C/min. Due to the complex nature of the manure, a high hydrogen peroxide dosage of 4 mL for a 30 mL sample volume was used, which was approximately 2 mL per 1% TS for a 30 mL sample. Three replicates were used for each set.

Part 2

This section is divided in two sub-sections. The first sub-section was dealt with the untreated samples (used to obtain the initial concentrations), while the second sub-section was with microwave treated samples. Though the primary reason to do this investigation was orthophosphate, analyses of SCOD and VFA were also verified in some cases.

Several dairy manure samples were generated with different percentage TS to be used in this part of the study. The same samples were also diluted by adding distilled water to obtain 0.5% TS. Since, previous research has proved that the most accurate results could be achieved at 0.5% TS, it was used as a basis for comparison (Chapuis-lardy *et al.*, 2003; Pan, 2005). Four sets of TS were used viz., 0.5, 1.6, 3.4 and 6.1% and the analysis were repeated. The samples had an approximate pH of 8. Three replicates were used.

Table 3 shows the experimental design for the second section of the part 2.

Two sets of TS of 0.5 and 3.4%were used in this part of the study. The samples were treated at 120°C, which can be able to compare with results from part 1.

Table 3. Experimental design for part 2 (microwave treated samples).

Set No.	TS (%)	pH	Temp (°C)	H$_2$O$_2$ dosage (mL)	Sample volume (mL)	Heating time (mins)
1a	0.5	4	120	1	31	10
1b	1.6	3.5	120	4	34	10
1c	3.4	4	120	6	36	10
2a	0.5	4	70	1	31	10
2b	3.4	3.7	70	6	36	10

There were also operated at a low temperature of 70°C to examine if the pattern varied with temperature. All the samples were acidified to an approximate pH of 4 by adding few drops 30% conc. sulfuric acid. The heating time was maintained at 10 minutes, and the ramp time was set as 20°C/min. The hydrogen peroxide dosage was kept as approximately 2 mL per 1% TS for a 30 mL sample. Set 1 (120°C) was analyzed for orthophosphate, soluble chemical oxygen demand (SCOD), and volatile fatty acids (VFA), while Set 2 (70°C) was analyzed only for orthophosphate and SCOD. Three replicates were used for each set.

Analytical procedures

After microwave treatment, the samples were allowed to cool down to the room temperature. The samples were then centrifuged at 3500 rpm for 15 minutes; the supernatant was filtered using 4.5 μm fiberglass filter. The filtered samples were analyzed for SCOD, orthophosphate, soluble ammonia, soluble nitrates and nitrites (NO$_x$) and VFA. A small portion of each treated sample was left unfiltered and used for the total chemical oxygen demand (TCOD) analysis. The initial untreated dairy manure samples were also centrifuged and filtered and analyzed for SCOD, orthophosphate, ammonia, soluble NO$_x$ and VFA, while the unfiltered samples were analyzed for TCOD, total phosphate (TP), total Kjeldahl nitrogen (TKN) and TS. The analytic procedures were in accordance with those enlisted in Standard Methods (1998).

RESULTS AND DISCUSSIONS

This study was an attempt to examine how effectiveness of this process to digest separated dairy manure for the release of useful soluble materials. The results of the nutrient release and solids disintegration from the MW/H$_2$O$_2$-AOP are expressed as SCOD, VFA, orthophosphate, soluble ammonia, and NO$_x$.

Part 1

Soluble phosphate

The results for orthophosphate are summarized in Table 4 and Figure 1. The dairy manure samples had a %TS of 1.6 with an initial orthophosphate concentration of 38.3 mgP/L and a total phosphate concentration of 84.3 mgP/L.

Table 4. Summary of results for part 1.

Set No.	pH	SCOD (gL^{-1})	TCOD (gL^{-1})	PO$_4$-P (mgL^{-1})	NH$_3$-N (mgL^{-1})	NO$_x$ (mgL^{-1})	VFA (mgL^{-1})
1	3.5	2.8 $\pm$ 0.6	9.8 $\pm$ 2.7	65.8 $\pm$ 0.98	208 $\pm$ 3.6	negligible	391 $\pm$ 31.3
2	7.3	2.75 $\pm$ 0.1	7.85 $\pm$ 1	9.7 $\pm$ 0.35	89.2 $\pm$ 4.4	0.6 $\pm$ 0.3	48.6 $\pm$ 0.91
3	12	2.5 $\pm$ 0.3	13.3 $\pm$ 5.5	10.4 $\pm$ 0.29	97.1 $\pm$ 3.2	0.17 $\pm$ 0.07	54.1 $\pm$ 13.4

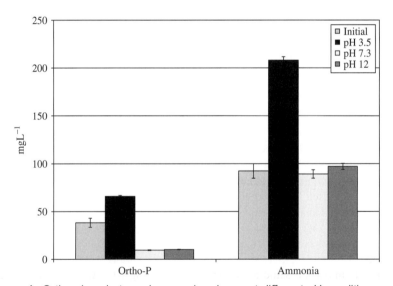

Figure 1. Ortho phosphate and ammonia release at different pH conditions.

At pH 3.5, the soluble phosphate concentration increased to 65.8mgP/L, an increase of 71.7%. A decrease of 74.7 and 72.9% was observed at pH 7.3 and pH 12, respectively. Thus, it could be clearly seen that acid addition was essential for releasing phosphorus from solid dairy manure. The results were similar to a previous study applying MW/H$_2$O$_2$-AOP on liquid dairy manure, wherein, the samples had to be acidified to obtain phosphorus release (Chan *et al.*, 2008). It is interesting to note that treating the sewage sludge, orthophosphate was released irrespective of the pH in the MW/H_2O_2-AOP (Chan *et al.*, 2007). The reason could be that sludge particles are mostly microbial cells and microwave radiation causes cell lysis resulting in release of orthophosphate from the cells into the solution. Whereas solid dairy manure contains a lot of undigested animal feed like straws along with organic matters and lipids. Phosphorus in dairy manure could not be solubilized without H_2SO_4. In fact, the reaction was reversed; it was converted to an insoluble form in the absence of acid as showed in Table 4. It might be due to the formation of struvite, or k-struvite (potassium magnesium phosphate) under neutral or basic condition, as a result, orthophosphate was precipitated out of the solution. This was similar to the previous study that less orthophosphate was obtained in the basic solution for dilute dairy manure (Pan, 2005).

Ammonia and NO_x

The results for soluble ammonia are presented in Table 4 and Figure 1. At TS of 1.6%, the solid dairy manure had an initial soluble ammonia concentration of 92.4 mgN/L and a TKN value of 570 mg/L.

Acid addition helps retain ammonia in the solution (Wong *et al.*, 2007). After microwave treatment at pH 3.5, the ammonia concentration had an increased of 125%. The treatment was found ineffective for samples with pH 7.3 and 12. At pH 7.3, it was of a 3.5% decrease of ammonia in the solution, and a mere 5% increase was observed at pH 12. It could be anticipated that in the absence of acid, high microwave temperature promoted volatilization of ammonia. The results indicate the benefit of using acid as a catalyst for the treatment of dairy manure. It released not only more orthophosphate, but also more ammonia in the solution.

The results for soluble NO_x are presented in Table 4. Initially the samples had a NO_x concentration of 0.15mgN/L. After treatment the acidified samples showed no presence of nitrates and nitrites, while, at pH 7.3, the NO_x concentration increased to 0.6 mgN/L. At pH 12, the NO_x concentration after treatment remained almost the same as the initial at 0.17 mgN/L. Ammonia could have got converted to NO_x when no acid or base was added to the solid dairy manure samples, thus, showing a high NO_x level at pH 7.3.

Volatile fatty acids

The results for VFA are summarized in Table 4 and Figure 2. The initial soluble VFA concentration was observed to be very low at 4.97 mg/L. The biodegradable organic matter in dairy manure will be transformed to VFA at high temperatures in the thermal process (Rico *et al.*, 2007). The VFA was present mainly in the form of acetic acid, however, the other volatile fatty acids such as propionic, butyric, valeric, hexanoic and heptanoic were also found along with acetic acid after the treatment.

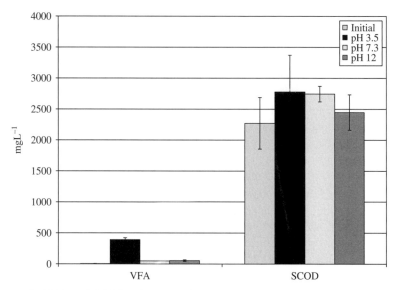

Figure 2. VFA and SCOD release at different pH conditions.

For pH 3.5, after treatment the VFA concentration increased to 391 mg/L, which was a remarkable increase from the initial value. At pH 7.3 and pH 12, the VFA concentration, in terms of acetic acid, increased to 48.6 and 54.1 mg/L, respectively. The results proved that VFA could be released into the solution irrespective of the pH. Acid treated samples released extensive quantities of VFA than those of neutral or base samples. Previous study using MW/H_2O_2-AOP on sewage sludge, also found that under acidic conditions the process produces more VFA Liao *et al.*, 2007).

Soluble COD

The results for soluble COD are summarized in Table 4 and Figure 2. The solid dairy manure samples had an initial soluble COD concentration of 2272 mg/L and initial TCOD concentration of 32.2 g/L.

At pH 3.5, a 22.3% increase was obtained in SCOD concentration after treatment. Increases of 21% and 7.7% were observed at pH 7.3 and 12, respectively.

Samples treated at pH 12 showed the least increase. It can be concluded that the soluble COD concentration was increased irrespective of the pH. After treatment, the TCOD concentrations dropped incredibly for all the three sets. Acid addition promoted more oxidation and gasification causing carbon dioxide to be released as a gas and thus reducing the TCOD value (Chan et al., 2007). In this case, at pH 3.5 and pH 7.3, the decrease in TCOD was the most; while at pH 12 the TCOD concentration was higher than the other two sets. It was suggested that at high pH, it might produce crude soap, therefore, SCOD increase was least among the treatments, and also retained the most TCOD.

The results indicated that the MW/Ht_2O_2-AOP was very effective to treat separated solid dairy manure under acid condition. More studies are underway in this laboratory. In order to obtain valuable-added products from dairy manure, other technologies commonly used in the sewage treatment, such as thermal, chemical, and thermal-oxidative, will also be studied (Camacho et al., 2004; Valo et al., 2004; Wen et al., 2004; Takashima, 2008).

Part 2

Due to interference problems with colorimetric measurement of phosphorus in soil, plant materials and manures, the resulting values might not be accurate (Pan, 2005; Chan et al., 2007). It was suggested that for dairy manure the ortho-phosphate analysis be performed at 0.5%TS, since least interference occurs at this point.

The results for part 2 are summarized in Tables 5a and 5b. Table 5a presents the results for untreated samples i.e. the initial concentrations of SCOD, PO_4 and VFA at TS 0.5, 1.6, 3.4 and 6.1%.

Table 5a. Summary of results for part 2 (untreated samples).

TS (%)	Initial SCOD (mgL^{-1})	Initial PO_4-P (mgL^{-1})	Initial VFA (mgL^{-1})
0.5	381 ± 4.4	12.1 ± 1.5	1.37 ± 1.2
1.6	2272 ± 416	8.8 ± 1	4.97 ± 1.8
3.4	2755 ± 504	31.6 ± 0.5	6.24 ± 1
0.5	489 ± 6.8	6.52 ± 0.05	1.2 ± 0.5
6.1	5520 ± 261	28.1 ± 1	10.5 ± 6.3

Considering that the results obtained at 0.5% TS were accurate for ortho-phosphate analysis. The values obtained at 1.6% of TS were compared to values obtained from 0.5% TS multiplied by 3.2 (ratio of 1.6 by 0.5). If the values obtained at 1.6% TS were close or within the standard deviation, then the analysis result was considered to be correct. For example: at 1.6% TS the initial orthophosphate concentration was 8.8 mg/L which was much below the data obtained for 0.5% TS (12.1 mg/L). Thus, the actual initial orthophosphate concentration for 1.6% TS should have been around 38 mg/L (12.1 mg/L multiplied by 3.2). Similar results were obtained at other TS (3.4 and 6.1%), which proves that the orthophosphate values obtained at higher TS were suppressed. The SCOD results seemed to be quite reliable, with the values obtained for both 3.4 and 6.1% TS being within the standard deviation. Figure 3 showed that the R^2 value was quite low for orthophosphate. Thus, it was clear that the orthophosphate concentration did not show a linear increase with TS. It was seen in Figure 3 that SCOD and VFA showed a linear increase with TS. The results indicate that only orthophosphate analysis seems to give incorrect measurements. To overcome this problem, the initial orthophosphate value for Part 1 (1.6% TS) was calculated using the data for 0.5% TS; the orthophosphate value for 0.5% TS was multiplied by 3.2.

Table 5b. Summary of results for part 2 (microwave treated samples).

Set No.	Treated SCOD (mgL^{-1})	Treated PO_4-P (mgL^{-1})	Treated VFA (mgL^{-1})
1a	686 ± 52.4	19.2 ± 0.5	15.1 ± 2.6
1b	3173 ± 600	74.6 ± 0.98	443 ± 31.3
1c	5694 ± 211	137 ± 7.8	424 ± 20.2
2a	695 ± 157	15.1 ± 0.4	–
2b	2878 ± 280	91.6 ± 6.4	–

Table 5b presents the data obtained from microwave treated samples. Here, set 1 represents the samples heated at 120°C while set 2 represents samples heated at 70°C. The comparison was done on similar basis as above.

For 1.6% TS, the data obtained at 0.5% TS was multiplied by 3.2, and if the values were within the standard deviation, the analysis are correct. At 120°C, when the data from 0.5% TS was compared with that obtained from 1.6 and 3.4% TS, both the SCOD and PO_4 values were within the standard deviation. While the VFA concentration obtained at both 1.6% and 3.4% TS were much higher than the one obtained at 0.5% TS. At 70°C, the orthophosphate results were found to be quite close when compared between 0.5 and 3.4% TS. However, the SCOD values at 3.4% TS for 70°C (Table 5b) were less when compared as above to 0.5% TS. It could be speculated that at 70°C,

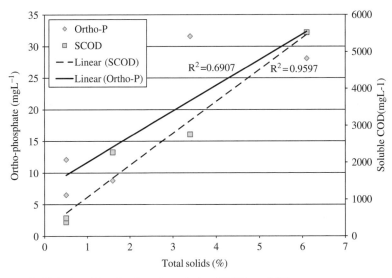

Figure 3. Relationship of ortho-phosphate, SCOD and TS.

the sample was not properly heated (cold spots were formed since sample was thick). The VFA analysis was not done in this case.

REFERENCES

Am. Public Health Assoc. (1998). Standard methods for the examination of water and wastewater, 20th ed., *Am. Public Health Association*, Washington, D.C., 1998.

Barnett, G.M. (1994). Phosphorus forms in animal manure. *Bioresource Technology*, **4**, 139–147.

Camacho, P., Delaeris, S., Geaugey, V., Ginestet, P. and Paul, E. (2004). A comparative study between mechanical, thermal and oxidative disintegration techniques of waste activated sludge. *Water Science and Technology*, **46**, 79–87.

Chan, W.I., Wong, W.T., Liao, P.H., Lo, K.V. (2007). Sewage sludge nutrient solubilization using a single-stage microwave treatment. *Journal of Environmental Science and Health*, Part A, **42**, 59–63.

Chan, WI., Lo, K.V., Liao, P.H. (2008). Nutrient release from fish silage using microwave enhanced advanced oxidation process. *J. of Environmental Science and Health*, Part B. (in press).

Chapuis-Lardy, L., Temminghoff, E.J.M., De Goede, R.G.M. (2003). Effects of different treatments of cattle slurry manure on water-extractable phosphorus. *Netherlands Journal of Agricultural Science*, **51**, 91–102.

Gungor, K. and Karthikeyan, K.G. (2005). Influence of anaerobic digestion on dairy manure phosphorus extractability. *Transactions of the ASAE*, **48**(4), 1497–1507.

Kleinman, P.J., Sharpley, A.N., Wolf, A.M., Beegle, D.B. and Moore, P.A. (2002). Measuring water-extractable phosphorus in manure as an indicator of phosphorus in runoff. *Am. Journal of Soil Sci. Soc.*, **66**, 2009–2015.

Kowalenko, C.G. and Babuin, D. (2007). Interference problems with phosphorantimonyl-molybdenum colorimetric measurement of phosphorus in soil and plant materials. *Communication in Soil Science and Plant Analysis*, **38**(9), 1299–1316.

Liao, P.H., Lo, K.V., Chan, W.I. and Wong, W.T. (2007). Sludge reduction and volatile fatty acid recovery using microwave advanced oxidation process. *J. Environ. Science and Health*, Part A **42**, 633–639.

Pan, S.H. (2005). Phosphorus release from dairy manure, MS thesis, University of British Columbia, Canada.

Rico, J.L., Garcia, H., Rico, C. and Tejero, I. (2007). Characterisation of solid and liquid fractions of dairy manurewith regard to their component distribution and methane production. *Bioresource Technology*, **98**, 971–979.

Takashima, M., and Tanaka, Y. (2008). Comparison of thermo-oxidative treatments for the anaerobic digestion of sewage sludge. *Jounal of Chemical Technology and Biotechnology*, **83**, 637–642.

Valo, A., Carrere, H. and Delegenes, P. (2004). Thermal, chemical and thermal-chemical pretreatment of waste activated sludge for anaerobic digestion. *Journal of Chemical Technology and Biotechnology*, **79**, 1197–1203.

Wen, Z., Liao, W. and Chen, S. (2004). Hydrolysis of animal manure lignocellulosic for reducing sugar production. *Bioresource Technology*, **91**, 31–39.

Wolf, A.M., Kleinman, P.A., Sharpley, A.N. and Beegle D.B. (2005). Development of a water-extractable phosphorus test for manure: An interlaboratory study. *Soil Sci. Soc. Am.*, **69**, 695–700.

Wong, W.T., Lo, K.V. and Liao, P.H. (2007). Factors affecting nutrient solubilization from sewage sludge in microwave advanced oxidation process. *J. Environ. Science and Health*, Part A, **42**(6), 825–829.

Profitable recovery of phosphorus from sewage sludge and meat & bone meal by the Mephrec process – a new means of thermal sludge and ash treatment

K. Scheidig, M. Schaaf and J. Mallon

Ingitec GmbH, Leipzig, Germany

Abstract A special metallurgical recycling process has been developed to simultaneously generate phosphate slag and gas for power generation within a single process. This would be the prerequisite for the phosphorus recycling at acceptable costs. The material to be processed might be sewage sludge and/or meat & bone meal as well as ashes from their mono-incineration. Depending upon market conditions this process can be operated for producing greater phosphorus containing slag while simultaneously generating less gas for power generation or for greater gas generation while simultaneously producing slag with lower phosphorus content.

The Mephrec process has attained this goal by performing a special metallurgical recycling method. The waste material is agglomerated by briquetting and subsequently smelted in a shaft furnace using limestone or dolomite for slag formation and coke for energy supply that simultaneously acts as a reducing agent. The liquid slag produced and a small liquid metallic phase will be tapped at about 1450°C and granulated within a jet of water or within a water bath.

Now ingitec reports on studies with the Mephrec process performed within a small shaft furnace situated at the hot test facility of the Freiberg Technical University (Bergakademie Freiberg), Saxony/Germany, representing a smelting capacity of 300 kg/h of briquetted material. The studies were performed using the above mentioned phosphorus containing wast materials.

PARTICULAR FEATURES OF THE MEPHREC PROCESS

Input material

The input material must be agglomerated (e.g. briquetted) as a prerequisite for using the Mephrec Process. There are well known technologies for briquetting fine grained materials such as ashes. But ingitec has developed a special method for using sewage sludge to convert mechanically dried (e.g. 25% dry matter) sludge into briquettes suitable for processing in shaft furnaces. The briquetting process involves the particular pure materials as well as the mixture of sewage sludge and ashes from the mono-incineration. A special technology has also been developed for the use of meat & bone meal.

Thus, a unique characteristic of the Mephrec Process is the use of very different phosphorus containing wastes, such as pulverized or even mushy material.

Equipment

The Mephrec process utilizes well known equipment and technologies (briquetting, smelting, shaft furnace, slag granulation and gas cleaning technologies). There are no complicated new apparatuses, which spells out reduced technological risk.

Investment

Compared to the 2-stage-process of phosphorus recycling from sewage sludge after mono-incineration the Mephrec process needs less capital investment and calls for lower operating costs.

ingitec has estimated the capital investment needed for the Mephrec process installation based on its experience derived from 2 shaft furnace projects for the smelting-gasification of waste materials. ingitec planned these projects by including the basic and detail engineering, supervision of construction and then put into operation. Representing throughput capacities of 10,000 t/a both of the shaft furnaces have been operated by ingitec for 12 and 10 months respectively in Erfurt and Leipzig, Germany.

The Mephrec process is expected to run profitably at capacities in the range of at least 40,000 t/a (throughput of briquetted material) because of decreasing costs at higher capacities; it has to be said that this depends on the input material and on local conditions.

Smelting technology

The agglomerated waste material will be smelted in a shaft furnace using lime-stone or dolomite for slag formation. Energy supply is given by metallurgical coke that is burned at temperatures up to 2000°C and that provides reducing conditions in the packed bed of the shaft furnace above the burning zone. The produced liquid slag and a small liquid metallic phase will be tapped at about 1450°C and granulated within a jet of water or within a water bath. The granulated slag can be separated easily from the lumped metal which represents a kind of pig iron.

The special smelting-gasification technology of the Mephrec process has been patented for ingitec.

RECENT RESULTS OF THE SMELTING-GASIFICATION WITH THE MEPHREC PROCESS

Slag formation

Basic slag components

The slag produced contains a calcium-phosphate with high phosphorus plant availability (Table 1). Finely ground it can be used as P-fertilizer as well as raw material for K, P-fertilizers or any other combi-fertilizer.

Table 1. Examples for the basic components of the Mephrec slag produced.

Slag components	CaO [%]	MgO [%]	SiO_2 [%]	Al_2O_3 [%]	P_2O_5 [%]	Plant availability of P_2O_5
Mephrec slag 1	49.0	3.46	21.0	17.4	4.64	92.7%
Mephrec slag 3	50.5	2.54	16.0	14.5	12.0	94.2%

P-contents of different Mephrec slags

Studies on the plant availability of P_2O_5 contained in the above mentioned waste materials showed very low figures, most of them less than 50%. Therefore, the object of the smelting-gasification of these materials was to find out how the plant availability could be changed by the Mephrec process. The results are shown in Table 2.

Table 2. Plant availability of P_2O_5 in the slag after smelting-gasification with the Mephrec process.

Components of the briquettes	Total P_2O_5 of the slag [mass-%]	Plant availability of P_2O_5 [mass-%]	Plant availability of P_2O_5 [% of total P_2O_5]
Mephrec slag 1 (100% sewage sludge)	4.64	4.3	92.7
Mephrec slag 2 (60% sewage sludge + 40% ash from sewage sludge-incineration)	10.49	9.49	90.5
Mephrec slag 3 (40% sewage sludge + 60% ash from meat & bone-meal-incineration)	12	11.3	94.2

In dependency on market conditions the process can be operated for the production of a slag containing larger amounts of phosphorus (Mephrec slags 2 and 3; see Table 2). Simultaneously, it generates less gas for power generation. On the other hand, the process can be operated for a higher gas generation and simultaneously produces a slag with lower phosphorus content (Mephrec slag 1; see Table 2).

Heavy metal content of Mephrec slags

The slag exhibits extraordinarily low heavy metal contents (Table 3). As the Mephrec process runs under reducing conditions, heavy metal oxides will be reduced and either alloyed with the liquid metal phase or evaporated depending on their vaporization point. Consequently, the heavy metal contents of the Mephrec slag are considerably lower than the proposed new limits for sewage sludge to be used as fertilizer, which is one more unique characteristic of the Mephrec Process.

Table 3. Comparison of legally allowed contents of heavy metals in the sewage sludge to be used as fertilizer, the proposed new limits and the data of the mephrec slag made from sewage sludge.

Parameter	Pb	Cd	Cr	Cu	Ni	Hg	Zn
Limited by law (valid since 1992 for use of sewage sludge)	900	10	900	800	200	8	2.500
Proposal 2007	120	2.5	100	700	60	1.6	1500
Mephrec slag 1 (made from sewage sludge)	<5.0	<0.4	68	123	13	<0.1	11

The results of the smelting-gasification of different phosphorus containing wastes have proven that the metallurgical recycling should provide a new way to make this recycling process profitable.

Empirical evaluation of nutrient recovery using Seaborne technology at the wastewater treatment plant Gifhorn

L.-C. Phan[a], D. Weichgrebe[a], I. Urban[a], K.-H. Rosenwinkel[a], L. Günther[b], T. Dockhorn[b], N. Dichtl[b], J. Müller[c] and N. Bayerle[d]

[a] Institute for Water Quality and Waste Management (ISAH), Leibniz University of Hannover, Welfengarten 1, 30167 Hannover, Germany, (E-mail: phan@isah.uni-hannover.de)
[b] Institute of Sanitary and Environmental Engineering, Technical University of Braunschweig, Pockelsstrasse 2a, 38106 Braunschweig, Germany
[c] PFI Consulting Engineers, Karl-Imhoff-Weg 4, 30165 Hanover, Germany
[d] Abwasser- und Strassenreinigungsbetrieb Stadt Gifhorn, Winkeler Str. 4a, 38518 Gifhorn, Germany

Abstract The Seaborne technology refers to combined chemical processes that not only eliminates heavy metals but also recovers phosphorus and ammonia from sewage sludge. Fertilizers are produced as an output of this technology. Phosphorus will be recovered in Nutrient Recovery System 1 (NRS 1) as struvite and phosphate salts. Ammonia will be partly recovered at NRS 1 like phosphorus and the rest will be stripped at NRS 2 as DiAmmonia Sulphate (DAS). A first full-scale plant with a modified Seaborne technology (Seaborne plant) was built on the wastewater treatment plant (WWTP) of Gifhorn in Germany in 2005. Since March 2007, the Seaborne plant has been investigated for the metal and heavy metal removals and nutrient recovery. Furthermore, the quality of struvite product from NRS 1 was evaluated. The product from struvite separation at Seaborne plant contained very low heavy metal concentrations that meet the EU standards. Phosphorus from sewage sludge was recovered successfully in NRS 1 (95% of phosphorus were removed from sewage sludge and 50% of phosphorus were recovered). However, this product consisted of not only struvite but also other phosphate salts of calcium and iron.

INTRODUCTION

In the past decades in Europe, the disposal and reuse of digested sewage sludge became more complicated. The disposal of digested sewage sludge at landfills have been either limited or completely prohibited. A complete thermal treatment (incineration) for digested sewage sludge is also not a favourable way because of the heavy metal emission to the environment and a high consumption of energy. For agricultural using, the EU standards for heavy metal concentrations in digested sewage sludge from WWTPs have become stricter. In the mean time, the natural phosphorus resource in the world for fertilizer production could be on a shortage due to the increasing demand on phosphorus-based fertilizers. Until now, the technologies like struvite crystallization with fluidized bed reactor are focusing on the phosphorus recovery from the liquid phase after digester (anaerobic digester supernatant) only. Because a large part of phosphorus still locates in the cells of organisms or solid phosphate salts in solid phase, the phosphorus recovery from anaerobic digester supernatant can reach only a limited efficiency (up to 30% of phosphorus in digested sewage sludge according to Battistoni *et al.* (2001). New technologies for digested sewage sludge treatment which can eliminate not only the heavy metal and organic pollutants but also can recover nutrient compounds are on demand. The Seaborne technology has been developed for these reasons. This technology refers to combined chemical processes for heavy metal elimination and nutrient recovery from sewage sludge. A part from removing heavy metals, the Seaborne technology reaches a high level of nutrient recovery efficiency thanks to the dissolving digested sewage sludge resulted from acidification.

SEABORNE TECHNOLOGY AT THE WASTEWATER TREATMENT PLANT GIFHORN

The WWTP Gifhorn is a conventional domestic WWTP (50,000 PE) with denitrication, nitrification and chemical phosphorus precipitation by iron chloride. The thickened waste activated sludge (WAS) was mixed with the primary sludge and external grease before they are anaerobically, mesophilically stabilized in sludge digesters for approx. 21 days. The digested sludge will be treated in the sewage sludge treatment plant with Seaborne technology (Seaborne plant, Figure 1).

The Seaborne technology, which applied at the WWTP Gifhorn, Germany, was modified and optimised from the original Seaborne technology concept. In comparison to the original concept, the Seaborne technology in Gifhorn does not have the RGU (**R**egenerative **G**as **U**pgrading), which cleans up the gas from sludge digester and increases the digester gas to the quality of natural gas.

At Seaborne plant, the anaerobically stabilized sludge is acidified by sulphuric acid in the extraction unit. Solids, which were not dissolved during acidification, are separated from the flow by a decanter (decanter 1) and a filter system. Afterwards, these separated solids are dried and incinerated. The ashes can be fed back to the acidification process. In the RoHM unit (**R**emoval of **H**eavy **M**etals), the sodium sulphide (Na_2S) is added to precipitate the heavy metals from the process flow. Again heavy metals are separated from the flow by using a filter system.

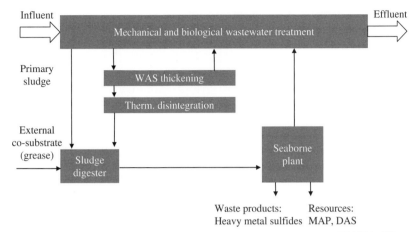

Figure 1. Simplified schema of the sewage sludge treatment at WWTP Gifhorn, according to Müller (2005).

In the following process steps, the nitrogen and phosphorous are recycled from the processed flow, which is now free of heavy metals. In the NRS 1 (**N**utrient **R**emoval **S**ystem) unit, the pH-value is increased up to 9 by using sodium hydroxide and magnesium hydroxide is added as precipitant in AEE step (equivalent adjustment step). The crystallisation of magnesium, ammonia and phosphorous to MAP (struvite) is then achieved. Afterwards, the MAP is separated from the flow by using a decanter (decanter 2) and a filter. Finally, the surplus ammonia, contained in the liquid after MAP precipitation, is separated as **D**i**A**mmonia **S**ulphate (DAS) in a stripping unit (NRS2). The remaining treated process stream is recirculated to the influent of the WWTP. The products of the Seaborne process MAP and DAS can be reused as fertilizer in agriculture (Günther *et al.*, 2008). An optional unit for calcium precipitation was installed for solving the operational trouble in NH_3-Stripping caused by the surplus calcium precipitation. This unit is only in operation when the surplus calcium concentration after NRS 1 unit is too high. The schema of Seaborne plant is shown in Figure 2.

EVALUATION METHODS

Different measurements have been taken to evaluate the efficiency of the Seaborne plant. Sample locations are shown in Figure 3.

For the evaluation, not only nutrient, metal parameters are considered but also organic compounds will be measured. The analysed parameters are therefore divided into 3 groups: general parameters, nutrient compound parameters, metal

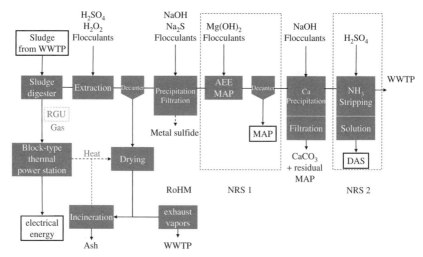

Figure 2. Schema of the sewage sludge treatment plant with the modified Seaborne technology (Seaborne plant).

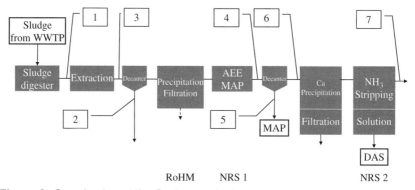

Figure 3. Sample plan at the Seaborne plant.

and heavy metal parameters. At the time the measurements were undertaken, the Ca-Precipitation and NH$_3$-Stripping units were not in operation. Hence, the output of Seaborne plant remains the solid phase after extraction unit (sample 2), the solid phase after NRS1 unit (sample 5) and the effluent Seaborne (liquid phase, sample 7).

The samples were collected from 29.09.2008 to 02.10.2008 during the operation of the Seaborne plant. The parameters were analysed by the Standard methods for water, wastewater and sewage sludge analysis (Table 1).

Table 1. Analysed parameters.

General parameters	Carbon parameter	Nutrient parameters	Metal parameters	Heavy metal parameters
TSS	TOC	TKN	Potassium (K)	Lead (Pb)
VSS	TIC	NH_4-N	Calcium (Ca)	Cadmium (Cd)
CODtot.	TC	NO_3-N	Magnesium (Mg)	Chromium (Cr)
CODfil	DOC	Ptot	Sodium (Na)	Copper (Cu)
	DIC	PO_4-P	Manganese (Mn)	Mercury (Hg)
	DC		Aluminium (Al)	Nickel (Ni)
			Iron (Fe)	Zinc (Zn)

Finally, a mass balance of the main compounds will be established by the mean values of measured concentrations from liquid, solid samples and the flow rate. Based on the mass balance, an empirical evaluation of the quality and quantity of products and the emission of Seaborne plant effluent to the WWTP Gifhorn can be done.

RESULTS AND DISCUSSION

As mentioned, the empirical evaluation of the Seaborne plant will follow three main criteria: i) the concentrations of heavy metals in fertilizer product and the effluent of Seaborne plant, ii) the emission impact of the Seaborne plant effluent on the WWTP Gifhorn, iii) the recovery efficiency of nitrogen and phosphorus compounds.

Metal and heavy metal distribution

According to the sample plan, there are 5 samples (sample 1, 3, 4, 6, 7) from the liquid phase and 2 samples from the solid phase (sample 2, 5). Figure 4 and Figure 5 show that apart from potassium and sodium, whose salts have a high dissolubility, the concentrations of other metals and heavy metals decrease significantly from influent (digested sludge) to effluent of the Seaborne plant. In particular, the concentrations of aluminium and iron in the effluent Seaborne samples are around 0.1 and 1 mg/l respectively. This means the metals and heavy metals were removed steps by steps through the Seaborne plant. While the heavy metals were mostly removed after the extraction unit, the other metals were removed by the NRS 1 unit.

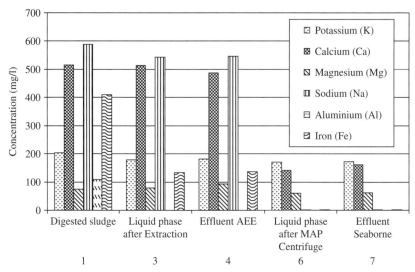

Figure 4. Metal concentrations in liquid phase at the Seaborne plant.

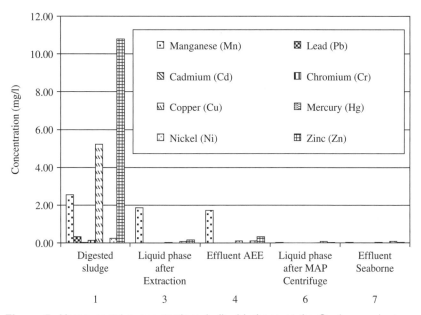

Figure 5. Heavy metal concentrations in liquid phase at the Seaborne plant.

As the solids after the extraction unit (incineration input, sample 2) and the solids after NRS1 unit (fertilizer product, sample 5) are the main outputs in the solid phase, the heavy metal concentrations of these samples will be compared to the EU Directive (RL 86/278/EWG, 2008), the German legal requirements (AbfKlärV, 1992) and the results from the German sewage sludge survey of the German Water Association, DWA (Durth *et al.*, 2003).

It can be seen from the Table 2 that the heavy metal concentrations of the solid outputs of Seaborne plant and the digested sewage sludge are far lower than the legal requirements. In comparison to the results from German sewage sludge survey, only the copper concentrations from the digested sewage sludge (sample 1) and the solids after extraction (sample 2) are greater than the average value from other German sewage sludge. The other heavy metal concentrations are lower than the average values. The high concentration of copper in sample 2 is not critical because it is just an immediate product for the next incineration step. On the other hand, the fertilizer product (MAP, sample 5) has very low concentrations of all heavy metals.

Table 2. Concentrations of heavy metal in input and outputs (solid phase) of Seaborne plant in comparison with German sewage sludge survey (Durth *et al.*, 2003), German legal requirements (AbfKlärV, 1992) and EU Directive (RL 86/278/ EWG, 2008).

Parameter	Digested sewage sludge (s. 1)	Solids after extraction (s. 2)	MAP (s. 5)	Results from DWA survey	German legal requirement	EU Directive
mg/kg TSS						
Lead (Pb)	25.92	32.00	9.00	61.7	900	750–1200
Cadmium (Cd)	1.33	1.18	0.10	1.52	10	20–40
Chromium (Cr)	11.80	27.20	4.82	60.5	900	–
Copper (Cu)	418.94	534.00	12.50	380.2	800	1000–1750
Mercury (Hg)	0.02	0.82	0.02	0.92	8	16–25
Nickel (Ni)	20.39	16.60	6.68	32.2	200	300–400
Zinc (Zn)	866.77	869.00	54.10	955.7	2500	2500–4000

Therefore, it can be said that the solid outputs of the Seaborne plant fulfilled the requirements on heavy metals according to the EU Directive and German standards.

COD, TOC distribution

The organic compounds along the process line can be presented by the COD and TOC concentrations from samples 1, 3, 4, 6, 7 (Figure 6). The results show that approximate 92% of organic compounds (COD and TOC) were removed by the extraction unit. The NRS 1 unit removed about 3% of total organic compounds. Totally, around 95% of total organic compounds were eliminated from the liquid phase of Seaborne plant. The rest 5% will be returned to WWTP. In order to investigate the impacts of the Seaborne effluent on the degradation of WWTP, two samples from the liquid phase after extraction (sample 3) and from Seaborne effluent (sample 7) were taken for the COD fraction.

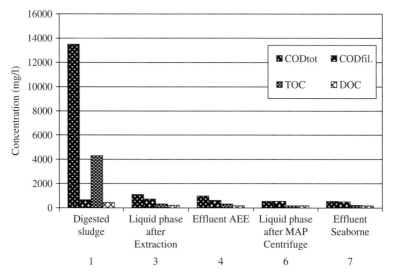

Figure 6. COD, TOC concentrations along the process line.

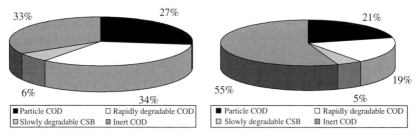

Figure 7. COD fraction of sample 3 and sample 7 (percentage of the total COD).

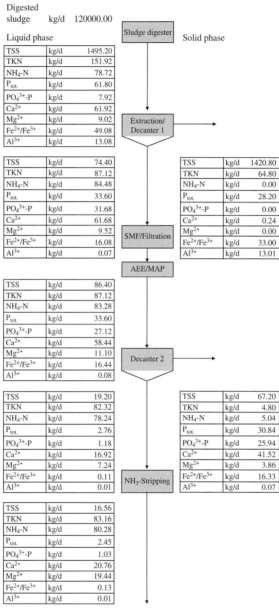

Figure 8. Mass balance for TSS, nitrogen, phosphorus compounds and ionic metals at the Seaborne plant.

The COD fraction shows that the inert COD increased from 33% in the liquid phase after extraction to 55% in the Seaborne effluent of total COD (Figure 7). The addition of chemicals caused such increase. However, the load of inert COD, which returned to WWTP, was only approximately 46 kg/d. Comparing to the COD loads of influent WWTP (4765 kg/d) and effluent WWTP (329 kg/d), the inert COD load from the Seaborne plant can be neglected.

Mass balance of the main compounds

Based on the concentrations and flow rate, a mass balance of TSS, TKN, NH4-N, $P_{tot.}$, PO_4^{3+}-P, Ca^{2+}, Mg^{2+}, Fe^{2+}/Fe^{3+} and Al^{3+} can be established. Because of the difficulty in determining directly the flow rate for the solid phase, the loads in the solid phase were estimated from the loads in the liquid phase.

In the decanter 1, after the extraction unit, 95% of TSS, 43% of TKN, 46% of $P_{tot.}$, 67% of iron and 99% of aluminium were removed from the main flow (Figure 8). Because this solid output will be incinerated, the TSS, TKN and $P_{tot.}$ loads in this output should be reduced in order to avoid the loss of nitrogen and phosphorus and to increase the nutrient recovery. The optimisation of this step can be done by a further reduction of pH at the extraction unit. However, a low pH (<2) at the extraction can disturb the separation in decanter 2 because at low pH, the polymers can lose their activities.

In the decanter 2, 50% of $P_{tot.}$, 3% of TKN, 6% of NH_4-N, 67% of calcium, 43% of magnesium, 33% of iron were recovered as fertilizer (Figure 8). The low ammonia ratio in the fertilizer product shows that this product contained not only struvite ($MgNH_4PO_4*6H_2O$) but also the phosphate salts of calcium and iron like $FePO_4*2H_2O$, $Fe_3(PO_4)_2*8H_2O$, $CaCO_3$, $Ca_{10}(PO_4)_6(OH)_2$. Despite the low ratio of struvite, the fertilizer product is suitable for sustainable agriculture thanks to their high percentage of phosphorus and low concentrations of heavy metals.

CONCLUSION

The Seaborne plant has removed most organic compounds (99% of TSS, 96% COD_{tot}, 95% TOC) from the digested sewage sludge. The inert COD fraction returned to WWTP is very low. There are small heavy metal concentrations in the Seaborne plant outputs at decanter 1 and 2 (fertilizer product) that meet the EU and German standards.

However, the extraction unit must be optimised by decreasing pH in order to achieve a lower ratio of phosphorus and nitrogen in the solid output at decanter 1. If the extraction unit is optimised, an efficiency of 96% for phosphorus recovery can be reached. On the other hand, a higher ratio of struvite in the NRS 1 output can increase the quality of fertilizers as Seaborne plant's product. Furthermore, if NH_3-Stripping is in operation, the NRS 2 can produce a DAS amount up to 80 kg NH_4-N/d.

To sum up, the Seaborne plant at WWTP Gifhorn at the evaluation period (29.09.2008 to 02.10.2008) can recover 50% of phosphorus from the digested sewage sludge and offer a fertilizer product with very low heavy metal concentrations. After the optimisation, a recovery of 95% of phosphorus and 85% of nitrogen should be reached.

REFERENCES

Abfallklärschlammverordnung (AbfKlärV) (1992). (German sewage sludge ordinance).

Battistoni, P., DeAngelis, A., Pavan, P., Prisciandaro, M. and Cecchi, F. (2001). Phosphorus removal from a real anaerobic supernatant by struvite crystallization. *Water Res.*, **35**(9), 2167–2178.

Durth, A., Schaum, Ch., Alessandro, M., Wagner, M., Hartmann, K.-H., Jardin, N., Kopp, J. and Otte-Witte, R. (2005). Ergebnisse der DWA-Klärschlammerhebung 2003 (German sewage sludge survey 2003).

Günther, L., Dockhorn, T., Dichtl, N., Müller, J., Urban, I., Phan, L.-C., Weichgrebe, D., Rosenwinkel, K.-H. and Bayerle, N. (2008). Technical and scientific monitoring of the large-scale seaborne technology at the WWTP Gifhorn. *Wat. Pra. Tech.*, **3**(1).

Müller, J. (2005). Umsetzung des Seaborne Verfahrens auf der Kläranlage Gifhorn (Implementation of the Seaborne process at the WWTP Gifhorn). Proceedings from the 75. Darmstädter Seminar.

PFI Planungsgemeinschaft, Institut für Siedlungswasserwirtschaft, TU Braunschweig im Auftrag des niedersächsischen Umweltministerium (2003). Machbarkeitsstudie zum Seaborne Verfahren (Feasibility study of the Seaborne process).

RL 86/278/EWG Klärschlammverwendung in der Landwirtschaft (2008). (EU Directive: RL 86/278/EWG, Using sewage sludge for the agriculture).

Schulze-Rettmer (1991). The simultaneous chemical precipitation of ammonium and phosphate in the form of Magnesium-ammonium-phosphate. *Wat. Sci. Tech.*, **23**, Kyoto, pp. 659–667.

Wittig E. (2006). Erste großtechnische Erfahrungen mit der Klärschlammaufbereitung-sanlage Gifhorn nach einem modifizierten Seaborne Verfahren (First large-scale experiences with the modified Seaborne process at the WWTP Gifhorn), DWA-Nord Klärschlammtagung.

Sewage treatment to remove ammonium ions by struvite precipitation

Sergey Lobanov

Department of Chemical Engineering, Perm State Technical University, Komsomolskij avenue 29a, 614990 Perm, Russia, E-mail: lobanov@daad-alumni.de

Abstract The process of struvite precipitation as applied to waste water purification to remove ammonium ions was studied. The influence on this process of different factors, such as pH, reagent flow mode, presence of impurities, was discovered. Crystal size distribution and crystal habit of struvite particles in suspensions obtained under different conditions were analyzed. Kinetic parameters of struvite reaction-crystallization processes were identified using Rojkowski hyperbolic size dependent growth model. Two different mechanisms of struvite crystallization were found. The two-step crystallization process has proved to be the most favorable process mode for struvite precipitation providing a high struvite recovery level as well as high-performance crystals.

INTRODUCTION

At present, a problem of sewage treatment to remove ammonium ions exists in many enterprises. The majority of them pollute the surface waters with waste products. Excess of ammonium nitrogen in natural reservoirs negatively influences human and animal health as well as microflora of reservoirs.

There is a number of physical, chemical and biological methods used for processing nitrogen-containing waste streams. However the majority of these methods may not be applied in purification processes of the high flow streams of sewage with the high contents of ammonium ions in it. Thus at the moment attention is focused on how to find a solution to the problem of sewage treatment to remove ammonium ions.

One of the most promising technologies of sewage treatment to remove ammonium ions for now is recovery of ammonium nitrogen as struvite. The main advantage of this method of purification is an opportunity of recycling of ammonium as a useful product NP-fertilizer.

Solubility of struvite had been thoroughly investigated. It is highly soluble at low pH and insoluble in alkaline media. Solubility product of struvite pK_{so} is 13.26 (Ohlinger *et al.*, 2001). However, the kinetics data of its formation are rather inconsistent. Only little research had been devoted to nucleation and precipitation of struvite. There is not much research carried out on studying the process of crystallization of struvite and influence on this process of different factors, such as pH, concentration of ammonium ions, reagent flow mode, etc. These investigations are rather important, because it is possible to obtain sediment with the given characteristics by managing the process of crystallization. This research will also enable us to find out optimal conditions for the achievement of a high purification level from ammonium ions.

The purpose of the research of the struvite crystallization process is establishing optimal conditions for obtaining of the sediment for the achievement of a high purification level to remove ammonium ions and for obtaining high-performance crystals.

RESULTS AND DISCUSSION

Several experiments were carried out to precipitate struvite under different conditions. Ammonium chloride, magnesium chloride hexahydrate, sodium hydrophosphate dihydrate, sodium hydroxide were used as initial substances. The precipitation occurred by the reaction:

$$Mg^{2+} + NH_4^+ + HPO_4^{2-} + 6H_2O = MgNH_4PO_4 \cdot 6H_2O + H^+ \qquad (1)$$

First, the influence of pH value on crystal dimensions and habit was studied. A solution of sodium hydroxide was used to adjust the pH value. The process of crystallization started immediately after adding magnesium chloride solution to a solution containing NH_4^+ and PO_4^{3-} ions. And the higher the pH value, the higher the struvite recovery level (as well as purification level from ammonium ions) and the more intensive the crystallization process was at the beginning. Struvite recovery level at pH 6.90, 8.72 and 11.64 was 37, 86 and 93% correspondingly. However crystallization proceeded slowly at high pH value 11.64 in comparison with experiment at pH 8.72. It can be connected with reducing of struvite solubility at excess amount of alkali. Struvite has the lowest solubility in water at pH 9.47 (Matýsek and Raclavská, 2001). At higher pH levels struvite becomes more soluble and that decreases its recovery level. These data are in accordance with previous research (Hultman et al., 1997; Lobanov and Poilov, 2006). Table 1 shows ion associates in solution in equilibrium with struvite at different pH levels. Dependence of wastewater purification level X on pH value is represented in Figure 1. As one can see, optimal precipitation, as well as purification level, is achieved between pH 8.0 and 10.5.

After the precipitation process the suspensions obtained underwent CSD analysis. Dimension of most crystals was between 20–80 μm. Struvite crystals

Table 1. Chemical reactions of struvite precipitation at different pH intervals.

Chemical reaction	pH interval	K_{eq}
$Mg^{2+} + NH_4^+ + H_2PO_4^- \leftrightarrow MgNH_4PO_4 + 2H^+$	<7.2	$3.16 \cdot 10^6$
$Mg^{2+} + NH_4^+ + HPO_4^{2-} \leftrightarrow MgNH_4PO_4 + H^+$	>7.2 и <9.2	$3.98 \cdot 10^0$
$Mg^{2+} + NH_3 + HPO_4^{2-} \leftrightarrow MgNH_4PO_4$	>9.2 и <12	$7.08 \cdot 10^9$
$Mg^{2+} + NH_3 + PO_4^{3-} + H_2O \leftrightarrow MgNH_4PO_4 + OH^-$	>12	$7.08 \cdot 10^7$

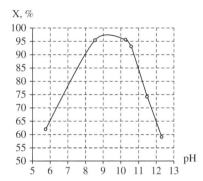

Figure 1. Wastewater purification level X as a function of the final pH level in precipitation of struvite.

are shown in Figure 2. At low pH values it could be observed that crystals had a lager size than at high pH. So, it is in accordance with the data, that low pH reduces supersaturation level. Low supersaturation causes slow nucleation and fast crystal growth, and thus, large crystals.

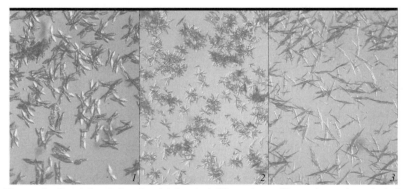

Figure 2. Crystals of struvite obtained at different conditions *(experiments 1, 2, 3)*.

Figure 2 also represents struvite crystal habit. At low pH value typical struvite crystals can be observed – twin crystals, or so called "macle". Such crystals were observed by Regy *et al.* (2002) and Suschka *et al.* (2003). These crystals are larger than needle-like crystals, precipitated at high pH level. Needle-like crystals are usually obtained in the presence of free ammonia excess, which can be observed at high pH level of mother liquid (Suschka *et al.*, 2003). At medium pH level one can observe some type of hybrid star-like crystals. Thus pH level and consequently also a supersaturation level significantly affect crystal size and habit of struvite. In all experiments large crystals as well as fines can be observed. That might confirm the existence of different nucleation and growth mechanisms during the process of crystallization.

Rojkowski hyperbolic size dependent growth model (RH SDG model), suggested by Koralewska *et al.* (2007), was applied to treat the data obtained and to define some crystallization parameters. The researchers found out that this mathematical model was the best suitable one of a number of empirical or semi-empirical equations to describe the overall size dependent growth kinetics of reaction-crystallization of struvite synthesis.

$$G = G_\infty - \left(\frac{G_\infty - G_0}{1 + aL}\right) = \frac{G_0 + aG_\infty L}{1 + aL} \tag{2}$$

$$n = n_0 \exp\left[-\left(\frac{1}{\tau}\frac{G_\infty - G_0}{aG_\infty^2}\ln\left(\frac{aG_\infty L + G_0}{G_0}\right) + \frac{1}{\tau}\frac{L}{G_\infty} + \ln\left(\frac{G_0 + aG_\infty L}{(1 + aL)G_0}\right)\right)\right] \tag{3}$$

From the RH SDG model, one can directly calculate nuclei population density, n_0, their (minimal) growth rate, G_0, as well as their maximal admissible growth rate (of the largest crystals), G_∞, assuming a selected suspension residence time, τ. Nucleation rate B can be calculated with the simple formula:

$$B = n_0 \cdot G_0 \tag{4}$$

Figure 3 represents dependence of population density of struvite crystals on their sizes. As one can see, the RH SDG model (solid lines) fits well to experimental data (points) and can be applied to the mathematical description of struvite reaction-crystallization process. However as was mentioned before, crystals could grow according to different mechanisms and finally have different shape and size. That was observed during the analysis of data obtained after experiment *2*. It was not possible to describe well the data using just one model curve. So it was suggested that the CSD curve in this case was polymodal. The RH SDG model was applied separately to two different ranges of crystal size.

As a result, the data were described with two model curves – one for large fraction and one for fines. These curves described the experimental data well. The suggestion that crystallization could proceed with two different mechanisms was also confirmed by crystal habit of struvite. As was mentioned before,

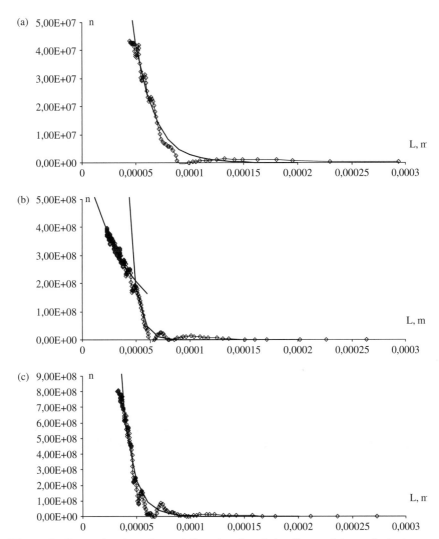

Figure 3. Dependencies of population density of struvite crystals on their sizes: *a – experiment 1; b – experiment 2; c – experiment 3.*

crystals obtained at low (6.90) as well as at high (11.64) pH levels had uniform shapes typical of struvite – twin crystals and needle-like crystals. However the crystals obtained at pH 8.72 (experiment *2*) had different hybrid forms and dimensions. That means that it grew according to different mechanisms. It can be observed when crystallization proceeds at different conditions during the whole process.

Parameters calculated using RH SDG model, are presented in Table 2. As was suggested before, nucleation rate B increased and maximal crystal growth rate G_∞ decreased with higher pH value (experiments *1* and *2-large fraction*) due to higher supersaturation level. In the case of experiment *2* the crystal formation proceeded according to two different mechanisms. Fine fraction is seen to be formed at low B and high G_∞, in comparison with the large fraction. It corresponds to crystallization mechanism at low supersaturation. The large fraction was formed at high B and low G_∞, according to crystallization mechanism at high supersaturation. Thus one can suggest that during the process of crystallization, supersaturation level changed; that led to different mechanisms of crystallization. That was due to the changing pH during the process. As was mentioned before, pH level significantly influences supersaturation of mother solution. In spite of the fact that fine crystals were formed at low supersaturation, that might lead to formation of large crystals, the supersaturation level favorable for low nucleation rate and high growth rate was adjusted for a short time. So the crystals couldn't grow further. The large fraction was formed at relatively high supersaturation level, which corresponded to high nucleation rate and low crystal growth rate. However this supersaturation level was adjusted during the longest time of whole crystallization process. So the crystals could grow during the whole crystallization process. Moreover, the large fraction crystals had dendritic shape that was favorable for crystal growth. These crystals can offer growth site to nuclei. Nuclei can stay attached on it. That suggestion was confirmed by dependencies of linear growth rates on their sizes – growth rate increases with a size of a particle for large fraction (experiment *2*). At the same time, growth rate is not changed significantly with the size for the fine fraction. That means that the large fraction was formed according to size dependent growth (SDG) mechanism. Growth rate of the fines didn't depend on crystal size that corresponds to size independent growth (SIG) mechanism.

During experiment *3* nucleation rate B was lower than B at experiment *2* (large fraction) and maximal growth rate G_∞ higher than the same parameter at experiment *2*. That could be explained by increasing of struvite solubility at alkali excess. As was mentioned before, excess amount of alkali increases struvite solubility and thus reduces supersaturation level that leads to low nucleation rate

Table 2. Parameters of RH SDG model.

	Precipitation mode			n_0, m⁻¹m⁻³	G_∞, m·s⁻¹	G_0, m·s⁻¹	B, m⁻³s⁻¹
1	Direct instant mode	$pH_{fin} = 6.9$		$4.57 \cdot 10^{10}$	$3.06 \cdot 10^{-8}$	$4.53 \cdot 10^{-10}$	21
2	(Mg^{2+} to mixture of PO_4^{3-}, NH_4^+ and NaOH)	$pH_{fin} = 8.72$	Large fraction	$3.77 \cdot 10^{11}$	$8.41 \cdot 10^{-9}$	$5.15 \cdot 10^{-9}$	1939
			Fine fraction	$1.08 \cdot 10^{11}$	$7.63 \cdot 10^{-8}$	$4.65 \cdot 10^{-10}$	50
3		$pH_{fin} = 11.64$		$2.43 \cdot 10^{11}$	$1.42 \cdot 10^{-8}$	$3.59 \cdot 10^{-9}$	874
4	Dosed feed of NaOH	0.5 ml·min⁻¹		$1.17 \cdot 10^{14}$	$3.05 \cdot 10^{-8}$	$1.03 \cdot 10^{-9}$	120555
5		2.5 ml·min⁻¹	Large fraction	$1.24 \cdot 10^{15}$	$1.35 \cdot 10^{-8}$	$5.84 \cdot 10^{-9}$	7216383
			Fine fraction	$7.50 \cdot 10^{13}$	$8.19 \cdot 10^{-8}$	$4.23 \cdot 10^{-10}$	31741

and high growth rate. Level of pH during experiments 1 and 3 was relatively constant so the crystals obtained were uniform and the crystallization process proceeded according to single mechanism. It can also be concluded that low supersaturation mainly causes SIG mechanism of struvite crystallization (experiments *1* and *2-large fraction*) and high supersaturation – mainly SDG mechanism (experiments *2-fine fraction* and *3*).

Thus it can be concluded that high-performance crystals can be obtained at low supersaturation level. However, the purification level was not high enough at low pH value and accordingly at low supersaturation. Experiments with dosage pumping of precipitating reagents into reactor were carried out at optimal pH level of about 8.5 to create low supersaturation and precipitate large struvite crystals. However these experiments mainly did not lead to obtaining of high performance crystals of struvite due to varying supersaturation level during the process of crystallization. Presence of fines was still observed. In the cases with precipitating of high performance crystals, struvite recovery level was not high enough due to low supersaturation and, as a result, high struvite solubility.

In this connection, it would be expedient to obtain large high performance crystals at low supersaturation in the first step and then to increase super-saturation level to force crystal growth and to increase the purification level in the second step. Thus the crystals obtained in the first step would be seeds for crystal growth on the second step. Moreover, twin crystals obtained at low supersaturation represents perfect seed material for further growth (Regy *et al.*, 2002). These crystals can offer growth site to nuclei.

So, experiments with gradual pH increasing were performed. Recovery levels are seen to be quite high for these experiments (86 and 89% for experiments *4* and *5* correspondingly). The crystals obtained are shown on Figure 4. The particles represent perfectly formed twin crystals of large dimensions. Only small amounts of fine crystals can be observed in the case of the faster rate of supersaturation creation. CSD analysis of the sediments showed that crystal size was no less than 50μm. The sediment obtained in experiment *4* contained slightly larger crystals than sediment obtained in experiment *5*. Crystal size achieved 200 μm. Crystallization process in experiment *4* proceeded according to single mechanism due to relatively constant supersaturation level on every step of the process. Due to the two-step crystallization process it was possible to achieve high growth rate as well as high nucleation rate (see Table 2). However, no fines were observed because of the fact that nucleation occurred on the surface of seed crystals. The rate of supersaturation creation in experiment *5* was relatively higher which led to a relatively significant change of supersaturation level and crystallization process proceeded according to two mechanisms. In this case, nucleation process occurred not only on the surface of seed crystals but also in mother liquor (primary nucleation). However, the amount of fine fraction in experiment 5 was not high and its dimensions were no less than 50 μm. Dependencies of linear growth rates on their sizes also shows that the crystals grew mostly according to SIG mechanism due to its formation at mostly low supersaturation.

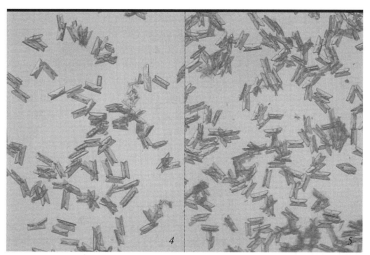

Figure 4. Crystals obtained in two-step crystallization process *(experiments 4 and 5).*

Thus two-step crystallization is the most favourable process mode for struvite precipitation. It enables a high struvite recovery level (hence high purification level) as well as high-performance crystals. However, the rate of pH increasing in the second step of the process should be rather slow to prevent primary nucleation and formation of fines.

In view of the fact that sewage is a multicomponent system, it is also necessary to take into account the effect of various impurities on sewage treat-ment to remove ammonium ions, because some of these substances can impair the efficiency of the process. The role of impurities in struvite precipitation is played by ions of potassium and calcium. At a high content of calcium ions in wastewater, part of phosphate ions is consumed for formation of insoluble calcium phosphates, which impairs the treatment efficiency (Regy *et al.*, 2002). In the presence of a large amount of potassium ions, they can substitute ammonium ions in struvite to form magnesium potassium phosphate (K-struvite) with identical structure and properties. In addition, the problem of sewage treatment to remove ammonium ions is topical for plants manufacturing urea, and, therefore, a study of the influence exerted by urea impurity on the sewage treatment process is of practical interest. It is known that urea can underwent hydrolysis in an alkaline medium to form ammonium ions, which may diminish the degree of sewage purification.

So, experiments on struvite precipitation in the presence of impurities were performed. In a study of the influence exerted by a urea admixture on the sewage treatment process, the simulated solution contained, in addition to ammonium ions, urea in a concentration of 30 $g \cdot l^{-1}$. The experiments performed demonstrated that urea is not noticeably hydrolyzed in alkaline solutions under the conditions specified during the time of the crystallization process. The urea admixture does not affect the efficiency of sewage treatment to remove ammonium ions and the process of struvite precipitation. An X-ray analysis of the precipitate obtained in the presence of urea demonstrated that the sediment was struvite free of any impurity phases.

Presence of a large amount of potassium ions in wastewater (68 $g \cdot l^{-1}$) slightly reduces the purification efficiency. The precipitate obtained in this case is only 89% composed of struvite. An X-ray analysis demonstrated that no new phases were present. Because magnesium potassium phosphate is formed under the same conditions as struvite, it may be assumed that magnesium potassium phosphate forms a solid solution with magnesium ammonium phosphate. K-struvite is isostructural with struvite, both substances crystallize in the orthorhombic crystal system, and the radius of the ammonium ion is close to that of the potassium ion.

The most significant influence on the process of sewage treatment to remove ammonium ions by precipitation of struvite is exerted by calcium ions. It has been found (Mitani et al., 2001) that struvite is formed mostly at ion concentration ratios $[Ca^{2+}] : [Mg^{2+}]$ lower than 0.25 : 1. For comparison, experiments were performed at $[Ca^{2+}] : [Mg^{2+}] = 1 : 1$. It was found that at $[Ca^{2+}] : [Mg^{2+}] = 0.25 : 1$ the degree of wastewater purification decreased, but, nevertheless, remained fairly high . An analysis of the precipitate formed in this case demonstrated that mostly struvite was formed (88%). At the ratio $[Ca^{2+}] : [Mg^{2+}] = 1 : 1$, the purification level decreased significantly (66%). An X-ray analysis demonstrated that the precipitates were mostly composed of struvite; however, no any phases associated with calcium phosphates were found. It may be assumed that an amorphous calcium phosphate $Ca_3(PO_4)_2$ was formed in addition to struvite. It is known (Regy et al., 2002) that hydroxylapatite, having the lowest solubility product ($pK_{SP} = 58.5$), is the most thermodynamically stable phase among all forms of calcium phosphate; however, intermediate phases, including amorphous calcium phosphate, can be formed in primary nucleation.

CONCLUSIONS

The influence of different parameters on the struvite crystallization process was studied in order to establish optimal conditions for obtaining of the sediment for the achievement of a high purification level to remove ammonium ions and for obtaining high-performance crystals. The RH SDG model was used to define the crystallization parameters of the process. The model described the experimental data well. It was found that the crystallization occurred according to two mechanisms in the case of significant changes of the supersaturation level during the process of crystallization. In this case, the large fractions of the sediment were formed at high nucleation rate and low crystal growth rate. The fine fractions were formed at low nucleation rate and high growth rate. It was also found that low supersaturation mainly causes SIG mechanism of struvite crystallization and high supersaturation – mainly SDG mechanism. The two-step crystallization process was found to be the most favorable process mode for struvite precipitation. It enables a high struvite recovery level as well as high-performance crystals.

ACKNOWLEDGEMENT

This work was supported by joint grant of German Academic Exchange Service (DAAD) and Russian Ministry of Science and Education and was carried out in collaboration with Chair of Separation Science and Technology of University Erlangen – Nuremberg, Germany.

REFERENCES

Hultman, B., Levlin, E., Löwén, M. and Mossakowska, A. (1997). Uthållig Slamhantering, Förstudie (Sustainable sludge handling, pre-study). Stockholm Water Co., R. Nr 23 Sept-97.

Koralewska, J., Piotrowski, K., Wierzbowska, B. and Matynia A. (2007). Reaction-crystallization of struvite in a continuous liquid jet-pump DTM MSMPR crystallizer with upward circulation of suspension in a mixing chamber – an SDG kinetic approach. *Chem. Eng. Technol.*, **30**(11), 1576–1583.

Lobanov, S.A. and Poilov, V.Z. (2006). Treatment of wastewater to remove ammonium ions by precipitation. *Russ. J. Appl. Chem.*, **79**(9), 1473–1477.

Matýsek, D. and Raclavská, H. (2001). Problems of dissolution and crystallization of struvite. In: *Proc. of the Second International Conference on Recovery of Phosphate from Sewage and Animal Wastes*, Amsterdam, Netherlands, 12–13 March.

Mitani, Y., Sakai, Y., Mishina, F. and Ishiduka, S. (2001). Struvite recovery from wastewater having low phosphate concentration. In: *Proc. of the Second International*

Conference on Recovery of Phosphate from Sewage and Animal Wastes, Amsterdam, Netherlands, 12–13 March.

Ohlinger, K.N., Young, T.M. and Schroeder, E.D. (2001). Kinetics and thermodynamics of struvite crystallization as it applies to phosphate recovery from municipal wastewater for agricultural fertilizer production. In: *Proc. of the Second International Conference on Recovery of Phosphate from Sewage and Animal Wastes*, Amsterdam, Netherlands, 12–13 March.

Regy, S., Mangin, D., Klein, J.P. and Lieto, J. (2002). Phosphate recovery by struvite precipitation in a stirred reactor. LAGEP report, Centre Europeen d'Etudes des Polyphosphates.

Suschka, J., Kowalski, E. and Poplawski, S. (2003). Study of the effects of the reactor hydraulics on struvite precipitation at municipal sewage works. Centre Europeen d'Etudes des Polyphosphates.

Full-scale plant test using sewage sludge ash as raw material for phosphorus production

W.J. Schipper[a] and L. Korving[b]

[a] Thermphos International B.V., Postbus 406, 4380 AK Vlissingen, The Netherlands, willem.schipper@thermphos.com
[b] N. V. Slibverwerking Noord-Brabant, Middenweg 38, 4782 PM Moerdijk, The Netherlands, korving@snb.nl

Abstract The use of sewage sludge incinerator ash as feedstock for white phosphorus production was investigated in a full-scale plant test. The iron content of the ash is critical for this application, as high Fe content in the feedstock gives rise to the formation of ferrophosphorus by-product. For this, a scan of sewage works in The Netherlands was performed to identify suitable sludges. Phosphate in sewage works was seen to be removed by either iron salt or aluminium salt dosage, or through the EBPR process. Dosage of iron salt contributes to the iron content of sewage sludge ash, whereas the aluminium or EBPR processes generate a sludge which can yield a suitable ash. These suitable sludges were incinerated separately which yielded a phosphate-rich ash low in iron. This ash was tried succesfully in multi-tonne quantities in the white phosphorus process, with little or no effects seen on the operational or environmental parameters of the process. It is concluded that this route allows the succesful recycling of large amounts of sewage phosphate. A comparison is made with alternative recovery routes of sewage phosphate and ways forward are suggested.

INTRODUCTION

Phosphate is a nutrient that plays an important role in our society. It is necessary for all life on earth and is an essential ingredient for the fertilizers used to maintain food production for an increasing world population. Phosphate reserves are limited, however, and are being used up, with the phosphate ending up being dispersed in soils, surface water or waste streams with little possibility to recover it. In addition the reserves of phosphate rock are concentrated in certain areas of the world (such as Morocco, China), which makes a large number of countries in the world dependent on the import of phosphate rock to meet their demand of phosphate. This is certainly the case for Europe which only has very limited phosphate reserves.

In the developed world, about 15–25% of these mined phosphates are present in the food chain or industrial products and end up in the sewer system. Since sewage needs to be treated before it is discharged, waste water is purified in waste water treatment plants (WWTPs), which also remove almost all of the phosphate burden in waste water. These phosphates are concentrated in the sewage sludge produced by these WWTPs. Excess sludge is continually removed.

Currently, sewage sludge ash is applied as low-grade fertilizer, but this use is restricted or prohibited in a growing number of EU countries. Therefore a growing amount of sludge is incinerated, with the phosphate in the ash being essentially lost to re-use, as the ash has no applications that re-use the phosphate.

In The Netherlands the total annual production of municipal sewage sludge amounts to circa 342 000 tons dry matter, containing 11 600 ton P, with a concentration of 34 g P/kg dry matter (Geraarts *et al.*, 2007). This is a large volume compared to the yearly use of phosphates in fertilizers in The Netherlands which amounts to 21 000 ton P/y and illustrates the importance to develop methods to recycle the phosphate that is collected in the sewage system, although it should be noted that the fertilizer use in The Netherlands is relatively low compared to other European countries because of the large manure production in the country.

This article introduces one such method to recycle the phosphates collected in the WWTPs. The advantage of this method is that it is based on minor modifications of existing technologies. The method uses the existing largely centralized systems for the treatment of sewage water and sewage sludge in The Netherlands. Both aspects ensure that the phosphates can be recycled in a cost effective manner with very little effort. With this method already 550 ton P/y is being recycled since mid-2008 and it is expected that this volume can increase considerably within five years.

PROCESS DESCRIPTION

The recycling of phosphate from sewage sludge has been achieved by a cooperation between Thermphos International and Slibverwerking Noord-Brabant (SNB).

Thermphos International is a producer of elemental phosphorus, phosphorus derivatives and defined phosphates. For the production of elemental phosphorus, mined apatite (phosphate rock) is granulated and reduced with coke and gravel in a submerged arc furnace at high temperatures. The process lends itself to the recycling of phosphates by replacing a part of the apatite by secondary phosphates (Schipper *et al.*, 2001). The concentrations of a range of impurities determine the suitability of secondary phosphates. Among these, iron is of notable importance, as this is reduced along with the phosphate rock to form ferrophosphorus by-product. This stream consists of an approximately equimolar mixture of FeP and Fe_2P. Apart from having a limited application potential, the formation of this byproduct also decreases the yield of elemental (white) phosphorus and requires additional raw materials and energy. Hence, the iron content of raw materials has a practical upper limit.

SNB operates four incineration lines for the treatment of sewage sludge from municipal WWTP's. On an annual basis more than 400 000 tons of wet sludge (95 000 t/y as dry matter) is incinerated in the installation. This amount represents 28% of the total sludge production in The Netherlands and is produced in more than 70 WWTP's representing a total capacity of approximately 8 million people equivalents. The incineration process completely mineralizes the sludge and transforms it into fly ash with a volume of 37 000 t/y.

The fly ash contains the complete phosphate load present in the sludge and this results in an average phosphate concentration in the ash of 80 g P/kg or 18 wt% P_2O_5. With this concentration the ash represents a low quality phosphate ore and could theoretically be used as an input material for white phosphorus production. Unfortunately the ash as produced under normal conditions, i.e. from sludges originating from all types of WWTP designs, also contains a large amount of iron, typically 10 to 30 times the iron content of common phosphate ores, which reduces the efficiency of the phosphorus production below acceptable limits, and results in unacceptable amounts of ferrophosphorus by-product.

However, it is known that the design and operational parameters of a WWTP determine the iron level of its sludge to a large extent. Sludges poor in iron are commonly encountered. If these are incinered separately, this might yield a suitable ash. Therefore, a study was launched to find suitable sludges for separate incineration.

INFLUENCES ON ASH QUALITY

For the production of white phosphorus, the Fe/P molar ratio should be as low as possible. The encountered Fe/P ratio of 0.6 in standard quality ash is too high to allow its use in the phosphorus process. In naturally occurring apatites, the Fe/P ratio is usually less than 0.05.

The Fe/P molar ratio in the different sludges was seen to vary considerably from 0.1 to values of 1.5 mol Fe/mol P. This immediately revealed that no sewage sludge ash would be able to meet the specifications of natural apatite. For this reason, the Fe/P molar ratio was tentatively limited to a maximum of 0.2. Based on this value, a scan was made of available sludges from various WWTPs or WWTP sludge pools to assess their iron and phosphate content.

The iron content in the sewage sludge is largely determined by the method of P-removal that is used in the WWTPs. Basically an operator of a WWTP can choose for three different methods to remove the phosphate from the wastewater:

(1) dosage of iron salts to precipitate the phosphate;
(2) dosage of aluminum salts to precipitate the phosphate;
(3) enhanced biological phosphorus removal (EBPR; "BioP").

The following table shows the relative use of the different P-removal technologies in The Netherlands (Geraarts *et al.*, 2007).

If iron salts are used to precipitate the phosphate, the iron level in the sludge will be too high to enable recycling of the sludge ash in phosphorus production. Therefore only sludges from WWTP's using aluminum salts or the EBPR process can potentially be used for the production of sewage sludge ash with a low iron level. Table 1 shows that 46% of the sludge incinerated by SNB would be potentially suitable for the production of an ash with a low iron content.

Table 1. Relative use of P-removal technologies in The Netherlands.

Type of P-removal	SNB	The Netherlands
Enhanced biological P removal (EBPR)	26%	25%
No additional P removal[1]	5%	11%
Precipitation with iron	49%	64%
Precipitation with aluminum	20%	

[1]normally about 50% of the influent P will be accumulated in the sludge without any additional measures.

In Figure 1 a cumulative distribution of sludge ash volumes at SNB is given as a function of the Fe/P ratio. It is seen that an unexpectedly low amount, less than 20% of the sludge incinerated by SNB, meets the Fe/P criterion of max. 0.2. This is due to the fact that iron from other sources than P removal may end up in the sludge. These sources include:

(1) Dosage of iron salts in sludge digesters to prevent H_2S formation in the biogas formed.
(2) Dosage of iron salts in sewer pumping stations to prevent odour formation.

(3) Discharges of iron containing sludges to the sewer from drinking water production.
(4) Leakage of iron-containing groundwater to the sewer system.

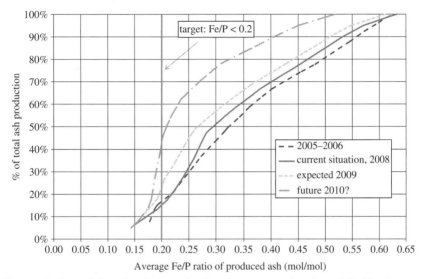

Figure 1. Cumulative distribution of ash volumes as a function of Fe/P ratio.

In addition, a percentage of suitable low-iron sludges is lost for reuse when they are dewatered in central dewatering locations together with sludges from iron-dosing WWTPs.

FULL SCALE TESTS

Two large scale tests have been performed by SNB and Thermphos in 2006 and 2007. In a first test, lasting two months, SNB produced 350 tons of low-iron ash by separating low-iron sludges in its sludge bunker. The produced low-iron ash was then buffered in an intermediate ash storage and consequently processed by Thermphos in a concentrated time span of 2 weeks, thus replacing 4% of the total input of phosphate rock to the plant.

This first test proved that it was feasible to separate the low-iron sludges in the sludge bunker of SNB and to produce ash with a low enough iron content to meet the criteria of the phosphorus process. The test showed that some changes were necessary to the process control of the sludge driers and furnace due to the higher caloric value of the low-iron sludges. These changes were possible within the constraints of the incineration process and did not lead to significant effects on emissions such as NO_x. Also at Thermphos no significant effects were observed. The material was processed without statistically significant influence on the process energy consumption, the capacity of the process, the quality of the end product, and the emissions to air. Additional amounts of ferrophosphorus were seen to fit the calculated mass balance closely; otherwise the phosphorus yield of the process was not affected by the use of the sewage sludge ash.

After the successful first test a larger test was initiated lasting over a period of 5 months from November 2006 until March 2007. In this period SNB produced a total volume of 2000 ton low-iron ash which was continuously processed by Thermphos, thus replacing on average 1 to 2% of the total input of phosphate rock.

The second test showed that it was possible to produce and process the low-iron sludge on a more continuous basis, while maintaining the required quality of the ash. With the second test the Fe/P ratio of the produced ash was slightly higher than with the first test, showing a need for continuous quality control of the ash production. Also during this period, no significant effects were seen in the phosphorus process other than a slightly elevated ferrophosphorus production.

Table 2 shows the composition of normal (standard grade) ash, and the quality of the iron poor ash for the two tests.

Table 2. Composition of normal ash and low-iron ash.

Component	Unit	Normal ash, 2007	Test 1, 350 ton	Test 2, 2 000 ton
Fe	g/kg	85	41	40
P	g/kg	78	116	92
Fe/P	mol/mol	0.61	0.20	0.24

CONCLUSIONS AND OUTLOOK

It has been demonstrated that it is possible to produce an iron poor sewage sludge ash as a feedstock for the phosphorus process, with little or no significant consequences for the operation of the sludge incineration and phosphorus production processes.

It should be noted that the implementation of this route has the advantage of using existing infrastructure. Expansion of suitable quantities can be implemented relatively easily by changes at WWTP level. Typically this will involve a change form iron salt dosage to aluminium salts as precipitation agents, as well as constructing all new WWTPs according to EBPR design. The consequences of aluminium salt dosage were investigated in (Geraarts et al., 2007); it is concluded that operationally, the consequences would be little whereas the economics depend strongly on the local availability of aluminium-containing by-products.

It is especially notheworthy that the recycling pathway described here has a significant advantage over more complicated side stream phosphate recovery methods. Typically these methods involve a separate sludge treatment step, where phosphate is released from excess sewage sludge into solution, and then precipitated in the form of fine particles or pellets (Loosdrecht et al., 1998; Brett et al., 1997). These methods all require a costly equipment extension at WWTP level and will typically only recover 50 to 75% of phosphate in the influent, with the remainder staying in the sludge, which is processed in the usual ways. The prohibitive cost of such modifications are best demonstrated by the fact that since the (experimental) implementation of side-stream phosphate recovery technologies in two out of approximately 370 WWTPs in The Netherlands in the 1990s, none have been upgraded since then. By taking the recycling pathway described here, these investments can be dispensed with. Additionally, the recovery-potential increases as the sewage sludge ash contains circa 90% of the P present in the influent to the WWTP.

Based on the experiences of the two tests, Thermphos and SNB have decided to continue their cooperation on a long-term basis. Iron poor ash will be continuously produced by SNB, and processed in the phosphorus process at Thermphos. From July 2008 onwards until the writing of this article (november 2008) SNB has been separating and incinerating low-iron sludge continuously on one incineration line. The produced ash quality is constant and meets the quality limit of a Fe/P ratio of 0.20.

Activities are now focused on increasing the volume of the available low-iron sludge. To achieve this, logistical improvements on the sludge incineration side have been made. More importantly, information has been spread to the WTTP operators to encourage the supply of low-iron sludge by switching to Al salt dosage, or by constructing new WWTPs according to the EBPR principle. A further encouragement is created by passing on the benefit of lower ash treatment costs to the suppliers of low-iron sludge.

Already two waterboards, which operate several WWTPs, have decided that they are prepared to change to the use of aluminium-salts if this will make their sludge meet the quality criteria. Other WWTP operators will start studies in 2009 to investigate their possibilities to improve the Fe/P-ratio in their sludge. Figure 2 represents the current amount of sludge ash to meet the Fe/P criterion, as well as future scenarios to increase this amount. It is seen that SNB can produce ash with an average Fe/P ratio of 0.2 with a volume corresponding to 17% of the total ash production. If the limit value of the Fe/P ratio would be relaxed to 0.25 the total ash volume that can be produced would increase to over 30% of the total ash production. It is expected that the iron content will further decrease in the coming years due to further optimization projects at the WWTP's. Based on the current knowledge it is expected that the low-iron ash production will increase from 17% to 26% in 2009 and can increase to as much as 45% in 2010. This increase will mainly be the result of changeovers from the dosage of iron salts to the dosage of aluminium salts. On longer term, also the increasing popularity of enhanced biological phosphorus removal (EBPR) will stimulate this pathway of phosphate recycling. Figure 2 illustrates the growing application of the EBPR-process in The Netherlands. Similar trends can be observed in other industrialized countries.

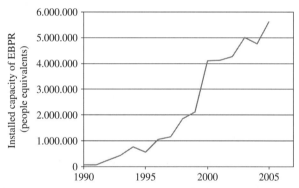

Figure 2. Growing application of the EBPR process for sewage treatment and phosphate removal in the Netherlands (Geraarts *et al.*, 2007).

REFERENCES

Brett, S., Guy, J., Morse, G.K. and Lester, J.N. (1997). Phosphorus removal and recovery techologies, Selper Ltd, London, UK.

Geraarts, B., Koetse, E., Loeffen, P., Reitsma, B. and Gaillard, A. (2007). Fosfaatterug-winning uit ijzerarmslib, STOWA, report 2007-31, ISBN 978.90.5773.380.2.

Loosdrecht, M.C.M. van, Brandse, F.A. and Vries, A.C. de. (1998). *Water Sci. Technol.*, **37**, 209–217.

Schipper, W.J, Klapwijk, A., Potjer, B., Rulkens, W.H., Temmink, B.G., Kiestra F.D.G. and Lijmbach, A.C.M. (2001). *Environmental Technology*, **22**, 1337.

Phosphorous recovery and nitrogen removal from wastewater using BioIronTech process

V. Ivanov[a], C.H. Guo[a], S.L. Kuang[a] and V. Stabnikov[b]

[a]School of Civil and Environmental Engineering, Nanyang Technological University, Singapore
[b]Institute of Municipal Activity, National Aviation University, Kiev, Ukraine

Abstract Two major phosphorous–containing streams on municipal wastewater treatment plant (MWWTP) are raw sewage with phosphorous concentration up to 20 g m^{-3} and reject water (return liquor, digester supernatant, sludge digester liquid), which is produced after the dewatering of biomass from anaerobic digester and containing phosphorous up to 200 g m^{-3}. Reject water is returned to the aeration tank contributing up 50% of phosphorus, nitrogen, and organics into the main stream of the MWWTP. Therefore, recovery of phosphorus from reject water can significantly reduce the phosphorus load to the aeration tank and improve quality of effluent. Recovery of phosphate, which is the major form of P in reject water, by precipitation with Mg, Ca, or Fe-containing reagents is commonly used on the MWWTPs. An innovative technology BioIronTech could diminish the cost of the reagents in 7–10 times because the iron salts in this technology are replaced with ferrous ions produced by iron-reducing bacteria (IRB) from the cheap iron ore. The aim of the research was to study the efficiency of phosphate precipitation from reject water using the bioreduction of iron ore by IRB. The ferrous production rate reciprocally depended on the size of iron ore particles. However, the phosphate recovery rate did not depend on the specific surface area of the iron ore particles when the size of iron ore particle was smaller than 7 mm. Iron-reducing bacteria ensured the production of ferrous ions from iron ore up to concentration of 550 g m^{-3} and the removal efficiency of phosphate from reject water up to 73%. The anaerobic/aerobic recovery of phosphate using BioIronTech process could be combined with the anaerobic denitrification and aerobic nitrification.

INTRODUCTION

The recovery of phosphorus from reject water

Phosphorus, discharged from municipal wastewater treatment plants (MWWTPs) to the aquatic systems, is the main factor causing their eutrophication. There are two major phosphorous – containing streams on the MWWTP: 1) raw sewage with concentration of P from 5 to 20 mg L^{-1}, and 2) reject water (other terms are return liquor, digester supernatant, sludge digester liquid) produced by the dewatering of anaerobically digested activated sludge and containing up to

200 mg P L^{-1}. Reject water is returned to the aeration tank and contributes from 10 to 50% of the nutrients and organics loads in the main stream of the MWWTPs (Ivanov *et al.*, 2005; van Loosdrecht and Salem, 2006). Therefore, removal or recovery of phosphorus from reject water can significantly reduce the phosphorus load to the aeration tank and improve quality of effluent.

The recovery of phosphate, which is major form of P in reject water, by precipitation with Mg, Ca, or Fe-containing reagents is commonly used for the the recovery of nutrients on the MWWTPs (Morse *et al.*, 1998; de-Bashan and Bashan, 2005; Takacs *et al.*, 2006). Ferric phosphate can be recovered by settling and used as a fertilizer (de-Bashan and Bashan, 2005). A disadvantage of the chemical precipitation methods is the high cost of reagents. For example the cost of 1 ton of Fe in $FeCl_3$ is approximately US$1015.

Another disadvantage of the ferric and ferrous salts applications is that they are stable only at low pH, so their addition to reject water requires also an addition of calcium or sodium hydroxides (Water Environment Federation, 2006). Therefore, the high costs of the ferric salts and lime are the main disadvantages of their applications for phosphate removal from wastewater.

The BiolronTech process

To diminish the expenses for the ferrous and ferric reagents, an innovative technology BioIronTech combining biological and chemical processes could be applied (Ivanov *et al.*, 2004; 2005; Tay *et al.*, 2008). The iron salts in this technology are replaced with the ferrous ions produced by iron-reducing bacteria (IRB) from the relatively cheap iron ore. An approximate cost of 1 ton of Fe in iron ore is US$140. That is 7 times cheaper than Fe of chemical reagents. The principles of BioIronTech process are described in the related US patent (Tay *et al.*, 2008). This process includes the anaerobic and aerobic steps. At the first step, the cheap source of Fe^{3+}, for example iron ore, is anaerobically reduced to Fe^{2+} by iron-reducing bacteria using organic substances present in wastewater:

$$4Fe_2O_3 + CH_3COO^- + 7H_2O \rightarrow 8Fe^{2+} + 2HCO_3^- + 15OH^- \qquad (1)$$

The ferrous ions, produced in the anaerobic bioreactor, precipitate phosphate from wastewater as well as urban stormwater, aqricultural drainwater, aquacultural recycled water:

$$Fe^{2+} + HPO_4^- \rightarrow FeHPO_4 \qquad (2)$$

At the second step of the process, ferrous phosphate is chemically or biologically oxidized by oxygen:

$$2FeHPO_4 + HPO_4^{2-} + 0.5O_2 + H_2O \rightarrow Fe_2(HPO_4)_3 + 2OH^- \qquad (3)$$

There are known some environmental engineering applications of iron-reducing bacteria (Fredrickson and Gorby, 1996; Lovley, 2000; Nielsen et al., 2002). However, the spectrum of the BioIronTech technologies wider than the known ones and includes such environmental engineering applications as the enhanced removal of phosphate from reject water (return liquor) of MWWP (it is described in this paper); enhancement of anaerobic digestion on MWWTP due to competitive inhibition of sulphate reduction by ferric reduction (Ivanov et al., 2004a,b; Ivanov et al., 2005; Stabnikov and Ivanov, 2006); treatment of food-processing wastes (Ivanov et al., 2004b); anaerobic treatment of fat-containing wastes due to precipitation of long-chain fatty acids with ferrous ions (Ivanov et al., 2002); anaerobic treatment of sulphate-containing wastes due to competitive inhibition of sulphate reduction by ferric reduction (Stabnikov and Ivanov, 2006); removal/ recovery of phosphate from stormwater, aquacultural and agricultural wastewaters (Ivanov et al., 1996; 1997; 1999; Stabnikov and Ivanov, 2004); biodegradation and complete removal of aromatic acids, chlorinated phenols and cyanides from industrial wastewaters (Tay et al., 2004; Ivanov et al., 2005).

Other potential applications of the BioIronTech process could be the biorestoration of eutrophicated lakes and ponds; the maintenance of water quality in aquacutural ponds; the production of ferrous coagulant for pre-desalination treatment of seawater, pre-treatment of industrial wastewater, and desilting of stormwater; the biocementation of the dams, slopes and land reclamation sites; the microbially-enhanced oil recovery; the treatment of landfill leachate; the removal of arsenic and other heavy metals as well as radionuclides from groundwater.

Almost all these technologies can be performed using the iron salts (Fytianos et al., 1998; Morse et al., 1998; Takacs et al., 2006). However, the ferrous ions can be produced in BioIronTech process from iron ore (Ivanov et al., 2002), which is significantly cheaper than the iron salts. Additionally, BioIronTech process does not require a mixer as well as the dosators of dissolved iron salt and lime that are required by any other process with the iron salt addition.

The potential application of BioIronTech process for the recovery of phosphate and removal of nitrogen from reject water

The most effective application of BioIronTech process is the recovery of phosphate and the removal of nitrogen from reject water of the MWWTP. Although the flow of recycled reject water is only from 2 to 4% of the total flow of raw sewage (Janus and Van der Roset, 1997; our data), it contributes from 10 to 40% of the nitrogen load and from 10% to 80% of phosphorous load on the activated sludge tank (van Loosdrecht and Salm, 2006; our data), due to high concentration of ammonium, from 200 to 1500 mg L^{-1} (Berend *et al.*, 2005; our data), and up to 200 mg P L^{-1} (Batistoni *et al.*, 1997; our data).

By our calculations, the annual supplies of nitrogen and phosphorous with reject water in Singapore with the population of 4 million are 5,700 thousand tons of P and 13,300 tons of N. Considering that world population, served with centralized sewage treatment, is 1.5 billion and the average domestic water consumption is the same as in Singapore, the annual supplies of phosphorous and nitrogen with reject water into main stream of the MWWTPs are approximately 2 and 5 million tons, respectively. Therefore, the recovery of phosphate and the removal of ammonium from reject water prior their recycling into main stream can significantly reduce the nutrient loads on MWWTPs, recover valuable phosphorous fertilizer, and diminish the concentrations of the nutrients in WWTP effluent, thus preventing eutrophication of the aquatic systems and deterioration of water quality.

The different combinations of nitrification, denitrification and anaerobic oxidation of ammonium had been reported for the nitrogen removal from reject water, such as bio-augmentation batch enhanced (BABE) technology (Salem *et al.*, 2004); the single reactor system high activity ammonium removal over nitrite (SHARON) process (Hellinga *et al.*, 1999); anaerobic ammonium oxidation (ANAMMOX) process (Strous *et al.*, 1997); SHARON process combined with ANAMMOX process (Van Dongen *et al.*, 2001); CANON process (van Benthum *et al.*, 1998); OLAND process (Kuai and Verstraete, 1998); aerobic/anoxic deammonification process (Gut *et al.*, 2006). However, only BioIronTech process could be hypothetically suited for the simultaneous removal of nitrogen and the recovery of phosphate from reject water of MWWTPs.

The aim of this research was to study the efficiency of BioIronTech process for the recovery of phosphate and partially for the removal of nitrogen from reject water of MWWTPs.

ANAEROBIC TREATMENT OF REJECT WATER USING THE BIORONTECH PROCESS

The components of the system

The major mineral of used iron ore was hematite (Fe_2O_3). The porosity of the iron ore particles was from 45% to 55% (v/v) depending on the size. The iron ore particles of different sizes were produced by the sieving of the crashed iron ore.

The content of reject water was as follows, mg L^{-1}: total organic carbon (TOC), 205 ± 4; total suspended solids (TSS), 460 ± 50; volatile suspended solids (VSS), 430 ± 50; total dissolved phosphate (TDP), 208 ± 2; total ferrous iron, 9.1 ± 0.1; dissolved NH_4^+-N, 690 ± 8; dissolved NO_3^--N, 0.68 ± 0.04; dissolved NO_2^--N, .05 ± 0.01; pH, 7.69 ± 0.02. All parameters have been measured using standard methods (APHA, 2005).

The volatile fatty acids comprise almost all TOC of the reject water. Empirical formula of organic matter of reject water was $CH_2O_{0.57}$ and average oxidation number of carbon in this organic matter was + 0.86, so the balanced equation, describing the reduction of ferric by organic matter of reject water, is as follows:

$$4.86Fe^{3+} + CH_2O_{0.57} + 1.43H_2O \rightarrow 4.86Fe^{2+} + CO_2 + 4.86H^+ \qquad (4)$$

The ferrous production from the iron ore

The ferrous production rate in anaerobic batch culture of iron-reducing bacteria in reject water with an addition of iron ore depended on the iron ore particle size. However, the phosphorus removal rate did not depend on the specific surface area of the iron ore particles when the size of iron ore particle was smaller than 7.6 mm (Table 1).

Table 1. The ferrous production and phosphate removal rates depending on the size of the iron ore particles.

Mean size of iron ore, mm	Total Fe^{2+} production rate, mg L^{-1} d^{-1}	Dissolved phosphate removal rate, mg P L^{-1} d^{-1}
35	0.37	1.27
17	0.72	1.46
7.6	1.27	2.18
2.4	1.51	2.18
0.6	5.99	2.19

The concentration of ferrous ions produced by IRB was up to 550 mg L^{-1}. The phosphorus concentration in reject water was diminished by 90% of the initial concentration during the bioreduction of iron ore. In another batch culture of IRB grown on the shaker in reject water with added iron ore particles with the size between 0.5 and 1 mm, the maximum rate of total ferrous production was 29 mg L^{-1} d^{-1}.

Organics and nitrogen removal from reject water using BioIronTech process

The maximum rate of TOC removal was 19 mg TOC L^{-1} d^{-1} in batch culture of IRB grown on the shaker in the reject water with the added iron ore particles with the size between 0.5 and 1 mm. According to the Equation 3, theoretical ratio of TOC/Fe is 0.044 g TOC oxidized g^{-1} Fe^{3+} reduced. Therefore, the quantity of organics oxidized by Fe^{3+} was 6.7% of total TOC removal. Probably, the major portion of TOC in reject water was removed by fermentation because the data on the content of VFA demonstrated the transformations of VFAs into mainly formic and isovaleric acids (Table 2).

Table 2. The changes of VFA contents in reject water after incubation with iron ore.

VFA	Formula	Content, molar %	
		before treatment	after treatment
formic acid	CH_2O_2	9.4	62.2
acetic acid	$C_2H_4O_2$	31.9	17.5
propionic acid	$C_3H_6O_2$	9.4	0.0
isobutyric acid	$C_4H_8O_2$	3.0	0.0
butyric acid	$C_4H_8O_2$	1.8	0.0
isovaleric acid	$C_5H_{10}O_2$	9.2	20.3
valeric acid	$C_5H_{10}O_2$	10.7	0.0
isocaproic acid	$C_6H_{12}O_2$	13.1	0.0
caproic acid	$C_6H_{12}O_2$	11.5	0.0

Hypothetically, the removal of phosphate during anaerobic stage of the BioIronTech process can be performed simultaneously with denitrification using VFAs of reject water by the equation:

$$4.86NO_3^- + 5CH_2O_{0.57} + 4.86H^+ \rightarrow 2.43N_2 + 5CO_2 + 7.43H_2O \qquad (5)$$

However, the calculations from this equation show that concentration of TOC in reject water is not sufficient for complete bioreduction of nitrate (produced from ammonium of reject water on aerobic stage) simultaneously with the bioreduction of iron ore and the recovery of phosphate by ferrous phosphate precipitation. Therefore, to perform simultaneous recovery of phosphate and denitrification on anaerobic stage of BioIronTech process, the additional donor of electrons must be added into reject water.

To ensure production of nitrate from ammonium in reject water, the aerobic stage of BioIronTech process must combine oxidation of ferrous and ammonium with the recycling of produced nitrate to anaerobic stage after recovery of ferric phosphate by sedimentation as shown in Figure 1.

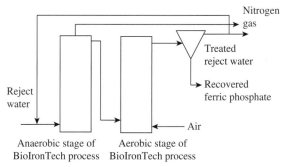

Figure 1. Schematics of the phosphate recovery from reject water in continuous BioIronTech process.

Effect of nitrate and organic acids on the phosphate recovery from reject water

When stoichiometrical ratio of acetate to nitrate in reject water corresponded to the following equation:

$$8NO_3^- + 5CH_3COOH + 8H^+ \rightarrow 4N_2 \uparrow + 10CO_2 + 14H_2O \qquad (6)$$

or was higher, the removal efficiency for nitrate was 99% and the recovery efficiency for dissolved phosphate, which was precipitated by the ferrous ions produced from an iron ore, was 73%.

The phosphate removal efficiency by the precipitation with biologically produced ferrous ions was slightly inhibited by nitrate and acetate additions (see Figure 2), provably due to the competition between nitrate, acetate, and phosphate anions reacting with ferrous cations.

CONCLUSIONS

1. Annual recovery of phosphorus from reject water of MWWTPs all over the world could reach 2 million tons.
2. Application of the BioIronTech process for the phosphate recovery can outcompete the conventional chemical precipitation of phosphate with iron, calcium, or magnesium reagents because of low cost of iron ore and simplicity of the process performance.
3. The recovery of phosphate from reject water of the MWWTR can be accompanied with the removal of nitrogen and organics.

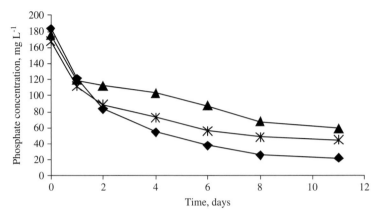

Figure 2. The removal of phosphate from reject water using iron ore and iron-reducing bacteria. ◆ Control 1 (no nitrate and acetate added); ∗ Control 2 (presence of nitrate but no addition of acetate); ▲ Experiment with addition of acetate and nitrate in stoichiometrical ratio corresponding to the following reaction: $8NO_3^- + 5CH_3COOH + 8H^+ \rightarrow 4N_2 \uparrow +10CO_2 + 14H_2O$.

REFERENCES

American Public Health Association (APHA), American Water Works Association and Water Environment Federation (2005). *Standard Methods for the Examination of Water and Wastewater*, 21st edition, Washington DC, USA.

Bashan, L.E. and Bashan, Y. (2005). Recent advances in removing phosphorus from wastewater and its future use as fertilizer (1997–2003). *Wat. Res.*, **38**(19), 4222–4246.

van Benthum, W.A.J., Garrido, J.M., Mathijssen, J.P.M., Sunde, J., van Loosdrecht, M.C.M. and Heijnen, J.J. (1998). Nitrogen removal in intermittently aerated biofilm airlift reactor. *J. Environ. Eng.*, **124**(3), 239–248.

Fredrickson, J.K. and Gorby, Y.A. (1996). Environmental processes mediated by iron-reducing bacteria. *Curr. Opin. Biotech.*, **7**(3), 287–294.

Fytianos, K., Voudrias, E. and Raikos, N. (1998). Modelling of phosphorus removal from aqueous and wastewater sample using ferric iron. *Environ. Pollut.*, **101**(1), 123–130.

Gut, L., Plaza, E., Trela, J., Hultman, B. and Bosander, J. (2006). Combined partial nitritation/Anammox system for treatment of digester supernatant. *Water Sci. Technol.*, **53**(12), 149–159.

Hellinga, C., van Loosdrecht, M.C.M. and Heijnen, J.J. (1999). Model based design of a novel process for nitrogen removal from concentrated flows. *Math. Comp. Modell. Dyn. Sys.*, **5**(4), 351–371.

Ivanov, V.N., Stabnikova, E.V. and Shirokih, V.O. (1997). Influence of ferrous oxidation on nitrification in aqueous and soil model ecosystems. *Microbiol.*, **66**(3), 428–433.

Ivanov, V.N., Sihanonth, P. and Stabnikova, E.V. (1999). Amensalism of nitrifying and iron-oxidizing bacteria in the biofilm of aquacultural biofilter. *J. Appl. Microbiol.*, **85**, 258S–259S.

Ivanov, V.N., Stabnikova, E.V., Stabnikov, V.P., Kim, I.S. and Zuber, A. (2002). Effects of iron compounds on the treatment of fat-containing wastewaters. *Appl. Biochem. Microbiol.*, **38**(3), 255–258.

Ivanov, V., Wang, J.Y., Stabnikov, V., Xing, Z.K. and Tay, J.H. (2004a). Improvement of sludge quality by iron-reducing bacteria. *J Resid. Sci. Technol.*, **1**(3), 165–168.

Ivanov, V., Wang, J.Y., Stabnikova, O., Krasinko, V., Stabnikov, V., Tay, S.T.L. and Tay, J.H. (2004b). Iron-mediated removal of ammonia from strong nitrogenous waste-water of food processing. *Wat. Sci. Technol.*, **49**(5–6), 421–431.

Ivanov, V., Stabnikov, V., Zhuang, W.Q., Tay, S.T.L. and Tay, J.H. (2005). Phosphate removal from return liquor of municipal wastewater treatment plant using iron-reducing bacteria. *J. Appl. Microbiol.*, **98**(5), 1152–1161.

Janus, H.M. and Van der Roset, H.F. (1997). Don't reject the idea of treating reject water. *Water Science and Technology.*, **35**(10), 27–34.

Kuai, L. and Verstraete, W. (1998). Ammonium removal by the oxygen-limited autotrophic nitrification-denitrification system. *Appl Environ. Microbiol.*, **64**(11), 4500–4506.

van Loosdrecht, M.C.M. and Salem, S. (2006). Biological treatment of sludge digester liquids. *Wat. Sci. Technol.*, **53**(12), 11–20.

Lovley, D.R. and Anderson, R.T. (2000). Influence of dissimilatory metal reduction on fate of organic and metal contaminants in the subsurface. *Hydrogeol. J.*, **8**(1), 77–88.

Morse, G.K., Brett, S.W., Guy, J.A. and Lester, J.N. (1998). Review: phosphorus removal and recovery technologies. *Sci. Total Environ.*, **212**(1), 69–81.

Nielsen, J.L., Juretschko, S., Wagner, M. and Nielsen, P.H. (2002). Abundance and phylo-genetic affiliation of iron reducers in activated sludge as assessed by fluorescence *in situ* hybridization and microautoradiography. *Appl. Environ. Microbiol.*, **68**(9), 4629–4636.

Salem, S., Berends, D.H.J.G., van der Roest, H.F., van der Kuij, R.J. and van Loosdrecht, M.C.M. (2004). Full-scale application of the BABE technology. *Wat. Sci.Technol.*, **50**(7), 87–96.

Stabnikov, V.P. and Ivanov, V.N. (2006). The effect of various iron hydroxide concen-trations on the anaerobic fermentation of sulfate-containing model wastewater. *Appl. Biochem. Microbiol.*, **42**(3), 284–288.

Stabnikov, V.P., Tay, S.T.L., Tay, J.H. and Ivanov, V.N. (2004). Effect of iron hydroxide on phosphate removal during anaerobic digestion of activated sludge. *Appl. Biochem. Microbiol.*, **40**(4), 376–380.

Strous, M., Van Gerven, E., Zheng, P., Kuenen, J.G. and Jetten M.S.M. (1997). Ammonium removal from concentrated waste stream with the anaerobic ammonium oxidation (Anammox) process in different reactor configurations. *Wat. Res.*, **31**(8), 1955–1962.

Takacs, I., Murthy, S., Smith, S. and McGrath, M. (2006). Chemical phosphorus removal to extremely low levels: Experience of two plants in the Washington, DC area. *Wat. Sci. Technol.*, **53**(12), 21–28.

Tay, J.H., Tay, S.T.L., Ivanov, V. and Hung, Y.T. (2004). Application of biotechnology for industrial waste treatment. In: *Handbook of Industrial Wastes Treatment.*,

2nd edn., (L.K. Wang, Y.T. Hung, H.H. Lo, C. Yapijakis, eds.). Marcel Dekker, N.Y., pp. 585–618.

Tay, J.H., Tay, S.T.L., Ivanov, V., Stabnikova, O. and Wang, J.Y. (2008). Compositions and methods for the treatment of wastewater and other waste. US Patent 7,393,452. Date of grant July 1, 2008. Date of filing 1 April 11, 2003.

Van Dongen, U., Jetten, M.S.M. and van Loosdrecht, M.C.M. (2001). The SHARON-Anammox process for treatment of ammonium rich wastewater. *Wat. Sci. Technol.*, **44**(1), 153–160.

Water Environment Federation (2006). *Biological nutrient removal (BNR) operation in wastewater treatment plants.* McGraw-Hill, USA.

Phosphorus speciation of sewage sludge ashes and potential for fertilizer production

S. Nanzer[a], M. Janousch[b], T. Huthwelker[b], U. Eggenberger[c], L. Hermann[d], A. Oberson[a] and E. Frossard[a]

[a]Institute of Plant Science, ETH Zurich, Zurich, Switzerland
[b]Swiss Light Source, Paul Scherrer Institute, Villigen, Switzerland
[c]Institute of Geological Sciences, University of Bern, Bern, Switzerland
[d]ASH DEC Umwelt AG, Vienna, Austria

Abstract In Switzerland, direct application of sewage sludge in agriculture has been forbidden, resulting in high quantities of phosphorus rich sewage sludge ashes. As sewage sludge ashes have high heavy metal contents, fertiliser production is only feasible after removal of heavy metals. ASH DEC Umwelt AG developed a thermo-chemical treatment to extract heavy metals from sewage sludge ashes. Swiss ashes have been collected and submitted to the thermo-chemical treatment. Chemical characteristics and phosphorus speciation of treated and untreated sewage sludge ashes have been analyzed. A pot experiment with ryegrass, comparing untreated sewage sludge ashes to water soluble phosphorus fertilizer and rock phosphate was performed. P-content of untreated and treated sewage sludge ashes ranged from 40 to 86 g kg^{-1}. Heavy metals were efficiently extracted with removal rates up to 98%. Mineralogical investigations revealed the presence of whitlockite and aluminum- and iron-phosphates in untreated sewage sludge ashes. However, after the thermo-chemical treatment phosphorus was present as chlorapatite. Investigated ashes enhanced plant growth significantly, which shows to potential of heavy metal free sewages sludge ashes for fertilizer development.

INTRODUCTION

Phosphorus (P) recycling from wastewater is of great interest and will become essential to ensure agricultural production. Since 2006 the direct agricultural use of sewage sludge (SS) is forbidden in Switzerland because of sanitary concerns. SS is either incinerated by co- or mono-combustion or enters cement production. The yearly incineration of approximate 4 million tons of Swiss SS results in 100 thousand tons of incineration ash, which is at present put into landfills (BAFU, 2006). Recycling P from these residues would allow substituting up to 100% of Swiss P imports.

Comparing fresh SS to sewage sludge ash (SSA), SSA is free of organic pollutants and pathogens and therefore a reasonable source for fertilizer development. But as SSA has high heavy metal contents, fertilizer production is only feasible after their removal. Based on the work of Wochele *et al.* (1999) ASH DEC Umwelt AG developed a thermo-chemical treatment to remove heavy metals from SSA, combining chloride-addition and high-temperature incineration. Before the production of an ash-based fertilizer can start in Switzerland, its advantages and risks must be studied. We currently investigate the elemental composition, P solubility and P speciation of SSA to assess the potential release of P from SSA to soil and plant.

MATERIALS AND METHODS

Ten of fourteen Swiss SSA mono-incineration plants have been sampled. The ashes were characterized for total contents of macronutrients and heavy metals (X-ray fluorescence spectroscopy and wet chemical total digestion), pH in water and P solubility in an acid, neutral and alkaline extractant. These parameters were used to select five ashes, which were submitted to the thermo-chemical treatment designed by ASH DEC Umwelt AG. The thermo-chemical treatment consisted of a $CaCl_2$-addition of 150 g/kg SSA. The additive-ash mixture remained 30 min in a rotary kiln at 950 or 1100°C. The samples were air cooled immediately. The treated SSA were characterized as the untreated. Additionally the P speciation was investigated by: X-ray powder diffraction (XRD), X-ray absorption P K-near edge spectroscopy (XANES) and extended X-ray absorption fine structure analysis (EXAFS) at the P K-edge. X-ray absorption spectroscopy was conducted at the Swiss Light Source, a facility of the Paul Scherrer Insitute (Villigen). In addition the SSA samples will be investigated by electron microscopy (scanning electron microscopy coupled to energy dispersive X-ray spectroscopy).

A growth trial with ryegrass (*Lolium mutliflorum*) applying untreated SSA has been carried out. Another pot experiment comparing thermo-chemically treated to untreated SSA is in progress.

RESULTS, PRELIMENARY DATA AND DISCUSSION

Total P-content of untreated and treated SSA ranged between 40 and 86 g kg^{-1} and decreased with increasing proportion of SS from industrial origin. Due to different P-precipitants addition during wastewater treatment, ashes could be grouped into aluminum, iron and calcium-rich SSA (Table 1). From each of these three groups at least one ash was selected (denoted Al-SSA, Fe-SSA, Ca-SSA) and submitted to the thermo-chemical treatment. More than 90% of cadmium, zinc and lead were removed. The removal of copper, chromium and nickel was less, with concentrations remaining in ranges considered not

Table 1. Characteristics of Swiss sewage sludge ashes (SSA) and sewage sludge (SS) source and incineration of the respective ash.

Incineration plant	1	2	3	4	5	6	7	8	9	10
SSA characteristics										
Type*	Fe	Al	Fe	Fe	Fe	Fe	Fe	Ca	Ca	Fe
P [g kg^{-1}]**	58	64	78	72	86	76	76	40	48	79
Al [g kg^{-1}]**	67	152	44	56	43	67	54	67	75	44
Ca [g kg^{-1}]**	116	58	123	142	97	118	97	232	209	114
Fe [g kg^{-1}]**	96	30	177	125	188	93	179	100	74	152
SS source										
Type	Fresh Ferm.	Fresh Ferm	Fresh Ferm.	Fresh	Ferm.	Fresh Ferm.	Ferm.	Fresh Ferm.	Fresh	Fresh Ferm.
Industrial [%]	15	<10	<10	<10	<10	20	<10	90	90	<10
SS incineration										
Furnace***	MH	FB	MH	MH	FB	RK	FB	MH	SF	FB
Temp. [°C]	875	900	850	800	900	875	875	850	1000	875

*Aluminum-, calcium- and iron-rich SSA composition (Al-, Ca- and Fe-SSA).
**Determined using X-ray fluorescence spectroscopy.
***Multi hearth (MH), fluidized bed (FB), rotary kiln (RK), stocker-fired (SF).

to be critical for plant nutrition. In general the complexion of heavy metals with chloride was more efficient at higher temperature.

Solubility of P in water was below 3% of total P and increased slightly due to the thermo-chemical treatment. P-solubility in citric acid significantly improved from average 52% of total P for untreated to 76% at 950°C and 62% at 1100°C for treated SSA.

X-ray powder diffraction showed the P-bearing phase whitlockite in untreated, and chlorapatite in thermo-chemically treated SSA. Changes in mineralogy are illustrated in Figure 1 showing XRD spectra of an untreated and thermo-chemically treated Fe-SSA. No Al- and Fe-phosphates could be identified. However, the detection of minerals was hampered due to poor cristallinity and interfering XRD spectra of the numerous phases. XANES data indicate the presence of Al- and Fe-phosphates in the untreated SSA and confirm apatite as a main mineral P-phase after the thermo-chemical treatment. Figure 2 shows X-ray absorption spectra of the untreated and treated Fe-SSA and hydroxylapatite (preliminary data).

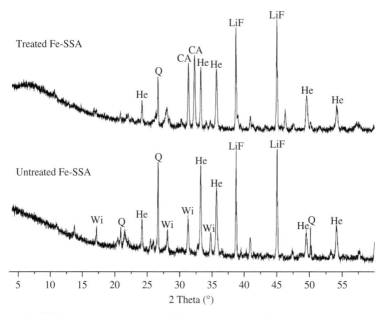

Figure 1. XRD spectra of an iron-rich untreated and thermo-chemically treated sewage sludge ash (Fe-SSA) containing chlorapatite (CA), hematite (He), lithiumfluoride (LiF) as reference, quartz (Q) and whitlockite (Wi).

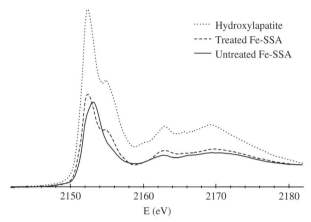

Figure 2. P K-edge of an iron-rich untreated and thermo-chemically treated sewage sludge ash (Fe-SSA) and hydroxylapatite.

Pot experiments with ryegrass under greenhouse conditions showed an enhanced performance – compared to unfertilized control – for untreated SSA. However, dry matter production of plants fertilized with untreated SSA was 30% and P-uptake 70% lower than for plants fertilized with water soluble P.

OUTLOOK

Growth trials to assess the potential of thermo-chemically treated SSA to be used as fertilizer are in progress. Phosphorus radioisotope techniques may deliver more information on the kinetics of P release from SSA to the soil solution. Finally, the use of SSA based fertilizers, as P source for crops will be tested in field experiments, with several crops.

REFERENCES

BAFU (2006). Abfall und Recycling 2006 im Überblick. Bundesamt für Umwelt BAFU, Bern.

Wochele, J., Ludwig, C., Stucki, S., Auer, P.O. and Schuler, A.J. (1999). Abfall- und Rückstandsbehandlung. Grundlagen zur Thermischen Schwermetallseparation, pp. 19–36 Forschung für eine nachhaltige Abfallwirtschaft. Ergebnisse des Integrierten Projektes Abfall Schwerpunktprogramm Umwelt des Schweizerischen Nationalfonds. Brandl, H. *et al.*, Villigen.

Savings from integration of centrate ammonia reduction with BNR operation: simulation of single-sludge and two-sludge plant operation

Morton Orentlicher[a], Alexander Fassbender[a] and Gary Grey[b]

[a]ThermoEnergy Corporation, 124 W. Capitol Avenue, Ste. 880 Little Rock AR 72201
[b]HydroQual. Inc, 1200 MacArthur Blvd.Mahwah, NJ 07430

Abstract Simulation studies of both single-sludge and two-sludge nutrient removal processes demonstrate that side-stream treatment of centrate with the patented Ammonia Recovery Process provides substantial savings in chemicals, energy and sludge production relative to conventional BNR. Generic models of the two major classes of nutrient reduction technology, single sludge and two sludge, were adapted to provide similar operation to those of New York City for the single sludge category and the Blue Plains plant of WASA for the two sludge category.

INTRODUCTION

The Ammonia Recovery Process (ARP) is a patented technology of the ThermoEnergy Corporation, and has been extensively studied at both the computational and pilot test level as a stand alone technology for the removal of ammonia from concentrated waste streams. ARP consists of a pre-treatment train followed by a vacuum distillation operation that removes over 80% of the influent ammonia, and a highly efficient ion exchange module that polishes the effluent to meet the required specifications. The applications studied in these simulations did not require the polishing stage since the ammonia level in the treated side stream was set at 100 ppm. Exhibit 1 displays typical data obtained with ARP vacuum distillation operation operating as modelled in this report.

Cost estimates for removal of about 90% of the centrate ammonia for the 85 mgd water pollution control plant (WPCP) operated by New York City DEP at 26th Ward indicate that a capital cost of $30 million for centrate treatment would accomplish the same reduction as would a capital cost of $115 million for BNR applied to the main stream. Estimates of operating costs similarly indicated major reductions in costs for energy, chemicals, and sludge disposal.

Operating benefits for side-stream treatment of centrate cannot be fully described in standard estimation worksheets due to the multiple interactions

© 2009 The Authors, *International Conference on Nutrient Recovery from Wastewater Streams.* Edited by Ken Ashley, Don Mavinic and Fred Koch. ISBN: 9781843392323. Published by IWA Publishing, London, UK.

between processes involved in nutrient reduction in a WPCP. Simulation studies using BioWin were conducted of the impact of centrate side-stream treatment with ARP on inputs needed for BNR. The simulation study provides the projected required process inputs to achieve a desired level of ammonia removal by the WPCP.

Exhibit 1. Ammonia removal from centrate by ARP.

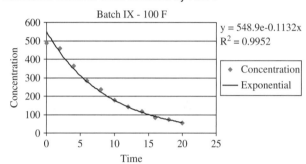

BACKGROUND AND CONDITIONS MODELED

Contemporary WPCP are multi-stage bio-chemical facilities, with complex interaction of the stages. Simple linear reasoning is inadequate to describe the ecology of the active organisms, which controls operation at aerobic secondary treatment, anoxic and aerated zones of nutrient reduction, and the anaerobic digesters. Consideration of the feedback of intermediate streams and the behavior of clarifiers and dewatering devices reinforces the conclusion that a simulation that accounts for these phenomena is needed for a realistic computational model. The most widely used simulation software is BioWin, developed by EnviroSim Associates and applied repeatedly to similar problems for more than a decade (Marsteller, *et al.*, 1994, Katehis, *et al.*, 1995). BioWin simulations to estimate the impact of ARP use on the inputs required for reduction of the ammonia effluent from a typical WPCP were performed for steady-state operation at 20 C.

Two broad categories of WPCP were studied: *single-sludge* plants, such as those of NYC DEP, in which modification of the secondary treatment is employed for nutrient reduction, and *two-sludge plants,* such as the Blue Plains plant of WASA, in which separate treatment trains are used for carbonaceous BOD removal and nutrient reduction. Generic 100 mgd models were used for each of these categories, and performance of an example of each category was used to adjust the model operation parameters to match typical centrate ammonia concentration. Parameters were specified in the generic BioWin model for each category using default values for rate constants, tank volumes, etc, and identical influent characteristics were used for the two categories of plant. Total nitrogen in plant effluent of 5 ppm was achieved for the single-sludge model and the 2 ppm for the two-sludge case. This is displayed for the single-sludge case in Exhibit 2, and for the two-sludge case in Exhibit 3.

ARP is itself a sequence of steps that are dependent on both the waste stream to be treated and the effluent specifications. A version of ARP to be modeled for centrate treatment in this study takes the effluent from the vacuum separation step as the return stream to the plant. This preserves the alkalinity of that stream to be used in the plant's BNR, while returning a stream with 100 ppm of ammonia-nitrogen. The effect of this application of ARP treatment of the centrate on chemical and energy inputs was projected for each category of plant. The BioWin model for each category was modified to include ARP as shown in Exhibits 2 and 3. The chemical (methanol and alkalinity) and oxygen inputs to the BNR process were adjusted in order to keep effluent nitrogen the same for ARP and non-ARP results. Oxygen demand was taken as a surrogate for energy use since the major energy benefit of ARP is the decrease in aeration required. A conservative estimate of caustic demand for ARP based on full conversion of centrate bicarbonate to carbonate was employed.

FINDINGS

Identical influent streams are assumed for the two plant categories studied. The impact of ARP on plant operations is measured by changes in the following set of variables for each category.

Input to model of wastewater plant	Definition
Total 'N' load	Total nitrogen input to BNR = Influent-N + Centrate-N
Sludge produced	Total sludge output from plant
Alkalinity	Total alkalinity required for nutrient reduction and maintaining an effluent pH of 7.3
Methanol	Methanol required for BNR
Oxygen	Oxygen input for carbonaceous BOD removal and nitrification

The N-load on the BNR process for either category of plant includes recycled centrate (and small amounts from other recycle streams) as well as influent nitrogen. Typically nitrogen in recycled centrate is about 20–40% of the total N load. Sludge produced by the plant is reduced by ARP treatment of the centrate, since nitrifier bacterial growth in the BNR is reduced due to the ammonia removed by ARP. Alkalinity demand is similarly reduced in the BNR process by the removal of ammonia by ARP, while half of the alkalinity added for ARP is available for use in the treated centrate that is recycled to the BNR process. Methanol demand is obviously reduced as well by the removal of ammonia by ARP, but in addition the soluble carbon made available by anaerobic digestion is not affected by ARP and is returned to the BNR as bio-available carbon replacing an equivalent amount of methanol. Last, the oxygen demand is lowered by the reduced ammonia nitrogen load caused by removal of ammonia by ARP.

Exhibit 2. BioWin model for single sludge model.

Flow: 100 MGD			
Centrate TN: 795 mg/L at 0.41 MGD			
Influent Quality			
Flow	MGD	100	
TKN	mg/L	25	
COD	mg/L	290	
CBOD	mg/L	142	
TSS	mg/L	138	
pH		7.0	
Effluent Quality		Without ARP	With ARP
NH_3-N	mg/L	0.56	0.58
NO_3-N	mg/L	2.54	2.28
TN	mg/L	5.3	5.1
COD	mg/L	33.5	32.5
CBOD	mg/L	8.66	8.28
pH		7.3	7.3

Single-sludge plant

The single-sludge plant, as shown in Exhibit 2, adapts secondary treatment to achieve nitrification and denitrification. This allows considerable capital cost savings in retrofit of existing equipment. The % reduction of sludge produced is less than the reduction in N load, since BNR is not the only source of bio-solids. Methanol reduction exceeds the reduction of nitrogen load since treated centrate and influent provide a source of bio-available carbon.

The generic model produced a centrate-N load of about 10% of influent TKN, whereas NYC DEP plants typically have centrate-N loads of about 20%. A calibrated site-specific model would be required for accurate estimation of savings to be achieved at a specific plant. However, the substantial % reductions in, methanol and alkalinity inputs and aeration are expected to be similar at an actual plant.

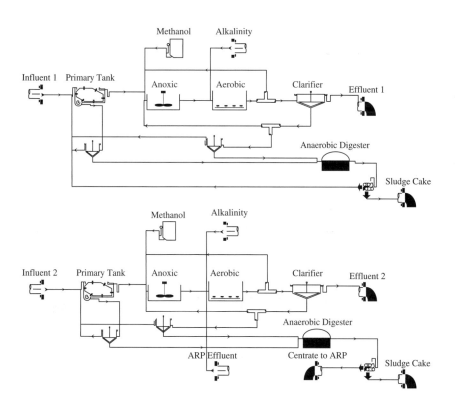

Two-sludge plant

The two-sludge plant dedicates a separate process to nutrient reduction, as shown in Exhibit 3. The generic model demonstrates that the two categories of plant have materially different benefits from the application of ARP. Influent carbon and BNR recycle in the single-sludge category lowers demand for addition of both alkali and methanol relative to the two-sludge category. While the reduction of methanol demand related to ARP is over 6000 kg/d for the two-sludge and about 5300 kg/d for the single-sludge model, the percent reduction calculated for the single-sludge model is much higher due to the lower overall demand for the single-sludge category. A similar result was obtained for the alkali demand.

Exhibit 3. BioWin model for two sludge model.

Flow: 100 MGD			
Centrate TN: 795 mg/L at 0.41 MGD			
Influent Quality			
Flow	MGD	100	
TKN	mg/L	25	
COD	mg/L	290	
CBOD	mg/L	142	
TSS	mg/L	138	
Effluent Quality		Without ARP	With ARP
NH$_3$-N	mg/L	0.06	0.08
NO$_3$-N	mg/L	0.47	0.46
TN	mg/L	1.96	1.91
COD	mg/L	22.7	22
CBOD	mg/L	3.1	2.9
pH		7.3	7.3

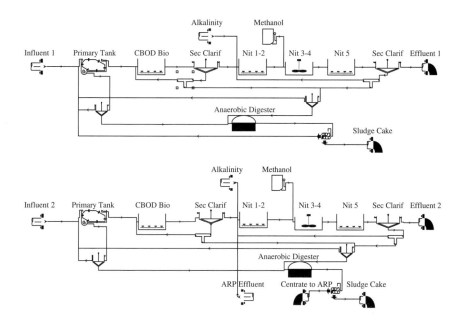

CONCLUSION

The generic BioWin model results are indicative of substantial savings in both operating and energy costs when ARP is integrated into the nutrient reduction design for either a single sludge or a two-sludge wastewater treatment plant. A more accurate estimate of cost savings for a specific plant requires use of actual plant parameters in the BioWin model and calibration of the model to match plant operational data, as well as unit prices for the process inputs and outputs for the plant location. The beneficial outputs include:

Bio win model results	Single-Sludge Plant	Two-Sludge Plant
Methanol reduction	38%	19%
Alkalinity reduction	10%	13%
Sludge reduction	6%	3%
Oxygen ("Energy") reduction	10%	13%

Reduction figures are calculated on a plant-wide basis.

REFERENCES

Dimitrios Katehis, Sudhir Murthy, Bernhard Wett, Edward Locke and Walter Bailey Nutrient Removal fromAnaerobic Digester Side-Stream at the Blue Plains AWTP Proceedings of the Water Environment Federation 79th Annual Technical Exhibition and Conference, Dallas, TX, USA, Oct 21–25, 2006.

Wilson, A.W. and Marstaller, T. Practical applications of biological nutrient removal models. Presented at the Preconference. Modeling and Simulation for Planning, Design, and Operation of Waste Water Systems for the Water Environment Federation 67th Annual Conference and Exposition, Chicago, Illinois, USA, Oct 15–19, 1994.

The use of phosphorus-saturated ochre as a fertiliser

S.T.D. Carr[a], K.E. Dobbie[a,b], K.V. Heal[a] and K.A. Smith[a]

[a]School of GeoSciences, The University of Edinburgh, Crew Building, West Mains Road, Edinburgh, EH9 3JN, UK
[b]Current affiliation: SEPA, Heriot Watt Research Park, Avenue North, Edinburgh, EH14 4AP, UK

Abstract Diffuse pollution is the major cause of water pollution worldwide. The long-term solution is a change in land-use practices and management. However, this will take time to implement and become effective and a short-term, low-cost treatment method is required. Ochre is formed in minewater settling lagoons as iron-rich precipitates which can be air-dried and used as a filter substrate. Ochre has a high adsorption capacity for P as it is comprised largely of $Fe(OH)_3$ and $FeO \cdot OH$ and contains other compounds known to adsorb P, such as aluminium oxides and calcium carbonates. When the P-adsorption capacity of the ochre filters has been reached, the substrate will offer a rich source of P (up to 30.5 mg P g^{-1}) which it is proposed could be used as a slow-release fertiliser. The studies reported here showed that the use of P-saturated ochre as a fertiliser compared to conventional fertilisers, such as K_2HPO_4, had a tendency to produce greater crop yields with no signs of stress, possibly due to the slow release of P from the ochre matrix. Concentrations of potentially toxic elements in the ochre-amended soil were within permissible standards and there was no evidence of soil contamination.

INTRODUCTION

After mines are abandoned, pumping of water ceases, often resulting in the rebound of the water table and the release of water containing potentially toxic elements (PTEs), especially iron (Fe). In order to treat these environmentally damaging discharges, mine water treatment plants (MWTPs) have been constructed across the UK. They are based around the oxidation of Fe(II) in solution to Fe(III), to form a precipitate known as "ochre". Ochre therefore requires removal from MWTPs and is currently disposed of to landfill as no alternative end-use is available. Ochre not only contains Fe, but, depending on the geochemistry, other contaminants such as Al and As. It is estimated that at least 3.7×10^4 tonnes of ochre are produced in the UK each year from coal MWTPs (Hancock, 2004), with associated processing and disposal costs between £35 to £100 per tonne.

Due to a high content of elements known to adsorb P, such as Fe, Al, Mg and Ca, ochre has been proposed as a material to remove phosphorus (P) from eutrophic and nutrient-enriched waters, with this potential confirmed in laboratory studies (Heal *et al.*, 2003). Parfitt (1989) states that the predominant mechanism of P sorption to goethite (noted by Heal *et al.* (2003) to be the major component of ochre) involves rapid ligand exchange with surface OH groups at very reactive sites and the formation of a binuclear bridging complex between a phosphate group and two surface atoms. Following the exhaustion of very reactive sites, weaker ligand exchanges occur at less reactive sites whilst over time there is a slow penetration of phosphates into the solid matrix through defect sites and pores.

After the capacity of ochre to adsorb P is exhausted it requires disposal. Upon saturation, ochre will offer a rich source of P (up to 30.5 mg P g^{-1}) which could be used as a slow release fertiliser. Dobbie *et al.* (2005) showed that the use of P-saturated ochre as a fertiliser resulted in higher yields at the same application rate as conventional fertiliser, with no signs of stress to the vegetation. Furthermore, there was no evidence of soil contamination since concentrations of PTEs in the ochre-amended soil were within permissible standards. Alternative methods of P recovery from P-saturated ochre are also under consideration, such as subjecting it to reducing conditions causing the release of P from the ochre matrix due to the reduction of Fe^{3+} to Fe^{2+}. A laboratory experiment showed that placing ochre in such conditions produced a leachate with a P concentration of 50 mg l^{-1} (Bozika, 2001). Hence the recovered P-rich solution rich in P could be concentrated or converted into a more conventional form for use in industry.

This paper examines the potential for using ochre to recover P from P-contaminated water. The potential of ochre for P removal is demonstrated and ongoing research is presented which aims to improve understanding of the processes affecting P-removal by ochre. Finally the use of P-saturated ochre as a slow release fertiliser is shown.

OCHRE AS A PHOSPHORUS ADSORBENT

Prior to experimentation and use, ochre is usually air-dried to transform it from sludge with up to 80–95% water content to a dry powder or granular form. Depending on the geochemistry of the mine water, the MWTP design and operation ochres from different sites (see Table 1) have different chemical compositions and particle size distributions and hence a varying ability to adsorb P. Ochres can therefore be selected for different P-removal applications based upon their characteristics. For example, fine-grained ochre from the Minto MWTP is suitable for dosing applications due to its fine particle size, whilst granular ochre from the Polkemmet MWTP, with a higher permeability, is more suitable for use in P filters.

Table 1. Chemical and physical properties of air-dried ochres from two MWTPs, Scotland (modified from Heal *et al.*, 2004).

	Polkemmet	Minto
pH (in distilled water)	7.2	6.9
%Fe[1]	65 ± 0.5	67.5 ± 3
%Al[1]	0.7 ± 0.02	0.1 ± 0.01
%Mg[1]	0.6 ± 0.01	0.8 ± 0.04
%Ca[1]	7.0 ± 0.1	11.8 ± 0.4
Dry bulk density (g cm^{-3})	1.8	0.8
Particle size range (mm)	0.25–10	<0.25–1
Saturated hydraulic conductivity (m day^{-1})[2]	26–32	0.7–1.7

[1]Mean of triplicate samples ± standard error. Determined by atomic adsorption spectrophotometry of acid digests (concentrated nitric and hydrochloric acid additions) of ashed samples.
[2]Determined in columns over 32 days using the falling head method and Darcy's Law.

Laboratory experiments

Laboratory batch experiments have demonstrated the capability of ochre to adsorb P from artificial solutions. The maximum P adsorption capacity of ochres from Polkemmet and Minto MWTPs was determined as 26 and 30.5 mg P g^{-1}, respectively (Heal *et al.*, 2003). Minto ochre has a smaller particle size than Polkemmet ochre and thus a larger reactive surface area and this is the likely reason for its higher P adsorption capacity. In these experiments the rate of the reaction was found to be rapid, with almost all P adsorbed within the first hour by Polkemmet ochre and within a few minutes by Minto ochre. The adsorption capacities of ochre are far greater than for other materials that have been investigated for P adsorption (Table 2). Ochres have also been shown to be efficient adsorbents of P in non-agitated mixtures. When P solution was added to Polkemmet ochre in a beaker P concentrations were reduced from 5 to <0.01 mg P l^{-1} within 8 minutes (Heal *et al.*, 2003, 2004).

Ochre as a filter substrate

The long term suitability of Polkemmet ochre to adsorb P was examined over a nine-month period by pumping a P solution (20 mg P l^{-1}) onto a gently angled trough packed with 10 kg ochre (Heal *et al.*, 2003). The solution was pumped at a flow rate of 1.2 l hr^{-1} ensuring a contact time of 4.5 hours. The reduction in P concentration during the experiment is shown in Figure 1. Passage through the trough reduced the P concentration of the solution by up to 99.8%, with the lowest efficiency being 95.2%. A similar experiment but on a far larger

Table 2. Maximum P adsorption capacities of different wetland substrates (Heal *et al.*, 2003).

Substrate	Adsorption capacity (mg P (g substrate)$^{-1}$)
Gravel	0.03/0.05
Bottom ash	0.06
Steel slag	0.38
Blast furnace slag	0.4–0.45
Fly ash	0.62
Shale	0.75
Laterite	0.75
Zeolite	1
Polkemmet ochre	26
Minto ochre	30.5

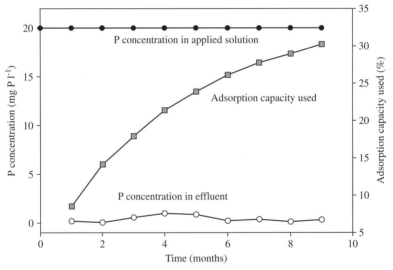

Figure 1. Phosphorus removal and % adsorption capacity used in Polkemmet ochre after nine months of trough experiment (Heal *et al.*, 2003).

scale was conducted by Dobbie *et al.* (accepted) using over one tonne of ochre from a variety of MWTPs to provide tertiary treatment of sewage effluent in southern Scotland. During the 27-month trial P concentrations were reduced by up to 80% in optimal flow conditions, with no detectable release of PTEs from the ochre. P removal rates declined over time, probably due to clogging of the filter unit.

Ongoing research on the use of ochre as a filter substrate

Current research is developing a process-based understanding of the factors affecting P removal by ochre in order to inform the design of a suite of ochre-based filters to treat water polluted with P over a range of flow conditions and P concentrations. The filter design will take into account the need for continued reactivity between water and filter matrix and the avoidance of clogging. Sorption-desorption between reactive surfaces and solution, dissolution-precipitation equilibria and kinetics will also be investigated. Ochres with different chemical and physical properties from seven sites in the British Isles are being investigated. The presence of competing ions and dissolved organic carbon (DOC) on the magnitude and rate of P adsorption by ochre will be assessed in batch experiments. An aspect of concern for the long term implementation of ochre filters, the potential release under anoxic conditions of PTEs, such as As, will be investigated through batch and column experiments.

A chemical model using the software ORCHESTRA (Meeussen, 2003) will be calibrated using the results of the batch and column experiments for the seven ochres and will form the basis for designing ochre-based filter units for field implementation. The development of a robust model will allow a range of scenarios to be tested which would otherwise not be possible, e.g. ten year filter implementation.

USE OF PHOSPHORUS-SATURATED OCHRE AS A FERTILISER

After P-saturation ochre removal from filter units and disposal will be required. Due to its high concentration of up to 30.5 mg P g^{-1} this ochre could be employed as a fertiliser. A similar approach was considered by Hylander and Simàn (2001) who investigated whether P adsorbed by blast furnace slag was plant-available. Other waste materials, such as steel slag without additionally adsorbed P, have been used as P fertilisers in agriculture (MacNaeidhe, 2001) and forestry (Jandl et al., 2003). However, even if waste products rich in P are effective fertilisers, application cannot proceed if it leads to soil contamination. Therefore two main issues require addressing for the successful use of P-saturated ochre as a fertiliser. Firstly, whether crop productivity resulting from application of P-saturated ochre matches that from conventional fertilisers, and secondly whether the application of P-saturated ochre leads to soil contamination from PTEs.

Materials and methods

Pot experiments

Research to address these issues was conducted by Dobbie *et al.* (2005) using ochre from Polkemmet MWTP which was air-dried, coarsely crushed and then saturated with P using solutions of KH_2PO_4. Pot experiments were conducted using agricultural soil collected in central Scotland. The soil was air-dried, sieved (4 mm) and mixed with sand to give it the texture of sandy loam. Available P, K and Mg were then measured in the mixture to determine the appropriate fertilizer application rates which were 85 kg P_2O_5 ha^{-1} and 90 kg K_2O ha^{-1}. Four litres of the soil-sand mixture were placed in each of 60 5-litre pots. The pots were sub-divided into two crop types, barley and grass, with five replicates of six different P treatments for each crop type. The six different P treatments used were: a control with no added P, a conventional P treatment using KH_2PO_4 at the recommended application rate, and four treatments using the P-saturated ochre at 0.5, 1, 2 and 5 times the recommended rate, as determined from the acetic acid extractable P content of the P-saturated ochre. Additionally 0.29 g K and 0.2 g N were added to all the pots in the form of K_2SO_4 and NH_4NO_3. A further 0.06 g of N was added in solution after three weeks in the form NH_4NO_3. The pots were sown to standard agricultural practice equivalents; for the barley 8 seeds (equivalent to 200 seeds m^{-2}) and the grass 0.16 g of seed (40 kg ha^{-1}).The pots were distributed randomly in an unheated greenhouse and redistributed every 2 weeks. Soil water content was determined gravimetrically and tap water added as required to maintain soil water content at approximately 80% field capacity. The experiment was conducted from July to October 2002, until the barley heads had ripened. The mass and total P content of the barley and grass at the end of the experiment were determined. In addition, at the start and end of the experiment. The soils were analysed for total and available P and total PTEs (Al, As [at end only], Cd [at start only], Cr, Cu, Fe, Mn, Ni, Pb and Zn).

Field trials

Field trials were conducted using the same crop types as in the pot experiments, for barley at a farm in central Scotland, and at a nearby acid grassland which had a low P status soil. At both sites four replicates of three P treatments were established: a control with no added P, a conventional P-application (triple superphosphate, TSP) and a P-saturated ochre amendment, which, as in the pot experiments, contained the same amount of acetic acid extractable P as the conventional P-treatment. For the barley trial, P was applied by broadcasting the conventional fertiliser and the P-saturated ochre by hand over their respective plots of 3 m $\times$ 2 m at an application rate of 85 kg P_2O_5 ha^{-1}. Following this the seedbed was cultivated to a depth of 15 cm before barley seeds were sown at a rate of 200 kg ha^{-1} prior to the plots being rolled. Additionally, at the start of the experiment 60 kg N ha^{-1} in the form of NH_4NO_3 and 60 kg K_2O ha^{-1} were applied to the plots, with 60 kg N ha^{-1} applied again 3 weeks into the experiment. P-saturated ochre and TSP was applied to the smaller grassland plots of 1 m^2 on an existing sward at 30 kg P_2O_5 ha^{-1} along with N at 62 kg N ha^{-1} in the form of NH_4NO_3. The plots were established in March 2003 and allowed to grow until August 2003. For the grassland, vegetation was sampled from an area of 25 cm $\times$ 25 cm in each plot in March prior to treatment and also in June and August. After each sampling the plots were mown and the remaining grass discarded. Barley was sampled from a 50 cm $\times$ 50 cm area in each barley plot in August. Vegetation samples were dried and weighed with total P determined. Total Cd and Pb concentrations were also measured in the barley grain. At the end of the trials, soil cores were taken in each plot from the top 15 cm (barley) and 7.5 cm (grass) and analysed for available and total P and for the same PTEs as described for soils in the pot experiments.

Results

Soil P concentration

The addition of P-saturated ochre in the pot and field experiments led to a significant ($P < 0.05$) increase in the amount of plant-available P as well as total P in the soil. Over the course of the experiments, the concentration of available soil P decreased in all treatments, with the largest reduction in the conventional fertiliser treatments as shown in Figure 2 for the barley pot experiments.

In the pot experiments, the uptake of P by vegetation in the control and all P-saturated ochre treatments substantially exceeded the depletion of available P in the soils. In the conventional fertiliser treatment, the P uptake by grass was slightly greater than the depletion of plant-available P in the soil, with the depletion greater than the amount of fertiliser added. In contrast, the barley from the conventional fertiliser treatment contained less P than was depleted from the soil, indicating a net fixation of P by the soil minerals.

The explanation for soil P depletion not being as high as plant P uptake in the P-saturated ochre treatments and control is the conversion of initially unavailable P in the ochre and soil into plant-available forms, possibly due to organic acids released from the root zone dissolving unavailable Fe, Ca and Al phosphates (Dakora and Phillips, 2002). In this way the P-saturated ochre acts as a slow release fertiliser throughout the growing season, resulting in larger amounts of residual P in the amended soils for the next growing season.

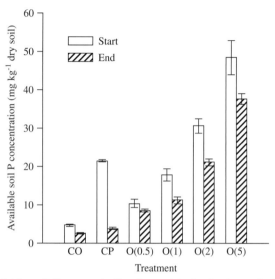

Figure 2. Available soil P concentration ($\pm$s.e.) in the barley soil at the start and end of the pot experiment. CO = unfertilised control; CP = conventional fertiliser; O(0.5)–O(5) = ochre applied at 0.5–5 times available P in the CP treatment. (adapted from Dobbie *et al.*, 2005).

This reduces the need for a repeat application of P-fertiliser in the following growing season. From the field trial results it is suggested that the addition of 40 t P-saturated ochre ha^{-1} would supply enough P for at least two growing seasons. The insolubility of the P bound to the ochre also means that it is less available in surface run-off than conventional water-soluble fertiliser.

Plant response

In the pot experiments crop yield from the ochre treatments was greater than from the conventional fertiliser and control treatments, although this was not always statistically significant. Yields from the P-saturated ochre treatment of half the recommended P application were higher (though not always significantly) than from the conventional fertiliser treatment, which initially contained twice the amount of available soil P. This confirms the conclusion that P was gradually released from ochre during the experiment, providing the vegetation with continuous nutrition. Results from the field trials also indicated that yields were greater in the P-saturated ochre treatment than in the conventional fertiliser treatment, though again these results were not statistically significant. It is therefore concluded that the use of P-saturated ochre in place of conventional P fertiliser has no adverse effect on crop yield and in fact there is a tendency towards an increase in yield with a less frequent application rate required. Furthermore P-saturated ochre could constitute a more sustainable source of P fertiliser than conventional fertilisers derived from finite mineral resources.

Potentially toxic elements

To assess whether ochre addition introduced contaminants to the soil in these experiments, PTE concentrations were measured in soil samples from the different treatments and compared with maximum permissible concentrations of PTEs in soils for the application of sewage sludge to agricultural land (MAFF, 1998). Soil PTE concentrations did not exceed the permissible concentrations with the exception of Ni in the pot experiments. However, the permissible concentrations for Ni were exceeded in soils from both the ochre and control treatments and were not significantly different, showing that exceedance was not due to the application of ochre. Plant uptake of Pb and Cd was examined by measuring the concentrations of these elements in the barley seed heads. The concentrations in barley grown in the ochre-treated soils were considerably lower than the maximum permissible levels in foodstuffs set by the European Commission (Commission Regulation (EC) No. 466/2001) and were not significantly different from those measured in the barley grown in the conventional fertiliser treatment. At the recommended application rate of 40 t ha^{-1}, ochre addition to this soil in the field would result in soil metal concentrations considerably below limits set out for the application of sewage sludge to agricultural land, with the exception of Ni. With respect to Fe, no maximum application rates are given within the UK (MAFF, 1998). Although the addition of Fe to soil in the field trials seems large (10.9 t Fe in 40 t ochre), it is only equivalent to raising the soil Fe concentration from 2.4 to 2.6% for the barley trial and 3.1 to 3.4% for the grassland trial, well within the usual range for this area of Scotland (Paterson et al., 2003).

CONCLUSIONS

The ability of ochre to rapidly adsorb P has been demonstrated through laboratory experiments and, with a high P-adsorption capacity (up to 30.5 mg P g for Minto ochre) it is an excellent candidate for use as a filter substrate. Sustained removal of P from sewage effluent by ochre has also been shown in large scale field trial. Dobbie et al. (2005) showed that ochre can be used as a slow-release fertiliser after P saturation has occurred, with no negative impact upon crop yield and soil quality. The P-saturated ochre lead to the continued release of P into the soil, with the potential for a single application to meet the needs of several growing seasons. Current research is focusing upon testing and developing ochre-based filters to adsorb P under a range of conditions with varying flow rates, P concentrations and competing ions. The environmental acceptability of ochre as a filter substrate will also be investigated by assessing the release of PTEs in laboratory experiments and through chemical modelling.

ACKNOWLEDGEMENTS

The UK Coal Authority funded preliminary experiments and supplied ochre. Some of the work reported here was conducted by Karen Dobbie who was funded by EPSRC Grants GR/R73539/01 and GR/R73522/01 and Enviresearch Ltd and supported by Scottish Water. Technical support from Andy Gray, Robert Howard, John Morman and Graham Walker (The University of Edinburgh) is acknowledged as is advice from Alison York regarding current legislation. Stephen Carr is funded by NERC and the Macaulay Institute, Aberdeen.

REFERENCES

Bozika, E. (2001). Phosphorus removal from wastewater using sludge from mine drainage treatment settling ponds. MSc Thesis, The University of Edinburgh.

Dakora, F.D. and Phillips, D.A. (2002). Root exudates as mediators of mineral acquisition in low nutrient environments. *Plant and Soil.*, **245**, 35–47.

Dobbie, K.E., Heal, K.V. and Smith, K.A. (2005). Assessing the performance and environmental acceptability of phosphorus-saturated ochre. *Soil Use and Management*, **21**, 231–239.

Dobbie, K.E., Aumônier, J., Heal, K.V., Smith, K.A., Johnston, A. and Younger, P.L. (accepted). Evaluation of iron ochre from mine water drainage treatment for removal of phosphorus from wastewater. *Chemosphere.*

Hancock, S. (2004). Ochre arisings and composition in the UK. MSc Thesis, Imperial College London.

Heal, K.V., Younger, P.L., Smith, K.A., Glendinning, S., Quinn, P. and Dobbie, K.E. (2003). Novel use of ochre from mine water treatment plants to reduce point and diffuse phosphorus pollution. *Land Contamination and Reclamation*, **11**, 145–152.

Heal, K.V., Smith, K.A., Younger, P.L., McHaffie, H. and Batty, L.C. (2004). Removing phosphorus from sewage effluent and agricultural runoff using recovered ochre. In: Valsami-Jones, E. (ed) Phosphorus in environmental technology: Principles and application, IWA publishing, 320–324.

Hylander, L.D. and Simán, G. (2001). Plant availability of phosphorus sorbed to potential wastewater treatment materials. *Biology and Fertility of Soils*, **34**, 42–48.

Jandl, R., Kopeszki, H., Bruckner, A. and Hager, H. (2003). Forest soil chemistry and mesofauna 20 years after an amelioration fertilization. *Restoration Ecology*, **11**, 239–246.

MacNaeidhe, F.S. (2001). Effect of application of basic slag and superphosphate on herbage yield and on soil and herbage concentrations of phosphorus in organic grassland. *Biological Agriculture and Horticulture*, **19**, 231–245.

MAFF (1998). The soil code: Code of good agricultural practice for the protection of the soil. MAFF/WOAD, MAFF (now DEFRA) publications.

Meeussen, J.C.L. (2003). ORCHESTRA: An object-orientated framework for implementing chemical equilibrium models. *Environmental Science and Technology*, **37**(6), 1178–1182.

Parfitt, R.L. (1989). Phosphate reactions with natural allophone, ferrihydrite and goethite. *Journal of Soil Science*, **40**, 359–369.

Paterson, E., Towers, W., Bacon, J.R. and Jones, M. (2003). Background levels of contaminants in Scottish soils. Report commissioned by SEPA. Available: http://www.sepa.org.uk/pdf/publications/reports4sepa/contaminants_scottish_soils.pdf [2008, 31 October].

Volatile Fatty Acid (VFA) and nutrient recovery from biomass fermentation

Q. Yuan, F. Zurzolo and J. Oleszkiewicz

Department of Civil Engineering, University of Manitoba, Winnipeg, Manitoba, R3T 5V6, Canada

Abstract A set of lab scale sequencing batch reactors consisting of a mother reactor and a fermenter were established to investigate nutrient recovery in acid-phase fermentation. Carbon recovery was achieved through post-fermentation VFA generation, whereas nutrient recovery was possible by struvite induced precipitation (aliquots of magnesium chloride). It was found that the acid-phase of biomass fermentation is an effective method to generate VFAs. The average specific VFA production rate was measured as 0.12 g VFA/g VSS per day. Observed biomass fermentation resulted in a 40% solids reduction. Significant release of phosphate and ammonia were recorded throughout the fermentation process, thereby facilitating the recovery of nutrients through struvite precipitation. Results suggest that struvite precipitation can remove high concentrations of phosphate and ammonia from wastewater via increases in pH, as well as magnesium dosing. Optimum struvite formation and magnesium chloride dosing were observed for a pH of 9, and 0.38 g/L, respectively. Nutrient recovery through struvite formation may be achieved without negative impacts on VFA generation.

INTRODUCTION

Phosphorus (P) removal from wastewater has recently become an essential part of wastewater treatment stemming from both domestic and industrial sources. Considered as the leading factor contributing to the eutrophication of lakes and rivers, sufficient phosphorus removal is critical to the viability and continued health of the world's waterways (Oldham and Rabinowitz, 2001).

Enhanced Biological phosphorus removal (EBPR) processes have been considered as one of the most effective and potentially least expensive ways to remove phosphorus from wastewater. The key to efficient EBPR performance rests in the presence of adequate volatile fatty acids (VFA) in the influent wastewater (Chu et al., 1994). Often, wastewater lacks sufficient naturally occurring VFA to impart suitable P removal, especially with a reduced chemical oxygen demand (COD) (Barajas et al., 2002). Therefore, the necessary introduction of a carbon source is considered an essential step in achieving adequate nutrient removal within wastewater treatment plants. However, the addition of a carbon source increases not only the operational cost but also the sludge production which ultimately raise the cost for sludge treatment.

© 2009 The Authors, *International Conference on Nutrient Recovery from Wastewater Streams.* Edited by Ken Ashley, Don Mavinic and Fred Koch. ISBN: 9781843392323. Published by IWA Publishing, London, UK.

As the cost for purchasing external carbon to facilitate biological nutrient removal continuously increases, process designers have began focusing their efforts to on-site VFA production. One widely used method of this design is to rely upon primary sludge for on-site production of VFA. Although the practice of using primary sludge fermentation is well established worldwide (Munch and Koch, 1999; Chanona *et al.*, 2006; Bouzas *et al.*, 2007), the actual process is difficult to control and the reliability of VFA generation is often not adequate, particularly in flat terrain large sewer systems such as in St Paul-Minneapolis MN, Winnipeg South MB or Gdansk PL. In addition, the volume of VFA produced from primary sludge fermentation is normally below the mass required to ensure efficient nutrient removal in all wastewater treatment plants; such as the Noosa BNR (Tomas *et al.*, 2003). As the result of these complications, recent research has focused its attentions towards generating VFA from fermentation of biomass or wasted activated sludge (WAS).

The biomass fermentation offers to not only supplement VFA but also to reduce sludge volume. These two major advantages are critical to the operation of WWTPs as both economically and environment friendly designs. However, a drawback of biomass fermentation is that the VFAs are generated in conjunction with ammonia and phosphate salts. If the fermented sludge liquor was used as a VFA source without any pre-treatment, the overall nutrient load to the system may offset the benefits of VFA produced. Alternatively, limited phosphorus resources have directed the research toward phosphorus recovery from sludge liquors. Phosphorus recovery through struvite formation is considered a cost-effective approach in biological nutrient removal plants equipped with anaerobic digestion facilities (Mavinic *et al.*, 2007). In addition to recovery, the controlled precipitation of struvite at one stage in the wastewater treatment process prevents clogging and scaling problems in downstream operations. This in-turn provides significant cost savings resulting from fewer maintenance and replacement operations and the value of struvite as a commercial fertiliser product.

The objective of this study was to investigate VFA generation in acid-phase fermentation as an internal carbon source for biological nutrient removal. Additionally, struvite formation was investigated as an alternative to recovering nutrients from wastewater treatment to achieve sustainable sludge management.

MATERIAL AND METHODS

Experiment approach

The biomass fermentation set up consisted of a bench scale mother reactor and fermenter combination. The mother reactor was seeded with the activated sludge from a lab sequencing batch rector (SBR) with stable phosphorus removal and was operated in an anaerobic\aerobic configuration. The mother reactor is a SBR which goes through three, eight hour cycles per day. In each cycle the reactor is fed with synthetic wastewater (COD of 300 mg/L, acetic acid of 150 mg/L) after which an anaerobic environment is created by bubbling nitrogen gas through the reactor. It is at the end of the anaerobic cycle that a portion of the biomass is pumped into the fermenter. Following the anaerobic cycle which lasts for 90 minutes, air is bubbled through the reactor to create aerobic conditions for 4.5 hours. Finally, the reactor is allowed to settle and the supernatant is decanted. Overall, an SRT of 10 days is achieved in the mother reactor with an HRT of 12 hours.

The fermenter receives biomass from the mother three times each day and is manually wasted once per day. The wasted liquor is then analysed, filtered, and the supernatant is used for the struvite precipitation tests. The fermenter is anaerobic, well mixed, and kept at a temperature of 35°C. Overall, the fermenter has an SRT and an HRT of 7 days. Although the fermenter is designed to model an in-line fermenter, the wasted liquor is not fed back into the mother reactor as the goal of this experiment is to determine the characteristics of the liquor (i.e. VFA, P, N concentrations) and its suitability for the precipitation of struvite.

Phosphorus recovery via struvite formation

The struvite precipitation tests are performed on the supernatant from the fermenter. Biomass is filtered from the liquor and the resulting supernatant is refrigerated until sufficient quantities are available for testing. Five tests were performed at pH of 7.5, 8.0, 8.5, 9.0, and 9.5 respectively. For each test, approximately 2 litres of supernatant is gently mixed together. The pH is raised to the desired level with a 40% w/w sodium hydroxide solution and the supernatant is divided into four beakers. Varying doses of magnesium (0.6 M magnesium chloride hydrate [$MgCl_2 \cdot 6(H_2O)$]) solution are added to each of the four beakers and are thoroughly mixed. Following a 24 hour period for struvite precipitation, the crystals are removed by filtration. The resulting supernatant is reserved for analysis and the crystals are dried and analysed.

Analytical procedure

Mixed liquor suspended solids (MLSS) and mixed liquor volatile suspended solids (MLVSS) measurements were performed according to Standard Methods (APHA, 1998). Dissolved phosphate and ammonium was measured by Lachat

Instrument Quik Chem 8500, followed Quik Chem Method orthophosphate 10-115-01-1-O and Quik Chem Method ammonia 10-107-06-1-I. Varian CP-3800 gas chromatography was used for VFA measurement. Magnesium and calcium were analyzed by a Varian VISTA-MPX. Struvite crystals were examined under a phase contrast microscope (Nikon). Powder X-ray diffraction analysis was carried out on a PANalytical X'Pert Pro diffractometer equipped with an X'Celerator detector using a $CuK\alpha_{1,2}$ ($\lambda = 1.540598$, 1.544426 Å) radiation source.

RESULTS AND DISCUSSION

VFA production and solids reduction

Figure 1 illustrates concentrations of VFA as well as phosphate and ammonium in the fermenter over time. Although the fermenter was operated using a low solids concentration (MLVSS of 0.9–1.3 g/L), an increase in VFA concentration is observed with an average concentration of 354 mg/L. VFA as high as 658 mg/L was measured, and the average specific VFA production rate was measured as 0.12 g VFA/g VSS per day. Acetic acid was the dominant VFA produced with an average range of 66.5%, followed by propionic acid (14.6%). The ratio of VFA to soluble COD was observed as 58%. Since approximately 8 mg of VFA is required to remove 1 mg P (Abu-ghararah and Randal, 1991), the fermentation of 1 g wasted biomass will yield sufficient VFA for the removal of 15 mg P. This correlation effectively translates to a significant reduction in the demand for an external carbon source.

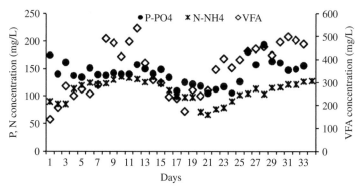

Figure 1. VFA, phosphate and ammonia concentration in the fermenter.

The large fluctuations of VFA concentrations were observed at approximately day 10, whereas phosphate and ammonium were fairly stable in this time frame (Figure 1). These fluctuations of VFA were most likely due to instrument malfunction. It was found at day 12 that the gas chromatograph flow rate regulator had malfunctioned, consequently the data recorded slightly before and after that day did not fit well with the rest of the recordings. The drop in VFA production around day 18 was probably a result of an accidental temperature increase of the fermenter to nearly 100°C (day 15). This increase of temperature definitely disrupted the fermentation process and caused a decrease in the VFA production. Nevertheless, it is apparent that the general trend of the graph describes an increase of VFA with time until a concentration of about 425 mg/L of VFA was reached, at which point the VFA production stabilised.

Another goal of acid fermentation is solids reduction. Figure 2 exemplifies that 40% solids destruction was achieved via biomass fermentation. This solids destruction will provide significant reduction for the cost of sludge handling, as the entire sludge treatment process accounts for nearly 60% of the total operating cost for the wastewater treatment plants (Horan, 1990).

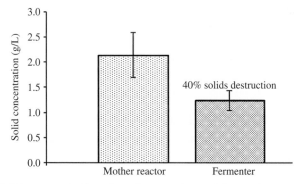

Figure 2. Solids concentration of mother reactor and fermenter.

Nutrient recovery

Solubilisation of P and N

The concentrations of phosphorus as phosphate and nitrogen (N) as ammonium were measured in the fermenter. Figure 1 illustrates the high concentrations of both elements in the fermented liquor with the average concentration of nitrogen measured to be 110 mg/L and the average concentration of phosphorus to be 142 mg/L.

The elevated concentrations of phosphate and ammonium in the fermenter, due to cell lysis, reinforce the suitability of a struvite precipitation process at this stage. If a fermented liquor rich in VFA, P and N would be returned to the mother reactor, the newly produced VFA would be completely depleted in the process of taking up the P that has been added by the liquor. If on average it takes at least 8 mg VFA for the removal of 1 mg P (Abu-ghararah and Randal, 1991), the observed 120 mg/L of P in the fermented liquor would require 960 mg/L of VFA, exceeding the total quality produced in the fermenter altogether. By combining the phosphorus and nitrogen together into struvite and precipitating it prior to adding the liquor back into the BNR system, establishes the benefit of added VFA for biological phosphorus removal, aiding the necessary reduction of nitrogen required further downstream.

The average released and solubilised ammonium and phosphate from the biomass fermentation was found to be 0.06 g NH_4-N/g VSS and 0.07 g PO_4-P/g VSS. One may argue that the amount of phosphorus released cannot be higher than that of ammonium. However, it should be noted that the biomass in the mother reactor was an enhanced culture containing significantly high population of PAO in the MLSS instead of the conventional activated sludge. This did not have an effect on the amount of VFA produced since the fermentation of other bacteria will produce the same results. However, lower P concentration does have an effect on the potential for struvite formation since high concentrations of the constituent components are required for precipitation. It is important to determine if there are sufficient amounts of P-PO_4 and N-NH_4 available in the fermenter prior to an attempt to precipitate struvite. It is hypothesized that a reason for this heightened release of phosphorus rather than that of ammonium can be that the biomass was withdrawn from the mother reactor at the end of its anaerobic cycle, at which time the phosphorus release has already happened within the mother reactor.

Figure 1 further indicates that during the fermentation stage, the ammonium and phosphorus production followed the same trend as VFA generation. It can be concluded that high ammonium and phosphorus levels may be associated with high VFA production; therefore, ammonia and phosphorus levels in the fermenter may be potentially used as an indicator in terms of the degree of biomass destruction.

Struvite formation test

Results suggest that the pH exhibits a stronger influence with phosphorus precipitation, rather than magnesium precipitation; this is illustrated in Figure 3. The rate of phosphorus removal increased while an increase in pH was experienced, as did the removal of ammonium and magnesium (Figure 4). It should be noted that the magnesium doses for a pH of 9 were different than those for the remaining trials. It was further hypothesized that the effects of pH could further influence the magnesium dosages for phosphorus removal. Initial studies began at a pH of 9 and appeared promising, further investigations were conducted to study if a wider pH range and magnesium dose still were able to provide beneficial results. However, results from the rest of the trials with different pH levels were in the agreement with the result from a pH of 9.0 using the magnesium dosage. This may be due to the over dosing of magnesium. It is suggested that lower magnesium dosages should be the focus of testing in future experimentation.

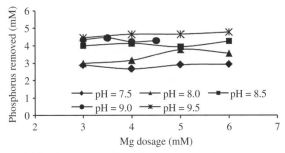

Figure 3. pH and Mg dosage on phosphorus removal.

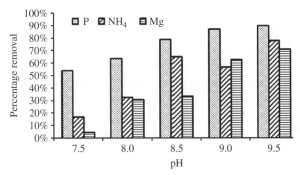

Figure 4. Phosphate, ammonia and megnisium removal with pH.

At a pH of 7.5, a small amount of crystallization was observed. It was found that much less magnesium was removed compared to ammonium or phosphate (data not show), and therefore not all the phosphate or ammonium was removed as struvite. Instead, it was thought that other reactions, such as calcium phosphate precipitation, were simultaneously competing for struvite formation, consequently forming other crystals containing phosphate and ammonia.

With an increase in pH levels of 8.0 to 8.5, both ammonia and magnesium removal were greatly improved compared to pH 7.5 (Figure 4); however, it still does not exhibit the 1:1:1 ratio characteristic of struvite precipitation. When the pH is increased to 9.0, a nearly perfect 1:1:1 ratio of removed ions was exhibited indicating struvite formation. The obtained precipitated crystals were examined via microscope; an identified orthorhombic shape was observed (Figure 5). Further analysis via powder x-ray defraction (XRD) techniques verified that the obtained defraction pattern was consistent with struvite, suggesting a minimum purity of 99%. This pH condition of 9.0 is considered the optimum condition for phosphorus removal through struvite precipitation. The results from this test suggest that the optimal dose for magnesium salt at pH 9 was 0.38 g/L, resulting in 84% of P-PO_4 and 60% N-NH_4 removal. At pH 9.5 significant ammonium removal was observed in the form of ammonia gas due to the high pH condition. The ammonium-ammonia equilibrium point occurs at pH 9.2 and so at pH levels higher than 9.2, ammonium converts to ammonia gas. Compared to the pH 9.0 trial, there were very similar quantities of struvite formed; however, the added production of ammonia gas at this higher pH leads to the conclusion that a pH 9.5 is less favourable.

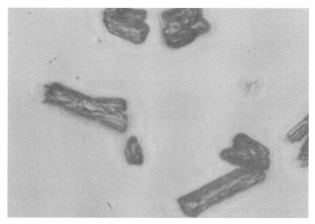

Figure 5. Picture of crystals from the fermentation liquor with $MgCl_2$ addition.

VFA concentration of the supernatant post struvite precipitation was analyzed. It was observed that only 5% VFA was lost in the bulk solution during the precipitation process. This suggests that the recovery of VFA and removal of phosphorus and ammonia can occur simultaneously without a negative synergistic impact.

CONCLUSION

In light of this preliminary study, the following conclusions can be made:

- Fermentation of waste activated sludge is an effective approach for VFA production, while reducing the sludge generation.
- Significant release of phosphate and ammonia occurs in the fermentation process.
- Struvite formation is an effective method by which phosphorus can be recovered from wastewater. Struvite precipitation can remove high concentrations of phosphate and ammonia from wastewater through increasing the pH and magnesium dosing. The best pH for struvite formation from this study is 9, with the optimal dose of magnesium chloride being 0.38 g/L.
- Nutrient recovery through struvite formation may be achieved without negative impacts on VFA generation.

ACKNOWLEDGEMENT

The work was sponsored by The City of Winnipeg Water and Waste Department and Manitoba Conservation Department's Sustainable Development Innovation Fund. J. Rak-Banville's technical support and editing is greatly appreciated.

REFERENCES

Abu-ghararah, Z.H. and Randall, C.W. (1991). The effect of organic compounds on biological phosphorus removal. *Wat. Sci. Technol.*, **23**(4–6), 585–594.

American Public Health Association (1998). *Standard Methods for Examination of Water and Wastewater*. 20th edn, American Public Health Association/American Water Works Association/Water Environment Federation, Washington DC, USA.

Barajas, M.G., Escalas, A. and Mujeriego, R. (2002). Fermentation of a low VFA wastewater in an activated primary tank. *Wat. SA.*, **28**(1), 89–98.

Bouzas, A., Ribes, J., Ferrer, J. and Seco, A. (2007). Fermentation and Elutriation of primary sludge: effect of SRT on process performance. *Wat. Res.*, **41**, 747–756.

Chanona, J., Ribes, J., Seco, A. and Ferrer, J. (2006). Optimum design and operation of primary sludge fermentation schemes for volatile fatty acids production. *Wat. Res.*, **40**(1), 53–60.

Chu, A., Mavinic, D.S., Kelly, H.G. and Ramey, W.D. (1994). Volatile fatty acid production in thermophilic aerobic digestion of sludge. *Wat. Res.*, **28**(7), 1513–1522.

Horan, N.J. (1990). *Biological Wastewater Treatment Systems*. Wiley, Chichester, UK.

Mavinic, D.S., Koch, F.A., Huang, H. and Lo, K.V. (2007). Phosphorus recovery from anaerobic digester supernatants using a pilot-scale struvite crystallization process. *J. Environ. Eng. Sci.*, **6**(5), 561–571.

Metcalf and Eddy (2003). *Wastewater Engineering Treatment and Reuse.*

Munch, E.V. and Koch, F.A. (1999). A survey of prefermenter design, operation and performance in Australia and Canada. *Wat. Sci. Technol.*, **39**, 105–112.

Oldham, K.W. and Rabinowitz, B. (2001). Development of biological nutrient removal technology in western Canada. *Can. J. Civ. Eng.*, **28** (Suppl. 1), 92–101.

Tomas, M., Wright, P., Blackall, L., Urbain, V. and Keller, J. (2003). Optimization of Noosa BNR plant to improve performance and reduce operating cost. *Wat. Sci. Technol.*, **47**, 141–148.

Phosphorus recovery from sewage sludge ash by a wet-chemical process

C. Dittrich[a], W. Rath[b], D. Montag[c] and J. Pinnekamp[c]

[a]MEAB Chemie Technik GmbH, D-52068 Aachen, Germany (E-Mail: carsten.dittrich@meab-mx.com)
[b]Aachen University of Applied Sciences, Department Applied Polymer Sciences (IAP), D-52074 Aachen, Germany
[c]Institute for Environmental Engineering, RWTH Aachen University, D-52056 Aachen, Germany

Abstract By the end of this century phosphate deposits with low concentrations of hazardous substances, primary cadmium and uranium, will be totally exploited. Phosphates cannot be substituted for plant nutrition by other substances. Thus, usage of secondary phosphate materials is becoming more and more important. Recovering phosphorus from ash enable the recovery of 80% of the inflow load to a municipal wastewater treatment plant (WWTP). The recovered phosphorus can be sold to the phosphate or fertilizer industry.

INTRODUCTION

At the Institute for Environmental Engineering (ISA) of RWTH Aachen University, several approaches to phosphorus recovery within the framework of wastewater treatment have been investigated and developed. The PASH process (Phosphorus Recovery of from Ash) enables phosphorus recovery from sewage sludge ash or ashes from the incineration of meat and bone meal.

The recovery of the phosphorus content starts with leaching the ash with diluted hydrochloric acid in a stirred tank at ambient temperature. Elevated temperature results in an increased iron concentration in the resulting leach solution and shall be avoided. The leach solution is filtered and the filter cake is washed with water and then carefully de-watered. The liquid filtrate (leach solution), containing a precipitate of phosphorus, calcium and metal compounds, is passed to a purification step for selective metal recovery followed by phosphate precipitation.

LEACHING THE ASH

Hydrometallurgical processing of apatite ores containing tri calcium phosphate $(Ca_3(PO_4)_2)$ has been applied commercially for many years. The traditional treatment route to achieve fertilizer grade phosphoric acid includes atmospheric sulphuric acid leaching of the apatite ores, followed by phosphorus gypsum separation and concentration to produce phosphoric acid to 52–54% P_2O_5. If a technical or food grade phosphoric acid is required, further process steps have to be included, such as solvent extraction purification, arsenic, fluoride and sulphate removal.

Characterisation of the ash material

Laboratory-scale experiments have been carried out with ashes from an industrial-scale sludge incineration plant that is fed with sewage sludge from a municipal wastewater treatment plant (WWTP). This plant is based on chemical phosphorus removal. The typical chemical analysis of the ash is summarised in Table 1. The pour density of the ash sample is 0.9 g/cm^3 and the average particle size is 75 μm.

Table 1. Typical chemical analysis of the ash from an industrial-scale sludge incineration plant (data referred to dry samples).

Parameter	Item	Concentration
Phosphorus	g P_2O_5/kg	180
Aluminium	g Al_2O_3/kg	95
Calcium	g CaO /kg	110
Silicon	g SiO_2/kg	340
Iron	g Fe_2O_3/kg	180
Magnesium	g MgO/kg	20
Copper	g Cu/kg	0.6
Zinc	g Zn/kg	3
Cadmium	mg Cd/kg	3.4
Chromium	mg Cd/kg	100

General leaching investigations

25 g of the ash was contacted for about 90 minutes at 30°C in a stirred mixing vessel with four different leaching solutions at a solid/liquid ratio of 1/5. After filtration, the resulting residue was washed with water, dried at 105°C and analysed. The following leaching solutions were investigated:

- Hydrochloric acid solution
- Sulphuric acid solution
- Phosphoric acid solution
- Sodium hydroxide

If hydrochloride acid is used, the equations below are representative of the leaching chemistry:

$$FePO_4 + 3HCL \Rightarrow H_3PO_4 + FeCl_3$$

$$FePO_4 + 4HCL \Rightarrow H_3PO_4 + FeCl_4^- + H^+$$

$$AlPO_4 + 3HCL \Rightarrow H_3PO_4 + Al^{3+} + 3Cl^-$$

$$Ca_3(PO_4)_2 + 6HCL \Rightarrow 2H_3PO_4 + 3Ca^{2+} + 6Cl^-$$

Results obtained from laboratory batch scale studies indicated that phosphorus leaching rates of $>90\%$ could be obtained using 8% hydrochloric acid solution at ambient temperature and in short time. Resulting leaching solutions were filtered and the solutions were saved. The leaching residue, containing mainly iron and silicon oxides, were washed with water and used for further experiments. Co-leaching rates of 57% for aluminium, 95% for calcium and 7% for iron were obtained. Based on these results it was decided to use 8% hydrochloric acid solutions in further test work.

Table 2 shows the summery of the leaching results for the dissolution of the phosphorus content using hydrochloric acid, phosphoric acid and sodium hydroxide.

Based on these results, further leaching was performed with 8% hydrochloric acid at some selected settings (Table 3).

Further optimisation of the leaching process was performed, focusing on the effect of residence time, iron and aluminium co-leaching. The process conditions are shown in Table 4 and the results are summarised in Table 5 and Table 6.

The next step involved the evaluation of the purification part of the leaching solution in the phosphorus recovery process. The testing programme contained laboratory scale experiments, mainly focused on phosphorus and heavy metals.

SOLVENT EXTRACTION

Solvent extraction principles

Solvent extraction is a selective separation procedure for isolating or concentrating a valuable substance from an aqueous solution with the aid of an organic solution. In the procedure the aqueous solution containing the

Table 2. Summery of the leaching results.

Operation conditions		HCl	H₂SO₄	H₃PO₄	NaOH
Concentration of leach solution	%	8	8	8	17
Solid/Liquid ratio	S/L	1/5	1/5	1/5	1/5
Temperature	°C	30	30	30	30
Time	min	90	90	90	90
Weigh-in	g	25	25	25	25
Weigh-out	g	15	22	20	24
Phosphorus leached	%	94	92	57	30

Table 3. Hydrochloric acid leaching.

Operation conditions				
Concentration of leach solution	% HCl	8	8	8
Solid/Liquid ratio	S/L	1/5	1/4	1/4
Temperature	°C	26	30	33
Time	h	1.5	1.5	1.5
Weigh-in	g	25	25	25
Weigh-out	g	15	15	15
Leaching rates				
Phosphorus	%	94	94	92
Iron	%	7	9	7
Aluminium	%	57	57	56
Calcium	%	95	94	92
Magnesium	%	65	66	66
Zinc	%	6	8	10

substance of interest, often at a low concentration and together with other dissolved substances, is mixed (extraction) with an organic solvent containing a reagent. The substance of interest reacts with the reagent to form a chemical compound, which is more soluble in the organic than in the aqueous solution. As a consequence, the substance of interest is transferred to the organic solution.

Table 4. Process conditions for hydrochloric acid leaching.

Concentration of HCl leach solution	%	8
Solid/Liquid ratio	S/L	1/5
Operation conditions for residence time		
Temperature	°C	33
Time	min	10, 30, 60, 75, 90
Operation conditions for temperature		
Time	min	30
Temperature	°C	30, 40, 60, 75, 90

Table 5. Summery of the leaching results: Effect of residence time.

Time	Min	10	30	60	75	90
Leaching rates						
Phosphorus	%	88	93	94	98	98
Iron	%	6	7	8	8	8
Aluminium	%	52	56	57	58	58

Table 6. Summery of the leaching results: Effect of temperature.

Temperature	°C	30	40	60	75	90
Leaching rates						
Phosphorus	%	90	93	99	100	100
Iron	%	6	7	11	19	23

Subsequently, in order to recover the extracted substance, the organic solution is mixed (stripping) with an aqueous solution that composition is such that the chemical compound between the substance and the reagent is split and, thus, the substance is recovered in the "new" aqueous solution. The concentration of the substance in the "new" aqueous solution may be increased, often to 10 to 100 times that of the original aqueous solution, through adjustment of the liquid flow rates. Freed from the substance of interest, the organic solution is returned for further extraction, either directly, after regeneration or after a fraction of it has been cleaned of impurities.

Chemical reactions involved using Alamine® 336

This reagent belongs to a class of reagents that is based on the principle of ion association. The active group contains basic nitrogen. They typically react with inorganic and organic acids to form amine salts, which undergoes ion exchange reactions with a variety of metal anions, such as iron(III) chloride, cadmium chloride, lead chloride and zinc chloride. A commercial reagent available of this type, Alamine® 336, was used for the experiments and its performance is shown in Figure 1. This reagent, manufactured by Cognis, is a reagent that was developed especially for the cobalt-nickel separation, vanadium extraction, platinum group metals separation and tungsten recovery. It is also used for the separation and purification of inorganic and organic acids from process and waste streams.

The general chemical reactions between the reagent and the iron chloride complex are shown below:

$$[R_3N]_{org} + [HA]_a \Rightarrow [R_3NH^+ + A^-]_{org}$$

$$[R_3NH^+ + A^-]_{org} + [FeCl_4^-]_{aq} \Rightarrow [R_3NH^+ + FeCl_4^-]_{org} + [A^-]_{aq}$$

As can be seen in the expression

- Alamine® 336 reacts in its amine salt form,
- Iron must occur in the aqueous solution as a iron chloride complex, and
- Alamine® 336 forms with iron chloride an ion-pair complex.

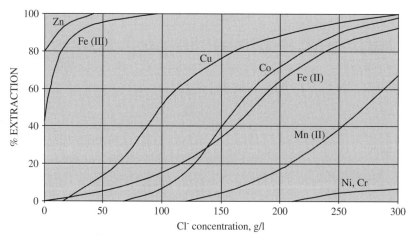

Figure 1. Alamine® 336 extraction from chloride solutions at 40°C and pH = 2 (Data MEAB Metallextraktion AB, Göteborg/Sweden).

Alamine® 336 can be stripped by a wide range of inorganic salt solutions such has sodium carbonate or ammonium carbonate. The general reaction between Alamine® 336 and an ammonium bicarbonate strip solution is shown below.

$$[R_3NH^+ + FeCl_4^-]_{org} + [4NH_4HCO_3]_{aq} \Rightarrow [R_3N]_{org} + [Fe(OH)_3 + 4NH_4Cl + 4CO_2 + H_2O]_{aq}$$

Laboratory-scale solvent extraction experiments

For the proposed purification process, the following functions are valid:

- Regeneration of the organic solvent to form the amine salt
- Extraction of iron, cadmium, copper, lead and zinc from the leach solution
- Scrubbing of co-extracted phosphoric acid from the loaded organic solvent
- Stripping of iron, cadmium, copper, lead and zinc from the loaded organic solvent with an alkali solution

Bench-scale experiments have been carried out with a downstream leaching solution provided from leaching tests. The solvent extraction operation was successfully tested in mixer-settler laboratory equipment, shown in Figure 2, and delivered and operated by MEAB Chemie Technik GmbH.

The MSU-2.5 laboratory and pilot plant mixer-settler units, available in PVDF, are traditionally designed pump mixer-settler units, with squared mixing chambers (mixers) and box-type settling compartments (settlers). The mixers are equipped with adjustable speed stirrers and the settlers with picket fences for the distribution of the dispersion over the whole cross-section of the settler.

An adjustable jack-leg for the outflow of the heavier solvent controls the phase boundary level at varying liquid densities. A heavier solvent recycle is integrated in the construction. Each unit is designed to operate hydro dynamically independent of other units, an advantage which enables all units to be placed horizontally, one after the other, thus imply flexibility in the experimental set-up. The flow (org + aq + recycle) range is up to 50 l/h at 2.5 m/h surface loading.

A general block diagram of the purification process is shown in the Figure 3.

The removal of iron, lead, cadmium, copper and zinc from the leach solution above is performed in a multi-stage solvent extraction operation. The leach solution, containing the metals in the form of chloride complexes, is mixed with the Alamine® 336 solution, containing kerosene. By the mixing the above-mentioned metals are transferred (extracted) to the Alamine® 336 solution. Phosphate, calcium and aluminium are left in the aqueous solution (raffinate). After separation, the aqueous solution is saved for the phosphate recovery.

Figure 2. Laboratory mixer-settler unit MSU-2.5.

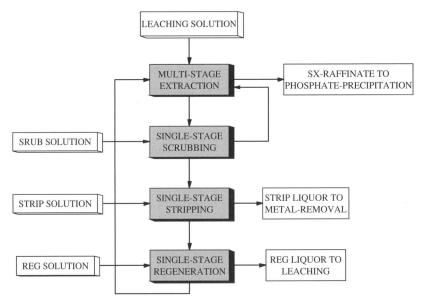

Figure 3. General block diagram of the purification process.

The organic solution, containing the metals, is scrubbed with a small amount of water to remove entrained leach solution (phosphate contamination). The stripping of the metals followed is performed with an ammoniacal solution. During the stripping, iron and lead form a hydroxide precipitate (solid). All other metals are soluble in the strip liquor. The organic solution is recycled after wash with diluted hydrochloric acid. The solvent extraction operation was success-fully tested in mixer-settler equipment (MSU-2.5) at the MEAB Chemie Technik GmbH in Aachen. The demonstration plant is shown in Figure 4.

Figure 4. Solvent extraction operation with MEABs MSU-2.5 in Aachen.

The composition of the leaching solution and the SX-raffinate is shown in Table 7.

PRECIPITATION

Precipitation of calcium (magnesium-) phosphate product and separation of aluminium

The third step in this phosphorus recovery process involves a precipitation of calcium phosphate or magnesium phosphate from the raffinate in the solvent extraction purification circuit. The aim of the laboratory bench scale

Table 7. Composition of the leaching solution and the SX-raffinate.

Parameter	Item	Leaching solution	SX-Raffinate
Phosphorus	mg/l	18,000	17,500
Calcium	mg/l	16,700	16,700
Magnesium	mg/l	2,200	2,200
Iron	mg/l	3,600	10
Aluminium	mg/l	7,800	7,800
Cadmium	mg/l	0.75	<0.05
Chromium	mg/l	4	4
Copper	mg/l	90	10
Nickel	mg/l	3	3
Lead	mg/l	29	1
Manganese	mg/l	1,500	1,500
Zinc	mg/l	250	1

testing carried out is to determine the initial operating conditions and the resulting precipitates.

Calcium and other metals form metal phosphates that are insoluble in water. The precipitation of metal phosphate and the re-release of aluminium are shown in the Figure 5.

The general block diagram of the calcium phosphate recovery process is shown in the Figure 6.

The acidic SX-raffinate from the solvent extraction operation, containing phosphoric acid, aluminium and calcium, is fed to a phosphate recovery vessel, where the pH value is raised to 5 using lime, limestone or sodium hydroxide. The operation is performed at room temperature and the residence time is 10–20 minutes. The main product from the neutralisation is a mixture of aluminium phosphate and calcium phosphate. If the aluminium content in the resulting SX-raffinate is too high, the pH is first adjusted to pH 2, were mainly aluminium phosphate is precipitated and removed. Then the resulting filtrate is fed to the calcium phosphate recovery section and the pH is raised to 5.

The general block diagram of the magnesium phosphate recovery process is shown in the Figure 7.

If magnesium phosphate is required as final product, calcium ions have to be removed by adding magnesium sulphate. The precipitation of calcium sulphate is shown below:

$$Ca^{2+} + Mg^{2+} + SO_4^{2-} + 2H_2O \Rightarrow Mg^{2+} + CaSO_4 * 2H_2O$$

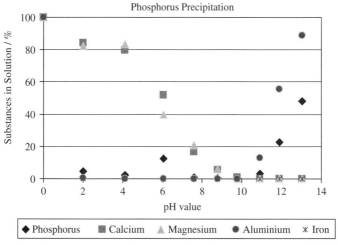

Figure 5. Precipitation of phosphates as a function of the pH value.

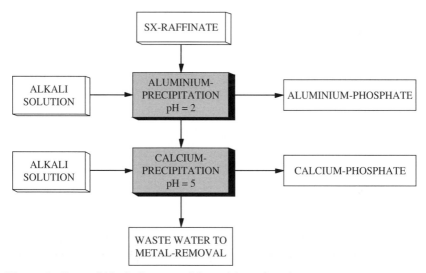

Figure 6. General block diagram of the calcium phosphate recovery process.

The raffinate from the solvent extraction operation is fed to a calcium sulphate removal circuit. By adding magnesium sulphate at room temperature and at a residence time of 20–30 minutes calcium sulphate is precipitated and the solid pulp is separated by filtration. The solid pulp is washed with water to

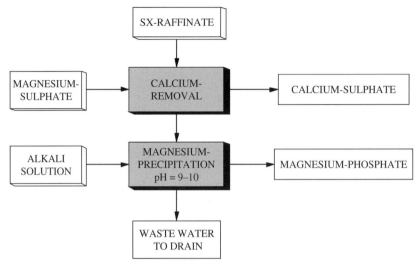

Figure 7. General block diagram of the magnesium phosphate recovery process.

remove phosphates from the calcium sulphate precipitate. The mother liquor from the filtration and the wash water are fed to the magnesium phosphate precipitation part. The pH is now raised to about 9–10 using sodium hydroxide or alternatives. The operation is performed at room temperature and a residence time of 10–20 min.

Preparation of the phosphorus product

Finally, the solid phosphate product from the precipitation is fed to a purification circuit, where clean water is used. The resulting filter cake (solid phosphate product) is dried, crushed and grinded. The mother liquor from the precipitation, containing the heavy metals and the wash water, is treated in a waste water plant.

The average content of two different phosphate precipitates is shown in Table 8 and Table 9.

Product quality

The investigation of harmful substances in the resulting products has shown that both products are within limits of fertilizer grade products. Table 10 shows the contamination of harmful substances. Numbers in extra bold letters are the values valid for fertilizer grade products in Germany.

Table 8. Average content in the calcium phosphate product (without aluminium removal).

Phosphorus	130 g/kg	Calcium	210 g/kg
Magnesium	25 g/kg	Aluminium	15 g/kg

Table 9. Average content in the magnesium phosphate product (after calcium sulphate removal).

Phosphorus	100 g/kg	Calcium	3 g/kg
Magnesium	130 g/kg	Aluminium	34 g/kg

Table 10. Contamination (heavy metals) in the phosphate product.

Cadmium	<0.5 mg/kg (1.5 mg/kg)	Lead	5 mg/kg (150 mg/kg)
Copper	5 mg/kg (70 mg/kg)	Nickel	60 mg/kg (80 mg/kg)
Mercury	<0.1 mg/kg (0.1 mg/kg)	Zinc	5 mg/kg (1,000 mg/kg)

The conceptual flow sheet of the described process is shown in Figure 8 (aluminium phosphate removal followed by calcium phosphate precipitation). In the final waste water treatment, heavy metals would be removed by addition of lime.

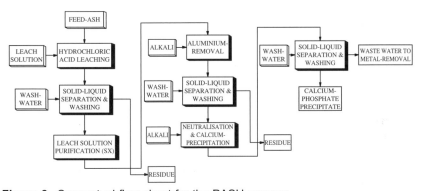

Figure 8. Conceptual flow sheet for the PASH process.

CONCLUSIONS

The PASH process is installed to recover the phosphorus from the ash after the incineration of municipal sewage sludge. All organic pollutants are eliminated in the incineration. Therefore, the recovery process has to deal with inorganic impurities only. The recovered product (calcium- or magnesium phosphate) contains only small amounts of metals and meets the requirements of the phosphate industry. Up to 85% of the phosphorus content in the inflow to a municipal wastewater treatment plant (WWTP) can be recovered by implementing the PASH process. Thus, non-renewable phosphate-ore deposits can be conserved and huge environmental impacts in the phosphate-exploiting countries can be avoided. Although the ashes used in the experiments were random samples, results proved to be repeatable.

ACKNOWLEDGEMENTS

This research and development project is funded within the framework of the German research cluster on phosphorus recycling which is supervised by the Project Management Agency Forschungszentrum Karlsruhe and funded by the Federal Ministry of Education and Research.

REFERENCES

Rydberg, J., Cox, M., Musikas, C. and Choppin, G. (2004). Solvent Extraction Principles and Practice. Second Edition. New York: Marcel Dekker, Inc.

Sole, K., Feather, A. and Cole, P. Solvent Extraction in Southern Africa: An Update of Some Recent Hydrometallurgical developments. In: Hydrometallurgy, 2005, 78, pp. 52–78 Internet http://www.meab-mx.se

Montag, D., Gethke, K. and Pinnekamp, J. (2007). Different Approaches for Prospective Sludge Management Incorporating Phosphorus Recovery. In: Filibeli, A., Sanin, F.D., Ayol, A., Sanin, S.L.: Facing Sludge Diversities: Challenges, Risks and Opportunities, pp. 289–296, IWA-Congress, March 28th–30th 2007, Antalya, Turkey, ISBN 978–975-441-238-3.

Phosphorus recovery from sewage sludge ash: possibilities and limitations of wet chemical technologies

Christian Schaum[a], Peter Cornel[b] and Norbert Jardin[c]

[a]Dr. Born - Dr. Ermel GmbH, Finienweg 7, D-28832 Achim, Germany, cs@born-ermel.de
[b]Technische Universität Darmstadt, Institut WAR, Petersenstraße 13, D-64287 Darmstadt, Germany
[c]Ruhrverband, Kronprinzenstraße 37, D-45128 Essen, Germany

Abstract In laboratory-scale experiments with ashes from different full-scale sludge incineration plants the elution behaviour of phosphorus and metals at different pH-values were examined. The elution of ashes with water alone did not cause any significant release of phosphorus, while depending on the ash origin and formation, a phosphorus release of maximum 30% was measured when using sodium hydroxide solution. With sulphuric acid, it was possible to release phosphorus and (heavy) metals quantitatively at pH-values less than 1.5. In addition to the elution experiments, precipitation and nanofiltration were investigated as methods for the separation of phosphorus and metals from the liquid phase. Sequential precipitation of phosphorus (SEPHOS Process) seems to be promising. The generated product, an 'aluminium phosphate', is a valuable raw material for the phosphorus industry. After alkaline treatment of the 'aluminium phosphate', it is possible to precipitate phosphorus as calcium phosphate (advanced SEPHOS Process). Following acidic elution of the ash, nanofiltration can also be used to separate phosphorus.

INTRODUCTION

In view of the limited resources of phosphorus and the rising concern about the sustainability of wastewater and sludge treatment, the recovery of phosphorus from wastewater, sewage sludge and sludge ash is a promising approach.

The phosphorus load into German municipal wastewater treatment plants (WWTP) has been decreasing continuously, mainly because of the use of phosphate-free detergents. At present, the phosphorus load amounts to between 1.6 and 2.0 g P/(PE · d), resulting in phosphorus influent concentration of 8 to 10 mg/l at 200 l per person and day. The effluent limits are 1–2 mg/l for most plants, depending on their size. This amounts to 0.2–0.4 g P/(PE · d). Thus, about 90% of the phosphorus must be eliminated in the WWTP.

Figure 1 shows the phosphorus balance for a typical wastewater treatment plant in Germany. Approximately 11% of the phosphorus is separated by the primary sludge (PS) and 28% by the secondary (surplus) sludge (SS) without any specific phosphorus removal. Thus, approximately 50% of the phosphorus has to be eliminated by means of enhanced biological phosphorus removal (EBPR), precipitation, or other phosphorus-removal techniques. With the generally applied EBPR and/or precipitation techniques, 90% of the incoming phosphorus ends in the sewage sludge. It seems worthwhile to develop methods for the separation of phosphorus from sludge, or, taking into account the future development towards sludge incineration, for extracting phosphorus from the ash.

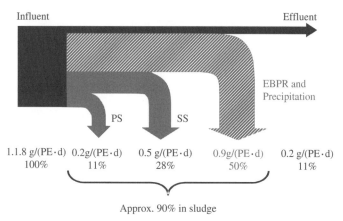

Figure 1. Phosphorus balance for a typical WWTP in Germany (PS: Primary Sludge; SS: Secondary Sludge (surplus sludge); PE: Population Equivalent).

MATERIAL AND METHODS

Origin and composition of the sewage sludge ashes

15 samples of sewage sludge ash were selected from mono incineration plants in Germany and European countries. All investigated samples were random samples, made anonymous as ash #A to #O.

Figure 2 shows the percentage distribution of the main components of the tested ashes in a box-plot diagram. The length of the boxes defines the range, to which 50% of all values belong. The average phosphorus content is about 16% P_2O_5. The comparison indicates that iron oxide and calcium oxide show a larger range within the 25 to 75 percentile than aluminium oxide.

Figure 3 presents a three-phase diagram of calcium oxide, iron oxide and aluminium oxide. The diagram shows the percentage distribution of these components in the ash samples, i.e. the sum of calcium oxide, iron oxide and aluminium oxide is 100% for each sample.

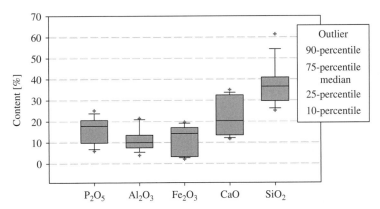

Figure 2. Distribution of the main components of sewage sludge ash as box-plot diagram.

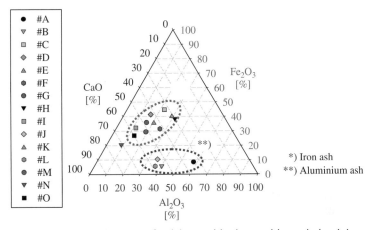

Figure 3. Three-phase diagram of calcium oxide, iron oxide and aluminium oxide in samples of sewage sludge ash, and their characteristic colouring (Schaum *et al.*, 2007).

Following the diagram, two types of ashes can be identified with reference to aluminium and iron concentrations:

– "aluminium ash" - This ash is defined by its high aluminium content (Al_2O_3 $>35\%$) combined with a low iron content ($Fe_2O_3 < 10\%$), as can also be seen in the greyish ash colour.

– "iron ash" – This ash is defined by its high iron content ($Fe_2O_3 > 25\%$), as can also be seen in the reddish brown ash colour.

In all ash samples, the content of calcium oxide (CaO) is between 30% and 70%, whereby iron oxide as well as aluminium oxide vary across this range.

Thermal treatment of mixtures of pure substances

In order to investigate the transfer of aluminium, phosphorus and calcium, the following mixtures of pure substances were heated in a muffle furnace at $T_{max} = 500$, 700 and 900°C:

– "mixture A" 10 g aluminium phosphate, 5 g calcium carbonate and 20 g silicon dioxide.
– "mixture B" 10 g aluminium phosphate, 10 g calcium carbonate and 20 g silicon dioxide.

By varying the calcium content, the influence of calcium on the transfer of phosphorus was to be determined. The reaction time at T_{max} was 2 h each. Following the thermal treatment, elution with hydrochloric acid and analyses of the dissolved substances were carried out.

Elution tests

In order to maximize the stability of the pH-value, elution tests with hydrochloric acid, cf. DIN 38414-4 1984, were carried out. The concentrations of the hydrochloric acid were 0–0.031–0.063–0.125–0.25–0.5–0.75–1.0–1.5–2.0 mol HCl/l, whereby deionised water was used for "zero concentration".

The ratio of ash to elution solvent was 1:10 (cf. DIN 38414-4 1984). In order to achieve a constant pH-value, the samples – transferred to plastic bottles – were treated in an overhead stirrer at room temperature overnight. Following sedimentation and filtration (0.45 µm Schleicher & Schuell ME 25), the pH-value was measured (Mettler Toledo Inlab 1003 electrode and WTW pH 197), and the dissolved substances were analysed via ICP-OES (DIN EN ISO 11855, 1998). In addition to the acidic elution, an alkaline elution was carried out, using a 1 mol/l sodium hydroxide solution.

Separation of phosphorus and metals via sequential increase of the pH-value

Following the acidic elution with 1 mol/l sulphuric acid, the pH-value is gradually increased by means of a 10 mol/l sodium hydroxide solution. The results are precipitation reactions. A fully-automatic, computer-controlled titration unit

(Metrohm, Type MPT-Titrino 798 with a Profitrode) was used to adjust the pH-value. After precipitate separation and filtration (0.45 μm Schleicher & Schuell ME 25) the dissolved substances were analysed via ICP-OES (DIN EN ISO 11855, 1998).

Separation of phosphorus and metals via nanofiltration

The separation of phosphorus from ash eluate was tested in a stirred cell, using a phosphorus selective solution-diffusion membrane (Membrane Desal 5 DK made by Osmonics). Sulphuric as well as nitric acid were used to produce the ash eluate. At regular intervals, permeate samples were taken, which were tested for dissolved substances via ICP-OES (DIN EN ISO 11855, 1998).

RESULTS AND DISCUSSION

Thermal treatment of mixtures of pure substances

Figure 4 shows the release rates of aluminium, calcium and phosphorus, with different incineration temperatures and different ratios of phosphorus to calcium (mixture A and B). By increasing the incineration temperature, the release of aluminium decreased, using acidic as well as alkaline elution. The release of phosphorus by temperature increase changed significantly only in case of alkaline elution. An increase of the percentage of calcium (mixture B) in combination with alkaline elution results in decreasing release rates of phosphorus, dropping to < 1% at 900°C. In contrast, with acidic elution the release of phosphorus is almost 100%, irrespective of temperature and calcium content.

Investigations of sewage sludge ashes by Schirmer (1998) using x-ray diffractometry showed, that through incineration the crystalline structures quartz,

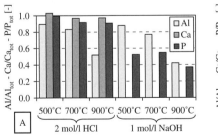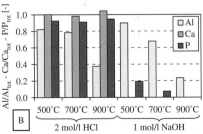

Figure 4. Release of aluminium, calcium, and phosphorus in mixtures of pure substances (aluminium phosphate, calcium carbonate, silicon dioxide), after thermal treatment at 500, 700, and 900°C; mixture A: $P_{tot}/Ca_{tot} \approx 1\ \frac{1}{2}$ [mol/mol]; mixture B: $P_{tot}/Ca_{tot} \approx \frac{3}{4}$ [mol/mol] (Schaum, 2007).

hematite (Fe_2O_3), whitlockite ($Ca_3(PO_4)_2$) and anorthite ($CaAl_2Si_2O_8$) dominate. The results of incineration tests with pure substances by Schirmer (1998), using a mixture of aluminium phosphate, calcium carbonate and quartz, confirmed the formation of calcium phosphate and anorthite. Through the incineration of a mixture of iron phosphate and calcium carbonate, he could also prove evidence of the formation of calcium phosphate and hematite. He identified whitlockite as well as hydroxylapatite ($Ca_5(PO_4)_3OH$) and assumed that incineration conditions affects the formation.

The change in solubility behaviour, which was shown in the experiments, confirmed the investigations by Schirmer (1998). Aluminium oxides hardly soluble in acids are formed during the thermal treatment of mixtures of aluminium phosphate, calcium carbonate and silicon dioxide. At the same time, the dissolution behaviour of phosphorus and calcium suggests that transfer of phosphorus occurs and calcium phosphate is formed. This phenomenon is affected by the incineration temperature and the ratio of phosphorus to calcium. It is important to note that calcium phosphate is soluble only in acids, cf. Stumm und Morgan (1996).

Elution tests

The results of elution tests with sewage sludge ash indicate that with pH-values < 1.5, it is possible to release phosphorus almost completely, irrespective of the type of ash (aluminium ash or iron ash) and despite the fact of limited release of aluminium and iron (cf. Figure 5). Thereby, the limited release rates of aluminium and iron confirm the assumption that oxides of low solubility are formed as well as calcium phosphate during sewage sludge incineration.

Figure 6 shows that with alkaline elution the release of phosphorus is limited. An average release rate of merely about 10% (P_{tot} based) was observed, ranging from <1 minimum to 30% maximum. Furthermore, ashes with low calcium concentrations, i.e. large ratio of P_{tot} to Ca_{tot}, lead to a higher release rate of phosphorus (cf. Figure 6). Analogue to the tests with pure substances, the results of alkaline elution of sewage sludge ashes show a higher release of phosphorus in those ashes with low calcium concentrations.

In order to summarise the results of phosphorus, calcium, aluminium and iron release, the process is described mathematically. The release process of phosphorus and calcium against the pH-value can be described by the general formula for the release of metal hydroxides according to Sposito (1984):

$$\text{According to Sposito (1984):} \quad \frac{P}{P_{tot}}; \frac{Ca}{Ca_{tot}} = \frac{1}{1 + e^{a+b \cdot pH}} \tag{1}$$

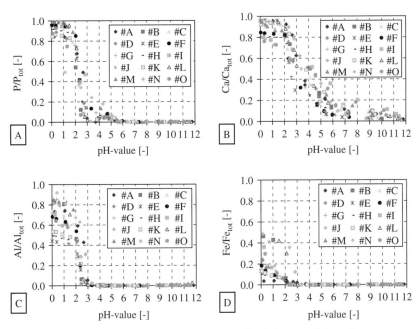

Figue 5. Release of phosphorus, calcium, aluminium and iron from sewage sludge ash at different pH-values; elution of the ashes #A – #O with HCl (Schaum, 2007).

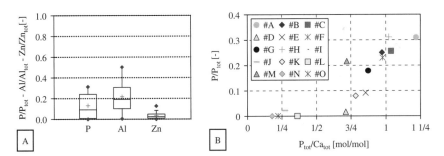

Figure 6. A: Release of phosphorus, aluminium and zinc after elution of sewage sludge ash (ashes #A – #O) with 1 mol/l sodium hydroxide solution, presentation as box-plot diagram; B: release of phosphorus as a function of the calcium content of the ash.

The non-linear regression showed that, using a correlation coefficient of $R^2 > 0.9$, Equation 1 is a good approach to describe the release behaviour of phosphorus and calcium. In order to map the significant increase in aluminium release between pH-values 3 and 2, Equation 1 was modified:

$$\text{Adaptation for the release of aluminium:} \quad \frac{Al}{Al_{tot}} = c \cdot \frac{1}{1 + e^{a+b \cdot pH}} \qquad (2)$$

In case of iron, no adaptation was possible neither with Equation 1 nor Equation 2 using a correlation coefficient of $R^2 > 0.9$. In both cases the correlation coefficient $R^2 = 0.5$. Figure 7 shows the release of phosphorus, calcium, aluminium and iron from sewage sludge ash vs. the pH-value. Thereby, the ash type (aluminium or iron ash) has hardly any influence on the release behaviour.

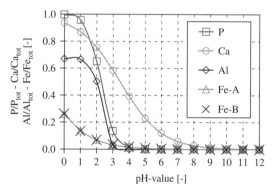

Figure 7. Release of phosphorus (Eq. 1), calcium (Eq. 1), aluminium (Eq. 2) and iron (Fe-A: Eq. 1, Fe-B: Eq. 2) from sewage sludge ash vs. the pH-value after non-linear regression.

Separation of phosphorus and metals via sequential increase of the pH-value

Following the acidic ash treatment (pH-value 1.5), the pH-value of the filtrate was increased step-wise by adding sodium hydroxide solution. Figure 8-A shows the concentration of aluminium, iron and phosphorus against the pH-value. Both aluminium and phosphorus dissolve in acidic as well as alkaline environment. Thereby, the run of the curves is very similar. It is assumed that by increasing the pH-value an amorphous aluminium phosphate precipitates at pH-values 3–4, which releases in alkaline environments.

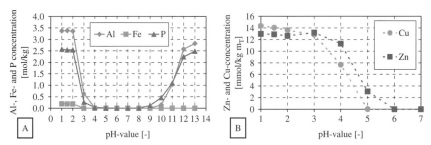

Figure 8. Sequential precipitation by increasing the pH-value (following the elution with 1 mol/l H_2SO_4), ash #A, A: aluminium, iron and phosphorus concentrations vs. pH-value, B: copper and zinc concentrations vs. pH-value.

Analogous to Figure 8-A, in Figure 8-B the concentrations of zinc and copper against the pH-value are illustrated. The run of the curves shows that the heavy metals remain mainly in solution at pH-values between 2 and 3. Due to this phenomenon, it is possible to produce a heavy-metal depleted precipitate by keeping the pH-value below 3.5.

Based on the investigations presented above, the so-called SEPHOS Process - Sequential Precipitation of Phosphorus was developed, cf. Cornel *et al.* (2004, 2006). First, sewage sludge ash is eluted with sulphuric acid. After removing the undissolved residues, the ph-value of the filtrate is increased, thus inducing precipitation reactions. Finally, the phosphate-rich and heavy-metal depleted precipitate can be reused elsewhere.

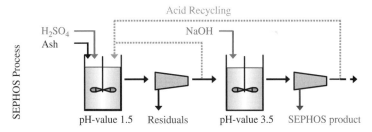

Figure 9. Flow diagram of the SEPHOS Process (Cornel *et al.* 2004, 2006).

Through alkaline treatment of the precipitate, phosphorus can be separated as calcium phosphate, and aluminium can be reused as precipitant, cf. Advanced SEPHOS Process Cornel *et al.* (2004, 2006).

Separation of phosphorus and metals via nanofiltration

The investigations using a stirred cell indicated that nanofiltration is a suitable method for phosphorus separation (Niewersch et al., 2007). In case of ash #C, about 50% of the phosphorus passed the membrane. Thereby, sulphuric acid as well as nitric acid were used for elution. With polyvalent ions, such as aluminium, iron, calcium and zinc more than 98% were retained by the membrane, cf. Table 1. Only sodium, i.e. as monovalent ion, passed the membrane. For charge balance, the anions chloride, sulphate, and nitrate, for the most part, could pass the membrane.

Table 1. Concentration of various components in feed and permeate, ash #C.

		P	Al	Ca	Cu	Fe	Mg	Na	Zn
Feed	mg/l	7.536	3.626	750	66	2.127	960	192	235
Permeate	mg/l	3.762	14	18	2	50	10	110	1
Retention	%	50	>99	98	97	98	99	43	>99

The results of the stirred-cell tests show, that it is possible to separate phosphorus selectively by nanofiltration. Thereby, it can be expected that advancements in membrane technology will increase the membrane permeability for phosphorus. In order to allow the implementation of nanofiltration as a suitable method for phosphorus recovery, it is necessary to remove the solids from the ash eluate, e.g. by pre-stage microsieving. An operational pressure of 60–70 bar was necessary to achieve permeate fluxes, thus leading to a high energy demand. As the phosphorus concentration of the permeate is relatively low and further anions, such as nitrate, sulphate and chloride are present (depending on the type of acid used for ash elution), direct reuse as phosphoric acid seems improbable. Thus, further treatment steps are necessary, e.g. precipitation as calcium phosphate or magnesium ammonium phosphate.

CONCLUSIONS

The mono incineration of sewage sludge leads to a transfer of phosphorus and thus to the formation of calcium phosphate and aluminium and iron oxides. By acidic elution at pH-values <1.5, phosphorus is released, as well as a large percentage of the metals. Due to the formation of calcium phosphate, a release in alkaline environments is limited and mainly dependent on the ash composition. The investigation of 15 sewage sludge ashes presented a maximum release rate of about 30% for phosphorus.

After removing the insoluble solids, sand in particular, phosphorus has to be separated from the metals. Hereby, two different processes were presented, i.e. sequential precipitation and nanofiltration.

Following the acidic elution of the ash, sequential precipitation of phosphorus (SEPHOS Process) is induced by increasing the pH-value systematically to approximately 3.5, thus leading to the precipitation of aluminium phosphate. As copper and zinc only precipitate at pH-values >3.5, separation of phosphorus from the heavy metals is possible. Aluminium phosphate can then be used in the electro-thermal phosphorus industry. Rough estimations show that the costs for chemicals needed to produce the precipitate "aluminium phosphate" via the SEPHOS Process are approximately 3–3 times higher than the current world price of raw phosphate. By optimizing elution and precipitation processes, the consumption of chemicals needed and thus the costs can be reduced. Furthermore, the prices of raw phosphate are expected to increase, following its decreasing quality and quantities.

Through alkaline treatment of the aluminium phosphate (Advanced SEPHOS Process), phosphorus as well as aluminium dissolves. By adding calcium, calcium phosphate precipitates, while aluminium stays in solution and can be recycled as precipitant.

The results of phosphorus separation by nanofiltration confirmed the general feasibility of this method for phosphorus recovery. It can be expected that advancements in membrane technology will increase the membrane permeability for phosphorus.

ACKNOWLEDGEMENT

The presented research activities were part of a PhD thesis, done at the Technische Universität Darmstadt, Institut WAR, Wastewater Technology (Schaum, 2007). The nanofiltration experiments were carried out in cooperation with RWTH Aachen University, Department of Chemical Engineering. Part of the investigation was conducted within a joint research project of the Ruhrverband and the Technische Universität Darmstadt, Institut WAR, Wastewater Technology, titled "Phosphorus recovery from sewage sludge ashes – concepts, strategies, economic efficiency". This project was funded by the Ministry for the Environment, Nature Conservation, Agriculture and Consumer Protection of North Rhine-Westphalia.

REFERENCES

Cornel, P., Jardin, N. and Schaum, C. (2004). Möglichkeiten einer Rückgewinnung von Phosphor aus Klärschlammasche, Teil 1: Ergebnisse von Laborversuchen zur Extraktion von Phosphor, GWF Wasser Abwasser, No. 9, Oldenbourg Industrieverlag GmbH, München, Germany.

Cornel, P., Jardin, N. and Schaum, C. (2006). Möglichkeiten einer Rückgewinnung von Phosphor aus Klärschlammasche, Teil 2: Ergebnisse von Laborversuchen zur Rückgewinnung von Phosphor, GWF Wasser Abwasser, No. 1, Oldenbourg Industrieverlag GmbH, München, Germany.

DIN 38414-4 (1984). Schlamm und Sedimente, Bestimmung der Eluierbarkeit mit Wasser, Beuth Verlage GmbH, Berlin, Germany.

DIN EN 13346 (2000). Bestimmung von Spurenelementen und Phosphor, Beuth Verlage GmbH, Berlin, Germany.

DIN EN ISO 11885 (1998). Bestimmung von 33 Elementen durch induktiv gekoppelte Plasma-Atom-Emissionsspektrometrie, Beuth Verlage GmbH, Berlin, Germany.

Niewersch, C., Koh, C.N., Wintgens, T., Melin, T., Schaum, C. and Cornel, P. (2007). Potentials of using nanofiltration to recover phosphorus from sewage sludge, proceedings, International Conference on Membranes for Water and Wastewater Treatment, 15–17 May 2007, Harrogate, UK.

Schaum, C. (2007). Verfahren für eine zukünftige Klärschlammbehandlung – Klärschlammkonditionierung und Rückgewinnung von Phosphor aus Klärschlammasche, Dissertation, Schriftenreihe WAR, No. 185, Darmstadt, Germany.

Schaum, C., Cornel, P. and Jardin, N. (2007). Phosphorus Recovery from Sewage Sludge Ash – A Wet Chemical Approach, proceedings, Wastewater Biosolids Sustainability: Techical, Managerial and Public Synergy, 24–27.06.2007, Moncton, Canada.

Schirmer, T. (1998). Mineralogische Untersuchungen an Klärschlammasche und an Ziegeln mit Zusatz von Klärschlammasche, Phasenbestand, Sinterprozesse und Schwermetallfixierung, Berichte aus der Umwelttechnik, Shaker Verlag, Aachen, Germany.

Sposito, G. (1984). The Surface Chemistry of Soils, Oxford University Press, New York, USA.

Stumm, W. und Morgan, J.J. (1996). Aquatic Chemistry, Wiley-Interscience publication, UK.

Phosphate adsorption from sewage sludge filtrate using Zinc-Aluminium layered double hydroxides

Xiang Cheng[a], Xinrui Huang[a], Xingzu Wang[a], Bingqing Zhao[b], Aiyan Chen[a] and Dezhi Sun[a,c]*

[a] School of Municipal and Environmental Engineering, Harbin Institute of Technology 202 Haihe Road, 150090 Harbin, China
[b] Research Center for Eco-Environmental Science, Chinese Academy of Sciences, 18 Shuangqing Road, 100085 Beijing, China
[c] College of Environmental Science and Engineering, Beijing Forestry University, 35 Tsinghua East Road, 100083 Beijing, China
*Corresponding author

Abstract A series of layered double hydroxides (LDHs) with different metal cations were synthesized to remove phosphate in excess sludge filtrate from municipal wastewater treatment plant for phosphorus recovery and eutrophication control as well. The highest phosphate adsorption capacity can be obtained by using Zn-Al-2-300, that is LDHs with Zn/Al molar ratio of 2 and calcined under 300°C for 4 h. Near-neutral and mild alkaline waters appeared suitable for the possible application of Zn-Al LDHs due to the amphoteric character of Aluminum hydroxide. Phosphate adsorption from sludge filtrate by the LDHs followed pseudo second-order kinetics equation, and the adsorption capacity at equilibrium was expected to be ~50 mg-P/g. Adsorption isotherms showed that phosphate uptake in this study was an endothermic process and had a good fit with Langmuir-type model. The absorbed phosphate can be effectively desorbed (more than 80%) from LDHs particles by 5 wt % NaOH solution.

INTRODUCTION

Various waste streams from industrial, agricultural and household activities contain considerable amount of soluble phosphorus, and its excessive input to the receiving water bodies could cause serious eutrophication of rivers, lakes and bays (Mulkerrins *et al.*, 2004). Biological nutrients removal process has been currently employed as the main method for phosphate separation from wastewater before discharge, it still, however, does not completely resolve the problem because of environmental risk and safety concerns in disposal of those excess sludge containing condensed phosphate, heavy metal and microorganism.

On the other hand, phosphorus was demonstrated essential for all forms of life, which plays an important role in DNA/RNA molecules and is also related to cellular energy transport via ATP. But phosphorus resource is nonrenewable and nonreplaceable. Its reserve under the present exploitive technology and economic condition was estimated to cover the demand for only around one century, including large amount of low-grade rock (Driver et al., 1999). Most of phosphorus in its cycle will end up in deep sediments in water, reentering the cycle probably after millions of years. Thus, it is urgent to protect, recover and reuse this limited resource for the sustainable development of the globe.

Chemical precipitation has been well documented as popular technology for phosphate recovery from wastewater both in pilot- and few full-scale (Jaffer et al., 2002; Pastor et al., 2008). It is of easy manipulation and will produce high-quality of phosphate precipitates, typically struvite, nevertheless chemical cost was added during the treatment and also large volume of chemical sludge may be produced, which is currently an acute disposable problem. More recently some researchers studied the feasibility of phosphate adsorption from phosphate-rich streams. The various adsorbents used include industrial materials and byproducts, natural or synthetic minerals, metal oxide/hydroxide and other materials.

Layered double hydroxides (LDHs), also called hydrotalcite-like compound or anionic clay, received wide attentions as an effective adsorbent. Its formula can be normally expressed as $[M^{2+}_{1-x}M^{3+}_{x}(OH)_2][A^{n-}]_{x/n} \cdot yH_2O$, where M^{2+} and M^{3+} are di- and trivalent metal cations in the octahedral positions of the positively charged brucite-like layers, A^{n-} is the incorporated anions in the interlayer space along with water molecules for charge neutrality and structure stability, and x normally ranges from 0.17 to 0.33 (Kwon et al., 1988). Due to the high charge density of the sheets from M^{2+} substitution by M^{3+} and the exchangeability of the interlayer anions as well, LDHs has been employed in several studies to remove phosphate in different environment, such as phosphate solutions (Badreddine et al., 1999; Das et al., 2006; Kuzawa et al., 2006; Ookubo et al., 1993), drain effluent (Seida and Nakano, 2002) and seawater (Chitrakar et al., 2005).

By secondary/enhanced biological process in municipal sewage works, various forms of phosphorus in water is terminally concentrated in excess sludge. When encountering anaerobic condition, the assimilated phosphorus tends to be released once again, leading to a higher concentration of soluble phosphate in sludge filtrate. Very few reports were as yet available on phosphate adsorption from sewage sludge filtrate by LDHs, probably because high levels of suspended solids in that stream easily make the adsorption column clogged. In the present study, we used magnetic particles of LDHs to remove phosphate in sludge filtrate from a sewage treatment plant. The particles were then effectively separated by magnetic-field for phosphate recovery. Furthermore, influencing factors of the adsorption process and recovery method were investigated and analyzed.

MATERIALS AND METHODS

Preparation of LDHs

A series of LDHs were synthesized by coprecipitation method using various divalent metal cations (Mg, Zn, Cu, Ni and Co) and trivalent ones (Al, Fe), and were abbreviated as M(II)-M(III). 200 ml of solution of mixed chlorides was prepared in a 500-ml beaker containing 0.5 M of M(II) and variable amount of M(III) (according to the M(II)/ M(II) ratio set). Fe_3O_4 powder was introduced as magnetic core for LDHs particle formation. Under vigorous mechanical stirring, 20% of NaOH solution was then added dropwise through a peristaltic pump to reach final pH of 9.0 ± 0.2. The synthesis was performed at 80°C in a water thermostat, followed by aging for 18 h at the same temperature. The resulting slurry was filtrated and rinsed with deionized water thoroughly until no chloride ion was detected. After dried at 80°C overnight, the precipitate was ground to ∼100-mm particles. A part of the obtained LDHs particles was calcined in an atmosphere of high-purity nitrogen for 4 h to study its effect on the adsorption capacity. All chemicals used were at least analytical reagents except the Fe_3O_4 powder being chemical grade.

Phosphate adsorption from sewage sludge filtrate

Adsorption assays were carried out in a thermostat shaker bath at room temperature. 0.02 g LDHs particles were added to 117-ml glass serum vials with 50 ml of sludge filtrate (from Taiping municipal wastewater treatment plant, Harbin, China) and sealed with butyl rubber stoppers. The water quality of the sludge filtrate is shown in Table 1. No pH (except the assays of pH effect) and other adjustments were conducted. After 24 h adsorption, the supernatant was immediately filtrated through 0.45-μm membrane for complete particles removal. The resulting liquid was used to determine the residual phosphate and then calculate the adsorption capacity of LDHs. LDHs-free vials were performed as control for non-adsorption degradation of phosphate in the water. All assays were conducted in triplicate. Phosphate adsorption kinetics was investigated similarly, except that 0.1 g of LDHs and 250 ml of the sludge filtrate were used, and the phosphate concentration in liquid was followed for 72 h.

Effect of pH on phosphate adsorption

The pH dependence of phosphate adsorption by LDHs was evaluated in batch assays with sludge filtrates from pH 2 to 12. The pH was adjusted using 1 M HCl/ NaOH solution. The phosphate residue in the solution was measured and the concentration of metal ions was analyzed as well to estimate the LDH dissolution.

Table 1. Characteristics of sludge filtrate from Taiping municipal wastewater treatment plant, Harbin, China.

Parameter	Sludge filtrate	Parameter	Sludge filtrate
PH	6.80 ± 0.00	Na^+, mg/L	83.65 ± 0.19
SS, mg/L	3.82 ± 0.00	K^+, mg/L	44.66 ± 0.05
COD, mg/L	514.62 ± 6.24	Al^{3+}, mg/L	0.06 ± 0.00
PO_4^3, mg-P/L	20 ± 4	Ca^{2+}, mg/L	105.8 ± 0.60
NO_2^-, mg/L	ND^a	Cu^{2+}, mg/L	ND^a
NO_3^-, mg/L	ND^a	Fe, mg/L	0.24 ± 0.00
SO_4^-, mg/L	112.53 ± 0.42	Mg^{2+}, mg/L	27.22 ± 0.09
Cl^-, mg/L	81.44 ± 0.50	Mn^{2+}, mg/L	1.34 ± 0.01
NH_4^+, mg/L	84.34 ± 4.82	Zn^{2+}, mg/L	0.01 ± 0.00

[a]ND: Non-detectable.

Study of adsorption isotherm

Adsorption isotherms were obtained by stirring known amounts of LDHs particles with 250 ml of sludge filtrates in 583-ml glass infusion bottles at fixed temperatures. Phosphate remainder was measured after 72 h when adsorption equilibrium was almost achieved based on the foregoing investigation. Isotherms of phosphate adsorption in KH_2PO_4 solutions were also tested for comparison.

Study of phosphate desorption

Phosphate-loaded LDHs was obtained by stirring Zn-Al-2-300 (Zn-Al LDHs with Zn/Al molar ratio of 2, calcined at $300°C$) with sludge filtrate at room temperature for 24 h. The particles were separated from the solution by imposing a magnet field, then washed in deionized water several times, and dried at $80°C$. Same amount of the recovered solids were added in three types of solutions with different concentrations (Table 2) for phosphate desorption. After 24 h of interaction at room temperature, the supernatant was sampled to determine the phosphate release.

Analysis

Characterization of LDHs

Metal ions were analyzed by ICP-OES (Optima 5300DV, PerkinElmer, USA). Crystal phases of LDHs were determined using an X-ray diffractometer (D/Max-RB, Rigaku, Japan) with CuKα radiation and operating at 45 kV and 45 mA. Samples were prepared as finely pressed LDHs powder in an aluminium holder. XRD patterns were collected over a 2θ from $10°$ to $90°$ in a scan rate

Table 2. Selection of solutions for phosphate desorption from LDHs particles.

Desorption solutions	Desorption rate (%)	Zn^{2+} (mg l^{-1})	Al^{3+} (mg l^{-1})
1 w/v% NaOH	20.58	23.90	16.75
2 w/v% NaOH	37.99	54.00	34.18
3 w/v% NaOH	52.27	65.40	33.70
5 w/v% NaOH	87.57	85.00	36.12
10 w/v% NaOH	89.34	333.10	61.17
20 w/v% NaOH	91.24	426.60	66.12
5 w/v% NaCl	7.56	3.20	30.54
10 w/v% NaCl	10.34	8.70	30.49
20 w/v% NaCl	21.17	9.50	30.19
5 w/v% NaCO₃	40.34	41.34	20.11
10 w/v% NaCO₃	84.22	100.35	43.06

of $5°$ min^{-1}. Thermogravimetry and differential thermal analysis (TG-DSC) was carried out on a STA449C instrument (NETZSCH, Germany). 4.675 mg of LDHs particles in an open Al_2O_3 crucible were heated from 35 to $1000°C$ ($10°C$ min^{-1}) in an Ar atmosphere at a flow rate of 30 ml min^{-1}.

Phosphate measurement

Water samples were filtrated through 0.45-μm membranes to remove suspended solids. The filtrate was then used to determine the concentration of soluble phosphate by molybdenum blue method (APHA *et al.*, 1998). No pH adjustment was conducted except in desorption assays the pH value of the samples was adjusted to 7.0 ± 0.1 due to their strong alkalinity.

RESULTS AND DISCUSSION

Parameters affecting phosphate adsorption from sewage sludge filtrate by LDHs

Metal compositon

Magnetic particles of LDHs prepared with various metal cations were subjected to phosphate adsorption assays. As shown in Figure 1, Zn-Al LDHs presents the highest adsorption capacity of up to 24.98 mg-P/g adsorbent in sewage sludge filtrate with initial phosphate concentration of 16.06 mg-P/L, followed closely by Mg-Fe and Zn-Fe particles. Low phosphate uptake was observed when Cu, Ni or Co was the divalent cation for LDHs synthesis. The result herein clearly differs from those in previous documents. Das *et al.* (Das *et al.*, 2006) reported that Mg-Al LDHs was the most efficient adsorbent for phosphate removal in potassium dihydrogen phosphate solution. Chitrakar and coworkers (Chitrakar *et al.*, 2005) suggested that Mg-Mn type and its heat-treated materials were more effective to adsorbing phosphate from seawater. In fact, phosphate uptake by layered double hydroxides in aquatic environment is as yet far from well understood; the difference in those results may be related to many specific conditions since in every study the synthesis method of LDHs could be quite different and the water environment tested and coexisting anions as well.

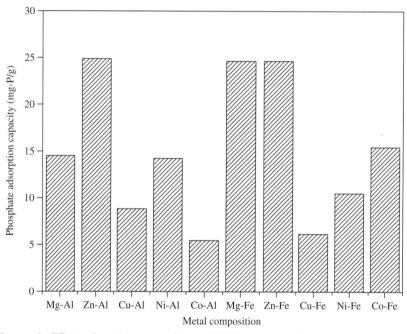

Figure 1. Effect of metal composition of LDHs on phosphate adsorption from sewage sludge filtrate.

Zn/Al molar ratio

In order to investigate the effect of Zn/Al molar ratio on the phosphate uptake, five different Zn-Al LDHs were used in adsorption assays. Zn/Al ratio of 2 appeared suitable for phosphate adsorption by LDHs; more Zn^{2+} replacement by Al^{3+} decreased the adsorption capacity instead of further improvement (Figure 2). Similar results have been reported elsewhere: Mg/Al of 2:1 (Das *et al.*, 2006), Mg/Mn of 3:1 (Chitrakar *et al.*, 2005) and Ni/Al of 3:1 (Wang and Gao, 2006). Phosphate adsorption activity of LDHs compounds results from the positively charged hydroxylated sheet, which is developed by partial substitution of trivalent for divalent cations. Introduction of large amount of Al^{3+} improved the net positive charge of the mixed metal hydroxide layers and subsequently the phosphate adsorption, on the other hand, however, may lead to structure disorder or even crystallinity loss (Xu and Zeng, 2001).

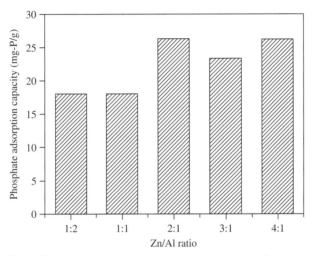

Figure 2. Effect of Zn/Al molar ratio on phosphate adsorption from sewage sludge filtrate.

Calcination

Calcined products of Zn-Al LDHs (CLDHs) generally had a higher adsorption capability as shown in Figure 3. The greatest phosphate uptake (40.77 mg-P/g) was observed when the adsorbent was calcinated under 300°C, which was 1.55-fold higher than that of the uncalcined. Beyond this temperature, the adsorption capacity of CLDHs slightly went down, with a sharp decrease after calcination under 600°C. The improvement in anions adsorption by LDH calcination has also been addressed previously (Das *et al.*, 2006, Wang and Gao, 2006).

Some thermally decomposed LDHs will undergo spontaneous rehydration and structural reconstruction, called "Memory Effect", when added in aqueous medium (Ferreira *et al.*, 2004). Anions incorporation occurs simultaneously in this process, increasing the anionic exchange capacity of this compound (Châtelet *et al.*, 1996). XRD pattern indicates that the LDHs phase completely disappeared and was replaced by single phase of metal oxides after calcination at 300°C for 4h (Figure 4). The result is in accordance with that of TG-DSC

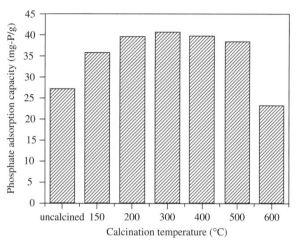

Figure 3. Effect of calcination temperature on phosphate adsorption by LDHs from sewage sludge filtrate.

analysis (data not shown), in which the first stage of mass loss of around 17% was observed with the temperature up to 300°C. This phase transformation was obviously responsible for the marked increase in phosphate adsorption by CLDHs compounds mainly based on the following three aspects: (1) higher surface area was obtained in calcined particles, (2) obstacle anions (mainly CO_3^{2-}) were released from the precursor, producing more active sites for phosphate adsorption and (3) phosphate incorporation occurred during the LDHs reconstruction. However, higher calcination temperature did not lead to further improvement in phosphate uptake. It can be explained by the fact that a mixture of ZnO and spinel-type $ZnAl_2O_4$ was formed under high temperatures instead of a metastable phase in the first stage, only which may reconstruct under appropriate condition (Figure 4).

pH

pH of solution generally plays an important role in the physicochemical reaction on water-solid interface. In Figure 5, buffering effect of pH and LDHs dissolution as well were observed in the range of acidic and highly alkaline due to the amphoteric character of aluminum hydroxide. Acidic environment with pH below 5 adversely affected phosphate uptake by the absorbent due to the serious LDH dissolution as indicated by sharp increase of Zn^{2+} in bulk water. Increased phosphate removal at pH 5, however, was attributed to coagulation process with Al^{3+} ions from partial dissolution of the LDHs compound.

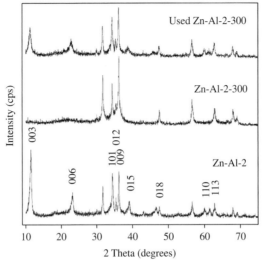

Figure 4. XRD pattern of Zn-Al-2, Zn-Al-2-300, and Zn-Al-2-300 after phosphate adsorption.

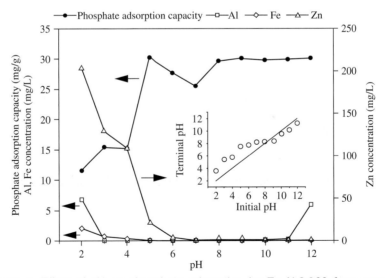

Figure 5. Effect of pH on phosphate adsorption by Zn-Al-2-300 from sewage sludge filtrate.

High adsorption capacity of ~30 mg-P/g was achieved when pH of the solution beyond 8, and a slight decrease in adsorption capacity of the LDHs in near neutral environment may result from the weakening of its electro-static interaction with phosphate, which transformed from divalent to mon-valent ion.

Adsorption kinetics

According to Figure 6, the majority of phosphate adsorption in sludge filtrate was completed in 24 h. The phosphate adsorption capacity of Zn-Al-2-300 at equil-ibrium could be almost obtained in 72 h and estimated close to 50 mg-P g-1. The kinetics data were analyzed using pseudo first-order, pseudo second-order and Elovich Equation (Figure 6). Though high correlation coefficient ($R^2 > 0.9$) is obtained in each model, the phosphate adsorption in present study obviously followed pseudo second-order equation, by which a very accurate estimation of qe is achieved as well as the highest correlation coefficient ($R^2 > 0.99$) (Table 3). Comparable results were observed in phosphate adsorption onto $ZnCl_2$ activated coir pith carbon (Namasivayam and Sangeetha, 2004), calcined electrocoagulated metal hydroxides sludge (Golder *et al.*, 2006) and amorphous zirconium hydroxide (Chitrakar *et al.*, 2006).

Adsorption isotherm

Adsorption isotherms of phosphate on Zn-Al CLDHs at different temperatures are presented in Figure 7. Under same equilibrium concentration of phosphate in sludge filtrate, adsorption capacity of the LDHs increased noticeably with the water temperature increasing from 25 to 30°C, and went back to that level in the 50°C test. This phenomenon may be because microorganisms existing in sludge filtrate exhibited higher activity and consumed part of phosphate in mesothermal environment, lowering the contribution of adsorption in phosphate removal. In KH_2PO_4 solution, however, only clear improvement in phosphate uptake at equilibrium was observed with the gradual temperature increase. The phosphate adsorption is therefore indicated as an endothermic process. Furthermore, much higher adsorption capacity obtained in pure phosphate solutions when temper-ature beyond 30°C demonstrated the enhancement of biological activity in the sludge filtrate.

Langmuir and Freundlich models were introduced to study the adsorption isotherms. The results of fitting the experimental data to the two models are presented in Table 4. Langmuir model clearly exhibits better fit for phosphate

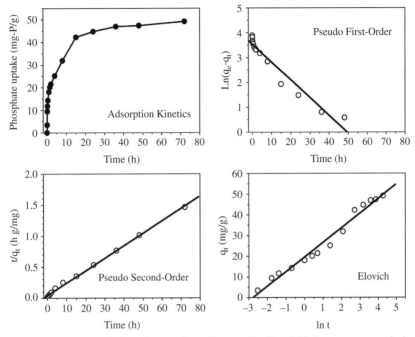

Figure 6. Kinetics of phosphate adsorption on Zn-Al-2-300 from sewage sludge filtrate and data fitting by pseudo first-order, pseudo second-order and Elovich equations.

Table 3. Kinetic models for phosphate adsorption on calcined Zn-Al LDHs from sludge filtrate and the calculated constants.

Model	Equation	Simplified form	Constant	R^2
Pseudo first-order	$\dfrac{dq_t}{dt} = k_1(q_e - q)$	$\ln(q_e - q_t) = \ln(q_e) - \dfrac{k_1}{2.303}t$	$q_e = 34.6$ $k_1 = 0.165$	0.945
Pseudo second-order	$\dfrac{dq_t}{dt} = k_2(q_e - q_t)^2$	$\dfrac{t}{q_t} = \dfrac{1}{k_2 q_e^2} + \dfrac{t}{q_e}$	$q_e = 50.0$ $k_2 = 0.00913$	0.997
Elovich	$\dfrac{dq_t}{dt} = \alpha \exp(-\beta q_t)$	$q_t = \frac{1}{\beta}\ln(\alpha\beta) + \frac{1}{\beta}\ln(t)$	$\alpha = 113$ $\beta = 0.142$	0.976

adsorption on the calcined LDHs compared to Freundlich one. In assays of both pure KH_2PO_4 solution and real sludge filtrate, very high correlation coefficients ($R^2 > 0.99$) were observed when using the former model.

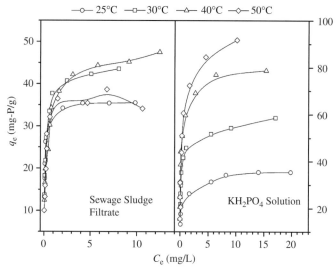

Figure 7. Isotherms of phosphate adsorption on Zn-Al-2-300 from sewage sludge filtrate and KH₂PO₄ solution.

Table 4. Langmuir and Freundlich constants for phosphate adsorption on calcined Zn-Al LDHs from sewage sludge filtrate.

	Langmuir equation				Freundlich equation		
Temp. (°C)	q_{max}	b	R_L	R^2	K_f	n	R^2
Sludge filtrate (Initial phosphate concentration: around 20 mg l⁻¹)							
25	35.81	13.64	0.003450	1.0000	28.32	6.436	0.8461
30	44.11	6.495	0.006947	0.9998	32.48	4.658	0.8796
40	47.82	3.114	0.01580	0.9988	30.08	4.239	0.9340
50	35.73	12.77	0.004855	0.9955	27.11	4.606	0.7928
KH₂PO₄ solution (Initial phosphate concentration: 20 mg l⁻¹)							
25	35.85	3.094	0.01590	0.9988	23.79	6.490	0.9123
30	58.22	3.993	0.01237	0.9980	40.28	6.362	0.9250
40	79.08	6.962	0.007131	0.9993	55.14	4.980	0.9293
50	92.58	4.250	0.01163	0.9972	58.37	4.059	0.8632

Phosphate desorption

Assays of phosphate desorption were conducted using three types of solutions as shown in Table 2. NaCl solution was not suitable for exchanging PO_4^{3-} from P-loaded Zn-Al LDHs even at high concentration (20 wt %) comparing with the other two adsorbents. Kuzawa and coworkers (Kuzawa *et al.*, 2006) used 30 wt % NaCl solution to elute phosphate from granular Mg-Al LDHs and the desorption rate was less (14%). Those results were expectable due to the relatively low affinity of Cl⁻ with LDHs compounds (Seida and Nakano, 2002). More than 80% of the phosphate on the LDHs can be desorbed by NaOH or Na_2CO_3 solution with concentration of 5 wt % and 10 wt %, respectively. Unfortunately, carbonate is the most serious obstacle anion against phosphate uptake by LDHs compound because of its high selectivity. Therefore, using Na_2CO_3 in the desorption step introduces a challenge to the regeneration of the used LDHs. On the other hand, CO_3^{2-} in desorption solution makes it impossible to recover the enriched phosphorus via calcium phosphate. Higher desorption rate of phosphate was achieved with the increasing concentration of NaOH solution as the desorbent, however, LDHs dissolution can not be overlooked under NaOH concentration beyond 5 wt %, as indicated by metal cations in liquid (Table 2). No report has as yet covered this point, therefore, the stability and activity of Al-contained LDHs compound in each step of phosphate adsorption/recovery needs further understanding.

CONCLUSIONS

Magnetic particles of layered double hydroxides were used in this study in order to separate and recovery phosphate in sludge filtrate from Taiping municipal wastewater treatment plant in Harbin, China. Zn-Al LDHs with Zn/Al molar ratio of 2 was demonstrated more efficient for phosphate adsorption comparing with other LDHs synthesized. Calcination remarkably improved phosphate adsorption by Zn-Al-2 because of the phosphate incorporation in the spontaneous rehydration and structural reconstruction of the calcined LDHs. After calcinations under 300°C for 4 h, a phosphate adsorption capacity of 40.77 mg-P/g adsorbent was achieved. Acidic or strongly alkaline environment adversely affected phosphate adsorption by Zn-Al-2-300 since aluminum hydroxide is amphoteric. Kinetics of phosphate adsorption from sewage sludge filtrate in this study followed pseudo second-order equation. Adsorption isotherms tested showed that phosphate uptake by Zn-Al-2-300 was an endothermic process and fitted Langmiur model well. Phosphate desorption from LDHs particles can be effectively realized by using 5 wt % of NaOH solution. Thus, despite of the complexity of the micro-process involved–there have been several theories currently–phosphate adsorption and recovery from sewage sludge filtrate by using optimized layered double hydroxides is promising.

ACKNOWLEDGEMENTS

This work was supported by High-Tech Research and Development Program of China (863 Program, Grant No. 2007AA06Z328) and National Eleventh Five-Year Research Program of China (Grant No. 2006BAD03A0201).

REFERENCES

APHA, AWWA and WEF (1998). Standard methods for the examination of water and wastewater, American Public Health Association, Washington, DC.

Badreddine, M., Legrouri, A., Barroug, A., De Roy, A. and Besse, J.P. (1999). Ion exchange of different phosphate ions into the zinc-aluminium-chloride layered double hydroxide. *Materials Letters*, **38**(6), 391–395.

Châtelet, L., Bottero, J.Y., Yvon, J. and Bouchelaghem, A. (1996). Competition between monovalent and divalent anions for calcined and uncalcined hydrotalcite: anion exchange and adsorption sites. *Colloids and Surfaces A: Physicochemical and Engineering Aspects*, **111**(3), 167–175.

Chitrakar, R., Tezuka, S., Sonoda, A., Sakane, K., Ooi, K. and Hirotsu, T. (2005). Adsorption of phosphate from seawater on calcined MgMn-layered double hydroxides. *Journal of Colloid and Interface Science*, **290**(1), 45–51.

Chitrakar, R., Tezuka, S., Sonoda, A., Sakane, K., Ooi, K. and Hirotsu, T. (2006). Selective adsorption of phosphate from seawater and wastewater by amorphous zirconium hydroxide. *Journal of Colloid and Interface Science*, **297**(2), 426–433.

Das, J., Patra, B.S., Baliarsingh, N. and Parida, K.M. (2006). Adsorption of phosphate by layered double hydroxides in aqueous solutions. *Applied Clay Science*, **32**(3–4), 252–260.

Driver, J., Lijmbach, D. and Steen, I. (1999). Why recover phosphorus for recycling, and how? *Environmental Technology*, **20**(7), 651–662.

Ferreira, O.P., Alves, O.L., Gouveia, D.X., Souza Filho, A.G., de Paiva, J.A.C. and Filho, J.M. (2004). Thermal decomposition and structural reconstruction effect on Mg-Fe-based hydrotalcite compounds. *Journal of Solid State Chemistry*, **177**(9), 3058–3069.

Golder, A.K., Samanta, A.N. and Ray, S. (2006). Removal of phosphate from aqueous solutions using calcined metal hydroxides sludge waste generated from electro-coagulation. *Separation and Purification Technology*, **52**(1), 102–109.

Jaffer, Y., Clark, T.A., Pearce, P. and Parsons, S.A. (2002). Potential phosphorus recovery by struvite formation. *Water Research*, **36**(7), 1834–1842.

Kuzawa, K., Jung, Y.J., Kiso, Y., Yamada, T., Nagai, M. and Lee, T.G. (2006). Phosphate removal and recovery with a synthetic hydrotalcite as an adsorbent. *Chemosphere*, **62**(1), 45–52.

Kwon, T., Tsigdinos, G.A. and Pinnavaia, T.J. (1988). Pillaring of layered double hydroxides (LDH's) by polyoxometalate anions. *J. Am. Chem. Soc.*, **110**(11), 3653–3654.

Mulkerrins, D., Dobson, A.D.W. and Colleran, E. (2004). Parameters affecting biological phosphate removal from wastewaters. *Environment International*, **30**(2), 249–259.

Namasivayam, C. and Sangeetha, D. (2004). Equilibrium and kinetic studies of adsorption of phosphate onto $ZnCl_2$ activated coir pith carbon. *Journal of Colloid and Interface Science*, **280**(2), 359–365.

Ookubo, A., Ooi, K. and Hayashi, H. (1993). Preparation and phosphate ion-exchange properties of a hydrotalcite-like compound. *Langmuir*, **9**(5), 1418–1422.

Pastor, L., Mangin, D., Barat, R. and Seco, A. (2008). A pilot-scale study of struvite precipitation in a stirred tank reactor: Conditions influencing the process. *Bioresource Technology*, **99**(14), 6285–6291.

Seida, Y. and Nakano, Y. (2002). Removal of phosphate by layered double hydroxides containing iron. *Water Research*, **36**(5), 1306–1312.

Wang, Y. and Gao, H. (2006). Compositional and structural control on anion sorption capability of layered double hydroxides (LDHs). *Journal of Colloid and Interface Science*, **301**(1), 19–26.

Xu, Z.P. and Zeng, H.C. (2001). Abrupt Structural Transformation in Hydrotalcite-like Compounds $Mg1-xAlx (OH)_2(NO_3)x \cdot nH_2O$ as a Continuous Function of Nitrate Anions. *J. Phys. Chem. B.*, **105**(9), 1743–1749.

Urine reuse as fertilizer for bamboo plantations

Jean Emmanuel Ndzana and Ralf Otterpohl

Institute of Wastewater Management and Water Protection, Hamburg University of Technology, Eißendorfer Strasse 42, D-21073, Hamburg, Germany (ndzana.emmanuel@tuhh.de)

Abstract It has never been reported on fertilization of bamboo plantations with urine derived from the Source Control Sanitation, the so called Ecological Sanitation (Ecosan), although urine has demonstrated its potential as a valuable resource for the agriculture. Basically urine is a pathogen-free mixture. In terms of nutrient elements nitrogen (N), phosphorus (P) and potassium (K) in wastewater streams, urine alone represents 87% (N), 50% (P) and 54% (K), making it the major component nutrient-rich of wastewater. Urine is in use for fertilizing a variety of crops ranging from vegetables to fruit trees. Usually application of urine on the soil only happens on a discontinuous basis; that means a couple of applications per year. This study is a pioneer one investigating the reuse of urine on bamboo plantations as fertilizer on a year-round basis. The optimum nutrient loading rate based on nitrogen feeding is researched. The nutrient uptake of the bamboo species used Phyllostachys *viridiglaucescens* (P. *viridiglaucescens*) is analysed. The effect of continuous feeding with urine on the biomass production is discussed. The experiments are conducted on a lab-scale plant under measured and controlled parameters. The year-round reuse of urine can be a method of choice for urine reuse where the storage of urine is not feasible.

INTRODUCTION

The fundamental idea of innovative and integrated water concepts is based on the principle of separating different flows of domestic wastewater according to their characteristics (Otterpohl *et al.*, 2002). This constitutes the basis of the Ecological Sanitation (Ecosan) where waste streams greywater, urine and faeces are collected and treated separately for the benefit of the agriculture. Urine is the major component nutrient-rich of municipal wastewater. In terms of nitrogen (N), phosphorus (P) and Potassium (K) human urine contains about 87% N, 50% P and 54% K. A complete picture of domestic wastewater constituents is given in Table 1. Urine has been and is currently used in agriculture as fertilizer for a diverse range of fruits and vegetables in many regions of the World. It is used in countries such as Mexico, Germany, USA, Sweden, and Denmark (Pearson *et al.*, 2007). In Zimbabwe a series of successful experiments were carried out on

© 2009 The Authors, *International Conference on Nutrient Recovery from Wastewater Streams.* Edited by Ken Ashley, Don Mavinic and Fred Koch. ISBN: 9781843392323. Published by IWA Publishing, London, UK.

tomato, covo, spinach and lettuce (Calvert *et al.*, 2004). In Sweden there is some experience with agricultural use of urine (Kärrrman *et al.*, 1999). However it has never been reported on bamboo plantations fertilization with urine derived from the Ecosan. Furthermore, the application of urine for fruit trees and vegetables occurs only on a discontinuous rhythm as urine is applied just a couple of times in the year at the beginning of the growing season (Calvert *et al.*, 2004).

Table 1. Typical characteristics of household wastewater components.

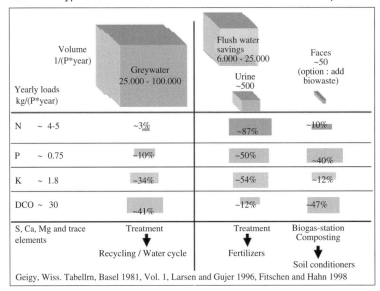

Volume 1/(P*year)		Flush water savings 6.000 - 25.000	Faces ~50 (option : add biowaste)
	Greywater 25.000 - 100.000	Urine ~500	
Yearly loads kg/(P*year)			
N ~ 4-5	~3%	~87%	~10%
P ~ 0.75	~10%	~50%	~40%
K ~ 1.8	~34%	~54%	~12%
DCO ~ 30	~41%	~12%	~47%
S, Ca, Mg and trace elements	Treatment ↓ Recycling / Water cycle	Treatment ↓ Fertilizers	Biogas-station Composting ↓ Soil conditioners

Geigy, Wiss. Tabellrn, Basel 1981, Vol. 1, Larsen and Gujer 1996, Fitschen and Hahn 1998

Bamboo as the fastest growing line vegetable species (Villegas, 1990) appears to be a very good candidate for urine application recipient. Bamboo reaches its full length within 2 to 4 months growth (Liese, 1985). The rest of the life of the plant, its culms – bamboo stems made of nodes and internodes – will harden through accumulation of nutrients in the culm tissue for them to come to maturity.

The rhizome system constitutes the structural foundation of the plant, in which nutrients are stored and through which they are transported (Liese, 1985). At the difference of the culm, the rhizome exhibits year-round non-stop growth. From this fact, a continuous feeding of bamboo plant with nutrients appears feasible.

At the Institute of Wastewater Management and Water Protection of the Hamburg University of Technology a pioneer research coupling bamboo plant and Ecosan has started. Figure 1 is an overview of the whole research concept. As entire part of it, this study aims to determine the effects and impacts of continuous direct reuse of urine derived from the Ecosan on bamboo plantations in terms of nutrients uptake, nutrient loading rate and biomass production.

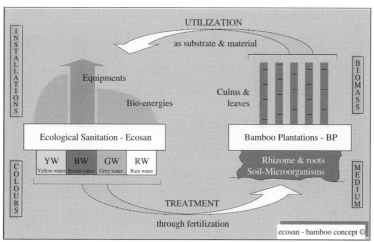

Figure 1. The Research concept Ecosan-Bamboo from which the current study derives.

MATERIALS AND METHODS

The experimental set up

Nine reactors were placed at the interior court of the Institute of Wastewater Management of the Hamburg University of Technology-TUHH. Each reactor is a huge PVC container of 160 l in which bamboo plant is planted.

The 53 cm height of the reactor allows efficient growth of bamboo; Li *et al.* (1998), An *et al.* (1995) and Widmer (1998) found that the effective root zones for several bamboo species is comprised between 0 and 30 cm of top soil. Thus efficient growth of bamboo is performed within the first 30 cm. Openings were made at the bottom part of each reactor to allow effluent collection that percolates only from the top part of the reactor; effluent collectors are sized and placed so that rainwater could not be collected. A schematic set-up of the experimental site is provided on Figure 2. Mature bamboo plants were transplanted from a living stand of the Botanic Garden of Hamburg. Culms height is comprised between 3.00 and 8.00 m. The bamboo species tested is

Phyllostachys (P) *viridiglaucescens,* a winter-hard plant that withstands temperatures up to −22°C. Bamboo roots balls were placed in the PVC container and filled with a self-made composition of soil. The soil in the container is of 2 layers; a drainage layer of 5 cm gravel at the bottom part and the rest upper part is soil. The soil composition is a mixture of flower earth (85%), sand (10%) and crushed tree barks (5%).

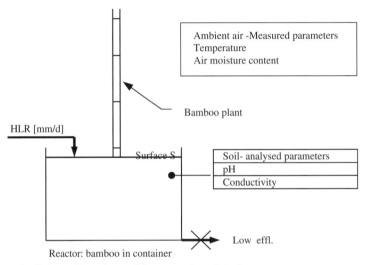

Figure 2. Schematic set-up of the experimental site.

The analytical method

Author's self-collected urine was used in this study to ensure no pathogen contamination and avoid possible side effects from pharmaceuticals (no tablets were taking during the whole study). Urine was mixed with tap water (substrate); the substrate was fed 3 times a week at the top of the reactor. The other 4 days of the week were fed with tap water when necessary to do not let the reactors to dry. The reactors were named S1 to S9 and fed with low hydraulic loading rate so that less effluent occurs, as the optimum nutrient uptake is one of the key issues researched in this study. As Midmore and Kleinhenz (2001) suggested a nutrient uptake of 350 kg N/ha/year, this value was taken as basis for the fertilization rate defined on Table 2. The argument of Toky and Ramakrishnan (1981) and Mailly *et al.* (1997) about less leaching of nutrients in bamboo stand sustains the defined rates.

Table 2. Fertilization rate of the 9 reactors.

Reactor	S1, S2, S3	S4, S5	S6, S7	S8, S9
Fertilization rate [kg N/ha/a]	400	800	1200	1400

Measured parameters and performed analyses

The air temperature and moisture content were measured by using a data-logger which records data each 30 minutes (results are not shown). Analyses were performed on the substrate and the soil. The bamboo biomass produced was analysed in terms of number of shoots, culms diameter and shoots growth pattern. For the analyses the German standard methods (DIN) for soil and substrate were applied. In the substrate, the Total Organic Carbon (TOC) and Total Nitrogen (TN) were analysed with a TOC/TN analyser. NH_4-N was measured by using a reflectometer. Dr. Lange cuvette tests were used to measure the Total Phosphorus (TP) and ortho-phosphate (ortho P-PO_4). The pH and electrical conductivity (EC) were measured by using electrodes incorporating temperature measurement (Microprocessor pH Meter pH 196 for pH and Conduktometer LF 191 for EC). For the soil, the German Norm *[VDLUFA, Methodenbuch Band I (1991)]* was used to analyse the pH and the electrical conductivity (EC). For the pH, the soil samples were collected at the first 10 cm of top soil. They were dried at ambient air, passed through a sieve of 1 mm and a solution of 0.01 molar $CaCl_2$ was added in a ratio of 1:10 before the measure is taken after 1 hour. The EC analysis followed the same procedure but with addition of deionised water and a waiting time of 2 hours.

RESULTS AND DISCUSSION

The results of this study are concerned with the biomass production, the TOC removal and nutrient uptake of bamboo in terms of N and P.

The biomass production

The first shoots emerged from S6 2 months later (May 2.) after potting. This shooting started within the normal shooting period of bamboo in Europe (the beginning of the Spring, when the soil starts to warm a bit and sunshine increases). By shooting at the year of planting, bamboo in the reactors shows very rapid rooting. Shooting was observed until the end of October. This also corresponds to the normal end of shooting season. The rate of production (average number of culm produced per reactor) was 5.77. This value is similar to the one in natural stands (Liese, 1985) (up to 10 shoots). The maximal shoot diameter (Ø) was exhibited by S3:2.5 cm. Table 3 summarized the biomass produced by the 9 reactors. Thus the continuous fertilization with urine did change neither the shooting season nor the rate of production. Except for reactor S3 and S7, all reactors showed some emerged shoots with a curved growth pattern at their basis. These are shoots that were primarily to establish as rhizome. Because of containment in container, they were converted into culms. The establishment of a bamboo stand starts with the development of a stable solid rhizome system.

Table 3. Biomass production of the 9 reactors.

Reactor	S1	S2	S3	S4	S5	S6	S7	S8	S9
Nber of shoots	7	8	1	9	4	7	2	9	5
Curved culms	2	1	0	3	2	3	0	3	1
Production	5.77								

TOC removal through urine application – soil pH and conductivity (salts concentration)

The mean value of TOC in raw urine is 2807 mg/l, the min value being 1080 mg/l and the max value being 6960 mg/l. The hydraulic retention time being 50 days, within which only 21 days are fed with the mixture tap water and urine, an average of 943 (for S1, S2, S3), 1886 (for S4, S5), 2829 (for S6, S7) and 3301 mg/l TOC (for S8, S9) was fed into the reactors, respectively to the fertilization rates 400, 800, 1200 and 1400 kg N/ha/year. Figure 3 gives TOC in the influent and effluent regarding the reactors and their corresponding fertilization rates. The TOC removal efficiency varied from 74.6, 87.1, 90.8 and 91.1 respectively for fertilization rate of 400, 800, 1200 and 1400 kg N/ha/a.

Independently to the fertilization rates the TOC concentration in the effluent of reactors remains almost at the same level varying between 239 and 295 mg/l.

The pH in the influent for each reactor remains almost at levels of which in the raw urine around 8.00 (detailed results are not shown). A slight increase of pH-value was encountered with the increase of nitrogen concentration: 8.44, 8.57, 8.61 and 8.62 respectively for 400, 800 1200 and 1400 kg N/ha/year.

Reactors presented effluent only when severe rain felled. S8 did not leak at all; no effluent was collected from this reactor.

Table 4 presents the results of soil pH and EC analyses. From these, the soil pH value at the start of the study was 3.91, with an EC value 201 µS/cm.

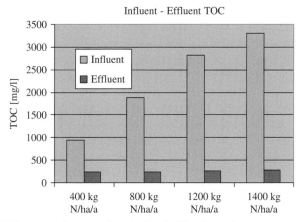

Figure 3. TOC removals according to the fertilization rates.

Table 4. Soil pH and electrical conductivity.

	Date	S1	S2	S3	S4	S5	S6	S7	S8	S9
pH [-]	04.08.08	4.12	4.36	4.23	4.13	4.33	4.47	4.18	4.18	4.15
	29.10.08	4.99	4.86	4.40	4.39	4.56	4.59	4.33	4.41	4.22
EC [μS/cm]	04.08.08	230	81	247	281	98	115	85	115	97
	29.10.08	66	68	84	71	63	96	77	88	78

In the course of the study, it slightly increases in all reactors to achieve 4.99 (value in S1). This is favourable to bamboo plants which prefer light acidic soils with pH between 5.5 and 6.5.

The decrease of EC in the soil consequently results to the decrease of salt content in the soil; this was not expectable since addition of urine on soil is usually known to increase it as Pearson *et al.* (2007) concluded their study based on some vegetables. The decrease of salts concentration in the top soil complies with the statement of Kleinhenz and Midmore (2001) that the top soil of bamboo stands is typically well aerated and natural mineralization of nutrients is usually quicker there than in deeper layers. Plant-available ions are effectively and almost immediately absorbed by the dense root system of bamboo plants in this horizon. This may explain the well-known statement that bamboo accumulate and sequester nutrients in its top layer.

Nutrient loading rate and nutrient uptake N and P

The average TN in raw urine is 5060 mg/l with a max value of 7390 mg/l and 2860 mg/l as min value. Taking into consideration the hydraulic retention time of 50 days, within which only 21 days were fed with substrate (the rest being tap water for watering the plants), and regarding the fertilization rates, the reactors were fed with 1700, 3400, 5100, 5951 mg/l TN. Similarly to TN, with consideration of the average of 424 TP in raw urine (the min value and max value of TP being 204 and 1050 mg/l respectively), the fed rates of TP were 142, 285, 427 and 499 mg/l TP for reactors fed with 400, 800, 1200 and 1400 kg N/ha/a respectively.

The effluent value both for TN and TP are so low that they are not noticeable on Figure 4 (TN and TP uptake by the system "soil-bamboo").

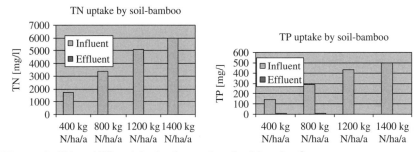

Figure 4. TN and TP uptake by the system "soil-bamboo".

The maximum levels of TN and TP in the effluent were found by reactor S6 and S1 and were 15.4 for TN and 7.97 mg/l for TP respectively. Almost 100% of nitrogen was taken up by the system "soil-bamboo" with an average of 99.8%. The same consideration is valid for phosphorus with a mean uptake of 97.9% by the system "soil-bamboo". N-NH$_4$ concentration in the effluent was considerably reduced. The maximum concentration is found by S6 and is 2.05 mg/l.

CONCLUSION

As a pioneer study investigating the fertilization of bamboo plantations with urine derived from the Ecosan, the results of this study show that human urine as a valuable nutrient can effectively be reused by stands of giant bamboos (in this case P. *viridiglaucescens*). From a wastewater point of view, bamboo plants constitute a remover of urine, while from an agricultural point of view; urine is reused as fertilizer for bamboo plantations. Thus the year-round application of urine through continuous feeding of bamboo plantations is feasible. Up to 1400 kg N/ha/year loading rate of the reactors, all bamboo plants remain healthy without any sign of disturbance in their growth. All of them operate well still. The biomass production did not change in comparison to natural stands of bamboo plantations, thus the continuous feeding of the reactors did affect neither the shooting season nor the rate of production. The continuous feeding of urine into the reactors at fertilization rate from 400 to 1400 kg N/ha/a resulted in to an efficient TOC removal; the efficiency increased from 74.6 to 91.1% regarding the considered fertilization rates although the concentrations of organic carbon is still high in the effluent. This study confirms the well known statement that bamboo plants absorb quite enormous quantity of nutrients at their top layer. In fact it was analysed that the EC of soil in its first 10 cm (top soil) decreased, while the expectations were turned to an increase as with most trials with vegetables and fruit trees. The pH of the soil slightly increased from 3.91 to 4.99 in the course of the study. The system soil-bamboo reacts very positively to continuous application of urine on land as the TN and TP are almost completely absorbed by it; about 100% for nitrogen and 97.9% for phosphorus.

REFERENCES

An, Q.-N., Wang, J.-P., Zhang, X.-M., Du, W.-Y., Wu, M.-H. and Hu, X.-Q. (1995). Study on fertilization in *Phyllostachys nidularia* forest. *J. Bamboo. Res.*, **14**, 73–80.

Calvert, P., Morgan, P., Rosemarin, A., Sawyer, R. and Xiao J. (2004). Ecological Sanitation, revised and enlarged edition. Stockholm Environment Institute, 2004.

Kärrrman, E., Jönsson, H., Gruberger, C., Dalemo, M. and Sonesson, U. (1999). Management of wastewater and organic waste – systems analysis. Paper presentation at *Managing the Wastewater Resource, 4th International Conference for Ecological engineering for Wastewater Treatment, Ås, Norway, June 7–11.*

Kleinhenz, V. and Midmore, D.J. (2001). Aspects of bamboo agronomy, Advances in Agronomy, Vol. 74, Academic Press, 2001.

Li, R., Werger, M.J.A., During, H.J. and Zhong, Z.-C. (1998). Carbon and nutrients dynamics in relation to growth rhythm in the giant bamboo *Phyllostachys pubescens*. *Plant Soil*, **201**, 113–123.

Liese, W. (1985). Bamboos – biology, silvics, properties, utilization. Deutsche Gesellschaft für Technische Zusammenarbeit (GTZ), Eschborn, Germany.

Mailly, D., Christanty, L. and Kimmins, J.P. (1997). Without bamboo, the Land dies: Nutrients cycling and biogeochemistry of a Javanese bamboo Talun-Kebun system, 1997.

Otterpohl, R., Braun, U. and Oldenburg, M. (2002). Innovative technologies for decentralised wastewater management in urban and peri-urban areas. *Keynote presentation at IWA Small2002, Istanbul, September 2002.*

Pearson, N.S., Mnkeni, Funso R. Kutu and Pardon, M. (2007). Evaluation of Human urine as source of nutrients for selected vegetables ad maize under tunnel house conditions in the Eastern Cape, South Africa – *Waste Management and Research*, 2008(**26**), 132–139.

Toky, O.P. and Ramakrishnan, P.S. (1981). Run-off and infiltration losses related to shifting agriculture (Jhum) in North-Estern India, 1981.

Villegas, M. (1990). Tropical Bamboo. Rozzoli New York, 1990.

Widmer, Y. (1998). Soil characteristics and *Chusquea* bamboos in the *Quercus* forests of the Cordillera de Talamanca, Costa Rica. *Bull. Geobot. Inst. Eth.*, **64**, 3–14.

Ammonium absorption in reject water using vermiculite

N. Åkerback[a]*, S. Engblom[a] and K. Sahlén[b]

[a]Novia University of Applied Sciences, R&D, P.O.Box 6, FIN-65201 Vasa, Finland
[b]Swedish University of Agricultural Sciences, Department of Forest Ecology and Management, S-90183 Umeå, Sweden
*Corresponding author

Abstract To be able to keep up with an enlarged demand for biomass for energy production the forest resources in the Nordic countries will be utilized to a far greater extent in the future. One way to meet this increased demand for tree biomass is to increase forest tree growth by supplying nutrients, of which nitrogen is the most important. Some of the waste products from the community have a chemical composition that makes them usable as nutrients for plants. Digestion of organic waste leaves a reject water which is rich in NH_4-N. If the nitrogen contained in the waste produced by mankind could be recovered and used as fertilizer in agriculture and forestry, it would be of great importance.

In this work we describe how vermiculite, pre-treated to increase its nitrogen absorption capacity, can be used to effectively remove ammonium nitrogen from reject water, and how the nitrogen can be used as a plant nutrient.

ABSORPTION EXPERIMENTS

Research on the ammonium nitrogen absorption by vermiculite was done using reject water coming from the Stormossen biogas plant near Vasa, Finland. Stormossen treats kitchen and household biowaste as well as sewage sludge. The hypothesis that the absorption of ammonium takes place during a short time span was tested in several laboratory experiments. In order to examine the reaction time of NH_4-N absorption by vermiculite from the reject water, the experiments were carried out at temperatures of 25°C, 50°C and 80°C and grain sizes from 125 to 500 μm. The amount of vermiculite was from 190 to 400 g and the volume of reject water was 2100 ml. Experiments were carried out with stirring and no stirring, and the reaction times used were from 24 hours to 13 days. The presence of microbes in the reject water is likely and the following microbes were analysed: Escherichia Coli, Coliform Bacteria, Enterococci Preliminary, Sulphate Reducing Anaerobs, Salmonella and Somatic Coliphages.

Analytical procedures

The NH_4-N and the NO_2/NO_3 were determined using a FIAstar 5000 Analyser. Conductivity was measured with a MeterLab CDM 210 and pH was measured with a Sentron pH metre. The total-N was analysed by the Kjeldahl method.

The Ion beam analyses by the PIGE (Particle Induced Gamma-ray Emission) method were done at the accelerator laboratory at Åbo Akademi University.

The effect of temperature

Experiments at temperatures of 25°C, 50°C and 80°C show that it is advantageous to maintain a higher temperature during the absorption process. Results show that there is a difference of 0.4% in ammonium content of vermiculite between absorption at 25°C and at 80°C. The NO_2/NO_3 in the reject water was not significantly affected. At the temperature of 50°C and above, the content of common microbes (e.g. Escherichia Coli) decreases.

The stirring experiment

This laboratory experiment was carried out in order to determine some characteristics of the future pilot plant, e.g. must the pilot plant container be air-tight and if stirring would be necessary. Furthermore, a comparison of the absorption of vermiculite pre-treated at the temperature recommended by Eklund et al. (2007) with that of vermiculite bought in a hardware store was done. The results from another experiment showed that higher temperature were favourable for the absorption of NH_4-N and also for limiting the growth of microbes. The experiment was carried out at the temperature of 50°C, grain sizes were 250–500 μm and the amount of vermiculite were 200 g and 400 g. The volume of the reject water sample was 2100 ml and stirring was carried out at 200 rpm, except for two of the samples that had no stirring for 23 hours.

Theoretically the amount of 400 g of vermiculite in the reject water should reduce the NH_4-N content to zero. The grain sizes in this experiment were 250–500 μm, because these sizes were more suitable for the green house experiment. Six of the bottles had air tight caps on and six of them were open bottles.

The bottles containing vermiculite and reject water were placed on magnetic stirrers. Samples of the aqueous phase were analysed for the content of NH_4-N after approximately 2 h, 3 h, 4 h, 5 h, 6 h, 7 h, 8 h and 23 h. Conductivity and pH were also measured in the aqueous phase.

Results

The ammonium nitrogen content in the reject water

The absorption of NH_4-N by vermiculite both pre-treated and the hardware store quality was very rapid. Within a few hours the NH_4-N content of the aqueous phase was halved with the vermiculite bought from a hardware store. For the vermiculite pre-treated at the temperature recommended by the group in Turku (Eklund *et al.* (2007)), the NH_4-N content had after about three hours decreased to approximately one third of the original NH_4-N content. A marked difference in the absorption of NH_4-N could thus be seen between the two differing vermiculites, Figure 1. Regarding open or closed bottles, no major difference could be seen with respect to the NH_4-N content in the aqueous phase.

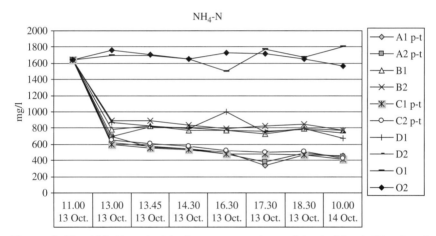

Figure 1. The NH_4-N content in the aqueous phase. A1 p-t, A2 p-t, C1 p-t and C2 p-t are samples containing pre-treated vermiculite. B1, B2, D1 and D2 are samples containing vermiculite bought from a hardware store. O1 and O2 are blank samples containing only reject water and no reduction can be observed. A1, A2, B1, B2, O1 are closed vessels, all other are open vessels.

The effect of stirring is shown in Figure 2, samples E1 (closed vessel) and E2 (open vessel) contained vermiculite pre-treated at the recommended temperature. After 23 hours without stirring the stirring was switch on for four hours. Without stirring the content of NH_4-N in the aqueous phase is down from 1646 mg/l to about 1200 mg/l after two hours. After four hours with stirring switched on the NH_4-N content went down from about 1000 mg/l to about 300 mg/l after three hours.

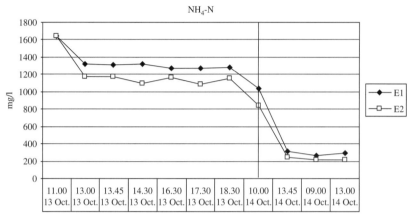

Figure 2. The effect of stirring on the NH$_4$-N content in the aqueous phase. E1 is a closed vessel and E2 is an open vessel.

The ammonium nitrogen content in the vermiculite

The effect of open or closed vessel is shown in Figure 3. There is a marked difference between the two. Closed vessels, samples A1 and B1, result in a higher absorption of NH$_4$-N by the vermiculite. As mentioned earlier there were no differences between open or closed vessels in the NH$_4$-N contents of the aqueous phase.

GREENHOUSE EXPERIMENT 1: AUGUST – DECEMBER 2004

Materials and methods

Growing conditions

Growing boxes of Panth model, with 121 containers were used. Four boxes constituted one m^2. The growing substrates were sand (grain size 0.9 mm $\pm$ 0.1 mm) and ammonium treated and untreated vermiculite (absorption substrate, AS), with grain size 0.125–0.25 mm. Water content of the substrates were determined at 105°C (16 h) to 6.1% (untreated) and 14.5% (ammonium treated) at the start of the growing experiment. The substrates were then mixed with water (175 ml/per litre of sand; 300 ml/per litre for untreated substrate; 275 ml/per litre for ammonium treated substrate). The following

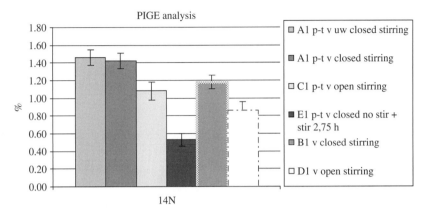

Figure 3. The nitrogen content in vermiculite. The sample "A1 p-t v uw closed stirring" represents pre-treated vermiculite, closed vessel and not rinsed before the analysis. The sample "A1 p-t v closed stirring" represents the same sample but the vermiculite was rinsed with water before the analysis. The rest of the vermiculite samples were rinsed before the PIGE analysis.

substrate mixtures were composed and were filled into 40 containers (60 ± 10 ml per container) per treatment: C0 = sand, C50 = 50/50 volume % sand/AS untreated substrate, D50 = 50/50 volume % sand/AS ammonium treated substrate, C67 = 33/67 volume % sand/AS untreated substrate, D67 = 33/67 volume % sand/AS ammonium treated substrate. Seeding was carried out on August 31, 2004 in greenhouse. The daily growing conditions were 18 h day (+20°C and light) and 6 h night (+15°C, no light). The growing boxes were watered at regular intervals. After five weeks, half the number of the boxes for each treatment, were irrigated with a solution of P, K, Mg, and B once a week. After 8 and 14 weeks, ten seedlings per treatment were randomly sampled and dried at +70°C for 24 h for determination of dry matter weight. The nutrient content in the sampled needles were determined by ICP-MS.

Results

Seedling weight

There were only small positive effects of ammonium treatment and nutrient irrigation after 8 weeks growing time for both pine and spruce (Figure 4). Six weeks later, the average dry weight of pine seedlings growing in untreated absorption substrate was almost unchanged and about 40 mg. Pine seedlings grown in ammonium treated substrate was about 25% heavier, and also heavier than seedlings growing in untreated substrate receiving nutrient irrigation.

Seedling size was somewhat greater for 67% admixture of absorption substrate than for 50% admixture. The heaviest seedlings (about 100 mg for pine and 60–70 mg for spruce) were found in ammonium treated substrate with additional nutrient irrigation.

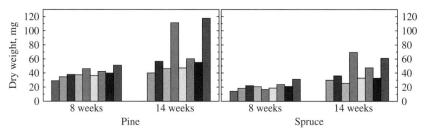

Figure 4. Seedling dry weight after 8 and 14 weeks growing time. For legend description, see Materials and methods.

Needle nutrient content

After 14 weeks, the concentration of nitrogen for plants growing in ammonium treated substrate approached 3,5% for pine and 2,5% for spruce (Figure 5).

For potassium, the concentrations were higher for nutrient irrigated substrates than for non irrigated substrate. The difference in calcium concentration between ammonium treated and not treated substrate was considerably higher for spruce than for pine. The differences in magnesium concentrations between treatments were small and with no evident treatment effect.

GREENHOUSE EXPERIMENT 2: APRIL – AUGUST 2005

Materials and Methods

Growing conditions

The same containers as above were used. The growing medium was absorption substrate (AS), with a grain size of 1–2 mm. Some of the substrate was previously treated by Novia's laboratory in Vasa with reject water from the biogas plant at Stormossen nearby Vasa, Finland, and then dried. Moisture content was determined at +105°C to 12 and 5 weight %, respectively, for treated and untreated substrate. The following substrate mixtures were composed :

C50 = 50/50 volume % sand/AS untreated (2 boxes)
D50 = 50/50 volume % sand/AS ammonium treated (2 boxes)
C100 = 100% untreated AS (2 boxes)
D100 = 100% reject water treated AS (2 boxes)

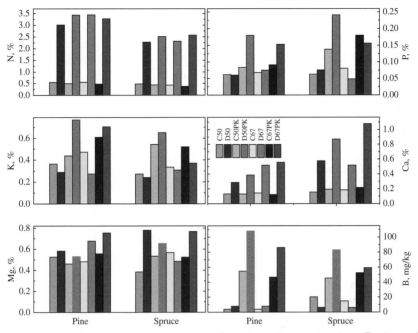

Figure 5. Needle nutrient concentrations after 14 weeks growing time. For legend description, see Material and methods.

Water was added to the untreated (300 ml/l) and treated substrate (230 ml/l) mixtures. A temperature increase from 20 to 65° was noticed during a couple of minutes, when water was added to the non-treated substrate. The substrate was filled into 30 containers (55–60 ml per container) per treatment. Seeding was carried out on April 4, 2005 with two seeds per container. Growing condition and nutrient addition was the same as in experiment 1.

On August 15, (after 19 weeks), 10 seedlings were randomly sampled from each treatment, cut at the substrate surface and individually photographed. After drying at 70°C for 24 hours the dry matter weight was determined.

Results

Seedling weight

Seedling weight in untreated absorption substrate was higher in the sand mixture (75 mg for pine and 39 mg for spruce) than in the pure substrate (61 mg for pine and 35 mg for spruce) (Figure 6). The weights were higher in untreated substrate, irrigated with nutrients (97 and 121 mg for pine and 54 and 52 mg for spruce, respectively, for sand mixed and pure substrate). Seedlings growing in reject water treated substrate without nutrient addition were heavier in pure substrate (131 mg for pine and 76 mg for spruce) than in sand mixture (118 mg for pine and 60 mg for spruce), and also heavier than seedlings growing in untreated substrate and untreated substrate with nutrient irrigation. The combination of reject water treated substrate and nutrient irrigation showed a considerably higher plant weight than all other treatments. For pine, the seedling weight was 350 mg in pure substrate and more than the double (763 mg) in mixed substrate, more than 10 times higher than in untreated substrate without nutrient irrigation. For spruce, the seedling weight was about 280 mg, with no difference between pure and mixed substrate.

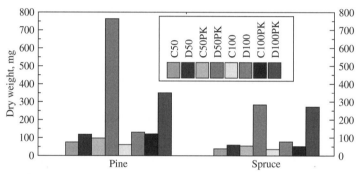

Figure 6. Seedling dry weight after 19 weeks growing in different substrates in greenhouse experiment 2. For legend description, see Material and methods.

Needle nutrient concentration

After 5 months growing, the nitrogen content of needles of seedlings growing in reject water treated substrate without addition of other nutrients was more than 3% for both pine and spruce (Figure 7 illustrates this for Scots pine). Nitrogen concentration was 2% for seedlings in pure reject water treated substrate with nutrient addition. For most other treatments, the nitrogen content was between 0.8 and 1.4%. Needle phosphorus concentration was 500–700 mg/kg for all treatments without nutrient addition, whereas for seedlings receiving additional nutrients, the concentration was much higher, between 1700 and 2500 mg/kg for pine and about 3500 mg/kg for spruce. Needle potassium content was 7000–8000 mg/kg for pine and 6000–8000 mg/kg for spruce, growing in all substrates without nutrient addition. The concentrations were higher for subatrates with nutrient addition, 12000–14000 mg/kg for both species in mixed substrate and somewhat lower in pure substrate. The calcium concentration was higher, 8000–10000 mg/kg for spruce and 6000–8000 mg/kg for pine, for seedlings in all reject treated substrates, than for non-treated substrates, 6000 mg/kg for spruce and 4000 mg/kg for spruce. Magnesium showed a similar trend as calcium with higher concentrations for seedlings growing in substrate without reject water treatment. For boron, the needle concentration was about 15 mg/kg for seedlings growing in substrates without nutrient addition for both pine and spruce. For substrates with nutrient addition, the concentrations were much higher, between 120 and 190 mg/kg for pine and 110–170 mg/kg for spruce.

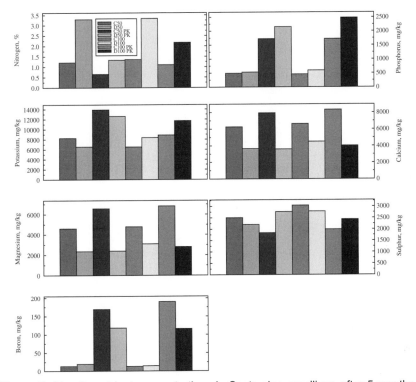

Figure 7. Needle nutrient concentrations in Scots pine seedlings after 5 months.

ACKNOWLEDGEMENT

The financial support by TEKES is gratefully acknowledged.

REFERENCE

Eklund, O., Toropainen, V., Shebanov, A., Åkerback, N. and Engblom, S. (2007). Patent Pending WO/2007/003689.

Alternating anoxic-aerobic process for nitrogen recovery from wastewater in a biofilm reactor

Mohamed F. Hamoda[a] and Rashid A. Bin-Fahad[b]

[a]Professor of Environmental Engineering, Dept. of Environmental Technology Management, CFW, Kuwait University, P.O. Box 5969, Safat 13060, Kuwait, e-mail: mfhamoda@yahoo.com, Tel. (+965)6606-0830, Fax (+965)2255-1157
[b] Ministry of Environment and Water Resources, Federal Government, Dubai, U.A.E., e-mail: binfahad@emirates.net.ae

Abstract In recognition of the importance of water reuse, the Dubai municipality in the United Arab Emirates has undertaken plans to improve the treated effluent quality and expand the treatment works at Al-Awir municipal wastewater treatment plant (WWTP), as to double its current design capacity of 130,000 m^3 d^{-1}. This study examined an anoxic/aerobic biofilm system to upgrade the exsisting high-rate activated sludge process. A pilot plant, utilizing a four-compartment biofilm reactor packed with Biolace media, was constructed and operated using both the aerobic submerged fixed film (ASFF) and the anoxic/aerobic submerged fixed film (A/ASFF) systems. Field experiments were conducted to examine the effect of loading on the oxidation of carbonaceous and nitrogenous matter was at loading rates in the range of 0.03 to 0.3 g BOD $\cdot$ g^{-1} BVS $\cdot$ d^{-1} and the range of 0.01 to 0.11 g $NH_3 \cdot g^{-1}$ BVS $\cdot d^{-1}$, corresponding to hydraulic retention times (HRT's) in the range of 0.7 to 8 h.

The results revealed that both the ASFF and the A/ASFF systems are viable options to upgrade the activated sludge process with minimal excess sludge production, but the A/ASFF system appears to be more capable of maintaining stable and efficient treatment at higher loading rates. Removal efficiencies of up to 98% for BOD, 75% for COD, and 97% for ammonia were obtained over a wide range of loading rates.. Substrate removal rate increased with organic loading and followed a first-order kinetic model. The denitrification process was able to eliminate about 3.35 mg BOD (or 6.6 g COD) versus 1 mg denitrified N-NO_3. This contributed to higher organic removal in the A/ASFF system. The denitification rate in the A/ASFF system reached up to 0.005 Kg TON $\cdot$ Kg^{-1} BVS. h^{-1}. On the other hand, high nitrification rates and nitrate production were observed in the ASFF system that could be optimized for nitrogen recovery.

INTRODUCTION

Currently, treated wastewater effluents in the Arabian Gulf countries are reused extensively in landscape and greenery irrigation and, to a lesser extent, in agricultural lands as reported by Hamoda (2004). In recognition of the importance of water reuse in the United Arab Emirates, the Dubai municipality has undertaken plans to improve the treated effluent quality and expand the treatment works in Dubai's main municipal wastewater treatment plant (WWTP) at Al-Awir. This plant was originally designed to treat 130,000 m^3 d^{-1} of municipal wastewater serving a population of 800,000 persons. The plant is currently overloaded and expansion works are underway to double its capacity and improve the quality of treated effluent produced to satisfy the requirements for water reuse in irrigation. Execution of such plans is very costly as the construction and installation of new treatment works is becoming more expensive. Another approach is to upgrade the exiting facilities to cope with increasing flows at a much reduced cost.

Biological suspended-growth systems, such as the activated sludge process, are commonly used for the secondary treatment of municipal wastewaters. This is the case with Al-Awir WWTP, where the high-rate activated sludge (HRAS) process is used for aerobic treatment of wastewater following grit removal and primary sedimentation. The secondary biological treatment stage is the focus of this study as it constitutes the backbone of the plant. In this stage, the high rate activated sludge (HRAS) is primarily designed for the reduction of carbonaceous material present in the primary settled wastewater. It consists of three rectangular aeration tanks with capacity of 4820 m^3 each, designed for a peak hydraulic retention time of 1.33 hours, and organic loading of 0.3 Kg BOD kg^{-1} dry solids. The HRAS operates at mixed liquor suspended solids (MLSS) of 5,600 mg l^{-1}. Up to the design capacity of the plant, the HRAS produced well-treated effluent with an overall BOD and COD removal efficiency of approximately 85–90% and 55–65%, respectively. However, the HRAS process has two main persistent operational difficulties since the facility was put in service in 1989. These are: (1) excessive sludge production reaching 0.85 to 1.0 kg sludge solids per kg BOD applied, and (2) poor sludge settleability and frequent rising sludge problems. It was, therefore, necessary to upgrade the biological treatment stage to overcome operational problems and to cope with increasing wastewater flow received at the plant. In this study, attempts were made to use an attached-growth system to upgrade the HRAS process since the attached-growth processes have the advantages of low sludge production, good sludge settleability and stable operation (Metcalf and Eddy, 1991; Lessel, 1994).

Attached-Growth (fixed-film) processes are biological systems in which the microorganisms responsible for biodegradation and stabilization of organic matter are attached or fixed to a solid inert media forming a biofilm. A number of innovative processes have been used such as the aerated submerged fixed-film (ASFF) process developed by Hamoda and Abd-El-Bary (1987). This process employs a four compartment-in-series reactor equipped with an array of submerged media (fixed ceramic plates) for biomass attachment that is maintained under continuous diffused aeration. Modification of the ASFF process (Hamoda, 1989) to operate in the anoxic-aerobic (A/ASFF) mode could have some advantages based on studies on the activated sludge process conducted by Hao and Huang (1996). There are certain advantages with the fixed-film processes depending on the system used for attached growth (Grady and Lim, 1980; Liu *et al.,* 1996; Chudoba and Pujol, 1998). These include simplicity, low sludge production, no foam or sludge bulking, long solids retention, stable operation and resistance against shock loads as compared to suspended-growth systems. Combining both attached growth and suspended growth of microorganisms has become a viable option to upgrade the activated sludge process (Rogalla *et al.,* 1989; Odegaard and Rusten, 1990; Su and Ouyang, 1996; Hamoda and Al-Sharekh, 2000).

   This study was conducted in order to investigate the feasibility of upgrading the high rate activated sludge process using the ASFF and A/ASFF systems and to examine the effect of increased loading rate on organics removal and nitrogen recovery.

MATERIALS AND METHODS

Description of the pilot-plant and the experimental set-up

The aerated submerged fixed-film (ASFF) bioreactor was used for conducting the pilot-scale experiments . This reactor is made of 6 mm thick plexiglass sheets, and divided into four, equal-size compartments connected in series. The length of the reactor is 72.5 cm, the width is 30 cm and the liquid depth is about 60 cm, providing a total liquid volume of 115 litres. A pilot plant was installed at the WWTP (Figure 1).

   The experimental program involved in-parallel testing of ASFF and A/ASFF reactors, each was packed with the "Biolace" support medium. Biolace is a structured medium of cross-linked textile fibres which is fixed vertically and stretched in a high-grade stainless steel cage. Five (5) sheets of the Biolace (each is 390 mm long and 230 mm wide), spaced at 25 mm were fixed in each cage, occupying approximately 44% of the compartment's volume. The biolace is manufactured by UTS, Germany. Each reactor was operated continuously at a preset feed flow rate. Different flow rates were tested in each reactor over a total period of nine months to obtain HRT's in the range of 0.7 to 8 hours. Aeration of the reactors was provided through medium-to-fine, tubular- membrane, air diffusers placed underneath the media and operated at a constant pressure of approximately 2–2.5 bar. The operational parameters applied in the ASFF and the A/ASFF bioreactors are summarized sin Table 1.

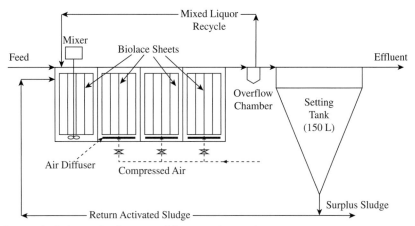

Figure 1. Schematic diagram of the experimental set-up.

Samples were collected daily from te influent and effluent of each reactor and were analyzed on the same day of collection. The samples were filtered using Whatman's Qualitative Filters size 4. The following parameters were determined on the filtrate: BOD_5, COD, Ammonia (NH_3-N), Nitrites (NO_2-N), Nitrates (NO_3-N), and Total Oxidized Nitrogen (TON). Unfiltered samples were used for other measurements such as Suspended Solids (SS) and Volatile Suspended Solids (VSS). The pH, dissolved oxygen concentration (DO) and temperature were measured on all samples collected. Compartmental attached biofilm mass was determined at the end of each experimental run. Representative compartmental samples of each medium were collected and oven dried at 105°C. The volatile (organic) fraction of the attached and suspended solids was determined by further burning the samples at 550°C. All laboratory analyses were performed according to Standard Methods (APHA, 1993).

RESULTS AMD DISCUSSION

Removal of organics

The organic removal profiles and removal efficiencies were determined for both the BOD and the COD parameters. Figure 2 shows mean (steady-state) BOD removal profiles, generated at all hydraulic retention times (HRT's). Similar patterns were observed in the ASFF bioreactor. Meanwhile, similar COD removal profiles were also observed in both bioreactors. It is clearly noticed that the majority of organic removal occurred in the 1st compartment and that organic organic removal followed a first-order kinetic pattern (Hamoda, 1989).

Table 1. Operational parameters applied to the pilot plant.

HRT[1]	FEED	NRS [2]	NML [3]	Total Inflow	Recycle	BVS[4]	Hyd. Load	BOD Load	COD Load	NH₃N Load
h	m³ d⁻¹	m³ d⁻¹	m³ d⁻¹	m³ d⁻¹	Ratio	g	m³.kg⁻¹ BVS.d⁻¹	g BOD.g⁻¹ BVS.d⁻¹	g COD.g⁻¹ BVS.d⁻¹	g NH₃N.g⁻¹ BVS.d⁻¹
The ASFF Bioreactor										
8	0.346	N/A	N/A	0.346	N/A	291.55	1.185	0.154	0.330	0.049
6	0.460	N/A	N/A	0.460	N/A	339.02	1.356	0.182	0.354	0.056
4	0.691	N/A	N/A	0.691	N/A	410.83	1.682	0.257	0.463	0.076
2	1.382	N/A	N/A	1.382	N/A	488.03	2.833	0.382	0.787	0.115
The A/ASFF Bioreactor										
8	0.173	0.173	N/A	0.346	1.0	694.69	0.497	0.033	0.076	0.009
6	0.229	0.229	N/A	0.458	1.0	574.46	0.797	0.056	0.990	0.015
4	0.389	0.389	N/A	0.691	0.8	544.33	1.270	0.109	0.186	0.034
2	0.821	0.821	N/A	1.382	0.7	580.23	2.382	0.218	0.379	0.064
1.5	0.864	0.864	N/A	1.814	1.1	464.02	3.910	0.290	0.525	0.084
1	0.950	1.037	0.691	2.678	1.8	617.60	4.337	0.222	0.369	0.065
0.7	1.382	1.382	1.210	3.974	1.9	601.35	6.609	0.290	0.526	0.099

[1]Hydraulic retention time.
[2]Nitrified return sludge.
[3]Nitrified mixed liquor.
[4]Biomass volatile solids.

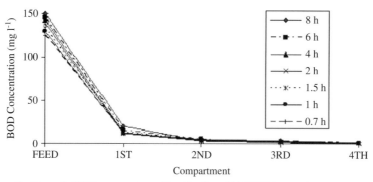

Figure 2. Overall BOD removal profiles for the A/ASFF bioreactor.

The A/ASFF process is classified as single-sludge predenitrification system. The required carbon source is essential to support the denitrification process, and since the experiment was on treatment of domestic sewage, internal carbon source was readily available through the utilization of the sewage's carbonaceous substrate; expressed as BOD and/or COD. Additional carbon source was internally provided through recycling of return sludge biosolids. First compartment's organic removals ranged from 85–92% and 60–67% for BOD and COD, respectively. No appreciable removal occurred in the remaining compartments as the substrate removal followed a first-order kinetic model. This is clearly demonstrated in Tables 2 and 3 for BOD and COD removal, respectively. By combining the organic removal in all compartments, the overall removal efficiency for BOD or COD was obtained in each case as shown in Tables 2 and 3, respectively. High "overall" removal percentages were obtained especially in the A/ASFF bioreactor which showed up to 98% for BOD removal and 75% for COD removal.

Comparison between the ASFF and the A/ASFF reactors' overall removal efficiency for organic matter (BOD) is shown in Tables 2 and 3. The superiority of the A/ASFF bioreactor is apparent in achieving exceptionally high carbonaceous removal at the excessive hydraulic loading rates applied. Although both reactors showed minimal effect due to variation in hydraulic loading, the A/ASFF bioreactor's BOD overall removal was slightly higher (96–98%) and dropped by approximately 1–2% going from HRT of 8 h down to 0.7 h. In contrast, the ASFF's BOD overall removal was generally between 92–97.5% for the applied loadings. Similarly, better COD removal was achieved with the A/ASFF bioreactor with an overall removal efficiency ranging from 73% to 80%. The ASFF bioreactor however, achieved COD removal efficiency in the

Table 2. Compartmental percentage BOD removal in the A/ASFF and ASFF bioreactors.

Reactor's Comp	A/ASFF bioreactor % BOD Removal HRT (h)							ASFF bioreactor % BOD Removal HRT (h)			
	8	6	4	2	1.5	1	0.7	8	6	4	2
1st	85.0	88.2	92.5	91.7	90.8	86.9	89.0	77.4	68.2	69.9	71.9
2nd	11.4	9.30	5.03	6.01	6.84	10.3	5.4	15.0	21.5	17.6	7.9
3rd	1.82	0.78	1.29	1.11	0.73	0.66	1.7	5.0	4.0	5.5	11.6
4th	0.55	0.47	0.31	0.1	0.4	0.74	0.4	0.4	0.4	3.0	1.1
overall	97.5	97.0	98.5	98.0	96.5	95.5	96.4	96.8	93.2	95.5	92.0

Table 3. Compartmental percentage COD removal in the A/ASFF and ASFF bioreactors.

Reactor's Comp	A/ASFF bioreactor % COD Removal HRT (h)							ASFF bioreactor % COD Removal HRT (h)			
	8	6	4	2	1.5	1	0.7	8	6	4	2
1st	66.5	60.9	70.1	66.8	61.1	65.8	62.0	46.2	37.8	43.3	35.6
2nd	8.95	8.38	1.79	4.75	5.73	8.25	7.1	11.4	14.9	9.5	8.8
3rd	2.34	2.03	0.96	4.70	2.03	1.47	2.0	3.4	9.2	12.7	16.3
4th	1.87	0.33	3.49	1.08	3.09	1.68	1.7	2.3	5.1	1.0	2.0
overall	78.5	72.1	77.5	78.2	72.5	78.0	74.5	67.1	67.5	67.3	63.0

range of 63% to 67%. This could be due to further utilization of the "hardly biodegradable" COD portion as a carbon source for the denitrification process in the anoxic stage of the A/ASFF bioreactor. In general, BOD removal was more stable than the COD removal due to fluctuations in the COD biodegradable fraction. Meanwhile, both the ASFF and the A/ASFF systems showed higher organics removal than the existing high- rate activated sludge (HRAS) system. An explanation for this increased organic biodegradation activity is a physio-logical difference between planktonic and biofilm microorganisms (Lazorova and Manem, 1995). Bacteria adhering to a surface establish strong relationship between them and are exposed to environmental conditions not found in the aqueous phase (Hallin et al., 2006).

Nitrogen transformations

In this study, the compartmental percentage ammonia removal and the total oxidized nitrogen (TON) production of the A/ASFF bioreactor were determined as illustrated in Table 4. Generally, the observed compartmental efficiency suggests the presence of active nitrifying microorganisms mainly in the 2nd and the 3rd compartments at all loading rates. This allowed the 4th compartment of the A/ASFF bioreactor to contribute to process stabilization and to act as a buffering zone to make-up for any fluctuations in performance. Therefore, the A/ASFF bioreactor would offer stable nitrification performance and quickly adapt to fluctuations in the organic and hydraulic loading rates applied on the system. In contrast, nitrification in the ASFF bioreactor was mainly accom-plished in the 3rd and 4th compartments.

Table 4. Compartmental mean (steady state) percentage NH3-N oxidized and TON producted in the A/ASFF bioreactor.

	2nd Comp.		3rd Comp.		4th Comp.	
HRT (h)	% NH_3-N Removal	% TON Prod.	% NH_3-N Removal	% TON Prod.	% NH_3-N Removal	% TON Prod.
8	31.40	52.52	17.68	43.31	2.55	−2.85
6	33.02	54.93	13.23	33.80	1.56	4.38
4	19.81	55.60	17.98	33.78	0.64	5.33
2	25.14	34.33	26.12	67.17	0.79	−11.10
1.5	19.00	26.20	17.00	57.50	2.60	10.60
1.0	20.00	11.60	19.00	66.00	2.20	4.07
0.7	12.00	18.60	13.00	51.50	4.50	14.10

Ammonia removal efficiencies were higher in the A/ASFF bioreactor as compared with those observed in the ASFF bioreactor especially at shorter

HRT's. This is illustrated in Figure 3 which shows percentage ammonia oxidation of up to 97% at HRT's of 6–8 h. Meanwhile, Figure 4 illustrates the mean concentrations of nitrates obtained in all compatments of the A/ASFF bioreactor at all HRT's. Presence of nitrates in the aerobic compartments (2nd, 3rd, and 4th) confirms the activity of the nitrifying bacteria in these compatments. In contrast, only traces of nitrates were observed in the anoxic compartment where dentirification takes place. The denitrification process was able to eliminate about 3.35 mg BOD (or 6.6 g COD) versus 1 mg denitrified $N-NO_3$.

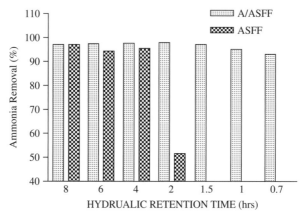

Figure 3. Mean percentages of overall ammonia oxidation in the ASFF and A/ASFF bioreactors.

Process application

Modification of the existing HRAS aeration tanks can be easily implemented by installing the Biolace media and necessary baffles at minimal cost, if compared with the construction of new tanks, to increase the capacity of Al-Awir WTTP with no delays. One of the benefits gained in adopting the ASFF bioreactor is apparent in securing steady ammonia oxidation and recovery of nitogen as nitrates at higher hydraulic loading rates than applied in the A/ASFF bioreactor which gives an edge to the ASFF bioreactor as a biological process for recovery of nitrates in the effluent that would be beneficial for effluent reuse in agriculture. On the other hand, the A/ASFF bioreactor could also be optimized for nitrogen recovery. A program is proposed for alternating ASFF and A/ASFF system operation as to optimize nitrogen recovery.

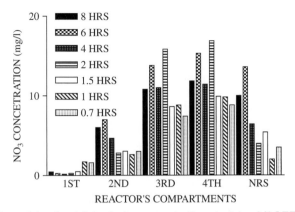

Figure 4. Mean (steady state) nitrate concentrations in tshe A/ASFF bioreactor.

CONCLUSIONS

Based on the experimental results obtained in this study, the following conclusions can be made:

1. Secondary treatment using either the ASFF or the A/ASFF process achieved high organic and nutrient removal efficiencies of up to 98% for BOD, 75% for COD, and 97% for ammonia. Oranic removal followed a first-order kinetic model.
2. Performance of the A/ASFF process was not adversely affected by a ten-fold increase in hydraulic/organic loading in the range of 0.5 to 6.6 m^3 kg^{-1} BVS d^{-1} (hydraulic) and in the range of 0.03 to 0.3 g BOD g^{-1} BVS d^{-1} (organic).
3. For nitrogen transformations, the A/ASFF process was superior to the ASFF process in ammonia oxidation by nitrification and nitrate reduction by denitrification, even at the shorter HRT's. On the other hand, the ASSF process can be better optimized for nitrogen recovery in the form of nitrates.
4. Success of the A/ASFF bioreactor in organic removal was mainly due to the efficient performance of the anoxic stage incorporated in the 1st compartment. More than 90% and 60% BOD and COD removal efficiencies were achieved in this stage, respectively.
5. The proposed system can be easily applied, with minimal required modifications, to the exiting aeration tanks as to double their capacities and achieve nitrogen recovery provided that the plant adopts an efficient aeration system and suitable media for biomass attachment.

REFERENCES

APHA (1993). Standard methods for the examination of water and wastewater. 18th ed., American Public Health Association, Washington, D.C.

Chudoba, P. and Pujol, R. (1998). A three-stage biofiltration process: performance of a pilot plant. *Wat. Sci. Tech.*, **38**(8–9), 257–265.

Grady, C. and Lim, H. (1980). Biological wastewater treatment: theory and applications. Marcel Dekker, Inc., New York, NY.

Hallin, S., Throback, I., Dicksved, J. and Pell, M. (2006). Metabolic profiles and genetic diversity of denitrifying communities in activated sludge after addition of methanol or ethanol. *Appl. and Environ. Microbiology*, **72**(8), 5445–5452.

Hamoda, M.F. (1989). Kinetic analysis of aerated submerged fixed-film (ASFF) bioreactor. *Wat. Res.*, **23**, 1147–1154.

Hamoda, M.F. (2004). Water strategies and potential of water reuse in the southern mediterranean countries. *Desalination*, **165**, 31–41.

Hamoda, M.F. and Abd-El-Bary, M.F. (1987). Operating characteristics of the aerated submerged fixed- film (ASFF) bioreactor. *Wat. Res.*, **21**, 939–947.

Hamoda, M.F. and Al-Sharekh, H. (2000). Performance of a combined biofilm-suspended growth system for wastewater treatment. *Wat. Sci. Tech.*, **41**(1),167–175.

Hao, O. and Huang, J. (1996). Alternating aerobic-anoxic process for nitrogen removal: process evaluation. *Water Res.*, **68**(1), 83–93.

Lazarova, V. and Manem, J. (1995). Biofilm characterization and activity analysis in water and wastewater treatment. *Wat. Res.*, **29**(10), 2227–2245.

Lessel, T. (1994). Upgrading and nitrification by submerged biofilm reactors – experience from a large scale plant,. *Wat. Sci. Tech.*, **29**(10/11), 167–174.

Liu, J., Groenestijn, J. and Doddema, H. (1996). Removal of nitrogen and phosphate using a new biofilm-activated sludge system. *Wat. Sci. Tech.*, **34**(1–2), 315–322.

Metcalf and Eddy, Inc. (1991). Wastewater engineering: treatment, disposal, reuse. McGraw-Hill Book Co., New York, NY.

Odegaard, H. and Rusten, B. (1990). Upgrading of small municipal wastewater treatment plants with heavy dairy loading by introduction of aerated submerged biological filters. *Wat. Sci. Tech.*, **22**, 191–198.

Rogalla, F., Bacquet, G. and Bonhomme, M. (1989). Fixed biomass carriers in activated sludge plants. *Wat. Sci. Tech.*, **21**, 1643–1646.

Su, J.L. and Ouyang, C.F. (1996). Nutrient removal using a combined process with activated sludge and fixed biofilm,. *Wat. Sci. Tec.*, **34**(1–2), 477–486.

Air stripping of ammonia from anaerobic digestate

Frank Wäger, Thomas Wirthensohn, Alberto Corcoba and Werner Fuchs

University of Natural Resources and Applied Life Sciences, Vienna, Dept. for Agrobiotechnology, IFA-Tulln, Institute for Environmental Biotechnology, Konrad Lorenz Strasse 20, 3430 Tulln, Austria

Abstract Effluents of anaerobic digestion plants generally show high content of nitrogen compounds. In many cases removal of ammonia is necessary to improve nutrient management e.g. in areas with high animal farming density. Air stripping is a promising technology not only to remove but also to recover the stripped ammonia as a valuable product. In this study the influence of typical pre-treatment technologies and the variation of the process parameters temperature and pH on stripping efficiency were investigated. The alkalinity of the substrate, i.e. the HCO_3^- concentration, was identified to be of critical importance. On the one hand it acts as a buffer system which makes elevation of the pH more difficult on the other hand the parallel desorption of CO_2 during NH_3 stripping stabilizes the pH. It was demonstrated that precipitation with lime milk ($Ca(OH)_2$) for removal of suspended solids and carbonates results in a strong decrease of alkalinity. Hence the pH in the course of the stripping process drops strongly which leads to a lower process efficiency. Therefore lime milk has to be overdosed to which leads an excessive chemical demand. As alternative suspended solids removal by means of microfiltration is proposed. In this case, only a moderate pH adjustment with caustic soda to pH 10 is suggested. To compensate, a high temperature, 80°C, should be applied during to the stripping process. Under such conditions elimination rates of 90% and even higher were achieved.

INTRODUCTION

Biogas technology has found great acceptance for power and heat production over the last years, not only because it is considered as a CO_2 neutral energy source but also as an appropriate technology for the treatment of organic waste. Nowadays, operators of biogas plants find themselves confronted with the challenge to find an appropriate technology that guarantees a low cost and long term assured possibility to treat the remaining digestate. In general, the remaining digestate is brought out to agricultural fields and used as a fertilizer. Nowadays biogas plants are often built in a larger scale – due to higher efficiency in the energy production – which leads to an increased use of supra-regional substrates. Hence, the required areas for bringing out the digestate are often not available in the vicinity of the production location. In such a case the legal limitation for bringing out nitrogen compounds requires transportation over long distances.

© 2009 The Authors, *International Conference on Nutrient Recovery from Wastewater Streams*. Edited by Ken Ashley, Don Mavinic and Fred Koch. ISBN: 9781843392323. Published by IWA Publishing, London, UK.

For that reason treatments for nitrogen removal and recovery have become processes of concern for improving waste management in areas with a nitrogen surplus. Air stripping in combination with absorption (typically sulphuric acid) is a promising technology to remove and recover ammonia from the liquid fraction of the anaerobic digestate (Bonmatí and Flotats, 2003; Rulkens et al., 1998).

Best air stripping efficiency can be achieved when a packed column is used, as it guarantees a huge surface and therefore optimal contact conditions of the gaseous and the liquid phase.

The pre-treatment of the anaerobic digestate effluent before entering the stripping column is a critical step, because it has a direct influence on the stripping efficiency and reliability of operation.

The parameters that have to be adjusted are:

- content of suspended solids (SS)
- temperature
- pH

pH and the temperature play the key role, as they influence the dissociation equilibrium of NH_4^+/NH_3 and the distribution between the aqueous and the gaseous phase. Furthermore dissolved solids, mainly inorganic ions like carbonates and phosphates, can cause severe problems in the process and may precipitate especially under conditions of high pH and/or high temperatures in the column or heat exchanger (Görisch and Helm, (2006)).

Besides the controlled precipitation of the mentioned substances in settling tanks with lime milk $(Ca(OH)_2)$, also stripping of CO_2 is a method often applied in practice.

Several alternative procedures are in use as a pre-treatment before stripping. All of them include a solid liquid-separation. Figure 1 shows two typical pathways for the pre-treatment of the anaerobic igestate. Pathway A includes a precipitation step. Frequently lime milk is added to the anaerobic digestate for precipitation, whereas at the same time the pH is lifted up to values from 10 to about 12 (Ozturk et al. (2003)). The disadvantage of precipitation with lime is the production of a huge amount of sludge (Kollbach et al. (1996)) which has to be removed in a subsequent settler or flotation. On the other hand lime precipitation removes carbonates which might lead to scaling or clogging of the stripping column.

Another pre-treatment option is shown in pathway B. In this case, the separation of the suspended solids from the liquid fraction is done by a centrifuge and subsequently a filtration step (microfiltration or ultrafiltration) eliminates all of the remaining SS. In many cases, carbonates are then eliminated via CO_2

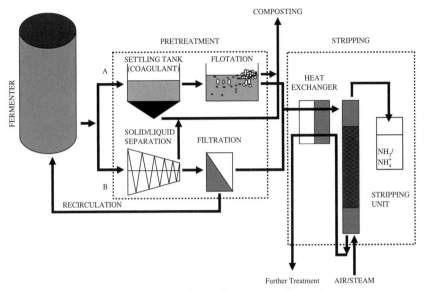

Figure 1. Schema for treatment options of anaerobic digestate.

stripping. Due to the elimination of carbon dioxide the pH ascends. Even higher pH values can afterwards be adjusted by the addition of base, e.g. caustic soda.

The purpose of the investigations described herein was to identify the best preconditioning method and the respective process parameters for a digestate from a local biogas plant.

MATERIALS AND METHODS

Anaerobic digestion plant

The material used for all experiments was taken from an anaerobic digestion plant in Styria/Austria. The biogas plant uses cow manure, kitchen garbage and other organic waste as substrate. Concentration of NH_4-N was from 3970 to 7.842 mg/l (Table 1). The anaerobic digestion effluent taken from the final storage and after microfiltration was stored at 4°C before usage for experiments.

Pre-treatment of the anaerobic digestate

Precipitation with lime milk

Different dosages of lime milk (10%) were added to 500 mL anaerobic digestate effluent in 1000 mL beakers. The samples were stirred for 30 min before the

Table 1. Characteristics of anaerobic digestion effluent.

Parameter	Unit	Final storage	After microfiltration
TS	%	4.81	1.40
pH		8.20	8.51
CSB	mg/L	51,493.5	10,264.5
TN	mg/L	9,183.5	8,209.8
NH_4-N	mg/L	6,732.5	6,522.3
PO_4-P	mg/L	1,008.2	428.0

resulting mixtures were then allowed to precipitate for 2 hours. Afterwards alkalinity and pH values were measured.

Elimination of CO_2 with sulphuric acid and pH adjustment

In order to investigate the influence of the presence/absence of carbonates, sulphuric acid with a concentration of 98% was added to the effluent after microfiltration until a final pH of 4.3. The liquid was stirred for 30 min before the pH was increased with caustic soda to the desired pH. For the pH lifting caustic soda with a concentration of 50% was added. After the addition of caustic soda the mixture was stirred until the pH got stable.

Stripping of carbon dioxide in bubble reactors

For stripping of carbon dioxide in batch scale, three two litre bottles were used, whereas the first two bottles were for air moistening and the third one contained the sample (Figure 2). In order to obtain the required temperature the incoming air was first mixed with steam. The temperature was controlled in the waterbath and in the inlet of the last bottle (with the anaerobic digestate effluent after microfiltration). A sample was taken every 15 min for a duration of 120 min, afterwards the pH, alkalinity and NH_4-N were determined.

Stripping of ammonia

Stripping of ammonia in bubble reactors

The experimental setup was the same as described in Stripping of carbon dioxide in bubble reactors. A sample was taken every 15 min, afterwards the pH, alkalinity and NH_4-N were determined.

Air stripping of ammonia with a stripping column

The schematic layout of the stripping device is presented in Figure 3. The column has a total height of 1.5 m whereas the filling bodies occupy 1.1 m.

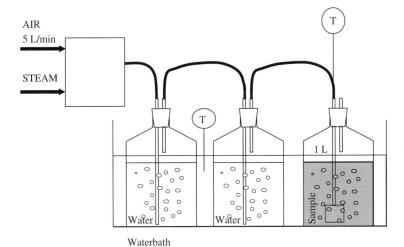

Figure 2. Experimental setup for stripping in bubble reactors.

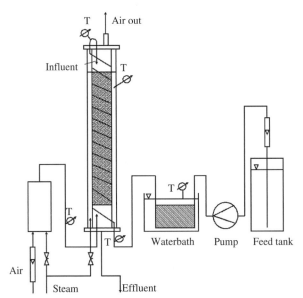

Figure 3. Stripping column with preheating of the liquid and the air inflow.

The used filling bodies are HIFLOW rings, with a specific surface of 313 m^2/m^3. The internal diameter of the inner column is 80 mm, the outer column serves as a mantle to guarantee the temperature conditions as adjusted for the experiments. Dimensioning of the column was done with the program RAPSODY (Version 2.18) from the company Rauschert Verfahrenstechnik. Air for stripping was mixed with steam before entering the column to enable saturation of the inflowing air with water and in the same way adjustment of the required temperature. Temperature was measured at all points of entrance and outlet to ensure stable conditions during the stripping process.

RESULTS AND DISCUSSION

Pretreatment

Precipitation with lime milk

Addition of lime milk is done to remove phosphate and carbonates and as well to increase the pH. Moreover SS are reduced significantly. Lei *et al.* (2007) found optimization on $Ca(OH)_2$ dosage at 27.5 g/L for an effectively removal of SS, turbidity NH_4-N, COD and PO_4-P. Furthermore, an addition of only 12.5 g/L was necessary to increase the pH to values $\geq$ 12. Addition of excess lime to prevent the pH from dropping was reported by Cheung *et al.* (1997) for landfill leachate. However, consumption of lime milk was in our case much higher, due to the high content of buffering substances. In order to increase the pH up to a value of 10, addition of 60 g/L lime was necessary (Figure 4). At this pH, 88.2% of the alkalinity was removed, which leaded to a diminished buffer capacity.

Equation 1 shows the connection of the addition of lime and the precipitation of $CaCO_3$.

$$2Ca^{2+} + 2OH^- + 2HCO_3 \leftrightarrow 2CaCO_3 + 2H_2O \qquad (1)$$

Stripping of carbon dioxide

The presence of carbonates in the anaerobic digestate affects the stripping process, on one hand, in terms of the reliability of operation due to the precipitation of lime, and on the other hand due to a high consumption of base for increasing the pH.

Lei *et al.* (2007) and Marttinen *et al.* (2001) described the increase of the pH when carbon dioxide is stripped at ambient temperature and at relatively moderate air flow rate. After one day of aeration with a flow rate of 2.5 L/Lmin the pH increased from 7.4 to 9.3.

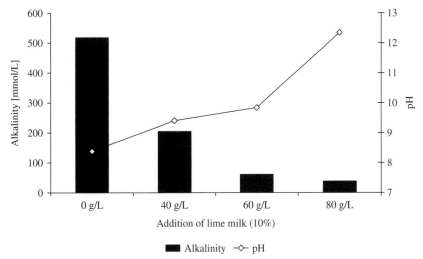

Figure 4. Removal of alkalinity and increase of pH due to precipitation with lime milk.

In our case this was suspected inefficient and therefore a higher temperature (80°C) and also an increased flow rate 5 (L/Lmin) were considered as convenient process conditions.

Figure 5 compares CO_2 stripping at 20°C and 80°C and demonstrates the much higher efficiency of elevated temperature. The elimination of alkalinity behaves oppositional to the pH value (Figure 6): the lower alkalinity gets, the higher the pH gets. Both curves come up to an equilibrium.

Adjustment of the pH

To obtain high efficiency of ammonia stripping at room temperature increasing of the pH up to 10.5–11.0 is inevitable (Liao *et al.* (1995)). The consumption of base is an important cost factor. Efforts, to keep the consumption of chemicals as low as possible and nevertheless achieve the required elimination efficiency of ammonia therefore often play a major role.

It is obvious that buffer systems in the digestate, primarily HCO_3^-/CO_3^{2-} and NH_4^+/NH_3, lead to a very high consumption of caustic soda resulting in high operation costs. The average pH value of the untreated anaerobic digestate was about 8.5. In order to increase the pH of the untreated digestate up to 10.16 mL/L have to be added. For adjustment to higher pH values the consumption of NaOH got drastically higher, e.g. 34.6 mL for pH 11 and 40 mL for pH 12.

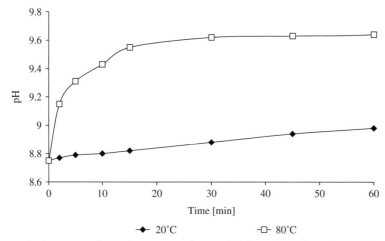

Figure 5. Increase of pH due to aeration at 20°C and 80°C and an airflow of 5 L/Lmin.

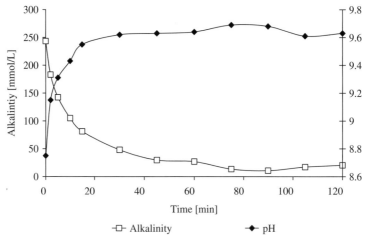

Figure 6. pH and alkalinity in dependence of time at an airflow of 5 L/Lmin and 80°C.

When the digestate first was CO_2 removed, in the range from pH 8.5 to pH 10 the progression of the curves was almost the same, whereas for pH values >10 the NaOH consumption was significantly lower (Figure 7).

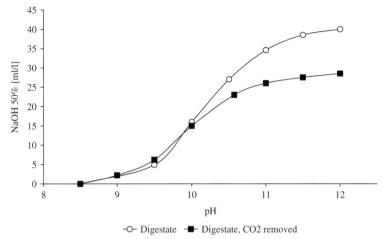

Figure 7. Consumption of caustic soda (50%) to digestate (untreated and after addition of sulphuric acid).

Stripping of ammonia

Stripping of ammonia in bottles

In order to investigate the influence of the presence of carbon dioxide on stripping efficiency of ammonia, experiments were carried out in bubble reactors. Two kinds of synthetic ammonia solutions (NH_4Cl and NH_4HCO_3) were compared with the anaerobic digestate effluent, untreated and CO_2 removed.

According to Equation 2, NH_3 is in equilibrium with NH_4^+. When NH_3 is desorbed to the gaseous phase the chemical force is driving the reaction to the right side and the protons generated lower the pH. During the stripping of the CO_2 the contrary effect occurs (Equation 3). Therefore the carbonate prevents a pH change during ammonia removal. Figure 8 illustrates this effect. Both for the microfiltrated digestate and the NH_4HCO_3 solution, the pH remained almost stable even at ammonia removal rates of nearly 100%. In contrast, if the carbonate buffer was removed the pH droped dramatically to levels as low as pH 7. The decline of the pH is due to the lacking buffer capacity. It can also be seen in Figure 8 that the pH of the CO_2 removed digestate is slightly more stable than of the synthetic NH_4Cl solution. The difference in behaviour is presumably due to a buffer capacitiy caused by other constitutents than carbonate, e.g. humic substances.

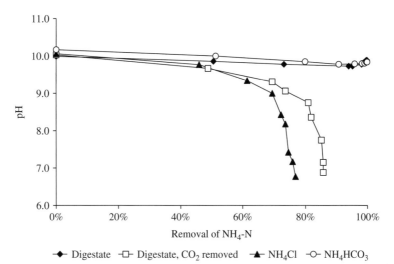

pH

Removal of NH$_4$-N

—◆— Digestate —◻— Digestate, CO$_2$ removed —▲— NH$_4$Cl —○— NH$_4$HCO$_3$

Figure 8. pH in dependence of removed ammonia of synthetic wastewater and anaerobic digestate at and 80°C and airflow of 5 L/Lmin.

With the ongoing decline of the pH the amount of free NH$_3$ present in the solution decreases until no further removal occurs.

$$NH_4^+ \leftrightarrow NH_3 + H^+ \tag{2}$$

$$HCO_3^- + H^+ \leftrightarrow H_2CO_3 \leftrightarrow CO_2 + H_2O \tag{3}$$

Figure 9 shows the progress of NH$_3$ removal with time. When the carbonates of the anaerobic digestate were eliminated by addition of sulphuric acid, total elimination of ammonia was not possible. The curve progression showed a similar behaviour as the NH$_4$Cl solution and stagnated at a remaining concentration of ammonia at 14% of the starting value during the last 30 minutes, whereas the remaining ammonium concentration of the not CO$_2$ removed anaerobic digestate was only 0.27%.

In comparison, the elimination rate of ammonia from the NH$_4$HCO$_3$ solution behaved the same as the not CO$_2$ removed effluent and showed much higher efficiency (Figure 9).

The course of alkalinity of the samples stripped at 80°C can be seen in Figure 10. With the ongoing process, alkalinity got more and more exhausted. The difference between the NH$_4$HCO$_3$ solution and the digestate is only due to different start concentrations of NH$_4$-N.

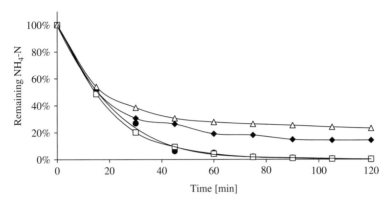

Figure 9. Remaining NH$_4$-N in dependence of time at 80°C and airflow 5 L/Lmin.

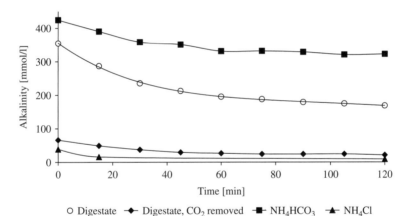

Figure 10. Alkalinity of synthetic wastewater and digestate (with and without CO$_2$ removal) in dependence of time at 80°C and airflow 5 L/Lmin.

Stripping efficiency was as also tested without pH adjustment.

A lower pH value favours the formation of undissolved CO$_2$ and therefore enhances its desorption (Figure 12). If desorption of CO$_2$ is higher than for NH$_3$, the pH is even going up. In Figure 11 it can be seen that the pH is converging to almost the similar value as the sample with a pH adjusted to 10.

In consequence, the effectiveness of the ammonia removal was in both cases almost the same.

Table 2 shows the air amount required for a certain percental reduction of ammonia.

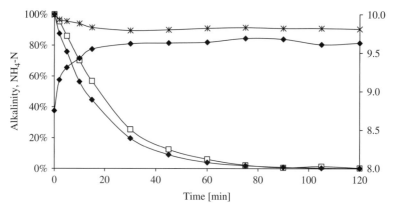

Figure 11. Remaining NH$_4$-N of digestate (untreated and after pH adjustment up to 10) in dependence of time, at 80°C and airflow of 5 L/Lmin.

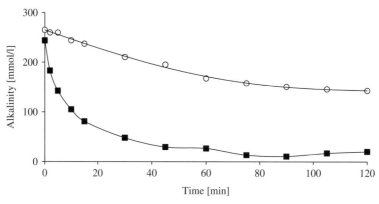

Figure 12. Remaining alkalinity of digestate (untreated and after pH adjustment up to 10) in dependence of time, at 80°C and airflow of 5 L/Lmin.

If 50% reduction is the goal, than the air demand is 1.34 fold the theoretical value for the pH adjusted digestate whereas it is 1.91 for the untreated one. Still this means that 43.5% more air is required if the pH is not changed. For higher elimination rates e.g. 90%, the multiple of the minimum airflow becomes even higher and increased to a value of 2.9 for the untreated digestate.

Thereby the difference between the untreated digestate and the pH increased sample show no big difference in respect to the multiple of the minimum airflow.

Table 2. Minimum airflow and multiple of minimum airflow for certain elimination rates at 80°C.

Elimination rate of NH$_4$-N [%]	Minimum airflow [L/L]	Applied airflow [L/L]		Multiple of minimum airflow	
		untreated	pH 10	untreated	pH 10
50	103	197.2	138.1	1.9	1.3
80	168	413.4	334.6	2.5	2.0
90	189	551.1	485.9	2.9	2.6

This small difference might be caused by the higher pH of the sample treated with caustic soda in comparison to the untreated digestate.

Air stripping of ammonia with a stripping column

Stripping in a packed column is much more efficient than stripping in a batch reactor due to the higher exchange surface. The actually required air demand in a stripping column is normally 1.2 to 1.5 higher than the theoretical minimum air demand (Perry and Green (1997)) and a more economic process to achieve higher NH$_4$ removal levels. For the further experiments an elimination rate of at least 80% was set as the removal goal.

Setting of flow conditions

To find out optimal liquid flow rate to achieve the required elimination rate, experiments with the stripping column were carried out at a pH of 10, a constant airflow of 20 m^3/h and temperatures of 60 and 80°C. Figure 13 shows that higher flow rates lead to reduced stripping efficiency. At a temperature of 80°C and a flow rate of 20 l/h only 9.63% of the starting concentration of NH$_4$-N remained in the effluent whereas the remaining concentration of NH$_4$-N at 60°C was 22.5%. For the following experiments the conditions of 80°C and a liquid flow rate of 20 l/h have been considered as appropriate.

A stripping column is mainly characterized by the parameters column height, column diameter, filling material, number of transfer units (NTU) and the height of transfer units (HTU).

Due to constant air and liquid flow in our experiments, column parameters were only maintained at a temperature of 20°C. Under these conditions the theoretical column height and the real column height were almost the same (Table 3, 1.1 m in comparison to 1.19 m).

For higher ammonia removal rates, the necessary number of transfer units increases. In our experiments the NTU value reached its maximum with 2.8 at 80°C. On the other hand, due to higher efficiency, the HTU value decreases

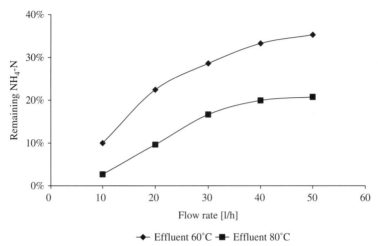

Figure 13. Remainig NH$_4$-N in the effluent of the stripping column in dependence of various liquid flow rates at 60 and 80°C and airflow of 20 m³/h.

Table 3. Theoretical column parameters and elimination of ammonia at pH 10.

Temp. [°C]	Remaining ammonia [%]	Remaining ammonia [mg/L]	NTU	HTU [m]	Theoretical column hight [m]	Deviation of the theoretical column hight
20	61.89	4,853.6	0.9	0.66	1.2	0.92
50	43.36	3,400.7	1.0	0.22	0.44	2.50
60	28.52	2,237.1	1.5	0.16	0.48	2.29
70	13.93	1,092.9	2.4	0.12	0.58	1.91
80	8.85	694.3	2.8	0.09	0.50	2.18

and the resulting column height is determined through H = NTU∗HTU$_{opt}$ (with HTU$_{opt}$ at least two times higher than HTU). To investigate the influence of the temperature and the pH value, we maintained the flow conditions constant, with the consequence that the air flow was higher than the determined minimum (5.21 times higher at 80°C). In the praxis the columns are generally built higher in place of higher air flow due to economical considerations. Nevertheless, the effective obtained elimination rate of ammonia matches well with the theoretical calculations. For the obtained elimination rates, the determined deviation with a maximum of 2.5 at 50°C was quite high, but considering the small column diameter and the thereby associated wall effects, the performance was considered as satisfactory.

Influence of temperature

The influence of temperature is a main factor for stripping efficiency. In Table 3 the remaining ammonia concentrations and percentages are listed. With increasing temperature, the rest concentration of ammonia in the effluent can be strongly diminished. Again, for reasons of comparison the presence/absence of HCO_3^- was investigated. Similar results as in the bubble reactor were obtained. While the pH of the effluent from the not treated feed did not fall lower than 9.5 at 80°C, the pH of the CO_2 removed sample decreased to a value of 8.5. A temperature increase did not much further contribute to the final removal rate (Figure 14).

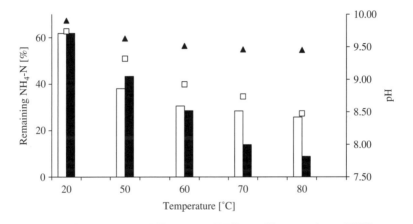

□ Effluent, CO_2 removed ■ Effluent -□- pH Effluent, CO_2 removed ▲ pH Effluent

Figure 14. Remaining NH_4-N and pH in the effluent after stripping at pH 10 and airflow of 20 m^3/h.

Influence of pH

As an air temperature of 80°C can be obtained using the waste heat from the gas engine, and it showed promising elimination results, further experiments were conducted at that temperature.

Figure 15 shows the results when varying the other main factor of influence, the pH.

At pH of 10 the residual concentration is already lower than 10%. Any further reduction causes a strong raise of the chemical costs which are required for increasing the pH (Figure 16).

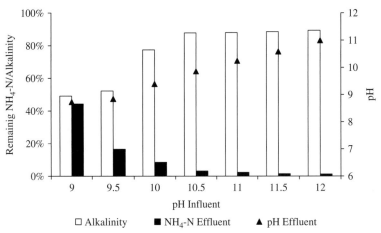

Figure 15. Remaining NH_4-N, alkalinity and pH in the effluent after stripping at 80°C and airflow of 20 m³/h in dependence of the pH.

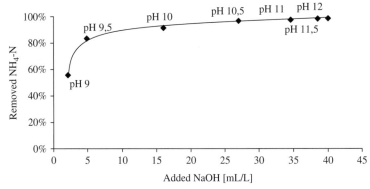

Figure 16. Removed NH_4-N and adjusted pH in dependence of NaOH added at 80°C and airflow of 20 m³/h.

CONCLUSION

To guarantee good operation reliability and high stripping efficiency the anaerobic digestate has to be pretreated. Precipitation with lime is advantageous due to the simultaneous suspended solid reduction, removal of carbonates and pH increase. However lime precipitation consumes a high portion of the alkalinity. This alkalinity keeps the pH at a constant level during the stripping process because of the simultaneous stripping of CO_2. Therefore the removal of alkalinity has to be compensated by the addition of a corresponding amount of base. E.g. an alkalinity of 400–500 mMol/L corresponds to 21 mL/L–26.2 mL/L NaOH (50%). Therefore lime is usually added in excessive amounts, which leads to high operational costs and problems with later neutralization.

To reduce chemical demand and sludge formation it is more convenient to opt for the second pathway proposed. In such a case the pH should not be raised above a value of 10. In order to still have efficient removal rates it is rather recommended to increase the temperature to 80°C. Under such conditions removal rates of up to 90% or even above can be obtained.

ACKNOWLEDGEMENT

This work was co-funded by the European Union: 6. Frame work research program – CRAFT Project: NIREC (Contract number: 033130)

REFERENCES

Bonmatí, A. and Flotats, X. (2003). Air stripping of ammonia from pig slurry: characterisation and feasibility as a pre- or post-treatment to mesophilic anaerobic digestion. *Waste Management*, **23**, 261–272.

Cheung, K.C., CHU, L.M. and Wong, M.H. (1997). Ammonia stripping as a pre-treatment for landfill leachate. *Water, Air and Soil Pollution*, **94**, 209–221.

Görich, U. and Helm, M. (2006). Biogasanalgen. Eugen Ullmer KG.

Kollbach, J.-St., Grömping, D. and Heinemeyer, L. (1996). Grundlagen zur Vorbehandlung von Prozeßwasser vor einer physikalischen Behandlung. Stickstoffrückbelastung – Stand der Technik 1996/97 – Zukünftige Entwicklungen.

Lei, X., Sugiura, N., Feng, C. and Maekawa, T. (2007). *Journal of Hazardous Materials*, **145**, 391–397.

Liao, P.H., Chen, A. and Lo, K.V. (1995). *Bioresource Technology*, **54**, (1995), 17–20.

Marttinen, S.K., Kettunen, R.H. Sormunen, K.M., Soimasuo, R.M. and Rintala, J.A. (2001). *Chemosphere*, **46**(2002), 851–858.

Ozturk, I., Altinbas, M., Koyuncu, I., Arikan, O. and Gomec-Yangin, C. (2003). Advanced physico-chemical experiences on young municipial landfill leachates. *Waste Management*, **23**, 441–446.

Perry, R.H. and Green, D.W. (1997). Perry's Chemical Engineers' Handbook, Seventh Edition, The McGraw-Hill Companies.

Rulkens, W.H., Klapwijk, A. and Willers, H.C. (1998). *Environmental Pollution*, **102**, S1, 727–735.

Effect of air temperature and air humidity on mass transfer coefficient for volume reduction and urine concentration

P.M. Masoom[a], R. Ito[b] and N. Funamizu[c]

[a]Dr. Engg Student, Environmental Engineering Department, Graduate School of Engineering, Hokkaido University, Japan, Kita-13 Nishi-8, Kita-Ku, Sapporo-060-8628, Email: mohammad_masoom@yahoo.com
[b]Ph.D, Assistant Professor, Environmental Engineering Department, Graduate School of Engineering, Hokkaido University, Japan, Kita-13 Nishi-8, Kita-Ku, Sapporo-060-8628, Email: ryuusei@eng.hokudai.ac.jp
[c]Dr. Engg., Professor, Environmental Engineering Department, Graduate School of Engineering, Hokkaido University, Japan, Kita-13 Nishi-8, Kita-Ku, Sapporo-060-8628, Email: funamizu@eng.hokudai.ac.jp

Abstract Urine contains nutrients such as Nitrogen, Phosphorus and Potassium. For application of urine fertilizer, its huge quantity is required to fulfil nutrient requirements for crops. For example Paddy requires about 134 kg-N fertilizer per hectare for which 14888 Litres of raw urine is to be transported to farmland located at 40 kilometres away from human settlements. Comparison of raw urine transportation cost with cost of 134 kg of commercially available N Fertilizer reveals that transportation cost is on higher side. Therefore, for 80% volume reduction, new onsite volume reduction system (OVRS) with application of atmospheric energy was tested. For a household of 10 family members 787-377 cm^2 and 1566-482 cm^2 sizes of vertical sheets would be required for various air temperature and air humidity conditions respectively.

INTRODUCTION

Urine contains high concentration of nutrients particularly Nitrogen, Phosphorus and Potassium. About 80% of Nitrogen and 50% of Phosphorus are found in human urine. Urine only represents just less than 1% by volume of domestic wastewater. The nutrient in urine, therefore are quite concentrated which can be readily available as fertilizer for crop cultivation.

Urine is pure nutrient solution containing low level of heavy metals and normally also of pathogenic organisms. Use of urine as fertilizer is problematic with regards to management, storage, and transportation (B.B Lind et al., 2001).

Estimation revealed that large quantity of urine is to be transported to farmland to fulfil nutrient requirements for particular crop. For example Paddy crop requires 134 kg-N fertilizer per hectare. For this purpose 14888 Litres of raw urine is required. Comparison of transportation cost of 14888 Litres of raw urine in Southern Pakistan with cost of commercially available N Fertilizer reveals that transportation cost for 40 kilometres distance is not feasible for farmers. Therefore, for 80% volume reduction, a new onsite volume reduction system (OVRS) was tested at laboratory scale (Masoom, Funamizu, Ito, 2008 Paper being submitted to Water Research Journal). Assessment of effect of air temperature and air humidity on mass transfer coefficient using OVRS was purpose of this research work. We used de-ionized water as a sample of urine because our objective is to determine mass transfer coefficient (MTC) of liquid through unit square area of a vertical cloth sheet. We understand that urine contains high concentration of salts which may create difficulty in making our assessment. Therefore, we decided to use the de-ionized water which is mainly free from salts.

CASE STUDY OF SOUTHERN PAKISTAN

Various crops are cultivated in Pakistan such as wheat, rice, sugarcane, cotton, maize and pulses using commercial fertilizer. Case Study of Southern Pakistan was carried out to compare cost of commercially available fertilizer with the transportation cost of equivalent quantity of urine.

National Fertilizer Development Centre of Government of Pakistan has recommended dose of N, P and K fertilizers per hectare for various crops. For example Paddy requires 134 kg-N Fertilizer, 67 kg-P Fertilizer and 50 kg-K Fertilizer.

Currently various commercially available fertilizers such as Urea, Calcium Ammonium Nitrate (CAN), Nitrophos (NP), Single Super Phosphate (SSP) and Di-Ammonium Phosphate (DAP), Sulphate of Potash (SoP) are used in Pakistan for cultivation of various crops. The variety wise sale rate of fertilizers is given in Table 1.

On an average estimates, farmland in Pakistan is located about 30–40 kilometres away from urban area, which is target distance for transportation Truck load freight rates charged in various countries is given in Table 2.

Market rates have been used for carrying out estimation of transportation cost of urine from human settlements to farmland.

Table 1. Variety-wise sale rate of fertilizers.

(Pakistan Rs. per bag of 50 kg)

Year	Urea 46% N	AN/CA N26% N	AS 21% N	NP 23:23	SSP(G) 18%	DAP 18:46	SoP 50% K	NPK 10:20:20	TSP (G)
1994–95	235	150	164	250	150	379	195	247	–
1995–96	267	172	172	320	183	479	331	–	–
1996–97	340	209	197	384	211	553	532	–	–
1997–98	344	222	209	412	196	574	538	–	–
1998–99	346	231	275	457	234	665	541	–	–
1999–00	327	231	286	464	298	649	543	–	–
2000–01	363	233	300	468	253	669	682	–	–
2001–02	394	268	308	519	280	710	765	–	–
2002–03	411	282	344	539	287	765	780	–	–
2003–04	421	298	373	622	316	913	809	–	–
2004–05	468	353	405	704	373	1001	996	–	801
2005–06	509	395	744	710	407	1079	1170	–	833
2005–06	Values in US Dollar as of June 2007								
	7.4	5.7	10.7	10.3	5.9	15.8	16.9	–	12
2008	850	600	925	900	1000	3200	2000	–	1000
	Values in US Dollar as of August 2008								
	11	8	12	11.5	12.9	41	25.8	–	12.9

– = Not Available
Source: National Fertilizer Development Centre (NFDC), Islamabad, Government of Pakistan

Comparison of cost of commercially available fertilizer with transportation cost of equivalent quantity of urine for paddy cultivation

Nitrogen fertilizer

Urine contains 7–9 g/L of Nitrogen (Guyton, 1986 and Krichman *et al.*, 1995). As per recommendations of Government of Pakistan, Paddy requires 134 kg-N Fertilizer/ha. Cost of 134 kg of commercially available Nitrogen Fertilizer is 30 US$. Keeping in view N concentration in urine, we require 14,888 Litres of raw urine containing 134 kg-N for one hectare of paddy field.

Figure 1 indicates transportation cost of 14888 L of raw urine proportionate to transportation distance, while it is equal to cost of commercially available N

Fertilizer at 14 km distance. Therefore, volume reduction is desired especially when farmland is located 40 kilometres away from residential area.

We conceived two proposals for 50% and 80% volume reduction of urine. When volume is reduced to 50% then required volume to transport gets 7444 Litres, whose transportation cost to 29 kilometres distance equals to cost of commercially available N Fertilizer. Therefore, this option is not feasible. When 80% volume is reduced then required volume becomes 2977 Litres whose transportation cost to the distance of 40 kilometres is estimated as 18 US$, whereas cost of equivalent quantity of commercially available N fertilizer is 30 US$. Therefore, farmers can reduce 12 US $ per hectare if they apply urine

Table 2. Truck load freight rates charged in various countries.

	Official Rates (2002)	Market Rates (2008) US$ per 1000 L km
Country	Average Cost per 1000 L km (US$)	
Pakistan	0.015–0.021	0.1374
India	0.019–0.027	Capacity of Truck:
Brazil	0.025–0.048	Type-I Big Trucks = 24,000 L
United States	0.025–0.050	Type-II Tractor Trolleys = 5000 L
Central Asian Republic	0.035–0.085	Type-III Small Van = 2600 L
Australia	0.036	Type-IV Donkey Carts = 800 L
China	0.040	

Road Transport Service Efficiency Study, India November 1, 2005, World Bank Report Pakistan Environmental Management Consultants

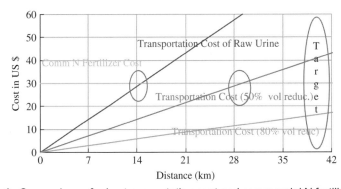

Figure 1. Comparison of urine transportation cost and commercial N fertilizer cost.

as fertilizer. Thus 80% volume reduction recovering 100% Nitrogen would be preferable to give financial incentive to farmers.

Phosphorus fertilizer

Urine contains 0.2–0.21 g/L of Phosphate (Krichmann and Petterson, 1995). As per recommendation of Government of Pakistan, paddy requires 67 kg of P Fertilizer/ha, while cost of commercially available P Fertilizer is 54 US$.

Keeping in view P concentration in urine, we require 335,000 Litres of raw urine containing 67 kg of urine based P Fertilizer to fulfil requirement of one hectare of paddy.

Figure 2 indicates transportation cost of 335,000 Litres of raw urine proportionate to transportation distance, while it is equal to the cost of commercially available P Fertilizer at 1 km distance. Thus volume reduction is desired. When volume is reduced to 50% then required volume to transport gets 167500 L, whose transportation cost to the distance of 2 kilometres is equal to cost of commercially available N Fertilizer. When 80% volume is reduced, then required volume to transport gets 67000 L, whose transportation cost to the distance of 5 kilometres is equal to cost of commercially available P Fertilizer. Therefore, it is not feasible to consider volume reduction target from 100% P recovery point of view for distance of 40 kilometres.

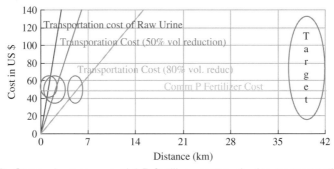

Figure 2. Comparison commercial P fertilizer cost and urine transportation cost.

Potassium fertilizer

Urine contains 0.9–1.1 g/L of Potassium (Krichmann and Petterson, 1995). As per recommendation of Government of Pakistan, paddy requires 50 kg of K Fertilizer/ha, while cost of commercially available K Fertilizer is 25.8 US$. About 50,000 L of raw urine is required to fulfil requirement of K fertilizer for one hectare of Paddy crop.

Figure 3 indicates transportation cost of 50,000 L of raw urine proportionate to transportation distance, while it is equal to the cost of commercially available K fertilizer at 4 km distance. Hence it is desired to reduce volume of urine. Two options 50% and 80% volume reduction were considered. When volume is reduced to 50% then required volume to transport gets 25,000 Litres, whose transportation cost to the distance of 8 kilometres is equal to cost of commercially available K fertilizer. When 80% volume is reduced, then required volume to transport gets 10,000 Litres, whose transportation cost to the distance of 18 kilometres is equal to the cost of commercially available K fertilizer. Therefore, it is not feasible to consider volume reduction target from 100% K fertilizer recovery point of view for a farmland located at 40 km.

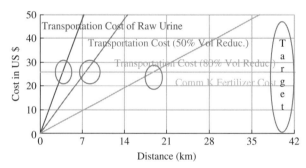

Figure 3. Comparison commercial K fertilizer cost and urine transportation cost.

Summary

We can not consider our target volume reduction on 100% Phosphate or Potassium recovery point of view. So we propose to reduce 80% volume of urine considering 100% Nitrogen recovery with additional recovery of 3 kg of P and 16 kg of K. Under present practices, farmers invests 110 US$ per ha on commercially available fertilizers for paddy. We can motivate farmers to use urine based fertilizer (100% N) plus deficit P (64 kg) and deficit K (34 kg), whose cost per ha for paddy becomes 87.2 US$, leading to 20% cost reduction per ha.

PREVIOUS VOLUME REDUCTION METHODS

There are various urine treatment and nutrient recovery techniques which includes hygienisation (storage), volume reduction (evaporation, freezing and thawing, reverse osmosis), stabilization (acidification, nitrification), P-recovery (MAP),

N-recovery (ion exchange, ammonia stripping, isobutylaldehyde-diurea (IBDU) precipitation, nutrient removal (anammox) and handling of micro pollutants (electro dialysis, nano-filtration, ozonization). However, all techniques require high energy and operation and maintenances cost. Moreover, Southern Pakistan has a hot and dry climate and poor piping system, so urine drying with natural energy like sun drying of laundry is expected in that area. Therefore, we decided to test a new onsite volume reduction system (OVRS) for urine treatment.

MATERIAL AND METHOD

Experimental set up can be seen in wind tunnel equipped with required material as shown in Figure 4.

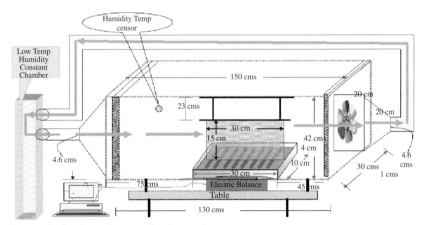

Figure 4. Onsite volume reduction unit.

Design of experiments

Basic Equation is Accumulation = q-Supply – q-Evaporation. Two types of experiments water supply rate and evaporation rate were considered for data evaluation, which are discussed in following paragraphs.

Water supply rate (WSR) experiments

Material

Wind tunnel, evaporation tank, vertical gauze sheet, electric fan (capacity 30 m³/min wind), constant temperature humidity chamber, humidity- temperature sensor, as shown in Figure 4, were used. Reason for using de-ionized water is that urine contains high concentration of salts which can create problem in assessment of water level in sheet due to salt attachment.

Method

The experiments were performed to assess how far and how fast does water move upward under capillarity in vertical sheet. This phenomenon largely depends upon character of sheet particularly pore size and type of thread of sheet.

We used Hagen-Poiseuille Equation to formulate q-Supply equation.

$$(W\ d)\ dH/dt = \sigma/\mu(H_{max}-H)/H\ (W\ d) \tag{1}$$

where W is width of vertical sheet (cm), d is depth of gauze sheet (cm), dH/dt is velocity of water molecules moving upward (cm/sec), σ is penetration factor (g/cm sec^2), μ is viscocity of air (g/cm sec), H_{max} is maximum water level (cm) and H is water level (cm).

H_{max} is capillary force, -H is gravity force for liquid, /H is pressure drop. As flow in pipes is affected by diameters and chractersitic lenght of pipes, therfore, in our case we consider this factor equivalent to sum of all diameters of small threads of cloth through which water flows.

Experimental conditions are shown in Table 3.

Table 3. Experimental conditions for experiment.

Air Humidity (%)	Air Temperature (°C)	Wind Velocity (m/s)
100%	25	0

Water Level in Sheet (H): H was measured with scale initially for every 10 minutes for first 1–2 hours, subsequently for every 3 hour.
Maximum Level of Water in Sheet (H_{max}): H_{max} was measured during 24 hours.
Penetration Factor (σ): Using measured water level, maximum water level, depth of gauze sheet (d), μ (from air temperature) and using Runge Kutta Method, we calculated σ.

Evaporation rate (ER) experiments

The purpose of these experiments is to evaluate ER and Mass Transfer Coefficient under various air conditions. Water Level in Sheet (H) was also measured for every 2 hours during evaporation. Virgin gauze sheet was used for every new experiment.

Material

Same as shown in Figure 4.

Method

Measurement of ER: Tank weight was measured every 10 minutes to evaluate ER of sample. Water level in sheet (H) was measured for every 2–3 hours. Mass Transfer Coefficient is calculated by following equation:

$$ER = M_{(Air)}Ky(Xi - X)A \tag{2}$$

where ER is evaporation rate (g/hr), Ky is Mass Transfer Coefficient (g mol/cm^{2}hr), $M_{(Air)}$ is molecular weight of dry air (g/g mol), Xi is saturated air humidity (g-water/g-dry air) which depends upon air temperature, X is humidity of supplied (g-water/g-dry air) which also depends upon climate conditions and A is effective drying area (cm^2), which is a design parameter for OVRS. A also specifies balance of q-Supply and q-Evaporation. A is calculated from sheet width (W) and water level (H).

Experimental conditions

Experimental conditions suitable for climate of Southern Pakistan as stated in Table 4 were selected. Each experiments was repeated thrice to confirm data.

Table 4. Experimental conditions for evaporation experiments.

Constant	Conditions	Variable Conditions				
Air Humidity (%)	Wind Velocity (m/s)	Air Temperature (°C)				
50	5	20	25	30	35	40
Air Temp (°C)	Wind Velocity (m/s)	Air Humidity (%)				
25	5	50	60	70	80	90

Model for design of vertical sheets for OVRS

Following fundamental equation was formulated for design of vertical sheet:
    Accumulation = q-Supply – q-Evaporation

Based on the above equation (P.M. Masoom, R. Ito, and N Funamizu, (2009) Design of onsite volume reduction system for source separated urine, submitted to Water Research), developed the following model:

$$dH/dt (d\ W) = \sigma/\mu(H_{max} - H)/H(d\ W) - M_{(air)}Ky(Xi - X)(W\ H) \tag{3}$$

Simplified form of equation (3) for steady state where q-Supply and q-Evaporation becomes equal thus $dH/dt = 0$, is as under, which can be used for calculating required H for evaporating a given volume of urine per day.

$$0 = \sigma/\mu(H_{max} - H)/H - M_{(air)}Ky(Xi - X)(H/d)   \qquad (4)$$

RESULTS AND DISCUSSION

WSR experiments

These experiments were performed to evaluate Penetration Factor (σ) and Maximum Water Level in the Sheet (H_{max}). Water Level in Sheet (H) was measured initially for every 10 minutes and subsequently for every 2–3 hours for 24 hours.

Figure 5 shows time course of water level in sheet (cm/min). This figure shows that rate of water rising initially was high owing to empty sheet during first 1–2 hours. Later speed of water movement slowed as water started accumulating and saturation was achieved during 20 hrs.

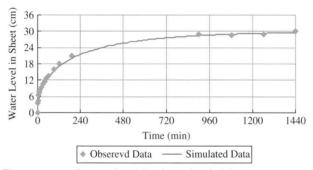

Figure 5. Time-course of water level in sheet (cm/min).

Runge Kutta Method was applied to simulate time-course of water level in sheet, based on which σ was found as 1.0×10^{-5} g/cm sec^2 and H_{max} 30 cm.

ER Experiments

These experiments were performed to assess effect of air conditions on mass transfer coefficient, as specified in the Table 4 were performed. The results are discussed in Figures 6–11.

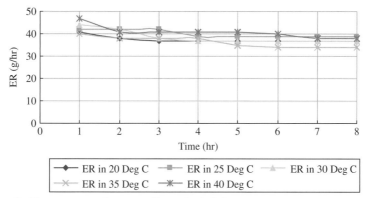

Figure 6. Time-course of evaporation rate (g/hr).

Effect of air temperature on mass transfer coefficient

Five experiments were performed for five air temperatures, as specified in Table 4. The findings are discussed in Figures 5–8.

Figures 6 indicates time-course of ER (g/hr) for air temperatures ranging from 20°C to 40°C, whereas relative air humidity (50%) and Wind Velocity (5 m/sec) were kept constant in each experiment.

Two points can be discussed with reference to this figure. ER in the beginning under all temperatures was found higher ranging between 50–40 g/hr. However, steady state of ER is observed after couple of hours. Thereafter no further decrease in ER was observed. Second point is that there is slight difference in ER while we compare all experiment data. ER was found 35–37 g/hr in steady state of all experiments, which shows no significant effect of air temperature. Reason is that narrow range of temperature was used.

Figure 7 shows time-course of water level in sheet H (cm). There are two different H in same experiment. In the beginning H is lower and subsequently H increased and attains steady state. This is because in beginning ER was high, so H was found low and in steady state ER, H increased a little and remains same. Another point is that H is slightly higher when air temperature is low as a result ER is slightly low. Small difference of H (1.9–2.2 cm), for air temperature 20–40°C is mainly because narrow range of air temperature was used.

Figure 6 indicates that there is minor difference in ER when air temperature is changed. Mostly ER remains same and steady state evaporation is observed within first 1–2 hrs. Therefore, H also changes slightly as shown in Figure 7.

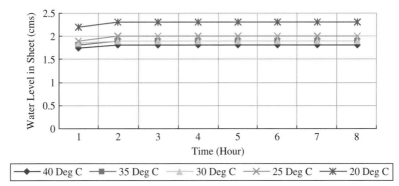

Figure 7. Time course of water level in sheet (cm).

Figure 8 shows time course of Mass Transfer Coefficient (MTC) while changing air temperature five times. This figure shows that there is minor difference in MTC which remains 1.75–2.1 g mol/cm^{2}hr. This is mainly because of narrow range of air temperature was used. However, since MTC is not same in all experiments which indicates that air temperature has some minor effect.

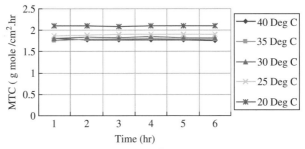

Figure 8. Time course of MTC (g mol/cm^2 hr).

Effect of air humidity on mass transfer coefficient:

Five experiments were performed by changing air humidity and keeping wind velocity and air temperature constant. Results are shown in Figures 9–11.

Figures 9 indicates time course of ER (g/hr). This figure reveals that ER was observed high in the beginning and steady state was observed after 1–2 hours and subsequently no change in ER was observed. Second point is that when air humidity was set at 50%, ER was observed as 40 g/hr which is the highest if we compare other cases. When air humidity is increased from 50% to 60%, we observed that ER decreased to 30 g/hr. This shows that dry air has capacity to evaporate much amount of water than wet air.

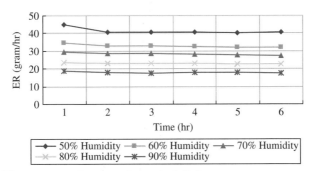

Figure 9. Time-course of evaporation rate (g/hr).

Figure 10 shows time-course of water level in sheet (cm). This figure shows that under low air humidity of 50%, H was observed lowest as 2.1 cm. Under the highest air humidity, H was observed as 3.8 cm which is the highest while comparing H in rest of experiments.

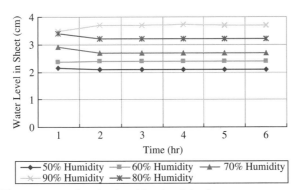

Figure 10. Time-course of water level in sheet (cm).

Another finding is that steady state H was observed after 1–2 hours when steady state ER was observed. Initially H was higher and in steady state, H was observed lower.

Figure 11 shows that value of MTC was found same (1.9 g mol/cm^2 hr) in all experiments.

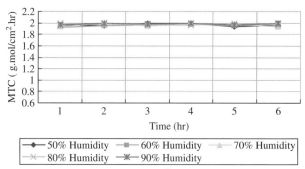

Figure 11. Time-course of MTC (g mol/cm² hr).

ESTIMATION OF VERTICAL SHEET FOR OVRS

For dry arid climate of Southern Pakistan, estimated size of vertical sheet for treating 80% volume of de-ionized water, based on 10 family members household is shown in Table 5.

Table 5. Estimated size of vertical sheet for various air temperature and air humidity conditions.

Given Data	Design Parameters			
Air Temperature (°C)	Required Dimensions of Vertical Drying Sheet			
	H (cm)	W (cm)	w = W/4 (cm)	Size (cm²)
20	2.3	342	86	787
25	2	235	59	470
30	1.9	255	64	485
35	1.9	239	60	454
40	1.8	209	52	377
Given Data	Design Parameters			
Air Humidity (°C)	Required Dimensions of Vertical Drying Sheet			
	H (cm)	W (cm)	w = W/4 (cm)	Size (cm²)
50	2	241	60	482
60	2.3	294	74	676
70	2.7	368	92	994
80	3.2	437	109	1399
90	3.5	447	112	1566

REFERENCES

Bo-Bertil Lind (2001). Volume reduction and concentration of nutrients in human urine. *Ecological Engineering*, **16**(4), 1, 561–566.

Carolina Schonning (2001). Evaluation of microbial health risks associated with the reuse of human urine. Swedish Institute of Infectious Diseases Control.

Carolina Schonning (2001). Urine diversion-hygienic risk and microbial guidelines for reuse, Department of Parasitology, Mycology and Environmental Mycology, Swedish Institute of Infectious Diseases Control.

Daniel Hellstrom, Erica Johansson and Kerstin Grennberg (1998). Storage of Human urine as a method to inhibit decomposition of urea. *Ecological Engineering*, **12**, 253–269.

Ek, M., Bergstrom, R., Bjurhem, J.-E., Bjorlenius, B. and Hellstrom, D. (2006). Concentration of nutrients from urine and reject water from an-aerobically digested sludge. *Water Science and Technology* **54**(11–12), pp. 437–444 (IWA publishing 2006).

Ellen M van Voorthuizen, Arie Zwijnenburg and Matthias Wessling (2005). Nutrient removal by NF and RO membrane in a decentralized sanitation system. *Water Research*, **39**, 3657–3667.

Food and Agriculture Organization of United Nations, Rome (2004). Fertilizer use by crops in Pakistan.

Guyton, A. (1986). Text Book of Medical Physiology. W. Saunders Co, Philadelphia, USA.

Hoglund, C., Strenstrom, T.A., Johnsson, H. and Sundin, A. (1998). Evaluation of Fecal contamination and microbial die off in urine separating sewerage System. *Water Science and technology*, **38**(6), 17–25.

Ito, R., Funamizu, N., Yokota, M. Energy analysis of composting toilets from full scale demonstration project on onsite differentiable treatment system for annual operation.

Krichmann, H. and Peterson, S. (1995). Human urine – chemical composition and fertilizer use efficiency. *Fertilizer Research*, **40**, 149–154.

Lind, B.B., Zsofia Ban and Stefan Byden (2005). Nutrient recovery from human urine by struvite crystallization with ammonia adsorption on zeolite and wollastonite. *Bioresource Technology*, **73**, 169–174.

Masoom, P.M., Funamizu, N. and Ito, R (2009). Design of onsite volume reduction system for source separated urine, being submitted to Water Research Journal.

Mats Johansson (2001). Urine separation – closing the nutrient cycle, Final Report on the R and D Project 'source separated human urine – a future source of fertilizer for agriculture in the Stockholm region.

Maurer, M., Pronk, W. and Larsen, T.A. (2006). Treatment processes for source separated urine. *Water Research*, **40**, 3151–3166.

Mourer, M., Schwegler, P. and Larsen, T. Nutrients in urine: energetic aspects of removal and recovery. EAWAG, Swiss Federal Institute of Environmental Science and Technology, Switzerland.

Naoyuki Funamizu (2002). Fractioning Gray-Water in the Differentiable Onsite Wastewater Treatment System. XXVIII congreso Interramexicano de Ingenieria Sanitariay Ambiental, Cancun, Mexico 27 al 31 de octobre.

Nasir, S.M. and Raza, S.M. (1993). Wind and Solar Energy in Pakistan. *Science Direct, Energy*, **18**(4), 397–399.

Nasir, S.M., Raza, S.M. and Yasmin Zahra Jafri (1991). Wind energy estimation at Quetta (Pakistan). *Renewable Energy*, **1**(2), 263–267.

Perry Green (1984). Perry's Chemical Engineers Handbook (McGraw Hill, 1984).

Pronk, W., Biewbow, M. and Boller, M. Assessment of processing alternatives for source separated urine. EAWAG, Swiss Federal Institute of Environmental Science and Technology, Switzerland.

Richard S. Brokaw (1960). Predicting transport properties of dilute gases. *J. Chem Phys.*, **32**(4), 1005–6, (1960) and **42**(4), 1140–6, (1965).

Tommer-Klank, L., Moller, J., Forslund, A. and Dalsgaard, A. (2006). Microbial assessment of compost toilet: in-situ measurements and laboratory studies on the survival of fecal microbial indicators using sentinel chambers. Waste Management.

Udert, K.M., Larsen, T.A and Gujer, W. Fate of Major Compounds in Source Separated Urine. EAWAG, Swiss Federal Institute of Environmental Science and Technology, Switzerland.

Warren L McCabe and Smith (1967). Unit Operations of Chemical Engineering (McGraw Hill 1967).

Wilsenach, J.A., Schuurbiers, C.A.H. and van Loosdrecht, M.C.M. (2007). Phosphate and Potassium recovery from source separated urine through struvite precipitation. *Water Research* **41**, 458–466.

World Bank (2005). Road Transport Service Efficiency Study.

Wouter Pronk, Martin Biebow and Markus Boller. Treatment of source separated urine by a combination of Bipolar Electro dialysis and gas transfer membrane. EAWAG, Swiss Federal Institute of Environmental Science and Technology, Switzerland.

Phosphorus cycling by using biomass ashes

Bettina Eichler-Loebermann and Silvia Bachmann

University of Rostock, Institute for Land Use, J. von Liebig Weg 6, D-18059 Rostock, Germany

Abstract The use of biomass as a source of energy will further increase during the next years. The reutilization of residues from the bioenergy processes in agriculture is an important issue in saving fertilizers and in realizing nutrient cycling. The phosphorus (P) fertilizing effect of biomass ashes was investigated in a pot experiment with a loamy sand for four different crops (Phacelia, Buckwheat, Ryegrass and Oil radish). As P source, poultry litter ash was compared with high-soluble mineral P (KH_2PO_4). Beside the plant P uptake different soil parameters were investigated. Soil P fractions were measured to follow the transformation process of ash-P in the soil. In general, a high P fertilizing effect of biomass ashes was found. Ash application resulted in an increase of plant P uptake and of high bio-available soil P fractions (resin-P, water-P, doublelactate-P) in comparison to the control. The degree of P saturation (DPS) increased when P was added, regardless if ash or KH_2PO_4 were given. The crops mainly influenced the readily available P fractions. Phacelia was found to increase of the highly available resin-P content. The results indicate that ashes may provide an adequate substitute for commercial P fertilizers.

INTRODUCTION

The use of biomass as a source of energy is estimated to triple during the next decade until 2020. This means that the amount of residues of the bio-energy processes will increase as well. The reutilization of these residues in crop production is an important matter to save fertilizers and to realize nutrient cycling in agriculture.

The residues of biomass combustion are the oldest mineral fertilizers in the world. They contain nearly all nutrients except nitrogen (N). This has special importance for phosphorus (P) since the P resources worldwide are limited and the prices for commercial P fertilizers continue to increase. Generally, biomass ashes have been evaluated positively regarding their effect on dry matter yield of crop plants (Mozaffari et al., 2002; Patterson et al., 2004). Concerning the effect of biomass ashes on plant P nutrition and plant available P in soil, results are indifferent. Codling et al. (2002) found a positive effect of poultry litter ash on plant P uptake and the Mehlich-3 extractable P of the soil. On the other hand, little or no effect of ash application on P uptake and plant available P has also been reported by others (Mozaffari et al., 2000; Mozaffari et al., 2002; Ohno and Erich, 1990). In a field trial a negative effect of wood ash on the plant available P in soil and the P supply of spruces (Picea abies) was found by Clarholm (1994). Besides being a source of nutrients itself, the application of biomass ashes may influence the form and availability of P, for example by increasing the pH of the soil (Ohno and Erich, 1990; Clapham and Zibilske, 1992; Muse and Mitchel, 1995).

So far, standard methods of soil examination have been used to characterize the effect of biomass ashes on soil P. But most of them concentrate on the estimation of readily bioavailable P, whereas the major part of the P compounds in the soil remains unidentified (Dalal, 1977). With the P fractionation method developed by Hedley et al. (1982) the effect of fertilization on soil P pools of lower availability can also be investigated. This contributes to a better understanding of the fate and the pathways of added ash P in the soil. Crop species may also influence the P fractions of the soil. The exudation of ions, organic acids or enzymes into the rhizosphere enables crops to aquire P of less available fractions. Especially catch crops are said to have access to strongly fixed P and their cultivation may also increase the P bioavailability (Horst et al., 2001; Bünemann et al., 2004; Eichler et al., 2008).

The aim of this study was to investigate the effect of biomass ashes on plant P nutrition in combination with catch crop cultivation. Next to standard soil tests, special attention was paid to the soil P-fractions.

MATERIALS AND METHODS

Four ashes remained from combustion processes of different biomasses (poultry litter, rape meal, cereal whole plants, and wood) were analyzed for the P contents with different extracting agents; water, double lactate, citric acid, and aqua regia.

Furthermore, a pot experiment with poultry litter ash was conducted. Different crops were cultivated in 6 kg soil in Mitscherlich pots for 2 month (see Table 1). Before sowing nutrients and the ash were given to the soil. For control no P was added. For the high soluble P treatment KH_2PO_4 was supplied and for the ash treatment P was added with poultry litter ash. This ash remained from combustion process in a special SFBC combustion plant (Small scale fluidized bubbling bed Combustion Plants). Water was supplied according the plant needs using deionized water.

Table 1. Tested factors in the pot experiment.

Factor A – Crop species	Factor B – P supply
a1 Phacelia	b1 Nutrient solution without P
(*Phacelia tanacetifolia*)	b2 Nutrient solution with P (KH_2PO_4),
a2 Buckwheat	0.23 g P per pot
(*Fagopyrum escultentum*)	b3 Nutrient solution poultry litter ash,
a3 Common ryegrass	0.4 g P per pot
(*Lolium multiflorum*)	
a4 Oil raddish	
(*Raphnus sativus oleiformis*)	

The P content in plant tissue was measured after dry ashing using the vanadate-molybdate method (Page *et al.*, 1982). For water extractable P (Pw) the method of Van der Paauw (1971) was used. P concentration in the extract was determined by the phosphomolybdate blue method applied to flow injection analysis with a Tecator FIAstar 5010 colorimeter. The double lactate extractable P (Pdl) content as well as pH ($CaCl_2$) were determined by routine methods as described by Blume *et al.* (2000). The oxalate soluble content of P, Al and Fe in soil (Pox, Alox, Feox) were analysed as described by Schwertmann (1964). Elemental concentrations were determined by inductively coupled plasma (ICP) spectrometry. After that the P sorption capacity (PSC [mmol kg^{-1}] = (Alox + Feox)/2) and the degree of P saturation (DPS [%] = Pox/PSC∗100) were calculated (Schoumans, 2000).

To determine the soil P fractions, the method described by Hedley *et al.* (1982) was used. The air dried soil (0.5 g; <0.01 mm) were at first extracted with an anion exchange resin (Anion-exchange membrane, 125∗125 mm strips, VWR International Ltd, Poole, BH151TD England) by shaking with 30 ml of deionized water for 18 hours. P sorbed on the resin-strip (Resin-P) is regarded to be the most biologically available. Subsequently, the soil residue was extracted with 30 ml of 0.5 M $NaHCO_3$ by shaking for 18 hours, centrifugation and decantation of the supernatant. These extraction steps were repeated with stronger extracting agents 0.1 M NaOH and 1 M H_2SO_4. P still remaining in the soil after the extraction steps is considered as residual-P. The residual-P content was determined by subtracting the amount of extracted P from total P (Pt) content. Total P was determined by aqua regia digestion in a Microwave Oven (MDS-2000, CEM- GmbH, Kamp-Lintfort, Germany). P concentrations in the supernatants were determined via ICP spectrometry. After the analyses of variance the least significant difference (LSD) test at p < 0.05 was used to compare means of soil and plant parameters (ANOVA).

RESULTS AND DISCUSSION

Nutrient content and P availability in different biomass ashes

The highest P content was found for cereal ash with more than 13% total P (aqua regia). For this ash even the water soluble P content was found to be high with more than 1% (Table 2). For all ashes a very high percentage of the total P was found to be soluble also in citric acid which indicates a relatively good availability to plants. Wood ash only had low contents of P compared to the other testes ashes.

Besides nutrients, secondary raw materials like biomass ashes may also contain harmful substances, mainly heavy metals. Those elements in waste materials have to be examined also to avoid environmental pollutions due to agricultural use of biomass ashes. For all ashes content of heavy metals was found to be below threshold values (data not shown).

Table 2. P contents in different types of ash in dependence of the extracting agent (mg 100 g^{-1}).

Ashes	P water	P double lactate	P citric acid	P aqua regia
poultry litter	0.05	1550	3600	3990
rape meal	0.12	4500	8259	9070
cereal	1284	5200	11851	13476
wood	0.84	22.0	123	128

Effect of ash application on dry matter yield and P uptake of tested catch crops

For further investigation we used the poultry litter ash in a pot experiment. The P supply affected the yield and P uptake of crops positively; independent if the P source was given as KH_2PO_4 or poultry litter ash (see Table 3). For Phacelia and Buckwheat no significant effect of fertilization on dry matter yield could be detected. In comparison to that, ryegrass and oil radish reacted with an increase of dry matter yield of about 17 to 27% on ash application compared to the control without P.

The different P supply became more obvious for plant P uptake (see Table 3). For each crop a significant increase of P uptake could be found after P application. A positive effect of poultry litter ash on plant growth was also found by Codling et al. (2002). On average, no differences were found between the ash treatment and the treatment with KH_2PO_4. But for oil radish the P uptake was highest when ash was added.

Table 3. Yields (g pot^{-1} DM) and P uptake (mg pot^{-1}) (shoot) in dependence of fertilizer treatment in a pot experiment (6 kg of P poor loamy sand, 5 weeks vegetation time).

	Treatment	Phacelia	Buckwheat	Ryegrass	Oil radish
	without P	13.8	16.5	11.4 a	18.8 a
Yield	KH$_2$PO$_4$	14.9	16.2	13.0 b	23.1 b
	ash	16.3	16.1	13.3 b	23.8 b
	without P	56.0 a	51.6 a	42.0 a	86.1 a
P uptake	KH$_2$PO$_4$	68.7 ab	64.7 b	63.7 b	129.4 b
	ash	87.1 b	68.7 b	60.8 b	149.1 c

Different letters label significant differences between the fertilization treatments (ANOVA, $\alpha = 0.05$).

Soil pH and soil P parameters as affected by ash application

With ash supply the average soil pH increased from 5.5 to 6.3 (see Table 4). Because of their high Ca content, the pH value of biomass ashes usually is between 10 and 13 (Obernberger, 1997; Patterson *et al.*, 2004). Hence, biomass ashes are proved to be an effective liming agent (Clapham and Zibilske, 1992; Muse and Mitchel, 1995). Besides the nutrient supply this liming effect can play an important role for increasing the plant available P, especially in acid soils. For all fertilizing treatments the cultivation of Phacelia resulted in an pH decrease. This may be induced by the high cation:anion ratio of the nutrient uptake, which was detected for Phacelia in former experiments (Eichler and Schnug, 2006).

Table 4. Soil pH, Pdl content in soil (mg kg $^{-1}$) and Pw content in soil (mg kg $^{-1}$) in dependence of fertilizer treatment for 4 different crops in a pot experiment (6 kg of P poor loamy sand, 5 weeks vegetation time).

	Treatment	Phacelia	Buckwheat	Ryegrass	Oil radish
	without P	5.1 a	5.7 a	5.8 a	5.7 a
pH	KH$_2$PO$_4$	5.2 a	5.7 a	5.7 a	5.8 a
	ash	6.3 b	6.5 b	6.5 b	6.6 b
	without P	35 a	36 a	38 a	33 a
P(dl)	KH$_2$PO$_4$	44 b	50 b	53 b	46 b
	ash	64 c	74 c	74 c	56 c
	without P	3.2 a	2.8 a	2.9 a	2.4 a
P(w)	KH$_2$PO$_4$	5.1 b	5.4 c	5.2 b	5.1 c
	ash	4.9 b	4.8 b	4.9 b	3.8 b

Different letters label significant differences between the fertilization treatments (ANOVA, $\alpha = 0.05$).

The fertilization had a significant effect on P parameters of the soil. The plant available P contents in soil increased when P was added as highly soluble P, as well as when added with ash. The P content soluble in double lactate (Pdl), which is considered as plant available P, was found to be highest in the ash treatment (see Table 4). In average in this treatment the Pdl content was with 67 mg kg^{-1} soil nearly double as high as in the treatment without P. Especially for Buckwheat and Grass high values were found. This might be due to mobilization processes or due to the lower soil P exhaustion because of the lower P uptakes of these crops. The P water (Pw) content in soil was found to be lower in the ash treatment than in the treatment with KH_2PO_4. However, in another experiment with different biomass ashes (wood, rape meal, cereal whole plant), high Pw contents were found when the ashes were supplied (data not shown).

For the P sorption capacity (PSC) and degree of P saturation (DPS) no significant differences were found. However, in average of all crop plants the DPS was slightly but significant higher when P was added (as KH_2PO_4 or ash). In average of all crops PSC was 32.4 for the control, 32.8 for KH_2PO_4 treatment and 32.6 for the ash treatment (all values in mmol/kg soil). The DPS was 38.5% for the control, 42.8% for the KH_2PO_4 treatment and 42.1% for the ash treatment.

The P fractionation of the soil showed, that the P fractions which are considered to be easily available for plants (Resin-P and $NaHCO_3$-P) amounted only 17% of total P in the soil (Figure 1). Most of the P was found in the moderate to hardly available P fractions (NaOH-P, H_2SO_4-P, Residual-P) which altogether amounted 83% of the total P. The highest percentage of P (41%) was extracted with NaOH. The NaOH-P fraction is regarded as moderate available P mainly sorbed to Al- and Fe-Oxides and humic acids or precipitated as Al- and Fe-Phosphates. This fraction usually is the biggest in highly weathered soils (Cross and Schlesinger, 1995).

Fertilization and catch crop cultivation mainly influenced the easily and moderate available P fractions. Like Pdl and Pw, the Resin-P fraction, which is considered as the most biologically available P (Hedley et al., 1982), became increased with fertilization, no matter if P was supplied as high soluble mineral fertilizer or ash. So, our results reflect a high availability of P supplied by ashes and are in consistence with the high P uptake of the plants from the ash amended soil.

The $NaHCO_3$ extractable P is regarded as P, labile sorbed to Al and Fe-Oxides or clay minerals and microbial P, still having a high availability to plants in short term (Hedley et al., 1982; Schlichting et al., 2002). In average of all crops, the $NaHCO_3$-P- as well as the NaOH-P content increased significantly when KH_2PO_4 was added to the soil. For ash lower values of these fractions

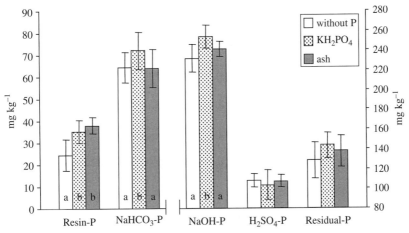

Figure 1. Soil P fractions as affected by application of poultry litter ash in a 6 weeks pot experiment on a loamy sand (average of all tested crops) (Duncan test, $\alpha = 0.05$).

were found. This is mainly due to the high depletion of both fractions by oil radish in the ash treatment (data not shown). According to its high P uptake, probably oil radish took the major part of its P uptake in the control and ash amended soil from this pool. This is also in accordance with findings of Zhang *et al.* (1997) who found a high ability of oil radish to aquire P sorbed to Al- and Fe-compounds.

The H_2SO_4 extractable P is mainly present as Ca bound P (Ca-P) in the soil. Like the Residual-P, H_2SO_4-P is not available for plants in short term. It was was not affected by fertilizer amendment nor by crop cultivation. Regarding the relatively high amounts of Ca in biomass ashes and the high amount of P being present as Ca-P (Obernberger, 1997; Codling, 2006), an enrichment of ash P in the H_2SO_4-P fraction could be expected. This result indicate either a high solubility of P in the tested poultry litter ash or its rapid transformation into other P forms making it better available.

CONCLUSIONS

The use of secondary raw materials (such as biomass ashes) in agriculture provides an opportunity for waste recycling and an adequate substitute for commercial P fertilizers. Provided the ashes are not loaded with harmful substances, the usage of biomass ashes in crop husbandry is an important method for the recirculation of nutrients in agriculture and saving of nutrient resources. The high plant P uptake and the high P content in soil in the ash treatment showed, that ashes may have a high P availability to plants. Regarding the limited P resources in the word and the increasing prices for P fertilizers the use of residues for fertilization should be promoted.

REFERENCES

Blume, H.P., Deller, B., Leschber, R., Paetz, A., Schmidt, S. and Wilke, B.M. (2000). Handbuch der Bodenuntersuchung. Terminologie, Verfahrensvorschriften und Datenblätter. Physikalische, chemische und biologische Untersuchungsverfahren. Gesetzliche Regelwerke, Wiley-VCH, Weinheim.

Bünemann, E., Smithson, P.C., Jama, B., Frossard, E. and Oberson, A. (2004). Maize productivity and nutrient dynamics in maize-fallow rotations in western Kenya. *Plant Soil*, **264**, 195–208.

Clapham, W.M. and Zibilske, L.M. (1992). Wood ash as liming amendment. *Commun. Soil Sci. Plant Anal.*, **23**, 1209–1227.

Codling, E.E. (2006). Laboratory Characterization of extrable phosphorus in poultry litter and poultry litter ash. *Soil Sci.*, **11**, 858–864.

Codling, E.E., Chaney, R.L. and Sherwell, J. (2002). Poultry litter ash as a potential phosphorus source for agricultural crops. *J. Environ. Qual.*, **31**, 954–961.

Cross, A.F. and Schlesinger, W.H. (1995). A literature review and evaluation of the Hedley fractionation. Applications to the biochemical cycle of soil phosphorus in natural ecosystems. *Geoderma*, **64**, 197–214.

Dalal, R.C. (1997). Soil organic phosphorus. *Adv. Agron.*, **29**, 83–113.

Eichler-Löbermann, B., Köhne, S., Kowalski, B. and Schnug, E. (2008). Effect of catch cropping on phosphorus bioavailability in comparison to organic and inorganic fertilization. *J. Plant Nutr. Soil Sci.*, **31**, 659–676.

Eichler-Löbermann, B. and Schnug, E. (2006). Crop plants and the availability of phosphorus in soil. In: *Encyclopedia of Soil Science*, R. Lal (ed.), Marcel Dekker, pp. 348–350.

Hedley, J.M., Stewart, J.W.B. and Chauhan, B.S. (1982). Changes in inorganic and organic soil phosphorus fractions by cultivation practices and by laboratory incubations. *Soil Sci. Soc. Am. J.*, **46**, 970–976.

Horst, W.J., Manh, M., Jibrin, J.M. and Chude, V.O. (2001). Agronomic measures for increasing P availability to crops. *Plant Soil*, **237**, 211–223.

Mozaffari, M., Rosen, C.J., Russelle, M.P. and Nater, E.A. (2000). Corn and soil response to application of ash generated from gasified alfalfa stems. *Soil Sci. Soc. Am. J.*, **165**, 896–907.

Mozaffari, M., Russelle, M.P., Rosen, C.J. and Nater, E.A. (2002). Nutrient supply and neutralizing value of alfalfa stem gasification ssh. *Soil Sci. Soc. Am. J.*, **66**, 171–178.

Muse, J.K. and Mitchel, C.C. (1995). Paper mill boiler ash an d lime by-products as soil liming materials. *Agron. J.*, **87**, 432–438.

Obernberger, I. (1997). Aschen aus Biomassefeuerungen. Zusammensetzung und Verwertung: Thermische Biomassenutzung. Technik und Realisierung ; Tagung

Salzburg, 23. und 24. April 1997. Düsseldorf: VDI-Verl. (VDI-Berichte, 1319), 199–222.

Ohno, T. and Erich, M.S. (1990). Effect of wood ash on soil pH and soil test nutrient levels. *Agric. Ecosyst. Environ.*, **32**, 223–239.

Page, A.L., Miller, R.H. and Keeney, D.R. (1982). Methods of soil analysis. 2. *Chemical and Microbial Properties*. 2nd ed. Madison, USA: American Society of Agronomy.

Patterson, S.J., Acharya, S.N., Thomas, J.E., Bertschi, A.B. and Rothwell, R.L. (2004). Barley biomass and grain yield and canola seed yield response to land application of wood *Ash. Agron. J.*, **96**, 971–977.

Schlichting, A., Leinweber, P., Meissner, R. and Altermann, M. (2002). Sequentially extracted phosphorus fractions in peat-derived soils. *Journal of Plant Nutrition and Soil Science*, **162**, 290–298.

Schoumans, O.F. (2000). Determination of the degree of phosphate saturation in non-calcareous soils. In: *Methods for Phosphorus Analysis for Soils, Sediments, Residuals, and Water,* G.M. Pierzynski (ed.), Southern Cooperative Series Bulletin 396, North Carolina, pp. 31–34.

Zhang, F.S., Ma, J. and Cao, Y. P. (1997). Phosphorus deficiency enhances root exudation of low-molecular weight organic acids and utilization of sparingly soluble inorganic phosphates by radish (Raghanus sativus L.) and rape (Brassica napus L.) plants. *Plant Soil*, **196**, 261–264.

Phosphorus recovery from high-phosphorus containing excess sludge in an anaerobic-oxic-anoxic process by using the combination of ozonation and phosphorus adsorbent

T. Kondo[a]*, Y. Ebie[a], S. Tsuneda[b], Y. Inamori[c] and K-Q. Xu[a]

[a] Research Center for Material Cycles and Waste Management, National Institute for Environmental Studies, 16-2 Onogawa, Tsukuba-shi, Ibaraki 305-8506, Japan
[b] Department of Life Science and Medical Bio-Science, Waseda University, 2-2 Wakamatsucho, Shinjuku-ku, Tokyo 162-8430, Japan
[c] Faculty of Symbiotic Systems Science, Fukushima University, 1 Kanayagawa, Fukushima-shi, Fukushima 960-1296, Japan
*(Corresponding author) (E-mail: kondo.takashi@nies.go.jp)

Abstract To meet increasingly stringent requirements for environmental protection and phosphorus resource recovery, we propose a continuous A/O/A (anaerobic/oxic/anoxic) process combined with sludge reduction process by microbubble ozonation and phosphorus adsorption process by a zirconium-ferrite adsorbent. Under various ozonation conditions, nutrient removal efficiency and sludge reduction efficiency were evaluated. When the amount of sludge reduction was high (over 9.4% of total MLSS per day), effective sludge reduction was achieved while nitrification efficiency was deteriorated. The decrease in the sludge amount for ozonation (9.4% of total MLSS per day) resulted in efficient nitrogen removal. Under this condition, no excess sludge was withdrawn during at least 2 months operation while MLSS concentration was gradually increased. Phosphorus concentration profile indicated that phosphorus was removed not only by oxygen utilizing PAOs (polyphosphate-accumulating organisms) but nitrate/nitrate utilizing PAOs, which contributing to the reduction of sludge production efficiency. Phosphorus accumulated in excess sludge was effectively solubilized by microbubble ozonation and a large part of the solubilized phosphorus consisted of PO_4-P. Then, over 90% of PO_4-P was recovered in the phosphorus adsorption column.

INTRODUCTION

In wastewater treatment plants (WWTPs), treatment and disposal of excess sludge have been serious problems, and the treatment of excess sludge may account for from 25% to 65% of total plant operating costs (Liu, 2003). One of the methods to reduce excess sludge is the solubilization of excess sludge by using ozone which is a strong oxidant, and recirculation of the solubilized excess sludge, containing readily biodegradable carbon, into a biological treatment system (Chu *et al.*, 2008; Cui and Jahng, 2004; Kamiya and Hirotsuji, 1998;

Sakai *et al.*, 1997; Yasui *et al.*, 1996; Yasui and Shibata, 1994). Recently, Chu *et al.* (2008) reported that microbubble ozonation process, which generate bubbles with diameters less than several tens of micrometers, provided effective and low cost operation for sludge reduction.

It is also effective for the excess sludge reduction to control the sludge production potential in biological processes. In biological phosphorus removal process, less sludge production was confirmed by using denitrifying polyphosphate-accumulating organisms (DNPAOs), which are capable of utilizing nitrate/nitrite as electron acceptors unlike PAOs which utilize only oxygen as electron acceptor (Ahn *et al.*, 2002a, 2002b; Kuba *et al.*, 1996, 1997; Soejima *et al.*, 2006; Tsuneda *et al.*, 2006). Previously, Tsuneda *et al.* (2006) described an anaerobic/oxic/anoxic (A/O/A) process and succeeded in causing DNPAOs to take an active part in simultaneous nitrogen and phosphorus removal in an acetate-fed sequencing batch reactor (SBR) while the additional acetate was required for the inhibition of phosphorus uptake under an oxic period.

From these backgrounds, we previously proposed an advanced system, the continuous A/O/A process combined with sludge reduction process by ozonation (Suzuki *et al.*, 2006). In this system, excess sludge containing high amount of phosphorus is ozonated and the supernatant passes through the phosphorus adsorption column to recover the phosphorus. The phosphorus is absorbed by zirconium-ferrite, and then, both the effluent of phosphorus adsorption column and the residual solid after ozonation were transferred to the oxic tank to inhibit the oxic phosphorus accumulation. In this experiment, effective nutrient removal with no sludge production was achieved while TOC removal efficiency was deteriorated due to the circulation of ozonated sludge. The inhibition of oxic phosphorus uptake was also failed, and thus most phosphorus was accumulated by only PAOs.

In this study, to achieve the efficient nutrients removal and to induce DNPAOs to take a large part in phosphorus accumulation, the previous system was improved by introducing the microbubble ozonation system and by changing the recirculation lines of ozonated sludge. During 152 days operation, operational conditions were optimized by changing ozone concentration, ozone dosage, and return ratio of the supernatant of the physicochemical process (ozonation and phosphorus adsorption) to the anaerobic and oxic tanks.

MATERIALS AND METHODS

Reactor design and operation

A bench-scale continuous A/O/A process with working volume of 36 L (an anaerobic tank, 10.3 L; an oxic tank, 10.3 L; an anoxic tank, 15.4 L), the same as previous study (Suzuki *et al.*, 2006), was operated for 152 days. The microbubble ozonation system and the phosphorus adsorption column were introduced to the A/O/A process (Figure 1).

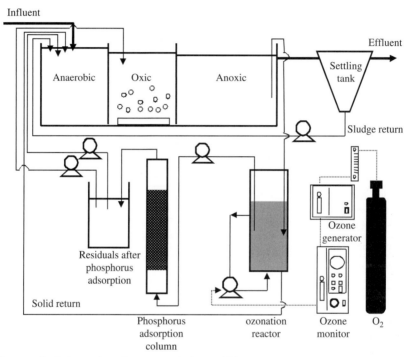

Figure 1. Schematic diagram of the A/O/A process with ozonation and phosphorus adsorption.

The microbubble ozonation system was a cylindrical reactor with an internal diameter of 0.2 m and a height of 0.8 m (effective volume of 20 L). The ozonation system received excess sludge continuously collected from the end of an anoxic tank daily. When the amount of excess sludge reached approximately 10 L (once in 1 to 3 days), sludge ozonation was conducted under specified ozonation conditions as shown in Table 1. Ozone gas was generated by an ozone generator (PO-10; Fuji Electric, Japan) and the applied ozone dose was monitored by a UV Ozone Monitor (PG-620HA; Ebara Jitsugyo, Japan). Then, ozone gas and excess sludge from the anoxic tank were mixed in turbulent flow by a current pump under high pressure (Nikuni swirling current pump M15NPD02S; Nikuni Co., Japan), and then the mixture was pumped out to the reactor. After ozonation, the supernatant with unsettled solids was flowed into the phosphorus adsorption column. The phosphorus adsorption column (internal diameter 75 mm; height 0.7 m; effective volume 2 L) was filled with 1.5 L of spherical zirconium-ferrite ($ZrFe_2(OH)_8$) adsorbent with an effective diameter of 0.7 mm (Japan Enviro Chemicals, Japan). Flow rate of the supernatant was set at 25 mL/min. After phosphorus adsorption, the adsorbent was backwashed with 15 L of tap water to remove residual SS. Then, the mixture of the effluent of phosphorus adsorption column and the backwash water were recirculated to the anaerobic tank and the oxic tank at an appropriate rate for around 17 h (Table 1).

Table 1. Operational conditions.

		Phase			
		1	2	3	4
Operating time [day]		0–22	23–54	55–83	84–152
Sludge amount for ozonation [L/day] (% of total MLSS in the reactoer)		8.4 (23)	5.6 (16)		3.4 (9.4)
Ozone concentration [mg/L]		40			
Ozone dose [mg-O_3/mg-MLSS]		20			30
Flow rate of the supernatant derived from the physicochemical process					
	Anaerobic tank [mL/min]	12.5		15	
	Oxic tank [mL/min]	12.5		10	
Flow rate of influent wastewater [mL/min]		60			
Flow rate of sludge return from the settling tank to the anaerobic tank [mL/min]		48			

The hydraulic retention time (HRT) of influent wastewater was adjusted to 10 h. Sludge return ratio from the settling tank to the anaerobic tank was controlled at 80%. The reactor was inoculated with activated sludge (initial MLSS was adjusted to 4,000 mg/L), which was collected from a WWTP (A/A/O process) with efficient biological phosphorus removal capacity. Raw rural wastewater, which collected from a rural sewage treatment plant daily, was flowed into the system. The characteristics of the influent wastewater were follows: 160–200 mg/L of SS, 55–80 mg/L of TOC, 45–55 mg/L of T-N and 4.0–5.5 mg/L of T-P. In this study, 4 different operational conditions were conducted to investigate the appropriate operational conditions (Table 1).

Analytical procedures

MLSS was measured following the standard methods (APHA 1995). To determine soluble TOC (S-TOC), NH_4-N, NO_{2+3}-N, NO_2-N, and PO_4-P, water samples were filtered using a glass-fiber filter (GF/C, Whatman Japan KK, Japan). TOC and S-TOC were measured using a SHIMADZU TOC-VSCH (Shimadzu, Japan). Total nitrogen (T-N), total phosphorus (T-P), soluble T-N (ST-N), soluble T-P (ST-P), NH_4-N, NO_{2+3}-N, NO_2-N, and PO_4-P were measured using a TRAACS 2000 (Bran+Luebbe, Japan).

RESULTS AND DISCUSSION

Reactor operation

The A/O/A (anaerobic/oxic/anoxic) process combined with microbubble ozonation system and phosphorus adsorption column was operated for 152 days. Figures 2 and 3 show effluent water quality and MLSS concentration in the A/O/A process, respectively. Effluent water quality and sludge production were affected by the operational conditions of ozonation (Figures 2 and 3). In Phase 1, MLSS in the A/O/A process dramatically decreased which indicating over sludge reduction. On day 23, the sludge amount for ozonation was reduced from 23% to 16% of total MLSS (Phase 2), and as a result, MLSS concentration was maintained around 3,000 mg/L. Although adequate control for sludge reduction was achieved, effluent T-N concentration increased to approximately 20 mg/L. Effluent T-N was mostly composed of NH_4-N, indicating that nitrification performance was deteriorated. It is well known that organic carbon induces the dissolved oxygen competition between nitrifying bacteria and heterotrophic bacteria, and thus it might results in the deterioration of nitrification performance. Then, to reduce the organic carbon loading to the oxic tank, the return ratio of the supernatant of the ozonated sludge was changed (Phase 3). However, nitrification performance was not improved. In Phase 4, to maintain the population density of slow-growing nitrifying bacteria, the sludge amount for ozonation was reduced by 3.4 L/day (9.4 % of total MLSS) (Table 1). As a result, efficient nitrogen removal was achieved over 2 months with the slight increase of MLSS concentration (Figure 3). Consequently, 87%, 76%, and 81% of TOC, T-N, and T-P in the influent wastewater were removed in Phase 4, respectively.

In previous study (Suzuki et al., 2006), effluent S-TOC was deteriorated by slowly biodegradable materials derived from ozonated sludge (Suzuki et al., 2006). In this study, this deterioration was not observed. This different indicates that some of the biorefractories and/or slowly biodegradable materials were

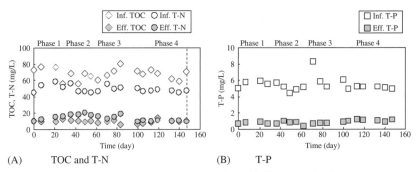

Figure 2. Time course of water quality profile of influent and effluent.

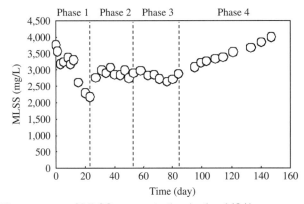

Figure 3. Time course of MLSS concentration in the A/O/A porcess.

oxidized to easily biodegradable materials by microbubble ozonation and/or these slowly biodegradable materials were degraded in the anaerobic tank.

Figure 4 shows the water quality profile in Phase 4 (day 147). In the anaerobic tank, phosphorus release by PAOs and DNPAOs was confirmed. Then, released phosphorus was then accumulated in sludge by PAOs and DNPAOs. Different from previous study (Suzuki *et al.*, 2006), phosphorus was not completely removed in the oxic tank. It suggested that oxic phosphorus uptake was inhibited and the remaining phosphorus was accumulated by DNPAOs in the anoxic tank.

Phosphorus recovery from excess sludge

Phosphorus concentrations in the influent and effluent of the phosphorus adsorption column are shown in Figure 5. During the operation, about 70% of

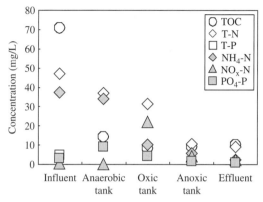

Figure 4. Water quality profile at day 147 (Phase 4).

the phosphorus in the ozonated sludge was solubilized by ozonation, and a large part of the solubilized phosphorus consisted of PO$_4$-P. Over 90% of the solubilized phosphorus was absorbed and effluent PO$_4$-P concentration was maintained at less than 1 mg/L until day 119 (Figure 5). After 119 days operation, effluent PO$_4$-P concentration reached 1 mg/L (Figure 5).

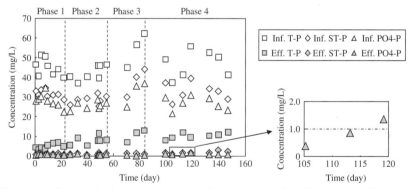

Figure 5. Concentration of each type of phosphorus in influent and effluent of phosphorus adsorption column. Right small figure shows the breakthrough point (PO4-P concentration: 1 mg/L) of the adsorbent at day 119.

Feasibility of the A/O/A system

As the amount of excess sludge increases, regulations for disposal have become increasingly stringent (Liu, 2003). Many attempts have been made to reduce excess sludge production. In this study, an ozonation system was added to the A/O/A process and no excess sludge was withdrawn during 152 days operation. Adequate sludge reduction was achieved in Phases 2 and 3; however, nitrification efficiency was deteriorated. Consequently, the reduction in the amount of ozonated sludge improved the nitrification efficiency in Phase 4 (9.4% of total MLSS in the reactor). Ozonated sludge would also act as a nutrient loading in the A/O/A process. Cui and Jahng (2004) reported that energy sources originating from solubilized sludge were not enough to reduce ammonia derived from solubilised sludge itself. In this study, although both nitrogen and organic carbon in the residuals were not analyzed, the deterioration of nutrient removal efficiency by this recirculation was not confirmed (Figure 4).

Reduction of sludge production potential is also important for excess sludge reduction. Kuba *et al.* (1996) reported that sludge production efficiency could be decreased by using DNPAOs. In this study, phosphorus was removed both in the oxic tank and anoxic tank. Therefore, DNPAOs contributed to both the nutrient removal and the sludge reduction. Meanwhile, sludge production potential should increase in the oxic tank by heterotrophic bacterial growth using the residuals of the physicochemical processes.

In this study, although slight increase in MLSS concentration was confirmed, both efficient nutrient removal and no excess sludge withdrawing was achieved during 2 months operation (Phase 4). Further studies are necessary to determine the sludge production potential. Changing ozonation conditions may be one of the most important parameters affecting nutrient removal because the solubilized sludge contains both readily- and slowly-biodegradable materials.

CONCLUSIONS

In order to meet increasingly stringent requirements for environmental protection and phosphorus resource recovery, an A/O/A process equipped with ozonation and phosphorus adsorption was tested under various ozonation conditions. Our main conclusions are as follows.

1. Under the optimum ozonation conditions (9.4% of total MLSS), both efficient nutrient removal and no excess sludge withdrawing was achieved during 2 months operation.
2. Phosphorus was not completely removed in the oxic tank different from previous study, suggesting that oxic phosphorus uptake was inhibited and the remaining phosphorus was accumulated by DNPAOs.
3. Most of phosphorus in excess sludge was solubilized by ozonation and a large part of solubilized phosphorus consisted of PO_4-P. Over 90% of PO_4-P was absorbed in the phosphorus adsorption column.

REFERENCES

Ahn, J., Daidou, T., Tsuneda, S. and Hirata, A. (2002a). Characterization of denitrifying phosphate-accumulating organisms cultivated under different electron acceptor conditions using polymerase chain reaction-denaturing gradient gel electrophoresis assay. *Water Res.*, **36**(2), 403–412.

Ahn, J., Daidou, T., Tsuneda, S. and Hirata, A. (2002b). Transformation of phosphorus and relevant intracellular compounds by a phosphorus-accumulating enrichment culture in the presence of both the electron acceptor and electron donor. *Biotechnol. Bioeng.*, **79**(1), 83–93.

APHA (1995). Standard Methods for the Examination of Water and Wastewater (19th ed), American Public Health Association, Washington, DC.

Chu, L.B., Yan, S.T., Xing, X.H., Yu, A.F., Sun, X.L. and Jurcik, B. (2008). Enhanced sludge solubilization by microbubble ozonation. *Chemosphere*, **72**(2), 205–212.

Cui, R. and Jahng, D. (2004). Nitrogen control in AO process with recirculation of solubilized excess sludge. *Water Res.*, **38**(5), 1159–1172.

Kamiya, T. and Hirotsuji, J. (1998). New combined system of biological process and intermittent ozonation for advanced wastewater treatment. *Water Sci. Technol.*, **38**(8–9), 145–153.

Kuba, T., van Loosdrecht, M.C.M. and Heijnen, J.J. (1996). Phosphorus and nitrogen removal with minimal cod requirement by integration of denitrifying dephosphatation and nitrification in a two-sludge system. *Water Res.*, **30**(7), 1702–1710.

Kuba, T., van Loosdrecht, M.C.M., Brandse, F.A. and Heijnen, J.J. (1997). Occurrence of denitrifying phosphorus removing bacteria in modified UCT-type wastewater treatment plants. *Water Res.*, **31**(4), 777–786.

Liu, Y. (2003). Chemically reduced excess sludge production in the activated sludge process. *Chemosphere*, **50**(1), 1–7.

Sakai, Y., Fukase, T., Yasui, H. and Shibata, M. (1997). An activated sludge process without excess sludge production. *Water Sci. Technol.*, **36**(11), 163–170.

Soejima, K., Oki, K., Terada, A., Tsuneda, S. and Hirata, A. (2006). Effects of acetate and nitrite addition on fraction of denitrifying phosphate-accumulating organisms and nutrient removal efficiency in anaerobic/aerobic/anoxic process. *Biopro. Biosys. Eng.*, **29**(5–6), 305–313.

Suzuki, Y., Kondo, T., Nakagawa, K., Tsuneda, S., Hirata, A., Shimizu, Y. and Inamori, Y. (2006). Evaluation of sludge reduction and phosphorus recovery efficiencies in a new advanced wastewater treatment system using denitrifying polyphosphate accumulating organisms. *Water Sci. Technol.*, **53**(6), 107–113.

Tsuneda, S., Ohno, T., Soejima, K. and Hirata, A. (2006). Simultaneous nitrogen and phosphorus removal using denitrifying phosphate-accumulating organisms in a sequencing batch reactor. *Biochem. Eng. J.*, **27**(3), 191–196.

Yasui, H. and Shibata, M. (1994). An innovative approach to reduce excess sludge production in the activated sludge process. *Water Sci. Technol.*, **30**(9), 11–20.

Yasui, H., Nakamura, K., Sakuma, S., Iwasaki, M. and Sakai, Y. (1996). A full-scale operation of a novel activated sludge process without excess sludge production. *Water Sci. Technol.*, **34**(3–4), 395–404.

Struvite control techniques in an enhanced biological phosphorus removal plant

R. Baur

Clean Water Services, Tigard, OR, USA

Abstract The Durham advanced Wastewater Treatment plants permit is 0.1 mg/l T-PO$_4$-P monthly median. To reduce chemical and sludge disposal costs the plant was converted to Enhanced Biological Phosphorus Removal process. Centrate storage was used to reduce the diurnal swings in nutrient loading. Struvite deposits soon plugged the centrate lines. Various pipe materials were tested and KYNAR^{TM} was found to prevent struvite from adhering to pipe wall. Several chemicals were tested to prevent the pH rise that causes struvite to precipitate. Ferric chloride was chosen for its pH lowering properties rather than its phosphorus removal properties. The Ostara struvite recovery process was tested to see if it could remove phosphorus and ammonia from the centrate and reduce the amount of nutrients recycled back to the plant for treatment. Adding magnesium to the reactor created struvite in 1 to 3 mm spheres that can be used as a slow release fertilizer. The reactor removed 90% of the phosphorus and 20% of the ammonia from the centrate. That translates to a 24% reduction in phosphorus and 6% ammonia entering the aeration basin for re-treatment reducing treatment costs and delaying the need for plant expansion. Start up of North America's first full plant Ostara struvite recovery facility is scheduled for spring 2009.

INTRODUCTION

The Clean Water Service's Durham Advanced Wastewater Treatment Plant located in Tigard, Oregon was the first plant in the U.S.A. to receive a Total Maximum Daily Load allocation for phosphorus in 1988. The TMDL was set at 0.070 mg/l l T-PO$_4$-P from May to November effective in 1994. The receiving stream, the Tualatin River, is a slow moving, warm, effluent dominated river. Previously, the Best Practicable Treatment for phosphorus was required resulting in effluent with a target of 2 mg/l T-PO$_4$-P using a tertiary alum dose. Pilot testing indicated that a dose of alum to the primaries and an additional dose in the tertiary clarifiers followed by mixed media filters could meet the 0.070 mg/l T-PO$_4$-P monthly median permit limit. Estimated dose was 170 mg/l of alum. Alum dosing systems were installed along with lime addition to maintain pH.

The full scale system worked better than the pilot and alum doses were soon optimized to 30 mg/l dose to the primaries targeted to provide secondary effluent of 0.5 mg/l S-PO$_4$-P followed by a tertiary dose of 30 mg/l. This resulted in 100% permit compliance for over 10 years. The permit limit has recently been increased to 0.10 mg/l T-PO$_4$-P after further river modelling and varies with river flow. The current average dry weather flow is 76 000 m^3/d (20 mgd).

Transition to biological phosphorus removal

Filamentous organisms were an ongoing problem at the plant, during an expansion the new train was designed to operate in several modes including Bardenpho and Modified University of Capetown modes utilizing their anoxic selectors to reduce filaments. The two other existing basins were retrofitted. The alum doses were maintained as there was not enough Volatile Fatty Acids (VFA) to do Enhanced Biological Phosphorus Removal (EBPR) and the primary alum addition suppressed EBPR by reducing the biologically available phosphorus. At the extremely low permit level, there was no room to experiment with EBPR so testing was initiated during the winter when phosphorus removal was not required. The plant was configured in Bardenpho mode and EBPR worked except in the train where the summer required nitrification was maintained. We needed additional VFA to remove the nitrate in the RAS.

A source of VFA would be needed to operate in the EBPR mode. Fermentation pilots resulted in patenting the Unified Fermentation and Thickening (UFAT) process. UFAT generates sufficient VFA for full EBPR which eliminates the primary alum dose and reduces the tertiary dose required. As a backup for EBPR, alum can be temporarily added to the secondary clarifier of a train that was not meeting the secondary target of 0.5 mg/l S-PO$_4$-P. Secondary effluent typically is less than 0.1 mg/l S-PO$_4$-P when EBPR is operating well.

Struvite appears

As part of the optimization of EBPR, anaerobic digested sludge dewatering centrate was determined to be a significant load of ammonia and phosphorus recycled back to the aeration basins. The centrifuges operated 24 hours per day 5 days per week. An existing tank was used to store centrate during the day and release it during the low flow evening and night periods in an attempt to shave the peaks and shift the nitrification load to the lower night time power rates.

After several weeks of operation, the centrate holding tank discharge pipes were plugged with struvite. The rubber lined flowmeter had much less struvite on it than the other pipes. The pipes were removed and the tank drained to a gutter which created a struvite stalagmite and plugged the gutter drains.

Material testing

Since the flowmeter did not accumulate as much struvite as the rest of the pipes, we tested different materials. Flanged elbows lined with rubber, polypropylene,

Teflon® and Kynar® (PVDF) were installed in series along with PVC pipe and tested until the line plugged. The Teflon® surface did develop a layer of struvite, but it was relatively easy to remove. There was no build up of struvite on the Kynar® lined piece; the microscopically smooth surface would not allow struvite to get a foothold. The other materials required hammer and chisel to break the struvite off. Kynar® lined pipe was approximately eight times as expensive as unlined pipe.

pH adjustment

The anaerobic digested sludge centrate is a saturated struvite solution and saturated with CO_2 at higher than atmospheric pressure, in our case 3740 pascals (15 inches of water column). When centrate is exposed to atmospheric pressure The CO_2 off gasses like a warm carbonated beverage and the pH rises (Figure 1). At the centrifuge, the centrate pressure drops to atmospheric pressure and is subject to violent agitation at the centrifuge discharge. As the pH rises the solution becomes supersaturated with respect to struvite and precipitation can occur. Any acid could be used to suppress the pH rise and prevent the supersaturation and precipitation.

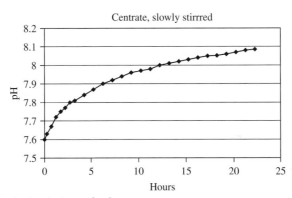

Figure 1. Centrate at atmospheric pressure.

We investigated several methods and chemicals to lower the pH.

- Acetic acid could be used and would supply some VFA to the aeration basin anaerobic zone to encourage EBPR when the centrate is recycled back to the aeration basin. Acetic acid is expensive and we have enough VFA from the UFAT process to maintain EBPR operation.

- Liquid CO_2 could be gasified and injected into the tank contents to lower the pH. Liquid CO_2 would require leasing a tank, installing eductors and purchasing liquid CO_2.
- The exhaust from the digester gas engine generator was tested and if injected into the centrate tank could be used to both mix the tank contents and lower pH due to the high concentration of CO_2 in the exhaust. There was enough CO_2 in the engine exhaust to lower the pH (Figure 2). A CO_2 compressor connected to the engine exhaust, and the boiler as a backup, and piping would be need to inject the exhaust into the centrate for pH adjustment and mixing. It is possible that NOx in the exhaust could be scrubbed by the centrate.

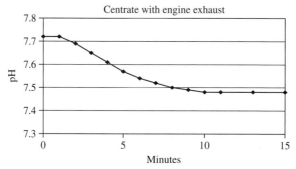

Figure 2. Injecting exhaust into centrate.

- Alum (aluminum sulfate) was considered as it would both lower the pH due to its acidic nature and remove some phosphorus. However, the sulphate could be reduced to hydrogen sulfide creating an odor and safety problem.
- Ferric chloride is stored near the centrate storage tanks and has an extremely low pH of minus 0.76! Low doses of ferric significantly reduced the pH (Figure 3). A reduction of just a few tenths of a pH unit is required. The ferric would also precipitate phosphorus and react with hydrogen sulfide reducing odor potential. Phosphorus is in great stoichiometric excess compared to the available magnesium so precipitating some phosphorus will not reduce the struvite potential as much as the pH reduction. The goal was to apply just enough ferric to suppress the pH rise due to off gassing, not to do biological phosphorus removal followed by chemical removal. The chemical option to remove phosphorus from the recycle stream by dosing centrate with ferric has removal efficiencies greater than in other less concentrated plant flows.

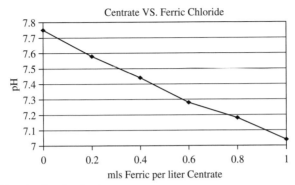

Figure 3. pH reduction by ferric.

Centrate storage system design

The dewatering system was redesigned with struvite in mind using what we learned from our experience and the pilot testing. The centrifuges were located directly above the centrate tank to minimize the distance the centrate had to travel. Previously we had problems with foamy centrate and long pipe runs. Rather than free falling to the bottom of the tank, the centrate is discharged into an open trough that slopes gently to the bottom to minimize turbulence and off gassing. Ferric was chosen to prevent the pH from rising and is added to the centrate in a hopper at the centrifuge. Ferric dosing was simple to install while it would have been more complex to install an exhaust blower and piping to inject engine generator exhaust. The pH of fresh centrate is compared to centrate leaving the holding tank and the ferric dose is adjusted so the discharge pH is slightly lower than the fresh centrate. As an added precaution, the holding tank discharge piping is Kynar® lined and pinch valves were used. To minimize the use of expensive Kynar®, after discharge from the holding tank the centrate is combined with the acidic VFA containing fermenter flow to lower the pH and dilute the centrate and the combined flow is piped to the aeration basins. This has been successful in eliminating struvite problems in the centrate system.

Digester struvite issues

Prior to EBPR and after 10 years of operation the digesters were inspected. A very small amount of grit was found, this was attributed to the efficient mixing from the mechanical mixer. The motor and gearbox mounted on top of the digester drive a water sealed shaft with impellers to keep sludge suspended and thoroughly mixed. Alum doses in the plant were reduced over the years as EBPR became more stable.

The lab noticed shiny crystals in the sludge. (Photo 1) Inspecting the holding tank revealed tons of loose 2 to 3 mm triangular struvite crystal piled up in the corners of the hexagon digester holding tank. No struvite was present on the tank walls or the mixer shaft or impellers. We believe that the efficient mixing keeps the surface supersaturation to a minimum with the struvite precipitation occurring on the billions of small struvite crystal circulating in the digester. Efficiently precipitating struvite on particles in the digesters should reduce struvite potential in the centrate by precipitating most of the available magnesium.

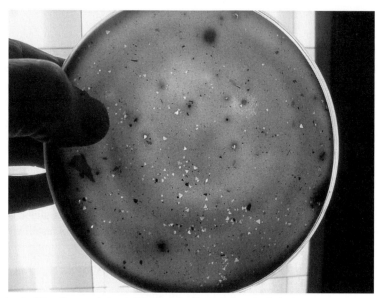

Photo 1. Struvite crystals visible in thin film of digested sludge.

Future plans for struvite recovery

The Durham plant piloted Ostara's www.ostara.com struvite reactor to see if we could reduce the levels of phosphorus and ammonia in the centrate recycle Prasad *et al.* (2008). By adding magnesium and creating struvite, it reduced the centrate phosphorus from 600 mg/l T-PO$_4$-P to 40 mg/l while reducing the ammonia from 1,200 mg/l NH$_3$-N to 1,000 mg/l. It produced struvite as a gray, hard round pellet that looks just like commercial fertilizer. CWS negotiated a contract with Ostara for the first full plant implementation of their system. CWS purchased the equipment and will sell the struvite to Ostara who will market it as Crystal Green™ to the local $1 billion container nursery industry located in our area. The 5% ammonia nitrogen, 27% phosphorus as P$_2$O$_5$, zero potassium and

10% magnesium (5-27-0 10% Mg) slow release fertilizer will be used in soil-less media to provide slow release nitrogen, phosphorus and magnesium for plants in containers. Construction started in October 2008. Startup of the three reactor system is scheduled for Spring 2009. Estimated production rate is 36 tonnes (40 US tons) per month.

CONCLUSION

In spite of a heavy load of EBPR waste activated sludge, our digesters do not have struvite build up on the wall, mixers or piping or in the centrate system. We believe the intense mechanical mixing prevents localized areas of super-saturation. The volume of loose struvite crystals in the digester does not significantly reduce the digestion volume. A significant amount of phosphorus leaves the plant as struvite crystals in the sludge cake. The rest of the phosphorus is in soluble and insoluble forms in the cake. The centrate is dosed with ferric only to suppress the pH rise from off gassing. Kynar® lined pipe is used sparingly until the centrate is diluted with acidic fermenter effluent to prevent scaling in the long pipe to the aeration basins. The Ostara system will reduce influent phosphorus by 24% and ammonia by 6%. This side stream treatment will further reduce nuisance struvite by reducing the influent load of phosphorus and turn the phosphorus into marketable struvite and produce revenue.

REFERENCE

Prasad, R., Baur, R. and Britton, A. (2008). Reducing Ammonia and Phosphorus Recycle Loads by Struvite Harvesting. *WEFTEC 2008 Proceedings* Session 83.

A novel waste sludge operation to minimize uncontrolled phosphorus precipitation and maximize the phosphorus recovery: a case study in Tarragona, Spain

R. Barat[a], M. Abella[b], P. Castella[b], J. Ferrer[a] and A. Seco[c]

[a]Instituto de Ingeniería del Agua y Medio Ambiente, Universidad Politécnica de Valencia, Valencia, Spain
[b]Empresa Municipal Mixta de Aguas de Tarragona (EMATSA), Tarragona, Spain
[c]Departamento de Ingeniería Química, Universidad de Valencia, Burjassot, Spain

Abstract The objective of this work is to evaluate by simulation a novel waste sludge operation in Tarragona WWTP aiming to minimize the precipitation problems and to obtain a stream enriched with phosphorus to recover. Previous studies showed that the increase in soluble P, K, Mg, NH4-N and Ca availability in the digester as well as the increase in pH are the main causes of precipitation problems in Tarragona WWTP. The proposed solution to minimize the precipitation problems in the digester consists in to perform a change in the configuration of the sludge line upstream the anaerobic digestion. This configuration enables to remove an important amount of phosphate from the system before the anaerobic digestion. The mixture of primary sludge (high volatile fatty acids concentration) with the secondary sludge (high intracellular poly-P concentration) allows to extract the internal poly-P and later to remove the soluble phosphate elutriating the mixed sludge in the gravity thickener. A new stream enriched with soluble phosphate can be obtained. Finally, the phosphate content in the supernatant stream could be recovered as struvite in a crystallizer. The proposed solution was simulated to evaluate its potential with the model BNRM1. Different simulations with different elutriation flow rates were performed. The results obtained showed that the higher elutriation flow the higher reduction of phosphate, ammonium and magnesium influent load to digestion and therefore, the higher concentration of this components in the supernatant stream to recover. However, the limitation of this solution lies on the capacity of the gravity thickener. According to the results obtained by simulation, a control system was designed in order to minimize the phosphate influent load to the anaerobic digestion and maximize its concentration in the supernatant stream.

INTRODUCTION AND BACKGROUND

The new regulations preventing eutrophication in natural ecosystems have reduced considerably the discharge level of phosphates in surface waters. Nowadays, biological phosphorus removal is often the preferred technology to achieve the effluent standards (Mino *et al.*, 1998). In this process the phosphate is removed from wastewater and accumulated inside the polyphosphate accumulating bacteria (PAO) as internal granules of polyphosphate. Therefore the fate of phosphate in a EBPR process is to be transferred from the water line to the sludge line.

At the same time, the concern about the limitation of the natural phosphate reservoir has arisen, increasing the amount of researches regarding the recovery of phosphate from wastewater. This recovery starts with the removal of phosphate from wastewater with the biological process above described. There are some researches (good review in Doyle and Parsons, 2002) that pointed out the possibility of recover phosphate on the supernatant stream obtained in the dewatering process of the digested sludge. This stream is highly enriched with phosphate, ammonium and metal cations. However, during the sludge treatment before the dewatering process, especially anaerobic digestion, the internal polyphosphates are released to the liquid phase (Wild et al., 1997) increasing significantly the phosphate and metal cations concentration. The increase on the concentrations of dissolved components and the high pH achieved during anaerobic digestion increase the precipitation potential in this stage of the treatment system. Different solids such as struvite and different calcium phosphates are likely to precipitate for high phosphate and metal ion concentrations and for high pH values (Nyberg et al., 1994). Precipitation problems lead to an important increase in the cost of the sludge management operations (Neethling and Benish, 2004).

This study has been performed in Tarragona Wastewater Treatment Plant (WWTP). The water line consists of an activated sludge process operated for biological phosphorus removal. The primary and secondary sludge are anaerobically digested after their separate thickening. Different precipitation problems have been found in the sludge line, especially during anaerobic digestion and post-digestion processes, which caused pipe blockage and accumulation on the surfaces of different devices of the sludge management. A previous study carried out in Tarragona WWTP (Barat et al., 2009) was focused on to study the causes of this precipitation and to quantify by means of mass balances the amount and extent of this precipitation. In this study approximately 100% of phosphate is fixed during the anaerobic digestion mainly as struvite. Also, an important release of phosphate during the sludge thickening was detected. However, there is no precipitation before the anaerobic digestion due to the low pH values achieve in the sludge thickening.

The objective of this work is to evaluate by simulation a novel waste sludge operation in Tarragona WWTP aiming to minimize the precipitation problems and to obtain a stream enriched with phosphorus to recover.

MATERIALS AND METHODS

This study has been performed in Tarragona WWTP in order to evaluate by simulation a novel sludge line operation that minimizes the uncontrolled precipitation and maximizes phosphate recovery. This work includes:

(1) To model the actual sludge line of Tarragona WWTP.
(2) To simulate a change in the sludge line configuration.
(3) To develop a process control.

Mathematical model

The mathematical model used is the Biological Nutrient Removal Model no. 1 (BNRM1, Seco *et al.*, 2004). This model considers the most important physical, chemical and biological processes taking place in a WWTP. The physical processes included are: settling and clarification processes (flocculated settling, hindered settling and thickening), Volatile Fatty Acids (VFA) elutriation and gas-liquid transfer. The chemical interactions included comprise acid-base processes, where equilibrium conditions are assumed, and phosphorus precipitation processes in the same way as in ASM2. The biological processes included are: organic matter, nitrogen and phosphorus removal; acidogenesis, acetogenesis and methanogenesis. The effect of temperature on the processes rates is considered in this model by using the general Arrhenius equation.

The settling processes model consists in a one-dimensional model based on the solids flux concept and the conservation of mass law. This model uses the settling velocity function proposed by Takács *et al.* (1991), which is corrected by a compression function in the lower layers. It has been linked to the biological model in order to consider biological processes taking place in primary and secondary settlers and gravity thickeners, i.e. they are considered as reactive elements. This model is described in detail by Ribes *et al.* (2002).

The simulation platform used for simulation was DESASS (DEsign and Simulation of Activated Sludge Systems, Ferrer *et al.*, 2008). DESASS is a software tool for WWTP design, simulation and optimisation. This software includes the model BNRM1 previously explained.

RESULTS AND DISCUSSION

Sludge line modelization

First of all, the model of the sludge line was performed and adjusted with experimental data. Due to the solution of the uncontrolled precipitation is based on the reduction of the phosphate load to the anaerobic digestion, only was simulated the gravity thickener, the DAF thickener and the mixture chamber before digestion (Figure 1).

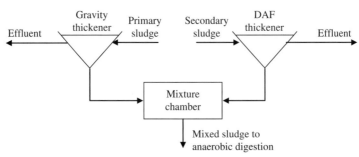

Figure 1. Tarragona WWTP sludge line configuration simulated with DESASS.

Table 1 and 2 show the experimental and simulated result obtained in the gravity thickener and DAF thickener. As can be seen, the simulated results showed that the model used was able to reproduce the physical and biological process that take place during the gravity and DAF thickening process. However some discrepancies exist between the experimental and simulated results. These discrepancies could be due to some problems associated to the sampling, which are punctual and in streams operated in a discontinuous manner.

As can be seen in Table 3, the model was able to reproduce the composition of the influent sludge to the anaerobic digestion. The measured pH value is lower than the simulated one. This difference could be due to the addition of $FeCl_3$ in the mixture chamber.

Furthermore, the model does not predict any precipitation upstream the anaerobic digestion. These results are in accordance with the low pH values achieved in the thickened and floated sludge which avoid the solid formation.

Proposal of a sludge management modification

Once adjusted the sludge line model of Tarragona WWTP with the experimental data, a change in the operation of the sludge line was evaluated by simulation aiming to minimize the phosphate, ammonium and metal cations load to anaerobic digestion.

The results obtained in previous studies in Tarragona WWTP (Barat *et al.*, 2009) demonstrated that the uncontrolled precipitation takes place mainly during

Table 1. Simulated and experimental results of the gravity thickener.

Parameter	Units	Experimental	s.d.	Simulated	Experimental	s.d.	Simulated
		Primary thickened sludge (line 3)			Supernatant (line 8)		
pH		6.2	±0.1	6.2	7.2	±0.0	7.2
TSS	mg/l	68425	±6524	68382	520	±60	411
VSS	mg/l	48233	±2572	48212	412	±58	294
COD tot.	mg O_2/l	115350	±6488	115142	994	±52	1021
COD sol.	mg O_2/l	3906	±1005	6132	326	±120	358
PO_4	mgP-PO_4/l	114.5	±66.1	69.5	15.5	±1.6	19.2
NH_4	mg N-NH_4/l	737.0	±132.5	169.1	55.5	±12.3	83.9
Ca tot	mg Ca/l	4676.7	±1028.9	4672.3	160.0	±7.1	163.7
Ca sol.	mg Ca/l	305.0	±128.5	138.1	132.2	±7.8	137.6
K tot	mg K/l	290.0	±57.2	343.8	44.6	±9.8	53.2
K sol.	mg K/l	107.8	±14.7	72.6	41.2	±6.8	51.5
Mg tot	mg Mg/l	447.5	±119.0	474.0	62.6	±3.2	67.4
Mg sol.	mg Mg/l	122.8	±22.0	94.1	55.2	±4.0	65.1
P tot	mg P/l	605.4	±130.6	831.3	70.6	±7.7	24.1
P sol.	mg P/l	127.2	±63.7	73.7	20.0	±1.9	19.4
N tot	mg N/l	2100.0	±418.3	2095	88.0	±11.0	96.8

Table 2. Simulated and experimental results of the DAF thickener.

Parameter	Units	Experimental	s.d.	Simulated	Experimental	s.d.	Simulated
		Secondary floated sludge (line 4)			Supernatant (line 7)		
pH		6.8	±0.1	6.5	7.1	±0.1	7.3
TSS	mg/l	23550	±3785	21713	2730	±458	2122
VSS	mg/l	19825	±1615	16063	2349	±373	1607
COD tot.	mg O_2/l	30482	±1666	32276	2736	±238	2855
COD sol.	mg O_2/l	138	±34	206	77	±17	58
PO_4	mgP-PO_4/l	174.1	±65.1	131.5	34.7	±7.5	4.2
NH_4	mg N-NH_4/l	520.6	±119.3	717.4	79.2	±27.0	134.7
Ca tot	mg Ca/l	568.0	±37.7	528.3	176.0	±16.7	162.3
Ca sol.	mg Ca/l	98.6	±23.3	174.6	122.6	±4.4	130.0
K tot	mg K/l	510.0	±40.6	427.0	86.0	±11.6	77.1
K sol.	mg K/l	154.8	±50.0	139.9	51.2	±7.8	45.8
Mg tot	mg Mg/l	344.0	±46.7	315.4	88.2	±9.1	72.8
Mg sol.	mg Mg/l	111.8	±39.1	73.9	57.2	±6.4	49.9
P tot	mg P/l	907.0	±181.9	1341.0	184.8	±23.5	118.3
N tot	mg N/l	2200.0	±273.9	1915.0	268.0	±70.1	274.5

the anaerobic digestion and the downstream processes. This is due to that during the anaerobic digestion the phosphorus from polyphosphates hydrolysis and organic matter degradation is released, increasing considerably the orthophosphate content in the system. The concentration of other ions such as

Table 3. Simulated and experimental results of the influent sludge to the anaerobic digestion (line 5).

Parameter	Units	Experimental	s.d.	Simulated
pH		6.0	±0.1	6.7
TSS	mg/l	48754	±10401	47115
VSS	mg/l	35780	±5586	33921
COD tot.	mg O_2/l	74994	±2943	78840
COD sol.	mg O_2/l	2254	±411	3937
PO_4	mgP-PO_4/l	339.3	±173.9	302.0
NH_4	mg N-NH_4/l	642.2	±85.0	444.1
Ca tot	mg Ca/l	2875.6	±532.6	2860.6
Ca sol.	mg Ca/l	214.5	±111.8	155.7
K tot	mg K/l	386.4	±56.3	381.1
K sol.	mg K/l	128.4	±34.1	167.2
Mg tot	mg Mg/l	402.1	±89.4	413.1
Mg sol.	mg Mg/l	118.0	±37.4	142.5
P tot	mg P/l	737.6	±196.4	1052.5
N tot	mg N/l	2144	±418.3	2016.3

ammonium, potassium and magnesium also increases during the digestion. The ammonium is released by decay and organic matter hydrolysis processes. The magnesium and potassium ere released by from polyphosphates hydrolysis (($K_{0.28}Mg_{0.36}PO_3$)n, Barat *et al.*, 2005) and organic matter degradation. All these factors with the addition of $FeCl_3$ to precipitate the sulphate produced and the increase of pH during the process (from 6.0 to 7.6) provoke the formation of different solids such as struvite, calcium phosphate and ferrous phosphate.

Therefore, the solution to minimize the uncontrolled precipitation problem will pass throw the minimization of the influent load of phosphate, ammonium and metal cations to digestion. This objective pushes towards changing the sludge line configuration before the anaerobic digestion.

In order to optimize the actual facilities and minimize the changes in the sludge line, a simple modification of the sludge streams is proposed as shown in Figure 2.

This new configuration consists on the mixture of the primary and secondary thickened sludge and this mixture is recirculated from the mixture chamber to the influent sludge in the gravity thickener. The mixed sludge is pumped to the anaerobic digestion from the bottom of the thickener.

In this new configuration we are mixing the primary sludge, with high volatile fatty acids concentration, (969 mg COD/l) with secondary sludge, enriched with PAO and therefore internal polyphosphate. The anaerobic

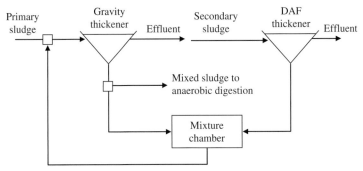

Figure 2. Proposed sludge line modification.

conditions and the presence of VFA induce the PAO release of polyphosphate as soluble phosphate. This phosphate release will take place in the mixture chamber and at the bottom of the gravity thickener. Elutriation of the soluble phosphate, ammonium and metal cations is carried out by the recirculation of the mixed thickened sludge from the thickener bottom to the inlet sludge zone. This elutriation lets to minimize digester influent load of phosphate, ammonium and metal cations and to obtain a supernatant stream enriched in these components.

Finally, the phosphate content in the supernatant stream could be recovered as struvite in a crystallizer, obtaining a product (struvite) that is a valuable slow-release fertiliser for agriculture and, hence, an economical benefit can be obtained.

The proposed sludge line modification showed in Figure 2 was simulated with DESASS aiming to optimize the operation process (elutriation flow) and quantify the maximum phosphate load reduction to anaerobic digestion and therefore the maximum phosphate concentration in the recovery stream. The different simulations carried out were subjected to some restrictions:

- To fix the sludge flow to anaerobic digestion (233 m^3/d) in order to maintain the sludge retention time in the digester around 22 days.
- To avoid the overflow in the gravity thickener.

Figure 3 shows the simulation results of anaerobic digester influent load reduction (%) and TSS in the supernatant at different elutriation flows. These simulations show that the higher elutriation flow the higher reduction of the influent load in the digester of phosphate, ammonium and magnesium. These reductions can achieve values of 54% of phosphate, 23% of total P, 30% of magnesium, 14.4% of total Mg, 46% of ammonium and 14% of total N.

Nevertheless, as can be seen in the evolution of the supernatant TSS concentration (see Figure 3), from a particular elutriation flow (250 m^3/d) does not increase the reduction and the thickener starts to be overflowed increasing significantly the TSS concentration in the supernatant.

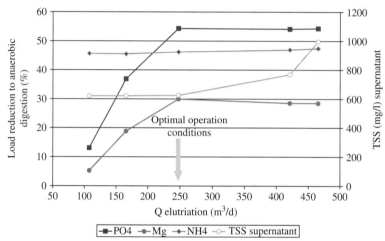

Figure 3. Simulated results (load reduction to anaerobic digestion and TSS in the supernatant) at different elutriation flows.

These results indicate the maximum load reduction to digester with the minor modification in the sludge line, only introducing the elutriation stream. Although this proposal does not guarantee the whole demise of the uncontrolled precipitation, these precipitations are significantly reduced and therefore minimized the operation problems associated.

Control system

The simulations carried out showed that the bottleneck point is the thickener overflow at high elutriation flows. In order to optimize the process it is proposed a control system based on the control of the sludge blanket height in the gravity thickener (see Figure 4). The information of the sludge blanket height (by means of an ultrasound sensor) and the flows in the different lines will be used by a fuzzy logic control system to act over the different pumps aiming to maintain the maximum elutriation flow without overflow.

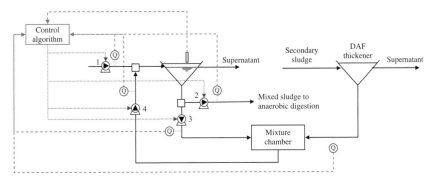

Figure 4. Control system.

CONCLUSIONS

In a first step the model BNRM1 was able to reproduce the physical and biological process that take place in the actual sludge line. Once adjusted the sludge line model with the experimental data, a change in the operation of the sludge line was evaluated by simulation aiming to minimize the phosphate, ammonium and metal cations load to anaerobic digestion. The simulation results show that the higher elutriation flow the higher reduction of the influent load in the digester of phosphate, ammonium and magnesium. These reductions can achieve values of 54% of phosphate, 23% of total P, 30% of magnesium, 14.4% of total Mg, 46% of ammonium and 14% of total N. These results indicate the maximum load reduction to digester with the minor modification in the sludge line, only introducing the elutriation stream.

The simulations carried out showed that the bottleneck point is the thickener overflow at high elutriation flows. In order to optimize the process, a control system based on the control of the sludge blanket height in the gravity thickener was proposed.

ACKNOWLEDGMENTS

This research work has been supported by the Empresa Municipal Mixta de Aguas de Tarragona (EMATSA) which is gratefully acknowledged.

REFERENCES

Barat, R., Montoya, T., Seco, A. and Ferrer, J. (2005). The role of potassium, magnesium and calcium in the enhanced biological phosphorus removal treatment plants. *Environ. Technol.*, **26**, 983–992.

Barat, R., Abella, M., Roig, J., Ferrer, J. and Seco, A. (2009). Study of uncontrolled precipitation problems in Tarragona WWTP (Spain), *This conference.*

Doyle, J.D. and Parsons, S.A. (2002). Struvite formation, control and recovery. *Water Res.*, **36**, 3925–3940.

Ferrer, J., Seco, A., Serralta, J., Ribes, J., Manga, J., Asensi, E., Morenilla, J.J. and Llavador, F. (2008). DESASS: A software tool for designing, simulating and optimising WWTPs, *Environ. Modell. Soft.*, **23**, 19–26.

Mino, T., van Loosdrecht, M.C.M. and Heijnen, J.J. (1998). Microbiology and biochemistry of the enhanced biological phosphate removal process. *Water Res.*, **32**(11), 3193–3207.

Neethling, J.B. and Benisch, M. (2004). Struvite control through process and facility design as well as operation strategy. *Water Sci. Tech.*, **49**(2), 191–199.

Nyberg, U., Aspergren, H., Andersson, B., Elberg Jorgensen, P. and la Cour Jansen, J., 1994. Circulation of phosphorus in a system with biological P-removal and sludge digestion. *Water Sci. Tech.*, **30**(6), 293–302.

Ribes, J., Ferrer, J., Bouzas, A. and Seco, A. (2002). Modelling of an activated primary settling tank including the fermentation process and VFA elutriation. *Environ. Technol.*, **23**, 1147–1156.

Seco, A., Ribes, J., Serralta, J. and Ferrer, J. (2004). Biological Nutrient Removal Model No.1 (BNRM1). *Water Sci. Tech.*, **50**(6), 69–78.

Takács I., Patry G.G. and Nolasco D. (1991). A dynamic model of the clarification-thickening process. *Water Res.*, **25**(10), 1263–1271.

Wild, D., Kisliakova, A. and Siegrist, H. (1997). Prediction of recycle phosphorus loads from anaerobic digestion. *Water Res.*, **31**(9), 2300–2308.

Study of uncontrolled precipitation problems in Tarragona WWTP (Spain)

R. Barat[a], M. Abella[b], J. Roig[b], J. Ferrer[a] and A. Seco[c]

[a]Instituto de Ingeniería del Agua y Medio Ambiente, Universidad Politécnica de Valencia, Valencia, Spain
[b]Empresa Municipal Mixta de Aguas de Tarragona (EMATSA), Tarragona, Spain
[c]Departamento de Ingeniería Química, Universidad de Valencia, Burjassot, Spain

Abstract Tarragona WWTP has a capacity of 35 000 m^3/d. The water line consists of an activated sludge process operated for biological nitrogen and phosphorus removal. The plant operators have continuously found precipitation problems in the sludge line, especially during anaerobic digestion and post-digestion processes. This work studies the precipitation problems in Tarragona WWTP in order to locate the precipitation sources and its causes from an exhaustive mass balance analysis. This work is the first step before the proposal of solutions trying to minimize the uncontrolled precipitation. The DAF thickener and anaerobic digester mass balances suggest that polyphosphate is released during the excess sludge thickening. Despite the high concentrations achieved in the thickened sludge, precipitation does not occur in this point due to the low pH. This study confirms that the precipitation problems were mainly found in the digestion stage. The results show that 99.5% of the available phosphate was fixed in the digester of which 42.8% precipitates as ammonium struvite, 4.2% precipitates as hydroxyapatite, and the remaining 53% was adsorbed on the surface of the solids or precipitated with iron. The increase in soluble PO_4^{3-}, K, Mg, NH_4^+ and Ca availability in the digester as well as the increase in pH were pointed out as the main causes of precipitation problems.

INTRODUCTION

Phosphate is the limiting component for growth in most ecosystems. The discharge of phosphates in surface waters may lead to eutrophication and blooming of algae. Therefore it is essential to control the phosphate emissions. Nowadays, biological phosphorus removal is often the preferred technology to achieve the effluent standards (typically in the range of 0.5–1 gP/m^3).

In biological nutrient removal processes (BNR), phosphates and metal cations are taken up and stored as polyphosphates (Poly-P) inside the bacterial cells. These polyphosphates are removed from the system in the excess sludge. During the sludge treatment, especially anaerobic digestion, these polyphosphates are released to the liquid phase (Wild *et al.*, 1997) increasing significantly the phosphate and metal cations concentration. The increase on the concentrations of dissolved components and the high pH achieved during anaerobic digestion increase the precipitation potential in this stage of the treatment system.

Different solids such as struvite and different calcium phosphates are likely to precipitate for high phosphate and metal ion concentrations and for high pH values (Nyberg *et al.*, 1994).

Dewatering of the anaerobic digested sludge by centrifugation was found to be a critical stage for precipitation in many wastewater treatment plants in Germany (Heinzmann and Engel, 2005). The solid formed was mainly struvite and small portions of different calcium phosphate compounds. Downstream of the centrifugation of the sludge, precipitation was so significant that the outgoing pipes were fully blocked.

Struvite precipitation is not simply a problem of BNR treatment works (Parsons and Doyle, 2004). However, Enhanced Biological Phosphorus Removal (EBPR) processes, which produce a Poly-P rich waste sludge, exacerbate struvite and calcium phosphate deposit problems in anaerobic sludge digestion. Therefore, in these cases, special attention must be paid to control the formation of these deposits.

Precipitation problems lead to an important increase in the cost of the sludge management operations (Neethling and Benish, 2004). Furthermore, these precipitation processes can be very important in the recycled streams composition from the sludge treatment causing variations in the dissolved phosphorus concentration that might affect the efficiency of the EBPR process.

The aim of this work is to study the precipitation problems in a large scale wastewater treatment plant in order to locate the precipitation sources and its causes from an exhaustive mass balance analysis. This work is the first step before the proposal of solutions trying to minimize the uncontrolled precipitation.

MATERIALS AND METHODS

This study has been performed in Tarragona Wastewater Treatment Plant (WWTP) in order to determine the type and extent of phosphate fixation in the waste activated sludge line.

Tarragona WWTP

Tarragona WWTP was established in 1993 with a capacity of 35000 m^3/d (140000 P.E.). Figure 1 shows the plant configuration. The water line consists of an activated sludge process operated for biological nitrogen removal following preliminary treatment and primary sedimentation. However, due to the high anoxic zone and low nitrification, this zone works mainly as anaerobic stimulating the biological phosphorus removal process. The water line consists of two identical lines divided in anoxic and aerobic zones each one, with a total volume of 8928 m^3 (18% anoxic and 82% aerobic).

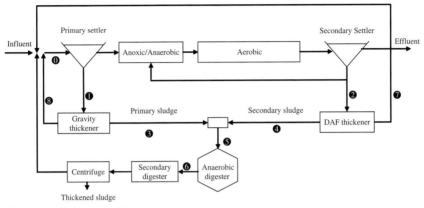

Figure 1. Tarragona WWTP flow chart and sampling points.

The primary and the secondary excess sludge are concentrated in one gravity thickener and one dissolved air flotation (DAF) thickener respectively. Once thickened, both sludge are mixed before being anaerobically digested. FeCl$_3$ is added in the mixture chamber to prevent the sulphate formation during digestion. The anaerobic digestion is carried out in one digester of 5129 m^3. The digested sludge is stored in a secondary digester and finally dewatered by centrifugation. The effluents from centrifuges and sludge thickeners are recycled to the water line before the primary settler.

The plant operators have continuously found precipitation problems in the sludge line, especially during anaerobic digestion and post-digestion processes, which caused pipe blockage and accumulation on the surfaces of different devices of the sludge management system as centrifuge, pumps and digested sludge pipes or supernatant pipes (Figure 2).

Analytical campaign

To identify the precipitation problems an intensive analytical campaign was carried out under normal operation conditions in different points of the sludge line. Figure 1 shows the WWTP configuration with the sampling points marked with black dots. Each line was sampled five days consecutively. The parameters analysed were: total suspended solids (TSS), volatile suspended solids (VSS), total chemical oxygen demand (tot. COD), soluble chemical oxygen demand (sol. COD), total phosphorus, phosphate, ammonium, pH and total and soluble

(a)   (b)

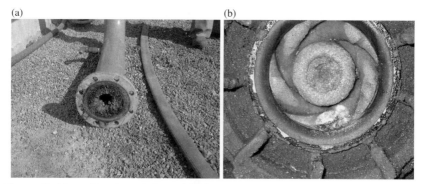

Figure 2. Deposits of precipitate: (a) digested sludge pipe; (b) centrifuge.

concentrations of calcium, magnesium and potassium. All the analyses were carried out according to the Standard Methods (APHA, 1998). Furthermore, the flow rates were measured in all the sludge lines sampled.

Mass balance

Phosphorus, potassium, magnesium and calcium mass balances can be used to understand the chemical fixation mechanisms in the sludge treatment system. However, in order to perform these balances, several assumptions must be taking into account. The mass balance equations and the assumptions considered in this paper are deeply explained in Barat *et al.* (2008).

As pointed out by Barat *et al.* (2008), organic matter degradation produce a release of phosphorus (P), potassium (K), magnesium (Mg), and calcium (Ca) associated with the content of each element in the organic matter. The amount of P, K, and Mg in the organic fraction of the primary sludge was estimated from experimental analysis (line 1, Table 2) assuming that suspended P, K, and Mg were mainly associated to organic matter. This assumption can be made because the presence of P, K and Mg precipitates in the influent wastewater is negligible and no precipitation of P, K and Mg is expected in the primary settler. Nevertheless, an accurately determination of the calcium content in the organic matter of primary sludge was not possible due to the high calcium carbonate concentration in the influent wastewater which is accumulated in the primary sludge.

The obtained values for primary sludge (Table 1) were used to perform the mass balances in the gravity thickening units. For the secondary sludge it is not possible to easily distinguish between the organic P, Mg, and K content and that associated with Poly-P structure. Hence, reported values for biomass composition were assumed. These values (Table 1) were used to carry out the mass balances in the DAF thickening stage.

Table 1. Organic P, K, Mg and Ca in primary and secondary sludge.

	Primary sludge[a]	Biomass[b]
P (mgP/mgCOD)	0.009	0.020
K (mgK/mgCOD)	0.0035	0.007
Mg (mgMg/mgCOD)	0.005	0.0035
Ca (mgCa/mgCOD)	–	0.0035

[a]Experimental data
[b]Metcalf and Eddy (2003)

RESULTS AND DISCUSSION

Table 2 shows the average values of the parameters analysed and the average flow rate for each line. Standard deviations are also provided.

Gravity thickening

As Table 2 shows, the soluble phosphate, ammonium, calcium, magnesium and potassium concentrations increased during the thickening process (lines 1, 3 and 8). The results show a significant hydrolysis of the suspended organic matter and latter fermentation at the bottom of the thickener. The mass balance shows a soluble COD increase of 429.4 kg/d in the thickener. These processes are associated with a significant release of phosphate and ammonium (from 19.1 mg P-PO_4/l and 83.8 mg N-NH_4/l in the influent to 114.5 mg P-PO_4/l and 737 mg N-NH_4/l in the thickened sludge. Furthermore this hydrolysis is responsible of an important increase of soluble cations concentration. The pH decrease in the thickened sludge (from 7.2 to 6.2) confirms the volatile fatty acids formation due to fermentation processes.

Despite the high phosphate, ammonium, magnesium and calcium concentration in the thickened sludge, the precipitation of struvite and calcium phosphate were not expected due to the low pH values achieved by fermentation.

DAF thickening

Previous studies confirmed that the first anoxic zone of the activated sludge reactor (18% of the total volume) works as an anaerobic zone due to the low nitrate concentration in the recirculation stream. Therefore, polyphosphate accumulating organisms (PAO) could growth internally accumulating phosphate as polyphosphate (Poly-P).

Regarding the phosphorus dynamic in the DAF thickener, a significant phosphate release took place in the thickened sludge. As can be seen in Table 2,

Table 2. Average analytical measurements and flow rates.

		Stream							
		1	2	3	4	5	6	7	8
TSS	mg/l	12730	5013	68425	23550	48754	30725	2730	520
	s.d.	570.8	70.5	6524.0	3784.6	10400.8	1767.1	457.7	59.9
VSS	mg/l	9083	4285	48233	19825	35780	14075	2349	412
	s.d.	384.4	403.2	2571.6	1615.3	5585.6	3455.8	372.5	57.5
Tot.COD	mg/l	21106	6614	115350	23288	74994	28300	2736	994
	s.d.	352	564	6488	1666	2943	1803	238	52
Sol.COD	mg/l	348	56	3906	138	2254	256	77	326
	s.d.	79	16	1005	34	411	66	17	120
Ammon.	mgNH$_4$-N/l	83.8	90.2	737.0	520.6	642.2	1293.8	79.2	55.5
	s.d.	24.8	24.9	132.5	119.3	85.0	115.2	27.0	12.3
Total P	mgP/l	165.9	299.2	605.4	907.0	737.6	509.4	184.8	70.6
	s.d.	65.2	15.8	130.6	181.9	196.4	93.7	23.5	7.7
Phos.	mgPO$_4$-P/l	19.1	30.6	114.5	174.1	339.3	3.1	34.7	15.5
	s.d.	10.5	14.1	66.1	65.1	173.9	0.8	7.5	1.6
pH		7.2	7.0	6.2	6.8	6.0	7.6	7.1	7.2
	s.d.	0.1	0.0	0.1	0.1	0.1	0.1	0.1	0.0
Sol. Ca	mg/l	137.5	126.8	305.0	98.6	214.5	161.2	122.6	132.2
	s.d.	9.7	9.0	128.5	23.3	111.8	9.7	4.4	7.8
Tot. Ca	mg/l	956.6	204.0	4676.7	568.0	2875.6	2574.0	176.0	160.0
	s.d.	113.0	15.2	1028.9	37.7	532.6	199.3	16.7	7.1
Sol. Mg	mg/l	65.0	58.2	122.8	111.8	118.0	77.2	57.2	55.2
	s.d.	2.4	9.7	22.0	39.1	37.4	5.8	6.4	4.0
Tot. Mg	mg/l	131.4	104.8	447.5	344.0	402.1	310.0	88.2	62.6
	s.d.	16.1	7.4	119.0	46.7	89.4	17.3	9.1	3.2
Sol. K	mg/l	51.5	52.2	107.8	154.8	128.4	187.0	51.2	41.2
	s.d.	0.7	16.8	14.7	50.0	34.1	1.9	7.8	6.8
Tot. K	mg/l	88.5	118.0	290.0	510.0	386.4	268.0	86.0	44.6
	s.d.	17.9	13.0	57.2	40.6	56.3	8.4	11.6	9.8
Q	m^3/d	700	826	123	96	219	219	716	590

the phosphate concentration increased from 30.6 mgP/l in the excess sludge to 174.1 mgP/l in the floated sludge. The amount of phosphate released in the DAF thickener was 32.7 kgP/d.

This phosphate release can be attributed to the formation of anaerobic zones inside the thickener, also observed in other WWTP (Barat et al., 2008). Under anaerobic conditions and in the presence of volatile fatty acids (VFA), PAO release the internal Poly-P to the bulk solution. These VFA are produced by the organic matter fermentation.

The precipitation of struvite and calcium phosphate were not expected alike the gravity thickener due to the low pH values achieved in the floated sludge.

Anaerobic digestion

Table 3 shows the results of the mass balances carried out in the anaerobic digester. The results obtained showed a lower release of phosphorus due to Poly-P hydrolysis (P_{libPAO}) than due to organic solids degradation (P_{libORG}). This result confirmed an important Poly-P hydrolysis before the anaerobic digestion. Considering that all the remaining Poly-P was released during the digestion process, the amount of Poly-P released in the digester (3.1 kgP/d) was lower than the amount of Poly-P released during the excess sludge thickening (23.7 kgP/d).

Table 3. Results of the anaerobic digester mass balances.

K_{TOTrel}	12.8	kgK/d
K_{ORGrel}	11.7	kgK/d
K_{PAOrel}	1.1	kgK/d
P_{PAOrel}	3.1	kgP/d
P_{ORGrel}	95.1	kgP/d
P_{fix}	128.3	kgP/d
$\%P_{fix}$	99.5	%
Mg_{PAOrel}	0.9	kgMg/d
Mg_{ORGrel}	33.3	kgMg/d
Mg_{prec}	43.1	kgMg/d
$\%Mg_{prec}$	71.8	%
Ca_{prec}	11.7	kgCa/d
$\%Ca_{prec}$	24.9	%
%P-MAP	42.8	%
%P-HAP	4.2	%
$\%P_{ads}$	53	%

The mass balance of phosphate showed an important precipitation of phosphate. The 99.5% of the available phosphate in the digester was fixed. Moreover, magnesium and calcium precipitation was detected. These results suggested that the conditions reached in the digestion process enhanced the precipitation. The high concentration of ammonium (1293.8 mgNH$_4$-N/l in the digested sludge versus 642.2 mgNH$_4$-N/l before the digestion) and the increase in the pH value (from 6.0 to 7.6) during the anaerobic digestion process gave rise to the formation of struvite and in that stage of the sludge treatment line.

Distribution of precipitates in the digester

Different phosphate magnesium salts can precipitate from a solution containing Mg^{2+}, NH$_4^+$, and PO$_4^{3-}$. Nevertheless, struvite is the one that precipitates in the pH values achieved in anaerobic digesters: neutral and higher pH values (Musvoto et al., 2000). Moreover, struvite deposits are quite common and actually found at just about every municipal WWTP where anaerobic digestion is carried out (Neethling and Benish, 2004). According to this, magnesium was considered to precipitate in the digester mainly as struvite, which made it possible to estimate the amount of phosphate precipitated as struvite (%P-MAP) from its stoichiometry.

Calcite precipitation was not considered due to the high phosphate concentrations in the digester. Several authors have reported that the presence of phosphates inhibits the calcite growth due to the adsorption of phosphates on the calcite surface enabling the formation of calcium phosphates (Plant and House, 2002; Lin and Singer, 2005). Therefore, the calcium precipitated in the digester was considered to be in the form of calcium phosphates. According to the Ostwald rule of stages, calcium phosphate formation takes place as a two-stage process. In that process, the phases that are thermodynamically less stable (i.e. precursors) are formed before the most stable phase, which is hydroxy-apatite (Ca$_5$(PO$_4$)$_3$OH). Hence, hydroxyapatite (HAP) was considered as the calcium phosphate formed in this work due to the high process retention time, which let the precursors transform into HAP. Then, assuming that calcium precipitates with phosphate as HAP, it was calculated the amount of phosphate precipitated with calcium (%P-HAP) using a molar ratio Ca/P of 1.67. The rest of the phosphorus fixed in the digester was considered to be adsorbed on the surface of solids or precipitated with iron (P$_{ads}$). Although phosphate adsorption during anaerobic digester has not been completely understood, it is considered as a possible mechanism for phosphorus fixation (Jardin and Pöpel, 1994; Wild et al., 1997).

According to this, 42.8% of the phosphate precipitates as struvite, 4.2% as hydroxiapatite and the remaining 53% was precipitated with iron or adsorbed on solids surface.

CONCLUSIONS

Analytical determinations of soluble P, K, Mg and Ca and mass balances were used to easily identify precipitation problems throughout the sludge treatment line of Tarragona WWTP.

The influent wastewater characterization showed the hardness of water and its potential to spontaneously precipitate different solids through the waste sludge treatment.

The mass balances carried out in the DAF thickener showed that a significant amount of polyphosphate was released during the excess sludge thickening probably due to the formation of anaerobic zones inside the thickener.

Precipitation problems were mainly found in the digestion stage. The 99.5% of the available phosphate was fixed in the digester: 42.8% precipitates as ammonium struvite, 4.2% precipitates as hydroxyapatite and 53% was precipitated with iron or adsorbed on the solids surface. The increase in soluble P, K, Mg, NH_4-N and Ca availability in the digester as well as the pH increase were pointed out as the main factors causing precipitation problems.

This study confirmed the important precipitation problems in the sludge management of a wastewater treatment plant with EBPR.

ACKNOWLEDGMENTS

This research work has been supported by the Empresa Municipal Mixta de Aguas de Tarragona (EMATSA) which is gratefully acknowledged.

REFERENCES

American Public Health Association, American Water Works Association and Water Environmental Federation, 1998. Standard methods for the examination of water and wastewater, 20th edn, Washington DC, USA.

Barat, R., Bouzas, A., Martí, N., Ferrer, J. and Seco, A. (2008). Precipitation assessment in wastewater treatment plants operated for biological nutrient removal: a case study in Murcia, Spain. *J. Environ. Man.*, in press. doi:10.1016/j. jenvman.2008.02.001.

Heinzmann, B. and Engel, G. (2005). Induced magnesium ammonia phosphate precipitation to prevent incrustations and measures for phosphate recovery. In *Proceedings Nutrient Management in Wastewater Treatment Processes and Recycle Streams*, Krakow.

Jardin, N. and Pöpel, H.J. (1994). Phosphate release of sludges from enhanced biological P-removal during digestion. *Water Sci. Technol.*, **20**(6), 281–292.

Lin Y.P. and Singer P.C. (2005). Inhibition of calcite crystal growth by polyphosphates. *Water Res.*, **39**, 4835–4843.

Metcalf and Eddy (2003). Wastewater Engineering: Treatment and reuse. 4th Ed. McGraw Hill, New York, USA.

Musvoto, E.V., Wentzel, M.C. and Ekama, G.A. (2000). Integrated chemical-physical processes modelling-II. Simulating aeration treatment of anaerobic digester supernatants. *Water Res.*, **34**, 1868–1880.

Neethling, J.B. and Benisch, M. (2004). Struvite control through process and facility design as well as operation strategy. *Water Sci. Tech.,* **49**(2), 191–199.

Nyberg, U., Aspergren, H., Andersson, B., Elberg Jorgensen, P. and la Cour Jansen, J., 1994. Circulation of phosphorus in a system with biological P-removal and sludge digestion. *Water Sci. Tech.*, **30**(6), 293–302.

Parsons, S.A. and Doyle, J.D. (2004). Struvite scale formation and control. *Water Sci. Technol.*, **49**(2), 177–182.

Plant, L.J. and House, W.A. (2002). Precipitation of calcite in the presence of inorganic phosphate. *Colloid Surface A*, 203, 143–153.

Wild, D., Kisliakova, A. and Siegrist, H. (1997). Prediction of recycle phosphorus loads from anaerobic digestion. *Water Res.*, **31**(9), 2300–2308.

Phosphorus recovery in EBPR systems by struvite crystallization

N. Martí[a], L. Pastor[b], A. Bouzas[a], J. Ferrer[b] and A. Seco[a]

[a]Dpto. Ingeniería Química. Universidad de Valencia. Doctor Moliner 50. 46100. Burjassot. Valencia. Spain (E-mail: nuria.marti@uv.es; alberto.bouzas@uv.es; aurora.seco@uv.es)
[b]Instituto de Ingeniería del Agua y Medio Ambiente (IIAMA). Universidad Politécnica de Valencia. Camino de Vera s/n. 46022. Valencia. Spain (E-mail: jferrer@hma.upv.es; laupasal@doctor.upv.es)

Abstract Phosphorus recovery by struvite crystallization is a suitable method for treating sludge rejected liquors especially in wastewater treatment plants with enhanced biological phosphorus removal (EBPR). In this paper, global phosphorus recovery efficiency by struvite crystallization has been evaluated following the fate of phosphorus in each stage of the sludge treatment process. The results obtained show that the main loss of phosphorus occurred in the anaerobic digester due to precipitation. By the other side, the major quantity of phosphorus available to be used in the crystallizer was present in the thickener supernatant stream. In the crystallization process, high phosphorus precipitation and recovery efficiencies of 83.0% and 69.9% were achieved, respectively. However, a low struvite percentage was obtained due to the high Ca/Mg molar ratio in the crystallizer inlet stream, which favoured the calcium phosphate formation. Finally, the total phosphorus mass balance (i.e. sludge line + crystallizer) showed that the global phosphorus recovery efficiency was around 40%, being 15% for the struvite recovery.

NOMENCLATURE

$\%P_{MAP}$	Percentage of phosphorus recovered as struvite
COD_T	Total chemical oxygen demand
NH_4-N	Ammonia nitrogen
PO_4-P	Orthophosphate
P_T	Total phosphorus
Q	Volumetric flowrate
TS	Total solids
TVS	Total volatile solids
w_P	Mass of phosphorus per mass of treated sludge

© 2009 The Authors, *International Conference on Nutrient Recovery from Wastewater Streams.* Edited by Ken Ashley, Don Mavinic and Fred Koch. ISBN: 9781843392323. Published by IWA Publishing, London, UK.

Subscripts

av	available
fix	fixed
lost	lost
ORGrel	release from organic matter degradation
PAOrel	release from polyphosphate hydrolysis
prec	precipitated
rec	recovered
TOTrel	total release

INTRODUCTION

Phosphorus recovery by struvite ($MgNH_4PO_4 \cdot 6H_2O$) crystallization is one of the most widely recommended technologies for treating sludge digester liquors (Battistoni *et al.*, 2002; Parsons and Doyle, 2004; von Münch *et al.*, 2001) especially in wastewater treatments plants with enhanced biological phosphorus removal (EBPR).

In EBPR process, phosphates and other ions (i.e. Mg^{2+}, K^{2+}) are taken up and stored as polyphosphates (Poly-P) inside the bacterial cells. During the anaerobic digestion process, these polyphosphates are released to the liquid phase (Wild *et al.*, 1997) increasing notably the phosphate, magnesium and potassium concentration. Moreover, ammonium concentration increases significantly as proteins are degraded and dissolved magnesium, phosphorus, calcium and potassium concentrations increase due to the cell lysis. Therefore, the rejected liquors from digested sludge dewatering show high phosphorus, ammonium and magnesium concentration which makes these streams very appropriate to recover phosphorus as struvite in a crystallization process. The thickener supernatant could also be used in the crystallizer since Poly-P hydrolysis can take place in the gravity thickener.

Moreover, other studies have shown that the product obtained by struvite crystallization can be used as an effective slow release fertilizer in agriculture (de-Bashan *et al.*, 2004; Shu *et al.*, 2006) and hence an economical benefit can be obtained. In order to guarantee phosphorus rich streams and thereby to obtain a high struvite production in the crystallization process, the uncontrolled phosphorus precipitation in the digester should be reduced. Other factors, such as calcium concentration in the crystallizer influent stream, can significantly affect the production of struvite (Pastor *et al.*, 2008a).

In order to enhance struvite production in the crystallization process, the sludge treatment scheme should be designed to ensure high levels of phosphorus in the sludge liquors. In this paper, a strategy for the sludge line operation has been tested following the fate of phosphorus in each stage of the sludge treatment process. This study will enable to evaluate the global phosphorus recovery efficiency in the system by struvite crystallization.

MATERIALS AND METHODS

Pilot plants description

Four pilot plants were used in this work (Figure 1). These pilot plants were operated for fermentation and elutriation of primary sludge (Plant 1), biological nutrient removal (Plant 2), anaerobic digestion of primary and secondary waste sludge (Plant 3) and struvite crystallization (Plant 4). All theses plants, except Plant 4 (which is placed in the Environmental Technologies Laboratory of the University of Valencia), are located in Carraixet Wastewater Treatment Plant (Valencia). Plant 1 and 2 are described in Bouzas *et al.* (2007) and García-Usach *et al.* (2006), respectively.

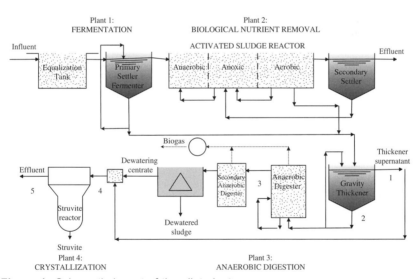

Figure 1. Schematic layout of the pilot plants.

The anaerobic digestion pilot plant consists of a gravity thickener, an anaerobic digester and a secondary digester. The anaerobic digester is a completely mixed type reactor of 160 litres in effective volume. The reactor is equipped with pH and temperature electrodes. A detailed description of the anaerobic digestion pilot plant can be found in Martí *et al.* (2008a).

The crystallization pilot plant is composed of the crystallization reactor, three stainless steel injection tubes for the influent and reactants, two peristaltic pumps, one membrane pump, and two balances. The reactor is a stirred tank reactor that is composed of two parts: a reaction zone, and a settling zone to prevent fine particles from being driven out with the effluent. The crystallization pilot plant and the reactor are described in detail in Pastor *et al.* (2008b).

Experimental procedure

As Figure 1 shows, primary sludge from the primary settler/fermenter and EBPR sludge from the aerobic reactor of the BNR system were pumped to the gravity thickener, where both were mixed thickened. Elutriation of the thickened sludge was also carried out by its recirculation from the thickener bottom to the inlet sludge zone. Moreover, the EBPR sludge from the aerobic reactor was used as a "washing-stream" that elutriates the thickened sludge to a greater extent. The thickened sludge (stream 2) was digested at mesophilic conditions $(T = 35 \pm 1°C)$ and at a solids retention time of 20 days. The digested sludge (stream 3) was dewatered by centrifugation. The thickener supernatant (stream 1) and the dewatering centrate were mixed (stream 4) and stored to be used in the phosphorus recovery process by struvite crystallization.

The crystallizer was operated in continuous mode for the liquid phase and in batchwise for the solid phase. The resulting hydraulic retention time in the reaction zone was 2.5 hours. The struvite precipitation process leads to a decrease in the pH value so NaOH was added in order to maintain the pH in the desired value. Software based on fuzzy logic control was used to maintain the pH at the set value (Chanona *et al.*, 2006).

Analytical methods

Total phosphorus (P_T), total chemical oxygen demand (COD_T), total solids (TS) and total volatile solids (TVS) analyses were performed in accordance with Standard Methods (APHA, 2005). PO_4-P, NH_4-N and soluble calcium, magnesium and potassium were analysed by ion chromatography (Metrohm IC, Switzerland).

Precipitates obtained in the crystallizer were analyzed by X-Ray diffraction (XRD) in order to check whether struvite crystals were formed. The XRD measurements were performed on a BRUKER AXS D5005 powder diffractometer.

Calculation of phosphorus precipitation in digestion and crystallization processes

Phosphorus precipitation assessment in the digester

PO_4-P, Ca^{2+}, Mg^{2+}, and K^+ mass balances were carried out to assess phosphorus precipitation in the digester using the methodology developed in Martí *et al.* (2008a). As explained in that work, PO_4-P, K^+ and Mg^{2+} are released during sludge anaerobic digestion due to polyphosphate hydrolysis and organic matter degradation, whereas Ca^{2+} is considered to be released only due to organic matter degradation. The organic content of phosphorus, potassium, magnesium and calcium in the mixed influent sludge was calculated using the proportion of primary (45%) and secondary sludge (55%) in the total sludge fed to the digester and the content of each element in the organic matter (Martí *et al.*, 2008b).

Phosphorus precipitation assessment in the crystallization process

Phosphorus precipitation in the crystallizer was assessed taking into account two types of efficiencies: recovery efficiency (Equation 1) and precipitation efficiency (Equation 2). The precipitation efficiency represents the process efficiency from a thermodynamic point of view provided that the supersaturation can be almost completely consumed, which should be the case with sufficient residence time. The recovery efficiency takes into account both the precipitation and crystal growth efficiency.

$$RecoveryEffiency = \frac{P_{Tin} - P_{Tef}}{P_{Tin}} \cdot 100 \tag{1}$$

$$PrecipitationEfficiency = \frac{PO_4\text{-}P_{in} - PO_4\text{-}P_{ef}}{PO_4\text{-}P_{in}} \cdot 100 \tag{2}$$

The percentage of precipitated phosphorus as struvite (%P_{MAP}) has been calculated by (Equation 3), considering that all the magnesium precipitated has lead to the struvite formation.

$$\%P_{MAP} = \frac{Mg^{2+}_{prec}}{PO_4 - P_{prec}} \cdot 100 \tag{3}$$

RESULTS AND DISCUSSION

Experimental characteristics of the main streams involved in the process (Figure 1) are summarized in Table 1.

Table 1. Experimental data of the pilot plant main streams.

Parameter		1	2	3	4	5
				Stream		
Q	(l/d)	38.5	4.3	4.3	41.3	41.3
PO_4-P	(mg/l)	46.5	98.6	67.6	43.0	7.3
NH_4-N	(mg/l)	24.8	110.6	796.0	129.0	123.7
Mg^{2+}	(mg/l)	57.2	72.3	56.4	49.8	39.4
Ca^{2+}	(mg/l)	157.5	124.7	57.0	129.9	71.6
K^+	(mg/l)	24.9	65.9	132.6	35.3	35.9
TS	(mg/l)	NA	29583	16850	NA	NA
TVS	(mg/l)	NA	18637	7545	NA	NA
COD_T	(mg/l)	NA	32650	11076	NA	NA
P_T	(mg/l)	NA	577	433	46.4	14.0
pH	(mg/l)	NA	6.2	7.2	NA	8.7

Phosphorus precipitation in the digester

The results obtained in the anaerobic digestion process confirmed a TVS removal (TVS_{rem}) of 60%, which represent a reference value for a well operated digester. Specific biogas production was 1 l/gTVS$_{rem}$, with CH_4 and CO_2 percentages of 63% and 32%, respectively.

In order to assess the precipitation of phosphorus in the digester mass balances were applied using the experimental concentrations of the influent and effluent digester streams (streams 2 and 3, Table 1). Due to the low pH in the gravity thickener, precipitation in this stage of the sludge treatment line was considered negligible. The results obtained are shown in Table 2.

Table 2. Precipitation results in the anaerobic digester.

K_{TOTrel} (mg/l)	K_{ORGrel} (mg/l)	K_{PAOrel} (mg/l)	P_{PAOrel} (mg/l)	P_{ORGrel} (mg/l)	P_{TOTrel} (mg/l)	P_{fix} (mg/l)
66.7	62.8	3.9	11.1	186.9	198.0	229.0

As can be observed, 229 mg/l of phosphorus were fixed in the digester. Although this amount seems to be a high phosphorus precipitation, other studies have confirmed that this sludge line management minimizes phosphorus precipitation in the digester (Martí *et al.*, 2008b). The mixed thickening of primary and secondary sludge enhances Poly-P hydrolysis in the thickener due to the presence of significant volatile fatty acid (VFA) concentrations from primary sludge. Moreover, the elutriation system using an external stream of high flowrate reduces the available phosphorus in the digester giving rise to a decrease in the precipitation. Hence, operational costs associated to the formation of deposits can be reduced and the phosphorus recovery in the crystallization processes can be enhanced.

Phosphorus recovery in the crystallization process

According to Equation 2 and Equation 3 and using the crystallizer influent and effluent PO_4-P and P_T concentrations (streams 4 and 5, Table 1), the recovery and precipitation efficiencies were calculated. Table 3 shows that precipitation and recovery efficiencies of 83.0% and 69.9% were obtained, respectively. Therefore, the agitated crystallizer has been shown to obtain satisfactory results for the struvite precipitation. Moreover, the process allowed achieving low P concentration (7.3 mg/l) in the effluent of the reactor (Table 1).

Table 3. Crystallization process results.

Phosporus removal efficiencies	
Precipitation efficiency (%)	83.0
Recovery efficiency (%)	69.9
Struvite obtained	
%P$_{MAP}$	37.0

According to the literature (Musvoto *et al.*, 2000), at the crystallizer conditions it can be considered that all magnesium precipitated as struvite (Equation 3), which allowed us to estimate the amount of phosphate precipitated as struvite (%P-MAP), 58%. No potassium struvite ($MgHPO_4 \cdot 6H_2O$) was formed in the crystallizer, in agreement with other authors (Schuiling and Andrade, 1999; Wilsenach *et al.*, 2006) that observed that potassium struvite ($KMgPO_4 \cdot 6H_2O$) could precipitate instead of struvite only in the case of low ammonium concentrations. Table 3 shows that a low percentage of phosphorus had formed struvite (37.0%). The low struvite percentage obtained was attributed to the high Ca/Mg molar ratio in the crystallizer inlet stream which favoured the calcium phosphate formation (Pastor *et al.*, 2008a).

The struvite formation has been confirmed by XRD analyses. Figure 2 shows the X-Ray difractogram of the solids recovered in the experiment together with the struvite pattern. The good correlation between the peaks confirms that the solids obtained were struvite. X-Ray analysis also showed that no crystalline calcium phosphates were recovered (data not shown), which suggests that the calcium phosphate precipitated was an amorphous calcium phosphate.

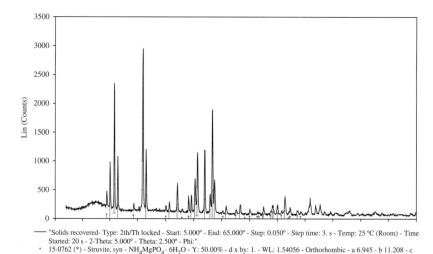

——— "Solids recovered- Type: 2th/Th locked - Start: 5.000° - End: 65.000° - Step: 0.050° - Step time: 3. s - Temp: 25 °C (Room) - Time
    Started: 20 s - 2-Theta: 5.000° - Theta: 2.500° - Phi:"
  *   15-0762 (*) - Struvite, syn - NH$_4$MgPO$_4$ · 6H$_2$O - Y: 50.00% - d x by: 1. - WL: 1.54056 - Orthorhombic - a 6.945 - b 11.208 - c
    6.1355 - alpha 90.000 - beta 90.000 - gamma 90.000 - Primitive - Pm21n(31) - 2 - 477.585 - I/Ic

Figure 2. X-Ray difractogram of the solids recovered (Struvite pattern).

Evaluation of global phosphorus recovery

Two P-rich streams can be used for phosphorus recovery by struvite crystallization: the centrate from the digested sludge dewatering and the thickener
supernatant.

In order to evaluate the phosphorus recovery in the system, the loss of PO$_4$-P
in any stage of the sludge treatment line and the available PO$_4$-P in the
supernatant and centrate produced were both calculated. The results were
calculated considering the mass flowrate of sludge treated.

In the sludge treatment line, phosphorus losses may occur in the anaerobic
digester due to precipitation as different phosphate salts, and always occurs in
the dewatering process where part of the soluble phosphorus remains in the
dewatered sludge.

Phosphorus losses in the sludge treatment line and phosphorus available in
the rejected liquors, both calculated per kilogram of sludge treated, are shown
in Table 4.

As Table 4 shows, the main loss of phosphorus (w$_{P\ LOST}$) is due to
precipitation in the anaerobic digester, which indicates the high importance
of reducing this precipitation in order to increase the potential phosphorus
available for the crystallization process. On the other hand, the main availability
of phosphorus (w$_{P\ AV}$) is due to the thickener supernatant stream. The great
phosphorus available in this stream can be attributed to the high supernatant

Table 4. Global phosphorus recovery.

Loss of phosphorus in the sludge line (gP/kg sludge)	
Precipitation in the digester	5.9
Dewatered sludge	0.6
Total ($w_{P\ LOST}$)	6.4
Available phosphorus to be recovered (gP/kg sludge)	
Thickener supernatant	12.3
Dewatered centrate	1.2
Total ($w_{P\ AV}$)	13.4
$w_{P\ TOT}$ **(gP/kgfango)** $= w_{P\ LOST} + w_{P\ AV}$	19.9
Phosphorus recovery in the crystallization process (gP/kg sludge)	
Total available ($w_{P\ AV'}$)	10.7
w_{Pprec} (gP/kgfango)	8.9
w_{Prec} (gP/kgfango)	8.1
$P_{rec\text{-}MAP}$ (gP/kgfango)	3.0

flowrate produced as a consequence of using the EBPR sludge (i.e. not settled) as a "washing-stream". Table 4 shows that, if no phosphorus had been lost, around 20 g of phosphorus per kg of treated sludge could have been recovered ($w_{P\ TOT}$). However, the loss of phosphorus in the sludge treatment line, mainly by precipitation in the digester, reduced this availability to 13.4 g of phosphorus per kg of treated sludge ($w_{P\ AV}$).

In the phosphorus recovery process, as shown in Table 4, only 10.7 g of phosphorus per kg of treated sludge ($w_{P\ AV}$) were fed to the crystallizer. The difference with $w_{P\ AV}$ can be attributed to a phosphorus precipitation in the storage tank. The results obtained show that 8 g of phosphorus per kg of treated sludge were recovered, 3 g of them as struvite. These results mean that the global efficiency in the system was around 40% for the phosphorus recovery, being 15% for the struvite recovery. Future investigations will focus on the study of different sludge line management strategies in order to optimise the global phosphorus recovery by struvite crystallization.

CONCLUSIONS

In this paper, a sludge line operation scheme has been tested in order to evaluate the global phosphorus recovery efficiency in the system by struvite crystallization. The main conclusions are shown below:

- A great amount of phosphorus is lost due to precipitation in the anaerobic digester, which points out the importance of reducing this precipitation to increase the available phosphorus in the crystallization process.

- The thickener supernatant stream is a phosphorus rich stream that provides a high amount of available phosphorus to the crystallizer. This is due to the high supernatant flowrate produced when the EBPR sludge is used as a "washing-stream" to enhance elutriation in the thickener.
- High phosphorus precipitation and recovery efficiencies of 83.0% and 69.9% were achieved, respectively, in the crystallization process. Only 37% of the recovered phosphorus in the crystallizer was struvite since the presence of calcium favoured the calcium phosphate formation.
- The global phosphorus recovery efficiency, which takes into account the total phosphorus in the system (sludge line + crystallizer), was around 40%, being 15% for the struvite recovery.

REFERENCES

APHA, 2005. Standard Methods for the Examination of Water and Wastewater. 21st ed. American Public Health Association, American Water Works Association and Water Environment Federation, Washington DC, USA.

Battistoni, P., De Angelis, A., Prisciandaro, M., Boccadoro, R. and Bolzonella, D. (2002). P removal from anaerobic supernatants by struvite crystallization: long term validation and process modelling. *Water Res.*, **36**, 1927–1938.

Bouzas, A., Ribes, J., Ferrer, J. and Seco, A. (2007). Fermentation and elutriation of primary sludge: Effect of SRT on process performance. *Water Res.*, **41**, 747–756.

Chanona, J., Pastor, L., Borrás, L. and Seco, A. (2006). Application of a fuzzy algorithm for pH control in a struvite crystallization reactor. *Water Sci. Technol.*, **53**(12), 161–168.

de-Bashan, L.E. and Bashan, Y. (2004). Recent advances in removing phosphorus from wastewater and its future use as fertilizer (1997–2003). *Water Res.*, **38**(19), 4222–4246.

García-Usach, F., Ferrer, J., Bouzas, A. and Seco, A. (2006). Calibration and simulation of ASM2d at different temperatures in a phosphorus removal pilot plant. *Water Sci. Technol.*, **53**(12), 199–206.

Martí, N., Bouzas, A., Seco, A. and Ferrer, J. (2008a). Struvite precipitation assessment in anaerobic digestion processes. *Chem Eng J.*, **141**, 67–74.

Martí, N., Ferrer, J., Seco, A. and Bouzas, A. (2008b). Optimisation of sludge line management to enhance phosphorus recovery in WWTP. *Water Res.*, **42**, 4609–4618.

Musvoto, E.V., Wentzel, M.C. and Ekama, G.A. (2000). Integrated chemical-physical processes modelling-II. Simulating aeration treatment of anaerobic digester supernatants. *Water Res.*, **34**, 1868–1880.

Parsons, S.A., and Doyle, J.D. (2004). Struvite scale formation and control. *Water Sci. Technol.*, **49**(2), 177–182.

Pastor, L., Martí, N., Bouzas, A. and Seco, A. (2008a). Sewage sludge management for phosphorus recovery as struvite in EBPR wastewater treatment plants. *Bioresour. Technol.*, **99**(11), 4817–4824.

Pastor, L., Mangin, D., Barat, R. and Seco, A. (2008b). A pilot-scale study of struvite precipitation in a stirred tank reactor: Conditions influencing the process. *Bioresour. Technol.*, **99**(14), 6285–6291

Shu, L., Schneider, P., Jegatheesan, V. and Johnson, J. (2006). An economic evaluation of phosphorus recovery as struvite from digester supernatant. *Bioresour. Technol.*, **97**(17), 2211–2216.

von Münch, E. and Barr, K. (2001). Controlled struvite crystallization for removing phosphorus from anaerobic digester sidestreams. *Water Res.*, **35**, 151–159.

Wild, D., Kisliakova, A. and Siegrist, H. (1997). Prediction of recycle phosphorus loads from anaerobic digestion. *Water Res.*, **31**(9), 2300–2308.

Index

A
acetate 500
acetic acid 349, 491, 638, 775
Acidithiobacillus ferrooxidans 504–506, 508
Acidithiobacillus spec. strains 503–508
Acidithiobacillus thiooxidans 504–506, 508
Acinetobacter calcoaceticus 479–481
activated sludge
    Design and Simulation of Activated Sludge Systems (DESASS) 783, 787
    high-rate (HRAS) 708, 715
    micro-organisms in 480, 482–483
    phosphates concentration in 482–483
    phosphorus concentration in 481–482
    return (RAS) 480, 484–485
        SCOD/PO$_4$ ratio for 484–485
    waste (WAS) 303–305, 309–313, 568
adsorption isotherms 674, 680–682
advanced SEPHOS process 659, 667, 669
advanced wastewater treatment plants (AWWTPs) 206, 209–210
aerated submerged fixed-film (ASFF) process 708–710, 712–716
AFP *see* anti-filament polymer
agricultural efficiency 30–31
agro-fuels production 49, 53
agro-industry, phosphate removal in 245
    anaerobic effluent characteristics 247
    analytical methods 247–248
    full scale unit 250–252
    pilot plant description 246–247
    pilot-scale tests 248–250
airlift reactor
    in batch mode 181–188

in continuous operation 183, 188–190
    lab-scale reactor 91, 93
    MAP particle size 183–184
    for phosphorus recovery
        induced struvite precipitation 179–191
    in phosphorus removal
        from industrial wastewater 89–96
Alamine® 336, 650–651
algae 3, 4, 6
    as bio-fuels 16–20
alum 351, 773, 776
aluminium ash 661
aluminium sulfate 480
ammonia 556–557
ammonia recovery process (ARP) 615
    saving from integration with BNR operation
        background and conditions modelled 616–617
        findings 617–621
ammonia recycle, reduction by struvite harvesting 351–359
Ammonia-Selective Electrode Method 495
ammonia stripping
    air stripping 722–724
    in bottles 727–731
    in bubble reactors 722
    influence of temperature 733
    pH influence 733–734
    setting of flow conditions 731–732
    with a stripping column 731–734
ammonium
    absorption by vermiculite
        absorption experiments 697–700
        greenhouse experiments 700–705

concentration 802
ions, removal by struvite precipitation
    579–589
ammonium nitrogen
    removal 498
    variation in crystallizer influent and
        effluent 498
anaerobic digestate, air stripping of
        ammonia from 719
    ammonia stripping 722–724, 727–734
    anaerobic digestion plant 721
    precipitation 798–799, 806–808
    pretreatment 721–722, 724–727
    of effluent before stripping 720
anaerobic digested sludge 775
anaerobic digestion 65, 302–303, 782,
        797–798, 802–804
    phosphorus precipitation calculation in
        805
anaerobic effluent characteristics 247–248
anaerobic phase
    phosphate release during 499–500
anaerobic/oxic/anoxic (A/O/A) process
        764–765
    feasibility 769–770
animal slurry separation, in growing cereal
        crops 317
    draught force 321–325
    measuring 318–320
    NH$_3$ 320
    pressure and flow 320–321
animal wastewater, phosphorus recovered
        as calcium phosphate precipitate
    process applications 462
        manure treatment systems without
            lagoon 464–467
        phosphorus extraction from digested
            swine lagoon effluents 462–464
    process chemistry 460–462
    process configuration 460
anoxic-aerobic (A/ASFF) process 709,
        712–716
anti-filament polymer (AFP) 481
AQUASIM platform 372
ASH DEC 406–408, 412–413
    phosphorus speciation of sewage sludge
        ashes 609–613

ashes
    aluminium ash 661
    biomass ashes 753–855
    eluates preparation 391–392
    FA ash 431, 436
    fraction 433
    HA ash 431, 436
    iron ash 662
    KO ash 431, 436
    leaching 645–649
        characterization of ash material 646
        general investigations 646–647
    quality, influences of 593–595
    sewage sludge ashes
        phosphorus recovery 163–164,
            405–414, 417, 419–424
        phosphorus speciation of 609–613
ASM2-Delft metabolic Bio-P model 372,
        376
ATAD see Autothermal Thermophilic
        Aerobic Digestion
attached-growth processes see fixed-film
        processes
automatic process control system 260–261
Autothermal Thermophilic Aerobic
        Digestion (ATAD) System, for
        sludge digestion 489, 493
    ammonium nitrogen concentration and
        removal 498
    batch test results 499–500
    crystallizer operation 492–494
        feed 492
        magnesium concentration 494
        pH control 494
    description 490
    key issues 490–491
    magnesium concentration 498
    monitoring 494–495
    objectives 491
    orthophosphate concentration and
        removal 496–498
    pH value 498–499
    reactor operation 495
    temperature 498–499
    total suspended solids concentration
        499
    volatile fatty acid content 495–497

AWWTPs *see* advanced wastewater
    treatment plant

B
backcasting 24
BAM 406
bamboo plantations, urine reuse as fertilizer
    analytical method 690–691
    biomass production 691–692
    experimental set up 689–690
    measured parameters and performed
        analyses 691
    nutrient loading rate and uptake N and P
        694
    rhizome system 688
    TOC removal through urine application
        692–693
bioaccumulation
    of phosphorus and heavy metals 504,
        508–509
bio-fuels 16–20
biogas technology 719
BioIronTech process 600–601
    anaerobic treatment of reject water
        components of the system 602–603
        ferrous production from iron ore 603
        nitrate and organic acid effects on
            phosphate recovery 605
        organics and nitrogen removal
            603–605
    applications 601
    for phosphate recovery and nitrogen
        removal 601–602
Biolace media 709
bioleaching
    of phosphorus and heavy metals 504,
        506–508
    reactor 505
biological conversion 456
biological nutrient removal processes
    (BNR) 783, 791
biological phosphorus removal process
    764, 781
biological P-remobilization, from sewage
    sludge 270–271, 273–274
biological suspended-growth systems 708
biomass fermentation 635

analytical procedure 637–638
    experiment approach 636–637
    importance 636
    merits and demerits 636
    nutrient recovery 639–642
    phosphorus recovery via struvite
        formation 637
    VFA production and solids reduction
        638–638
BioWin model 616–617
    for single sludge model 618–619
    for two sludge model 620
BOD removal 616, 710, 712–713, 715
brushite 522, 527, 529

C
calcination 677–678
calcite precipitation 798
calcium
    composition 472
    removal
        HRT and pH values for 473–474
calcium-acetate-lactate extractable
        phosphate (CAL-P) 217–221
calcium-phosphate-phases (Ca-P-phases)
        crystallization 521–528
    crystallization experiments 522–523
    objective 522
calcium carbonate 436
calcium oxide 436
calcium phosphate 301
    amorphous 807
    crystallization process 121
    recovery process 654–655
calcium phosphate precipitation
    phosphorus recovery from animal
        wastewater 459–467
    from wastewater treatment plants,
        P recovery
        experiment 294
        pH adjustment 294–297
        principle 292
        raw water 292–294
        reuse feasibility 298
calcium silicate hydrate compounds (CSH)
    301, 303, 522–523
    in EBPR, P-load 306–314

economical considerations 314
efficiency 527
of grain sizes 523–524
specification of 523
XRD patterns of 526
CAL-P *see* calcium-acetate-lactate
    extractable phosphate
carbon dioxide 776
    elimination with sulphuric acid and pH
        adjustment 722
    stripping 263–265, 720, 724–725
    in bubble reactors 722
carbonaceous substrate 712
CBOD removal 59–60
CFD *see* Computational Fluid Dynamics
changing diets 29
Clean Water Services (CWS) 351, 358,
    773, 778
coagulation 236
COD removal 56, 91, 558–559, 574, 576,
    710, 712, 714–715
colorimetry 218, 247, 471, 552, 559
completely stirred reactor (CSTR) 471
Computational Fluid Dynamics (CFD)
    132–133, 135
    modelling 133–135
coproducts, of corn processing 443,
    445–447, 451, 455
corn processing nutrients 443
    characterization 446
    dry grind 450–451
    wet milling 447–450
    extraction, conversion and use 455–456
    membrane separations 451
        gluten filtration 451–454
        thin stillage filtration 454–455
crop residues 36
Crystal Green™ 356–358, 778
crystallization
    of calcium phosphate 521–529
    phosphorus precipitation calculation
        805
    and phosphorus recovery 802, 807–808
    of struvite
        RH SDG model 582–588
        SIG mechanism 583–584, 586
crystallizer reactor 374–375

CSH *see* calcium silicate hydrate
    compounds
CWS *see* Clean Water Services

D
Danfoss Magflo® 318
Davies activity coefficient 122, 124
denitrification process 712, 716
denitrifying polyphosphate-accumulating
    organisms (DNPAOs) 764, 768
Design and Simulation of Activated Sludge
    Systems (DESASS) 783, 787
dewatering 777
    of anaerobic digested sludge 792
differential algebraic equations (DAEs)
    364–365
digested sludge 151, 793
dissolved air flotation (DAF) 793
    thickening 795–797
draught requirement
    for nozzles
        pointing backward 322–325
        pointing downward 322–324
        pointing forward 321–324
dry grind process 443, 445, 450–451
dry matter
    separation efficiencies of 543–547
Durham Advanced Wastewater Treatment
    Plant 351, 357, 773

E
Ecological Sanitation (Ecosan) 687
ecological testing of products
    phosphorus recovery processes 225–233
        test strategy 227–228
        trace metals 231
electrical conductivity 282
    analyses 283–284
    and ionic strength, relationship 283,
        286–289
    temperature dependence 283–286
elutriation 787, 804, 806
energy efficient nutrient recovery, from
    household wastewater 511,
        516–518
    chemical analysis 515–516
    greenhouse gas reduction and 519

pilot plant 512–515
  sampling 515
enhanced biological phosphorus removal
    (EBPR) 160, 301, 351, 594, 598,
    635, 774–775, 777, 792
  analytical methods 303
  economical aspects 314
  long-term experiments 303–309
    with EBPR sludges 305–306,
    309–314
equilibrium model
  model formulation
    assumptions 113–114
    equations 114–115
    prediction comparison 116–117
Eulerian continuum descriptions
    133–134
eutrophication 2–4
evaporation rate (ER) experiments
    744–750
  mass transfer coefficient
    air humidity effect on 748–750
    air temperature effect on 747–748
extended X-ray absorption fine structure
    analysis (EXAFS) 610

F
FA ash 431, 436
fed-batch suspension crystallizer 364,
    366–367
feedback controller model program
    development 262–263
ferric chlorine 480
ferric dosing 776–777
ferrous sulfate 480–481
fertilizer
  ash-fertilizers 127, 432
  future demand
    energy plants 49
    farming technologies 48–49
    nutrition 48
    population growth 48
  granulation 409–410
  "green" fertilizer 529
  manure 33–34
    treatment systems without lagoon
    464–467

manure phosphorus
    strategy for separation 235–242
  nitrogen fertilizer 739–741
    recycling 215–223
  PhosKraft® NPK fertilizer granules 410
  phosphorus fertilizer 407–409, 427,
    460, 741
  phosphorus-saturated ochre as 627
    field trials 628–629
    plant response 630–631
    Pot experiments 628
    potentially toxic elements 631
    soil P concentration 629–630
  PK-fertilizers 425, 427
  potassium fertilizer 741–742
  relative fertilizer effect (RFE) 222
  Thomas-fertilizers 407
  urine as, in bamboo plantation
    analytical method 690–691
    biomass production 691–692
    experimental set up 689–690
    measured parameters and performed
    analyses 691
    nutrient loading rate and uptake N
    and P 694
    rhizome system 688
    TOC removal through urine
    application 692–693
FIAstar 5000 Analyser 698
filamentous bacteria
  involvement in phosphorus recovery
    cycle 479–487
filtration
  gluten filtration 451–454
  nanofiltration 389, 395–402
  sludge resistance to filtration 471–472,
    476–477
  tests 392–394
  thin stillage filtration 454–455
  ultrafiltration 391–394, 399–400
Fixed Growth Reactor (FGR) 489
fixed-film processes 708–709
flowsheet, for wastewater treatment
  modelling 63–66
  nitrogen recovery 63
  technical and economic hurdles 66
  treatment mechanisms 56–58

treatment steps and process systems
58–59
aerobic MBR 59–61
anaerobic MBR digestive system
61–62
phosphorus and nitrogen removal
62–63
food chain 2–4
efficiency 30, 35
food waste 35–36
Freundlich model of phosphate adsorption
680–682
full scale Ostara reactor system 356–358
full scale struvite plants 250–252
future fertilizer demand
energy plants 49
farming technologies 48–49
nutrition 48
population growth 48

G
Genetic Algorithm 122
German Federal Research Centre for
Cultivated Plants 408
global phosphorus demand 27–28
demand reduction measures 28–29
agricultural efficiency 30–31
changing diets 29
food chain efficiency 30
gluten filtration and nutrient separations
451–454
Goshen Ridge Farm 465–466
gPROMS (general process Modelling
System) 362
Graphical User Interface (GUI)
for reactor operation 265
for stripper model 266
gravity thickening 795
"green" fertilizer 529
Gulf-Anorexia 3–4

H
HA ash 431, 436
Hach Lange Cuvette Tests 303
Hagen-Poiseuille Equation 744
Hampton Roads Sanitation District
(HRSD) 193–202

heavy metal
bioaccumulation of 508–509
bioleaching of 504, 506–508
content of Mephrec slags 566
recycling of 425–426
HIFLOW rings 724
high-pressure injection tines 318–325
sand bin 318–322
soil bin 320, 323
high-rate activated sludge (HRAS) 708,
715
homogeneous struvite precipitation
operational conditions 469–477
household
energy system of 515, 518
wastewater flows in 513–515
water components 688
human excreta 34–35
hydraulic retention times (HRT) 471, 710,
716
hydroxyapatite (HAP) 436, 439, 798
hydroxylapatite-like (Hap-like)
Ca-P-phase 522, 526–527

I
ICP see inductively coupled plasma
incineration 645–646
see also ashes
induced struvite precipitation 179–191
inductively coupled plasma (ICP) method
552, 755
industrial manufacturing plant 412–414
influent wastewater characteristics 374
inland water 5
Institute for Plant Nutrition 408
investigated model system 181
ion activity product (IAP) 362
ion chromatography 506
ion speciations 362–363
ionic strength and electrical conductivity,
relationship 283–289
iron ash 662
iron-reducing bacteria (IRB) 600, 603

K
K-struvite 587–588
KO ash 431, 436

Koljern™ isolation technique 512, 515, 518–519
Kynar® 775, 777, 779

L
lab-scale reactor 91, 93, 134
Lagrangian descriptions 133–134
Langmuir model of phosphate adsorption 680–682
layered double hydroxides (LDHs) 672
  characterization 674–675
  parameters affecting phosphate adsorption
    calcinations 677–678
    metal composition 675–676
    pH 678–680
    Zn/Al molar ratio 676–677
  preparation 673
leaching the ash 645–649
  characterization of ash material 646
  general investigations 646–647
  see also bioleaching
livestock wastewater see animal wastewater
LIX® 418, 420
Lulu Island Wastewater Treatment Plant (LIWWTP) 261

M
magnesium 798
  composition 472
  concentration, during sludge digestion 494, 498
  removal
    HRT and pH values for 474
magnesium ammonium phosphate see MAP
magnesium phosphate recovery 654–656
manure see fertilizers
MAP crystallization process, swine wastewater with 328
  analytical methods and instrumentation 330–331
  bench scale experiments 328–329, 331–334
    calcium and carbonate 332–334
    magnesium 332
    pH value 331

pilot scale experiments 329–330, 334–337
  continuous flow 336–337
  sequencing batch 334–336
MAP crystallizers 149, 371–372, 486–487
  calculative value of 149
  precipitation costs 151–153
  effectiveness 156–157
MAP particle size 183–184
MAP struvite crystallization 121–122
MAP supersaturation 123
mass balance 794–795
  for ammonium 123
  for magnesium 123
mass transfer coefficient, air temperature and air humidity effect on 737
  evaporation rate (ER) experiments 744–750
    air humidity effect on mass transfer coefficient 748–750
    air temperature effect on mass transfer coefficient 747–748
    experimental conditions 745
    OVRS, model for vertical sheets design 745–746
    previous volume reduction methods 742–743
    Southern Pakistan case study 738–742
    vertical sheet for OVRS estimation 750
    water supply rate (WSR) experiments 743–744, 746
MathWorks™ 263
Matlab software 263
meat-and-bone meal
  filed experiments 216–219
  pot experiments 217–221
membrane EBPR (MEBPR)
  for phosphorus removal and recovery 371–386
membrane separations, in corn processing 451
  extraction, conversion and use 455–456
  gluten filtration 451–454
  thin stillage filtration 454–455
Mephrec process
  equipment 564
  input material 563–564

investment 564
smelting-gasification technology of 564
    basic slag components 565
    heavy metal content of slags 566
    P-contents of slags 565–566
MeterLab CDM 210, 698
MgNH$_4$PO$_4$·6H$_2$O *see* struvite
microbubble ozonation process 764–765
microorganisms 505
    in activated sludge 480, 482–483
Microthricx parvicella 479–481
microwave-enhanced advanced oxidation
        process (MW/H$_2$O$_2$-AOP) 551–561
Milestone Ethos TC microwave oven
        digestion system 553
Millennium Development Goals 14
mine water treatment plants (MWTPs)
        623–624, 628
MINTEQ 365
Minto ochre 625
mixed liquor suspended solids (MLSS)
        708
molecular weight cut-off (MWCO) 390
monocalcium phosphate (MCP) 223
MSU-2.5 651–653
MW/H$_2$O$_2$-AOP *see* microwave-enhanced
        advanced oxidation process

N
nanofiltration, in phosphorus recovery
        processes 389, 395–399
    analytics 394
    ash eluates preparation 391–392
    economical aspects 401–402
    filtration experiments 392–394
    model solutions preparation 394–395
    retention 393
    theoretical background 390–391
    ultrafiltration 399–400
Nansemond Treatment Plant (NTP)
        193–195
National Fertilizer Development Centre of
        Government of Pakistan 738
Newton-Raphson method 122
nitrification 715
nitrogen
    separation efficiencies of 543–547

nitrogen fertilizer 739–741
nitrogen recovery 13–14
    from wastewater, alternating anorexic-
        aerobic process for 707
    nitrogen transformations 715–716
    organics removal 710, 712–715
    pilot-plant description and
        experimental set-up 709–710
    process application 716
nitrogen removal 8–9, 601–605
Nocardia amarae 479–481
nucleation 72–73
NuReSys$^®$ 251
nutrient recovery
    from biomass fermentation
        solubilization of phosphorus and
            nitrogen 639–640
        struvite formation test 640–642
    COD distribution 574, 576
    empirical evaluation of 567–577
    energy efficient 511, 516–518
        chemical analysis 515–516
        greenhouse gas reduction and 519
        pilot plant 512–515
        sampling 515
    heavy metal distribution 571–573
    ionic metals, mass balance for 575–576
    metal distribution 571–573
    nitrogen, mass balance for 575–576
    phosphorus compounds, mass balance
        for 575–576
    TOC distribution 574, 576
    TSS, mass balance for 575–576
nutrient release 556
nutrient uptake 690, 694

O
ochre
    as filter substrate 625–626
        ongoing research 627
    importance 623–624
    Minto ochre 625
    as phosphorus absorbent 624–625
        laboratory experiments 625
    Polkemmet ochre 625, 628
    see also phosphorus-saturated ochre, as
        fertilizer

online monitoring, process control system
for 257
analytical methods 265–266
carbon dioxide stripping model
263–265
description 258
dominant technique 261–262
expectations 266
feedback controller model program
development 262–263
GUI 265–266
instrumentation and process monitoring
260–261
objective 258
terminology 259
controlled variable 260
manipulated variable 260
set point 260
struvite solubility product 259–260
supersaturation ratio (SSR) 259
Oregon State University Extension Service
357
organic matter
degradation 794–795
as dissolved chemical oxygen demand
484–485
orthophosphate 447, 449
analysis
release at different pH conditions
556
reliability 559–561
removal 496–498
variation in crystallizer influent and
effluent 496–498
see also phosphate
ortho-phosphorus concentration 199
osmotic pressure, on transmembrane flux
during concentration of PFS manure
531, 533–535
resistance and concentration
polarization 539–541
and transmembrane flux 535–538
Ostara process 196, 352, 492, 778–779
reactor effluent 356
reactor operation 352–355
reactor products 356
ozonation process 764–765

P
Particle Induced Gamma-ray Emission
(PIGE) method 698
PASH process see phosphorus recovery,
from sewage sludge ash
patch flocculation 236
effect of 237–241
strategy for 241–242
pH value 720
adjustments 725–727, 775–777
with aeration 296–297
with $Ca(OH)_2$ 294–295
with NaOH 296
control, in crystallizer influent and
effluent 494, 498–499
in dependence of removed ammonia
728
development during calcium phosphate
crystallization 527
influence 733–734
on crystal dimensions and habit
580–582
and wastewater treatment 460–463, 467
Phacelia 757
PhosKraft® NPK fertilizer granules 410
phosphate
impact of supply and demand 45–53
P consumption, demand 47–50
P reserves, supply 46–47
price development 50–53
release during anaerobic phase 499–500
uses 591
see also orthophosphate
phosphate accumulating organisms
(PAOs) 499–500
phosphate adsorption using zinc-
aluminium layered double
hydroxides 671
adsorption from sewage filtrate 673
adsorption isotherms study 674,
680–682
adsorption kinetics 680–681
characterization of LDHs 674–675
LDHs preparation 673
parameters affecting adsorption 675–680
pH effects on adsorption 673
phosphate adsorption study 674

phosphate desorption 674, 682–683
phosphate measurement 675
phosphate conversion 127
phosphate desorption 674, 682–683
phosphate price development
   current and historical development
      50–51
   future development 51–53
phosphate recovery
   BioIronTech process 60–602
   sidestream flow system for see
      membrane EBPR (MEBPR)
   technologies for 212
   from wastewater 782
phosphate recycling 149
phosphate release 160–162
phosphate removal 245–253, 328,
   335–336, 605
phosphate rock 32–33
   business-as-usual scenario 52
   reserves and heavy metal contents 47
phosphomolybdate blue method 755
phosphorus 215–216, 470
   accumulation by filamentous bacteria
      480
   adsorption column 765–766
   bioaccumulation of 508–509
   bioleaching of 504, 506–508
   composition of 472
   consumption 47–50
   cycling, using biomass ashes 753–855
      ash application effect on dry matter
         yield 756–757
      nutrient content and P availability, in
         different biomass ashes 756
      soil pH and soil P parameters
         757–759
   elimination 523–528
   essential potentials quantification
      99–110
   fertilizer 407–409, 427, 460, 741
      recycling 215–223
   fractionation method 754, 758
   needs, framework
      classification matrix 26–27
      demand options 26–28
      demand reduction measures 28–31
      global food demand 24–26
      global phosphorus demand 27–28
      historical and future scenarios 37
      institutional challenges and policy
         implications 37–39
      supply options 26–27, 31–37
   particle size distribution of 548–549
   removal
      HRT and pH values for 473–474
   reserves supply 46–47
      in 2030 47
   separation efficiencies of 543–548
   volutins, in filamentous organisms
      482–483
   in waste water treatment 302–303
   see also phosphorus removal and
      recovery, from sewage sludge
phosphorus recovery 10–13, 599–600
   from BioIronTech process 601–602
   different fractions 102
   process descriptions
      comparison and outlook 165–166
      economics 164–165
      post precipitation, in WWTP effluent
         160
      PRISA process 160–162
      RPA process 163–164
   via struvite formation 636–637, 640–642
   technologies and costs see phosphorus
      recovery, economy of
phosphorus recovery, by calcium
      phosphate precipitation
   from wastewater treatment plants
      experiment 294
      pH adjustment 294–297
      principle 292
      raw water 292–294
      reuse feasibility 298
phosphorus recovery, by nanofiltration
      389, 395–399
   analytics 394
   ash eluates preparation 391–392
   economical aspects 401–402
   filtration experiments 392–394
   model solutions preparation 394–395
   theoretical background 390–391
   ultrafiltration 399–400

phosphorus recovery, by thermochemical
    treatment of sewage sludge ash
    417, 419–424
  heavy metal recycling 425–426
  pot trials 425, 427–429
phosphorus recovery, economy of 145
  costs 150–156
    component process 150–151
    MAP precipitation costs 151–153
    phosphate remobilization costs
      154–155
    supplies input, reducing 155–156
    technologies and 159–166
  resource economy of phosphorus
    146–149
    nutrients value 148–149
    phosphorus price 147
phosphorus recovery, from animal
    wastewater 459
  process applications 462
    manure treatment systems without
      lagoon 464–467
    phosphorus extraction from digested
      swine lagoon effluents 462–464
  process chemistry 460–462
  process configuration 460
phosphorus recovery, from ash 163–164
phosphorus recovery, from excess sludge
    768–769
  in anaerobic-oxic-anoxic process 763
  analytical procedures 767
  feasibility of A/O/A system 769–770
  reactor design and operation 764–766
  reactor operation 767–768
phosphorus recovery, from sewage sludge
    ash 405, 503, 505–506
  conceptual flow sheet 657
  elution tests 662, 664–666
  leaching the ash 645–647
    characterization of ash material 646
    general investigations 646–647
  origin and composition of sewage
    sludge ashes 660–662
  phosphorus and heavy metals
    bioaccumulation of 508–509
    bioleaching of 506–508
  precipitation

calcium (magnesium-) phosphate
    product and separation of
    aluminium 653–656
  phosphorus product preparation 656
  product quality 656–657
  process 406–407
  product 407–410
  separation of phosphorus and metals
    via nanofiltration 663, 668
    via pH-value increase 662–663,
      666–667
  solvent extraction
    chemical reaction using Alamine®
      336, 650–651
    laboratory-scale experiments 651–653
    principles 647–649
  technology transfer 410
    industrial manufacturing plant
      412–414
    prototype manufacturing plant
      410–412
  thermal treatment of mixture of pure
    substances 662–664
phosphorus recovery cycle
  filamentous bacteria involvement in
    479–487
  practical implications 486–487
phosphorus recovery in EBPR systems, by
    struvite crystallization 801
  anaerobic pilot plants description
    803–804
  analytical methods 804
  experimental procedure 804
  global phosphorus recovery evaluation
    808–809
  phosphorus precipitation calculation
    in anaerobic digestion 805
    in crystallization process 805
  phosphorus precipitation in anaerobic
    digester 806–807
  phosphorus recovery in crystallization
    process 807–808
phosphorus recycle reduction by struvite
    harvesting 351–357
phosphorus remediation, from animal
    slurry 431
  ash fraction determination 433

dry matter determination 433
extraction 434
phosphorus analysis 434, 439
thermal gravimetric analysis 433–435
volatile solids 433
X-ray diffraction spectroscopy 433,
 436–439
phosphorus removal 7–8
modelling approach 104–106
 model enhancement 107–108
 module integration 108–109
methods 99
phosphorus-saturated ochre, as fertilizer
 627
field trials 628–629
plant response 630–631
pot experiments 628
potentially toxic elements 631
soil P concentration 629–630
see also ochre
phosphorus solubility 230–231
phosphorus speciation
of sewage sludge ashes 609–613
phosphorus supply options 31–32
crop residues 36
food waste 35–36
human excreta 34–35
manure 33–34
other options 37
phosphate rock 32–33
Phostrip process 102, 372
PHREEQC model 208
Phyllostachys viridiglaucescens 689–690
piggery wastewater 339
biochemical analyses 341
precipitation runs 341–348
solid analyses 342
supernatants 340–341
pilot plant 512–515
energy system, of house 515, 518
wastewater flows, in house 513–515
pilot-scale reactor
dimensions 134
phase properties and simulation settings
 136
simulated average solids volume
 fraction 140

volume fractions of solid phases
 137–139
pilot-scale struvite crystallizer 366
pilot-scale struvite plants 252–254
anaerobic effluent characteristics 247
analytical methods 247–248
description 246–247
full-scale unit 250–252
pilot-scale tests 248–250
PK-fertilizers 425, 427
Polkemmet ochre 625, 628
pollution 1, 24, 35, 121, 159, 327, 364,
 411–412, 414, 627
pollutants 3, 159, 226, 302, 389–390,
 406–407, 418–419, 657, 743
poly-β-hydroxybutyrate (PHB) 160
polymer bridging 236
polyphosphate (poly-P) 791–792, 795
polyphosphate (poly-P) accumulating
 organisms (PAOs) 504, 506, 508,
 764, 768, 782, 795
hydrolysis 797, 802, 806
potassium fertilizer 741–742
precipitation 720
of calcite 798
of calcium phosphate 292–298, 459–467
experiments from sewage sludge
 potash wastewater 271–272, 276–277
 seawater 272–275
 upflow reactor 272–273, 276–278
with lime milk 721–722, 724
of MAP 151–153
of phosphorus precipitation 653–657,
 805
pilot-scale reactor 272–273
problems in WWTP 782, 791–793
anaerobic digestion 797–798
analytical campaign 793–794
DAF thickening 795–797
gravity thickening 795
mass balance 794–795
precipitates distribution in digester
 798–799
of struvite 798–799, 804, 806–808
for ammonium ions removal
 579–581, 587–588
chemical equilibrium model 122–126

homogeneous 469–477
induced 179–191
precipitation potential 472–473
thermochemical approach, modelling
from wastewater 121–128
upflow reactor 276–278
prefiltrated swine (PFS) manure
characteristics of 533
conductivity 539–541
PRISA process 160–162
process control system, for online
monitoring 257
analytical methods 265–266
carbon dioxide stripping model 263–265
description 258
dominant technique 261–262
expectations 266
feedback controller model program
development 262–263
GUI 265–266
instrumentation and process monitoring
260–261
objective 258
terminology 259
controlled variable 260
manipulated variable 260
set point 260
struvite solubility product 259–260
supersaturation ratio (SSR) 259
process systems engineering, application
of 361, 367–370
modelling 362
chemistry and thermodynamics
362–363
description 364
kinetics 363–364
model simulation and thermodynamic
validation 364–365
parameter estimation 366–367
P-RoC-technology 521–522, 524, 529
protein 5, 802
recovery 9
prototype manufacturing plant 410–412

R
RAPSODY program 724
RAS see return activated sludge

Rauschert Verfahrenstechnik Air 724
reactor
airlift reactor 89–96, 179–191, 722
bioleaching reactor 505
bubble reactor 722
completely stirred reactor (CSTR) 471
crystallizer reactor 374–375
fixed growth reactor (FGR) 489
full scale Ostara reactor system
356–358
lab-scale reactor 91, 93, 134
operation 265, 352–355, 495, 764–768
pilot-scale reactor 134, 136–140,
272–273
products 356
reactor design 764–766
reactor effluent 356
suspended growth reactor (SGR) 489
upflow reactor 272–273, 276–278
reject water
BioIronTech process
anaerobic treatment using 602–605
application for phosphate recovery
and nitrogen removal 601–602
phosphorus recovery 599–600
see also urine; wastewater
relative agronomic effectiveness (RAE)
222
relative fertilizer effect (RFE) 222
relative supersaturation 363
relevant flows 100–101
and fractions 101–102
resistance-in-series model 539
resource recovery
composting 13
eutrophication 2–4
limiting nutrients, concept 4–5
nitrogen
recovery see nitrogen recovery
removal see nitrogen removal
sources 7
nutrients surplus 7–9
phosphate
recovery see phosphate recovery
removal see phosphate removal
phosphorus
recovery see phosphorus recovery

removal *see* phosphorus removal
  sources 6
protein recovery 9
urine separation 14–16
wastewater conversion 5
see also other recovery and removal
  processes
retentate stream 454–455
return activated sludge (RAS) 480
  SCOD/PO$_4$ ratio for 484–485
reverse osmosis (RO) 532
  transmembranes, osmotic pressure on
    531–541
Rojkowski hyperbolic size dependent
  growth model (RH SDG model)
    582–588
  parameters 585
RPA process 163–164
Runge Kutta Method 746

S
sand bin experiment 318–322
SCOD 556, 558–559
scum
  micrograph of 482
  phosphates concentration in 482–484
  phosphorus concentration in 481–482
  SCOD/PO$_4$ ratio for 484–485
    disintegration time 485–486
Seaborne technology, at WWTP Gifhorn
    567–579
  COD distribution 574, 576
  evaluation methods 569–571
  heavy metal distribution 571–573
  ionic metals, mass balance for 575–576
  metal distribution 571–573
  nitrogen, mass balance for 575–576
  phosphorus compounds, mass balance
    for 575–576
  sample plan 570
  schema of 569–570
  TOC distribution 574, 576
  TSS, mass balance for 575–576
Sentron pH meter 698
separation
  animal slurry separation, in growing
    cereal crops 317

draught force 321–325
NH$_3$ 320
pressure and flow 320–321
nutrient separations
  extraction, conversion and use
    455–456
  gluten filtration 451–454
  thin stillage filtration 454–455
  of phosphorus and metals
    via nanofiltration 663, 668
    via pH-value increase 662–663,
      666–667
  separation efficiencies
    of dry matter 543–547
    of nitrogen 543–547
    of phosphorus 543–548
  slurry separation 544
  solid-liquid separation 235–236
  urine separation 14–16
Sequential Precipitation of Phosphorus
    (SEPHOS) process 667
sewage sludge ash (SSA) 609
sewage sludge ash, P-recovery from
  process 406–407
  product 407–410
  technology transfer 410
    industrial manufacturing plant
      412–414
    prototype manufacturing plant
      410–412
sewage sludge ash, phosphorus recovery
    from 389, 417
  leaching the ash 645–647
  by nanofiltration 395–399
    analytics 394
    ash eluates preparation 391–392
    economical aspects 401–403
    filtration experiments 392–394
    model solution preparation
      394–395
    theoretical background 390–391
  precipitation 653–657
  solvent extraction 647–653
  thermochemical treatment 419–429
  by ultrafiltraion 399–401
sewage sludge ash, phosphorus speciation
    610–613

sewage sludge, precipitants and
    P-remobilization from 269
  biological 270–271, 273–274
  pilot-scale precipitation reactor 272–273
  seawater and wastewater from potash
      production as magnesium sources
      271–272
  seawater as precipitant 274–275
  upflow precipitation reactor 276–278
  wastewater from potash production as
      precipitant 276–277
sewage sludge incinerator ash, for
    phosphate production
  recycling of phosphate
    ash quality, influences 593–595
    full scale tests 595–596
    process description 592–593
sidestream flow system, for phosphate
    recovery see membrane EBPR
    (MEBPR)
sidestream treatment options 352
sidestream wasting 376
SIG see size independent growth
    mechanism, of struvite
    crystallization
Simulink software 263
single-sludge plants 616
  BioWin model for 618–619
single-sludge predenitrification system 712
Single Super Phosphate (SSP) 427
size independent growth (SIG)
    mechanism, of struvite
    crystallization 583–584, 586
Slibverwerking Noord-Brabant (SNB)
  full scale tests 595–596
  P-removal technologies 594
  recycling of phosphate 593
sludge 763
  activated see activated sludge
  anaerobic digested sludge 775, 792
  BioWin model 616–620
  digestion 151, 489–499, 793
  EBPR sludges 305–306, 309–314
  line modelization 783–784
  management modification proposal
      784–788
  microorganisms in 480, 482–483

production, potential reduction 770
  sewage sludge 270–271, 273–274
  sewage sludge ashes 163–164, 405–414,
      417, 419–424, 609–613
  sludge resistance to filtration (SRF)
      471–472, 476–477
  SVI (sludge volume index) 471, 476
  see also other individual entries
slurry separation 544
smelting-gasification technology
  of Mephrec process 564
    basic slag components 565
    heavy metal content of slags 566
    P-contents of slags 565–566
soil bin experiment 320, 323
solid dairy manure 551
  initial characteristics of 553
  MW/H$_2$O$_2$-AOP effectiveness on
      551–561
    ammonia 556–557
    analytical procedures 555
    apparatus and substrate 553
    experimental design 553–555
    NO$_x$ 556–557
    orthophosphate analysis, reliability of
        559–561
    SCOD 556, 558–559
    soluble phosphate 556–557
    volatile fatty acids 556–558
solid-liquid separation 235–236
soluble phosphate 556–557
solution thermodynamics 69–70
  supersaturation 71–72
  thermodynamic equilibria 70–71
  thermodynamic solvers 72
solvent extraction, for phosphorus
    recovery
  chemical reaction using Alamine® 336,
      650–651
  laboratory-scale experiments 651–653
  principles 647–649
speciation
  of ion 361–363
  model 209–212
  of phosphorus 609–613
SplitBox technique 512–513, 518
  chemical analysis 515–516

energy saving and greenhouse gas
    reduction 519
installation 514
nutrient recovery 518–519
sampling 515
wastewater flows in 514
SRF *see under* sludge
SSR *see* super-saturation ratio
stripping 181, 425
    air stripping 92, 246, 249, 719–724,
        731–734
    of ammonia 275, 336, 473, 722–724,
        731–734
        in bottles 727
        in bubble reactor 722
    of carbon dioxide 181, 184–186, 191,
        263–265, 724–725
        in bubble reactor 722
    of carbonate 334
    with preheating of liquid and air inflow
        723
struvite 111–113, 170, 469
    chemistry 205
    content, calculation 81–82, 85–86
    control techniques, in enhanced
        biological phosphorus removal
        plant 773
        appearance of struvite 774
        centrate storage system design 777
        digester struvite issues 777–778
        future plans for struvite recovery
            778–779
        material testing 774–775
        pH adjustment 775–777
        transition to biological phosphorus
            removal 774
    crystallizer 374–375
    elemental analyses 81, 84–85
    formation 469–470
    image analysis. 81
    phosphate-based precipitates 79–87
    precipitation potential (PP) of 472–473
    properties 90
    single crystal formation 80
struvite crystallization 69, 81, 160–162
    phosphorus recovery in EBPR systems
        801, 804

phosphorus removal from industrial
    wastewater 89–96
struvite crystals
    microphotography 95
struvite formation 636–637, 640–642
struvite harvesting 212
    analytical techniques 207
    modelling techniques
        formation model 207–208
        model comparison 210–212
        PHREEQC model 208
        solubility curve 209
        speciation model 209
    to reduce ammonia and phosphorus
        recycle 351
        full scale struvite recovery analysis
            356–358
        Ostara process, piloting 352–356
        sidestream treatment options 352
struvite nucleation studies 69–77
struvite precipitation 512–513, 515,
    579–588
    ammonium ions removal by 579
        chemical reactions of 580–581
        impurities, role of 587–588
    chemical equilibrium model
        model formulation 122–124
        solution strategy 124–126
    homogeneous 469–477
    induced 179–191
    thermochemical approach, modelling
        from wastewater 121–128
struvite prills 354, 356
struvite recovery, dewatering centrate
    pilot testing and economic evaluation
        193–202
        cost-benefit analysis 201
        pilot system 198
struvite recovery, from urine
    social and economic feasibility 169–177
        currently available fertilizers 173
        fertilizer prices 175
        financial feasibility 172–176
        regression estimates for nutrient
            values 173–174
        regression estimates for struvite
            processing estimates 176

struvite recovery process
  description 195–196
  growth rate parameter estimation
    366–367
  process systems engineering,
    application of 361, 367–369
    chemistry and thermodynamics
      362–363
    description 364
    kinetics 363–364
    model simulation and thermodynamic
      validation 364–365
struvite solubility product 259–260
  standardization 203–212
substrate-related resistance, on
    transmembrane flux 531
  as function of PFS manure conductivity
    539–541
supersaturation 71–72
supersaturation ratio (SSR) 259, 362, 375,
    492–493
SUSAN (Sustainable and Safe Re-use of
    Municipal Sewage Sludge for
    Nutrient Recovery) 406, 417
suspended growth reactor (SGR) 489
SVI see under sludge
swine lagoon liquid, phosphorus removal
    from 462–464
swine wastewater, with MAP
    crystallization 327
  analytical methods and instrumentation
    330–331
  bench scale experiments 328–329,
    331–334
    calcium and carbonate 332–334
    magnesium 332
    pH value 331
  pilot scale experiments 329–330,
    334–337
    continuous flow 336–337
    sequencing batch 334–336

T
Tarragona WWTP case study
  uncontrolled precipitation problems
    791
  waste sludge operation 781

TCOD 556, 559
Teflon® 775
temperature 720
thermal gravimetric analysis (TGA)
    433–435
thermodynamic equilibria 70–71
thermodynamic solvers 72
Thermphos International
  full scale tests 595–596
  recycling of phosphate 592–593
thin stillage filtration and nutrient
    separations 454–455
Thomas-fertilizers 407
total ammonia nitrogen (TAN)
  composition 472
  removal
    HRT and pH values for 473–474
    see also ammonia
total Kjeldahl N 545, 698
Total Maximum Daily Load (TMDL) 773
Total Organic Carbon (TOC) removal
    692–693
total suspended solids (TSS) 495
  variation in crystallizer influent and
    effluent 499
trace metals 231
triple superphosphate 215–216
TSS see total suspended solids
two-sludge plants 616
  BioWin model for 620

U
UBC MAP fluidized bed crystallizer
  heterodynamics of 131–143
    CFD modelling 133–135
    crystallization process 133
    schematic diagram 134
ultrafiltration, in phosphorus recovery
    391–394, 399–400
United Fermentation and Thickening
    (UFAT) 774–775
upflow anaerobic sludge bed (UASB)
    effluent characteristics 247–248
urine 737
  as fertilizer
    Ecosan-bamboo research concept 689
    importance 687–688

quality 171–172
urine-diverting dry toilets (UDDTs) 171

V
vanadate-molybdate method 755
vermiculite, ammonium absorption 697
    absorption experiments 697–700
    greenhouse experiments 700–705
VFA *see* volatile fatty acid
volatile fatty acid (VFA) 494, 556–558,
        774–775, 777, 806
    production, from biomass fermentation
        638–639
    variation in crystallizer influent and
        effluent 495–497

W
waste activated sludge (WAS) 303–305,
        309–313, 568
waste sludge operation, Tarragona case
        study 781
    control system 788–789
    mathematical model 783
    sludge line modelization 783–784
    sludge management modification
        proposal 784–788
wastewater treatment
    green development 55–67
    new flowsheet 56–66
wastewater treatment plant 159, 763
    in Al-Awir 708
    P recovery, by calcium phosphate
        precipitation from

experiment 294
    pH adjustment 294–297
    principle 292
    raw water 292–294
    reuse feasibility 298
    in Tarragona 782
water extractable phosphorus (WEP), in
        dairy manure 552
water supply rate (WSR) experiments
        743–744, 746
WEP *see* water extractable phosphorus, in
        dairy manure
wet milling 443–444, 447–450
WWTP *see* wastewater treatment plant

X
X-ray absorption P K-near edge
        spectroscopy (XANES) 610, 612
X-ray diffraction (XRD) 79–85, 87, 433,
        435–439, 523, 610, 612, 804
    patterns of CSH substrate 526

Y
yellow water 155–156

Z
zeolite clinoptilolite adsorption 512–513,
        515–517, 519
zinc 231, 425, 429
Zinc-Aluminium LDH, phosphate
        adsorption 671–683